SHIPS FOR VICTORY

SHIPS FOR VICTORY

A History of Shipbuilding under the U. S. Maritime Commission in World War II

by
FREDERIC C. LANE

with the collaboration of
Blanche D. Coll · Gerald J. Fischer · David B. Tyler
Charts by Joseph T. Reynolds

with a new preface by Arthur Donovan

THE JOHNS HOPKINS UNIVERSITY PRESS
BALTIMORE AND LONDON

© 1951, 2001 The Johns Hopkins University Press
All rights reserved
Printed in the United States of America on acid-free paper

Originally published by The Johns Hopkins Press, 1951
Johns Hopkins Paperbacks edition, 2001
9 8 7 6 5 4 3 2 1

The Johns Hopkins University Press
2715 North Charles Street
Baltimore, Maryland 21218-4363
www.press.jhu.edu

Library of Congress Cataloging-in-Publication Data

Lane, Frederic Chapin, 1900–
 Ships for victory : a history of shipbuilding under the U.S. Maritime Commission in World War II / by Frederic C. Lane, with the collaboration of Blanche D. Coll, Gerald J. Fischer, David B. Tyler ; charts by Joseph T. Reynolds.—Johns Hopkins paperbacks ed.
  p. cm.
 "Originally published by Johns Hopkins Press, 1951"—T.p. verso.
 Includes bibliographical references and index.
 ISBN 0-8018-6752-5 (alk. paper)
 1. Shipbuilding—United States—History—20th century. 2. World War, 1939–1945—United States. 3. United States. Maritime Commission—History. I. United States. Maritime Commission. II. Title.
VM23 .L3 2001
338.4′762382′0097309044—dc21

2001018657

A catalog record for this book is available from the British Library.

To
KENT ROBERTS GREENFIELD

CONTENTS

	PAGE
PREFACE TO THE 2001 EDITION	xi
PREFACE TO THE ORIGINAL EDITION	xvii
LIST OF FIGURES	xxiii
LIST OF TABLES	xxv
LIST OF PLATES	xxvii

CHAPTER 1. THE COMMISSION AND THE SHIPBUILDING INDUSTRY
- I. A General View of Wartime Shipbuilding 1
- II. The Maritime Commission Before the War 10
- III. The Long-Range Program 21
- IV. Standard Types of Cargo Vessels 27
- V. The Shipyards of the Country in 1940 32

CHAPTER 2. EMERGENCY SHIPBUILDING BEFORE THE DECLARATION OF WAR
- I. The Inception of the Emergency Program 40
- II. Selecting the Sites and Managers 46
- III. The Second Wave of Expansion 55
- IV. The Third Wave 60
- V. Popularizing the Program 66

CHAPTER 3. DESIGN AND INITIAL PROCUREMENT FOR THE LIBERTY SHIP
- I. The Basic Design Decision and the Working Plans . . . 72
- II. Modifications in Design 80
- III. Purchasing Through a Central Agent 89
- IV. The Break with Gibbs & Cox 97

CHAPTER 4. CONTRACTS WITH SHIPBUILDERS AND THEIR SUPERVISION
- I. Lump-Sum Contracts 101
- II. Facilities Contracts 107
- III. Manhour Contracts 117
- IV. Price-Minus Contracts 123
- V. The Commission's Supervisory Staff 127

CHAPTER 5. EXPANSION AND REORGANIZATION AFTER PEARL HARBOR
- I. The Fourth Wave of Expansion 138
- II. The Fifth and Final Wave 143
- III. Politics and Administrative Methods in the Selection of Shipyard Sites 150
- IV. The Commission Under the War Powers 161
- V. Decentralization of Supervision 168

CONTENTS

		PAGE
CHAPTER 6.	EXCESS CAPACITY AND THE CANCELLATION OF THE HIGGINS CONTRACT	
I.	The Inclusion of Military Types and the Signs of Overexpansion	173
II.	Plans and Progress at the Higgins Yard	184
III.	The Decision to Cancel	190
IV.	Inquest Post-Mortem	194

CHAPTER 7. SPEED AND PRODUCTIVITY IN MULTIPLE PRODUCTION
  I. Standardization and Subassembly 202
  II. The Plans of the New Shipyards 214
  III. Assembly-Line Methods 224
  IV. The Productivity of Labor 230

CHAPTER 8. BUILDING THE LABOR FORCE
  I. Job Breakdown and Union Rules 236
  II. Wages and Recruitment 244
  III. Negroes and Women in Shipyards 252
  IV. Training Programs 258

CHAPTER 9. COLLECTIVE BARGAINING
  I. The Formation of the Shipbuilding Stabilization Committee 268
  II. The Zone Conferences 276
  III. Strikes and Loafing 287
  IV. Inflation and the Stabilization Committee 306

CHAPTER 10. THE BATTLE FOR STEEL
  I. Opening Engagements, 1941 311
  II. "Taking It in the Teeth" 322
  III. Regaining Some Lost Ground 335
  IV. The Outcome: Production and Inventory 343

CHAPTER 11. GUIDING THE FLOW OF MATERIALS
  I. Priorities 353
  II. Placing Orders 359
  III. Scheduling Ship Construction 362
  IV. Scheduling Components 365
  V. Scheduling Materials, PRP and CMP 369
  VI. The Completed System in Operation 379
  VII. Expediting 387
  VIII. The Causes of Large Inventories 390

CHAPTER 12. INCREASING THE SUPPLIES OF COMPONENTS
  I. Propulsion Machinery 396
  II. Financing Suppliers 399
  III. The Case for Small Business 404
  IV. Inspection and Conservation 408

CONTENTS

PAGE

CHAPTER 13. STABILIZATION AND MORALE IN THE LABOR FORCE
- I. The Problem 411
- II. Policing Wages 415
- III. Federal Aid in Transport and Housing 427
- IV. Shipyard Communities 437
- V. Health and Safety 446
- VI. Worker Participation 451

CHAPTER 14. MANAGING MANAGEMENTS
- I. Varied Talents and Planned Competition 456
- II. Freedom of Management vs. Bureaucratic Controls . . . 471
- III. The Selective-Price Contract 487

CHAPTER 15. CHANGING MANAGEMENTS
- I. Tampa and Savannah 493
- II. South Portland, the Local Scene 503
- III. South Portland, the Interregnum 513
- IV. South Portland, A New Team 526
- V. Minor Instances and Conclusions 535

CHAPTER 16. CRACKS IN WELDED SHIPS
- I. Crises, Criticisms, and Correctives 544
- II. Efforts to Improve Welding 555
- III. Research and Retrospect 565

CHAPTER 17. THE VICTORY SHIP
- I. The Desire for Faster Cargo Carriers 574
- II. The Hull Design of the Victory Ship 577
- III. The Search for Engines 583
- IV. Collision with the Controller of Shipbuilding 587
- V. Resolving the Conflict with WPB 601

CHAPTER 18. MILITARY AND MINOR TYPES
- I. Landing Ships and Escort Vessels 608
- II. Transports 617
- III. Concrete Barges and Other Minor Types 627

CHAPTER 19. THE CONTRAST BETWEEN 1943 AND 1944
- I. Programs and Production in 1943 and 1944 637
- II. Supplying Components in 1944 647

CHAPTER 20. THE MANPOWER AND MANAGERIAL CRISIS
- I. The New Manpower Problem of the Shipyards 654
- II. Controls of the War Manpower Commission 662
- III. Fumbling with Functions of Management 670
- IV. The Lump-Sum Contracts in 1944 and 1945 679

CONTENTS

PAGE

CHAPTER 21. ADMINISTRATIVE PROBLEMS: (A) THE REGIONAL OFFICES
    I. The Four Regions 689
    II. Relations with Washington 692
    III. Coordination Within the Regions 703

CHAPTER 22. ADMINISTRATIVE PROBLEMS: (B) THE FLOW OF MONEY
    I. Appropriations 712
    II. Disbursements 721
    III. Auditing Construction Costs 728
    IV. The Accounts 739

CHAPTER 23. ADMINISTRATIVE PROBLEMS: (C) THE COMMISSION AND THE WAR SHIPPING ADMINISTRATION
    I. Expansion of Personnel 750
    II. WSA and the Joint Services 754
    III. Reorganizations in the Finance Division 765
    IV. Commission Action 773
    V. The Disruption of the Wartime Commission 787

CHAPTER 24. ADVENTURES IN HINDSIGHT
    I. Renegotiation and Investigations of Profits 798
    II. Analyses of Costs 819
    III. Conclusions 829

BIBLIOGRAPHICAL NOTE 834
INDEX 835

PREFACE TO THE 2001 EDITION

The Second World War ended nearly two generations ago, and the number of living veterans who participated in this enormous global conflict, both those who served in the armed forces and those who worked and worried on the home front, is rapidly declining. While this epochal war is in no danger of being forgotten, subsequent generations will have to rediscover and re-imagine the war for themselves when they can no longer ask the people who lived through it about their war experiences. And when these younger citizens ask about the enormous wartime shipbuilding program that enabled the United States to supply its allies and support its troops on the far sides of the Atlantic and Pacific oceans, they will come to appreciate Frederic Lane's *Ships for Victory*.

Lane's book, like all great histories, succeeds on several levels. Comprehensive and based on a careful examination of the relevant records, *Ships for Victory* provides a richness of detail and the kind of scholarly apparatus one expects from a definitive historical study while maintaining its focus throughout. Yet the book remains accessible and useful and is not to be considered just another dust-gathering official history. Its author has presented his vast and complicated subject in a clear, informative narrative, producing a book to be read as well as consulted and presuming no special expertise on the reader's part. Filled with stories of interesting individuals and dramatic situations, as well as descriptions of ships, shipyards, and administrative assignments, this is the book with which to start if you are interested in U.S. merchant shipbuilding during World War II. As you read Lane's narrative you will be learning from a master storyteller as well as from a scholar with an unrivaled command of his subject.

Frederic Lane was a prominent history professor at the Johns Hopkins University when the war began. An economic historian who specialized in Venice in the age of the Renaissance, Lane was also a distinguished maritime historian like his notable contemporaries Samuel Eliot Morison and Robert G. Albion. The Johns Hopkins University Press published Lane's classic and often reprinted *Venetian Ships and Shipbuilding of the Renaissance* in 1934, and his study of Andrea Barbarigo, a Venetian merchant of the fifteenth century, in 1944. Again like Morison and Albion, Lane was committed to making a broad view of history interesting and meaningful to general readers. With two collaborators he wrote a textbook called *The World's History*, the first edition of which appeared in 1950, one year before *Ships for Victory* was published. Thus, following the end of the war, when President Roosevelt directed every federal agency to prepare "a final report that will sum up both what was accomplished and how the job was done," it was entirely appropriate that the U.S. Maritime Commission invited Lane to take on this task in its behalf.

In the original preface Lane tells us that during 1946 and the first half of 1947 he worked on this project as an employee of the Maritime Commission, but when the Commission had to terminate his contract due to a lack of funds, the Johns Hopkins University and the Rockefeller Foundation provided financial support that enabled him to finish his work. The book was written for a series of "Historical Reports on War Administration," but it is not the official wartime history of the Maritime Commission. Lane was, however, given complete access to the Commission's records, and he also received invaluable assistance, which he generously acknowledged, from many members of the Commission's staff. The Maritime Commission imposed no restrictions on what he saw or said, nor did they review or endorse his conclusions. Perhaps such ministerial poise when subjected to officially-mandated historical scrutiny is only possible after one has emerged victorious from a demanding but morally unambiguous war.

PREFACE TO THE 2001 EDITION

Lane's book tells the story of the greatest shipbuilding program in American history. As war raged in Europe, and German submarines decimated merchant shipping in the Atlantic, the United States remained officially neutral. Well before the Japanese attack on Pearl Harbor forced America into war, however, President Roosevelt was looking for ways to aid Great Britain as she struggled to carry on alone against the seemingly invincible German forces. One solution was to build merchant ships for the British, an undertaking that enabled the Maritime Commission to begin planning for an expanded shipbuilding program once the United States entered the war. When the United States was no longer on the sidelines, German submarines began sinking U.S. merchant ships at a rate that far exceeded the shipyards' ability to replace them. The fate of the American allies and forces in Europe looked grim. The Maritime Commission responded to this crisis by expanding the nation's shipbuilding capacity again and again. By the end of the war U.S. shipyards working under the Maritime Commission's direction had built a total of 5,777 cargo ships, 2,708 of which were the "ugly duckling" Liberty ships built to a British design. *Ships for Victory* recounts how this was accomplished through a massive national effort.

Lane's story also includes accounts of Admirals Land and Vickery in Washington, of Henry Kaiser and the other industrialists who built new shipyards on both coasts for the Commission, of the women who learned how to cut and weld steel, and of the Louisiana African Americans who flocked to the Higgins shipyard in New Orleans to learn skills denied them in peacetime. Urgency encouraged innovation, and new techniques of pre-assembly and assembly-line production were developed to turn out ships in record time. Only a nation fully mobilized for war could have successfully mounted a heroic effort of this sort. The combination of improved anti-submarine warfare and rapidly increasing ship production kept supplies flowing to the allies and armed forces overseas. Ships were built and victory was achieved.

Anyone reading Lane's account of this historic achievement half a century after the event will want to know more about related topics that are not covered in his book. The story of the brave men who sailed these hastily-built ships in dangerous, submarine-infested seas was as much a peak experience for American seafarers as building the ships was for American shipbuilders. Readers interested in this story will enjoy John Bunker's *Heroes in Dungarees: The Story of the American Merchant Marine in World War II* (Naval Institute Press, 1995). The story of American shipbuilding and the American merchant marine after the war is, by contrast, a story of long-term decline rather than enduring achievement. As far as commercial operations were concerned, the United States became complacent after winning the war and failed to meet the challenges posed by its battered allies and vanquished enemies, who soon began building more efficient cargo ships and reclaiming a growing share of the world's trade. When the war ended, America's emergency shipyards were closed and the wartime techniques used to produce ships swiftly and inexpensively were abandoned, at least until U.S. competitors abroad began using them in their own commercial yards. A decade after the end of the war America's position as the dominant carrier of the world's trade was also diminishing, and today the U.S. flag has nearly disappeared from the world's commercial fleet engaged in international trade. It is a long, discouraging story of repeated failure to respond adequately to commercial challenges from abroad. Readers interested in knowing more about the attitudes and policies that guided the U.S. maritime industries in this period will find them described in Andrew Gibson and Arthur Donovan, *The Abandoned Ocean: A History of U.S. Maritime Policy* (Univ. of South Carolina Press, 2000).

Students of twentieth-century American history will also be curious about the relationship between the World War II emergency shipbuilding program and the similar program initiated in World War I. Lane does not treat this topic systematically, but he does make a number of illuminating comparisons (see pp.

312, 829–30). Safety considerations aside, the federal government's involvement in maritime affairs in the twentieth century has been overwhelmingly driven by military, foreign trade, and employment concerns, and not by a determination to foster a competitive maritime industry. The Maritime Commission that managed the emergency shipbuilding program in the Second World War was the direct descendent of the United States Shipping Board created in 1916. When the European great powers went to war in 1914, the foreign ships that had routinely brought imports to America and carried its exports overseas were redeployed to support their nations' war efforts and were no longer calling at U.S. ports. With the products of America's farms and factories rotting and rusting on its docks, President Wilson and Congress agreed to create and fund a new federal agency charged with building a fleet of cargo ships capable of carrying America's foreign trade to its markets abroad and of supporting the nation's armed forces when deployed overseas. The program Lane describes was thus the second great war-impelled shipbuilding program of the twentieth century. Twice as many ships were built to support the Second World War as were built for World War I; and the Second World War was the first time the United States had to support its armed forces in a two-front war fought on opposite sides of the world. There were, therefore, interesting differences as well as similarities in how the two programs were organized and managed, and surplus ships were disposed of in different ways when the wars had ended. Readers interested in further details will find useful the essays in Robert A. Kilmarx, ed., *America's Maritime Legacy: A History of the U.S. Merchant Marine and Shipbuilding Industry* (Westview Press, 1979).

The memory of the shipbuilding program Lane describes lives on in a number of artifacts as well as in the pages of his book. While almost all the Liberty ships have been scrapped, two have been lovingly preserved and continue to sail from their homeports in San Francisco and Baltimore. To go aboard them and see the breakbulk cargo-handling gear, the crew's quarters, the minimally equipped bridge, and the antique but still functioning boilers and triple-expansion engines is to travel back in time. A World War II Victory ship also still sails on excursions from the West Coast. In Washington, D.C., the Maritime Collections of the Smithsonian Institution's National Museum of American History have archived and preserved a large number of the ship plans discarded by the Maritime Administration, the successor agency to the Maritime Commission. These artifacts, together with a number of oral and visual histories of maritime activities in World War II, can help enliven and expand on the stories so comprehensively and authoritatively told in Lane's *Ships for Victory*.

Will we ever see such a shipbuilding program again? As unlikely as it seems, it is not impossible. America's armed forces are still committed to being prepared to simultaneously fight two major overseas wars. The Korean, Vietnam, and Gulf Wars have all demonstrated that fighting a land war overseas still requires massive sealift, even in the age of airpower and ballistic missiles. The challenges Admiral Land and the Maritime Commission faced when the United States entered World War II are thus not unlike challenges we could be forced to address again. The shipbuilding program Lane describes should not, therefore, be treated as a chapter out of a distant and buried past. The story is too good and its contemporary relevance is too great for that. However, if we are able to avoid another conflict like World War II, it is not likely there will ever be another war-impelled shipbuilding program like the one described in *Ships for Victory*. Let us hope that will be the case, and in the meantime, let us take the time to enjoy and learn from Lane's wonderful telling of one of the great stories of World War II.

ARTHUR DONOVAN

United States Merchant Marine Academy
January 2001

PREFACE TO THE ORIGINAL EDITION

WHILE commenting on the draft of this book, Admiral Land, Chairman of the U. S. Maritime Commission during World War II, recalled a remark he made to a committee of the Senate: "If you want fast ships, fast shipbuilding, fast women, or fast horses, you pay through the nose." That epitomizes one side of the story. The other side is expressed in the book's title: the shipbuilding herein described achieved its purpose in the victories of 1945.

A history composed so soon after the event has special advantages and disadvantages in its sources and in its perspective. Ours will be better understood if we recount how this book came to be written. In the spring of 1946 I was granted leave of absence by The Johns Hopkins University in order to accept employment under contract with the U. S. Maritime Commission as Historical Officer to prepare a narrative account of the wartime activities of the Commission. President Roosevelt had directed that "soon after the war each agency should have ready a final report that will sum up both what was accomplished and how the job was done." The suggestion that I undertake this work for the Maritime Commission came from Clarence H. Danhof, then attached to the Committee on Records of War Administration which had been set up in the Bureau of the Budget to assist the execution of the President's directive. Although this book has been brought to completion independently, by private efforts, it was planned as part of the series, Historical Reports on War Administration, fostered by that committee.

The decision of the Maritime Commission to employ me was due to Commissioner John Michael Carmody, and I am glad of this opportunity to pay tribute to his appreciation of the need of a historian, in order to meet the standards of his profession, not to be limited to recording the views of administrative superiors, but to be free to express his own point of view. Through the goodwill of Mr. Carmody, this understanding was placed in the record at the time of my engagement and the Maritime Commission placed no restrictions on our investigation or writing.

On June 24, 1947, the Maritime Commission informed me that my contract of employment would be terminated at the close of that fiscal year because the appropriation to the Commission was being cut by Congress. During the previous fifteen months, with the aid of the staff provided by the Commission, we had completed about half of the research on which this volume is based. Hating to see so much hard work go down the drain, I undertook as a private individual in July 1947, at the same time that I returned to The Johns Hopkins University, to carry to completion the volume now published. I received essential assistance in this project by a grant-in-aid from the Rockefeller Foundation. This enabled Blanche D. Coll and Gerald J. Fischer to give many months to this work in 1947-1948, and The Johns Hopkins University gave me a partial leave of absence in the spring of 1948. It is a pleasure to have this opportunity to record my sincere gratitude for the aid received from the Rockefeller Foundation and The Johns Hopkins University.

In continuing the work I also received essential assistance from the Maritime Commission in that the Commission continued to give me unrestricted access to their files and arranged for Joseph T. Reynolds to complete the drafting of way charts for each major yard and for the compilation of the statistical record which Gerald J. Fischer had organized and supervised. This *Statistical Summary of Shipbuilding* was put in final form for publication by Russell M. Brown, Florence C. Bryant, and Violet B. Torney and issued as Historical Reports of War Administration, United States Maritime Commission, No. 2. These projects and my own work were very much aided by the friendly interest which the Commission, especially Commissioner Richard Parkhurst, took in the completion of the historical record.

Most members of the Commission and its staff who appear as principals in this volume had left the Commission before we began work on its history, but many of them made their knowledge available to us through frequent interviews. It is a matter of keen regret that I was not able to talk to Admiral Vickery before his death, and spoke only briefly with Captain Macauley before he left Washington. The other three wartime Commissioners, E. S. Land, T. M. Woodward, and J. M. Carmody were available in Washington; and as we worked we frequently went to them for information, and also to many others who had taken part in the events we were recording. I am especially indebted to Admiral Land for the quickness and keenness with which he read and commented on all the chapters. First to last, the following also helped us in our research, and many of them read chapters in various stages of drafting. To them all we wish to express our appreciation of their assistance: Eugene J. Ackerson, Robert Earle Anderson, Joseph F. Barnes, James L. Bates, W. H. Blakeman, S. M. Buffett, R. J. Carroll, B. F. Carter, Henry Z. Carter, Percy Chubb, Wesley Clark, John G. Conkey, Captain Granville Conway, Henry L. Deimel, Edward Scott Dillon, Margaret E. Dowden, P. J. Duff, William G. Esmond, Norman R. Farmer, E. W. Frederick, Ward B. Freeman, J. T. Gallagher, E. Stanley Glines, J. E. P. Grant, J. F. Harrell, I. M. Heine, Harvey H. Hile, A. C. Himmler, Leroy G. Kell, J. G. Kendrick, James Fitzpatrick, William U. Kirsch, Harvey Klemmer, Irene Long, F. L. Lynch, Mary E. Mangan, C. D. Marshall, E. E. Martinsky, Clyde L. Miller, Edison G. Montgomery, Huntington T. Morse, Magdelene Napier, L. P. Nickell, Alan Osbourne, Paul D. Page, Jr., John R. Paull, Carl Carroll Perry, John M. Quinn, T. H. Reavis, Daniel S. Ring, L. R. Sanford, Samuel Duvall Schell, O. H. Schulze, T. E. Shaw, W. L. Slattery, W. E. Spofford, Elizabeth Sprott, T. E. Stakem, Harold E. Steffes, H. P. Strople, Arthur L. Swift, III, Edward J. Tracy, Francis H. Van Riper, Eleanor H. Van Valey, C. D. Vassar, John L. Vassar, John Vasta, O. B. Vogel, A. C. Waller, W. Walsh, Ivan J. Wanless, William A. Weber, E. E. White, A. J. Williams, J. L. Wilson, L. M. Wuertele, all employees of the Commission during the war; and also to many not connected with the Maritime Commission: to Robert G. Albion, Jennie Pope Albion, Don Q. Crowther, Horace B. Drury, Edward M. Gordon, Frank

A. Hodge, Albert H. Huntington, R. M. Leighton, Daniel M. Mack-Forlist, Jean Maxwell, Maurice Nicholls, and H. Gerrish Smith. The nature of their contributions is indicated as far as possible in the footnotes; and far more than footnotes could indicate we are indebted to the aid received from those in charge of the Maritime Commission's records, especially: Barbara J. Boardman, R. A. Chandler, Stephen C. Manning, W. L. Nesbitt, Jr., William J. Turner, George A. Viehmann, and Annie S. Wright.

The photographs are from the files of the Maritime Commission and its successor agency, the Maritime Administration in the U. S. Department of Commerce, and were arranged by Joseph T. Reynolds.

The statements in this volume have not been endorsed by the Maritime Commission or Maritime Administration; although much help was received from the Commission and from persons connected with it, neither they nor the Commission were asked to approve the views here expressed.

I welcome this opportunity to thank also those who worked with me in the Office of the Historian, U. S. Maritime Commission: Pearl M. Amidon, Hymen E. Cohen, Chester L. Guthrie, Sarah C. Kieffer, Helen E. Knuth, Nicholas R. Lederle, Allen M. Ross, Violet B. Torney, Marie D. Werner, and John Worth. Their contributions have also been indicated, often inadequately, in the footnotes, and even special mention here cannot fully express our appreciation of the help of Sarah C. Kieffer, not in secretarial labors alone, but in dependability and friendliness.

Mr. Reynolds and my other collaborators whose names appear on the title page were also part of the staff working with me in the Maritime Commission. They have joined me in acting as private individuals to bring the volume to completion. Many chapters are based on the research of my associates and the essays in which they presented their findings In that sense, chapters 8, 9, 13, and much of 20 are Miss Coll's; chapters 10, 11 in part, and the last section of 7 are Mr. Fischer's; chapters 3, 16, and 17 are Mr. Tyler's. Their essays—which are cited in the notes and like other works cited are entered in the index—are two or three times as long, however, as the corresponding sections of this volume. In condensing, I revised freely, sought additional information on some points, and am responsible for this text; but many passages in those chapters are in their words. Moreover, their part in making this book has not been limited to those chapters. I have depended on Mr. Fischer, who was head of the Statistical Analysis Branch in the Production Division of the Maritime Commission, 1942-1944, for knowledge of programs and problems of ship production, as well as for statistical analysis, and on Miss Coll for research on many points as well as analysis of problems connected with labor. All these associates went over the general problems of organizing the book with me many times in 1946-1950, read my manuscript in whole or in very large part, shared in the proof reading, and made helpful suggestions about other chapters besides those for which their research provided the foundations.

I wish finally to express my thanks to Lilly Lavarello, Anne G. Draper, Katherine Van Eerde, and Jonathan P. Lane, for their aid with the copy, proof, and index, and to my wife for her large share in bringing another joint venture, full of the unexpected, happily to completion.

FREDERIC C. LANE

The Johns Hopkins University
March 31, 1951

FIGURES

FIGURE		PAGE
1.	Total Tonnages of Principal Types	5
2.	Shipbuilding in the U. S. During Two World Wars	7
3.	Location of Shipbuilding, 1941-1945	9
4.	Yards Building Large Merchant Ships in 1940	35
5.	Programs Formulated in 1940 and 1941	41
6.	Merchant Shipyards of 1941	59
7.	Losses vs. Construction, 1940–1942	64
8.	Shipyards in the Maritime Commission Program, 1942-1944	142
9.	Location of Maritime Commission Shipbuilding Facilities	153
10.	Construction Time of Liberty Ships, Average of All Yards	174
11.	Construction Time of Liberty Ships in Selected Yards	175
12.	Actual Speed of Production Compared with Scheduled Speed	177
13.	Programs Formulated in January–July 1942	181
14.	Losses vs. Construction, December 1941–June 1944	203
15.	Time on the Ways of Liberty Ships, in Selected Yards	208
16.	Manhours Per Ship for Liberty Ships, By Rounds	209
17.	Timing of Fabrication, Assembly, and Erection of Steel	211
18.	Layouts of Shipyards for Multiple Production	221
19.	Productivity in the Construction of Liberty Ships	231
20.	Manhours Per Ship for Vessels Built in New Yards, Selected Types	234
21.	Weekly Earnings in Shipbuilding and in Other Industries	245
22.	Workers in Maritime Commission Shipyards	251
23.	Shipbuilding's Share of Steel Shipments, 1940 and 1943	312
24.	Steel-Plate Requirements, Allocations, and Shipments	334
25.	Programs Formulated in June–December 1942	336

FIGURES

Figure		Page
26.	New Shipyard Facilities Authorized by the Maritime Commission	341
27.	Actual Inventories Compared with Minimum Standards	345
28.	Steel Plate Being Used in Ship Construction	348
29.	Inventory Ratios of Steel Plate	351
30.	The Maritime Commission's Production Division	367
31.	Straight-Time Hourly Earnings	425
32.	Gross Hourly Earnings	426
33.	Admiral Vickery's Chart Room in 1943	461
34.	Managerial Groups Building Ships for the Maritime Commission, 1941–1945	470
35.	Railroads and Facilities at South Portland, Maine	507
36.	Crack Arrestors	549
37.	Programs Formulated in 1943	597
38.	Class II Shipyards That Delivered Coastal, Concrete, and Military Vessels	630
39.	Types of Ships in Construction on the Ways in Class I Yards	639
40.	Deliveries Compared with Shipsworth Produced	641
41.	Indexes of Production, Manhours, and Productivity	643
42.	Labor Productivity in Maritime Commission Shipyards	644
43.	Effects of Changes in Types on Productivity of Labor	645
44.	Programs Formulated in 1944	651
45.	Hires vs. Separations in Maritime Commission Shipyards	655
46.	Hours Worked Weekly Per Employee	660
47.	The Maritime Commission and War Shipping Administration	764
48.	Kaiser Shipyards and Some Corporate Affiliates	809
49.	Total Direct Cost of Principal Types	820
50.	Distribution of Total Cost of Maritime Commission Shipbuilding	823

TABLES

Table		Page
1.	Vessels Delivered in U. S. Maritime Commission Program, 1939–1945	4
2.	Characteristics of Some Principal Types Built by Maritime Commission	28
3.	Percentages of Working Time Lost Due to Strikes, 1941–1945	305
4.	Strikes and Man–Days Lost in Strikes in Merchant Shipyards	305
5.	New Facilities Authorized for Shipyards and for Production of Ship Components	397
6.	Main Propulsion Turbine Production for Maritime Commission Vessels, 1944	399
7.	Average Hours and Earnings for Selected Maritime Commission Shipyards, By Areas, 1941–1945	419
8.	Actual Way Time and Contract Estimates, Awards of December 1942	458
9.	Anticipated Cost and Actual Cost of Selected Shipyards	472
10.	Actual Manhours and Contract Estimates, Liberty Ships	475
11.	Concrete Barge Compared to Liberty Ship in Cost and Capacity	634
12.	Way Time and Number of Employees in Selected Yards, January–May 1945	683
13.	Deliveries, Building Berths, and Employees by Regions	690
14.	Construction Fund of U. S. Maritime Commission	717
15.	Types of Activity of Administrative Personnel of the Maritime Commission and War Shipping Administration, Fiscal Years 1940–1947	755
16.	Personnel in Organizational Units of Maritime Commission and War Shipping Administration at End of Fiscal Years 1940–1946	758

TABLE		PAGE
17.	Results of Renegotiation to March 31, 1947	802
18.	Profits Before Taxes on Various Enterprises	806
19.	Profits on Selective Price Contracts	817
20.	Average Costs of Selected Types	819
21.	Total Cost of Maritime Commission Shipbuilding	822
22.	Cost of Liberty Ships	826
23.	Production, Employment, Manhours, and Productivity, 1941–1945	828

PLATES

PLATE		FACING PAGE
I.	Admiral Land. Side Launching of a C1. C2 Cargo Vessel .	42
II.	Standard Types: C3 Cargo, C1 Cargo, T3 Tanker . . .	43
III.	British and American Libertys. H. J. Kaiser and W. S. Newell	74
IV.	Keel Laying. Launching. Sponsor, Marian Anderson . .	75
V.	Maritime Commissioners Woodward, Land, Macauley, Carmody, and Vickery. Assembly Platens behind the Ways at Richmond, California	106
VI.	J. E. Schmeltzer. Workers in the Mold Loft. Steel Storage Yard	107
VII.	Laying a Keel. Half-finished Hull. Marinship at dusk . .	138
VIII.	Oregon Ship at Night. Lifting a Cofferdam. Welding on a Stern Casting	139
IX.	Workers: Riveters, Rivet Passer, Welders, Shapers . . .	234
X.	Aerial Views of Shipyards: Jones-Brunswick and Marinship	235
XI.	Workers: Blacksmiths and Apprentice Welders	266
XII.	Training: Shipfitters, Lady Welders. Changing Shifts at North Carolina Ship	267
XIII.	Preassemblies. Lifting a Deckhouse. T2 Tanker	330
XIV.	Preassemblies at Houston and Bethlehem-Fairfield. Automatic Burning Machine	331
XV.	Division Directors Bates, Anderson, Rockwell, and MacLean. Launching an Escort Carrier	362
XVI.	T2 Tanker. Liberty Tanker. Liberty Collier	363
XVII.	C2 as Naval Auxiliary. C2 Cargo Ship. Subassemblies at Oregon Ship	490
XVIII.	Admiral Vickery. Transferring a double-bottom. Kaiser-Swan Island in Snow	491
XIX.	Aerial Views of Cushing Point, South Portland, Maine . .	522

PLATES

Plate		Facing Page
XX.	Fractured Ships: *Esso Manhattan, Schenectady, Valery Chkalov*	523
XXI.	Victory Ships: In Outfitting, Cargo Carrier, Transport	586
XXII.	Military Types: LST and BB3. Aerial View of Kaiser-Vancouver	587
XXIII.	Military Types: P2 Transport, C4 Transport, Frigate	618
XXIV.	Minor Types: Coastal Cargo, Concrete Ship, Ocean-going Tug	619
XXV.	Regional Directors Spofford, McInnis, Sanford, and Flesher. Aerial View of Calship	746
XXVI.	Shipyard Scenes	747
XXVII.	Navy Attack Cargo Ship. Propeller. Yard Scene. Concrete Barge	778
XXVIII.	Workers Leaving Bethlehem-Fairfield. A Liberty Rides at Anchor	779

SHIPS FOR VICTORY

Chapter 1

THE COMMISSION AND THE SHIPBUILDING INDUSTRY

I. A General View of Wartime Shipbuilding

THE COMBINATION of government regulation and private enterprise through which American industry operated during World War II was a strikingly decentralized kind of administered economy. In shipbuilding all the major problems of production—management, labor supply, capital, and materials—were tackled directly by both private corporations and government agencies. Both acted to draw managerial talent from other fields of activity. Both placed orders for materials and components. Both campaigned to add a million to the labor force in the shipyards, and by joint action they created the conditions that made the workers want to come. The distribution of materials, the financing of operations, and the application of efficient engineering became problems both for corporation managers and government officials. Consequently the history of shipbuilding in World War II is at once a history of private enterprise and of governmental activity.

Plans were made on a national scale and were carried out in such a way as to strengthen rather than weaken private corporations. This policy had the solid foundation of being in accord with the prevailing ideal of free enterprise and of utilizing the momentum of going concerns and the incentives of the profit system. Under the strain of war, it proved to have both advantages and difficulties. On the whole the advantages are most evident in the first years of the war, in the outpouring of energy, the rapid technical achievement, and the soaring output; the difficulties are more apparent in the later years, after victory seemed assured and it became more necessary to count the cost.

The terms on which government and private enterprise worked together were shaped by the character of the shipbuilding industry. In time of peace shipbuilding was one of those parts of the American economy in which production was in the hands of a small number of companies having a substantial community of interest. The number of firms was restricted by the large overhead required. Large-scale shipbuilding called for an extensive area of valuable waterfront; for heavy machinery such as furnaces, rollers, and cranes to shape and move the steel; for a specially equipped machine shop; for many warehouses and storage areas; and for a spacious mold loft in which the templates used as patterns in shaping steel were laid out with mathematical accuracy. Above all, shipbuilding required an effective engineering organization containing a small army of draftsmen and headed by engineers of long training and experience. The technical leaders of the industry, and to a large extent its business leaders also, were naval architects and marine engineers. This was a small and select group. They had the standards, fellow feeling, and pride of skill that are usual in a highly trained profession.

This "shipbuilding fraternity," as Admiral Land often called it, was accustomed to working together for many purposes. It cooperated, along with ship operators and underwriters, to support the American Bureau of Shipping, which set standards for workmanship, strength of materials, and seaworthiness for merchant ships. Its interests were presented to the public and Congress by the National Council of American Shipbuilders.

The shipbuilding industry was also accustomed to working with the government—on the one hand seeking support from the government, and on the other accepting various kinds of supervision. To assure safety at sea, all merchant ships were subject to inspection by the Bureau of Marine Inspection and Navigation in the Department of Commerce. The prosperity of shipyards had for a long time depended on government subsidies and government orders, which entailed additional government inspection.

Closer integration of the industry with government agencies in a nationally planned program began a year before Pearl Harbor and was intensified after the declaration of war. Full wartime integration was carried out through two sets of controls. Like the rest of American industry, shipbuilding was subject to the rules and regulations issued by special agencies such as the War Production Board and the War Manpower Commission created to act only during the war. More direct and firm, however, were the controls resulting from the fact that the government became the industry's only customer, a customer ready to buy all that the industry could produce, and more. Government orders fixed the production goals of the shipyards, and government contracts formulated rules stipulating how corporations should operate.

These procurement activities of the government were administered for shipbuilding by two agencies, the Navy Department and the United States Maritime Commission. Moreover, they became the agencies through which many general wartime controls were applied to shipbuilding. Subject to the more general war plans and production goals defined by the President and his chief advisers, the Navy and the Maritime Commission each directed and coordinated a section of the shipbuilding industry. Consequently the history of American shipbuilding in World War II falls into two parts.

For the shipbuilders under the Maritime Commission the first main task was to build merchant ships faster than they were being sunk. Without merchant shipping, the United States could not make its will felt beyond this continent either to bring help to its friends or to carry war to its enemies. In 1941 and 1942 shipping losses exceeded new construction. The Axis powers were not only winning the war on land, they were winning the war at sea, for if they continued to sink more tonnage than was built they would succeed in cutting their enemies' lines of supply. The perilous and feverish race between sinkings and construction was not definitely won until 1943.

After triumph over the submarine menace, the need for cargo ships continued to be intense, for again as in World War I a " bridge of ships " was essential to the American Army. The shipyards under the Maritime Commission provided the bridge, delivering in 1939-1945 roughly 50 million deadweight tons of large cargo carriers and tankers.[1] In 1943 alone they built 18 million tons, more than the total merchant fleet under the flag of the United States in 1939.[2] See Figure 1.

Merchant ships were not the only kind built under the Maritime

---

[1] Rounded off totals for "Major Types" from Table B-3 in Gerald J. Fischer, *Statistical Summary of Shipbuilding under the U. S. Maritime Commission during World War II* [Historical Reports of War Administration, U. S. Maritime Commission, No. 2] (Government Printing Office, 1949).

[2] *Ibid.* The merchant fleet of the United States in 1939 totaled 12 million tons; that of the world, 81 million.—U. S. Maritime Commission, *Report to Congress* for the Fiscal Year ended June 30, 1948 (Government Printing Office, 1949), p. 64.

Commission. Since more transports and naval auxiliaries were needed than could be built in yards under contract to the Navy, they were built in Maritime Commission yards. In the fall of 1944 most of its shipyards on the West coast were building vessels classified by the Commission as "military" types. Some of these vessels classified as "military" were designed purely for military use—for example, the tank-carrying ships for effecting landings on

TABLE 1

VESSELS DELIVERED IN U. S. MARITIME COMMISSION PROGRAM
1939-1945

	Number	Tonnage in thousands of tons*		
		Displacement (Light)	Deadweight	Gross
Grand Total	5,777	22,218	56,292	39,920
By Type of Contract				
Maritime Commission	5,601	21,478	54,102	38,490
Private	111	539	1,581	997
Foreign	65	201	608	433
By Type of Ship				
Standard Cargo†	541	2,496	5,349	3,834
Emergency Cargo (Liberty)	2,708	9,412	29,182	19,447
Victory Cargo	414	1,851	4,492	3,151
Tankers	705	3,707	11,365	7,061
Minor Types	727	1,562	2,601	1,980
Military Types	682	3,188	3,303	4,452

* Displacement tonnage (the weight of the ship) and deadweight tons (the weight of the cargo) are both measures of weight based on the long ton of 2,240 pounds. Gross tonnage, in contrast, is a measure of the volume of the closed-in space on the ship. It is the cubic carrying capacity measured in cubic feet (100 cubic feet to the ton). An average freighter carrying 10,000 deadweight tons has a displacement, light, of about 4,000 tons, and cubic capacity of about 6,000 gross tons. Passenger liners have large gross tonnage compared to their displacement or deadweight.
† Includes passenger-and-cargo types.
Source: Fischer, *Statistical Summary*, Tables B-3 and B-4. Figures rounded off so that columns may not add.

enemy beaches (the LST's) or the aircraft carrier escorts, commonly known as "baby flattops." But most of the vessels built by the Maritime Commission and classified by it as "military" were transports, attack cargo ships, or oilers. Their designs were basically those of merchant vessels modified to meet military needs. Although these ships could be converted into merchant vessels and were counted as merchant vessels by the American Bureau of Shipping (Figure 2), they were built for delivery to the Army or Navy and were considered military types by the Commission. In terms of cost, military types constituted 23 per cent of the Commission's shipbuilding for 1939-1945 (Figure 49).

TOTAL TONNAGES OF PRINCIPAL TYPES
Of Vessels Delivered In Maritime Commission Program, 1939-1945

▩ MILITARY ▦ MINOR ▢ TANKER ▦ VICTORY CARGO ▩ LIBERTY CARGO ▨ STANDARD CARGO

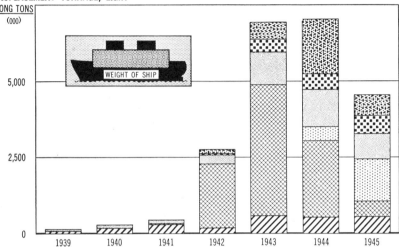

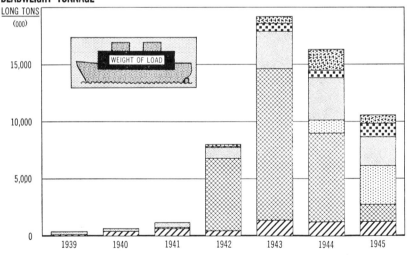

FIGURE 1. Source: Fischer, *Statistical Summary*, Tables B-3 and B-4. Vessels built in the United States for foreign governments and all vessels over 2,000 gross tons constructed in the United States for private concerns, as well as deliveries to the U. S. Maritime Commission, are included because all this construction was programmed by the Maritime Commission.

6 THE COMMISSION AND THE INDUSTRY

The building under the Maritime Commission taken as a whole was therefore a complex of construction of many different types. In analyzing the importance of these different types, it is necessary to distinguish between deadweight tonnage and displacement tonnage. Deadweight tonnage is the weight of the cargo that a ship can carry without overloading. Displacement tonnage, light, is the weight of the ship itself, before it has been loaded. The addition of the deadweight tonnage to the displacement tonnage, light, gives the total displacement, namely the weight of water displaced by the fully loaded ship. For the study of industrial activity, displacement tonnage, light, is the best indication of a ship's size. Where " displacement tons " is used hereafter in this study, the reference is to displacement tons, light. Analysis of deliveries in terms of these displacement tons is the best way of indicating by tonnage the amount of productive effort expended on various types of ships. In these terms, the much-talked-of Liberty ship, the dry cargo vessel which earned the title of " workhorse of the fleet " and was built in far larger numbers than any other type, formed only 42 per cent of the total output (Table 1).

Production goals were set in terms of deadweight tons, however, because the construction programs were directly related to the need for carrying capacity. For many purposes it is necessary to describe both the program and the output in terms of deadweight tons. In these terms the Liberty ship dominates the picture until it was gradually replaced in 1944 and 1945 by a new type of emergency cargo carrier, the Victory ship (Figure 1).

The totals for the whole Maritime Commission program are impressive no matter how they are measured. Under Maritime Commission contracts 5,601 vessels were delivered, and additional deliveries of merchant vessels in 1939-1945 for private companies and foreign governments bring the grand total up to 5,777 vessels. Their construction consumed a total of about 25 million tons of carbon steel and engaged the labor of 640,300 workers at the peak of employment.[3]

The magnitude of the shipbuilding task under the Maritime Commission can also be seen in perspective by comparison with the only other similar efforts in the rapid construction of ships. British production of merchant ships during 1941-1945 measured

---

[3] Fischer, *Statistical Summary*, Tables B-3, B-4, and G-2, and estimate by Fischer for steel consumption on basis of sources he cites in *ibid.*, Section F, notes.

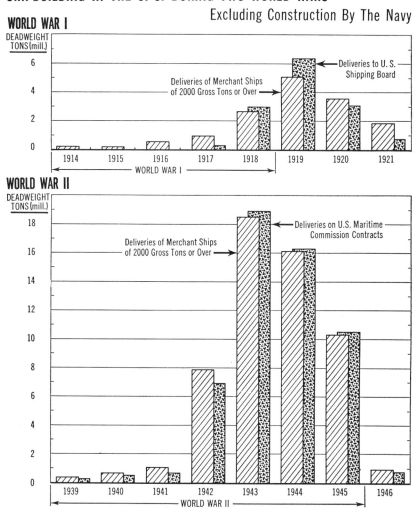

FIGURE 2. Sources: U. S. Shipping Board, *Sixth Annual Report,* June 30, 1922, pp. 266-67; Fischer, *Statistical Summary,* Table B-3; Vessel Cargo Data Branch of MC (for deliveries to MC in 1946); and American Bureau of Shipping, *Bulletin,* January 1949 (for deliveries of merchant ships).

in deadweight tons was less than a sixth that of the United States building under the Maritime Commission, for the two governments agreed that the yards in Britain would concentrate on combat vessels,[4] and although the British did build merchant vessels also they were largely special types for war transportation.[5] The one instance in America of similar emergency production was in World War I. During the five years 1916-1920 the output of cargo ships under the Shipping Board was one-third of that under the Maritime Commission in the five years 1941-1945 when measured in number of ships, and less than one-fourth when measured in deadweight tonnage. Shipbuilding also expanded more rapidly and more promptly in World War II; in World War I the peak of production came only after the war was over, and in World War II the program was well underway when war was declared and reached its peak in the fall of 1943 (Figure 2). Both 1943 and 1944 were record-breaking years in which 1,849 and 1,786 vessels were delivered, with a total displacement tonnage of 6.9 and 7.0 million tons, respectively (Figure 1).

Comparison with the shipbuilding being done at the same time for the Navy is difficult because ordnance and armament are large items in the total cost for the Navy, and the greater complexity of warships makes a comparison in terms of displacement tons inadequate. The final products are hardly more comparable than a barracks and a fortress. The direct cost of ships delivered in 1941-1945 was about $18 billion for the Navy, exclusive of ordnance, about $13 billion for the Maritime Commission.

A striking contrast between the Navy expenditures and those of the Maritime Commission is in the way they were distributed by shipbuilding areas (Figure 3). Actually the money spent flowed through the shipyards, which were the assembly points, back all over the country to the steel mills and to manufacturers who made engines, boilers, cables, and innumerable other components. But the records compiled enable us to follow the money only as far as the shipyards. The biggest concentration of Navy building was in yards on rivers and estuaries of the Northeast coast which were the traditional centers of shipbuilding—Boston

[4] Letter of Franklin D. Roosevelt to Winston Churchill published in *The Times* (London), Aug. 4, 1943, p. 8. Cf. below, chaps. 17 and 23.

[5] Sir Amos L. Ayre, " Merchant Shipbuilding during the War," in Institution of Naval Architects, *Transactions*, vol. 87 (London, 1945), pp. 3-28.

Bay, New York Bay, the Delaware River, and the Chesapeake Bay. The Maritime Commission expenditures were heaviest in yards around San Francisco Bay, and near the mouth of the Columbia River System, and were extensive also in southern California, and scattered points along the South Atlantic coast and the Gulf. But the Maritime Commission also did substantial building on the Delaware and the Chesapeake; and the Navy built much in San Francisco Bay, in Puget Sound, and on the Texas coast. Conse-

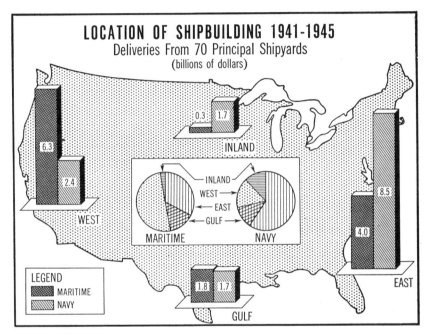

FIGURE 3. Source: Fischer, *Statistical Summary*, Table H-7.

quently the total expenditures for ship construction by both Navy and Maritime in San Francisco Bay amounted to $3.9 billion, as against $2.6 billion on the Delaware, a similar amount on the Columbia, and $2.5 billion on the Chesapeake. These four areas stand out above the others.[6]

[6] All these previous figures except for sinkings are from Fischer, *Statistical Summary*. See especially note to Table H-7. For sinkings see William Chaikin and Charles H. Coleman, *Shipbuilding Policies of the War Production Board, January 1942-November 1945* [Historical Reports on War Administration: War Production Board Special Study No. 26] (Civilian Production Administration, Bureau of

Of the total munitions production of the United States, July 1, 1940 to August 31, 1945, ships were 22.5 per cent. They formed a larger percentage than any other single item except aircraft as measured by the War Production Board in "standard 1945 munitions dollars."[7] The ships produced under the Maritime Commission were 38.2 per cent of the total value of ship construction and conversions, and 7.5 per cent of the total munitions production.[8]

II. THE MARITIME COMMISSION BEFORE THE WAR

If Congress or the President had set up a new agency for the purpose of executing a shipbuilding program of the magnitude just described, they would probably have centered power over the agency in one man. The wartime agencies were all so organized. Even when they were called a Board or Commission, such as the War Production Board, or the War Manpower Commission, the Executive Orders by which they were created gave all the power to the Chairman.[1] Practically the situation was much the same in the Maritime Commission during the war but in law it was very different—each of the five Commissioners had an equal vote. By act of Congress power was lodged not with the Chairman but with the Commission; for the Commission had been established to perform very different functions from that which became its main task during the war years.[2]

The U. S. Maritime Commission was created by the Merchant Marine Act of 1936 in order to modernize the American merchant marine. It was expected to achieve this aim by a judicious distribution of subsidies among applicant shipbuilding and ship-

Demobilization, Multigraphed, Apr. 15, 1947), p. 7; Samuel Eliot Morison, *History of United States Naval Operations in World War II*, Vol. I, *The Battle of the Atlantic* (Boston, 1947), especially pp. 404, 410-414; *The United States at War: Development and Administration of the War Program by the Federal Government* [Historical Reports on War Administration, Bureau of the Budget, No. 1] (War Records Section, Bureau of the Budget, Government Printing Office, n. d.), p. 137.

[7] Bureau of Demobilization, Civilian Production Administration, *Industrial Mobilization for War, History of the War Production Board and Predecessor Agencies 1940-1945* [Historical Reports of War Administration, War Production Board General Study No. 1] (Government Printing Office, 1947), I, 962.

[8] Percentages computed by Gerald J. Fischer from U. S. Civilian Production Administration, *The Production Statement*.

[1] *The United States at War*, pp. 104-06.

[2] Merchant Marine Act, 1936, 49 Stat. 1985, Title II, Sec. 201.

operating companies. These subsidies to American private enterprise were intended to make up for the differential between their costs and the costs of their European competitors. As these differentials in costs were of two kinds, so were the subsidies. Companies operating passenger ships or freighters could apply for a subsidy to compensate for their higher costs and enable them to operate in competition with foreign ships sailing the same routes. The Commission was empowered to grant such subsidies if it found that the route served was essential and if the operating company had an acceptable plan for replacing its old ships with modern ones. The cost of the new ships would be higher if they were built in American yards than if built in foreign yards, and this differential was the basis for the second kind of subsidy which the Commission was empowered to grant, a construction subsidy. If an operating company was ready to make a contract with a shipbuilding company for a ship the design of which was approved by the Commission and the Navy, the Commission was empowered to pay the builder a subsidy equal to the difference between his costs and foreign costs of building.[3] The conception of those who drafted the Merchant Marine Act of 1936 was that the Maritime Commission would be a quasi-judicial body, for it would pass on applications for differential subsidies.

It had judicial functions too in its power to review the agreements on rates made by the various shipping pools and in this respect could be compared to the Interstate Commerce Commission.[4] It was also instructed to operate schools for merchant seamen and officers, and was here enjoined to take the initiative. And it was expected to accomplish its main function—the replacement of over-age ships by new ones of approved designs—by acting on the applications for construction and operating subsidies submitted by interested companies. Passing judgment on financial questions of this kind and laying down general regulations were functions which it was natural to entrust to a Commission. Moreover, shipping subsidies had a malodorous reputation, spiced by recent scandals, so that the burden of granting future subsidies seemed more appropriately placed not on one man but on a bipartisan board of five.

From the very first, however, the Chairman of the Commission

[3] *Ibid.*, Titles I, V and VI.
[4] Statements by ex-Commissioner Thomas M. Woodward to Frederic C. Lane, June 25, 1946, Notes in MC, Historian's Collection, Research file 105.

assumed an outstanding role which made him more than first among equals. Joseph P. Kennedy was made the first Chairman with the general understanding that he would give personal attention to winding up the mail subsidies which were considered a political scandal. This required action very different from judicial procedure. When Mr. Kennedy resigned in February 1938, Rear Admiral (later Vice Admiral) Emory Scott Land, USN (Retired), already a member of the Commission, became its Chairman. Under his leadership the Commission and its staff set to work to get modern freighters built for the American merchant marine without waiting for applications from the industry for construction subsidies. The Commission, not the private companies, took the initiative; and consequently it functioned more like an operative agency, less like a judicial agency, than had been contemplated in the Merchant Marine Act of 1936. The Commission drew plans for modern fast cargo ships, invited bids from shipyards, played newcomers against old-timers to beat down prices, and then awarded contracts—all without any assurance of being able to sell the ships to operators. This procedure was authorized as a last resort by Title VII of the Act of 1936, which required that such construction directly by the Commission be subject to the express approval of the President.[5] In general, Commissions are thought of as instruments of Congress in contrast to Departments which are instruments of the President, but the reliance of the Maritime Commission on Title VII made it in large measure an agency of the Executive.

The active or operating role assumed by the Commission was one factor which tended to cause more power to concentrate in the Chairman. Another reason for the dominating position which Admiral Land held as the Commission's chairman from 1938 to 1946 was the close personal relationship between him and President Roosevelt. Their friendship had begun in the Navy Department some twenty-five years earlier when Roosevelt was Assistant Secretary from 1913 to 1920. During six of those years, "Jerry" Land was a Lieutenant or Lieutenant Commander in the Bureau of Construction and Repair. Their friendship continued in the twenties when both were associated with the Daniel Guggenheim Fund for the Promotion of Aeronautics, and was renewed when

---

[5] U. S. Maritime Commission, *Report to Congress* for the Period ended October 25, 1939 (Government Printing Office, 1940), pp. 4-5, and below, chap. 4.

President Roosevelt in March 1933 nominated Rear Admiral Land to be Chief Constructor of the Navy and Chief of the Bureau of Construction and Repair.[6] Since Admiral Land was a naval constructor, not a line officer, this post placed him at the top of his profession. He was detached from it and at his own request placed on the retired list in April 1937 when he was then only fifty-eight years old and as agile and tireless as ever. His "retirement" caused some wonder, but its meaning became clear when President Roosevelt appointed him that same month to the Maritime Commission and then, sending Mr. Kennedy to London as Ambassador, made Admiral Land the Chairman of the Commission.

Admiral Land's personality contained many qualities likely to appeal to Franklin Roosevelt and qualities important in explaining the performance of the Commission under his chairmanship. He set a very high value on personal loyalty—loyalty to one's superior, to friends, and to subordinates who were loyal in turn. He worked, moved, and thought with startling rapidity. In his office he went through a host of papers, dictating or revising memoranda and leaving his characteristic " L " on innumerable reports. Even across a desk, however, his cold, bold eyes and quick sallies conjured away the wrapping of bureaucracy and recalled instead the dangerous swordplay of Viking ancestors. A flair for salty phrases led him occasionally to make impolitic pungent remarks in after-dinner speeches, but shone to his advantage before Congressional committees where his footwork was just as agile as it had been on the football field in 1900 when he went through the Army line on a trick play to score the Navy's winning touchdown.[7] He was not inclined to belittle in the least his own importance, but he knew how to disarm critics by down-to-earth disclaimers of infallibility.

" I got my start in life by working from sunrise to sunset, with my hands, for $10.00 a month and keep," [8] asserted Admiral Land frequently, insisting he was no " brass hat." From schooling in

[6] Memorandum by Land to Frederic C. Lane, June 23, 1948, in MC, Historian's Collection, Research file 118.1.1.

[7] This biographical sketch is based mainly on the " Biography " prepared Sept. 30, 1944, by the MC, Division of Public Information, and on Donald E. Keyhoe, "You Can't Sink This Admiral," *American Magazine*, Mar. 1940, pp. 21, 68, whose version of the Army-Navy game of 1900 is even more dramatic than that in *The Sun*, Baltimore, Monday morning, Dec. 3, 1900, which I have followed.

[8] E. g., Senate, *Maritime Commission Nominations, Hearings*, before a Subcommittee of the Committee on Commerce, 75th Cong., 1st sess., Apr. 10, 1937.

Canon City, Colorado, he went to Wyoming as a cowpuncher and student at the university, and from Wyoming was appointed to the Naval Academy. When he graduated in 1902 he was tops in athletics and sixth in his studies. After a post-graduate course in naval architecture at Massachusetts Institute of Technology, service with Admiral Sims in London in 1918, many years in the Construction Corps of the Navy, a brief period of work in aeronautics (he took up flying when fifty and qualified for a pilot's license), and the four years as chief of the bureau in charge of the Navy shipbuilding, he was a proud member of the " shipbuilding fraternity," and accepted, as its highest honor, election in 1940 to the presidency of the Society of Naval Architects and Marine Engineers. In Admiral Land, the Maritime Commission had a chairman who was highly esteemed and thoroughly at home in the Navy and in all the shipbuilding industry.

In view of his career, it is not surprising that Admiral Land preferred bureaus to commissions. Speaking informally at the end of the festivities when he was installed as President by the Society of Naval Architects and Marine Engineers, and after some obviously joking remarks, Admiral Land said: "You know, I don't believe in the commission form of government; I think you ought to have a one-man show and shoot him at sunrise if he doesn't run it right." [9] Although these were his sentiments, Admiral Land was a realist who worked with men and institutions as he found them in order to attain what he thought was the main objective. Before the war was over he did business with profiteers among the capitalists and with trouble makers among the labor leaders, both of which he no doubt disliked much more intensely than he disliked commissions. In accord with this instinct for using the means at hand, he could work with and through a commission-form of government, at least as long as Roosevelt was in the White House.

How much the Commission would act as a unit and how much the Commissioners would divide up its functions and assign one sphere to one Commissioner and another sphere to another Commissioner was constantly an issue, however, and was solved in different ways from time to time. As a rule, the Navy officers who were members of the Commission favored the bureaucratic

---

[9] Society of Naval Architects and Marine Engineers, *Transactions*, XLVIII (1940), 276, and on his biography, *ibid.*, p. 61, and the sources above cited.

division of functions, the civilian members opposed it. At the early meetings this problem was discussed and it was agreed that the Commission would act together as a unit as much as possible;[10] but either force of circumstances, or the diversity in the interest and the capacity of the Commissioners, or the dominance of the Navy element led to a considerable amount of division of functions among the Commissioners. In 1937 the division was roughly as follows: Chairman Kennedy had responsibility for financial, accounting, personnel, and other general functions; the training of ships' personnel was given to Admiral H. A. Wiley; Commissioner Thomas M. Woodward, a Pennsylvania lawyer familiar with the Shipping Board and the ICC, occupied himself particularly with hearings concerning rates and regulations; and Admiral Land took charge of ship design and construction.

Certain features of this functional specialization persisted a long time. Mr. Woodward served until 1945 and kept a special interest in rates, regulations and semi-judicial hearings.[11] After Admiral Wiley left the Commission in 1940, another Navy officer, Captain Edward Macauley (Retired) was appointed a Commissioner, in April 1941, with the understanding that he would supervise seaman training and labor relations.[12] When Admiral Land became Chairman in 1938 he gave primary attention to the external relations of the Commission—its contacts with the White House, with Congress, with other agencies—and to its program and public relations as a whole; but as early as May 1937 he had made arrangements to have Commander Howard L. Vickery brought over from the Navy's Bureau of Construction and Repair to act as his assistant.[13] On September 26, 1940, Howard Vickery, then Commander, later Vice Admiral, became a Commissioner, having been appointed on Admiral Land's recommendation; so there continued to be one Commissioner in charge of construction.[14]

[10] Interview with ex-Commissioner Thomas M. Woodward by Chester L. Guthrie and Allen M. Ross, June 7, 1946, in MC, Historian's Collection, Research file 105.
[11] Interviews with Admiral Land and ex-Commissioner Woodward. An organization chart dated Apr. 28, 1937 carries the functional division much further than was ever formally approved.
[12] Land to the President, May 6, 1942, in MC, Historian's Collection, Land file.
[13] Kennedy to Claude A. Swanson, Secretary of the Navy, May 3, 1937, in Land reading file.
[14] Land to Schuyler Otis Bland, Aug. 29, 1940; Land to Senator Josiah W. Bailey, Aug. 29, 1940, in MC, Historian's Collection, Land file.

No such continuity is to be found in regard to the financial and general business affairs which were the special competence of Joseph Kennedy, the first Chairman. Admiral Land did not feel particularly qualified in the field of finance. When John J. Dempsey was serving as Commissioner from January to July of 1941, Admiral Land spoke of him as the " best businessman on the Commission." On Mr. Dempsey's resignation, the Chairman asked the President to appoint someone who would strengthen the Commission in this respect, mentioned a number of persons, and recommended one of them.[15] The President appointed John M. Carmody, who served throughout the war. Mr. Carmody was not among the men Admiral Land had suggested but he had had some business experience, especially in industrial production and labor relations, before entering government service in 1933 at the beginning of the New Deal. He had just served as Administrator of the Federal Works Agency and was much interested in systems of efficient administration and scientific management.[16] For financial matters, however, the most active representative of the Commission throughout the war was Robert Earle Anderson, whom Admiral Land selected to become the Commission's Director of Finance in August 1938.[17] Mr. Anderson had worked as naval architect in the Navy's Bureau of Construction and Repair in 1904-1913, been comptroller, treasurer, and vice president of the Winchester Repeating Arms Co., 1915-1924, and financial vice president of Paramount Pictures in 1935-1936. Not only was he in charge of accounting, auditing, and disbursement as Director of the Division of Finance; he negotiated with companies concerning their financing and he advised the Commission on the financial side of contracts and in many matters involving business judgment.[18]

Under the Commissioners, the highest official of the Commission was the Executive Director who was responsible for the general execution of the Commission's orders. The post was first filled by Mr. Kennedy's associate, Joseph R. Sheehan, for whom it was

---

[15] Land to the President, July 1, 1941, in Land reading file.

[16] Hymen Ezra Cohen, "Commissioner John Michel Carmody of the United States Maritime Commission 1941-1946," (typescript, Nov. 18, 1946), p. 7, in MC, Historian's Collection, Research file 118. 3.

[17] Land to Anderson, June 24, 1938 and July 20, 1938, in Land reading file.

[18] Mr. Anderson's activities are reflected in his reading file and in the Minutes of the Commission.

created. After Mr. Kennedy and Mr. Sheehan left, the post was given to one of the many officials who came to the Maritime Commission from the U. S. Shipping Board Bureau of the Department of Commerce, Samuel Duvall Schell. He had been Chairman of the Board of Trustees and President of the U. S. Shipping Board Merchant Fleet Corporation and so had been in direct charge of the operation and maintenance of all the government-owned merchant ships and ship terminal properties before the creation of the Maritime Commission. Both a lawyer by training and an experienced auditor, Mr. Schell had been for more than twenty years connected with the merchant marine, mainly in government service. At the creation of the Commission he served for a while as Special Expert in connection with training and operations.[19] When he became Executive Director, Mr. Schell was particularly concerned with the handling of personnel and with coordinating the activities of the various divisions. For example, he wrote or edited administrative orders, okayed press releases, and went over with the budget officer the details of the budget. All recommendations for appointments and promotions were routed through him and given his close personal attention.[20] Possible conflicts of jurisdiction among divisions were often settled without any administrative order, for Mr. Schell depended largely on personal conferences with the directors concerned.[21]

Most of the staff with which the Commission began its work were like Mr. Schell transferred from the Shipping Board Bureau of the Department of Commerce, which was the continuation of the Shipping Board of World War I. Many of the directors of divisions were new, but the section chiefs were very largely seasoned employees of the Shipping Board, as were for example all three of the section chiefs under Mr. Anderson in the Finance Division.[22] The Commission's offices were located where those of the Shipping Board Bureau had been, mainly on the fourth floor of the Commerce Building. The personnel, numbering 973

---

[19] Biography furnished by the Division of Information, in MC, Historian's Collection, Research file 116.1.

[20] MC, Admin. Order No. 44; statements by E. G. Montgomery to Frederic C. Lane, Mar. 26, 1948.

[21] Statements by Samuel D. Schell to Hymen E. Cohen, Sept. 25 and Oct. 1, 1946, notes in MC, Historian's Collection, Research file 105.

[22] Information from Personnel Division concerning William U. Kirsch, Joseph M. Quinn, and Joseph A. Honsick.

on May 1, 1937, was reorganized into seven divisions.[23] By the middle of 1940 the personnel had expanded to 1,754 and the divisions numbered sixteen.[24] During the reorganizations, with many new persons coming in at all levels, Mr. Schell worked to keep up the morale of the old timers concerned about their future. He told the story of what happens if you put little stones and big stones in a bottle and shake. The big stones come to the top. In the shake-up of the old bureau, those who were sure they were big stones need not worry about rising to the top, so Mr. Schell strove to reassure them, asking: "Big stone or little stone?"[25]

The sixteen divisions into which the personnel was organized at the end of 1940 may be divided for our purposes into three groups: divisions concerned with construction, divisions concerned with operations, and divisions rendering general services. The divisions which rendered service to the activity of the Commission as a whole were at the end of 1940 the Secretary's Office, Legal Division, Finance Division, Research Division,[26] Maritime Promotion and Information Division, and Personnel Division.[27] Together with the 37 special assistants and secretaries attached directly to the Office of the Commissioners, the total personnel amounted to 638. About the same number were attached to the seven divisions concerned with operations and maintenance of ships: Regulations, Operations, Insurance, Maritime Personnel, Examining, Training, and Maintenance and Repair. Under the last mentioned division was the care of old Hog Islanders and other ships tied up in the reserve fleet.[28] The divisions concerned with operations expanded their activities enormously during World

---

[23] House, *Third Deficiency Appropriation Bill for 1937, Hearings* before the Subcommittee of the Committee on Appropriations, 75th Cong., 1st sess., June 15, 1937, p. 251; U. S. Maritime Commission *Report to Congress* for the Period of Oct. 26 to Dec. 31, 1936, p. 2.

[24] Number of Employees by Divisions Annually from June 30, 1939 through June 30, 1945, in MC, Historian's Collection, Research file 116, Organization Chart, Apr. 4, 1941.

[25] Interview with Samuel D. Schell, Oct. 1, 1946.

[26] In addition to the Research Division there was for a while a Section of Special Studies. The two were merged and became the Division of Economics and Statistics on Mar. 28, 1941. MC Admin. Order No. 37, Suppl. No. 30.

[27] The Purchase and Supply Division might be added because it did the general buying of office supplies, but its main activity was buying the "allowance list" for the *America* and other Commission ships.

[28] The functional chart dated Apr. 4, 1941 is confirmed by an analysis of administrative orders to be a summary of the situation at that time. See Table 16.

THE COMMISSION BEFORE THE WAR

War II, as did those concerned with construction, and will not receive here the attention they deserve, because this is only a history of shipbuilding under the Maritime Commission, not a history of all its activities.

From the point of view of the production of ships, the heart of the Commission was the Technical Division.[29] It had a special status because it was directly supervised first by Commander Vickery as special assistant to Admiral Land, and later by Admiral Vickery as Commissioner. He was placed by Commission action in charge of all matters related to construction.[30] Very few of the Technical Division's staff in 1940 had come from the Shipping Board Bureau, and the two most important individuals to be so transferred, J. E. Schmeltzer and William G. Esmond, had been with the Shipping Board Bureau only since 1934. James L. Bates who became Director of the Technical Division on January 1, 1939 had been for more than thirty years a civilian employee of the Navy Department. Before he was formally named Director of the Technical Division he spent his mornings at the Navy on one side of the Ellipse and his afternoons at the Maritime Commission in the Commerce Building on the other side. Although the Commission at times found it so hard to secure draftsmen that they drew some from the Department of Agriculture, the Navy was naturally looked to as the obvious source of supply for the kind of skilled labor needed in the design work of the Technical Division.[31] So much personnel was obtained by the Maritime Commission from the Navy at a time when the Navy also was expanding construction that the Secretary of the Navy registered objection.[32] The Navy tint in the Technical Division was heightened of course by the close attention it received from Admirals Land and Vickery, but a large part of the growing staff was recruited outside the government from private industry.

[29] MC Admin. Order No. 20, Jan. 5, 1937. At first a division with that name was charged with maintenance and repair as well as new construction, but the two functions were soon separated. Maintenance and Repair became a separate division, leaving in Technical the staff which had meanwhile been formed to design new ships and inspect their construction.

[30] MC Admin. Order No. 41, Dec. 28, 1937. He was made Rear Admiral Jan. 22, 1942 (confirmed by Senate, Apr. 9, 1942) and Vice Admiral, Oct. 24, 1944 (confirmed by Senate, Nov. 27, 1944). See below, p. 459.

[31] See below, in this chapter, sect. III.

[32] Land to Claude H. Swanson, June 7, 1938 and Sept. 22, 1938; Land to Richard B. Wigglesworth, June 21, 1938; Land to Assistant Secretary of Navy, July 23, 1938; Land to Admiral Sexton, Nov. 22, 1938, all in Land reading file.

Although all sixteen divisions looked equal on an organization chart, there was actually considerable difference in the extent to which directors of divisions worked immediately with the Commission or reported to it through the Executive Director. Admiral Vickery carried the affairs of the Technical Division directly to the Commission, and Mr. Anderson also dealt directly with the Commission, at least in advising it concerning contracts and other business matters. On the other hand, the Executive Director, Mr. Schell, who was regularly present at Commission meetings, spoke with authority on many subjects and especially in regard to the handling of the staff and the organization of its work.

In spite of the extent to which individuals had taken over specialized functions, on the whole the Commission acted as a five-man Commission in supervising its staff and registering basic decisions. It held two regular Commission meetings a week to pass on a long docket full of routine matters, and held also many special meetings. Admiral Land's leadership was recognized and became more secure with the retirement from the Commission of Admiral Wiley, and E. C. Moran, Jr., in 1940 and 1941 and the appointment to the Commission of Admiral Vickery, Mr. Dempsey, and Captain Macauley. Divided votes were rare in 1940-1941 and not a fixed pattern. There were sometimes long discussions, occasionally heated, but the Commissioners made a deliberate effort to avoid recording dissents and splitting the Commission into regularly opposing groups. The commission-form still had vitality on the eve of the war.[33]

What gave the Commission vitality was its representative or balanced membership. This representative quality was not to be found in the geographical distribution of the membership; nor indeed in its political composition.[34] By law not more than three of the five Commissioners could be of the same party, but there

---

[33] These conclusions are based on interviews and general study of the Minutes.

[34] Admiral Land wrote to George E. Martin, editor and publisher of *Pacific Shipper*, on Apr. 3, 1941: "To the best of my knowledge and belief (and I think I know what I am talking about) there has not been a single regional appointment on this Commission since the law was enacted."—Land reading file. The appointment of Captain Macauley on almost that same day might be said to give representation to the West coast, although there were other reasons for his appointment. Admiral Land himself was born in Colorado; Mr. Dempsey had been a Representative from New Mexico; Admiral Vickery was from Ohio; Mr. Woodward from Philadelphia.

were no active Republicans, for the Navy officers were considered nonpartisans and the civilian members were usually men who had been long in government service.[35] The balance sought in the membership of the Commission seems to have been rather a balance of abilities. Admiral Land when first appointed was outstanding as a naval constructor and as Chairman proved highly gifted in handling public relations and political contacts. Admiral Vickery was a production man. Captain Macauley was skilled in maritime labor relations. Mr. Woodward was a lawyer. When these four composed the Commission, a man of business experience was sought to round out the combination. If its members could so work together as to pool their capacities such a commission might administer an agency better than one man. But the Commission form was to be subjected to a severe strain by the war.

III. THE LONG-RANGE PROGRAM

The undisputed leadership which Admiral Land held in the Commission in 1941 was due in part to his success in carrying through a program for new construction. This prewar construction was very small compared to the production of the war years, but the building of 1938-1940 was extremely important in preparing for the work done in 1941-1945. It laid the foundations in shipyards, in designs, and in forming a trained staff for the Commission.

Merchant shipbuilding had come to a very low ebb in the United States when the boom after World War I collapsed, and it made only a slight revival under the Merchant Marine Act of 1928. Most of the fleet, having been built in 1918-1922, would in 1942 be twenty years old, which was generally considered obsolescent. Out of a total of 1,422 ocean-going vessels (2,000 gross tons and over) registered in 1942 under the American flag, 91.8 per cent would be twenty years old, as would all the 225 government-owned ships, most of them in moth balls from World War I. Most of the dry-cargo vessels were of 10- to 11-knot speed. Although twenty-nine large combination passenger-and-cargo vessels had been built under the subsidy provisions of the Merchant Marine Act of 1928, the United States had no liners to rival the

---

[35] The only party affiliation given in *Who's Who* (1946-47) was Vickery, Republican.

British "Queens," and 60 per cent of the combination passenger-freight vessels were near obsolescence.[1]

The first ship contracted for through the Maritime Commission was a symbol, the *America*, the largest liner ever built in the United States. It was not nearly so big as the huge luxury liners of the French and British, the *Normandie* and the *Queen Mary*, which were the record makers of the North Atlantic traffic, but in some features such as fireproofing it excelled even these famous ships, and it was a notable advance in American passenger vessel construction. The contract was awarded September 30, 1937, and the *America* was intended for operation by the United States Lines to give the American merchant marine a representation of which it could be proud in the most publicized of all trade routes.[2]

More needed than the highly publicized liners were dry-cargo carriers. Speedy tankers too were wanted as auxiliaries for the Navy. Subsidies to increase the speed of tankers of the *Cimarron* class to 18 knots were among the first subsidies to be paid by the Commission,[3] for this extra speed was reckoned a "defense feature." But American oil companies had kept up their tanker fleets; the glaring deficiency in the American merchant marine was in dry-cargo vessels. The main problem of the Commission was to build speedy, efficient freighters.

Although everyone favored the building of new freighters, practical agreement on what to build was another matter. The Navy, the Commissioners, and the prospective operators found difficulty in agreeing. The Navy was given an important voice in the matter by the Merchant Marine Act of 1936. The possible use of the vessels by the Navy in time of war was a main reason why the national government was interested in new construction.

---

[1] *Economic Survey of the American Merchant Marine* (Government Printing Office, 1937), pp. 36-39. *The Use and Disposition of Ships and Shipyards at the End of World War II, A Report prepared for the United States Navy Department and the United States Maritime Commission by the Graduate School of Business Administration, Harvard University, June, 1945*, printed for the use of the Committee on the Merchant Marine and Fisheries, Document No. 48 (Government Printing Office, 1945, cited hereafter as *Harvard Report*), p. 174.

[2] Robert Earle Anderson, *The Merchant Marine and World Frontiers* (New York, 1945), p. 71; MC Minutes, Sept. 30, 1937, pp. 1772-74. A general view of the long-range program is Emory S. Land, "Building an American Merchant Marine," the *Annals* of the American Academy of Political and Social Science, CCXI (Sept., 1940), 41-48.

[3] MC Minutes, Dec. 21, 1937, p. 2498; *Marine Engineering and Shipping Review*, Mar., 1939, p. 103.

All contract plans and specifications for the proposed vessels were to be submitted by the Commission to the Navy Department for approval, so that the Navy could urge any changes it deemed necessary to make possible the speedy and economical conversion of ships into naval auxiliaries. On this basis the Commission assumed the cost of all national defense features.[4] The Navy's conception of the kind of ships desired was naturally one-sided, but it was an important side and had to be considered.[5] At the other extreme was the point of view of the ship-operating companies. Whereas the Navy wished as speedy ships as possible for tactical reasons, the shipping companies wished economy in operation even at the sacrifice of some speed. Even within the Commission there was sharp divergence of views as to the way in which the Commission should proceed and the type of ships that should be built.

How these divergences of view were reconciled or overridden is another story. Important for the shipbuilding of World War II is the fact that they had been overcome by positive action in which the Commission took the initiative under Admiral Land's leadership. Merchant shipbuilding was booming as war drew nearer, and the revival of the industry was along the lines laid down by the Maritime Commission in its "long-range program." It was called a long-range program because it was the answer to the Congressional injunction that the Commission make a definite plan for replacing obsolete vessels in sufficient volume to serve essential trade routes. It called for the building of 50 ships a year for ten years.[6] Although that does not seem a large program when in retrospect we compare it to the production during the war years, it seemed highly ambitious in 1938 when the first contracts in the long-range program were awarded.

After World War II began in Europe the demand for shipping increased, as it usually has in time of war. With a more favorable market for the ships and with their potential use as naval auxilia-

---

[4] Merchant Marine Act, 1936, 49 Stat. 1985, Secs. 101, 501 and 502.

[5] Kennedy to Swanson, June 10, 1937, in MC general files (cited hereafter as gf) 107-5; Land Memorandum for files, Aug. 9, 1937, in MC, Historian's Collection, Land file.

[6] Land to the Commission, Dec. 20, 1938, in Land file, and Gerald J. Fischer, "The Programming of Ship Construction by the U. S. Maritime Commission 1938-1945" (typescript, Nov. 27, 1946) p. 8, in MC, Historian's Collection, Research file 202. 1.

ries growing more urgent, the long-range program of the Maritime Commission was accelerated. On August 27, 1940 the Commission voted to speed up that program so that 200 ships would be contracted for prior to July 1941. Contract authorization was raised $50 million by the Second Deficiency Appropriation Act of 1940. Before the end of October 1940, 47 ships had been delivered and contracts had been awarded for 130 more. The long-range program was proceeding well ahead of schedule.[7]

In launching the long-range program, the Commission developed a technical staff of the kind needed for planning and supervising an extensive program of ship construction. The very much larger staff which undertook the planning and supervision of the wartime program had as its nucleus the personnel recruited in connection with the long-range program. Placing a wartime program in the hands of a group previously organized for a peacetime purpose is in theory debatable. It can be argued that absorption in the particular objective set for them in peace, in this case the building of ships of high quality for competitive use in international trade, will interfere with acceptance of wartime objectives, such as maximum speed in multiple production. To some extent it is inevitable that procedures adapted to conditions of peace should be continued by force of routine or tradition into the changed conditions of war. On the other hand it is hard to see how the planning and supervision of the emergency shipbuilding program could have been organized at all with anything like the speed with which it was organized if there had not been a core of officials already concerned with supervising merchant shipbuilding.

The group of men who were the guiding spirits determining the design of the ships for the long-range program came together even before the formation of the Maritime Commission. Acting on a suggestion from the Navy, the Shipping Board Bureau of the Department of Commerce in 1934 engaged George G. Sharp, consulting naval architect, together with J. E. Schmeltzer and William G. Esmond of his staff, to work on the design of standard types of merchant vessels acceptable to the Navy.[8] The Navy's

[7] U. S. Maritime Commission, *Report to Congress* for the Period Ended October 25, 1940, pp. 6-8, gives the misleading impression that it was expected that their construction would be completed July 1941. But see Land to Harold D. Smith, Oct. 23, 1940 in MC gf 506-1, and Fischer, *Statistical Summary*, Table A-2.

[8] J. W. Barnett, Acting Chief of Division of Loans and Sales, U. S. Shipping Board Bureau, to Acting Director, Aug. 7, 1934. The plan for engaging a naval

THE LONG-RANGE PROGRAM 25

representatives on the joint board which met with these architects included the then Commander Howard L. Vickery, and occasionally James L. Bates.[9] When shipbuilding began to pick up, Mr. Sharp returned to the affairs of his firm of naval architects,[10] but Messrs. Schmeltzer and Esmond continued with the Shipping Board Bureau until they were transferred with the rest of its staff to the Maritime Commission. The organization of the new Technical Division was completed in 1939 by making James L. Bates Director, Mr. Esmond the head of the Hull Section and Mr. Schmeltzer of the Engineering Section. In July 1940, after the standard cargo vessels had proved brilliantly successful, Mr. Schmeltzer was made Associate Director of the Division in recognition of his primary responsibility for designs of propulsion equipment.[11]

As the long-range program materialized in more and more ships to be designed and inspected, additional units were formed within the Technical Division. A Preliminary Design Branch investigated the possibilities of various types of ships, developed initial plans, tested models, and worked up tentative drawings and calculations. Both for the hull and for the engines, separate sections were formed to turn these preliminary plans into exact specifications, and other sections checked the detailed working plans which were prepared by the constructor's naval architect. A Materials Section investigated materials and equipment of many kinds. There was an Interior and Styling group also, and a Clerical Section. To find what shipyards were capable of building the vessels being planned, an Investigation Section was formed. Later it was absorbed by the Construction Section which sent inspectors into the yards to report on the way the work was progressing and see that it was being done according to specifications. A Trial Board reporting directly to the Commission was formed to pass on acceptance of vessels. In 1940, after the ships of the long-range program had

architect for the work is described in the letter of Mar. 16, 1934, same to same. Both in MC gf 505-1.

[9] Memoranda on meetings of the "Committee on Standard Merchant Vessels" in MC gf 505-1. The notes on the meeting of Aug. 27, 1934 contain the earliest use I have seen of the well-known designations T2, C2, and P2.

[10] J. M. Johnson, Acting Secretary of Commerce, to Claude A. Swanson, Secretary of the Navy, July 3, 1936, in MC gf 505-1. See *Harvard Report*, pp. 174-80, on the gradual revival of both Navy and merchant shipbuilding.

[11] J. M. Chambers (approved by Vickery) to J. L. Spilman, Classification Division, C. S. C., Mar. 30, 1940, in MC, Personnel Division files.

been in use long enough, a Performance Section within Technical was charged with drawing lessons from their operation, especially from any breakdowns.[12] Thus various units in the technical staff of the Commission were planning, supervising, and replanning the long-range program as it grew in volume.

Recruiting the technical staff was a serious problem even before the war. From next to nothing at the formation of the Commission, it grew to 161 in July 1939 and to 369 in July 1940. An addition of nearly 300 was planned for 1941.[13] This growth was all the more difficult because at the same time the Commission was constantly losing many good men to the new shipyards which were being formed to construct ships for the long-range program. A steel fabricator or some other entrepreneur moving into the shipbuilding business who had his first contract from the Commission and his initial guidance from its staff was likely to look to its personnel for recruits to his staff. When in 1940 the shipbuilding revival became a boom, members of the Commission's staff received many fine offers, so the Director of the Division warned in asking increased salaries. He listed twenty persons lost in this way.[14] Since men with proper training were difficult to get, plans were made for taking college graduates with degrees in mechanical or electrical engineering, classifying them as P-1 trainees, and having the section chiefs provide them with training in marine matters. Certain members of the Division conducted night classes for such personnel.[15] In recruiting its own staff, as well as in its relations to the shipbuilding companies, the Technical Division became an educational institution giving important training in naval architecture and marine engineering.

[12] Administrative Orders and Quarterly Reports of the Technical Division analyzed in Blanche D. Coll, " Administration of Functions Assigned to the Technical Division, 1937-1947 " (typescript, Mar. 10, 1947) in MC, Historian's Collection, Research file 106.

[13] Director, Technical Division to Budget Officer, Aug. 27, 1940, in files of the office of the Director of the Technical Division, " Budget " (Mary Mangan's files) ; Vickery to Director, Division of Finance, Nov. 25, 1940, in MC gf 201-2-11; statement of W. J. Turner to F. C. Lane in interview, Feb. 25, 1948. Cf. Technical Division Quarterly Report, July 1, 1939 to Sept. 30, 1939.

[14] Assistant Director, Technical Division to Director, Division of Personnel Supervision and Management, Sept. 18, 1940; Bates to Commission, Oct. 22, 1940, both in MC gf 201-2-11; Memorandum, Bates to Budget Officer, Oct. 2, 1940, in files in the office of the Director of the Technical Division.

[15] Director, Technical Division to Chiefs of all Sections, June 19, 1940, in MC gf 201-2-11. Statement by Montgomery.

Alongside the development of the technical staff which planned and inspected new construction grew an auditing staff in the Finance Division. The Merchant Marine Act of 1936 limited the profits of the builders to 10 per cent of the contract price; consequently the Commission's representatives checked over a company's statement of cost to see whether some of the price stated in the contract could be recaptured by the Commission as excess profit. For this purpose resident auditors were sent to all yards working on the Commission contracts. In January 1940, there were at Sun Shipbuilding, for example, a Principal Construction Cost Auditor, a Junior Construction Auditor, and three clerks. About fifteen yards had somewhat similar units,[16] and in Washington various units in the Construction Audit Section of the Finance Division guided their work. The basic rules for them to follow, especially in the important matter of allocating overhead, were laid down in a booklet, *Regulations Prescribing Methods of Determining Profit*.[17] These rules and these trained auditors were the foundations from which the Commission built its system for auditing the wartime contracts.

IV. STANDARD TYPES OF CARGO VESSELS

The ships designed before the war for the long-range program continued during the war to be the Commission's ideal of what they would like to build. They were commonly called " standard types " to distinguish them from the other main groups of types— emergency, military, and minor. Even under the pressure of war the Commission succeeded in building so many of the standard types that its " long-range " program of completing 500 of these vessels in ten years had been exceeded by 1946. The standard dry-cargo carriers were called C-types. They were of three sorts, C1, C2, and C3, the letter indicating that they were cargo ships and the number showing the relative size. See Plates I, II, XVII. Their main characteristics are shown in Table 2.

Compared to the freighters built before 1939, all of the C-types

---

[16] Organization chart dated Jan. 18, 1940, giving names of employees and distinguishing those which were merely " proposed."

[17] Adopted May 4, 1939, by the U. S. Maritime Commission, including amendments and annotations to Sept. 1, 1941, p. 9. Copy in MC, Historian's Collection, Research file 24. 1.

were remarkable for their speed. The *Challenge*, one of the first of the C2's, completed her maiden voyage from Boston to Cork, 2,742 nautical miles, in 6 days, 18 hours, and 38 minutes—an average speed of 16.82 knots, although the design speed of the C2 was only 15.5 knots.[1] Because of their better lines and improved propulsion machinery, these 15.5-knot ships had about the fuel consumption per nautical mile of the earlier 11-knot ship. At its sea trials in June 1939 the *Challenge* "chalked up a new world's record for fuel economy with a fuel consumption of 0.552 pounds per shaft horsepower per hour for all purposes." Its

TABLE 2

CHARACTERISTICS OF SOME PRINCIPAL TYPES BUILT BY MARITIME COMMISSION*

	Liberty	C1	C2	C3	Victory
Length, overall	441' 6"	417' 9"	459' 6"	492'	455' 3"
Beam, moulded	56' 11"	60'	63'	69' 6"	62'
Draft, loaded	27' 8"	27' 6"	25' 9"	28' 6"	28' 6"
Deadweight tonnage	10,419	9,075	8,794	12,500	10,734
Bale capacity	500,245	452,420	536,828	732,140	453,210
Speed, knots	11	14	15.5	16.5	16.5

* For costs and displacement tonnage of these types see Table 20.
Source: Booklets prepared by the MC Division of Vessel Disposal and Government Aids and "Liberty Ship Stowage and Capacity Booklet," prepared for the Commission by the Bruce Engineering Co., New York.

high speed cross compound turbines developed 6,000 shaft horsepower and drove the propeller shaft at 92 revolutions per minute through double reduction gears.[2] While their speed made these ships attractive to the Navy as potential auxiliaries, their combination of speed and economy enabled them to meet the needs of competitive commercial operators.

Most of the C-types were powered with turbines, and the basic reason why they could combine speed and economy was the progress made in building double reduction gearing which reduced the very rapid revolutions necessary for an efficient turbine to the relatively slow revolutions necessary in an efficient propeller. But Diesel engines, which did not require double reduction

[1] *Marine Engineering and Shipping Review*, Oct. 1939, p. 469.
[2] *Ibid.*, July 1939, p. 305; and H. Gerrish Smith, "Events of 1939 Prove Wisdom of U. S. Shipbuilding Policy," in *ibid.*, Jan., 1940, p. 50. See also J. E. Schmeltzer, "Engineering Features of the Maritime Commission's Program," in *Transactions* of the Society of Naval Architects and Marine Engineers, XLVIII (1940), 332-70.

gearing, were installed on many C2's.[3] Few Diesels had been installed in American merchant vessels before the Maritime Commission launched its program, and Admiral Land considered recognition of the feasibility of the Diesel to be one of the main contributions of the Maritime Commission to the technical development of the shipbuilding industry.[4]

The other major contribution to which he pointed with pride in 1939 was the high pressure, high temperature steam turbine power plant being installed on C3's. Indeed, the C3 was an extremely fast ship for a cargo carrier. The design speed was 16.5 knots, but one of the first group built, the *Sea Fox*, on her official trials in March 1940 attained a speed of 19.5 knots and showed a fuel consumption rate of 0.563. These ships had a shaft hp. of 8500. A modified form of the basic C3 design providing for 95 to 192 passengers was also built. One was the round-the-world passenger liner *President Jackson* delivered in 1940.[5]

Much less ambitious were the C1's. Having a speed of only 14 knots, they were not up to the general standard required for Navy auxiliaries, but were believed suitable for some particular trade routes. Five were delivered in 1940, twenty-nine in 1941, powered partly by turbines and partly with Diesel engines. But the chief source of pride to the technical staff of the Commission when the war began was the excellent performance of the C2's and C3's.[6]

Also a source of pride but a dream which existed only on drawing boards was the trans-Pacific passenger liner with the design designation P4-P. To compete with rumored Japanese liners, it was to have a speed of 24 knots. It would be 760 feet long with a beam of 98 feet and be propelled by two turbines, each of 29,000 shp. operating on reheat regenerative cycles. Since its speed and size made the P4-P potentially an airplane carrier, that possibility was kept in mind in its design. For example, smoke stacks were placed on one side so as to leave room for a flight deck. Other P (passenger ship) designs of various sizes

[3] Most of the "C" Diesel ships had single reduction gearing between the two or four Diesel engines and the line (propeller) shaft. The ships with low speed Sun Doxford engines, like the C2-SU's, had no reduction gearing.
[4] *Marine Engineering and Shipping Review*, Aug. 1939, pp. 355-56.
[5] *Ibid.*, Nov. 1940, pp. 92, 108.
[6] E. S. Land, "Some Policies of the U. S. Maritime Commission," in *Transactions* of the Society of Naval Architects and Marine Engineers, XLVIII (1940), 258-76.

(P1, P2, and P3) were included in Admiral Land's long-range program. Although none of these was built before the war, the technical staff of the Commission had worked on the basic problems which arose when the construction of such ships for transports was decided on during the war.[7]

For tankers similarly, design designations (T1, T2, and T3) were used to indicate various sizes, but before 1941 the only tankers built with Maritime Commission aid were those of the high speed *Cimarron* class, the T3-S2-A1, which were built with designs furnished by the Standard Oil Company of New Jersey,[8] and for which the Maritime Commission paid only the cost of defense features.

All dry-cargo and passenger ships built under Maritime Commission contracts embodied a large number of improvements in addition to their more efficient propulsion machinery. Ever since the catastrophic burning of the *Morro Castle* within sight of the Jersey coast in September 1934, American naval architects, led by George G. Sharp, had concentrated on designing fireproof ships. Non-combustible materials were used wherever possible. Metal furniture replaced wood in officers' and crew's quarters as well as in passengers' accommodations. Insulation of bulkheads and decks partitioned the ship into zones so that there would be no clear sweep of flames as in the *Morro Castle* tragedy. New aids to navigation and better life-saving equipment also served to make the C-ships among the safest vessels afloat.

Since the Maritime Commission was charged with rebuilding morale among American seamen as well as replacing obsolete and inefficient vessels, a notable feature of all its vessels was better accommodations for officers and crew. The crew were given cabins in the house amidships instead of being bunked altogether in the traditional forecastle. Proper mess rooms were provided, as were hot and cold running water, and many more ventilator fans. Modern refrigeration and other equipment were provided for the ship's galley. Also a system of rat proofing was added to the improvements which raised the standards of health and comfort for seamen.

From the point of view of commercial operation, an essential

---

[7] *Ibid.*, and J. E. Schmeltzer, "The Commission's P4-P Design," in *Marine Engineering and Shipping Review*, Feb. 1941, p. 75.

[8] Statements by E. S. Dillon, Feb. 28, 1949.

feature of the C-types was their equipment for handling cargo. On most, this consisted of 14 or 15 five-ton booms and one thirty-ton boom. All were worked by electric winches. They were rigged to 10 king posts which formed so prominent a part of the profile characteristics of the C-types.[9]

The methods used in constructing the C-types anticipated some features of the wartime program. Welding was just coming into extensive general use on large merchant vessels. In the Commission's design riveting was replaced by welding to such an extent as to save a substantial amount in the weight of the hull and so secure greater deadweight carrying capacity.[10] The proportions of riveting and welding varied from yard to yard. The *Exchequer*, a C3 built by the Ingalls Shipbuilding Corp. at Pascagoula, was hailed in November 1940 as being " the first all-welded ship to be completed under the Maritime Commission shipbuilding program," and consequently 600 tons lighter than other C3's.[11] With Commission encouragement, the trend was towards more and more welding.[12]

More definitely connected with Commission policy was the trend towards standardization and the effort to apply methods used in mass production. Traditionally, operators determined the design of each ship according to their particular ideas of the special needs of the ports and routes in which they intended to operate the ship. The C-types, in contrast, were designed initially by the Maritime Commission after consultation with many operating companies and with trade associations whose suggestions were compromised and combined as far as possible so as to produce a standard ship adapted to a variety of needs. Modifications were then made to meet special needs of individual operators, but many components and items of equipment were the same. Consequently, it was hoped that this application of the American system

---

[9] General descriptions in the article by Admiral Land, " Some Policies," above cited and in Robert Earle Anderson, *The Merchant Marine and World Frontiers*. Detailed descriptions of specific ships in many articles in the *Marine Engineering and Shipping Review* in 1939 and 1940.

[10] H. Gerrish Smith, in *Marine Engineering and Shipping Review*, Jan. 1940, p. 50.

[11] *Marine Engineering and Shipping Review*, Nov. 1940, p. 112.

[12] David B. Tyler, " Welding Before World War II," (typescript in MC, Historian's Collection, Research file 204. 3), p. 14, citing D. Arnott, " Some Observations on Ship Welding," in *Transactions* of the Society of Naval Architects and Marine Engineers, L (1942), 327-28.

of standardizing parts would lead to some economy in production. *The Economic Survey of the American Merchant Marine* issued by the Commission in 1937 and used as a basis of policy said: " The practice of American manufacturers, such as the automobile makers, of having parts manufactured at central points for assembly elsewhere might be emulated in the construction of ships." [13] Then it would not be necessary for every shipyard to have all the equipment for fabricating in the yard the various materials going into a ship. There would be less overhead and lower costs.

Another form of saving through standardization was by awarding contracts, not for one ship at a time, but for groups of four to six identical ships. This simplified the preparation of working plans, the ordering of materials, and the control of construction in a yard.[14] The type awarded to a yard was adapted to its equipment and most of the smaller, newer yards were standardized on one type.[15] Although American shipbuilding in 1940 was far from the technique of multiple production developed during the war, it was already moving in that direction in building the Maritime Commission's standard cargo vessels.

V. The Shipyards of the Country in 1940

The value in 1941 of this long-range program lay not only in the ships it had produced but at least equally in the shipyards which as a result were then operating. The slump in shipbuilding after the First World War had threatened to cripple the industry. This was the more serious because naval architecture was an intricate science. Dominating the clatter and seeming random dispersion of a shipyard, guiding the work of its huge cranes, heavy presses, and other special equipment were the calculations of highly trained engineers and experienced managers. Their practical and theoretical knowledge had to be kept employed if the industry was to survive and have in it the possibility of sudden growth to meet a new emergency.

In the lean years from 1922 to 1938 only the very strong shipbuilding companies were able to keep going. The strongest

---

[13] P. 21.
[14] Land, " Some Policies," pp. 266-67, and H. Gerrish Smith in *Marine Engineering and Shipping Review*, Jan. 1940, p. 49.
[15] Fischer, *Statistical Summary*, Table C.

survivors may be called the Big Five: Newport News, Federal, New York, Sun, and Bethlehem. Because of their equipment, the high reputation of the engineers, the location of their yards, and their corporate or banking affiliations, they were the best placed for securing the scanty orders for tankers, ore carriers, and Navy vessels which offered some nourishment during the barren period. The Newport News Shipbuilding and Dry Dock Company was the oldest and perhaps the most experienced in the country. It had been founded by the railroad magnate, C. P. Huntington, in 1890 and since that time had built ships of all kinds, including battleships. In 1940 it was outfitting the *America*. The Federal Shipbuilding and Dry Dock Co. of Kearney, New Jersey, was a subsidiary of U. S. Steel. The New York Shipbuilding Co. at Camden, New Jersey, founded in 1899, had built battleships, cruisers, and passenger liners. During World War II, it built entirely for the Navy; from 1936 to 1946 it had no Maritime Commission contracts. The Sun Shipbuilding and Dry Dock Company, Chester, Pennsylvania, was a subsidiary of Sun Oil Company and constructed many tankers for that concern and other oil companies.[1]

Bethlehem's position was unique in that it held a leading position in the three fields of steel manufacture, shipbuilding, and ship repair. In all three of these fields, its main investments were in the Northeast. Steel mills at Sparrows Point near Baltimore, Maryland, and in Pennsylvania at Lackawanna, Bethlehem, Cambria, and Steelton, had a steel-ingot producing capacity of more than 11 million net tons. Bethlehem Steel Company's biggest yard was the Fore River yard at Quincy, Massachusetts. Also important was the Sparrows Point yard and there was a smaller yard at Staten Island, New York. But Bethlehem's repair facilities included not only docks at East Boston, Brooklyn, Baltimore, and Hoboken, but also facilities on the West coast in San Francisco Bay and at San Pedro. There were small steel plants on the West coast also, at Seattle, South San Francisco, and at Vernon, Los Angeles, although none of these compared in capacity to the big Eastern mills.[2] Bethlehem reopened the

[1] *Harvard Report*, pp. 167-74.

[2] Testimony of A. B. Homer, Vice President, Bethlehem Steel Company, in Senate, *Investigation of the National Defense Program, Hearings* before a Special Committee Investigating the National Defense Program, 77th Cong., 1st sess., pt. 6, July 15, 1941, pp. 1550-53. Hereafter cited as Truman Committee, *Hearings*.

old Union Iron Works yard in San Francisco Bay so as to join in the Maritime Commission's long-range shipbuilding program. Altogether Bethlehem was building ships in four yards in 1940: at Fore River, Massachusetts; Staten Island, New York; Sparrows Point near Baltimore, Maryland; and at San Francisco, California.[3]

In addition to the Big Five, two other shipbuilders, both of primary importance to the Navy, were active even during the depth of the shipbuilders' depression: the Electric Boat Company at New London, Connecticut, which built submarines, and the Bath Iron Works, at Bath, Maine, which built destroyers.[4] The Bath Company was drawn into the Maritime Commission program in 1940 when it contracted to build four C2 freighters. When the emergency came, men from Bath, as well as from Federal, Newport News, Bethlehem, and Sun, were better able to take part in a large shipbuilding program because previous orders from the Maritime Commission had given them more experience, more equipment, and a more skilled staff. But they probably would have survived anyhow.

When war began in Europe and shipping boomed, new companies entered the field of shipbuilding, and in 1940 Admiral Land boasted that seven practically new yards had been developed by the Maritime Commission.[5] Three of these were on the Gulf and four on the West coast, whereas the Big Five were all concentrated in the Northeast, except for the Bethlehem San Francisco yard building for Maritime. The new yards were small, but they were seeds from which came establishments of importance during the war and they were scattered, as was desirable from the point of view of national defense and the labor supply. At the end of 1940, they were: [6]

(1) Tampa Shipbuilding and Engineering Co., at Tampa, Florida, which was building Diesel-powered C2's on three ways.

[3] *Ibid.*
[4] *Harvard Report*, pp. 172-73.
[5] House, *Independent Offices Appropriation Bill for 1942, Hearings* before the House Subcommittee of the Committee on Appropriations, 77th Cong., 1st sess., pt. 1, Dec. 19, 1940, p. 442.
[6] A basic source for all this survey of the state of the shipyards in 1940-41 is the set of "Way Charts" prepared by Joseph T. Reynolds under the direction of Gerald J. Fischer and F. C. Lane in the office of the Historian, U. S. Maritime Commission, 1946-47. They are based on the records kept by the U. S. Maritime Commission Production Division and by the Navy, Bureau of Ships, supplemented by letters from shipyards. Cited hereafter as Reynolds' Way Charts.

(2) The Ingalls Shipbuilding Corp. at Pascagoula, Mississippi, which was building C3's on four ways in a yard which was celebrated as the first constructed in the United States " exclusively for the building of welded ships." It was an offshoot of the Ingalls Iron Works of Birmingham, Alabama.[7]

(3) Pennsylvania Shipyards, Inc., at Beaumont, Texas, had laid the keel for a C1. Its two ways were designed to have room for four such ships, since ships were launched sideways. Another

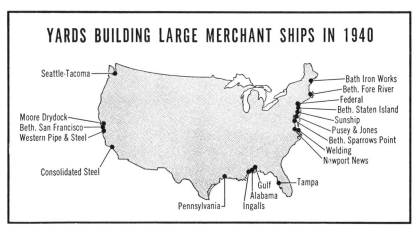

FIGURE 4. Class I yards, that is, those building ocean-going vessels over 400 feet in length. Source: Fischer, *Statistical Summary*, Table E-1, and Reynolds' Way Charts.

special feature of its launchings was the use of bananas to grease the ways.

(4) Consolidated Steel Corp. Ltd., at Long Beach, California, was building two C1's in the one long way in their recently acquired Craig yard. Here also side launchings were used.[8]

(5) Western Pipe and Steel Co. was also working on two C1's in a new yard in San Francisco using side launchings.

(6) Moore Dry Dock Co. of Oakland, California, primarily a repair company with three ways, was engaged in outfitting three C3's.

[7] *Marine Engineering and Shipping Review*, Jan. 1940, p. 35; and Nov. 1943, p. 166; Dec. 1943, p. 180; Jan. 1944, p. 196; *Fortune*, July 1943, p. 218.
[8] *Consolidated Steel Corp., Annual Report for 1944.*

(7) Seattle-Tacoma Shipbuilding Corp., in the new yard it was building at Tacoma, Washington, had two C1's on the ways and three in outfitting as the year 1940 ended.[9]

In addition (8) The Pusey and Jones Corp. at Wilmington, Delaware, was building a C1; (9) Gulf Shipbuilding Corp. at Chickasaw, Alabama, was starting two C2's, of which it laid the keels in 1941; and (10) Alabama Dry Dock and Shipbuilding Co. at Mobile was at work on a private tanker, as was (11) Welding Shipyards, Inc. of Norfolk, Virginia. Altogether the yards working on large merchant ships at the end of 1940 totaled 19.[10] To the eleven already mentioned should be added: (12) Newport News Shipbuilding and Dry Dock Co., building C3's and privage cargo ships on six of its ways; (13) Federal Shipbuilding and Dry Dock Co., about half full of Navy work but building merchant vessels under private contracts on five of its ways; (14) the eight ways of Sun Ship—as the Chester yard was commony called—which were filled half with private tankers, half with C-types. In the Bethlehem yards, C-types or tankers were being built on seven ways at Sparrows Point (15), four ways at Fore River (16), two ways at Staten Island (17), and two ways at the San Francisco yard (18). Finally (19) the Bath Iron Works had C2's under construction on two ways. Since in some yards, particularly those which launched sideways, there was more than one building berth on a single way, the total number of building berths occupied with ocean-going merchant shipbuilding on January 1, 1941, was 53.[11]

None of these shipyards, neither the yards of the Big Five nor of the smaller new companies, had any idle ways in the fall of 1940. What was not filled by the Commission's long-range program or by the small amount of unsubsidized private building was filled by the Navy's program. After years in which new construction was very slight, a substantial amount of Navy building was approved in 1934. Then came in 1938 the Twenty-Percent Expansion Act and on top of it in June 1940 the Eleven-Percent Expansion Act. This construction seemed like a lot until France fell and Germany controlled all the coasts from the Arctic to

---

[9] *Marine Engineering and Shipping Review*, Mar. 1941, p. 54.

[10] That is, building ocean-going vessels over 400 feet in length.

[11] Fischer, *Statistical Summary*, Table E-1, but the figure for building berths there given, 51, does not include the two ways at Bath Iron Works because those ways were shorter than the standard used in that table to determine building berths.

Portugal. Immediately, in July 1940, a Seventy-Percent Expansion Act increased the authorized tonnage of the Navy from 1,724,480 tons to 3,049,480 tons, a two-ocean Navy. Consequently all yards were fully occupied before the Maritime Commission started its emergency program, and mostly with Navy contracts.[12]

The expansion of the Navy's program in July not only required the full use of existing shipyards but called for the building of new yards. Some abandoned yards, such as the Cramp Shipbuilding Company of Philadelphia, where the cranes had rusted away, were provided by the Navy with money to re-equip the yards and resume construction. Small yards such as the Gulf Shipbuilding Corporation, were given contracts for destroyers and the facilities with which to build them. Some of the Navy building, especially of auxiliaries, was placed in Gulf or West coast yards, but most of it was on the northeast seaboard where the yards were best equipped and their labor most experienced.[13]

The building of warships began to crowd out merchant ships in some yards. Especially heavy were the demands of the Navy on the Big Five. Newport News took so many Navy contracts in 1941 that it became absorbed, as New York Shipbuilding had been all along, in work on battleships, carriers, and cruisers. Federal took contracts for 5 cruisers and 37 destroyers, leaving only a little room for merchant building.[14] Three of Bethlehem's four yards were assigned to do Navy work entirely—Fore River, which was the largest, Staten Island, and the San Francisco yard. The latter which had been revived by the Maritime Commission as a modest 3-way yard, added more ways and appropriate machine, plate, pipe, and sheet-metal shops to take on Navy work, as did also Fore River, Staten Island, and the repair yard at San Pedro. All these yards were destined for Navy work as soon as ships under contract for the Maritime Commission and on the ways were finished.[15] Only Sparrows Point remained free for merchant shipbuilding. Sun Ship, which had done no building of naval types, was induced to take contracts for auxiliaries, seaplane

[12] *Harvard Report*, pp. 176-77.
[13] House, *Navy Department Appropriation Bill for 1942, Hearings* before the Subcommittee of the Committee on Appropriations, 77th Cong., 1st sess., Feb. 1941, pp. 668-72, 690-91, 722-28.
[14] *Ibid.*, p. 666.
[15] Testimony of A. B. Homer before the Truman Committee, *Hearings*, pt. 6, p. 1558; House, *Navy Department Appropriation Bill for 1942, Hearings*, p. 668.

tenders and destroyer tenders.[16] Even the smaller shipbuilding companies which had been brought to life by the Maritime Commission, such as Tampa and Ingalls, were drawn into the Navy's plans. The Consolidated Steel Corporation, which was building for the Maritime Commission at Los Angeles, agreed to build at Orange, Texas, a plant for the construction of destroyers.[17] Moore Dry Dock and Seattle-Tacoma added new facilities for the same purpose.[18] These contracts did not directly interfere with the work on ships previously ordered by the Commission for its long-range program, but they were a drain on the skilled managers and engineers of even those shipbuilding companies which the Maritime Commission thought of as peculiarly its own.

Compared with the capacity or equipment of the American shipbuilding industry the Navy's program was enormous. The estimated cost of vessels under contract was $5 billion.[19] On June 1, 1940 only 6 private yards were working on Navy orders. By February of the next year there were 68. The bulk of these contracts were let in September of 1940 as soon as Congress appropriated the money.[20] From that time on, not only were the ways filled, but also the shops which manufactured gears, turbines, and all the other machinery essential to a ship.

Making new ship ways was relatively easy and the steel output was still large enough to handle the demand. The greatest difficulty was in obtaining machinery because of the scarcity of machine tools. In terms of shipbuilding that meant a scarcity of cranes, gear hobbers, gears and turbines. To meet that situation the Navy was given priority by the Vinson Act of June 28, 1940. By issuing a certificate of priority, the Secretary of the Navy could reserve for Navy building any scarce component, and he did reserve accordingly those items of machinery most difficult to obtain for the building of fast merchant ships.[21]

Compared with the Navy's building, the Maritime Commission's

---

[16] *Ibid.*, pp. 671-72, 723, 726.
[17] *Ibid.*, p. 723; *Consolidated Steel Corp. Annual Report for 1944*
[18] House, *Navy Department Appropriation Bill for 1942, Hearings*, p. 723.
[19] *Ibid.*, p. 668.
[20] *Ibid.*, pp. 238-39. The figure 68 includes builders of small craft.
[21] Charles H. Coleman, *Shipbuilding Activities of the National Defense Advisory Commission and Office of Production Management, July 1940 to December 1941*, (Report No. 18 of the Policy Analysis and Records Branch, Office of Executive Secretary, War Production Board, dated July 25, 1945), pp. 52-62. Cited hereafter as Coleman, *Shipbuilding Activities*.

long-range program was but a drop in the bucket. The 177 ships which had been contracted for prior to October 15, 1940 were estimated to cost $438,200,000.[22] Between October 1940 and July 1, 1941, the Commission planned to award contracts for 23 more vessels estimated to cost $101,100,000.[23] Trifles compared to the Navy's $5,000,000,000! But small as was the merchant shipbuilding program, it was being seriously interfered with by the Navy's priorities, especially on turbines and gears. Congestion was already intense in the American shipbuilding industry when in October of 1940 a British Merchant Shipping Mission assembled in New York to plan an emergency program.

[22] MC, *Report to Congress* for period ended Oct. 25, 1940, p. 7.
[23] Land to Harold D. Smith, Oct. 23, 1940 in MC gf 506-1.

Chapter 2

EMERGENCY SHIPBUILDING BEFORE THE DECLARATION OF WAR

I. The Inception of the Emergency Program

IN THE WINTER of 1940-1941, Britain was straining every moral and material resource to resist the kind of lightning war by which Hitler had overwhelmed France during the previous summer. The Nazis controlled all the coast of Europe from Lapland to the Pyrenees. German aircraft and submarines, operating from much better bases than they had had in World War I, seemed likely to strangle Britain by destroying its shipping. The ships of the United States were forbidden to enter the combat area by the Neutrality Act which had been passed in November 1939, but President Roosevelt was pushing various methods of aiding Britain while at the same time strengthening the defenses of the United States in the western hemisphere. The exchange of fifty destroyers to the British in return for the lease of naval bases in Newfoundland and other British territory in America was arranged in September 1940. Other forms of aid to Britain were to follow before long. Although the United States was making preparation for war (Congress passed the Selective Service Act in September 1940), it was not yet at war. Preparations were referred to as " the defense effort."

Under these conditions three emergency shipbuilding programs were laid on top of the accelerated long-range program during 1941. They form three waves of expansion. The first wave consisted of 60 ships contracted for by the British and the 200 which were announced on January 3, 1941, for the United States. The second wave came in April after Congress passed the act providing for transferring merchant ships and other forms of aid to Britain under the formula of " lend-lease." The third rolled in gradually during the rest of the year, until on December 7, 1941, the Japanese attack opened the flood gates for two even bigger waves.

PROGRAMS FORMULATED IN 1940 AND 1941
For Tonnage To Be Delivered In 1942

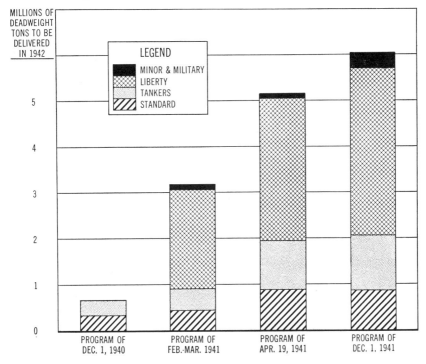

FIGURE 5. Private and British shipbuilding in the United States are included as well as ships contracted for by the U. S. Maritime Commission. Source: Fischer, *Statistical Summary*, Table A-2.

In each wave of expansion before the declaration of war, the officials of the Maritime Commission were called on to make decisions that later had important effects on the way the subsequent wartime programs were carried out. The atmosphere in which decisions were made and also the other agencies with which they were worked out kept changing. The very first wave was clearly a part of aid to Britain at a time when no general formula for giving aid, such as lend-lease, had yet been worked out or received legislative approval. Coordination of the national economy for the nation's defense was still a long way off; the war organization of the economy existed only in embryo, under the National Defense

Advisory Commission (NDAC). Under the chairmanship of William S. Knudsen the NDAC was drawing leading industrialists to Washington and giving the "New Deal" administration a general view of how to mobilize industrial resources. Shipbuilding plans were routed through the NDAC to be fitted into the general effort. Admiral Land was in charge of the shipbuilding section of the NDAC as well as Chairman of the Maritime Commission, and acted in both capacities in preparing plans for the President or in advising the British.[1]

Although there was general concern over Britain's ability to resist the Nazis, opinion in Washington was divided during the winter of 1940-1941. On the one side were the "all-outers," who felt that American economic life must be converted as fast as possible to a war economy either to aid Britain in her fight or to be ready for her fall. On the other side were the "not-quite-outers," who did not want to upset the American economy so much.[2] Emotions ran high in discussions of how much, how generously, or how selfishly the United States should aid Britain, but there was no doubt some aid would be given, and the British Merchant Shipping Mission which had assembled in the United States in October 1940 was assured that they could do some building in American yards.

The British were prepared to back an extensive program involving the construction of new yards. They recognized that new yards could hardly be expected to produce any ships before 1942, and they discovered at once that they could not have fast ships, for there were not enough of the needed turbines and gears, not even enough to supply the Navy and the Maritime Commission's long-range program. But they obtained permission to build 11-knot 10,000 ton freighters.[3] The British mission worked in consultation with Admirals Land and Vickery who on November 12 approved its general plan.[4] Official approval of the locations

---

[1] Coleman, *Shipbuilding Activities of NDAC and OPM*, pp. 1-2; *The United States at War*, pp. 23-52.

[2] The phrases are from Donald Nelson's *Arsenal of Democracy* (New York, 1946), p. 126.

[3] The British plans were politically approved on Oct 30, 1941 by Secretary of the Treasury Morgenthau, who was then the clearing house for aid to Britain. *Journal of Commerce*, Nov. 1, 1940.

[4] C. T. Ballantyne, British Purchasing Commission, to Philip Young of the President's Liaison Committee, Nov. 13, 1940, and Knudsen to Young, Nov. 18,

PLATE I. Above, side launching of a C1 at Consolidated Steel's Long Beach yard. Lower left, a C2-F, one of the fast freighters built by the Maritime Commission during the Chairmanship of E. S. Land, Vice Admiral, USN (Ret.) (lower right).

PLATE II. Above, a C3, largest and speediest of the standard types built by the Maritime Commission before the war. Center, a smaller freighter, the C1-B. Below, a tanker (T3 S A1) built at Bethlehem's Sparrows Point yard.

selected for new yards by the British was recorded by the Commission on December 11, at a time when emergency construction by the United States was being planned also.[5] Workmen were just clearing the sites where the British ships were to be built when President Roosevelt announced at a press conference on January 3, 1941 that the United States would build 200 similar ships.[6] The British had been urging action of this kind. " Looking into the future," wrote Prime Minister Churchill to President Roosevelt on December 8, 1940, " it would seem that production on a scale comparable to that of the Hog Island scheme of the last war ought to be faced for 1942." [7]

The idea that the United States should build 11-knot freighters such as the British would have to build was not relished by Admiral Land. What was later to be called the Liberty Fleet was in 1940 conceived of as an evil, perhaps a necessary evil, but an evil to be avoided if possible. The energies of the Maritime Commission were focused on the standard types. The decision that there must be additional construction to meet the emergency was made at the White House, and as a consequence the Commission accepted as a necessity late in December of 1940 the diversion of its energies from the goal which had been set up in peaceful times. Since early November, Admiral Land had been advising the White House how a program for the production of ships in quantities should be arranged, what types of ships were possibilities, what size of shipyard, and where the shipyards should be. Although at first he referred to emergency building as if it was a British program only, the arguments could apply and were applied to American building.

Admiral Land's advisory memoranda kept the Commission's long-range program in the picture as long as possible. In his first, on November 8, he was careful to dissociate the simple, slow type of ship which he proposed for the British from the " United States " types, by which he meant those of the Maritime Commis-

---

1940, in MC, Historian's Collection, Vickery file " 1940—Shipyards and Construction Data," which is a collection of the series of memoranda of Nov. and Dec. 1940, cited below.

[5] MC Minutes, pp. 15336-37.

[6] *Journal of Commerce*, Dec. 27, 1940; Samuel I. Rosenman, ed., *The Public Papers and Addresses of Franklin D. Roosevelt*, Vol. IX, *War—and Aid to Democracies* (New York, 1941).

[7] Winston S. Churchill, *Their Finest Hour* (Boston, 1949), p. 564.

sion's long-range program.[8] On November 18 he wrote: "In my judgment we are not interested in the type of ship proposed by the British, which type is for emergency use only. If it is decided to augment our own program, we should build ships for 20-years life and have an eye on the future. Therefore build ships to our standard designs."[9] On the twenty-ninth he set down that it would be a mistake for the United States to build emergency type ships and lease or charter them to the British. If we did build that type, "we should sell the ships to the British and be entirely clear of this design of vessel which is suitable for their purposes but would not be suitable for ours." "Furthermore, if our emergency becomes equal to or greater than that of the British, we can always commandeer the vessels. [Apparently he expected that possibility to be clarified within the year, before the ships were delivered.] The last thing I want to do is to repeat the mistakes of the last war and have a lot of obsolete vessels on our hands unless the emergency is so great as to make this an absolute necessity."[10] But he did not propose to do nothing, he proposed to build more of the standard types. He wrote in long hand at the bottom of a memorandum for the President: "If we don't watch our step the Merchant Marine will be 'The Forgotten Man' in the national defense picture."[11] He was thinking of the long-range program and how excellent cargo ships already under construction were being delayed by Navy priorities. If only those priorities could be shaken, the Maritime Commission would jump at the chance to build more of the svelte speedy freighters which would give the United States the world's best merchant marine.

A month later the emergency had grown and American building of 200 simple cargo ships, in addition to the British program of

---

[8] Land to Knudsen, Nov. 8, 1940. Carbon in MC gf 506-1, original in WPB file 324, 1041, in National Archives, Washington.

[9] Memorandum, "Fabricated Ships" by Land, Nov. 18, 1940, in MC gf 506-1. A letter from Knudsen to the President, Nov. 19, 1940 in Vickery files, "1940—Shipyards and Construction Data" is based on this memorandum of Admiral Land's but omits the paragraph quoted above. Knudsen's letter does not oppose building the simple type for the United States, and makes no mention of the British building which Knudsen had approved on the 18th.

[10] Land to the President, Nov. 29, 1940, in MC gf 500-3, also in MC, Historian's Collection, Vickery file, "1940—Shipyards and Construction Data."

[11] Memorandum, Land to the President, Dec. 2, 1940, in MC gf 500-3.

THE INCEPTION OF THE EMERGENCY PROGRAM 45

60, was an accepted objective of the Maritime Commission. In spite of Admiral Land's efforts, standard types had no place in the first wave of expansion. Indeed, according to the principles he himself had advanced, there was no room for them at the moment in a program attempting something like a mass production of ships. "The proper procedure for the production of ships in quantities," he wrote November 8, "is to settle on the *type* of ship, which in quantity production for quick delivery, (considering the excessive load on the auxiliary building and turbine building capacity of this country) is a simple type of cargo ship with reciprocating engines, boiler pressure of 200 to 220 lbs., with all steam auxiliaries. The quantity of ships desired and the times at which deliveries are required should be established." [12] After Christmas he was busy applying these principles to an American building program, but a program developed because of Britain's immediate peril.[13] No one could say for sure what would be done with the ships once built. That was an "iffy" question, as President Roosevelt said, a question which would not have to be answered until after a year had passed, but many people thought of the 200 ships announced by the President on January 3 as destined for the British and not really part of the American merchant marine.[14]

---

[12] Land to Knudsen, Nov. 8, 1940, in gf 506-1.

[13] An American emergency program for building 200 ships is assumed by Admiral Land's Memorandum for the President, Dec. 26, 1940 in gf 507-1. The President authorized the program to get under way by a memorandum to Knudsen, Dec. 28, 1940, telling him to have negotiations for the facilities started (WPB files 324. 1041 and Coleman, *Shipbuilding Activities of NDAC and OPM*, p. 33, and by allocating to the Maritime Commission from the Emergency Fund for the President $500,000 in cash and $36,000,000 in contract authorizations (Harold Smith to Land, Dec. 31, 1940, in MC gf 301-6). Congress added to this by H. J. Res. 77 which became Public Law No. 5, 77th Cong., 1st sess., on Feb. 6, 1941, creating a total fund of $350,000,000 to be known as the "Emergency Ship Construction Fund, United States Maritime Commission."

[14] *The Public Papers and Addresses of Franklin D. Roosevelt*, IX, 645. Testifying in favor of the appropriation to build these ships, Admiral Land said he did not know what would happen in a year or two.—Senate, *Emergency Cargo Ship Construction, Hearings* before the subcommittee of the Committee on Appropriations, 77th Cong., 1st sess., Jan. 29, 1941, H. J. Res. 77, p. 9. Later Commissioner Dempsey testifying before the Truman Committee spoke as if they were to be chartered under lend-lease. *Hearings*, pt. 5, p. 1296.

II. Selecting the Sites and Managers

Once a program of quantity production was decided on, the selection of shipyard sites became a pressing issue. In the survey which had been going on since early November the two basic problems of size and location had been well explored, and special attention had been given to the experience at Hog Island in World War I.[1] That was the largest and most talked about of the emergency yards used in America's previous experience with building a host of standardized ships in a hurry. The conclusions were that any yard with as many as 50 ways created an inefficiently excessive concentration of men and materials, and that it was particularly important not to have such a concentration near the cities of the Northeast. At the other extreme was the possibility that there should be "hundreds of ways" in a large number of small yards all over the country. This solution was rejected by Admiral Land on the ground that it was too hard to find persons with knowledge of shipbuilding to manage all these yards, or to find the machine tools to equip them. He repeated again and again that the most dangerous bottle-neck was the shortage of men able or experienced in the running of shipyards. The slight experience of this kind which there was in the nation could best be used by having as few shipyards as possible and yet turn out the ships.[2]

In the middle ground, studies were made to see what was involved in building and operating 28-way shipyards. Among the plans brought over by the British was one for a yard of that size.[3] Within the Technical Division of the Maritime Commission, plans were drawn and an estimate made of the cost of a 28-way yard.[4]

---

[1] An extensive canvass of possibilities was made in memoranda of Nov. 8 and other memoranda cited above which discussed ostensibly the British program.

[2] Knudsen to the President, Nov. 19, 1940, and Land to the President, Nov. 29, 1940 (in MC, Vickery file, "1940—Shipyards and Construction Data") put the argument as follows: "The congested condition of the machine tool and equipment market" is such that "a multiplicity of small yards increases the problem of getting the yards in production and increases the cost involved. Ships will be obtained more quickly from three or four assembly yards with centralized management properly located with regard to steel fabricators and built to the needs of the program rather than from a considerable number of small yards widely distributed geographically."

[3] Liverpool *Journal of Commerce*, Nov. 21, 1940.

[4] In the MC Technical Division files is a blueprint dated Nov. 14, 1940 and

SELECTING THE SITES AND MANAGERS 47

President Roosevelt had given instructions that plans be made for 100 ways. While pointing out that no such large number of ways was needed, Mr. Knudsen and Admiral Land mentioned 28-way shipyards producing 100 ships a year among the alternatives to be considered, but nothing that large was recommended.[5] At one time Admiral Land suggested that 4 yards of 14 ways each would be the best solution because of the difficulty of finding machine tools or first-class managers for any larger number.[6] Later a 6-way yard was considered about the smallest efficient operating unit. In the concluding months of 1940, when nearly all the factors in the problem—even the number and type of ship to be built—were variables, shipyards ranging in size from 4 to 14 ways were considered good possibilities.[7]

Connected with the size and number of new yards was the problem of their location, a question full of political dynamite. Each region was optimistically conscious of its possibilities and would feel slighted if not given what it considered its share. Within each region many more sites were advocated than could be used. Later, a flood of letters came from Senators, Representatives and other political leaders, urging consideration of proposals from their constituents. But the initial selection, which laid down the lines followed in future growth, was made when the political pressure was less than it became after the program was announced and its potentialities were appreciated.

The main considerations were the adequacy of transportation and the availability of labor and management. Among the many lists of sites to be considered the longest was the first, set forth on November 8, 1940. It mentioned 14 yards, included places remembered for their shipbuilding in World War I such as Bristol,

---

entitled " Typical Layout for 28-way Fabricated Shipbuilding Plant." The estimate, avowedly based on Hog Island data, was $33,600,000 for 28 ways.

[5] Knudsen to the President, Nov. 19, 1940, in MC, Historian's Collection, Vickery file, " 1940—Shipyards and Construction Data."

[6] Land to the President, Dec. 2, 1940, says: " The greatest material bottleneck in the national defense program today is the machine tool situation and this situation led us to the conclusion that the number of yards for this British program should not exceed four. With the type of ship proposed (Type A [11 knots]), the number of ways is of much less importance than formerly as the time a ship is on the ways can be materially reduced, particularly as we expect to do a great deal of welding rather than riveting. This means that a maximum of four yards with fourteen ways per yard as the probable maximum number of ways will take care of the situation."

[7] Land to Knudsen, Nov. 8, 1940, in MC gf 506-1.

Pennsylvania; Groton, Connecticut; and Newburgh, New York; but these World War I sites were quickly eliminated from consideration because they were not going concerns with experienced staffs. They were abandoned yards where the machinery had been removed or rusted away. They were no better than new sites, and against them was the fact that they were in the area of industrial concentration where the naval program and other industrial activity was putting a strain on the labor supply and transportation.[8]

Great weight was given to the managerial associations of the sites proposed, although the Maritime Commission had a large part in bringing site and management together in many instances. First to be definitely agreed to were those sites which Todd Shipyards included in its plans. South Portland, Maine, and Richmond, on San Francisco Bay in California, were approved as sites for British building on December 11, 1940; and three other widely scattered yards, at Los Angeles, at Houston, and at Portland, Oregon, were approved for the American program on January 10, 1941.[9] Although precise locations were selected by members of the Todd group, all these yards were in places which had been included in Admiral Land's first list, that of November 8.[10] Three being on the West coast, one on the Gulf and one at the other extreme in Maine, they were all located where new supplies of labor could be found without competing directly with the already busy shipbuilding centers of the Northeast.

Concern with the labor supply and with developing the South led to the selection also of Mobile and New Orleans. Franklin D. Roosevelt had ever since World War I been enthusiastic for shipyards in the South.[11] The Maritime Commission had already built up shipbuilding in the Gulf area by giving contracts for standard freighters to Tampa Shipbuilding and Engineering Co., to Ingalls

---

[8] Notes by F. C. Lane of interview with P. J. Duff, Sept. 20, 1946. About this time, Mr. Duff, of the Technical Division of the Maritime Commission, went over many of the old yards to see how difficult it would be to reopen them. He found them pretty completely rusted away. On the congestion in the Northeast, and the uselessness of the old yards also, see Admiral Land's testimony, Jan. 29, 1941, in Senate, *Emergency Cargo Ship Construction, Hearings*, p. 11, and Admiral S. M. Robinson's testimony, Feb. 1941, in the *Navy Department Appropriation Bill for 1942, Hearings*, pp. 731, 743, and before the Truman Committee, *Hearings*, pt. 5, pp. 1421-22.

[9] MC Minutes, pp. 15753-15753-A, Jan. 10, 1941.

[10] Except for Richmond instead of Alameda and So. Portland instead of Portland, they were the same cities.

[11] Notes of interview of F. C. Lane with Admiral Land on Sept. 30, 1946.

SELECTING THE SITES AND MANAGERS 49

Shipbuilding Co., to the Pennsylvania Shipyards, Inc. at Beaumont, Texas, and by approval of private contracts to the Gulf Shipbuilding Corp. at Chickasaw, Ala., but Admiral Land was not vastly encouraged by the results.[12] He set forth his attitude in a memorandum to the President December 2, 1940 thus: " It is recognized that serious consideration should be given to locating these plants in the South in order to distribute the burden and to utilize to best advantage the labor available. The Maritime Commission's experience at Tampa, Florida, has been quite unhappy on account of labor troubles and unfortunately some K.K.K. difficulties which have only recently been ironed out. Theoretically the shipbuilding labor situation in the South looks attractive but practically this is not the case. Many difficulties are already being experienced with existing plants in the South which have contracts for the Maritime Commission and for the Navy program. ' Scamping ' shipbuilding labor in the South is already a difficult problem." [13] The South lacked not only management but also " engineers, naval architects, draftsmen, managers, foremen, quartermen, and leading men," all kinds of " white collar brains." [14]

The arguments on the other side were set forth by the National Resources Planning Board of which the Chairman was Frederic A. Delano, the uncle of the President. The Northeast had the advantage of better facilities with which to train new workers, but satisfactory new workers were becoming scarce. " In the southern states, on the other hand, there are large numbers of workers who would be available for unskilled work or for training. Other industries are not competing keenly for labor." On the West coast the obvious centers—Seattle, Portland, San Francisco, and Los Angeles—were crowded with naval activity or aircraft industries, and a steel industry to support shipbuilding was lacking. " No ship plates are produced in this area." The Great Lakes could not build large ships because large ships could not be got out through the locks. By these comparisons the possibilities of the Gulf for shipbuilding stood out all the brighter. It had a favorable climate and was protected from attack. It was " a low

[12] Awards in MC Minutes, to Tampa, pp. 3691, 8949, 9455; Ingalls, pp. 6272, 9454, 14042, 14218; Pennsylvania, p. 9594; Gulf, p. 15280.
[13] In MC, Vickery file, " 1940—Shipyards and Construction Data."
[14] Land to Glenn E. McLaughlin, Chief, Industrial Location Section of the National Resources Planning Board, Mar. 5, 1941, in MC gf 503-1.

income area " which was " in need of supplementary employment opportunity." Although its labor was unskilled, yet, the Planning Board argued, " with the aid of key skilled workmen supplied from existing yards, a shipbuilding program could be effectively introduced into additional ports, particularly those engaged in shipbuilding during the World War. . . . The South is handicapped somewhat by the absence of mills making ship plates and by the scarcity of machine shops, but it possesses adequate supplies of lumber and certain iron and steel materials." [15]

Arguments such as these, backed as they were by the White House, made it necessary to take the gamble of putting some yards in the South. New Orleans was a natural choice since there were railroad connections and no other yards nearby, but Mobile was for a long time left off the lists since expansion of shipbuilding there would draw labor away from the nearby yards which were building standard-type cargo ships—the Gulf yard at Chickasaw, further up Mobile Bay, and the Ingalls yard at nearby Pascagoula, Mississippi.[16] There was talk of a tie-up between Todd interests and the Alabama Dry Dock and Shipbuilding Co. In the end the management of a new emergency yard was supplied by Alabama Dry Dock and Shipbuilding Co., which did repair work primarily but was building two tankers.[17] Mobile, New Orleans, and Houston together gave the South a substantial place in the plans for emergency shipbuilding, and it gained an even larger place later.

Along the East coast, Baltimore was considered so well situated that it was included in every list, although the management which would develop a site there was treated as doubtful until the very last moment.[18] Wilmington, North Carolina, on the other hand, was not particularly favored until the Newport News Shipbuilding & Dry Dock Co. decided to develop a yard there. On the earlier list it had been bracketed on equal terms with Norfolk, Virginia, and Moorehead City, North Carolina. The latter site was strongly backed by the Governor of North Carolina, and other political

---

[15] McLaughlin to Land, Feb. 20, 1941, in gf 503-1.

[16] On Nov. 19, 1940, Knudsen wrote, " The Maritime Commission and the Navy do not approve of Mobile, Alabama, because of interference with ships now under contract." On Nov. 29, Land wrote, " Mobile has been dropped for the time being. Permanently, I trust." See MC, Vickery file " 1940—Shipyards and Construction Data."

[17] Reynolds' Way Charts.

[18] Land to the President, Dec. 26, 1940, in Land reading file.

leaders in the state, but when the leaders of the shipbuilding industry picked Wilmington, other sites in North Carolina were passed over.[19]

At the opening of 1941, nine yards with a total of 65 ways were approved to build in two years the desired total of 260 ships at the following locations: [20]

	Ways
South Portland, Me.—the Todd-Bath Iron Shipbuilding Corp. for the British	7
Baltimore, Md.—Bethlehem-Fairfield Yard	13
Wilmington, N. C.—North Carolina Shipbuilding Co.	6
Mobile, Ala.—Alabama Dry Dock and Shipbuilding Co.	4
New Orleans, La.—Delta Shipbuilding Co.	6
Houston, Texas—Houston Shipbuilding Corp.	6
Los Angeles (Terminal Island), Cal.—California Shipbuilding Corp.	8
Richmond, Cal.—Todd-California Shipbuilding Corp. for British	7
Portland, Oregon—Oregon Shipbuilding Corp.	8

First and last the main concern in launching the emergency shipbuilding program was to enlist the leaders of the shipbuilding industry. The Big Five, who had most of the "shipbuilding brains" on their payrolls, were not eager to join in or were prevented from doing so by the Navy's claim that its work should have priority. Admiral Land and Admiral Vickery talked to Joseph N. Pew, Jr., of Sun Ship on December 24 and Mr. Pew worked out a plan by which his yard would launch three cargo ships a year from each of seven of its ways; but Navy wanted the Sun yard to build tenders for destroyers and submarines, and Mr. Pew said they would have to be postponed if the merchant ships were to be turned out.[21] Admiral Land sent the proposal along to Secretary Knox, saying on the routing slip: "Putting this bluntly and brutally, it resolves itself into a question of priority, i. e.: which is more important under present conditions, 30 to 40 merchant vessels or 6 tenders for destroyers and seaplanes."[22] The Navy answered that it would utilize all of the facilities of Sun continuously during the rest of the war, and even asked Maritime to stay

[19] Land to Thomas F. Little, Jan. 17, 1941, in Land reading file.
[20] Press Release 804, Jan. 8, 1941, (in MC gf 105-2) said that the sites for building the 200 ships had been selected. For dates of Commission action, Jan. 10-17, see Fischer, *Statistical Summary*, Table E-5.
[21] Joseph N. Pew, Jr. to Land, Dec. 30, 1940, in MC gf 507-1.
[22] Land to Knox, Jan. 2, 1941. The quoted passage is on the routing slip, in MC gf 507-1.

out of the whole neighborhood because it was in the middle of the area where naval building was concentrated and any diversion of skilled labor to building merchant ships would interfere with the Navy's plans.[23] Just another " question of priority "!

Newport News, another shipbuilding leader, had so many Navy contracts there was no hope of getting any more merchant building into that yard, but Admiral Land worked hard to interest the company in undertaking the management of a new yard. As late as December 26, 1940, Newport News Shipbuilding was noted along with New York Shipbuilding as among the major firms " not interested to date." [24] New York Shipbuilding never was interested, but Admiral Land persuaded Homer L. Ferguson and Captain Roger Williams of the Newport News Company. By many a long distance telephone call he put it to them that " this was a job that with their capacity they could afford to do." [25] On January 10 they announced that they would build emergency ships at a new plant at Wilmington, North Carolina.[26]

Bethlehem Steel Co. was the natural firm to turn to for leadership in developing the yard at Baltimore, since it was operating a big yard adjacent to its steel mill at Sparrows Point southeast of the city and had a repair yard also in Baltimore harbor. Most Bethlehem yards were committed for naval work, and as late as December 26 the J. G. White engineering firm as well as Bethlehem was listed as the possible operator of a new Baltimore yard, but in the second half of January 1941 terms were finally arranged for a Bethlehem subsidiary to operate a new emergency yard named, because of the suburb in which it was located, Bethlehem-Fairfield.[27]

Federal, the remaining member of the Big Five, was considered as one source from which managerial talent might be drawn for the emergency yard planned in New Orleans.[28] It was very busy

[23] Knox to Land, Jan. 9, 1941, in MC gf 507-1.

[24] Land to the President, Land reading file; statements by Land, June 17, 1949.

[25] House, *Emergency Cargo Ship Construction, Hearings*, before the subcommittee of the Committee on Appropriations, 77th Cong., 1st sess., H. J. 77, Jan. 18, 1941, p. 13.

[26] *Journal of Commerce*, Jan. 10, 1941, p. 22. The contract was not approved until Jan. 17.

[27] Land to the President, Dec. 26, 1940, Land reading file; Minutes, pp. 15834-15834-A (Jan. 17, 1941).

[28] The memorandum of Dec. 26 couples Federal and New Orleans. It mentioned the American Ship Building Co. as among those to be considered without linking it to any site.

SELECTING THE SITES AND MANAGERS   53

already, however, working for the Maritime Commission and the Navy, and its managerial talents might be stretched to the limit by a future expansion of its existing yards. Accordingly the Maritime Commission turned to a leading shipbuilding company of the Great Lakes, the American Ship Building Co. Since the Liberty ships could not be built on the Lakes (being too long to pass the locks) the shipbuilding talent of the Lakes region was drawn upon by arranging with American Ship Building for the operation of the yard at New Orleans. A new company called the Delta Shipbuilding Co. was formed for that purpose.[29]

Thus two of the nine new emergency yards were operated by old-line companies, Bethlehem and Newport News. One was in the hands of a company transplanted from the Great Lakes, and one, at Mobile, was in the hands of managers of a repair company. All the other five, the five first approved, were under the combine headed by Todd Shipyards Corp., a new team. With the old-line yards busy with naval building and unable to supply experienced management for many new yards, the Commission had to turn to a relatively inexperienced group.

Under the Todd name had been gathered together a wide variety of men whose managerial talents had been displayed in various fields. Todd Shipyards itself had one of the very largest ship repair organizations with yards on all coasts and strong financial backing. It had built ships during World War I and led in the formation of the new Seattle-Tacoma shipyard. The ability of its president, John D. Reilly, as a man of bold decisions was spotlighted by newspapers and by *Fortune*, "the Magazine of Business," under the title: " Biggest Splash in Emergency Construction will be made by Todd." A second figure in the combination, also ranking high in the shipbuilding fraternity, was " Pete " (William S.) Newell, head of the Bath Iron Works. *Fortune* said, " Of Newell's know-how there can be no doubt." Towards the third element in the group, the Kaiser construction companies, *Fortune* expressed the general attitude by saying in July 1941:

---

[29] MC Minutes, pp. 15834-15834-A, Jan. 17, 1941; E. B. Williams, " The Delta Shipyard," *Marine Engineering and Shipping Review*, Apr. 1943; Outline of the Plan for Construction and Operation of a Shipyard, and the Building of Ships in It, at New Orleans, and for the U. S. Maritime Commission, in MC gf 507-3-5. The urging which induced the American Ship Building Company to join reluctantly is set forth by W. M. Gerhauser in House, *Investigation of Shipyard Profits, Hearings* before the Committee on the Merchant Marine and Fisheries, 79th Cong., 2d sess., 1946, pp. 518-20, 267-68.

"Whether Kaiser and his coterie of dam builders have bitten off more than they can chew remains to be seen, although the presumption is that they can get away with it." [30] Kaiser's only previous connection with shipbuilding was joint ownership with Todd in the Seattle-Tacoma Shipbuilding Corp. Kaiser would be useful, it was understood, in getting new yards built quickly—he had cranes, bulldozers, and construction gangs at his command—but his prominence in the shipbuilding of World War II was still in the unknown future.

Actually the "Kaiser" part of the Todd group consisted of seven or eight West coast construction firms known, oddly enough, as the Six Companies.[31] Henry J. Kaiser began business with a photographer's shop in New York, switched to a sand and gravel business, built up a reputation on the West coast contracting for all kinds of construction, and then brought together other construction firms to join in the achievements which won him fame as "Fabulous Kaiser." His group built the Hoover, Bonneville, and Grand Coulee dams, and the Bay Bridge from San Francisco across to Oakland, his headquarters. He was an organizer and a salesman, full of ideas and readiness to tackle the "impossible." Working for him were "six of the smartest young men in the country," men barely over 30, who, it was said, "keep the promises that Henry makes." [32]

In shipbuilding the kingpins among Henry Kaiser's young men were Edgar Kaiser, his son, and Clay Bedford. The former became head of the emergency shipyard at Portland, Oregon; the latter became general manager at Richmond, California, where the yard known originally as Todd-Cal, later as Richmond No. 1, was located, and where other yards were later built for the Maritime Commission. The third emergency yard on the West coast, that of the California Shipbuilding Co., at Los Angeles, (commonly known as Calship), was managed by associates of Kaiser in the Six Companies, at first by the McCone-Bechtel organization, later

---

[30] *Fortune*, July 1941, p. 124, and *New York Times*, Dec. 20, 1940, p. 1; Dec. 21, 1940, p. 5; Feb. 9, 1941, p. 25.

[31] In addition to Henry J. Kaiser Co. and The Kaiser Company, they were W. A. Bechtel Company, Bechtel-McCone-Parsons, General Construction Co., Morrison-Knudsen Co., MacDonald & Kahn, Inc., and Pacific Bridge Co. House, *Investigation of Shipyard Profits, Hearings*, charts following p. 374.

[32] *Fortune*, Oct. 1943, p. 147 ff.; see also *Fortune*, July 1941, pp. 124 ff.; Frank J. Taylor, "Builder No. 1," *Saturday Evening Post*, June 7, 1941.

by J. A. McCone. On the East coast, William S. Newell was in charge of the emergency yards at South Portland, Maine. The fifth yard of the Todd group, that at Houston, Texas, was most directly under Todd management. Ownership in all seven yards (counting the two building for the British) was divided between Todd Shipyards Corp. and the Kaiser group, with Bath Iron Works Corp. owning a part also in the Maine yards.[33] The interlocking ownership created a presumption of common Todd management at the very top. In January 1941 the later importance of Kaiser-managed yards could hardly be foreseen, for in shipbuilding circles Kaiser was still overshadowed by the Todd name, and it was not realized that later expansion would make the first wave of the emergency seem like a mere ripple.

In the perspectives of early 1941, an emergency program of 260 merchant ships, on top of the Navy's huge program and on top of the accelerated long-range program, seemed very large indeed. Admiral Land advised the President that to increase further the building of merchant ships would bring grave difficulties and that if any more were planned they should be of the standard designs and be built in existing yards.[34]

III. The Second Wave of Expansion

The second wave of expansion was directly linked to Congressional approval of the lend-lease program. It contained provision for more Liberty ships, for more C-type freighters, and also for government-built tankers.

After lend-lease was provided for by the Defense Aid Supplemental Appropriation Act, approved March 27, 1941, Admiral Land was called to advise how the funds should be allocated for about 200 more ships. He successfully urged on President Roosevelt that at least half the 200 be C-types.[1] Construction of tankers

---

[33] Corporate arrangements and management are fully discussed in House, *Investigation of Shipyard Profits, Hearings*, above cited.

[34] In the "Land file" collected from Admiral Land's desk are the original and many carbons of a memorandum dated Jan. 27, 1941, apparently intended for the President but never sent in that form. The statement cited is in the Jan. 27 memorandum and in one of Feb. 5 (in the same file and the reading file) which apparently was sent, and is addressed to the President via Mr. Knudsen.

[1] Land to the President, Mar. 27, 1941 and Apr. 2, 1941, both in MC gf 506-1. Also Land to General J. H. Burns, Mar. 28, 1941 in Land reading file.

was decided on in addition, and between the President's initial approval in general and the final authorization of contracts by the Commission, the number of ships was increased. The program was referred to the Office of Production Management (OPM) which was created in January 1941 to continue the work of the NDAC in directing the "defense effort." On April 17 the OPM approved a total of 306 new ships, and they were incorporated in the Commission's program of April 19, 1941, namely: 112 Emergency ships, 24 C1's, 46 C2's, 52 C3's, and 72 tankers.[2]

This program of April 19 included so many standard-type cargo ships, counting the 72 tankers, that the Commission had to fall back for authority on Title VII of the Merchant Marine Act of 1936.[3] The Commission was using its authorizations to the limit in an effort to build as much as possible and especially as many of the standard type as possible.

The inclusion of tankers of a commercial type for which the government would contract directly was a new element in the Commission's shipbuilding activities, and one which was to become of major importance during the war. Hitherto, an adequate number of tankers had been provided in the American merchant marine by the initiative of the oil companies. Aside from the Navy's construction of oilers, the government had entered into tanker construction only to pay for "defense features," which meant among other things the extra speed given to those of the *Cimarron* class. The tankers planned for in April 1941, on the other hand, were a standard commercial design which the Sun Shipbuilding and Dry Dock Company had been building for the Standard Oil Company of New Jersey. A design providing for electric drive was selected because of the shortage of gears and was given the designation of T2-SE-A1.[4] With speed of 14 to 15

[2] MC gf 506-1, Apr. 16, 1941. The staff of the Maritime Commission had been working on plans since March 22 but there were signs of last minute haste in the final compilations dated Apr. 16. In the list for the OPM there was an error of addition which Admiral Land detected and corrected. In the memorandum submitted to the Commission one item (of 24 ships) was skipped, and had to be acted on later. See MC Minutes, pp. 16959-60, 17072 (Apr. 24, 1941). A much commented-on copy is in gf 507-3-1, as well as preparatory material on costs.

[3] MC Minutes, p. 16960. But the Minutes fail to give the reference to the subsequent approvals under that title. Just how many ships were being built at a particular time under Title VII depends on how many of those previously begun under Title VII had subsequently been shifted to Title V.

[4] Statement by James L. Bates, Apr. 1941.

THE SECOND WAVE OF EXPANSION

knots and a deadweight tonnage of 16,735, these tankers were considerably bigger and faster than the emergency cargo ships, and became a very substantial element in the output of the Commission's yards.[5]

At first glance this April program adding 306 ships seems a reversal of the attitude Admiral Land was presenting to the President just two months earlier. In February 1941 he wrote: " Further dilution of shipbuilding brains . . . . will only result in gross waste and inefficiency without accomplishing the ultimate result desired, namely the delivery of finished ships." A survey had shown that the shipbuilding capacity of the country was fully occupied. But such reports, which were made again and again, although they reflected less confidence in newcomers to shipbuilding, such as Kaiser, than would have been accorded a year later, did not mean that Admiral Land thought no further expansion was possible; they meant merely that no further expansion was recommended *unless*—unless some of the brains and machinery pre-empted by the Navy were released by modifying priorities on Navy vessels not scheduled for delivery until 1944, 1945, or 1946, and by assigning old-line yards to the Commission's program.[6]

Since the White House was resolved to expand the building of merchant ships, Admiral Land's insistence that it could not be done without upsetting Navy priorities bore fruit. The Priorities Committee of the Army and Navy Munitions Board was instructed by the Office of Production Management that merchant ships should be treated on the same basis as ships for the Army and Navy.[7] Of greater immediate effect, however, was a new assignment of existing facilities and shipbuilding organizations. The first move towards the second wave of expansion was to pry the Navy out of one of the Big Five establishments and destine its facilities and " shipbuilding brains " for the Maritime Commission program. By a Presidential conference on March 14, 1941, it was agreed that the Navy would withdraw from the big Sun Yard at Chester, Pennsylvania. The division made that spring in the yards of the Big Five endured throughout the war. Navy was to place no orders with Sun Ship. Maritime was to place no contracts with Newport News or New York Ship, nor in the

[5] Fischer, *Statistical Summary*, Tables A-1, B-3.
[6] Land to the President, Feb. 5, 1941, Land reading file.
[7] Coleman, *Shipbuilding Activities of NDAC and OPM*, pp. 65-66.

Bethlehem yards at Fore River, San Francisco (Union Iron works), and Staten Island. Of the Bethlehem yards, that left only Sparrows Point in operation for merchant ships. Federal and a number of smaller yards, such as Western Pipe & Steel, Moore Drydock, and Ingalls were working for both the Maritime Commission and for the Navy, and as a part of the arrangement by which the Navy withdrew from Sun, Western Pipe & Steel increased the amount it was constructing for the Navy. In many smaller yards the Maritime Commission and the Navy continued to struggle for space, way by way.[8]

The assigning of Sun Ship to Maritime Commission contracts was the basis of the tanker program. Its eight ways were already committed to tanker construction on private accounts, but the company had the staff with which to manage a larger yard, and a twelve-way addition paid for and owned by the government was built along the Delaware River to handle the government contracts.

More shipways were needed also to build the added standard and emergency cargo ships, and they were provided by similarly adding to yards in existence, or at least yards of existing companies. To build C-types, 5 ways were added at Seattle-Tacoma, 3 ways at Pennsylvania Shipyards, 2 ways (added or renovated and enlarged) at Western Pipe and Steel, 1 way at Moore Dry Dock and 1 way at Pusey and Jones.[9] The nearest to a new yard was constructed at Wilmington, California, in the port of Los Angeles for operation by the Consolidated Steel Corp., Ltd. This steel-fabricating company was building C1's in the small Craig yard nearby.[10] To enable them to build more C1's, the Wilmington yard was built in another part of the port of Los Angeles, but Admiral Land called it " extending an existing facility " because no new management was involved.[11] Thus the Commission strengthened the companies which had participated in its long-range program and might form part of the nation's postwar shipbuilding potential.

[8] Land to Knox, Mar. 14, 1941; Memorandum for the Bureau of Ships, Apr. 7, 1941, Land to Bernard Baruch, Apr. 28, 1941; in Land reading file.
[9] Fischer, *Statistical Summary*, Table E-5.
[10] *Annual Report of Consolidated Steel Company*, 1944.
[11] Land to Senator Downey, July 1, 1941, in Land reading file. Only 4 ways were authorized in June 1941; 4 more in August.

THE SECOND WAVE OF EXPANSION 59

At the same time, the emergency yards which were just starting were told to increase the number of ways: Bethlehem-Fairfield from 13 to 16, Calship from 8 to 14, Oregon Ship from 8 to 11, Todd-Houston from 6 to 9, and North Carolina from 6 to 9. The two managements building yards for the British were given contracts to construct new yards next to those already being built: at

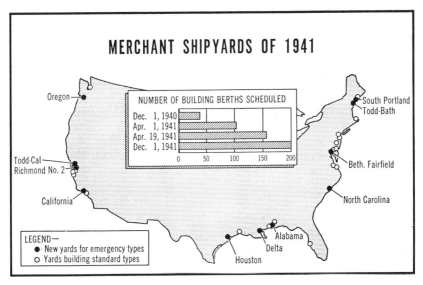

FIGURE 6. Class I yards. Names of the yards building standard types, many of which were much expanded in 1941, are given on Figure 4.

South Portland, Maine, a new 4-way yard under a new corporation, the South Portland Shipbuilding Corp.; and at Richmond, California, a new 6-way yard, commonly known as Richmond No. 2 under the Richmond Shipbuilding Corp., a subsidiary of Kaiser's Permanente Metals Corp.[12] In each case the new yard was under the same management as the old so that all the additions in the second wave were in that sense an expansion of existing yards and in accordance with the Commission's policy of not starting any more new yards.

[12] Fischer, *Statistical Summary*, Table E-5.

IV. THE THIRD WAVE

The first two waves of expansion were clear-cut moves; each was programmed and announced as a whole. For each there was one decisive Congressional action and Presidential authorization. The third wave was a composite and dragged out from spring until winter of 1941. It was composed of four elements at least: (1) small vessels designed for transfer to the British, (2) ore carriers, (3) sea-going tugs and concrete barges, and (4) major types such as had been the exclusive concern of the two previous waves of expansion. Each of these elements in the program was a response to a different and distinct aspect of the general shortage of shipping.

The program of April 19, 1941 had hardly been formulated before Sir Arthur Salter of the British Mission pointed out to Admirals Land and Vickery in detail that more was needed. Assuming the current rate of sinkings and the adequacy of the April program, when combined with British building, to make up for losses once that program was fully underway, Sir Arthur stressed the very urgent need for tonnage during the first half of 1942. He suggested that this be provided for by drawing on the shipbuilding potentialities of the Great Lakes for harbor tugs, small cargo ships, and small tankers which could be delivered early in 1942. Prompt action, he said, was essential. The Technical Division made an extensive and detailed report to Vickery within 16 days and funds were provided under lend-lease, but Commission approval came only on June 20, 1941, no satisfactory contracts were negotiated until August 1941,[1] and none of these ships was delivered before September 1942, when they could serve not in the anticipated crisis but in another one.[2]

The shipping situation in 1941 was directly responsible also for expanding in June and July the program for Libertys and C-types.

---

[1] MC, Historian's Collection, Land file, Apr. 25, 1941, Salter to Land with an aide-memoire of Apr. 22 attached; Vickery file, " Program of Construction as recommended in Sir Arthur Salter's letter to Captain Vickery on Apr. 24, 1941 "; MC Minutes, p. 17806 (June 20, 1941); " U. S. M. C. Ship Construction Program, 1938-1945," Part I: a collection of directives compiled by the Secretary of the MC. Cited hereafter as Program Collection.

[2] *U. S. M. C. Official Construction Record—Vessels Delivered 1939 through 1945,* House Committee Print No. 106 of the Committee on the Merchant Marine and Fisheries. Cited hereafter as *Official Construction Record.*

On May 27, 1941 the President declared a state of Unlimited National Emergency. Next day Admiral Land sent him a memorandum explaining the possibility of expanding American shipbuilding 50 per cent after August through speeding up the work in existing yards. If there were enough gears, steel, skilled labor, and so on, output per way per year could be counted at the rate of 3 or $3\frac{1}{2}$ ships instead of 2. But British shipping losses for the first week of June were at a startlingly high figure,[3] and when Admiral Land put his recommendations for expansion in more specific form he recommended adding 48 more shipways, about half for the minor types and half for more Libertys or C-types.[4] In reply the President expressed his appreciation of the splendid progress made but directed the Commission to " use every possible means, even to re-writing the contracts to get these ships built at an earlier date than is now contemplated." [5] Responding to this injunction, the Commission wired all the managers of emergency shipyards asking how many they could possibly deliver by the end of the year,[6] and Admiral Land in a widely circularized speech appealed to the workmen for greater efforts.[7] The result was a new schedule accelerating and expanding deliveries so that in June 1941 the total of Liberty ships planned for delivery in 1941 was raised from 1 to 19 and the total for 1942 was raised from 234 to 267. As these figures show, the acceleration of the program in June was mainly in the speed with which the yards were expected to get into production.[8]

Writing contracts that would give effect to the new schedule was a slower process. Presidential approval was given in June but a supplemental appropriation of over one billion dollars had to be obtained from Congress and that took time. In July Admiral Vickery explained the need for this appropriation. " Sinkings during the first five months," he said, " were at the rate of 6,600,000 dead-weight tons per year. After our proposed expansion comes

[3] Sir Arthur Salter to Land, June 11, 1941, in MC, Historian's Collection, Land file.
[4] MC gf 107-1, Memoranda to President from Land, May 28 and June 11, 1941.
[5] The President to Land, June 16, 1941, in gf 107-1.
[6] J. E. Schmeltzer to Todd-Bath Iron Shipbuilding Corp., June 17, 1941, in gf 507-4-14, and similar wires in files of other yards.
[7] Land to the President, July 1, 1941, in gf 506-1.
[8] Program tabulations of Apr. 19, 1941, June 12, 1941, and July 1, 1941. MC, Historian's Collection, Research file 117, Statistics, containing many papers from the files of the Statistical Analysis Branch, Production Division.

into full operation, ship construction in the United States will begin to keep up with this rate of losses." [9] The appropriation became law on August 25, 1941, with a clause permitting some of the vessels constructed with this money to be made available to the British under the Lend-Lease Act.[10]

A notable feature of this third wave of expansion, insofar as it affected the major types, was the change that occurred in the way the program was conceived. In the first and second waves the objective had been stated first in terms of total number of ships, and then calculations made as to how long it would take to build them. During the third wave the Commission's planning began to emphasize more the rate of production. Taking a year or two as the unit of study, it calculated how many ships could be built with all yards working at full capacity. Admiral Vickery said to the Appropriations Committee in July 1941, " Our aim is to load the shipbuilding industry with all the ships it can absorb as fast as it can absorb them, as long as this war lasts."[11]

The new facilities for the third wave involved no new yards, except those for building minor types. " Our theory has been to stretch the management across the plant rather than create new facilities, because when you start a new company you have to take the personnel from some other place and put them at the new site," said Admiral Vickery. " So our theory has been to favor expansion rather than the creation of new facilities." [12] More production of C-types was provided by strengthening the long-range yards, two new ways being built at Bethlehem's Sparrows Point yard, and two at Ingalls in Pascagoula, Mississippi. The new yard of Consolidated Steel at Wilmington, California, was doubled in size, from four to eight ways. In emergency yards two ways were added at the Delta yard in New Orleans, two at the yard of the South Portland Shipbuilding Co. and three at Richmond No. 2. Alabama Shipbuilding and Dry Dock Co. in Mobile which had

---

[9] House, *First Supplemental National Defense Appropriation Bill for 1942, Hearings* before subcommittee of the Committee on Appropriations, 77th Cong., 1st sess., July 17, 1941, p. 409.

[10] 55 Stat. 669, First Supplemental National Defense Appropriation Act, 1942, Title III.

[11] House, *First Supplemental National Defense Appropriation Bill for 1942, Hearing*, p. 410.

[12] *Ibid.*, p. 424.

begun four ways for Liberty ships added eight ways to be used for building tankers.[13]

In some places these changes made it necessary to alter the plan of a yard once or twice after construction had begun, especially since contractual arrangements for the third wave of expansion were not completed until September, after Congress had passed the appropriation. Late changes in plan had serious consequences, for example, at South Portland, Maine, where the original plan for the four ways added in the second wave called for a relatively tight yard; adding half as many more ways led to very serious overcrowding at the head of the ways.[14] Adjustments to changing plans had to be made in practically all the shipyards. That was a consequence of the gradual and cumulative way in which the emergency shipbuilding program developed.

The other two elements in the third wave of expansion were both connected with difficulties in the steel supply. To maintain the flow of iron ore from Lake Superior to the foundries, the construction of 16 ore carriers was declared essential.[15] Ore boats already under construction on private contracts were nearing completion so that ways would be vacant for new construction. New building was discussed with the Lake Carriers Association, but the members of the Association failed to present an acceptable agreement by private initiative. Accordingly the Commission authorized Admiral Vickery to make plans and negotiate with the builders. Contracts were awarded in October 1941.[16]

While building ore carriers to maintain the steel supply, the Maritime Commission also decided to build concrete barges to make up for the lack of steel. A proposal for building sea-going tugs and steel barges had been put forward in June 1941 by Mr. Bates, the Director of the Technical Division, as the most practical way of obtaining additional cargo-carrying capacity early in 1942. In that report he wrote: "The availability of steel has been something of a mystery."[17] By August, he foresaw a con-

---

[13] Fischer, *Statistical Summary*, Table E-5 and Reynolds' Way Charts.
[14] Truman Committee, *Hearings*, pp. 6017-19.
[15] MC Minutes, p. 19237 (Oct. 9, 1941).
[16] MC Minutes, pp. 19297, 19314, 20033.
[17] "The Possibility of Additional Cargo Carrying Tonnage for Use Early in the Calendar Year 1942," a memorandum by J. L. Bates, June 28, 1941, p. 6, in MC gf 506-1, exhibits.
Mr. Dunn's report of Feb. 22, 1941 to the President was the basis of recommen-

64 BEFORE THE DECLARATION OF WAR

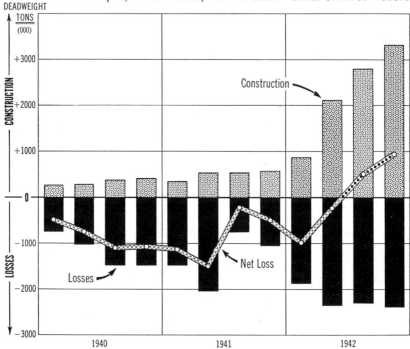

FIGURE 7. Source: *WSA Shipping Summary*, vol. II, no. 7 (July 1945), p. 22.

tinued steel shortage, and on that ground recommended that the vessels be built of reinforced concrete. Before the end of 1941, contracts were let for 15 concrete barges at three new shipyards, and for 25 sea-going tugs. What happened to the concrete barge program will be told in Chapter 18.

The complexity which distinguishes this third wave of expansion from the first two was a result of the varied elements which entered into it and also of the long period over which it was drawn out. The delays and hesitations which characterized it— from Sir Arthur Salter's letters on April 24 and 25 urging the need of more expansion to the letting of contracts in November

dations to Admiral Vickery from the Technical Division, May 9, 1941. " Program of Construction as recommended in Sir Arthur Salter's letter to Captain Vickery on Apr. 24, 1941," p. 1 and encl. A, in Vickery's file.

THE THIRD WAVE

and December for some of the very ships he had suggested in the spring—can be traced to a number of causes. Of basic importance, even when in the background, was the fluctuation of ship losses during this period. In April almost 800,000 deadweight tons were lost and during May and June the losses continued to be very high. During the first week of June they were reported to be at an even higher level.[18] June was the month in which the Great Lakes program, the barge program, and the expansion of the major types were all being pushed. Then reported sinkings fell off sharply, until in July they were less than the new tonnage being built,[19] and in August new construction of dry cargo vessels exceeded losses. In September losses went up again, to over 400,000 tons.[20]

Less fluctuating than the shipping losses but also an uncertain element was public opinion. Not all the country was persuaded that aid to Britain was the best policy. War openly declared gives a singleness of purpose from which comes extra productive capacity, but shipbuilding could not yet count on that reserve of energy.

Moreover, the shipbuilding capacities of the country were, by all ordinary standards, already exhausted before the third wave of expansion had even started. In April 1941 the construction authorized called for a labor force six times that normally employed in shipyards. Although the program of the Maritime Commission was large, that of the Navy was almost four times as big.[21] As Admiral Edward L. Cochrane said later:

> The Navy, of necessity preempted the services of the handful of experienced shipyards in the country which were in operating condition at the time.... Admiral Land has never complained of the Navy's action in preempting the services of the country's experienced shipbuilders, although we know that by this action, in order to meet the enormous production requirements which had been placed upon his shoulders as the Chairman of the Maritime Commission, he was forced to recruit shipbuilders from among agencies which had demonstrated outstanding ability in the field of general construction but which had limited, if any, experience in the shipbuilding game.[22]

---

[18] Salter to Land, June 11, 1941, Land file, said 168,000 gross tons were lost the week ending June 8.

[19] Stark to Land, Sept. 2, 1941, Land file.

[20] *WSA Shipping Summary*, prepared by the Division of Statistics and Research, WSA, Vol. II, No. 1, p. 18.

[21] In the speech cited in note 23 below Admiral Land said construction of merchant vessels for the Maritime Commission and private account was two billion dollars in value, that for the Navy seven and one half.

[22] "Shipbuilders Meet the War Challenge," in *Marine Engineering and Shipping Review*, December 1944, p. 149.

While the Maritime Commission was going ahead in face of these handicaps to meet the submarine menace by building, its Chairman sometimes wondered whether some other ways might be found for meeting that menace besides asking the impossible from the shipbuilders. In a radio talk on April 8, 1941, Admiral Land compared sinkings by Nazi submarines to a fire, and compared the building of more and more ships to feeding the fire with fuel, whereas we should aid the British to put out the fire. After announcing the lend-lease program of April 1941 (the second wave), he had said:

> The United States can do the jobs assigned so far and the expansion I have just outlined, but to extend or enlarge this program greatly will defeat the all important objective of obtaining completed ships, of obtaining destroyers and torpedo boats and submarines for the Navy, as well as cargo ships. . . . Any further dilution of shipbuilding brains, whether the skilled or unskilled labor in the yards or the managerial and professional talent in the office, will result in gross waste and inefficiency.[23]

Having stated this warning, he later went ahead with expansion as it was forced upon him, realizing that some waste and inefficiency was inevitable.

Before the declaration of war the Maritime Commission had programmed, in addition to the British building and a little private building, almost 5 million tons for 1942 and a productive capacity of over 7 million tons for 1943. Quite a contrast to the 341,219 tons which were the total deliveries of 1939.

V. Popularizing the Program

Not only was the emergency shipbuilding well underway before Pearl Harbor, but basic steps had been taken to present the program to the general public as a national enterprise in which they could join with pride. President Roosevelt himself led off by explaining in his fireside chats and news conferences the need for ships to defend America by aiding Britain.[1] Speeches by the Commissioners explained why the new type of ship was being built, the plans for new shipyards, and their importance in national policy.[2] The Commission's Division of Maritime Promotion and

---

[23] Press Release 881, MC gf 105-2, Apr. 9, 1941.

[1] *New York Times*, Dec. 30, 1940; Jan. 4, 1941; May 1, 1941; and May 28, 1941.

[2] See texts of addresses by Vickery, Feb. 7, 1941; Woodward, June 21, 1941; and Macauley, Sept. 4, 1941, in MC gf 105-5.

Information expanded its staff to assist in the preparation and dissemination of speeches and articles, and directed its efforts especially at making shipyard workers feel that their labor was not just effort they had to put out in order to get a pay check but was also a contribution to the nation.[3] The division undertook a poster campaign designed to do two jobs: "First, identify the shipyard worker with his job. Make him realize that he is building ships for the survival of democracy. If the ship on which he is working is not finished quickly, Britain may fall, and if Britain falls, the United States is in great danger and his wife and children are in danger. . . . Second, persuade him to work all the time." The first required serious posters with emotional power; the second humorous posters to ridicule loafing.[4] In such campaigns the Maritime Commission was moving in the usual channels of publicity. The keynote reiterated on banners, broadcasts, insignia, and awards was SHIPS FOR VICTORY. It has seemed appropriate to use this slogan as the title for a history of the program.

Out of the nature of the shipbuilding industry arose two special forms of publicity: the naming of ships and the launching ceremony. These were both matters of symbolic value, touched with emotion and inherently important therefore in public relations.

Naming the product is half the battle in any advertising campaign. The ship which was to win popular interest under the generic name "Liberty ship" started off under the severe handicap of being known as the "ugly duckling" because the President in his January announcement of the emergency building program referred to it as a "dreadful looking object," and *Time* magazine reported it under the heading "Ugly Ducklings."[5] That name went unchallenged until the middle of April when Admiral Land received a letter from J. A. Farrell protesting on the grounds that the ships were really good looking. Admiral Land circulated the letter through the Commission with the note: "I agree. We can do our bit by calling them 'Emergency Ships.'"[6] The Executive Director recommended on May 2, 1941 a press release saying that the emergency ships had been designated as the EC2, their

---

[3] See below p. 760, and, on the size of the division, Table 16, p. 758.
[4] MC Minutes, pp. 17403-04 (May 22, 1941).
[5] Press Conference of Jan. 3, 1941 in *The Public Papers and Addresses of Franklin D. Roosevelt* (New York, 1941), IX, 647; *Time*, Jan. 13, 1941, p. 14.
[6] J. A. Farrell, Pres., India House, New York, to Land, Apr. 16, 1941 with attachments, in gf 507-1.

design classification, and should be so referred to.[7] Whether they were really to be American ships or were to be turned over to the British was still not clear, but the press release added that those that were to operate under the American flag would be known as the " Liberty fleet." [8] At Congressional committee hearings in June and July they were still commonly referred to as " ugly ducklings," but Commissioner Dempsey insisted on the name " Liberty fleet." [9]

The new name, modified finally into " Liberty ships," was fixed in the public eye and ear by the widely publicized festivities celebrating the launching of the first of the type at the Bethlehem-Fairfield yard on September 27, 1941. It was proclaimed Liberty Fleet Day, and on Admiral Land's urging the President personally took part. The ceremony was eminently successful in catching public attention; and by the time the talk was over and newspaper articles had been written all over the country, " Liberty fleet " had become " Liberty ships." [10]

Partly by virtue of this name, ships of emergency type, as they continued to be called in the Commission's official statistics, became a kind of national symbol, although none of the standard or C-types of which the Commission was so proud ever won a similar place. There is nothing glamorous about the name C2 or T2. Traditionally the name of a ship is the affair of its owner or operator, not of its builder. Most of the Commission's standard types were destined to pass to a private company. If they acquired a popular name as a type it was the name with which the operating company advertised its fleet or its line. The naming of the Liberty fleet, popularized as Liberty ships, reflected the change from building for private enterprise to building for the use of the government. It aroused public consciousness that the Commission's emergency program was public business.

Ceremonies can be as important as names in catching public

---

[7] MC Minutes, p. 17184.

[8] PR E-16, May 2, 1941, in gf 507-1.

[9] Truman Committee, *Hearings*, pt. 5, pp. 1293-94 where Dempsey urges " Liberty fleet," and House, *First Supplemental National Defense Appropriation Bill for 1942, Hearings*, p. 414, where they are still referred to even by Admiral Vickery as " ugly ducklings."

[10] PR 1033, Sept. 24, 1941 and PR 1032, Sept. 25, 1941, in MC gf 105-2. Commissioner Macauley referred to them as " Liberty Ships " in an address on Sept. 4, 1941. In gf 105-5.

interest or shaping public opinion. The nature of shipbuilding created a stirring ceremony well fixed by tradition, the launching. It too changed in character as shipbuilding became so largely public business. Traditionally the company for which the ship was being built dominated the launchings, selecting the sponsor who would christen the vessel by breaking the bottle of champagne over her bow and providing the presents and entertainment. In the new yards, however, launchings were ceremonies for the builders, for the workmen as well as for the management, and for the public. The very first launching of a Liberty ship was made a sort of national holiday or festival as has been described. When the yard at Portland, Oregon, was ready to launch its first ship, *The Star of Oregon*, it planned to serve Coca-Cola and other similar refreshments to the workers in a general open-house celebration. In wiring the Commission for assurance that the cost would be reimbursed to the company as part of the cost of the ship, Mr. Kaiser said, " This is a goodwill program and will stimulate labor-employer relations and is not a private party for a few select friends of the builders." [11] Other new yards planned similar extensive and popular celebrations of their first launchings. In later launchings, workmen's wives were chosen to sponsor many ships. After production was in full swing and ships were being built so fast that one was launched every other day, as they were during the war, launching ceremonies were very much abbreviated, but more persons than ever before had a personal interest in launchings.[12]

A traditional feature of launching ceremonies which became a source of embarrassment to the Commission was the custom of giving a present to the lady chosen as sponsor. When a private shipbuilding company was building a ship for a private operating company, the size of the present and choice of the sponsor to receive it was a private, personal matter. When building was done for the Commission under contracts by which the Commission supervised all the costs of the ship, the situation was different, but the immemorial customs of the sea and of the shipbuilding fraternity are hard to kill. The Commission at once decided against reimbursing for launching gifts and set a limitation of $500

[11] MC Minutes, p. 18977.
[12] MC Minutes, p. 18916 (Sept. 12, 1941); p. 20100 (Dec. 12, 1941); p. 20158 (Dec. 16, 1941).

on the total that would be reimbursed for launching ceremonies after the first. After Pearl Harbor even this allowance was eliminated.[13] Most of the shipyards adopted a standard gift of moderate cost such as an engraved silver plate, a box containing the broken bottle used in the ceremonies, and pictures of the event. On the other hand a few of the old-line shipbuilders clung to older customs. Newspaper reports of launchings appeared featuring the champagne, fancy lunches, and diamond-studded wrist watches, and were re-echoed critically by members of Congress. Rumors of this sort did not add to the effectiveness of the launching ceremonies in symbolizing common efforts in a national undertaking. Admiral Land answered Congressional criticism in December 1941 by emphasizing that none of these was being paid for by the Maritime Commission and declaring: " The Maritime Commission has consistently discouraged the practice of presenting gifts to sponsors of merchant ships launched in private shipyards in the United States." [14] But the occasional expensive launching gifts hurt the reputation of the Maritime Commissioners and prevented the full symbolic transformation of the launching from a private party to a public ceremony.[15]

Although launching arrangements were handled by the yard managements, the selection of the name of individual Liberty ships was done by the Maritime Commission through a Ship Naming Committee. Names of individuals—patriots, scientists, journalists, educators, artists, and industrialists—were used, and had to be investigated to determine their suitability. In each ship there was installed a brief biography of the person whose name it bore.[16]

[13] MC Minutes, (Sept. 12, 1941) ; p. 20100 (Dec. 12, 1941) ; p. 20158 (Dec. 16, 1941).

[14] House, *Independent Offices Appropriation Bill for 1943, Hearings,* 77th Cong., 2d sess., pp. 296-97; House, *Independent Offices Appropriation Bill for 1946, Hearings,* 79th Cong., 1st sess., Jan. 15, 1945, pp. 590-94.

[15] The way in which this practice hurt the war effort is illustrated by the letter from a father whose boy was in action in the Pacific at $61 a month, protesting what he had read about the gifts of $2000 bracelets to government officials and saying, " If this is true, I have purchased my last war bond." Attached to Land to Hassett, Oct. 10, 1944, in Land reading file. The list of sponsors' gifts, which included $2000 wrist watches, was compiled and made public by the Commission during the Congressional Investigations of 1947, being available in the press room of the Division of Information at that time.

[16] H. E. Cohen, " Commissioner John Michael Carmody," pp. 94-108. The committee, of which Mr. Carmody was chairman, 1943-1944, also named the Victory ships, using in the main the names of towns and cities.

Like the rest of the Commission's activity, the effort to put patriotic drives back of the shipbuilding program was intensified after Pearl Harbor, but before that attack solidified the nation and deepened its danger many of the main lines of policy had been determined. Basic decisions had been made in public relations and in the selection of shipyards and their managements. No less important were the decisions already made in regard to ship design and the forms of contracts.

CHAPTER 3

DESIGN AND INITIAL PROCUREMENT FOR THE LIBERTY SHIP

I. The Basic Design Decision and the Working Plans

STANDARDIZATION of the type ship to be built was the first essential for the success of the emergency program. Speed and economy within a particular yard depended on having that one yard build one design. In the planning which the British had done for wartime building within Britain, they did not attempt to impose any single standard on all the British yards; they settled on one design for each yard.[1] But in the United States, where new yards were being built, standardization was attempted on a nationwide basis. Nationwide standardization gave two major advantages: specifications and drawings to guide ship construction, once prepared, could be rapidly reproduced for use in additional yards; and the procurement of components could be organized in a steady, flexible flow from a number of vendors supplying interchangeable articles to a number of shipyards.

These advantages of nationwide standardization were sought in many parts of the Commission's program. Their widest application came of course after Pearl Harbor when production reached its peak; but they were most successfully applied in what came to be known as the Liberty ship, and the basic decisions concerning the standardized production of this emergency-type ship were made in the period before Pearl Harbor.

In fact, the most basic decisions favorable to standardization were made at the very start of the program. Once plans for the shipyards were settled, the most pressing problems were to supply contractors with detailed drawings and to place with engine makers

[1] Robert C. Thompson and Harry Hunter, "The British Merchant Shipbuilding Programme in North America 1940-42," North East Coast Institution of Engineers and Shipbuilders, *Transactions*, Vol. 59, with discussion in pp. D47-D64. This article has been reprinted in *The Shipping World*, December 2, 9, 16, 1942. See also *The Times Record of British War Production, 1939-1945*, p. 14.

THE BASIC DESIGN

and other vendors the orders for components. While the Commission was still negotiating its contracts with the shipbuilders basic decisions were made about plans and procurement.

These decisions were implemented with celerity through a private enterprise under William Francis Gibbs, the head of Gibbs & Cox, Inc. This firm and that of George Sharp were the foremost naval architects in the United States. An administrator and negotiator of marked ability, William Francis Gibbs had supervised the construction of the *America* and done much awe-inspiring and secret work for the Navy. Foreseeing war and the need in a world conflict for multiple ship production, he had been building the necessary organization to service shipbuilders in such an emergency. At the end of 1940 his firm employed about a thousand persons and was soon to issue 700 acres of blue prints per year.[2] When the Todd-Newell-Kaiser alliance contracted in December 1940 to build 60 ships for the British, they arranged for Gibbs & Cox to be their naval architects. The biggest yards had their own naval architects and draftsmen, but smaller, newer shipbuilding companies in the Commission's program customarily made contracts with a firm of naval architects to supply them with detailed working plans, draw up precise specifications for materials and parts, and invite bids.[3] George G. Sharp performed these services for the new Tampa yard, for example. When the Maritime Commission launched its emergency program, Gibbs & Cox was already engaged to take care of procurement and prepare working plans for the British ships built in this country.[4] William Francis Gibbs was prepared to play a similar big role in the American program. This will be made clear as we examine first the story of the Liberty ship design and second the initial system of procurement.

Out of the 5 million tons scheduled in 1941 for delivery in

---

[2] House, *Investigation of the Progress of the War Effort, Hearings* before the Committee on Naval Affairs, 78th Cong., 2d sess., pursuant to H. Res. 30 (Vol. 5) May 8 and 9, 1944, p. 3970.

[3] *Ibid.*, pp. 3970 ff., 4026.

[4] The "Office Memoranda" of the Technical Division, Maritime Commission, define the usual relations of the Maritime Commission, the shipyard and the Marine Architect. Precedents are also discussed in the long "Memorandum" by R. E. Anderson, May 22, 1942, in gf 507-1-4. A copy of the contract of the Maritime Commission with Gibbs & Cox (MCc-ESP-15) is in gf 507-1-4, Feb. 7, 1941. An itemized list of the services of Gibbs & Cox is filed under Apr. 5, 1941 in Emergency Construction Div. file QM8-A3-2.

1942, 3 million tons were to be of the new type originally referred to as " ugly ducklings " and later famous as the Liberty ships. As the emergency increased, so did their proportionate role in production, until the " Libertys " became four or five times as numerous as any other type in the American merchant marine.

The history of the design which was to hold this dominant place in shipbuilding during the first part of World War II is very different from that of the designs of other Maritime Commission ships. Hitherto designs of ships built under the Maritime Commission had been developed in the Commission's Technical Division, which was accustomed to working on the designs long before any contracts were made with builders.[5] When the emergency program was decided on, there was no time to do more than choose among existing designs and then to make some modifications to meet special needs.

On the day after the President's broadcast concerning the emergency program, January 4, 1941, James L. Bates, Director of the Technical Division, held a conference with Messrs. Esmond, A. MacPhedran, E. P. Rowell, and I. Wanless, members of the Division who were directly concerned with matters of design.[6] He explained that the question of design of the emergency ship would be threshed out on January 8 when William F. Gibbs would arrive for a conference. The main characteristics were to be like those of the 60 ships the British were building: length about 440 feet, speed 11 knots, and weight-carrying capacity about 10,000 tons. Like the British they would have reciprocating engines, the only kind obtainable. Two changes from the British specifications had been decided on during the evening of January 3 after the President's announcement. In consultation with Mr. Bates, and with Mr. Schmeltzer, the Associate Director of the Division, Admiral Vickery decided that construction would be facilitated by the adoption of water-tube instead of Scotch boilers and oil firing instead of coal.[7]

---

[5] Office Order No. 43, Series of 1939, Sept. 21, 1939, in Technical Div. files; Admin Order No. 37, Suppl. No. 1, Dec. 13, 1938; Office Order No. 18, Series of 1938, May 24, 1938, in Technical Div. files; Quarterly Reports, Technical Division, July 1, 1939-Dec. 31, 1940.

[6] Memorandum, " Design C2-S-C1," Jan. 6, 1941, in Wanless File, EC2-S-C1, Correspondence.

[7] This account of discussion of details of the new design is based on David B. Tyler, " Construction and Operation of Liberty Ships from the Technical Point of View," Oct. 23, 1946, in Historian's Collection, Research file 111-1a.

PLATE III.

Henry J. Kaiser (left) and W. S. (Pete) Newell (right) planned new shipyards to build freighters like the *Ocean Courage* (above) for the British and to build for the Maritime Commission a modified form of the same design, the Liberty ship (below). Vessels of the British Ocean class were often called British Libertys.

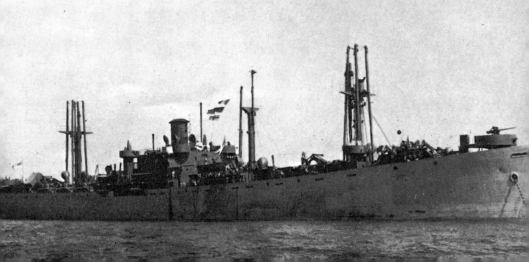

PLATE IV.

Above, the keel of a Liberty is laid at the Todd-Houston yard although the construction of the yard is still far from finished. Center, the Oregon Shipbuilding Corp. celebrates the launching of its first ship. Lower left, Marian Anderson sponsors a Liberty ship, the *Booker T. Washington*.

THE BASIC DESIGN

Any alternate to the designs being used by the British Merchant Shipping Mission would have to be ready for presentation on January 8. The crux of the matter, as Mr. Bates saw it, was whether Gibbs & Cox, who were already busy revising the British plans for use in American shipyards, would be willing to undertake in addition the preparation of another entirely different set of plans.

Working plans had to be obtained quickly and in large numbers for distribution to the several yards, and Gibbs & Cox was the only organization capable, at that moment, of handling this particular mass production of blue prints. In spite of the very short time before the scheduled meeting with Mr. Gibbs, it was decided to go ahead with the preparation of an alternate design for a ship with the same deadweight and of the same cost.[8]

At this time two engineers in the Technical Division, Alan Osbourne and F. E. Reed, made a search of the files to see if there was not an earlier American design that could be utilized in this emergency. The most promising plans they unearthed were those of the so-called "Los Angeles" class of vessel of which some 238 were built by mass production at the same time that 110 Hog Islanders were being put together. The "Los Angeles" type vessels had about the same deadweight as the British ships but somewhat smaller dimensions.[9] They had used triple-expansion engines, water-tube boilers, and had a designed speed of $10\frac{1}{2}$ knots. At first glance this type vessel appeared to be exactly what was needed, but, on further consideration, its suitability was called into question. Although the plans were complete and had the advantage of being third angle projection instead of first angle projection as in the case of the British plans, they were not in very good condition. They would have to be revised for welding. It was recalled that these vessels had developed certain defects in operation such as a weak skeg (the after part of the keel), a badly located air pump, and insufficiently rigid engine foundation. The water-tube boiler used was excessively heavy and expensive by modern standards, and since the engine-builder was no longer in business he would have no pattern nor be able to give the benefit

---

[8] *Ibid.*

[9] There is a picture of SS *Greylock*, a "10,950 dead-weight ton steel cargo-ship, built by the Los Angeles Shipbuilding and Dry Dock Company" in Edward N. Hurley, *The Bridge to France* (Philadelphia and London, 1927), p. 298.

of his experience in large production. These disadvantages were not insurmountable but, taking into consideration the great pressure to get these ships built in the quickest possible time, the up-to-date British plans, already being prepared for American use, looked like a better solution to the problem.[10]

The decision to explore the possibilities of an alternate design arose partly from criticisms of the British design. One of these criticisms was that the form of the hull was not that best suited for rapid mass production. During the two-day interval, from January 6 to 8, an effort was made to develop a straight-line form which would be easy to construct and would have speed or sea-going qualities equal to the British vessel. Studies of straight-line forms had been under consideration during World War I. The object was to eliminate, as far as possible, double turns or twists in individual plates, which would require furnacing or power presses for shaping plates. A simplified form would facilitate fabrication.[11] A midship house, usual with Maritime Commission vessels, was thought more suitable for North Atlantic crossings than the British two-house design because crew's quarters would be kept drier and it would be a means of economizing on piping and heating. Therefore another design was considered based on that of a T2-type tanker, originally prepared by George G. Sharp for the Shipping Board some time before.[12] Mr. Esmond had worked on this design with Mr. Sharp. It was essentially a straight-lined form with curved bilges and some curved framing at the after quarterpoint. Although it had tested very favorably in the model basin, no such vessel had ever been built. This was one of the chief counts against it now, for it was risky to adopt a form whose sea-going qualities were untried.[13]

At the meeting on January 8 several representatives of shipbuilding companies were present, as well as William Francis Gibbs. Admiral Vickery presided and opened the meeting by

[10] Memorandum, A. Osbourne to W. E. Spofford, Jan. 31, 1946, in Historian's files. F. E. Reed to David B. Tyler, July 2, 1946, in MC, Historian's Collection; Notes by David B. Tyler of interviews with J. L. Bates, June 12, 1946, and A. Osbourne, June 28, 1946.
[11] Notes by David B. Tyler of interview with W. G. Esmond, June 1946.
[12] Memorandum, " M. C. Design C2-S-D1," Jan. 7, 1941, in Wanless file, EC2-S-C1, Correspondence; Notes by David B. Tyler on interview with W. G. Esmond.
[13] Ibid. The general characteristics of this design were the following: L. W. L. 420 ft.; beam 58 ft.; displacement 13,490 tons; draft 26′ 3″; deadweight 10,090 tons; cruising radius at 11 K. 8,000 miles.

THE BASIC DESIGN

explaining that an alternate design had been prepared with the idea of seeing if the same deadweight and speed could be embodied in a vessel which would be cheaper and simpler to construct than the British vessel. Line drawings of the British and Commission designs were exhibited and compared. The latter appeared to be simpler to construct because of greater emphasis on straight-lines. Mr. Gibbs, however, pointed out that its adoption would delay the beginning of construction. In reply to this it was argued that this initial delay would be more than counterbalanced by the speeding up of later deliveries. Bethlehem agreed to prepare a plating model according to the Commission's design. The representatives of Todd and Newport News indicated their readiness to consider anything that would expedite the program.[14]

No final decision was reached at this time but the thing that impressed Admiral Vickery and Mr. Bates particularly was the readiness of the shipbuilders to undertake construction of vessels using the British design. No ship had as yet been built from that design but it was one that had been developed out of long experience so that its general characteristics had proved their worth.[15]

Admiral Vickery went over the whole question again in New York on January 13. The plating model made by the Bethlehem people indicated that the Commission design would give a minimum of furnaced and rolled plates. Estimates for deadweight and cubic indicated that this design would compare favorably with the British one. There was no time for model basin tests, so no conclusive opinion could be reached with regard to comparative speed and power of the two designs. Some decision had to be made, and the responsibility fell primarily on Admirals Land and Vickery.

Admiral Land decided in favor of the British design.[16] The fabricating required for this design could be managed in mass production with only a slight increase in manhours over that required for the Commission design. The latter was untried, as compared with the former. Moreover, work could begin much

---

[14] I. Wanless, "History of Emergency Ship Design," Mar. 28, 1941, in Technical Division files EC2-S-C1, Pt. I. Admiral Vickery had seen to it that certain of these representatives of shipbuilders had a chance to look at the British plans. See Vickery to James B. Hunter, Dec. 31, 1940, in gf 500-3.

[15] Notes by David B. Tyler of Interview with J. L. Bates, June 12, in MC, Historian's Collection, Research file 105.

[16] Note made by Admiral Land on a draft of this chapter.

sooner if the British design were used, since Gibbs & Cox were already preparing drawings for American shipyard use.[17]

Adoption of the British design meant that Gibbs & Cox would perform the functions of marine architect for all the emergency yards. To cooperate with Gibbs and with the companies building the emergency ships the Maritime Commission created a whole new division, the Division of Emergency Ship Construction, which was carved out of the existing Technical Division with the idea that, as Admiral Vickery explained, "... when the job is done—we will wash the whole division right out." [18] It was created to operate on an emergency basis and get results fast, even at the sacrifice of some of the painstaking procedures which were part of the usual practice of the Technical Division.

Except for the approval of shipyard plans, which will be discussed later, the most pressing task of the new division was to approve the working plans which were being prepared by Gibbs & Cox. Heretofore approval of working plans for use in the yards participating in the long-range program had been done in relatively leisurely fashion. The usual procedure was for the marine architects of the yards to send copies to the Technical Division where they were checked in the Hull Plan Approval and the Engineering Plan Approval Sections.[19] William G. Esmond and John E. P. Grant were now transferred from these sections and became chiefs of the corresponding sections in the newly created Emergency Ship Construction Division. To speed up the work they moved to New York in February and established an office at 21 West Street, in the same building with Gibbs & Cox. They made it a practice to consult with the agent's draftsmen while plans were being drawn, interpreting specifications according to the Commission's needs. This became known informally as " bed-

[17] Wanless, " History of Emergency Ship Design "; J. E. Schmeltzer to Chairman, May 15, 1941, in gf 507-1-4; notes by David B. Tyler on interview with Mr. Bates; MC Minutes, pp. 18447-49. Commissioner Dempsey's prepared statement in the Truman Committee, *Hearings*, pt. 5, June 3, 1941, p. 1466, said: By adopting the design and following the plans which Gibbs & Cox " had prepared " for the 60 ships " under construction for the British . . . it was possible to effect a material saving in the main essential, time," and to take advantage of what Gibbs & Cox had already done in procurement.

[18] House, *Emergency Cargo Ship Construction, Hearings* before the Subcommittee of the Committee on Appropriations, 77th Cong., 1st sess., on H. J. Res. 77, Jan. 18, 1941, p. 16.

[19] Quarterly Reports, Technical Division, July 1, 1939-December 31, 1940; Office Memoranda No. 1 and 2 in Technical Division files.

THE BASIC DESIGN

side " approval. It had the obvious advantage of doing away with the necessity of sending plans back and forth and meant that very little time needed to be spent on examination of the finished plans.[20]

The extent of the work that had to be done by Gibbs & Cox in preparing working plans was not realized at first. There was the somewhat easy assumption that since the British had prepared plans nothing much had to be done except to reproduce them in quantity and make the changes desired by the Commission. The drawings had, however, to be put in third angle projection and many details added which in British practice were omitted. In spite of their large staff and wide experience, Gibbs & Cox found they had bitten off almost more than they could chew. This was reflected in delays and misunderstandings. Their wheels did not always mesh with those of the Commission which employed them. For example, the Audit Branch in the Finance Division of the Maritime Commission found that Gibbs & Cox " pretty generally " ignored all communications emanating from that office with the result that the Branch sent memoranda to Mr. Grant with the request that he consult with Gibbs & Cox on the matter in question.[21]

Even at the bedside there were sometimes difficulties. On one occasion Mr. Grant had great difficulty securing drawings for condensers. A delay of this character meant that no manufacturer could proceed in the manufacture of condensers without risking rejection, and the program was delayed by that much.[22]

It should not be forgotten that the whole program did proceed at a rapid pace, however, and a mass of design work was accomplished in a short time. On the whole Messrs. Grant and Esmond got along very well with the Gibbs & Cox staff. If there was some confusion, there was also ability and drive, which in the end resulted in achievement.[23]

---

[20] W. G. Esmond, Chief of Hull Section, Emergency Ship Construction Division to Director, Emergency Ship Construction Division, Apr. 21, 1941, in Production Div. files QM8-A3-0-1.
[21] Interoffice Memo, R. H. Mohler, Acting Chief, Emergency Construction Audit Branch, Construction Audit Section, Division of Finance, to J. A. Honsick, July 21, 1941 in gf 507-4-1.
[22] J. E. Schmeltzer to Gibbs & Cox, Att. W. F. Gibbs, July 14, 1941 in gf 507-4-1.
[23] Interviews with Messrs. Grant and Esmond, June and July 1947 in MC, Historian's Collection.

II. Modifications in Design

Many changes distinguished the American " Liberty Ship " from its British prototype. The larger changes, such as the single central deckhouse, were decided on in the Commission and embodied in the specifications, but many others originated in the drawing of the working plans. Although seemingly small, some of these features turned out to be important, as, for example, the insufficiently reinforced square hatch corners and the cut in the sheer strake for the accommodation ladder, features which were the starting points of so many fractures. When these features were approved, there had been no experience with welded ships under wartime conditions of heavy deck loads, poor ballasting, and unusual stresses. Other changes were made which were indeed without structural importance but which made the American Libertys different from their British sisters.

The story of these variations naturally starts with the plans brought to the United States by the British Technical Merchant Shipbuilding Mission.[1] They were for a vessel which was representative of a type of 10,000 ton tramp for which similar designs were developed in various shipyards in England during the depression years. There has been some doubt as to which was the actual progenitor of the vessels built in the United States. As a matter of fact, no English-built ships had been based exactly on the plans used for the Liberty ship, for Sir Amos L. Ayre, Director of Merchant Shipping in the Admiralty,[2] altered slightly the plans he brought from England to America, in the interest of quick construction. He sought to avoid furnaced plates by giving a slight amount of curvature to the whole ship, thereby eliminating most of the double curves at the bow and stern.[3]

---

[1] This Mission was headed by Robert C. Thompson, Managing Director of Joseph L. Thompson & Sons, Ltd., Shipbuilders at Sunderland. The other members included: Harry Hunter, a marine engineer connected with Swan, Hunter, & Wigham Richardson, a shipbuilding and engineering firm at Wallsend-on-Tyne, Northumberland; William Bennett, Principal Surveyor of Lloyd's Register of Shipping for the U. S. A. and Canada; J. S. Heck, Principal Engineer Surveyor of Lloyd's Register, New York; and R. R. Powell, Assistant Secretary, Admiralty, as well as Secretary of the Mission.

[2] Chairman of the Burntisland Shipbuilding Co., Ltd., Burntisland, Fife, before joining the Ministry of Shipping in 1939.

[3] Notes by David B. Tyler on interview with J. L. Bates, July 1946. His object was also to prepare the plans so that the ships could be welded. The plating model

MODIFICATIONS IN DESIGN

Before World War II the British had laid plans for standardizing wartime production, yard by yard. Each yard had prepared a design which was to be standard for that particular yard. The plans brought to America were, before Sir Amos Ayre modified them, the plans being used by Joseph L. Thompson & Sons of Sunderland. The first of these standard vessels was the *Empire Liberty*, launched August 23, 1941 in that company's North Sands Shipbuilding Yard.[4]

The aim of the British designers of this type was to produce a vessel that would have the maximum deadweight and speed compatible with economy of fuel consumption. The result was a blunt-bowed type of tramp which some shipping people referred to as mere " powered scows." The news that a large quantity of these vessels would be built was greeted with little enthusiasm by British shipbuilders. They could not forget that after the First World War they had been saddled with a lot of slow and inefficient tonnage. They protested that the slowness of the emergency ship made it unsafe in convoy and unusable for postwar competitive operation.[5] The Liverpool *Journal of Commerce* sought to remind them that this type of ship was the result of shipyard research during the depression years and had proved itself an economical and efficient cargo carrier.[6]

The British Mission contracted for thirty ships each from Todd-Bath and Todd-California, on December 20, 1940. These vessels, which came to be known as the " Ocean " class, were powered with reciprocating engines, and were welded, except that the frames were riveted to the shells. According to Thompson and Hunter's account, they were welded " to ensure a good supply of labour, to facilitate production, and to get the best value for money." [7] The first built in America, the *Ocean Vanguard*, went down the ways at Todd-California, on October 15, 1941.[8] She was christened by Mrs. Emory S. Land at the invitation of the British.

---

indicated that only two plates on each side of the forefoot would have to be " furnaced," Esmond to Schmeltzer, Feb. 1, 1941 in gf 507-4-1.

[4] R. C. Thompson and Harry Hunter, "The British Merchant Shipbuilding Programme," North East Coast Institution of Engineers and Shipbuilders, *Transactions*, Vol. 59, pp. D47-D64.

[5] Liverpool *Journal of Commerce*, Nov. 21, 1940.

[6] *Ibid.*, Feb. 27, 1941.

[7] Thompson and Hunter, "The British Merchant Shipbuilding Programme in North America 1940-42," p. 64.

[8] The *Ocean Vanguard* and the *Empire Liberty* are compared in " British Proto-

An additional contract for 26 ships was placed with Burrard, Canadian Vickers, and Davie Shipbuilding, in Canada. These ships, known as the "Fort" class, were riveted, since they were built in established yards accustomed to following British plans closely.[9]

The British Mission found that the lack of detail in the plans they brought from England presented a big problem for American shipbuilders and called for considerable interpretation and amplification. The British practice was to make only about 30 per cent as many drawings as was customary in American yards, because their workmen were accustomed to doing more bench work, that is, hand fitting and filing. The designer often left details to be decided by the pattern-maker. For example, contrary to American practice, the designer did not put down clearances between crankshaft and bearing.[10] As another example, the propeller was fitted tightly to the shaft without the use of a gland sealing ring.[11] Other difficulties arose from differences in pipe flange standards, timber sizes, hemp and wire rope sizes, use of the Standard Whitworth thread, etc.[12] It became the job of the members of the Mission to cover these points, which they accomplished, in part, by having representatives of the shipbuilders inspect British ships of the same type, when in port.[13]

The British prototype of the American Liberty ship was, therefore, the "Ocean" type cargo vessel of which sixty were built in the United States, half at Todd-Bath and half at Todd-California. These welded vessels were the same as the twenty-six "Fort" class

type of the Liberty Ship," *Marine Engineering and Shipping Review*, Apr. 1942, pp. 168-71. Comparative measurements are given on p. 64. Other tramp types sometimes referred to as the prototype of the Liberty Ship are the *Scottish Monarch*, 29,300 deadweight ton vessel built by the Caledon Shipbuilding Co., in 1938 and mentioned in *Fairplay*, Feb. 20, 1941, and a standard motorship tramp of 10,200 deadweight tons, known as the "Economy Ship," built by the Burntisland Shipbuilding Co., Ltd., and mentioned in *Shipbuilding and Shipping Record*, Nov. 27, 1941.

[9] Thompson and Hunter, p. 64.

[10] Notes by David B. Tyler of Interview with J. E. P. Grant, Aug. 1946.

[11] This tight fit is not easily achieved under mass production conditions and if it is not achieved corrosion may result where the steel shaft and bronze casing come together. This was noted and pointed out by A. Osbourne in 1941, but it was not felt necessary to make a change. Notes by David B. Tyler of interview with Mr. Osbourne, Jan. 1947. This appears to have been the origin of propeller troubles suffered by several Libertys after they had been in use for a considerable length of time.

[12] Thompson and Hunter, p. 65.

[13] *Ibid.*

MODIFICATIONS IN DESIGN

vessels built in Canada, except that the latter were riveted. Both of these groups were built from the slightly modified design of the *Empire Liberty* type being produced at North Sands, Sunderland, England, which, in turn, was similar to the " economic " tramp ship developed in other British shipyards during the stringent prewar period. Thus, it can be said that the design was one that had evolved over a period of years and represented an effort to produce a ship that would be inexpensive to construct and to operate but would have a large carrying capacity.

When the Maritime Commission adopted the British design for the 200 emergency ships, it was designated the " EC-2 " to indicate that the vessels would be about the size of the standard C2 but different from it in being an " Emergency " type. The emergency character of this vessel was impressed on Congress. This is indicated in the report of the House Committee in January 1941 supporting the special appropriation. There the emergency ship is described as " what might be termed a ' 5-year ' vessel."

> It is slow and seaworthy and has the longevity of a modern steel ship, but for the demands of normal commerce in foreign trade it could not compete in speed, equipment, and general serviceability with up-to-date cargo vessels. The design is the best that can be devised for an emergency product to be quickly, cheaply, and simply built. They will be constructed for the emergency and whether they have any utility afterward will have to be determined then. The coastal trade may offer some possibilities in that direction.[14]

As previously noted, the Commission had decided upon certain changes even before it decided upon the adoption of the British design. The most important of these included the use of oil fuel, water-tube boilers, and a single midship house. Beginning in January 1941, representatives of the Commission and of Gibbs & Cox held a series of conferences for the purpose of settling details. The Commission was represented by Mr. Schmeltzer, Chief of the new Division of Emergency Ship Construction, and Messers. Esmond and Grant who handled questions involving hull and engineering, respectively. The agent's principal representatives were William Francis Gibbs and his chief assistants, C. A. Ward and Henry C. E. Meyer.

One of the first matters requiring adjustment had to do with

---

[14] House, *Appropriation to U. S. M. C. for Emergency Cargo Ship Construction, Report 10,* 77th Cong., 1st sess., p. 4.

the extent of welding on the frame and shell. The sixty British ships were designed to have the frame riveted to the shell and this was done in all American yards except at Bethlehem-Fairfield which preferred to weld the frame to the shell and to rivet the seams of the shell.

It was also decided, in the interest of economy and speed of construction, to have no camber on the weather deck between hatches and to have a straight camber from the outside of the hatches to the side of the ship. Although wood was to be used for hatch covers and for furniture, wooden decks were to be eliminated wherever possible. Wooden hatch covers were preferred since they could also serve as rafts, in case the vessel sank.[15]

The changeover to oil fuel was in line with American practice and logical where oil was easily obtainable. This change brought in its train several minor changes. Fuel tanks replaced fixed ballast in the double bottom and deep tanks were added in No. 1 hold to provide for salt-water ballast forward. Coal bunkers were eliminated and arrangements for oil fueling and for smothering fires with CO_2 were installed. The single midship house and the elimination of coal bunkers permitted the use of masts instead of kingposts and the lengthening of No. 3 hatch.

Certain other differences were less noticeable. They included such things as the change from chain rails to bulwarks at the sides of the weather deck so as to give better protection for deck cargo; higher bridge bulwarks in place of canvas wind dodgers; addition of non-slip deck covering in way of guns; addition of ladders to provide access to upper and lower holds without removing hatches, a dangerous practice in bad weather; and addition of such things as ratproofing, after steering station, 12" searchlights, more refrigerated store space, running water in officers' staterooms, cooled water tap in midship house, and slop chest for crew.[16]

It will be noted that many of these changes tended to make the ship more comfortable for the crew than was the case with the British "Ocean" type. At the same time many items were omitted which had become standard on American vessels. These omissions were decided upon with the idea that this was an emergency ship

---

[15] Chief, Hull Section Emergency Construction to Director, Emergency Construction Division, Feb. 1, 1941 in MC gf 507-4-1.

[16] A. MacPhedran to Vickery, Feb. 17, 1941, in Wanless file, EC2-S-C1, Correspondence.

and that cost should be made as low as possible without sacrificing seaworthiness or serviceability. After consultation with representatives of the Bureau of Marine Inspection and Navigation, it was decided to substitute round bar davits with operating gear in place of the more easily operated but more costly mechanical davits,[17] to use composition deck covering of noncombustible aggregate supplied by manufacturers not on the approved list, to reduce the heating range in crews' quarters, to eliminate emergency Diesel generator, spare bower anchor, and to reduce anchor chain from 300 to 240 fathoms.[18]

In addition there were about thirty-five deviations from the Commission's own standards for fire-proofing and crew comfort. Of these the following are worth mentioning: failure to use noncombustible material on all bulkheads, ceilings, linings, and furniture; omission of mechanical ventilation in holds, engine and boiler rooms, officers' and crew's quarters; reduction in size of crew's lockers and rooms; oil instead of electricity for galley and for accommodation lights; no engine room entrance into shaft tunnel; no fire detection system; hand instead of electric deep-sea sounding machines and no mechanical gear for skylights or ventilator cowls; cement instead of tile in crews' toilet spaces and no heat insulation on shell plating in their quarters; also, omission of certain refinements and navigation aids such as gyrocompass, radio direction finder,[19] radio and searchlight in motor lifeboat, stainless steel galley equipment, and shutter type blinker and flag bag and rack with halyards to a triatic stay.[20]

Two other important changes from the British " Ocean " type were adopted because of wartime operating conditions. One was the contra-rudder, developed by the Goldschmidt Corporation of New York, which gave a small increase in speed and maneuverability and at the same time cost 40 per cent less than the one of British design.[21] The other change was the addition of a fatho-

[17] Before the first ships were finished mechanical davits were specified.

[18] Vickery to Commander R. S. Field, Director, Bureau of Marine Inspection and Navigation, Apr. 14, 1941 in MC gf 507-4-1. There is a pencil notation on this letter, " Not sent out," but the decisions mentioned above are in accord with U. S. M. C. " Specifications for the Construction of a Single Screw Cargo Vessel, Design EC2-S-C1," Edition of Apr. 1941.

[19] These items, at first omitted, were added later.

[20] Vickery to Field, Apr. 14, 1941, in gf 407-4-1. These notations check with Specifications, Edition of Apr. 1941.

[21] Wanless, " History of Emergency Ship Design," p. 5.

meter. This became necessary because degaussing seriously affected the magnetic compass and gyrocompasses were not obtainable.[22]

A number of changes were made to simplify the process of construction or to eliminate " bugs " which were likely to develop during hasty construction in new yards. One of these improvements was a new form of hawsepipe designed by W. G. Esmond, which was so shaped that even if there was some variation in the length of the hawsepipe it could still be fitted to the shell in a way that would assure a four-point landing of the anchor. Without this improvement there would have been danger of losing anchors in heavy weather.[23]

Since all the 200 emergency ships were to be built to one design, the design had to be " faired " only once.[24] This was done in the mold loft of the Newport News Shipbuilding and Dry Dock Company. Offsets of the frames were then made and distributed to the other shipyards making vessels of the EC2 design. The lines were made from a parent set sent from England and were identical except that some slight changes were made to the waterline endings at the stern post and at the bow where a small radius stem casting was used to which the paravane skeg was attached.[25] In connection with lines it should be mentioned that the American Bureau of Shipping, after careful study of the original midship section and scantling sections which were based on the requirements of Lloyd's Register of Shipping, found that their own requirements for a maximum draft of 26'-10" were exceeded and, consequently, agreed to certify the ship for a draft of 27'-7" which gave an additional deadweight of 430 tons.[26]

Before actual construction had progressed very far a few additional changes were made. Mr. Harry Hunter and Mr. Robson, of the British Supply Council, met with Messers. Grant and Esmond, of the Commission, and C. A. Ward, of Gibbs & Cox, to consider changes which marine experience indicated were

[22] Memorandum, A. Osbourne to W. E. Spofford, Jan. 31, 1946.

[23] " Liberty Ships. Historical Data " by W. G. Esmond, folder in MC, Historian's Collection, Research file 111.

[24] " Fairing " means to lay down or draw the lines in the mold loft so as to check the correctness and harmoniousness of the lines and so as to establish definitely the dimensions of frames.

[25] Historical Notes prepared by W. G. Esmond in W. E. Spofford, " History of the Design, Construction and Operation of the EC-2 Emergency (Liberty) Ships," p. 32, in MC, Historian's Collection, Research file 111.

[26] Wanless " History of Emergency Ship Design," p. 6.

desirable. The most important was the substitution of $\frac{1}{4}''$ steel in conjunction with plastic armor instead of $\frac{3}{4}''$ steel as protection for the chart and wheelhouse and radio room. It had been found that the steel alone was not sufficient protection. This change also achieved a saving in the amount of steel required. It involved, however, the procurement of a considerable amount of special equipment at the shipyards. Other suggestions made at this time, which were adopted, included the addition of ventilators for the machinery space, the substitution of mechanically operated davits for manually operated ones, and the addition of direction finders.[27]

Some changes were dictated by scarcities, especially the scarcity of steel. Mr. Esmond found several ways of economizing on the use of steel. For example, the conventional " eye " beam used to support wooden hatch covers was replaced by a lighter beam in the latter part of 1942,[28] and deliveries of steel were speeded up by reducing the number of thicknesses used. In December 1941, Mr. Esmond prepared and secured the approval of the American Bureau of Shipping for a schedule which reduced the number of gauges for steel plate from 75 to 27.[29] After Pearl Harbor also, as will be explained in chapter 10, steps were taken to use the narrower plates which could be rolled by strip mills.

Another result of steel scarcity was the progressive reduction of anchor chain supplied the Liberty ships. At first the peacetime requirement of 300 fathoms was reduced to 240 fathoms, and later reduced again to 210 fathoms, the chain being divided so that one anchor had 135 fathoms and the other only 75. Even so, many ships went out with only one anchor.[30] Wartime requirements brought about the opposite result in the case of cargo booms. The original plans called for five-ton booms only, but the need to load tanks and other heavy equipment necessitated the installation of fifteen- then thirty-, and finally fifty-ton booms.[31]

The new type of door from the engine room to the shaft alley is an example of a change made to meet the needs of the crew and wartime experience. A vertical sliding door had been used in

[27] Vickery to R. R. Powell, British Supply Council in North America, June 16, 1941, in MC gf 507-4-1.
[28] Historical Notes by W. G. Esmond, in W. E. Spofford, "History," p. 41. This beam was designed by John C. Temple.
[29] *Ibid.*, pp. 36-38.
[30] *Ibid.*, p. 35.
[31] *Ibid.*, pp. 92-98.

this location but it was likely to jam and become immovable so that it was usually left open, contrary to the safety rules of the regulatory bodies. In the case of several sinkings it had apparently been the means of flooding the engine room; consequently the British decided to eliminate it. The first American Liberty ships were also designed without this door. As a result, oilers had to go periodically up to the weather deck and down again to the shaft alley and then up again after making their inspection. Upon the complaint of the seamen's unions, the matter was reconsidered and a small, quick acting, hinged door was installed.

The minor changes just enumerated did not affect the vessel's appearance. It was the single midship house in place of the fore and aft houses, the different cargo-handling equipment, and the solid weather-deck bulwark instead of chain rails that made the Liberty ship look different from the " Ocean " class vessel. See Plate III. The crews were most concerned about the presence or absence of such things as running water and mechanical ventilation. British crews were accustomed to more austere and comfortless quarters than had been enjoyed by American crews since the establishment of the Commission in 1936.

The Liberty ships and their British prototypes were sisters under the skin, so to speak. They were spawned of the same need— the need for a ship with maximum carrying capacity combined with simplicity of design and of operation. In the case of the British, this need came in the struggle of the depression years when a ship was wanted that would not cost too much to build and operate. In the case of the United States, the need grew out of the wartime pressure for more ships which could be built quickly and be easily operated. Cost, with us, was a secondary consideration. So was it also with the British when they came to build their " Ocean " and " Fort " class vessels, for by that time it had become a question of ships and more ships at whatever cost. And these new ships were needed in a great hurry. This urgency dictated the further simplification of an already simplified design and the extensive use of welding instead of the more thoroughly tested but slower process of riveting.

Since the all-important consideration with regard to the Liberty ship was its ability to move goods, its carrying capacity is of special interest. A typical general cargo Liberty ship (as distinct from special types designed as boxed-airplane transports, barrel tankers or bulk cargo vessels) had a deadweight tonnage of 10,419. Cargo was stored in five holds. No. 1 hold, in the bow, had the least capacity, 76,077 cu. ft., due to the presence of two deep tanks in this locality. These deep tanks were used for dry cargo or ballast and, together with a third tank aft of the engine-room, had a capacity of 41,135 cu. ft. No. 2 hold was the largest, reaching to the ship's floor and measuring 134,638 cu. ft. No. 3 was just as deep but was narrowed to make room for the engine so that its cubic was only 83,697. Holds 4 and 5 , being aft of the engine room, were interrupted by the shaft tunnel and had cubics of 82,263 and 82,435 respectively. The Liberty's total bale or cubic capacity was, therefore, 500,245 cu. ft.

This does not take into account the deck space which became an important consideration. Except for the midship housing and fore and aft gun turrets, there were no important deck obstructions for the whole of the Liberty's 441' 6" overall length. Hatches were all 19' 10" wide and those for holds No. 2, No. 4, and No. 5 were 34' 10" long while No. 1 was shorter by about a foot and No. 3 was only as long as it was wide.[32]

III. Purchasing Through a Central Agent

The purchase of components and materials for the Liberty ships was closely connected with the preparation of working plans because the same naval architects had an important role in both operations. To secure the advantages of large-scale purchasing and flexibility in deliveries, the Commission undertook to furnish the builders of Libertys with steel, engines, and nearly all other materials and components instead of leaving each shipyard individually to buy its own material as was done by the yards in the long-range program. As early as December 1940, Admiral Land advised the President that there should be a central control through which material would be " purchased and apportioned to the contracting yards as the working schedules require." This was the root of the wartime system in which the Procurement Division of the Maritime Commission bought the materials and components for Liberty ships (and many others) and the Production Division

---

[32] See note to Table 2, p. 28 above.

allotted these materials to the yards. But in December 1940 Admiral Land showed no desire to build up a bureaucracy to perform these functions, although that became necessary later. Instead, he proposed that purchasing and allocating be done for the government by a private firm under contract to act as agent.[1] It was the quickest way to get started. Accordingly the system of centralized procurement and government-furnished materials was introduced at the outset of the program, but through a private company acting as agent for the government.

If procurement was to be centralized under a private firm, Gibbs & Cox was the obvious choice. In spite of the slight differences between the American Liberty, the British Ocean type, and the Canadian Forts, the standardization was sufficient so that very many components would fit interchangeably into any of these ships. From the point of view of broad industrial planning, British and American emergency shipbuilding was all one operation. Politically and legally the British building and that of the Maritime Commission were of course quite distinct, and the Commission was careful not to assume responsibility for the terms of the arrangements made between the British Mission and private American corporations. But as a problem in scheduling, the two programs were one. Orders for components for the British ships were, on the whole, placed first, but long before they were ready for delivery the American building was under way and producers of components were told to deliver to whatever shipyard needed them at the moment. Thirty of the engines ordered originally for the British ended up in American Liberty ships.[2]

As naval architects for the companies building for the British, Gibbs & Cox had already arranged to handle procurement for the Todd-Bath and Todd-California yards. It was usual for a shipbuilder to have his naval architect draw up the precise specifications of the materials and parts he would need (these specifications are called requisitions) and invite bids from vendors. The actual purchasing was usually left to the shipyard—it decided which bids to accept—but for the 60 British ships Gibbs & Cox were to do the buying also and to arrange deliveries so that everything would

---

[1] Land to the President, Dec. 26, 1940, in MC gf 507-1.
[2] R. C. Thompson and Harry Hunter, " British Merchant Shipbuilding, Programme in North America," (Andrew Laing Lecture before the North East Coast Institution of Engineers and Shipbuilders) printed in *The Shipping World*, Dec. 2, 1942, p. 502.

reach each shipyard at the right time. Having placed the orders, they would tell the vendors exactly when and where to deliver; in short, they would schedule the flow of materials.[3] By January 29, 1941, the Commission had in effect made up its mind that Gibbs & Cox would do the ordering also for the 200 American Libertys.[4] In April, 112 more ships were added so that in the summer Gibbs & Cox was buying and setting tentative delivery dates on the basic materials for 372 ships, more than ten times the total national output two years before.

But Gibbs & Cox did not have nearly so free a hand in buying for the Maritime Commission as it did in buying as an agent for the yards. W. F. Gibbs would have preferred that his firm render all its services as an agent of the shipbuilding companies and that was his expectation when he started work on the designs. Only with reluctance was he persuaded to make his contract with the United States Government and act as its agent in supplying the yards.[5] His contract made his awards to vendors subject to the Commission's approval, which was exercised through the Division of Emergency Ship Construction. The procedure was as follows: Gibbs & Cox issued inquiries to vendors, checked proposals from vendors, and awarded commitments for the Commission. After approval of the commitment by the Director of the Emergency Ship Construction Division for the Commission, Gibbs & Cox prepared a formal purchase order which was in turn subject to final approval by the Commission's staff.[6] Gibbs & Cox also prepared the requisitions on which the inquiries and purchase orders were based, and these were approved by the same method in which working plans were being approved—by stationing in the New York offices the

[3] W. F. Gibbs describes his arrangements with the British in his testimony, House, *Navy Department Appropriation Bill for 1945, Hearings,* 78th Cong., 2d sess., p. 3870, and in a letter to Admiral Land, May 7, 1941, in MC gf 507-1-4, par. 11.

[4] At that date began a flood of letters telling engine manufacturers and others that the contracts for components would be let through Gibbs & Cox. MC gf 507-1 and 507-1-1.

[5] W. F. Gibbs to Land, May 7, 1941, in gf 507-1-4.

[6] Contract MCc-ESP-15, Article 2 (b) in gf 507-1-4; Admin. Order No. 56, May 1, 1941; Office Order No. 2 of the Emergency Ship Construction Division, in gf 202-4-1; Gibbs & Cox, Inc. to Director, Emergency Ship Construction Division, Apr. 7, 1941, Encl. A, par. 30, in gf 507-1-4. Major purchases were approved by the Director or his Assistant, usually over the phone, and minor purchases were approved by Messrs. Esmond and Grant.—Statement of Van Riper to G. J. Fischer, Sept. 22, 1947.

chiefs of the appropriate sections of the Division of Emergency Ship Construction.[7]

In exercising its power to approve, the Commission emphasized two principles and worked with Gibbs & Cox in joint efforts to carry them into effect. One was that the work should be spread widely in order to insure a good supply, since many of the large producers were already working to capacity for the Navy or for the Commission itself on its C-type program. This was also a desirable policy from a political and economical point of view at a time when war conditions were upsetting the usual course of business. The other principle was that parts should be interchangeable. This applied more particularly to the engines and boilers. These two principles were not easy to apply. Gibbs & Cox usually found it more convenient to award orders to one or two large suppliers. It could always be argued that time and paper work would be saved by limiting the field. When it came to making interchangeable engine parts, where exact reproduction was essential, there was the difficulty of inducing manufacturers accustomed to competitive business practice to work in close cooperation.

On March 31, 1941 Mr. Gibbs presided at a meeting in New York with representatives of the shipyards. Besides members of his own organization and of the Commission there were present representatives of the Bethlehem, Newport News, and American Ship Building companies. The purpose of the meeting was to discuss ways of purchasing and expediting materials and to decide in as much detail as possible just what would be done by the Maritime Commission, by Gibbs & Cox, and by the individual shipyard. It was agreed that the Commission would do the expediting. When the matter of subcontracting came up, American Ship Building stated they planned to subcontract for deck coverings and insulation and Bethlehem stated their intention to subcontract insulation, brickwork, and pipe fabrication. A matter brought up by Mr. Gibbs had to do with the loan of personnel to his organization to work on procurement and production. Federal Shipbuilding had already loaned two men and Newport News one, but more were needed. When each company responded that they had no experienced men to spare, Mr. Gibbs asked them

---

[7] Admin. Order No. 37, Suppl. 22, Jan. 7, 1941.

to think the matter over, since the yards themselves stood to gain by such a transfer of personnel.[8]

Problems typical of the whole organization for mass production arose in connection with the engine. The basic design used on all the 372 ships for which Gibbs & Cox was ordering was British,[9] and the plans had to be redrawn to adapt them to American industrial practice. British marine engine builders and other manufacturers in general left many details of the drawings to shop practice, an arrangement which worked out very well when the staff had grown up on the job, and where the whole unit was usually cast, machined, assembled, erected, installed, and delivered under one roof. For use in the United States, however, it was necessary to amplify the British plans and re-dimension them in respect to tolerances, fits, clearances, etc. As a result, 80 British working plans had to be expanded into about 550 plans.

Placing the orders became more difficult as the program expanded. All the engines for the 60 British ships were to be supplied by the one manufacturer which Gibbs & Cox judged best qualified, the General Machinery Corp. of Hamilton, Ohio.[10] To supply the American Libertys ten other engine makers were also given orders. A conference to bring about standardization and efficiency among these many producers was held on June 19 and 20, 1941. It was presided over by Henry C. E. Meyer of Gibbs & Cox and was attended by representatives of the companies, of Lloyd's Register of Shipping, the American Bureau of Shipping, the Office of Production Management and by John E. P. Grant of the Maritime Commission and Harry Hunter of the British Merchant Shipping Mission.

Mr. Meyer explained that the purpose of the conference was threefold, viz: (1) to afford an opportunity for examination of one engine nearly completed and several in various stages of construction; (2) to answer questions in connection with design, now that all problems had been met and solved in the completion of the first engine; (3) and to attempt to dispose of all major questions of builders so as to reduce future correspondence to a minimum. Mr. Meyer was very appreciative of the cooperation of Mr. Gregor,

[8] Minutes of Meeting held Mar. 31, 1941 in Emergency Ship Construction Div. file, QM8 A-3-2.
[9] They were designed by the North Eastern Marine Engineering Co., Ltd., a Division of Richardson, Westgarth & Co., Wallsend-on-Tyne.
[10] Thompson and Hunter, pp. 74-76.

chief engineer for the General Machinery Corp., who had prepared detailed plans and gone to the trouble of keeping all builders informed of corrections as they came along.

Mr. Meyer was supported by Messrs. Hunter and Grant in emphasizing that a standardized, reliable product that could be built in a minimum amount of time was what was wanted. It was not doubted that this engine could be improved upon, but improvements were not wanted. The only changes in design that would be considered were those which the General Machinery Corp. had found advantageous in constructing the first engine and those which might be found necessary in cases where limitations of facilities or materials might make them desirable. The design was a well-proven and tried one and should not be departed from, and might have a bearing on the outcome of the war if the need " for more and still more ships becomes more pressing." The need was for identical engines with interchangeable parts and, in order to obtain them, methods of manufacture and quality of workmanship should be as nearly identical as was humanly practicable.

Certain detailed technical questions were then discussed, such, for example, as the use of steel shims for lining up the centers of the cylinders with the centers of column and crank pins. It was agreed that there could be two alternate designs for pistons, one the original British design to which General Machinery Corp. was now working and the other a British design adopted by the Worthington Pump & Machinery Corp. It was also decided that, although all parts must be properly marked for identification during assembly, it was not practical to outline a standard procedure for assemblage.

After adjournment, the work in progress in the erecting shop was observed and the rugged appearance of the engine was remarked upon. On the following day there was further discussion of details in the light of this inspection. It was brought out that the only difference between the British and American engines was in the use of different gauge boards. On the afternoon of the second day the visitors watched while one complete throw of crankshaft was shrunk together, an operation which took only about 15 minutes. The General Machinery people stated, with a good deal of satisfaction, that, " when these crankshafts were put in a

PURCHASING THROUGH A CENTRAL AGENT 95

lathe for final finishing, the maximum error they had been able to detect was .007." [11]

Interchangeability was also an important factor in the building of boilers. In this case the design used on the British ships was not being followed on the American, which used a water-tube boiler designed by Babcock & Wilcox Co. It was placed athwartship with the firing aisle fore-and-aft, instead of transversely as in the case of the British coal-fired Scotch boilers. With this arrangement it was possible to make them interchangeable by having all connections to the " same hand," that is, by placing the safety valve, vent and cleaning connections, main steam nozzles, water level gauges, etc. symmetrically about the centerline of the boiler and drum. The " dog house " containing the superheater was also arranged symmetrically, so that the superheater could be installed on either the port or starboard side.[12] Boilers for the first 200 emergency ships were supplied by Babcock & Wilcox, Combustion Engineering, and Foster-Wheeler, at the rate of nine, six, and three per week, respectively. These companies were selected by Gibbs & Cox and the Maritime Commission because they had extensive works and were experienced manufacturers of boilers and boiler parts. When the program was enlarged to 312 ships, four more boiler manufacturing companies were given contracts.[13]

The usual problems in connection with subcontracting developed while arrangements were being made to procure the boilers. In their purchase commitment with Babcock & Wilcox, Gibbs & Cox had specified the use of Diamond soot blowers, as being preferred by that manufacturer and as insuring interchangeability. But one of the other companies, Combustion Engineering, proposed to use less expensive blowers made by the Bayer Co. The Commission inquired of Gibbs & Cox why the more expensive had been specified.[14] Upon the insistence of the Commission, blower contracts were divided between the Diamond, Vulcan,

---

[11] Gibbs & Cox to MC, June 21, 1941, in Emergency Ship Construction Division file, QM22, enclosing Minutes of Meetings of June 19 and 20.

[12] Gibbs & Cox to MC, Feb. 21, 1941, enclosing Minutes of Meeting held Feb. 21, 1941, in Emergency Ship Construction Division file, QM 8, S51.

[13] Minutes of Meeting, Feb. 21, 1941; J. E. Schmeltzer to Gibbs & Cox, Att. W. F. Gibbs, May 29, 1941 in Emergency Ship Construction Division file, QM8 S51.

[14] Joseph V. Santry of the Combustion Engineering Company to Schmeltzer, Apr. 15, 1941; Schmeltzer to Gibbs & Cox, March 22, 1941, in MC gf 507-1-1.

and Bayer companies. Representatives of these companies consulted among themselves and agreed that there would be prohibitive delays if they all built to any one of their designs or to a new design, but that each could make the four component parts of a blower, namely, head, element, wall box, and bearings, so as to be interchangeable and usable on any boiler or piping system.[15]

As the example of the soot blowers indicates, the placing of orders by the architect was closely supervised by the Commission. Frequently the Commission's staff sent Gibbs & Cox the names of would-be suppliers with instructions to the architect to include the names on the list of those from whom they asked bids. In some important cases, Gibbs & Cox asked by telegram specific approval from the Commission of a particular order.[16] As a general rule, bids were tabulated and evaluated by Gibbs & Cox and the information given to the Construction Division, together with the recommendation of the agent, for ratification. Usually there was a tentative ratification over the telephone in order to expedite the commitment. Detailed information followed by letter.[17] In standardizing and spreading the work the staffs of the Commission and of the architect worked together with the common aim of harnessing quickly many various industries into a shipbuilding team.

Political pressure operated in favor of the wider spreading of contracts. Manufacturers who felt that Gibbs & Cox was discriminating against them turned to their Congressman or Senator or wrote to the President. In regard to the soot blowers, the Bayer Company wrote directly to the President. Their representative, after seeing Mr. Schmeltzer, apologized for sending a copy of the letter at the same time to Senator Harry S. Truman, but made clear that he was ". . . looking to the U. S. Government [not to Gibbs & Cox] for justice."[18] Their claim to a part of the business was considered quite well founded, as has been explained, and they obtained a share. In most cases the company mentioned by a Congressman was at least called to the attention of the architect.

---

[15] Minutes of Meeting, Feb. 21, 1941; Gibbs & Cox to MC, Att. Mr. Schmeltzer, Apr. 9, 1941 in Emergency Ship Construction Division file, QM8 S51.

[16] Gibbs & Cox to Schmeltzer, March 21, 1941 and Schmeltzer to Gibbs & Cox, March 26, 1941, in MC gf 507-1-1.

[17] MC Minutes, pp. 18447-48 (Aug. 12, 1941).

[18] F. W. Linaker, Vice President, The Bayer Company, to Schmeltzer, March 31, 1941, in gf 507-1-1.

Much of the political pressure was against the overconcentration of the work in the Northeast. Senator Monrad C. Wallgren of Washington asked why all the boilers for the Liberty ships were being made in the East, although many of the ships were being built on the West coast.[19] Congressman Albert Thomas urged the consideration of Houston firms, which were his constituents, in buying components for the ships which were to be built at Houston.[20] A flood of inquiries seeking business were directed at the Commission and were answered by referring the industrialist or his Congressman either to the shipyard, which was buying its yard equipment, or to Gibbs & Cox, or sometimes to both.

IV. THE BREAK WITH GIBBS & COX

Throughout the first half of 1941, Gibbs & Cox and the staff of the Maritime Commission worked very closely together, almost as one organization, in the preparation of plans and the placing of orders. Then they parted company and Gibbs & Cox dropped out of the emergency shipbuilding program, which they had done more to organize than had any other one firm. In July and August they were finishing the last of the plans. At the same time the Commission was reorganizing its staff to take over procurement.[1] Gibbs & Cox had placed the orders for the 60 British and 312 Liberty ships which were authorized in the first two waves of expansion in the Maritime Commission's program. For the additional ships of the third wave, ships to be paid for by the appropriation bill which passed Congress in August 1941, the Commission was preparing itself to do the buying.

Reasons for the termination of the contract with Gibbs & Cox are not hard to find, although it is difficult to evaluate their relative importance. Basic, probably, was the growth in the size of the program. As the nation moved closer to war it became more and more apparent that procurement for the emergency shipyards would become so large an operation, and so closely connected with priorities and allocations, that the government would have to take direct control. Even in 1941 the concentration of so much power in the hands of a private firm was criticized by industrialists who

[19] Senator Wallgren to Land, March 24, 1941, in gf 507-1-1.
[20] Congressman Thomas to Land, Apr. 16, 1941, in gf 507-1-1.
[1] Vickery to Commission, July 25, 1941, with Admin. Order No. 37, Suppl. 42.

felt excluded and by politicians who felt that the private agency was too distant from their influence.[2] The General Accounting Office at a later date questioned the propriety of the contract. Referring to the power given Gibbs & Cox to prepare requisitions and purchase materials, Mr. McGinn of the General Accounting Office wrote: "It appears that under ordinary circumstances, such authority is inherent in the members of the Commission as officials of the Federal Government and may not be delegated by contractual agreement to a party not an official of the government, whose responsibility and duty is no more than contractual." [3]

Although the very growth of the program would of itself have forced a change, probably, sooner or later, quite different considerations appear on the surface in connection with the Commission's decision in June 1941 not to continue the arrangement with Gibbs & Cox. William Francis Gibbs was not satisfied with what the Commission was willing to pay.

The issue involved was how much the fee should go down as the volume went up. Gibbs & Cox worked on contracts giving them costs plus fixed fees. When the contracts were made for the 60 British ships, the compensation of Gibbs & Cox was fixed at $10,000 per ship. Their fee was $6\frac{1}{4}$ per cent of the shipbuilders' fee (of $160,000) and about $\frac{2}{3}$ of 1 per cent of the estimated cost of the ships. When asked by the Todd interests to perform similar services for the Libertys, Gibbs & Cox offered to do so for $6,875 per vessel, which was $6\frac{1}{4}$ per cent of the shipbuilders' base fee. The contract was not made with Todd, however, but with the Maritime Commission. It was made for all the 200 ships of the emergency ship program, and Gibbs & Cox agreed to cut the fee down to $3,750 per ship (a total of $750,000), which was $6\frac{1}{4}$ per cent of the shipbuilders' minimum fee and about $\frac{1}{4}$ of 1 per cent of the estimated cost of the ships. When the number of ships was increased to 312, the Commission proposed to keep the total fee the same, which would cut the rate down to $2,403.84 per ship. But this was too low for Gibbs & Cox. They held out for $3,076.92 per ship. While that question was still open, work got under way and the Commission's staff discovered to its surprise that the plans

[2] There is a hint of Congressional disapproval in Admiral Land's account of a telephone conversation with Congressman Adolph J. Sabath, Memorandum for J. E. Schmeltzer and F. E. Hickey, Sept. 3, 1941, in Land reading file.
[3] B. A. McGinn to Chairman, MC, Feb. 19, 1942, in MC gf 507-1-4.

for the British had not been previously prepared but that the plans for both the British and the Commission's account were being produced simultaneously and for all practical purposes were the same plans. Since it was really all one job, and since the British were paying $600,000 on it, the Commission decided that $500,000 more would be enough, and on April 25, 1941 made a firm offer of that amount. William Francis Gibbs protested indignantly that this was only $1,602.56 per ship which was only 11/100 of 1 per cent of the cost of a ship, only $12.33 per ship per week, and concluded: "We have no precedents for such low compensation. . . ."

The Commission stuck to its offer of a fee of $500,000. Since the expenses incurred in performing the contract were being paid for anyhow under the cost-plus-fixed fee contract, the half million was really payment for the services of William Francis Gibbs himself and his twelve top men. Their salaries were not included among the expenses, and the Commission's advisers admitted that William Francis Gibbs, the genius of the firm, rated a salary as high as any paid in industry. But he and his twelve top men were handling work for the Navy which totaled far more than the merchant shipbuilding they were supervising for the British and for the Maritime Commission. Whatever fee they received from the two latter sources would be payment for part-time services for about two and a half years. The Commission's advisers felt it should pay enough so that William Francis Gibbs earned at least $200,000 a year, but that was fully provided for, they argued, in the fee of $500,000, considering all the other fees his firm was collecting. Mr. Gibbs rejected all these arguments and persistently emphasized how little he received per ship.[4] Finally, still protesting, he accepted the Commission's terms and sent in the signed contract on June 24, 1941 after the actual work of drawing plans and placing orders was far advanced.[5]

The next December Admiral Land summed up for an Appropriation Committee the story of the stiff bargaining, "We felt

---

[4] Mr. Gibbs reviewed how his fee, when figured on a percentage basis, had been progressively lowered in stating his side of the case in a long letter to Admiral Land, May 7, 1941. Reasons for considering the $500,000 adequate were set forth in memoranda by J. E. Schmeltzer on May 15, 1941 and R. E. Anderson on May 22, 1941. These are in MC gf 507-1-4. For Commission action see Minutes, pp. 17077-78 (Apr. 8, 1941), and p. 17782 (June 19, 1941).

[5] W. F. Gibbs to Land, June 24, 1941, in gf 507-1-4.

that we paid plenty for it and they felt that we did not pay enough, so we just split business with them and said, ' Good-bye boys,' and kissed each other on both cheeks, and we are running our own show now." [6] Gibbs & Cox kept very busy with work for the Navy, which was turning out destroyers with great rapidity and paying the usual fee on each.[7]

The role of Gibbs & Cox in the emergency program was characteristic of the transitional period, that is of the first half of 1941, when the nation was not at war, when it was politic to refer not to preparations for war but to the " defense effort," and when preparations were governed by the intent of interfering as little as possible with the peacetime system of business. After mobilization of the nation was vigorously taken in hand by government bureaus, Gibbs & Cox remained important in the preparation of designs and working plans. During the transition period, Gibbs & Cox also provided the instrument for quickly centralizing procurement, a centralization which was a necessary consequence of the decision to have the Commission supply materials and components for the Libertys.

[6] House, *Independent Offices Appropriation Bill for 1943, Hearings* before the Subcommittee of the Committee on Appropriations, 77th Cong., 2d sess., on the Independent Offices Appropriation Bill for 1943, Dec. 9, 1941, p. 277.

[7] House, *Navy Department Appropriation Bill for 1945, Hearings*, 1944, p. 4028. Mr. Gibbs' testimony before that Committee, pp. 3986-4027, is a vigorous defense of his fees.

Chapter 4

CONTRACTS WITH SHIPBUILDERS AND THEIR SUPERVISION

I. Lump-sum Contracts

WHILE the Commission was choosing sites for new shipyards and settling on a new ship design, its staff was also shaping up new forms of contracts; while the technical staff worked on specifications, the legal and financial divisions drew clauses defining in new terms the Commission's relations with shipbuilding companies.

Most wartime contracts provided that the shipbuilder would be paid the costs of construction plus a fee. Literally they were, therefore, cost-plus contracts, but they were not "cost-plus" in the sense in which that term had come to be commonly used after World War I. The term had become associated with contracts that assured the contractor his costs plus a percentage of the cost. Those cost-plus-a-percentage-of-cost contracts were invitations to extravagant spending because the more a contractor spent the more he received as fees and the higher was his profit. The evil effect of those contracts in World War I caused them to be carefully avoided. When it was found necessary to use contracts in World War II that provided for reimbursement of costs, the terms of the contract were so arranged that higher costs would result in a lower fee for the contractor. Contracts of this sort are often called cost-plus-fixed-fee contracts because fixed maximum and minimum fees are specified in the contract, but they may more accurately be called cost-plus-variable-fee contracts.[1]

The use of cost-plus contracts, even in this improved form, seems at first glance a revolutionary change from prewar practice of awarding contracts which stated a lump sum as the price of the ship. The Merchant Marine Act of 1936 required competitive

---

[1] John Perry Miller, *Pricing of Military Procurement* (New Haven, 1949), p. 124.

bidding.[2] In simplest form this was illustrated by the letting of the contract for the *America*. The Newport News Shipbuilding Co. made the lowest bid and received the contract at that price.[3] But as soon as the Commission launched its program for standardized types of ships, it began a system of allocating a few ships to each of a number of yards. Then Congress authorized the use of wide discretion in negotiating these awards, for the Commission was empowered to take into consideration regional differences in costs, and the needs of defense and of relieving unemployment. Earnings of the companies were limited by the Merchant Marine Act of 1936 to 10 per cent of the contract price. In most cases that clause fixed the amount of profit received by the shipbuilders. A review of the Commission's lump-sum contracts under the long-range program will show, therefore, that the wartime practice of negotiating cost-plus-variable-fee contracts, although an important change, was not so substantial a change in substance as it was in form.

The first group of contracts awarded by the Maritime Commission were those in 1937 for 12 high-speed tankers with special national defense features. The Sun Shipbuilding Co. was the low bidder. The other bidders—Federal, Bethlehem, and Newport News—were then informed that if they would meet Sun's figure they might have three ships each. They lowered their bids and the work was spread out over the four yards.[4] That same winter bids were invited on the construction of the Commission's new design for standard cargo ships, C2's. The Commission stated its intention to have at least twelve started at once, if possible; but under the terms of the bids no bidder was to be awarded more than four.[5]

Such a practice of allocating might cause the shipbuilders to keep their bids high, assuming that even if they were underbid at first they could negotiate later and get a considerable share of whatever work was arranged for. This was the more likely to happen in case the only bidders were three or four of the Big Five. When the Commission asked bids on the standard cargo ships,

---

[2] Secs. 502a and 703a and 703c.

[3] MC Minutes, pp. 1623, 1772-77.

[4] This is the story as commonly told.—*Congressional Record*, 75th Cong., 3d sess., vol. 83, art. 6, p. 5939, Apr. 28, 1938. Strictly speaking, the contracts to build were given by the Standard Oil Company of New Jersey; MC Minutes, p. 2498.

[5] MC Minutes, p. 2425.

Sun, Bethlehem, and Newport News were among the bidders, and all their bids were much higher than the estimate of cost prepared by the Technical Division of the Maritime Commission. Competition came from some new bidders, however, including a little known concern whose yard had been burned over and which was close to bankruptcy, the Tampa Shipbuilding and Engineering Co. Whether Tampa could be considered a " responsible bidder " was a subject of considerable investigation and debate.[6] Another concern which bid about the same price, the Hess Company, was ruled out on the grounds that it had no plant, had hardly a skeleton organization, and was without working capital. But Tampa, it was decided, had the essentials of a plant and an organization which had built ships during the First World War and could borrow working capital from the Reconstruction Finance Corporation. Its price was reasonable. Accordingly the Commission awarded it a contract for four ships and negotiated with Sun and Newport News in hopes of lowering their prices to near the same level. As the specifications were changed a little during these negotiations, it was found legally necessary to issue a new invitation to bid. The outcome was that in the spring of 1938, Tampa obtained from the RFC the loans it needed to keep going, and four ships each were allocated also to Sun, Federal, and Newport News at prices subtantially below the bids originally submitted by the big builders of the Northeast.[7]

During the spring of 1938 Congressional action recognized and elaborated this system of allocations. Mr. Kennedy had suggested to the House Committee on the Merchant Marine in December of 1937 that it would be wise in the interest of national defense to allocate a certain percentage of shipbuilding to the West coast and the South, and the Commission would be ready to do this if

[6] A " Memorandum for Maritime Commission " by Admiral W. P. Robert in MC Land file, Sept. 25, 1937 says of it: " all the superstructures of ways were destroyed by fire some years ago." The Comptroller General declared later that at the time of the contract the company was " practically insolvent " and in need of at least $1,000,000 operating capital to engage in shipbuilding—House, *Tampa Shipbuilding & Engineering Co., Hearings on H. Res. 281*, before the Committee on the Merchant Marine and Fisheries, 77th Cong., 2d sess., Dec. 10-12, 1942, p. 9. (Hereafter cited as House, *Tampa Hearings.*) Analysis by Vickery and John Slacks in MC Minutes, pp. 3656-67 gives bids of all the yards, and they are in the Hearings just cited, pp. 199-200.

[7] *Ibid.*, pp. 9 and 28; MC Minutes, pp. 3074, 3412, 3462, 3656-3667, 3777-85, 3690-91; and MC, *Report to Congress, 1938*, p. 41.

the law were amended to " give some discretion " to the Commission.[8] The House Committee accordingly proposed and carried through an amendment which permitted the Commission, with the consent of the President, to award a contract at a price in excess of the lowest bid, if they believed it fair and reasonable, taking into consideration the conditions of unemployment and the needs of national defense.[9] The main purpose was to permit the awarding of contracts to West coast yards at higher prices than those bid by East coast yards,[10] but in the Congressional debates and reports it was recognized that the Commission was being given a very free hand " to trade with the shipbuilders." [11]

The kind of trading that followed may be indicated by one incident in January 1941, just as the emergency program was getting under way. By that time Western yards had been drawn by negotiation into the Maritime Commission program and were able, as the incident shows, to compete with Eastern yards. Bids had been asked on sixteen C2's. On the basis of the bids received, four C2's with Diesel propulsion were awarded to Tampa (subject to arrangements with the Navy and the Reconstruction Finance Corporation), six with turbine propulsion to Moore Dry Dock of Oakland, California, and six to Federal, *provided* Federal would lower its bid to match that of Moore Dry Dock. Federal protested violently and threatened to work only for the Navy and to eliminate itself entirely from bidding for Maritime Commission contracts. Admiral Land for the Commission reiterated that the bid from Federal had been quite out of line, being higher than that of a West coast bidder whereas the legal differential was in the opposite direction, and closed his telegram: " Commission deeply regrets your decision to eliminate your company from Maritime Construction contracts." Two weeks later a compromise was reached by which Federal received more than Moore Dry Dock, but less than they had bid.[12]

[8] House, *Hearings* before the Committee on Merchant Marine and Fisheries, 75th Cong., 2d sess., on H. R. 8532, Dec. 2, 1937, p. 36.
[9] House *Report*, No. 2582, 75th Cong., 3d sess., Conference Report June 3, 1938, pp. 4-5.
[10] House *Report*, No. 2168, 75th Cong., 3d sess., printed Apr. 20, 1938, p. 11. The amendment provided that the added costs in these contracts should be paid by the Commission as a part of national defense, and should not be counted in figuring the construction differential subsidy. Chap. 600, Public Law No. 705, 52 Stat. 957.
[11] *Congressional Record*, vol. 83, p. 5939.
[12] MC Minutes, pp. 15931-34 for bids; p. 16053 (Jan. 30, 1941), p. 16182 (Feb. 11); Land to Korndorff, Land reading file, Jan. 24, 1941.

To call this bargaining mere shadow boxing would be quite unfair, but with the advantages of hindsight we can see that it had little effect on the cost of ships to the government. That price was being determined mainly by the costs of building as found by cost accountants. According to the Merchant Marine Act of 1936, the books of the builders were to be open at all times to the scrutiny of auditors sent by the Maritime Commission; the builders were to prepare statements of the actual costs of construction, the charges for overhead, and the profits; and these were to be checked by the Commission's auditors. If a builder's profits were more than 10 per cent of the contract price, all profit in excess of 10 per cent was to be recaptured by the Commission.[13] Actually, on every contract authorized by the Commission in 1937 and 1938, except that to Tampa, there was profit which could be claimed under the recapture clause. The recaptures averaged 6.3 per cent of the contract price, and ranged from 1.3 per cent to 14.1 per cent. As affairs worked out, all the big firms received their cost plus 10 per cent of the contract price.[14] To be sure, lowering the contract price saved the Commission a little, but the main factor in determining the cost to the Commission of these ships was the cost of construction as found by the auditors.

Apparently what happened was that these yards, equipped to build ships more complicated that the C2's, allowed in their bidding for carrying charges on their expensive equipment,[15] and that these charges were later scaled down because when the Navy program began the companies had other construction to which to charge their overhead. Moreover the builders may have been afraid that the Maritime Commission would be just as exacting in its standards as was the Navy.[16] Whatever its cause, this situation could be summarized at the beginning of 1941 by saying that the bidders on contracts all added to their costs 10 per cent for profit and then 5 per cent more as a margin of safety.[17] Later, referring to other situations, Admiral Land gave the following general statement on why bids were too high:

[13] Sec. 505 (b).
[14] Figures on recaptures supplied by Henry Z. Carter, General Auditor, MC, Division of Finance, Construction.
[15] Slack's memo in Land file, 1938, not dated.
[16] Statement by Joseph Barnes Nov. 20, 1946.
[17] "Justification, Proposed Joint Resolution Authorizing Negotiated Contracts," Jan. 3, 1941, unsigned in MC gf 120-10.

"Let us assume, for example, that I am the bidder and I make my estimates for my material and my estimates for labor; and I add to that my 10 per cent profit; and then depending upon how much guts I have, I add 10 or 20 per cent for fear. Now I do not know what that fear means, but I do know that when it enters into the bids they are too high." [18]

One kind of fear likely to raise bids on fixed-price contracts was concern lest the price of labor and materials go up. In an effort to remove this fear, "escalator" clauses were introduced. They provided for the adjustments upward of the contract price (a) if there was a general rise in wage rates as shown by the index of the Bureau of Labor Statistics for average hourly earnings in the durable goods group of manufacturing industries, and (b) if there was a general rise in material costs as shown by the Bureau's index of wholesale prices for metals and metal products.[19] Contracts which contained these clauses were at first called adjusted-price contracts, in contrast to the fixed-price contracts which had no such clauses. The name lump-sum contract covers both forms. In 1938 bidders were asked to name figures to apply to each form, but thereafter escalator clauses were included in nearly all Maritime Commission contracts. A lump-sum contract was almost always an adjusted price contract.[20] Naturally, every contract price was adjusted also to allow for changes in the specifications. Although the ships were begun by the Maritime Commission on standard plans, before they were completed they were contracted for by shipping companies which then arranged for modifications to meet their special needs. Changes in specifications were therefore numerous. All these adjustments meant more work for the cost accountants.

In view of the escalator clauses in the contracts, of the tendency of most yards to overbid, and of the likelihood of changing specifications during construction, the bid had only limited relation to the final cost of the ships. Advertising for bids might, under some conditions, have led to sufficiently keen competition among builders so that prices would be driven down close to cost or below. But those were not the conditions governing lump-sum

---

[18] House, *Second Supplemental National Defense Appropriation Bill for 1942, Hearings*, 77th Cong., 1st sess., Sept. 26, 1941, p. 273.

[19] For example, Contract MCc 502, Federal.

[20] MC *Report to Congress*, 1940, pp. 35-42.

PLATE V. Above, left to right. Commissioners Woodward, Land, Macauley, and Carmody congratulate Vickery on his promotion to Rear Admiral. Below, platens at Richmond showing roofs that gave weather protection for workers on subassemblies.

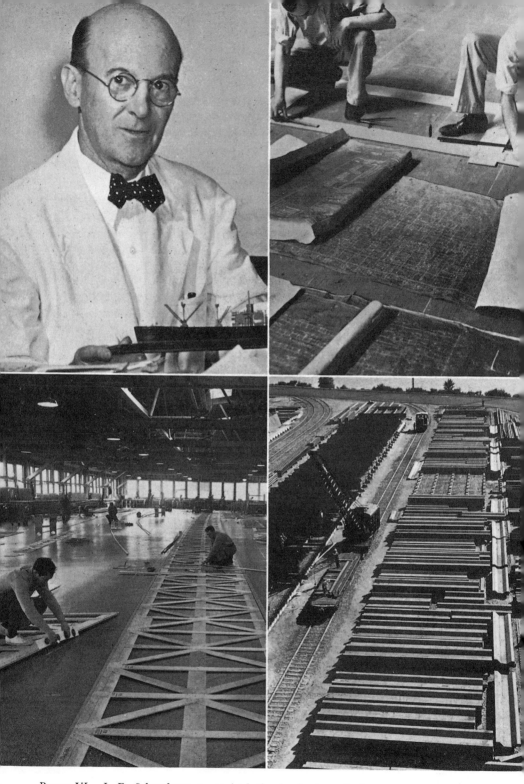

PLATE VI. J. E. Schmeltzer supervised the beginning of the emergency program. Wooden templates measured from blueprints in the mold loft served as patterns for shaping the steel arriving in the storage yards.

contracts which had been awarded by the Commission before 1941. In nearly all cases the final price had been cost plus 10 per cent of the contract price. Competitive bidding was sure to be even less effective under the conditions produced in 1941 by the emergency programs of the Navy and Maritime Commission. For several reasons new types of contracts were required.

II. FACILITIES CONTRACTS

The first new kind of contract required for the emergency program was that providing for the construction of shipyards. The same companies which received contracts for ships undertook to build the shipyards in which they were to operate.

When these facilities contracts were drawn there was a discussion going on in agencies like the National Defense Advisory Commission as to how the industrial expansion needed for the defense program should be financed. Should it be financed by the government or by private capital? That question applied not only to shipyards but to factories building airplanes, tanks, and so on. One proposal was that the government be the owner of the new facilities and pay directly the cost of their construction. It was obvious that this method would have to be used for munition plants which would have absolutely no commercial use. But there was a strong desire that this method should not be used in airplane factories, engine works, or other plants that might be readily converted to peacetime production. Government ownership of plants of this kind seemed like a step towards socialism. There was fear of what the government might do with them in the postwar period, and the commercial banks wanted to handle the business of financing industrial expansion in the defense effort as they had handled the past industrial expansion of the nation. Since the postwar use of the new plants was problematical, however, the banks hesitated to loan money on them. Consequently provision was made for rapid amortization, and a form of contract worked out, the EPF contract, by which the government would pay to the contractor over several years the amount he spent on new facilities, and so ultimately take the facilities off his hands if he did not desire them.[1] Contracts of this latter type

---

[1] Gerald T. White, "Financing Industrial Expansion for War: The Origin of the Defense Plant Corporation Leases," *Journal of Economic History*, IX (Nov.

did not prove of much practical importance but they show the atmosphere in which the Maritime Commission drew its facilities contract, and they help explain one of its most troublesome clauses.

Although yards for the construction of merchant ships might seem superficially to be facilities of a commercial type, the Commission decided to finance the emergency yards as if they were arsenals. It was believed that they would have little or no postwar value. In January 1941 the emergency shipyards were thought of as waterfront assembly points which would not have extensive equipment and would be dismantled as soon as the program was finished.[2] In accordance with this conception the Commission's facilities contracts provided that the government pay all the cost of the facilities and own them, although the Commission did not seek title to the land on which the facilities were to be erected.

The contractor who undertook to build the shipyard contracted also to maintain it and, as long as he had ship construction contracts requiring its use, he had the right to operate it, subject to clauses concerning default and cancellation. He was reimbursed for his construction and maintenance costs every 15 days, or at shorter intervals on presentation of bills paid in excess of $300,000. He received no fee and paid no rental;[3] therefore the only profit made directly from constructing the facilities went to the subcontractors, but the prospect of profits on shipbuilding contracts gave the prime contractors incentive to get the yards built well and rapidly. There was no similar incentive for being economical, since the government was paying all proper costs. Contractors were required to submit yard plans, specifications, leases, and subcontracts to the Commission for approval.[4] A limit to the cost was fixed in the contract, but the Commission could and repeatedly did approve expenditures above these limits.

The question was sometimes raised whether companies were

1949), 156-79; Ethan P. Allen, *Policies Governing Private Financing of Emergency Facilities*, May 1940 to June 1942 [Historical Reports of War Administration: War Production Board, Special Study No. 12 Civilian Production Administration, 1946]; Robert H. Connery, *The Navy and the Industrial Mobilization for World War II* (Princeton, N.J.: Princeton University Press, 1951), pp. 344-351.

[2] See below, chap. 7, sect. I, and Table 9.

[3] Form of Government-Owned Facilities Contract, and e. g., Contract MCc-1284, in Historian's Collection, Research file 112. 1.

[4] *Ibid.*; Office Order No. 1 [Emergency Ship Construction Division] Jan. 28, 1941 in gf 507-1.

interested in building shipyards because they hoped to take them over after the war. A postwar slump in shipbuilding was anticipated, however, so that only those yards which could also be used as repair yards seemed likely to be in postwar operation. The main material inducements offered to a company which built facilities were in the shipbuilding contracts which followed.

By June 1941, the Maritime Commission had awarded facilities contracts of this type to sixteen shipbuilding companies to build facilities that it was then estimated would cost about $50 million. Later waves of expansion lifted the total cost of new facilities for Maritime Commission shipyards to about $600 million by 1945.[5] Nearly all of it was spent under the type of contract just described, but some use was made also of two other types. Some of the Class II yards brought into the Maritime Commission program during the third wave of expansion to build minor types made use of the kind of contract developed by the Defense Plant Corporation (DPC), a subsidary of the Reconstruction Finance Corporation. Under this arrangement the DPC supplied the funds and owned the plant and the land and leased it to a shipbuilder sponsored by the Maritime Commission. In this way ten shipyards constructed and operated facilities costing $7 million.[6] These DPC contracts, which came to be used very extensively in building plants of the commercial type for suppliers, are explained more fully below in Chapter 12. Some other contracts, more like the Maritime Commission's standard facilities contract but differentiated by special clauses, were made to expand the facilities of existing shipyards, for example those with Ingalls. In these cases the shipbuilder paid rent for the use of the government-owned facilities under certain circumstances, and there were special provisions concerning maintenance and option to purchase.[7]

[5] Estimates for shipyards planned are in House, *First Supplemental National Defense Appropriation Bill for 1942, Hearings,* p. 423. Lists of the final costs of shipyards are in House, *Independent Offices Appropriation Bill for 1946, Hearings,* 79th Cong., 1st sess., pp. 546-57 and House *Independent Offices Appropriation Bill for 1948, Hearings,* pt. 2, 80th Cong., 1st sess., pp. 611-17. See also total in Chaikin and Coleman, p. 58.

[6] The DPC leases sponsored by the Maritime Commission are listed in House, *Independent Offices Appropriation Bill for 1946, Hearings,* pp. 584-86; and in the compilation by C. D. Vassar, "Defense Plant Corporation Projects" (Plancors), photostat in MC, Historian's Collection, Research file 207.6.2.

[7] Ingalls contract MCc 1674. Concerning such rentals see the recommendations from the Construction Division to the Commission, Aug. 18, 1941 in gf 503-17;

The real-estate problems connected with the shipyards are so closely linked to the facilities contracts that it seems best to consider them at this point even though they carry us ahead of our story. The facilities contracts approved by the Commission in January 1941 provided that the land on which the new shipyards were to be built should be owned or leased by the contractor. When the shipyard was no longer needed for national defense, it might be sold or leased to the contractor. But if no agreement was reached for such a sale or lease, the Commission could be required by the contractor to " enter upon such real estate and within 90 days thereafter demolish or remove any facilities installed thereon not so purchased by the Contractor: Provided, That the Real Estate on which shall be located the Facilities or such thereof as shall be demolished or removed shall be restored so as to leave the same in as good condition as immediately prior to the acquisition, construction, or installation of the Facilities thereon. . . ." This clause, Article 13, placed the Commission in a disadvantageous position in arranging after the war to dispose of the facilities in emergency yards such as those at Baltimore and Los Angeles.

In explaining why the policy of not owning the land was adopted Admiral Land testified in January 1941, " We are purposely avoiding owning the land on account of getting mixed up in real estate deals which are always provocative and difficult." [8] The following considerations help interpret that statement. In January 1941 the country was not at war and was far from being united in regard to policy toward the war. The Commission therefore had reason to proceed in ways which would not arouse opposition. Acquisition of waterfront properties by the Federal Government might be resisted and resented by localities which feared the withdrawal of the land from local taxation. Fast action was needed to acquire lease or title and it was believed this could be done in quickest fashion by leaving it up to the contractors. Perhaps the shipbuilding companies preferred that they be the owners or

Anderson to the Commission, Aug. 18, 1941, Appendix B to the report to the Commission of the Committee on Awards, March 27, 1945 in Weber's files, C3, Ingalls. But the facility contracts with Western Pipe and Steel, May 1, 1941 (MCc 1950) and that with Consolidated Steel Corp., Aug. 26, 1941 (MCc 1675) provide for no rentals and are quite similar to the facilities contracts for the new emergency yards.

[8] Senate, *Emergency Cargo Ship Construction, Hearings,* p. 13.

lessors of the land. During World War I the government had paid the cost of facilities erected on land belonging to shipbuilding companies, and some companies had acquired the facilities after the war at a small fraction of their cost.[9] Since the Commission was having difficulty in drawing experienced shipbuilders into its program in January 1941, it had reason to consider their preferences.

Admiral Land gave a further justification for Article 13 in the facilities contract by saying that it had been worked out with the National Defense Advisory Council.[10] The EPF contract, which that Council sponsored, provided that " in the case of intermingled facilities, . . . the contractor . . . may require the government to remove the facilities and restore his property to its original state,"[11] and it assumed that the land could be owned by the contractor.[12] The first lease concluded by the Defense Plant Corporation, with Packard Motor Car Co. on September 6, 1940, provided that the government-owned machinery and equipment should be removed by the government if Packard did not desire to buy.[13] The Navy also built government-owned facilities on land not owned by the government, relying on its ability to take the land by condemnation proceedings if it wanted.[14]

Whatever the reasons for not taking title to the land originally, it became more evident, as the government increased its investment by supplying more equipment than had been originally foreseen, that the government should have title. Partly because it was decided that the shipyards would have to fabricate their own steel, partly because the costs of constructing shipyards proved larger than predicted, the investment in facilities turned out to

[9] Senate, *Munitions Industry, Naval Shipbuilding, Preliminary Report of the Special Committee on Investigation of the Munitions Industry*, 74th Cong., 1st sess., Report 944 (Pt. I), pp. 346, 350.

[10] Senate, *H. J. Res 77, Hearings*, p. 13.

[11] Allen, *Policies Governing Private Financing*, p. 36; White, " Financing Industrial Expansion," p. 173.

[12] Allen, *Policies Governing Private Financing*, p. 32.

[13] White, " Financing Industrial Expansion," p. 170. The Defense Plant Corporation shortly changed its policy and required government ownership or long-term lease of the land.—*Minutes of the Advisory Committee to the Council of National Defense, June 12, 1940 to October 22, 1941* [Historical Reports on War Administration: War Production Board, Documentary Publication No. 1] (Government Printing Office, 1946), p. 108.

[14] Testimony of Admiral Robinson, Truman Committee, *Hearings*, pt. 5, p. 1453.

be five or ten times as much per way as had been estimated.¹ Heavier investment made it more desirable that the government own the land.[16]

Accordingly the Commission changed its policy and began to acquire title to the land on which its shipyards were built. To do so they frequently condemned and took land under the power of eminent domain. The potentialities of this power were little appreciated by the Commission when the emergency program began, but it was appealed to in April of 1941 when acquisition at a reasonable price by other means proved impossible in the case of the Eddystone property which adjoined the Sun yard at Chester and was considered essential for the planned expansion of that yard. The owners asked $2,594,000; the Commissioners thought $1,000,000 was just compensation. Although under political pressure the Commission compromised and by a split vote approved paying the owners $1,650,000, the case marked the beginning of appeals to the power of eminent domain.[17]

By 1942 the Commission as a general rule sought ownership of the land to be used for government-owned facilities, and condemnation proceedings were frequently resorted to in order to acquire title. In some cases this delayed taking possession;[18] in other cases it speeded the process. The normal peacetime procedure would have been for the Commission to call on the Department of Justice to handle condemnation proceedings; but because of the need for speed the legal staff of the Maritime Commission drew up the papers, obtained the signatures of the Federal district attorneys, took the papers to the judge, and made sure the matter went through the court. Paul D. Page, Jr., Solicitor of the Commission, who was in charge of such matters, described in a letter to the Department of Justice how the Commission obtained some of the land needed for the new Richmond yard begun just after Pearl Harbor. He was informed late in the afternoon of Thursday,

[15] See below Table 9.

[16] Statements by Paul D. Page, Jr., Apr. 25, 1947; and memorandum on ownership and rentals of land by R. E. Anderson, Nov. 2, 1942 in Anderson reading file.

[17] MC Minutes, pp. 16937 (Apr. 15, 1941), 16966, 17239-40, 17158, and for the compromise, pp. 17822 and 17988 (July 7, 1941), and final payment, p. 18454. Statements by T. Woodward. The importance of the case as starting appeals to the power of eminent domain was emphasized by Paul D. Page, Jr., in interviews Apr. 25, 1947 and Nov. 22, 1949.

[18] Testimony of K. K. Bechtel, House, *Production in Shipbuilding Plants, Hearings*, p. 693, June 22, 1943.

January 15, 1942 that it would be necessary to condemn a property belonging to the Richfield Oil Corp. When a representative of Kaiser had tried to obtain the land, a prominent official of the oil company had terminated the interview, according to the way Mr. Page heard the story, " by settling back in his overstuffed chair, letting his eyes rove around a few hundred linear feet of mahogany paneling, flicking the ash from his Corona, and delivering himself as follows:

> There will be no condemnation. We have influence in Washington. We will not sell. If you desire to lease our property you may submit an offer to our General Counsel. When in due course I receive a report from him, I will take the matter up with my Board and advise you as to our decision.

Aroused by this picture, Mr. Page started by plane next day for San Francisco equipped with necessary papers. In conference with real estate appraisers and shipyard officials on the West coast he found that the appraisal with which he had been furnished was too high and the metes-and-bounds description inaccurate. A new appraisal and new survey were made within twenty-four hours; hurried typists hammered out new copies of the Complaint, the Judgment, and the Declaration of Taking; and the Commission was informed in a telegram " approximately eight feet long." Monday morning Mr. Page explained the situation to the local attorneys in the Land Division of the Department of Justice and again consulted the Commission by teletype. He received " the green light " from Washington at 2:40 p. m., immediately secured a check for the revised amount of just compensation from the regional disbursing officer, and reached the Federal Building a few minutes before the Judge was to leave. Judgment on the Declaration of Taking was entered at 4:07 p. m. that same Monday. " Ordinarily, of course, we would have awaited preparation of a careful and detailed appraisal of the property before proceeding. The exigencies of war, however, made it essential for us to secure control of the property at the earliest possible moment. It simply was not possible to wait." [19]

Once aware of how much could be done under the government's wartime powers of eminent domain, the Maritime Commission

[19] The quotations are from Paul D. Page, Jr., to Robert J. Hayes, General Assistant, Land Division, Department of Justice, Jan. 22, 1942. Copy in MC, Historian's Collection, Research file 210. 1. See also MC Minutes, p. 20638 (Jan. 17, 1942), p. 20645 (Jan. 22, 1942), and pp. 26045-46 (Aug. 24, 1943).

used condemnation proceedings extensively. They took in this way not only industrial property but also private homes of erstwhile residential districts; they condemned land not only for shipyards but land for housing, for access highways, and parking lots.[20] Altogether the Commission acquired fee simple title to not less than 5,072 acres of land in 18 states at a cost of about $14 million. There were 27 direct purchases and 106 condemnation cases. Some of the latter involved the acquisition of leasehold interest, for during 1943 it was decided that court action to secure the right to use and occupy was in most cases the better way to get land for housing. Leases were acquired to 835 acres. Appraisal and title costs of the entire program were about $118,000.[21] In determining just compensation, use was made of local appraisers and also of experts loaned by the Bureau of Valuation of the Interstate Commerce Commission.[22]

Although condemnation was speedily applied in instances where it seemed necessary to prevent delays which would interfere with production, there were many cases in which it was not used. Indeed, at the end of the war the Commission lacked title to the land under many of the largest emergency yards. Even two of the big yards begun in 1942, Swan Island and Vancouver, were built on real estate which the Maritime Commission did not own. In the 1941 emergency yards, the land was in some cases acquired later by condemnation, but in other cases the leases were renewed. Since the shipbuilding companies were already in possession, it made no difference to production whether or not the government acquired title. The issue was merely one of cost and good local relations.

In the case of Swan Island and Vancouver, the land used for the shipyards belonged to the local port authorities. They charged low or nominal rentals. The lease from the Port of Vancouver, concluded in January 1942, required the premises to be restored to as good condition as prior to its leasing,[23] but the contract

---

[20] The Minutes of the Commission for March 1943 are full of records of condemnations. On its use at South Portland see chap. 15.

[21] Paul D. Page, Jr., to Commissioner Carmody, Sept. 9, 1946, copy in MC, Historian's Collection, Research file 210.1; MC Minutes, p. 24558 (Mar. 11, 1943), on condemnation for leasehold interest on the example of the FPHA.

[22] MC Minutes, pp. 23934-35 (Dec. 31, 1942); p. 24795 (Apr. 6, 1943).

[23] The lease dated Jan. 27, 1942 is filed with contract MC 2049, Jan. 9, 1949 in MC Finance Division. Rental was $300 a year.

leasing Swan Island from the Port of Portland, concluded in April 1942, modified this obligation by relieving the Commission from obligation to restore ground levels which had been regraded.[24] At Los Angeles also the land used for the emergency shipyards—those operated by Calship and Consolidated Steel—belonged to the local port authority.[25] Another reason for not seeking to take title there was that a court might place very high value on some of the land in view of the possibility that there was oil beneath it.[26] Large obligations for restoration of real estate were assumed in the port of Los Angeles and at the end of the war California Shipbuilding Co. was paid $2.5 million and given title to facilities which had cost about $25 million and also surplus materials on the facilities in a settlement of claims in which they undertook to assume the government's obligation to restore the real estate to the good condition required by the terms under which it had been leased.[27]

A somewhat similar situation developed out of different conditions at the Bethlehem Fairfield yard in Baltimore. To start this emergency yard as rapidly as possible in the spring of 1941, the Bethlehem company had leased industrial properties at high rentals from the Union Ship Co., the Pullman Company, and other owners at rentals which were approved by the Commission at the time but which were in 1943 judged to be excessive.[28] In negotiation with the owners the rentals were reduced on some portions of the land.[29] When the Pullman Company refused to come down to what the Commission considered a reasonable price, condemnation was used.[30] In regard to the property of the Union Ship Co., the heart of the shipyard, the owners asked

[24] The lease is contract MCc 2484, filed with contract MCc 2393, March 4, 1942 in MC Finance Division. Rental was $248,000 paid at the time of execution of the contract and $10 a year thereafter. The lease was for three years, renewable for 7 years.

[25] House, *Navy Department Appropriation Bill for 1947, Hearings*, 79th Cong., 2d sess., pp. 67-75. House, *Independent Offices Appropriation Bill for 1948, Hearings*, pp. 749-50.

[26] Statement by Paul D. Page, Jr., Nov. 22, 1949.

[27] House, *Navy Department Appropriation Bill for 1947, Hearings*, pp. 70-75. For the Commission action and Carmody's vote of "nay," see MC Minutes, pp. 33918-23 (Nov. 27, 1945), and below chap. 23.

[28] MC Minutes, pp. 24723-29 (Mar. 30, 1943). Also Anderson to Vickery, Jan. 10, 1941, in Anderson reading file.

[29] MC Minutes, pp. 26898-26902, 26932, 26973.

[30] MC Minutes, pp. 26874-77 (Nov. 23, 1943), and p. 26969.

$3 million, although the government's appraisers considered $1.5 million fair. But the Commission's Solicitor warned that in condemnation proceedings the Commission would very likely have to pay about $2.5 million since the owners could adduce in favor of their valuation an offer (hard to prove not bona fide) of $3 million, and since the Commission itself had earlier approved a yearly rental at first of $280,000 and then of $230,000. Threat of condemnation proceedings was used to obtain modifications of the lease which reduced the probable cost of the Commission's obligation to " restore " from about $2,000,000 to about $650,000. The Commission voted 3 to 1 in favor of renewing the lease on these terms, Commissioner Woodward voting " nay " because he believed the land should have been condemned.[31] At the end of the war the Bethlehem-Fairfield Co. was paid $1,473,812 and given title to facilities which had cost about $30 million (excluding the Pullman and Geis properties, which were retained by the government) and to surplus materials in a settlement of claims in which they assumed the government's obligation to the owners from whom land had been leased.[32]

Congressional criticism of this situation was summed up by Congressman Albert Thomas, who said:

> But you did buy a lot of land at some place and none at others. Where you were dealing with a local government agency I can see some reason for not buying, but in case of a private independent concern, where you could have gone in and condemned the land I do not see why that procedure was not followed, since you are going to have to go in and spend millions to undo what was done.[33]
>
> After you put that money into it nobody in the world, you must have known, was going to be interested in buying it except these people.[34]

Before thus disposing of the Calship and Beth-Fairfield yards the Commission tentatively estimated the cost of the restorations to which it was obligated by the terms of its leases and facilities contracts as $44 million.[35] There are some reasons for believing that this was a gross overestimate since it was presented to Congress

---

[31] MC Minutes, pp. 27685-91 (Feb. 29, 1944).
[32] House, *Navy Department Appropriation Bill for 1947, Hearings*, pp. 55-60. For the Commission action and Carmody's " nay " vote see MC Minutes, pp. 33916-19 (Nov. 27, 1945), and below in chap. 23.
[33] House, *Navy Department Appropriation Bill for 1947, Hearings*, p. 355.
[34] *Ibid.*, p. 61.
[35] *Ibid.*, pp. 352-55.

as part of a request for appropriations, and since, after circulating bids, a contract was awarded for demolition and restoration of the Consolidated Steel shipyard at Wilmington, California, amounting to $392,613,[36] whereas the cost of restorations at that yard was estimated in 1945 at $1,118,700.[37]

Obligations to restore the condition of the real estate formed only one of the difficulties, however, in the disposal of shipyards after the war. Shipyards were a drug on the market, even those to which the Commission had a clear title, land and all. Plans were made to keep four of the yards as standby yards ready for use in another emergency—namely: Wilmington, North Carolina; Alameda, California; Richmond No. 3, also in San Francisco Bay; and the Vancouver yard near Portland, Oregon.[38] All the rest were up for sale and only a few found purchasers willing to pay even 12 per cent of what the yards cost.[39]

Whether the total cost to the government would have been less if the Commission had taken title by condemnation in all cases, even those in which it would have had to pay a high price, can hardly be decided with assurance on the basis of the data available; but it does seem clear that the arrangements concerning real estate in the original emergency yards were a mistake. It proved to be a costly mistake since, for reasons readily understandable, these arrangements were made without any prescience of the amount the government was ultimately to invest in such facilities.

III. Manhour Contracts

Contracts for ship construction presented other problems. A change from lump-sum contracts to any form of cost-plus was sure to be criticized because of the unsavory reputation of the cost-plus-a-percentage-of-cost contracts of World War I. But obtaining reasonable prices in lump-sum contracts depends on ability to predict costs and on competitive bidding. Prices were

[36] MC, *Report to Congress, 1948*, p. 23.

[37] House, *Navy Department Appropriation Bill for 1947, Hearings*, p. 354. See also House, *Independent Offices Appropriation Bill for 1948, Hearings*, pp. 746-50.

[38] House, *Navy Department Appropriation Bill for 1947, Hearings*, pp. 18-19; *Independent Offices Appropriation Bill, 1948, Hearings*, p. 751; U. S. Maritime Commission *Report to Congress, 1947*, p. 9; U. S. Maritime Commission *Report to Congress, 1948*, p. 23.

[39] House, *Independent Offices Appropriation Bill for 1948, Hearings*, pp. 611-17.

going up in January 1941, especially the price of machinery, and shipbuilders feared that they could not get any firm bids from subcontractors. Since the Navy was asserting priorities over the scarcest items, deliveries were uncertain. The companies could not figure costs closely on the basis of previous construction, for the emergency ships were of a new type and were to be built in new yards. And on top of all this was the fact that shipbuilding companies already had all the business they could handle. Competitive bidding under these conditions would have been quite useless.

Back of the faith in competitive bidding lay the assumption that there would be a number of reliable concerns interested in building ships and the more efficient builder would send in the lowest bid and so obtain the work. But with all the established concerns already having all the work they could handle, they either would not bid on additional construction, or would make a relatively high bid, being sure to allow for a profit and for all kinds of extra expenses. On the other hand, some fly-by-night concern, consisting perhaps only of a few lawyers in Washington, a couple of foreign marine architects of lofty and mysterious reputation, and a lease on an abandoned shipyard of World War I, could optimistically turn in a low bid. With competitive bidding, as Admiral Land said, " you would get in—I am sorry to say this—but it is true—you would get in a lot of jackanape cock-eyed bids; you would get a lot of irresponsible persons putting bids in there and get a lot of people who had their offices in their hats." [1] What chance was there that they would be able to execute the contract? And what kind of a showing would the Maritime Commission make if the builders to whom it gave contracts were unable to execute them? Under the conditions existing in 1941, the period in which the country was not yet at war but the demand for shipbuilders was of an intensity like that of war, the more reliable shipbuilders were likely to be the highest bidders.[2] The

---

[1] House, *Emergency Cargo Ship Construction, Hearings* before the Subcommittee of the Committee on Appropriations, 77th Cong., 1st sess., Jan. 18, 1941, on H. J. Res. 77, p. 20.

[2] For Admiral Land's defenses of the contract see House, *Emergency Cargo Ship Construction, Hearings*; House *Foreign Construction Costs, Negotiation for Construction and Charter, Overtime Pay, Executive Hearings* before the Committee on Merchant Marine and Fisheries, 77th Cong., 1st sess., on H. R. 3252, Feb. 28, 1941; House, *Independent Offices Appropriation Bill for 1943, Hearings*, 77th Cong., 2d

question was not who would promise to build cheapest; the question was how to enlist, even if at high cost, the efforts of those who would really do the job.

As a result of this situation, the Maritime Commission decided that some kind of cost-plus contract had to be used but sought a form which, unlike the cost-plus-a-percentage-of-cost contracts, would not pay higher fees to the more wasteful contractors, but which would instead pay higher fees to those contractors who reduced cost. One answer was the " manhour " contract, which was used for all Liberty ships. It provided for reimbursement to the contractor of all his costs in ship construction, although specifying that he would act " as an independent contractor, and not as agent." [3] At the same time it aimed to make the size of the fee depend on economy in the use of manhours, and on rapid delivery.

Since no authority to negotiate contracts of this kind was given the Commission by the Merchant Marine Act of 1936, there was prepared and presented to the Commission on January 3, 1941 by its Committee on Legislation, the text of a Congressional joint resolution giving the Commission full powers to negotiate, such as the Navy had received.[4] This proposal was not acted on, however; the Commission decided to go one step at a time. For the emergency program of 200 ships, the authorizing resolution waived the need of competitive bidding.[5] On the rest of the program, it was waived May 2, 1941 when Public Law 46 conferred on the Maritime Commission authority to negotiate contracts, as well as a number of other exceptional powers, for the duration of the national emergency.[6]

sess., pp. 261-62; and the formal request for authority to negotiate under Public Law 46, 77th Cong., 1st sess., Land (by Walston S. Brown) to the Director, Bureau of the Budget, June 19, 1941 in MC gf 506-1.

[3] Article 1 in " Form of Contract for the Construction of Emergency Cargo Ships " in MC, Historian's Collection, Research file 209.5. In World War I, an " agency form of contract," as well as lump-sum and cost plus, was used. *Report of Director General Charles Piez to the Board of Trustees of the United States Shipping Board Emergency Fleet Corporation*, Apr. 30, 1919 (Government Printing Office, 1919), pp. 14, 23.

[4] Memo of Committee on Legislation to the Maritime Commission, Jan. 3, 1941, in gf 120-10. The Navy had received such authority June 28, 1940, C440, 54 Stat. 676; Connery, *The Navy and the Industrial Mobilization*, p. 202.

[5] See 55 Stat. 5, H. J. Res. 77, Public Law 5, Feb. 6, 1941, 77th Cong., 1st sess.

[6] 55 Stat. 148, H. R. 3252, Public Law 46, 77th Cong., 1st esss. The authority was extended by the First Supplemental National Defense Appropriation Act, 1942, ap-

In testifying before Congressional committees concerning the proposed new contracts, Admiral Land laid stress on penalizing the less efficient yards. Expressing it in percentages, Admiral Land said that he thought the standard fee should be 7 per cent of the estimated price. A very efficient yard should be rewarded with 3 per cent more (which would bring it up to the established legal maximum of 10 per cent), and the inefficient should be penalized by getting only 5 per cent. One Congressman expressed doubt whether builders would accept the possibility of being penalized for inefficiency to the extent of having their fees cut from 7 per cent to 5 per cent. Admiral Land said he was confident they would. " Yes indeed. If there is not enough patriotism among the shipbuilders of the United States I will come here and tell you about it." [7] In February 1942 Admiral Land boasted: " I think we have the toughest contract put out by any branch of the government." [8]

In these contracts to build Libertys, efforts to state an estimated price were abandoned. What was needed was a common standard of performance that could be applied to all the yards of the country. There were many reasons for not stating that standard in dollars. (1) Wage-rates differed in different localities so that the same standard of efficiency in using labor would have to be expressed differently in dollars from region to region. (2) Wage rates appeared to be going up and any figure set would have to be continually revised upwards to allow for this increase. (3) Most of the materials were to be supplied by the Commission so that labor would be the chief item of cost. (4) Costs in dollars, either of labor or of materials supplied by the builder, might not be known until sometime after the construction. A standard was needed which could be applied day by day as the work progressed, so that the shipbuilder could see how near he was to the standard. For these reasons the standard—the bogie, as it came to be called—was set in manhours.[9]

proved Aug. 25, 1941 (55 Stat. 681, Public Law 247, 77th Cong., 1st sess.) and by the President under the First War Powers Act, Ex. Order 9001, Dec. 27, 1941, 6 F. R. 6787. Cf. Miller, p. 52.

[7] House, *Foreign Construction Costs, Negotiation for Construction, Executive Hearings,* Feb. 28, 1941, pp. 10-11.

[8] House, *Fifth Supplemental National Defense Appropriation Bill for 1942, Hearings,* 77th Cong., 2d sess., p. 101.

[9] These reasons are a consensus of the reasons given to me by various members

To set the standard it was necessary to start from some past experience with ships not too different in type. The figures supplied by resident auditors concerning the construction of C2's were taken as a starting point. The figures were in dollars, but were broken down into material costs and labor costs for various parts of the work, and the manhours could readily be determined by dividing by the wage rates at a particular yard. The Liberty was a simpler type of ship, but by comparing the amounts of work of various kinds—for example, the weight of the piping on the two types—figures could be estimated, item by item, for each kind or part of the work on the EC2's. The total thus determined gave a standard for manhours directly employed in building the ship. After compromise with the builders the figure of 510,000 manhours was established. Then allowance was made also for the labor used in the general management of the yards, an allowance which was called indirect manhours, and which was made slightly more per ship for the smaller yards. Direct and indirect manhours together provided a means of putting the yards on a comparable basis so that the more efficient could be rewarded more, the less efficient given less.[10]

In addition to the rewards for building with fewer manhours and the penalties for using too many, there were bonuses for speed and penalties for delay. The standard of speed set up when the first contracts were drawn was about 150 days from keel laying to delivery.[11] Examples of fast building at the time when this figure was set were the completion of a C3 cargo ship at Federal in 189 days,[12] and the completion of a 14,000 ton private

of the staff of the Maritime Commission and especially by Joseph F. Barnes, Nov. 20, 1946 and by J. T. Gallagher on Nov. 21, 1946.

[10] J. T. Gallagher in an interview on Nov. 21, 1946 described how he made an estimate of the direct manhours needed. On the compromise, see testimony of Admiral Vickery in House, *Production in Shipbuilding Plants, Executive Hearings*, part 3, p. 907. The indirect manhours were fixed as 27.5 per cent of the direct manhours for a yard with 4 to 6 ways, and 23.1 per cent for a yard with 13 ways, and for the others proportionally. See letter of Vickery to J. V. Hayes, Counsel of the House Committee on Merchant Marine. Appendix B., Aug. 20, 1942 in MC gf 507-3-14.

[11] The figure of 150 days applies to the third or fourth round of the ways. More than six months were allowed for the first round, See the way chart prepared in the Construction Section of the Technical Division, Feb. 14, 1941. Now in MC, Historian's Collection, Research File 117, Statistics.

[12] MC, "Permanent Report of Completed Ship Construction Contracts." It was delivered in Sept. 1940.

tanker at Sun in 184 days.[13] The Liberty ships were smaller and simpler, to be sure, but they were to be built by new organizations in new yards. It would seem slow two years later, but in the spring of 1941, delivering a 10,000 ton ship five months after keel laying seemed fast work.

Those shipbuilders who exactly met the standard in manhours and in speed received what was called the base fee, $110,000 per ship. Those that fell behind were penalized $400 a day for each day's delay and $33\frac{1}{3}$ cents for each manhour employed above the bogie, except that in no case was their fee to be less than $60,000 per ship. Those that did better than the standard performance were to have their fees increased by $400 a day for delivering ahead of schedule, and by 50 cents for each extra manhour saved, except that in no case was their fee to exceed $140,000.

The actual average number of manhours was obtained by treating all the vessels built under one contract as a unit and dividing the total actual manhours worked by the number of ships. This system of fees for the construction of Liberty ships was worked out through conferences with the shipbuilders and was approved by the Commission March 11, 1941.[14]

The size of the fees was well under the legal limits permitted by statute. Public Law 46 of May 2, 1941 specified that fixed fees should not exceed 7 per cent and that the aggregate of such fixed fees plus bonuses to secure maximum performance should not exceed 10 per cent.[15] Since the cost of a Liberty was estimated in March 1941 as $1,750,000,[16] it would have been legal to set the the maximum at $175,000. The percentages actually used were explained by Admiral Vickery as follows: " We estimated what the cost of the ship was and stated the lower limit would be 3 percent and the upper limit would be 8 percent." In practice, he said, the percentages would be less than that, since labor and materials had risen.[17] When the yards began producing in quantity, cost fell, and there was an argument for reducing the percentage of

---

[13] Card file of private tankers kept by the Statistical Unit of the Production Division. It was delivered in December 1940.

[14] MC Minutes, pp. 16200, 16545, 16594 and contract forms on file in the Legal Division.

[15] 55 Stat. 148, sec. 2, and see below chap. 24 on profit limitations.

[16] Land to the President, March 27, 1941 in MC gf 506-1.

[17] House, *Fifth Supplemental National Defense Appropriation Bill for 1942, Hearings,* Feb. 12, 1942, pp. 116-17.

profit as the volume increased. Mr. Anderson suggested in March 1942 that recent contracts be renegotiated and lower fees put in new contracts,[18] but the contract awarded to Oregon Shipbuilding Corp. on May 8, 1942 still fixed the maximum at $140,000.[19] Fees were cut in half, however, to $70,000 maximum, $55,000 base fee, and $30,000 minimum, in the set of new contracts awarded December 24, 1942,[20] and cut again, to $60,000, $45,000 and $20,000 in April 1943.[21] By that time most of the yards were working on new types for which there was a different scale of fees, and under the Renegotiation Act the Price Adjustment Boards were adjusting all fees downward.[22]

The name for contracts of this type is not well established. In Maritime Commission records they are generally called cost-plus-variable-fee contracts, but some general discussions of wartime procurement include them under the name of cost-plus-fixed-fee (CPFF) contracts, for although they contained incentive payments the fees were limited by a fixed minimum and maximum.[23] Commission officials later called them manhour contracts to emphasize the manhour bogies which distinguished them from other ship construction contracts.[24]

IV. PRICE-MINUS CONTRACTS.

The cost-plus-variable-fee or manhour contracts, which have just been described, were very different in character from the cost-plus-a-percentage-of-cost contracts that had such a bad reputation following their use in World War I. But their name was easily abbreviated into the same unsavory tag, cost-plus. A different form of contract with a pleasant sounding name, price-minus, was used by the Maritime Commission in contracting for the tankers

[18] Anderson to Vickery, March 5, 1942, in Anderson reading file. See also the unsigned memorandum written by an "old shipbuilding friend" of Admiral Land and filed in gf 120-10 under March 31, 1941.
[19] Minutes, p. 21882 and contract MC 7950.
[20] MC Minutes, pp. 23865-68 (Dec. 24, 1942).
[21] MC Minutes, p. 24910 (Apr. 20, 1943).
[22] See below in chap. 24. On reduction of fees because of increased costs of facilties, see chap. 14.
[23] Miller, *Pricing of Militray Procurement*, pp. 125, 127.
[24] R. Earle Anderson, "United States Maritime Commission Procedure," section 2 of "Proposals and Contracts," in F. G. Fassett, Jr., ed., *The Shipbuilding Business in the United States of America* (2 vols., New York, 1948), II, 23.

and the standard-type cargo ships added to the program in April 1941.[1] A price-minus type of contract had been devised during World War I by Joseph W. Powell, then of the Bethlehem Steel Co., and in 1941 a special assistant to the Assistant Secretary of the Navy.[2] It was preferred by R. Earle Anderson, Director of Finance of the Maritime Commission, on the ground that it reduced the burden of auditing and was more effective than the manhour contract in throwing "the definite burden of costs reduction on the contractor."[3]

Under the price-minus contract the Commission paid the builder his costs, to be sure, plus a minimum fee, and offered a higher fee as an incentive to more efficient operation. In that sense it too was a cost-plus contract although it was not so called.

The distinctive feature of the price-minus contract was the method by which the incentive fee was figured. A price estimated to include cost and a certain amount of profit was named in the contract. If the actual costs of the contractor went over this sum, he was nevertheless to be paid his cost and a fee of about 3 per cent of the contract price.[4] That case was not expected to be typical. In the eventuality that the builder's costs were under the contract price, as they were expected to be, the builder received as fee half the difference between the contract price and the actual cost plus the minimum fee. The other half of the difference was "saved" by the Commission. Additional sums might be added to the fee as a reward for speed or subtracted from it for delay, but in no case was the total fee to be more than 10 per cent of the contract price, or more than a sum (less than 10 per cent) stated in the contract.[5]

Here is an example of how a price-minus contract would work.

---

[1] It was approved by the Commission May 8, 1941. Minutes, p. 17269 and form contracts in the Legal Division.

[2] *Supreme Court Reporter*, vol. 62, p. 594.

[3] Anderson to Chairman and Vice Chairman, MC, Nov. 25, 1942 in Anderson reading file.

[4] Art. 6, par. d. and Art. 15, Form Contract, No. DA-MCc, approved May 8, 1941, in MC gf 506-1.

[5] R. Earle Anderson, "'Price-Minus' Contracts Speed Shipbuilding," in *Purchasing*, April 1943. He explains the name by saying, "the Commission gets the ship for the contract price minus half the savings—hence the term 'price-minus.' It is the antithesis of cost-plus." The last sentence is true if by "cost-plus" is understood, as was common usage, cost-plus-a-percentage-of-cost. See also Anderson's article in Fassett, ed., *Shipbuilding Business*, II, 25-27.

Referring to a C3, Admiral Vickery testified: "The actual cost of building those ships with no profit in it, at the best estimate we can get at the present day, is about $2,900,000. We have set the price, including profit, at $3,200,000. [The "profit" here allowed for is about 10 percent.] We give them a $35,000 fee when the ship is launched and a $35,000 fee when the ship is delivered. [The minimum fee was then even less than 3 percent of the "actual cost," although in this same testimony Vickery referred to the fee as 3 per cent.] Now if the ship cost $3,000,000 we take the $3,200,000 and subtract $3,070,000 [cost plus minimum fee] and divide the remainder by two, give them $65,000 and take the other $65,000. So the final cost of the ship is not $3,200,000 but $3,135,000."[6] It might be added that the builder's profit was then $135,000, only 4.2 per cent of the contract price. If the builder cut the cost to $2,900,000, the figure the Maritime Commission started with as "actual cost," he made $185,000, which is 5.8 per cent. To make the statutory legal maximum of 10 per cent, he had to cut his costs down to $2,630,000. While it might be very hard to cut the cost that low, it might not be surprising if the builder cut it low enough to make 6 per cent of the contract price.

In estimating the cost in dollars and making the size of the fee depend on the saving in dollar costs, the price-minus contract was well adapted to the conditions surrounding the building of the standard type of cargo ships. In form it resembled a lump-sum contract; most of the clauses were the same in the two contracts, and the sharp contrast in the content of the two contracts occurred mainly in one article, number 15. The ships were to be constructed by going concerns which had been building for some years the type of ship to which the contract referred. The builders knew what their costs had been and could use that knowledge to estimate future costs, allowing for contingencies. Unlike the builders of Liberty ships, they were buying their own materials and components. They were protected against general rises in costs of material and labor by the escalator clauses, which were present in the price-minus as in the lump-sum contracts, but each yard was responsible for keeping its expenses for materials and labor relatively low in dollars, and the price-minus contract

[6] Admiral Vickery before an Appropriations Subcommittee, July 17, 1941, House, *First Supplemental National Defense Appropriation Bill for 1942, Hearings,* p. 438.

recognized this by varying the fee accordingly. To be sure, price-minus contracts did not provide for a common standard, such as the manhour bogie, by which a group of yards could be measured and placed on comparable terms. Such a common standard could have been applied to only a very few of the yards building tankers and C-types; for the long-range program contained four major types, and there were variations within a single type, for example the C3's, from one yard to another. Moreover, the manhour type of contract required elaborate auditing. For these reasons it was not wanted by yards building for the long-range program.

Neither did they in the main want to continue with lump-sum contracts under the new emergency conditions. But there were exceptions. Two of the nine yards approached in April or May 1941 preferred to stick to lump-sum contracts.[7] When the President of Ingalls read the form of the proposed price-minus contract, he wrote to Admiral Vickery protesting that he hoped that the Commission would not insist on terms so unlike those to which his company was accustomed.[8] The Commission acquiesced [9] but the various lump-sum contracts concluded thereafter with Ingalls included figures which limited the company's profits to less than the statutory 10 per cent.[10] Federal also adhered to the lump-sum form.

An extreme case of a company which objected to anything approaching cost-plus came up later when the Commission launched its tug-building program. Ira S. Bushey was willing to take a contract only at a fixed price and without the customary 10 per cent limitation on profits. In recommending the contract to the Commission Admiral Vickery said: " Obviously it is more advantageous for the Commission to have these contracts—the original price being the same in either instance—on a ' fixed price,' rather than upon an ' adjusted price ' basis. In addition, if the contractor is willing to assume the greater risk inherent in contracts of ' fixed price ' type, it would appear equitable on the part of the Commission not to limit the profit." [11]

[7] Land to Director, Bureau of the Budget, June 19, 1941, in MC gf 506-1, says only two were willing to make contracts on a lump-sum basis.
[8] Lanier to Vickery, May 22, 1941, in gf 506-1.
[9] MC Minutes, p. 18393 (Aug. 8, 1941).
[10] For example, the contract with Ingalls, Feb. 1, 1945 (MCc-35, 749) in Art. 16 limits the profit to $2,100,000 on seven ships priced at $23,800,000.
[11] MC Minutes, p. 18825 (Sept. 5, 1941).

THE SUPERVISORY STAFF

After the emergency program had been under way for some years, a fourth type of contract for ship construction, the selective price contract, was added. The nature of the selective price contract and the reasons for using it will be described in later chapters in which attention is given to how the contracts worked. In addition to these four regular forms of ship construction contracts—the selective-price, lump-sum, price-minus and manhour or cost-plus-variable-fee—there were some construction contracts which provided for payment of costs and a single fixed fee.[12] Fixed fees were usual in contracts for special services. Examples are the contracts with naval architects for working plans,[13] and the contract made with the Standard Oil Company of New Jersey to inspect for the Commission tankers built at Sun Ship.[14] Inspection was not usually delegated by contract, however; it was performed by a staff organized under Civil Service.

In the case of many contracts the actual execution of the contract occurred after work had begun and the contract was pre-dated to cover work already under way. In 1944, for example, while negotiations for new contracts were delayed, letters of intent were sent.[15] Generally speaking, the notification to a company that an award had been formally approved by Commission action was the signal to go ahead. It enabled the Commission to apply for priorities, but Commissioner Vickery said that until the contract was signed the Commission felt free to change the award.[16]

V. THE COMMISSION'S SUPERVISORY STAFF

The supervision of the new contracts was an occasion for expansion and reorganization in the staff of the Commission. In January 1941, an addition of 300 people in Washington and in the field was planned, 150 in the Emergency Construction Audit Section of the Finance Division and as many in a newly formed Division of Emergency Ship Construction.[1] At the time of Pearl Harbor

---

[12] See below, chap. 18 on military types.
[13] On that with Gibbs & Cox, see above, chap. 3. Examples with other firms: MC Minutes, p. 26462 (Oct. 7, 1943); pp. 26557-58 (Oct. 19, 1943).
[14] MC Minutes, p. 31099 (Feb. 20, 1945) and others there referred to.
[15] See chap. 20.
[16] House, *East Coast Shipyards, Inc., Hearings*, July 12, 1944, p. 397.
[1] House, *Emergency Cargo Ship Construction, Hearings*, Jan. 18, 1941, p. 16 and Table p. 18.

the number employed was 2,359.[2] The increase of about 600 in the previous 18 months was mainly to supervise construction.

The new types of contracts gave the auditors more voluminous duties than they had had under the lump-sum contracts used in the long-range program. Under the lump-sum contract the auditors checked the companies' records only to exclude " excessive or unreasonable " payments. In a " mixed " yard, such as Federal, working for Navy, Maritime, and private account, there were delicate problems in assigning items of overhead among the three; but in a yard working on Maritime contracts only, practically everything that went into the yard would be paid for by Maritime sooner or later and they could check the value of the item by the detailed estimates which the builders had submitted with their bids.[3] Under the manhour or price-minus forms of contract adopted for the emergency there was no doubt that all shipbuilding materials going into the yards would be paid for by Maritime sooner or later, but there was considerable difficulty in preventing overcharge and an enormous amount of paper to be examined.[4]

According to the new types of contract the contractor agreed to keep accounts " in conformance with a condensed chart of accounts which the Commission will furnish " and to back up its showings of actual costs by access at all times to " all books, records, correspondence, instructions, plans, drawings, receipts, vouchers, and memoranda . . . pertaining to said work." In fact all these accounts were the property of the Commission.[5] In accordance with this article a " Uniform Classification of Accounts for Shipbuilders, USMC, Emergency Ship Program," was issued by the Construction Audit Section of the Finance Division and resident auditors were sent out to new yards.[6]

More auditors per yard were needed than under the lump-sum contracts because far more papers had to be gone through in checking on the builders' vouchers. The auditors did not have,

---

[2] House, *Independent Offices Appropriation Bill for 1943, Hearings*, Dec. 9, 1941, p. 272.

[3] Statement by Leroy G. Kell, Mar. 19, 1947, in MC, Historian's Collection, Research file 105.

[4] Statement by Henry Z. Carter, Aug. 13, 1947, in MC Historian's Collection, Research file, 105.

[5] Article 9.

[6] Copy in Historian's Collection, Research file, 106.24.

as under the lump-sum contract, any detailed builder's estimate against which to check the bills later submitted or the orders which the builder wished to have approved. In the main, the purchase orders were approved even as to price by the engineering inspectors, who also represented the Commission, before they reached the resident auditor.[7] But even if all he had to do then was to match up the papers and make sure that what was ordered was according to the contract and what was delivered was signed for in due order, still in the rapidly growing and amorphous shipyards it was quite a task even to make sure that all the papers checked. Accordingly the Finance Division set out to double and then redouble its field force of construction auditors. Positions were set up and graded in cooperation with the Commission's Division of Personnel and as early as February 1941 appeal was made to the Civil Service Commission to approve the jobs and help in finding the men to fill them. At the 9-way shipyard being built at Wilmington, North Carolina, for example, an auditing staff of 14 was planned, headed by a Principal Construction Cost Auditor receiving a salary of $3,800.[8] Securing men for such posts became increasingly difficult so that in June the Director of Finance asked to have the requirements eased.[9] The expansion was gradual, to be sure, as one yard after another came into production, but breaking in new men placed an added burden on the experienced members of the Construction Audit Section.

To get the work going right at the new yards, experienced auditors from old yards were sent to nearby places and gradually moved up in this way to become district supervisors. For example, W. L. Marshall who had previously been stationed at Newport News, supervised also the auditing at North Carolina and later the yards in Georgia;[10] and William H. Quarg, who had been at Moore's, supervised new yards at Richmond.[11]

No general instructions concerning the auditing of the new type of contracts, the cost-plus-variable-fee contracts, were issued

[7] Interview with Leroy G. Kell, Aug. 19, 1947.

[8] Anderson to Guy E. Needham, Director Division of Personnel, Feb. 25, 1941, and Needham to the Commission, Mar. 7, 1941 (approved Mar. 11) in gf 201-2-5. Cf. organization chart in MC, Historian's Collection, Research file 106.23.10 dated in pencil "About April, 1941."

[9] Anderson to Director, Personnel, June 7, 1941, in gf 201-2-5.

[10] Various letters in gf 201-2-5, especially Honsick to Marshall, Feb. 10, 1942.

[11] He was made Regional Chief in May 1940 and District Construction Auditor in Dec. 1941. MC Personnel Division files.

at least for several months.[12] Of course, the *Regulations* adopted in 1939 covered many matters, and there was the " Uniform Classification of Accounts " newly issued to guide the emergency yards; but there were also many questions arising from the administration of the new type of contract. They were answered as they came up. The Chief of the Construction Audit Section shared the burden of deciding these matters and supervising the auditors by calling to Washington R. H. Mohler who had been Principal Construction Cost Auditor at the Bethlehem Fore River yard at Quincy, Massachusetts. He was placed in charge of not only the field force but also two auditing units, the Home Office Shipyard Unit and the Subcontract Audit Unit.[13] For a time, Mr. Mohler was considered Chief of an Emergency Construction Audit Branch.[14]

While the field force was growing, so was the Washington staff of the Construction Audit Section. Although the section performed in 1941 the function of examining vouchers sent in for payment, that was not its main task. Its main concerns were clarifying for the auditors the way they should exercise their duties, and using the information supplied by the auditors in calculations which assisted the Commission in performing its managerial functions.[15] Mr. Anderson, the Director of the Finance Division, was almost as active as Admiral Vickery in negotiating with shipyard managements,[16] and he was also primarily concerned with speedy production in the emergency. Under his direction, a Special Assignment or Special Studies Unit developed statistics by which to rate the yards in terms of how fast they were getting

[12] Statements by Henry Z. Carter, and Leroy G. Kell in the interviews above cited. An examination of the " Instructions to Auditors " showed that very few new instructions were issued in 1941.

[13] J. A. Honsick to W. H. Disbrow and eight other persons, Oct. 17, 1941, in gf 201-2-5. Mr. Mohler's previous position is shown on the organization chart dated July 2, 1941 in the MC Historian's Collection, Research file 106. 23. 10.

[14] There was no administrative order creating an Emergency Construction Audit Branch, but on Mar. 11, 1941 the Commission approved a memorandum from the Director, Division of Personnel, concerning the organization chart for the branch and the personnel.—MC Minutes, pp. 16541-42. In Jan. 1942, Mr. Mohler was signing as " Chief, Emergency Construction Audit Branch." See letters in gf 201-2-5.

[15] See below, chap. 22.

[16] Testimony by Vickery in House, *East Coast Shipyards, Inc., Hearings* before the Subcommittee on Production in Shipbuilding Plants of the Committee on the Merchant Marine and Fisheries, 78th Cong., 2d sess., July 12, 1944, pp. 398-404, and interviews with Wesley Clark, June 6, 1947, and Elizabeth Sprott.

on with contracts.[17] Other units at Washington of the Construction Audit Section, in addition to those already mentioned, were the Voucher Examining Unit, the Administrative Unit, the Report Review Unit, and the Cost Valuation Unit.[18]

In preparing its statistics for rating the yards, the Special Assignment or Special Studies Unit used both the reports of the auditors and those from the Commission's engineers. The function of the auditors did not extend to the actual physical inspection of construction and materials. Although there was considerable difference of opinion first and last as to whether auditors were concerned with the control of inventories and other checks on the relations between the papers and the realities they reported,[19] there was never any doubt that the Maritime Commission also needed to be represented at the yards by technical inspectors who passed on the quality of the materials received and the work performed. They were the responsibility of the Division of Emergency Ship Construction.

The director of this new division was J. E. Schmeltzer, hitherto Associate Director of the Technical Division; and its five sections— Hull Plan Approval, Engineering Plan Approval, Plant Engineering, Production Engineering, and Administrative—were all, except Plant Engineering, headed by former officials of Technical.[20] It was conceived as a temporary diversion of talent to perform for the Liberty ships the functions performed for all other types by the Technical Division and to be wiped out when the job was done.[21] Admiral Vickery later explained the creation of a parallel division by saying: " At the time of the creation of the Emergency Ship Construction Division it was considered advisable to divorce it entirely from any other activity because it was necessary for this new Division to immediately become organized and to negotiate contracts for new ships; to approve plant layouts; supervise the building of new shipyards; to approve the plans for the ships; to ratify the purchasing of practically all the machinery, equip-

---

[17] Mr. Anderson's personal activity in developing these statistics was stated by Wesley Clark and Frank R. Hunter in interview of June 6, 1947. The earliest such rating which I have found published, for Sept. 1 through Dec. 1, 1941, is in House, *Independent Offices Appropriation Bill for 1943, Hearings*, p. 279.

[18] Charts prepared by Marie D. Werner, in Historian's Collection, Research file, 106 (MC) 23.10.

[19] See below, chap. 22.

[20] Admin. Order 37, Suppl. 22, Jan. 7, 1941 and Suppl. 26, Jan. 16, 1941.

[21] House, *Emergency Cargo Ship Construction, Hearings*, Jan. 18, 1941, p. 16.

ment and material entering into the building of these ships." [22] Approval of working plans and purchasing, which has already been described, demanded immediate attention in January 1941 when the new division was formed,[23] but hardly less pressing was supervising of the construction of the new shipyards.

At the new emergency yards the first engineers needed to represent the Maritime Commission were the Resident Plant Engineers. These men had reason to feel that they were the top ranking representatives of the Maritime Commission in the locality. Generally they had a Civil Service grade which was above that of the auditors and of the other engineers.[24] The directive they received was of a kind to increase their sense of importance and encourage them to take the initiative. It began: " The Resident Plant Engineer shall be the official representative of the Maritime Commission at the site of the shipbuilding Contractor building ships under the Emergency Ship Construction Act, insofar as the plant facilities, construction and maintenance features are concerned." He was to " advise and direct the shipbuilder in connection with engineering, construction and maintenance activities pertaining to the shipbuilding facilities; " " to inspect all plans, . . . review all proposals and prices " and consult with the other engineers on the efficient flow of material.[25] Since building a shipyard involved in some localities substantial changes in the topography and in city planning, the Resident Plant Engineer was likely to become for a time at least the Maritime Commission's representative in the eyes of the community.

Yet none of the resident plant engineers was a previous employee of the Maritime Commission. They had none of the kind of tradition and professional pride which the Maritime Commission's resident auditors brought to their new assignments. Since the plant engineers were to fulfill a function not previously performed by the Commission, they had to be recruited from outside. The Civil Service Commission provided lists and some men

[22] MC Minutes, p. 18216 (July 25, 1941).

[23] Above, chap. 3.

[24] Their grade was P-5. At that time the other engineers and the auditors were CAF-11's at best (except on the West coast, where the Principal Hull and Machinery Inspectors were the equals of the Plant Engineers, rating CAF-12). Organization charts in MC, Historian's Collection, Research file 106. MC Minutes, p. 16287 (Feb. 18, 1941); Personnel Division files.

[25] Office Order No. 3, Mar. 1, 1941, of Emergency Ship Construction Division, in gf 202-4-1.

THE SUPERVISORY STAFF

sent in applications. Six or seven out of the first nine had been engineers in PWA or WPA; the rest had been employed in other government work supervising construction. A majority, but by no means all, had attended colleges or engineering schools. The best qualified men were already in their sixties. Previous experience with shipyard construction was generally in building the shipyards of World War I. In default of experience in shipyards, work in other kinds of waterfront construction enabled candidates to qualify. The section chief to whom they reported in Washington was also a new employee and was soon replaced by another new man. Although some of the resident plant engineers recruited hurriedly in the first months of 1941 proved successful and moved up into higher positions in the expanding programs, others proved unqualified or unworthy.[26]

The pressure was intense in January 1941 to get engineers on the spot so that dredging and filling and the delivery of equipment could start on the recently selected sites.[27] The instruction issued the plant engineers directed them to be available in the yard at all hours. They were to check over the yard plans before they were submitted to Washington for final approval. It was provided: " No purchase shall be made without written approval of the resident plant engineer." He was to receive copies of all invitations to bid, of all bids, and of all receipts of material. He could himself approve individual items of $5,000 or less; for larger items his approval was subject to action in Washington. Although the yard was used by the shipbuilding company, it belonged to the Commission, and it was the duty of the plant engineer to see that it was " rapidly and economically constructed and well maintained." [28]

Whereas plant engineers were a new sort of official in the Maritime Commission organization—and perhaps for that reason unrestrained by precedent—a staff of hull and machinery inspectors had been built up during the previous two years. Consequently there were experienced officials in Washington to supervise these inspectors, but so many new inspectors were

[26] MC Personnel Division files.
[27] Report Covering the Activities of the Emergency Ship Construction Division January-March 1941 said seven had been appointed and located.
[28] Office Orders, Division of Emergency Ship Construction, Order No. 1 (Jan. 28, 1941), and Order No. 3 (Mar. 1, 1941), in gf 202-4-1.

needed in the field that most of them were new employees of the Commission.[29] Many of its employees were lured away from the Commission by the shipbuilding companies, and the inspection force was understaffed throughout 1941.[30]

The duties of the inspector were more extensive in the new emergency yards than in the established shipyards. He had to help in forming effective organizations with men of whom few were experienced shipbuilders, and even fewer understood how to build welded ships of standardized design at maximum speed. In the emergency yards the inspectors needed to do more than pass judgment on finished work to see that it was up to the standard; they needed to pass on the plans and methods while they were being worked out. They received not only the same instructions in regard to standards to be met which had been issued to inspectors under the Technical Division,[31] but also from their section chief more detailed indication of how to perform their tasks in the yards.[32] They were told to maintain such relations with the contractors' representatives as would help in the interpretation of plans and specifications. They were to insure that construction was in keeping both with the specifications and with modern shipyard practice. Some inspectors took the latter phrase to mean that they should insist on use of the newly developing methods for speeding up the work. It seemed to imply that they should not only check on how welding had been done, for example, but on the way the work was being laid out for the welders. But the inspectors were specifically told that they " should at no time interfere with workmen in the carrying out of work, all comments should be made to the authorized representative of the Contractor." They were expected to settle on the spot controversies concerning unsatis-

[29] I have inquired concerning those whose names are on the organization charts of early 1941 both from personnel records and from the memories of James Kirkpatrick and Thomas Fraser and can not find that more than about six out of thirty had been with the Commission a year earlier.

[30] It contained 286 at the beginning of Oct. 1941 and 396 at the end of the year, but the chief of the section said that 70 more were needed in Washington and 800 in the field. Quarterly Report, Construction Division Oct. 1 to Dec. 31, 1941, in MC gf 212-8 and Production Division adm. file A9-3.

[31] Office Orders of the Emergency Ship Construction Division and Construction Division, in 202-4-1, repeat practically word for word the " Office Memoranda " of the Technical Division, in Technical Division files.

[32] Office Memoranda, " Instructions to Inspectors in Emergency Construction Yards." Copies in MC, Historian's Collection, Research file 203, and J. T. Gallagher to Principal Inspectors, June 4, 1941, in gf 507-3-1.

factory work, but to avoid questions of policy. Such matters should be referred to Washington.

In Washington the reports of the inspectors were handled by a hull coordinator and machinery coordinator. These officials were in the same section and the instructions to hull and machinery inspectors came from the same section chief. During the first half of 1941 this was J. T. Gallagher, Chief of the Production Engineering Section of the Emergency Ship Construction Division. But before building had gone far in many yards, a reorganization occurred, occasioned partly by the termination of the contract with Gibbs & Cox. It abolished the Emergency Ship Construction Division and created in its place the Construction Division which included all the inspectors, those of the old yards in the long-range program as well as those in the emergency yards. In the new Construction Division was created an Inspection Section headed by L. R. Sanford, who in the Technical Division had been in charge of inspectors in the long-range yards. To it were transferred all the resident hull and machinery inspectors.[33]

Viewed as a whole, the organization of auditors and inspectors initially set up to help and to supervise the new shipyards seems to have had three main weaknesses. (1) The staff was relatively inexperienced in the specific kind of duties which the emergency required. (2) There were four or five quite separate lines of authority and communication running between each shipyard and officials of the Commission in Washington. Reports were sent to Washington by the auditor, the plant engineer, the hull inspector, the machinery inspector and later from a purchase controller, each to a different superior.[34] (3) The inspectors and auditors were only $3,000-a-year or $4,000-a-year men, and were being sent out to deal authoritatively with $10,000-a-year company executives who were also in effect being paid by the Commission, since their salaries were included in the cost of the ships. Each of these weaknesses was at least in part unavoidable.

These difficulties were recognized and efforts were made to

[33] Admin. Order No. 37, Suppl. No. 42, July 25, 1941; Admin. Order No. 37, Suppl. No. 43, July 25, 1941; MC Minutes, pp. 18216 (July 25, 1941) and 18859 (Sept. 9, 1941) approving organization chart; and Quarterly Reports, Construction Division, July through Sept., 1941.

[34] On the purchase controller see Revision of Aug. 5, 1941 of Office Order No. 2, in gf 202-4-1, and "Instructions to Inspectors in Emergency Construction Yards," Aug. 11, 1941, cited in n. 32.

correct them. The inequality in company salaries and government salaries when both were being paid with government funds impressed John M. Carmody when he became a member of the Commission late in 1941. Admiral Vickery made a special plea to the Civil Service Commission for higher ratings.[35] The difficulty in having no one high-ranking representative at the shipyard was foreseen by Mr. Anderson. In March 1941 he addressed a memorandum to Admiral Land saying that the Commission should have at each emergency yard a man who was not merely an auditor, or not merely a technician, but was a man of business judgment who would pass on matters affecting speed and cost, and " whose personality will be such that he can deal firmly and wisely with the shipyard officials." At the bottom of the memorandum Admiral Land wrote, " I agree thoroughly and consider it of urgent importance." It was routed to Admiral Vickery and on the routing slip Executive Director Schell wrote that he thought appointment in positions exempt from Civil Service rules could be arranged " if you can pick the men you want." [36] But no such general representatives from the Maritime Commission to the new shipyards were appointed, partly because there was a question whether setting up such a general representative at each yard might not lessen the contractor's responsibility, partly because it was difficult to find the right men.[37] How useful the right kind of men could have been in such positions is clear from the story, told elsewhere, of what happened at the South Portland yards.

For the success of its shipbuilding program the Commission relied less on the staff it placed in the yards to supervise the contracts than on the executives of the shipbuilding companies. Enlisting managerial talent had been considered by Admiral Land the central problem in launching the program. In shaping the contracts a primary aim was to obtain a maximum exercise of managerial initiative and ability. Although the Commission sent out auditors and engineers to check fraud and inefficiency, the

[35] Statement by John M. Carmody, June 17, 1947; Vickery to Fleming in Kirkpatrick file in MC, Historian's Collection. The quarterly report of the Inspection Section of the Construction Division, Dec. 31, 1941, speaks of the scamping by yards of government personnel, and dissatisfaction among the inspectors at seeing men with less capacity and knowledge of shipbuilding receive higher salaries than inspectors.

[36] Anderson to Land with attachments, Mar. 26, 1941, in gf. 507-3-1.

[37] Anderson to F. C. Lane, May 3, 1950.

THE SUPERVISORY STAFF

Commission was not trying to give commands to the executives or workmen of the shipbuilding industry. It had no power to do so, and even if it had had the power it could not have told the shipbuilders in detail what to do. It did not know. Nobody knew. New problems were arising and the shipbuilding companies were expected to find the solutions, accepting from the Commission's staff the help they could give. The success or failure of the Commission's program depended on calling forth, through the incentives offered by the contracts and through all other means available, extraordinary efforts and achievements from the shipbuilding companies.

CHAPTER 5

EXPANSION AND REORGANIZATION AFTER PEARL HARBOR

I. THE FOURTH WAVE OF EXPANSION

THE JAPANESE attack and the declaration of war by Germany changed both the shipbuilding program and the conditions affecting its success. Sinking of merchant vessels took a pronounced turn for the worse because German submarines attacked the coastwise shipping lanes that run from the mouth of the St. Lawrence to the Gulf of Mexico and the Caribbean. Over all that area they found unarmed and unescorted freighters and tankers an easy prey in the first half of 1942. The huge loss in merchant vessels—6.4 million deadweight tons in the first half of 1942—made fast construction in the Maritime Commission yards ever more urgent. But building for the Navy was also desperately needed immediately after Pearl Harbor. At the same time Army commanders, looking ahead to the time when the men they were training would be brought into combat, demanded transports as well as cargo carriers. Under such pressure the Maritime Commission's program underwent two more waves of expansion, the fourth and the fifth.

Just before the attack on Pearl Harbor, the Maritime Commission scheduled 5 million deadweight tons for 1942 and 7 million for 1943.[1] These figures were in accord with the assumption that the yards would speed up production from two ships per way a year, as planned for their first year, and would produce thereafter at the rate of four ships per way a year.[2] With the declaration of war, the British building at Richmond and South Portland, as well as what little there was of private building, was included to bring the Commission's schedules on January 1, 1942

---

[1] Fischer, "Programming," Table 1. Tonnage figures in this chapter are in deadweight tons.

[2] House, *Independent Offices Appropriation Bill for 1943, Hearings,* Dec. 9, 1941, p. 269, testimony of Admiral Land.

PLATE VII. Above, shipfitters and welders laying the keel of a Liberty on one of the ways at Oregon Ship, and the half-finished hull. Below, Marinship at dusk as cranes bend to complete a Liberty ship ready for launching.

PLATE VIII. Shifts around the clock, big subassemblies, and welding speeded production. Above, Oregon Ship at night; lower left, forward cofferdam going into a tanker at Swan Island; lower right, welding on stern casting for a Liberty.

FOURTH WAVE OF EXPANSION 139

up to about 6 million in 1942 and 8 million in 1943.³ The incorporation in the Maritime Commission's program of the two yards which were building for the British had been planned the previous autumn, for on November 7, 1941 the Commission approved the leasing of the Todd-Cal yard by the Richmond Shipbuilding Corp. and the development of facilities having in view the use of the ways for the Maritime Commission when all the British ships had been launched.⁴ Contracts providing for the use of both British yards were made on January 16, 1942, and in the same month negotiations were begun with the British for the purchase of the yards.⁵ Practically speaking, the two British yards and the nine emergency yards of the Commission had been a joint plan long before Pearl Harbor. After December 7, 1941 their production was scheduled as a unit.

The declaration of war was also a signal for closer common action at the highest political levels, and before Christmas the Prime Minister of Britain crossed the ocean to confer at the White House concerning the objectives and materials of war. During January 1942, the shipbuilding goal was set as the delivery of 8 million tons in 1942, an increase of roughly a third. For 1943, 10 million tons were planned. Admiral Vickery, who was called on to say whether it could be done, replied that it would be no problem to get 10 million in 1943 if production in 1942 could be stepped up to 8 million.⁶

To increase production by a third in 1942, Admirals Land and Vickery depended mainly upon increasing the rate of production in existing yards and called on both management and labor to increase their efforts.⁷ It was believed they could speed up production from four Liberty ships per way per year and turn out five or even six per way per year. Speed was analyzed in terms of "rounds of the ways." The launching of one ship from each way in a yard completed one "round." The building time of Liberty ships delivered in 1941 was about 250 days, but they were

---

³ Fischer, "Programming," Table III; *Statistical Summary*, Table A-3.
⁴ MC Minutes, p. 19639 (Nov. 7, 1941).
⁵ MC Minutes, p. 20628 (Jan. 16, 1942). Land to Harry Hopkins, Jan. 6, 1942, MC, Historian's Collection, Land file. Records of the subsequent negotiations are in MC gf 507-3-14 and gf 507-3-11.
⁶ House, *Production in Shipbuilding Plants, Executive Hearings*, pt. 3, June 28, 1943, pp. 915-16.
⁷ Land reading file, Jan. 17 and 19, 1942.

the first round of the ways; and the original contracts called for the completion of later rounds in 150 days. Now, in January 1942, the shipbuilders were called on to contract for new Liberty ships on the basis of a building time of 105 days. Since that allowed 60 days on the ways, 45 days in outfitting, it would produce six ships per way per year.

In negotiating the new contracts Admiral Vickery first called in members of the Kaiser group from the West coast. When he told them that the new period for building a ship would be 105 days, they threw up their hands in protest. But after forty-eight hours they agreed to contracts on these terms.[8] In new contracts the delivery dates for new ships were based on the assumption that not only these ships but also the ships built for their first contract would be only 60 days from keel laying to launching, after the first round to come off the ways.[9]

Once Kaiser's men, the newcomers from the West, had said they would accept the speeded-up schedules as the basis for their contracts, Admiral Vickery invited the old-line shipbuilders of the East coast to do as well, but none of the Eastern yards had dates in their contracts which applied the 105-day standard to both the new and old contracts.[10] Bethlehem-Fairfield for example

[8] Vickery's account of the negotiations in his testmony House, *Higgins Contracts, Executive Hearings*, before a subcommittee of the Committee on Merchant Marine and Fisheries, 77th Cong., 2d sess., pt. 3, Aug. 20, 1942, pp. 251-52.

[9] The dates put in the contract were worked out by preparing charts showing how one ship followed another on the ways. Of course, the date at which the seventh ship, eighth ship, etc., would come off a way depended on how early the keel was laid and that depended in turn on how quickly the earlier ships on that way had been made ready for launching. Thus the delivery dates placed in the second contract called for a speeding up of the deliveries under the first contract. The way charts used in preparing the second group of contracts are in the folder "Way-chart Work Sheets for EC2 Contracts, January-May of 1942," in Historian's Collection, Research file 117, Statistics. Comparison of these charts with dates of delivery recorded in the MC Minutes shows that the dates worked out by Norman R. Farmer and Harvey H. Hile on the basis of 60 days on the ways and 45 days outfitting were the dates approved for the contracts with the Kaiser yards. For Calship the second and third round off the ways were on some ways allowed more time, but the new ships were allowed only 57 days on the ways, 42 days for outfitting.

[10] I have compared the contract delivery dates given in the MC Minutes with the dates on the work sheet schedules drawn up by Mr. Hile and Mr. Farmer for Admiral Vickery, which are dated at various days in Jan. 1942. Most of these sheets are marked "dates lifted," "used as basis for contract," but this is not true of the schedule for North Carolina Ship. The Schedule made out for North Carolina is like those made out for the Kaiser yards in allowing only 60 days on

submitted a schedule from which were taken the delivery dates for their contract. It allowed 85 days on the ways for the last ship of their first contract and an average of 64 days on the ways for the 110 ships of the new contract.[11] Yards that faced special handicaps were given contracts at slower schedules. The contract signed with Houston allowed time for the first ships of the new contract to be 81 days on the ways and 51 days outfitting.[12] In January 1942, when these contracts were made, Houston and many other companies had not finished their yards and had not launched a single ship. Even if they ultimately caught up with the speed of the Kaiser yards their average speed for 1942 was sure to be much slower.

For the program as a whole, the speed-up in rate of production was estimated as about 25 per cent, giving roughly five ships per way per year instead of four. Accordingly it was necessary to add a few new yards for major types to meet the boost in objectives of nearly a third which was ordered by the President in January 1942.[13]

Two new yards were built by a company headed by Henry Kaiser, who now organized new shipbuilding companies of his own in which Todd Shipyards had no shares. Kaiser and his associates also acquired full ownership at this time of the West coast yards they were managing and sold to Todd Shipyards their interest in the Todd-managed yards elsewhere. The new Kaiser company received contracts to build and operate for the Commission a basin shipyard known as Richmond No. 3, with five basins, located near the yards already being operated by Kaiser in Richmond under the immediate direction of Clay Bedford,

---

the ways for the ships after the first round. But the contract with North Carolina Ship did not use these dates; the contract dates call for deliveries on the second contract from three to four months later than those worked out by applying the standard of 60 days on the ways, 45 days in outfitting. William A. Weber's recollection of the conference of Admiral Vickery and Capt. Williams was that the latter said he was willing but acknowledged he could not come anywhere near fulfilling such a contract.

[11] A photostat of the schedule is in the folder "Work Sheets." There it is diagrammatically indicated on the schedule-chart submitted by Bethlehem-Fairfield that the average way-time for the 110 ships is 60 days, but study of the dates on that chart and in the contract does not confirm this.

[12] Work sheets cited and MC Minutes, p. 20624-25.

[13] Statement of Land in House, *Fifth Supplemental National Defense Appropriation Bill for 1942, Hearings*, Feb. 12, 1942, p. 100.

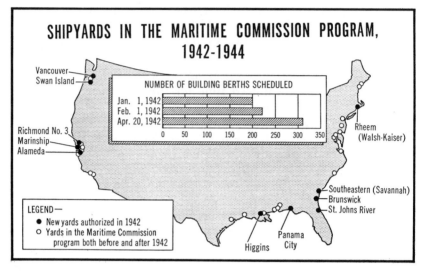

FIGURE 8. Class I yards. Names of yards built before 1942 are on Figures 4 and 6.

and to build and operate likewise a new yard at Vancouver, Washington, near the yard of the Oregon Shipbuilding Co. directed by Edgar Kaiser.[14]

Except for the two new Kaiser yards, there was only one other new yard planned to meet the Presidential directive of 18 million tons. That was the Alameda yard in San Francisco Bay which the Bethlehem Steel Co. proposed to develop from a repair plant into a four-way yard for the construction of transports.[15] A facilities contract was also given to complete a yard at Savannah, Georgia, but it was not a new yard. It was a yard which had been developed as a three-way yard by a private promoter to such a point that the Commission decided it was worth finishing as a six-way yard, considering the emergency.[16] Another addition at this time to the facilities available to the Commission was in the yards operated by the Federal Shipbuilding Co. The Navy

---

[14] MC Minutes, pp. 20502-04 (Jan. 13, 1942); p. 20491 (Jan. 10, 1942). On the ownership of the Kaiser companies see below, chap. 20.

[15] *Ibid.*, p. 20957 (Feb. 3, 1942).

[16] MC Minutes, pp. 20341-44 (Dec. 30, 1941); p. 20649 (Jan. 20, 1942); below, sec. III of this chapter.

agreed that two more ways in the Kearny plant were to be used for Maritime Commission work.[17]

In this fourth wave of expansion of the Maritime Commission's program, the increase in facilities was quite moderate, however. The yards building for the British had already been in operation. Therefore the increase in the fourth wave, during January 1942, was only 22 building berths, from 199 to 221, or slightly more than 10 per cent.[18] But it was possible to plan an increase of output of 30 per cent because existing yards were expected to speed up production by about 25 per cent.

II. THE FIFTH AND FINAL WAVE

After this fourth wave of expansion, Admiral Land reiterated his warning: "The shipbuilding cup is full to overflowing."[1] The Maritime Commission was under all kinds of pressure to build more shipyards and try new types of ships,[2] but like the Navy it felt that it had its hands full. At a conference in the office of Admiral S. M. Robinson, Chief of the Bureau of Ships, on January 15, 1942 with a representative of the Office of Production Management " it was definitely determined that any additional shipbuilding beyond that discussed at this conference was in excess of the capacity of the shipbuilding industry in the United States." They agreed not to build any more unless so ordered by the President of the United States.[3]

They were so ordered. Pressure from the Army demanded more and more ships. Sinkings were running over a half a million tons a month[4] and on the other hand the Army was training men

[17] House, *Fifth Supplemental National Defense Appropriation Bill for 1942, Hearings*, 77th Cong., 2d sess., Feb. 12, 1942, p. 113. Federal's way schedule in MC, Magdelene Napier's office.

[18] Because some yards had long shipways on which simultaneous erection of more than one vessel was the rule, the expression "building berth" is more accurate than "shipway." The figures which were the basis of the February 1942 program (according to the work sheets now in the Historian's Collection, Research file 117, Statistics) enumerated 193 building berths, but they need some correction for Federal, Pusey and Jones, and Consolidated Steel Co.

[1] House, *Fifth Supplemental National Defense Appropriation Bill for 1942, Hearings*, 77th Cong., 2d sess., Feb. 12, 1942, p. 95.

[2] See sec. III of this chapter.

[3] Land reading file, Jan. 17, 1942. Memo to Admiral Robinson and W. H. Harrison.

[4] *WSA Shipping Summary*, Vol. II, No. 1, p. 18 (Jan. 1945).

with the intent of making 1,800,000 ready for overseas service by the end of December 1942, and 3,500,000 ready by the end of 1943. Less than half that number could be maintained as a fighting force overseas, so General George Marshall wrote to the President on February 18, 1942, unless steps were taken " to increase the tempo of the shipbuilding program to a much higher figure. . . . The war effort of the United States, less what can be done by the Navy, will be measured by what can be transported overseas in troops and materiel."

The next day, February 19, Admiral Land had a conference in President Roosevelt's bedroom. The impossible must be done, the objective must be lifted again, from a total of 18 million tons in 1942-1943 to 24 million tons. The President asked that plans be made to build 9 million tons in 1942 and 15 million tons in 1943. " I realize," wrote President Roosevelt in the memorandum in which he confirmed this decision, " that this is a terrible directive on my part but I feel certain that in this very great emergency we can attain it." [5]

While Admiral Land was receiving in the President's bedroom the new directive, Admiral Vickery was on a tour of inspection in the South to see how much could be expected from the yards there. He had decided that the shipways already planned would not quite produce the 8 million tons in 1942 but that they would just about make the total of 18 million for the two years 1942-1943. He was at Mobile when Admiral Land called him and said they were now to try to make 9 million tons in 1942 and 24 million in the two years. He said, " You know it is impossible to get 9,000,000 tons in 1942." Admiral Land said, " Yes," and " I have so informed the authorities, and all I said was we would try." [6]

While the Commission was planning how to achieve the new

---

[5] MC, Program Collection, Feb. 21, 1942. The President to Admiral Land, with memorandum from Marshall to the President attached. At another conference at the White House, attended by representatives of the Army, the WPB and the MC but with no representative of the Navy present, the President reiterated on Feb. 23 the need for increasing the amount of merchant marine tonnage. Land reading file, Feb. 23, 1942.

[6] House, *Higgins Contracts, Executive Hearings*, pt. 3, pp. 252-53; Cf. similar accounts by Vickery in House, *Investigation of Rheem Manufacturing Co., Executive Hearings* before the Committee on Merchant Marine and Fisheries, 78th Cong., 1st and 2d sess., p. 61; and House, *Production in Shipbuilding Plants, Executive Hearings*, pt. 3, June 28, 1943, pp. 915-16.

FIFTH AND FINAL WAVE

"terrific" directive, the record of the yards was not sufficiently good so that the Commission felt justified in counting on any higher rate of production than that assumed in the schedules already made, namely, 6 Liberty ships per year per way.[7] The actual rate of production in February and early March 1942 was only about $2\frac{1}{2}$ ships per way per year. In order to try for the new objective it was decided to build more shipways. It was done by adding to existing yards and by creating wholly new yards at new locations. Established managements were available in the yards which were to be expanded, but untried combinations headed by men unskilled in shipbuilding had to be drawn on for the new yards.

Immediately upon return from the conference in the President's bedroom, Admiral Land called the Sun Shipbuilding Co. and placed with them an order for 50 vessels of a new type called "seatrains."[8] To enable Sun Ship to add this work to the tankers it was already building, it was awarded a facilities contract for the construction of an addition of 8 ways at the expense of the Commission.[9] They were adjacent to the 20 ways already being operated by Sun and may be considered therefore as part of a single 28-way yard, the largest single shipyard in the country. Although 28-way yards had been discarded in the planning of the Commission in the winter of 1940-1941, successive expansions pushed this one yard up to that size in 1942.

For other additions to existing yards, the Maritime Commission turned to the West coast. Again, as in the previous month, Admiral Vickery looked for aid to the Kaiser management because of its outstanding record at that date. The companies under old-line management had not yet shown what they could do: North Carolina had not yet launched any second-round ships; Bethlehem-Fairfield just one, on January 24, a hull which took 119 days from keel laying to launching. The yards on the Gulf were just getting started. But out on the West coast Oregon Ship was getting a ship ready to launch 71 days after keel laying, and at Richmond, yard No. 1, which worked for the British but was under the same management as nearby Richmond No. 2 and No.

[7] House, *Higgins Contracts, Executive Hearings*, pt. 3, p. 255. Cf. Figure 10-12.
[8] Land to the President, Feb. 21, 1942, in Land reading file. On the evolution of this order into an order for transports (C4's), see below, chap. 16.
[9] MC Minutes, pp. 21192-93 (Feb. 24, 1942); and p. 21271 (Mar. 3, 1942).

3, was finishing the second round off the ways in about 80 or 90 days for each ship.[10] After Admiral Land's call informed him of the new directive, Admiral Vickery sent for Edgar Kaiser to join him. At Mobile they discussed shipbuilding on the Willamette and the Columbia. It was decided to increase to 12 ways the Vancouver yard which had been begun just the previous month, and to build a new 10-way yard on Swan Island. Although Swan Island was later reduced to 8 ways, the total in all the yards of the Portland, Oregon, area was even then 31 ways. The yards were scattered some distance from each other but were all under the management of Edgar Kaiser.

While conferring in Mobile, Admiral Vickery wired to Richmond, California, that Richmond No. 2 should be expanded from 9 to 12 ways. That brought to a total of 24 ways the three Kaiser yards in Richmond. A little later 3 more ways were added in a sort of annex originally called No. 3A, later No. 4; and thus the Kaiser yards on San Francisco Bay, with a total of 27 ways, almost equalled in the number of their shipways the yards of Sun on the Delaware. Although the Richmond yards were not contiguous nor under a single corporation, they were quite close together and were all under the same general manager, Clay Bedford.[11]

An entirely new yard also was established on the West coast by enlisting part of the management previously associated with Calship. After Admiral Vickery returned to Washington, the Commission sent out on March 2, 1942 identical telegrams to all seven companies of the " Six Companies " group. Each was asked to submit a proposal for a shipyard somewhere on the West coast which would produce ships in 1942. Adequate production in 1943 was in sight, they were told; the paramount problem was production in 1942, and for that they were asked to contribute directly their top personnel. The telegram concluded: " The emergency demands all within your power to give your country ships." At 11 p. m. the same night S. D. Bechtel wired back, " We are studying the problem tonight and will give you our sincere best judgment tomorrow." Some other executives of the " Six Companies " worked up proposals but none so rapidly. Although the telegram was a surprise to him, Mr. Bechtel had his plans

---

[10] Vickery's bar charts, in MC, Historian's Collection.

[11] House, *Higgins Contracts, Hearings,* pp. 252-53; Fischer, *Statistical Summary,* Table E-5.

FIFTH AND FINAL WAVE

and a contract ten days later. He was able to enlist men from the allied construction companies and from Calship, and to fabricate the steel for his first ships in the facilities of that company, 400 miles away from his new yard at Sausalito, Marin County, across the Golden Gate bridge from San Francisco. Admiral Vickery approved verbally the site as one he knew well enough already and on the same day that Mr. Bechtel presented his plans told him to go ahead: " I am betting on you fellows; I expect you to produce." [12] The new yard, later called Marinship, did in fact deliver five ships in 1942.[13]

Marinship was a 6-way yard, one of five new 6-way yards which were begun to meet the President's new directive of February 19, 1942. The location of the other four, Admiral Vickery later testified, was decided by him mainly in consideration of where labor was available.[14] Of course the availability of good sites was also a factor. Congressmen from New York later claimed that there was unemployed labor in that big metropolis so that the Maritime Commission should have put one of its shipyards on or somewhere near Staten Island.[15] But Admiral Land emphasized the difficulty of finding in that neighborhood enough unoccupied land to serve as a site for a shipyard, especially when it was realized that World War II shipyards required relatively little waterfront but much depth compared to the shipyards of World War I.[16] So, either because of incorrect predictions concerning the labor situation in that area, or because of the lack of a good site, New York harbor was passed over at this point, and one of the 6-way yards was located in New England, the remaining three in the South.

Providence was selected in New England because it was believed that there was a good labor situation there. The Commission had

---

[12] Testimony by S. D. Bechtel in House, *Production in Shipbuilding Plants, Hearings*, June 22, 1943, pp. 691-708. MC Minutes, pp. 21362-64 (March 12, 1943). The telegrams and replies are in gf 507-1 under the dates cited.

[13] The initial contract for the yard and Liberty ships was awarded to W. A. Bechtel Co., a partnership. Late in 1942 the joint venture was changed to Marinship Corp. House, *Investigation of Shipyard Profits, Hearings*, 1946, pp. 249-51.

[14] House, *Investigation of Rheem, Hearings*, p. 61.

[15] House, *Walsh-Kaiser Co., Inc., Hearings*, before the Subcommittee on Production in Shipbuilding Plants of the Committee on Merchant Marine and Fisheries, 78th Cong., 2d sess., pp. 295-99.

[16] House, *Independent Offices Appropriation Bill for 1944, Hearings*, 78th Cong., 1st sess., p. 691.

only one yard building big ships in New England, at South Portland, Maine, so that a location on Narragansett Bay would even off the geographical distribution of yards. For a management, Admiral Vickery picked the Rheem Manufacturing Co., which was experienced in manufacturing water heaters, shells, cartridge cases and many other items, but not ships. It was highly recommended when he inquired about it from the Bethlehem Steel Co. Richard S. Rheem, president of the company, would have preferred to go to Stockton, California, but Admiral Vickery decided the West coast was now overloaded and persuaded him to take a site at Providence.[17]

The new 6-way yards in the south were at Brunswick, Georgia, Jacksonville, Florida, and Panama City, Florida. The latter yard and ultimately that at Brunswick were to be managed by the J. A. Jones Construction Co. Mr. Jones also had no experience in shipbuilding. He was a general contractor from North Carolina who had done construction of many kinds and was recommended to Admiral Vickery by W. H. Harrison of the War Production Board as a good manager. He wanted to go to a small town in South Carolina, but was persuaded instead to go to Panama City which Admiral Vickery considered a favorable site.[18] Jacksonville was urged by a local ship repair firm which allied with a New York firm of contractors to build the southernmost of the yards on the East coast.[19]

Besides the new 6-way yards, the increase in the yards and ways managed by Edgar Kaiser and Clay Bedford, and the addition to the Sun Shipbuilding Co., a new yard under Andrew J. Higgins at New Orleans was provided for to meet the President's "terrific" directive of February 1942. It was authorized as a 44-way yard, although it was of so original a design that conventional rating in terms of number of ways could really not be applied to it. But the story of the Higgins contract, the peak of the expansion, will be told in the next chapter. The new yards of 1942 are shown on Figure 8.

[17] House, *Investigation of Rheem, Hearings*, p. 61; *ibid.*, *Interim Report* (Report No. 2057), pp. 1-2; House, *Higgins Contracts, Executive Hearings*, p. 253.
[18] *Ibid.*, p. 255.
[19] House, *Investigation of Shipyard Profits, Hearings*, 1946, pp. 306-11. Land reading file, Feb. 21, 1942; and MC Minutes, p. 21271 (March 3, 1942), p. 21180 (Feb. 20, 1942), and p. 21288 (Mar. 3, 1942). Their contract called for delivery of 6 ships in the last half of December, and then none till April.

FIFTH AND FINAL WAVE

While adding blue prints of new shipyards, the Maritime Commission was still losing to the Navy some of the ways already in efficient operation. At the end of March 1942 the entire facilities of the Seattle-Tacoma Shipbuilding Co., 8 ways, a yard developed from its inception for Maritime Commission work, were allotted to the Navy, which wanted it for auxiliary aircraft carriers, constructed partly by conversion of the C3's already building in that yard. In return the Commission received a promise of future exclusive use of the building facilities of Moore Dry Dock and Ingalls,[20] a gain of 2 ways.

Even after all contracts for new yards had been awarded, in April of 1942, the Commission's building schedule did not come up to the President's directive of 9 million tons for 1942. The expansion in shipways was about 40 per cent,[21] but very little indeed could be expected off the new ways within the remaining months of 1942. The schedule of April 20, 1942 provided for a total of 7,715,800 tons in 1942. The hope of reaching that goal and if possible doing better depended mainly on speeding the work in existing yards, but in April the rate of production in most of the yards was disappointing.

For 1943 the Commission's schedules of April 1942 provided 15,413,600 tons, as much as the President had asked and something additional towards making up the deficit of 1942.[22] Actually, Admiral Vickery figured that he had enough shipways to build 16 million tons in 1943, " if everybody came through." As he said, he " put that much margin in it." [23] The President had raised the objective for 1943 by 50 per cent when he named 15 million tons; the Commission similarly expanded the number of shipways by roughly 50 per cent

This fifth wave of expansion in facilities was so large as to make it extremely unlikely that the number of shipways would be the limiting factor in the program. The bottleneck was more likely to be steel or engines. One member of the Commission's staff

---

[20] Land to General J. H. Burns, Munitions Assignment Board, Mar. 30, 1942 in Land reading file; Vickery to Harry Hopkins, June 24, 1942 in gf 107-1, MC Minutes, p. 22178. The gain at Moore and Ingalls was one way in each yard.

[21] Additions totalled 99 but the loss of Seattle-Tacoma made the net gain only 91. The total planned was thus raised from 221 building berths to 312, counting Higgins as 44.

[22] Fischer, *Statistical Summary*, Table A-3.

[23] House, *Higgins Contracts, Executive Hearings*, p. 255.

remarked in a committee assigned to study shipyards that it
" would be wise to build a lot of shipyards, whether we can expect
production soon or not. In this manner we might escape criticism
through placing the blame on the steel bottleneck." To this
vision of shipways without any hulls on them, a section chief in
the Construction Division added the image of hulls without
engines. He " ventured that there would probably be many ships
afloat within the next few months without machinery." These
remarks were made under such circumstances that they may not
have been intended in all seriousness, but there certainly was a
feeling that steel and engines were now the limiting factors.[24]
The Commission was making quite sure that its program would
not fall short of the President's goals because of lack of facilities
in 1943.

To make assurance doubly sure, the Commission was improving
the equipment of the shipyards begun a year or more earlier as
well as constructing new building ways. For example, Oregon Ship,
which was leading the Liberty yards in speed records, had no
room for more ways, but was putting vessels together so fast that it
needed to enlarge the plate shop and to add an outfitting berth.
For similar additions to its facilities an expenditure of about two
and a half million dollars was provided by the Commission.[25]
During April, May, and June of 1942, the Commission authorized
far more expenditure on facilities than was required simply to pay
for added shipways.[26] See Figure 26. One reason for improvements
of this kind was that some of their good effect would be felt even
in the totals for 1942.

III. Politics and Administrative Methods in the Selection of Shipyard Sites

Having narrated the expansion of the Maritime Commission's
program to its maximum, we may appropriately turn back a little
at this point, back even to the period before Pearl Harbor, to

[24] " Minutes of the Shipyard Site Planning Committee," Mar. 23, 1942, exhibit in MC gf 507-2. The nature of this committee is explained in the next section.
[25] MC Minutes, pp. 21901-02 (May 12, 1942). For the approval of $1,029,962 for Richmond No. 1, " to permit the handling and erection of the larger unit assembly under the latest method of prefabrication . . ." see MC Minutes, p. 21270 (Mar. 3, 1941).
[26] Fischer, *Statistical Summary*, Table E-7.

SELECTION OF SHIPYARD SITES 151

round out the story of how locations and managements for shipyards were decided on. Selections have been briefly described, but rejections from an even larger part of the story. Very many sites were rejected which their sponsors thought just as good as the ones chosen. The whole process of selecting and rejecting was done under heavy political pressure, which was most evident in some cases of rejection.

As it became clear that new shipyards were to be financed by the government a torrent of letters poured in to the Commission offering sites. Some were from senators, governors, and representatives; some came on fine-looking letterheads and referred more or less vaguely to substantial backing; and others were longhand efforts calling attention to a bit of waterfront on which ships had once been built and which might as well, some neighbor thought, be used again. In January 1942 Admiral Land, in answering negatively one such personal letter, said there were 139 applications in the Commission's files from persons who wished to start new yards at the government's expense.[1] And there were many more letters too vague to be called applications, and many which were in behalf not of a particular company or site but in support of a locality. Labor leaders as well as senators, governors, mayors, and Congressmen voiced local desires to have the government spend money for a shipyard in their neighborhood.[2]

While some political leaders advanced only the claims of their state or city, others urged the consideration of particular companies or particular waterfront properties, and many callers came to the Maritime Commission's offices seeking facilities contracts. If they boasted of what they had done in World War I, Admiral Vickery confronted them with the records, which he kept on his desk, of what yards had built during that war.[3] After one group had left, Admiral Land dictated a memorandum for the files, " It is of interest to note that some parts of their proposal represent a stock selling proposition [and] that some of their personnel are to be drawn from going concerns. . . ."[4] The main argument against most of the proposals was that they could not provide adequate management, except by weakening established

[1] Land to Thomas H. Latta, Jan. 9, 1942, Land reading file.
[2] MC gf 507-1 is full of such letters.
[3] House, *Production in Shipbuilding Plants,* Executive Hearings, pt. 3, p. 908. Vickery's charts are in MC, Historian's Collection.
[4] Land reading file, May 10, 1941.

yards. Only a small percentage needed to be taken seriously, for only a few were based on study of the railroad and highway connections and estimates of the labor supply. A host of the inquiries came from persons who did not even know what factors to consider.

Having to reject so many applications, the Commission was exposed to the charge of playing favorites. Individuals who were sent away empty handed felt they or their communities were being discriminated against. There was talk about not having the right "contacts."

An embarrassing feature of the Commission's position was that it announced after each wave of expansion that no new shipyards were to be financed. These general grounds were given as the reason for rejections. Then when a new Presidential directive unexpectedly made necessary the financing of new yards, possible sites and managerial teams were immediately combed through again. When announcement was made that new yards were to be built it stimulated a new wave of applications which was again met with rejections on grounds that no more shipyards were contemplated.

An opportunity to air the feeling that the Maritime Commission was distributing shipyards with arbitrary discrimination was offered just after the second wave of expansion by the hearings of the Truman Committee on June 3, and July 9, 1941.[5] Three particular cases were discussed: the Cargill yard at Albany, the yard at Newburgh, New York, which had been used in World War I, and the plant of the Groton Iron Works at Groton, Connecticut. Newburgh and Groton had already been urged on the Commission and rejected because on the one hand they would have to be supplied with new equipment almost as much as would a new yard, and on the other hand they were in the area in which the Navy's building was concentrated.[6] The Navy said that for the Maritime Commission to go into Groton would simply draw workers from the work being done for the Navy at the Electric Boat Co.[7] In regard to Cargill, Commissioner Dempsey, who was testifying, had to admit the need to investigate.

[5] Truman Committee, *Hearings,* pt. 5, Commissioner Dempsey testified for the Commission June 3, pp. 1277-97. Unused shipyards are also discussed in Admiral Robinson's testimony on pp. 1420-24.
[6] *Ibid.,* pp. 1421-23.
[7] *Ibid.,* p. 1288. See also Admiral Land's correspondence in his reading file Mar.

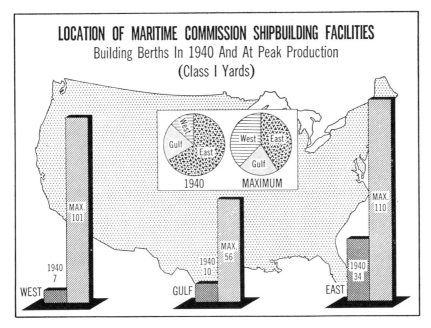

FIGURE 9. Number of building berths utilized for Maritime Commission shipbuilding on January 1, 1941 and maximum number. Source: Fischer, *Statistical Summary*, Table E-1.

Cargill, Inc., were grain merchants of Minneapolis. Their shipway was next to their grain elevator in Albany and had been begun for the purpose of constructing a vessel which would be particularly suitable for the carriage of bulk grain, but in the spring of 1941 they were finishing it as a 16-knot, 12,500 ton tanker. When Commissioner Dempsey referred the matter to Admiral Vickery, he at once sent inspectors to Albany and had the financial condition of Cargill looked into also. As a result of these investigations, Admiral Vickery asked them to make a definite bid on tanker construction. But at this point Cargill withdrew. They declared the labor situation was too uncertain and that they had been " thoroughly disappointed in their efforts to secure promises of early deliveries of machinery." The inter-

6, 14, 31, Apr. 10, and Aug. 2, 1941. In rejecting proposals regarding Groton Iron Works, Admiral Land referred to a list presented by Congressman W. J. Fitzgerald as " a phony list . . . at least 75% phony."

ested Senator and Congressman from New York were immediately informed of Cargill's withdrawal.[8]

One of the advantages often claimed for a Commission over a single administrator is that where responsibility is solidly placed on a Commission as a whole, there will be less suspicion of individual prejudice or favoritism. In the selection of shipyards some ill feeling was almost sure to be aroused by the rejection of so many projects and the approval of only a few. When some of the few went sour, the rejected could say that they could have done better. Accordingly it would be advantageous for the Commission to be able to say that all applicants had received due and impartial consideration. A committee reporting to the Commission would be a protection against charges that everything depended on having "contacts" with the right person.

A couple of weeks after the discussion in the Truman Committee of shipyard selections, the Maritime Commission created a Shipyard Site Planning Committee with Commissioner Woodward as chairman. The other members were L. R. Sanford, of the Construction Division, P. J. Duff, of the Plant Engineering Section, R. K. Chase, of the Legal Division, later replaced by Paul D. Page, Jr., John Slacks, special assistant to Admiral Land, and later Carl Carroll Perry, who handled much of the correspondence of the Committee. According to the administrative order creating it, the Committee was to review all projects and "determine the suitability of projects from the standpoint of geographical position, availability of labor, power and transportation, and the financial and technical ability, and experience of the applicants," and no projects were to come before the Commission unless previously reviewed by this Committee.[9] At its first meeting it agreed on a plan for routing all applications to one man and drew up a list of questions to be answered by applicants. During the next thirteen months the committee met nearly every Monday, Wednesday, and Thursday at 9 a. m., discussed 260 shipyards and to nearly all sent letters over Commissioner Woodward's signature explaining why they were rejected.[10] But not every proposal came to this Committee before

---

[8] MC gf 503-45, a relatively small file, gives this story on Cargill.

[9] MC Minutes, p. 17766 (June 19, 1941).

[10] The minutes of the committee meetings, from June 19, 1941 to May 14, 1942 are an exhibit back of gf 507-2. They record letters and memoranda approved and

SELECTION OF SHIPYARD SITES 155

going to the Commission. The proposals from Higgins and Bechtel, which have been mentioned, were invited by Admiral Vickery and negotiated and recommended by him directly to the Commission. Gradually it became clear that responsibility for the shipbuilding program was so concentrated on Admiral Vickery that decisions of the Shipyard Site Planning Committee were, practically speaking, recommendations to him.[11]

This subordinate role—less than was implied by the administrative order creating it—was not always accepted by the Committee without protest. In September 1941 it asked without success for a clarification of its position,[12] and two cases came up that month in which the view of the Committee, or at least of its chairman, did not prevail when matters were brought to a vote in the meeting of the Commission. Commissioner Woodward objected to a contract with the Leathem Smith yard at Sturgeon Bay, Wisconsin, because of the terms on which it was to be financed through the Defense Plant Corporation, a legal or financial question not central to the deliberations of the Committee.[13] He objected to the award of a contract to Savannah Shipyards on the general ground, which was central to the role of the Shipyard Site Planning Committee, that if contracts were to be given to a shipyard at Savannah, proposals of other concerns should have been considered.[14]

Savannah Shipyards, Inc., was indeed one of those doubtful and complicated decisions which was to cause trouble both politically and industrially. It was organized as an offshoot of the Empire Ordnance which was a company formed by Frank Cohen, previously a promoter of insurance companies in New York City, to take war contracts.[15] Mr. Cohen and his technical adviser had seen Admiral Vickery on March 27, 1941. There

opinions expressed by members. The record of opinions expressed is somewhat disjointed and does not inspire complete confidence. There is no evidence the minutes were read and approved. Commissioner Woodward summarized the Committee's work in a report to the Commission—Minutes, p. 22669 (Aug. 6, 1942).

[11] In some cases the Committee's action was, in form also, a recommendation to Commissioner Vickery.
[12] MC Minutes, pp. 18876-77 (Sept. 9, 1941).
[13] *Ibid.*, pp. 18793-95 (Sept. 4, 1941).
[14] *Ibid.*, p. 18796 (Sept. 4, 1941), but the memo is dated Aug. 22, 1941.
[15] Senate, *Investigation of the National Defense Program, Additional Report* of the Special Committee Investigating the National Defense Program, 77th Cong., 2d sess., No. 480, pt. 5, pp. 74-82. Cited hereafter as Truman Committee, *Additional Report.*

was not then and never was any question of the Commission awarding them a facilities contract so that Savannah Shipyards could build their plant with government money. Mr. Cohen's was one of the few proposals which did not call for government financing immediately. Instead, Mr. Cohen planned to finance his yard through the Savannah Port Authority. It owned the prospective site and it would issue bonds and use the money to pay Savannah Shipyards for the improvements they would be making on the leased property. All that was asked of Admiral Vickery in the interview of March 27, 1941 was that he assure the Savannah Port Authority that the site was from a technical point of view sound, and that " he was not opposed to the proposed personnel for the new yard." Perhaps, in view of what was to happen later, Admiral Vickery was to regret that he did not somehow stop Mr. Cohen's project in its tracks then and there. His reply was not that abrupt. As he had it carefully recorded, he answered that " as far as he could tell the proposals made were feasible and that we had no objection to Savannah Shipyards going ahead with arrangements and construction of a shipyard, although we are in no position to offer any assurance of a contract at this time." [16] It is noteworthy that Admiral Vickery did not specifically endorse the proposed personnel, and that the chief technical adviser of Mr. Cohen in this conference dropped out of the project shortly after.[17] But Mr. Cohen was able to start building his shipyard. Although he had no assurance of a contract, he counted on the view, generally agreed to at the conference, that no shipyard available in the near future was likely to be " a ghost town."

In August Mr. Cohen came back to the Commission seeking its aid in raising the money necessary to complete the shipyard, in which he said he had then invested about $600,000. He negotiated with Mr. Anderson, Mr. Schmeltzer, and Admiral Vickery.[18] Thereupon the Shipyard Site Planning Committee sent to the

[16] Memorandum, David E. Scoll to Commissioner H. L. Vickery, Mar. 28, 1941 in MC gf 503-40. In the same file under Mar. 27 is a report on Cohen by Scoll.
 This memorandum is the only case of which I have yet found record in which a careful summary of one of Admiral Vickery's many interviews was written up immediately afterwards.
[17] W. R. Crowley, president, Savannah Shipyards, to Land, Apr. 25, 1941 in gf. 503-40.
[18] Crowley to U. S. Maritime Commission Aug. 15, 1941 in gf 503-40, and various memoranda by R. E. Anderson, Aug. 14, through Sept. 12, 1941 in gf. 507-4-15.

SELECTION OF SHIPYARD SITES   157

Commission a memorandum setting forth its belief that contact should be made also with other groups interested in building a shipyard in the Savannah area. It said other persons had been told that no Federal funds would be available for such a shipyard. " It does not appear that the work which the above named company [Savannah] has done should place it in a different category from such other groups with respect to the application of this declared Commission policy." Accordingly, it was urged, the Commission should canvass carefully to see that it was dealing in the Savannah area with the " best possible available shipbuilding organization and would not run the risk of being manipulated into contracting with the above named company by reason of its having proceeded in apparent disregard of policy statements previously made." [19]

Admiral Vickery wanted the Savannah yard, however, because other yards were not available to fill the program for which Congress was appropriating that August. After a trip of inspection, Mr. Duff reported that he was much impressed with the progress made at Savannah since he had last visited it six weeks before.[20] The fact that they had actually spent some of their own money to start a yard seemed to Admiral Vickery and Mr. Anderson to put Savannah Shipyards in a different category from the other " interests " mentioned. The first proposal which they made to the Commission was withdrawn (made September 22, withdrawn September 26) [21] but a new proposal was approved on September 30. It required the Savannah Shipyards to complete the facilities before the Maritime Commission was obligated to make any payments.

So stringent were the clauses placed in the contract for the protection of the Commission that it was not accepted and signed by Savannah Shipyards until November 25, 1941, after it had been faced with a deadline. But it provided that the Commission pay a rent of 15 per cent a year by which the company would

---

[19] MC Minutes, p. 18796 (Sept. 4, 1941). The memorandum, which was dated Aug. 22, 1941, was merely " noted " by the Commission. For the previous discussion in the Shipyard Site Planning Committee see its minutes (cited above) for Aug. 20 and 22.

[20] *Ibid.*, minutes for Sept. 5 and Sept. 26.

[21] I have not found any text of this proposal. It is mentioned in the memorandum returning it. Secretary to the Committee on Awards, Sept. 29, 1941, gf 503-40.

gradually get back its investment in the plant, if it delivered ships; and Commissioner Woodward voted against it.[22]

Shortly after this, Empire Ordnance, Savannah Shipyards, and the whole network of Frank Cohen's companies were scrutinized by the Truman Committee in its investigation of lobbying. The charge had been made that Thomas Corcoran, intimate of the White House, was in the employ of Cohen and had used improper influence to get him the contract with the Maritime Commission. Admiral Vickery at once sent to the chief counsel of the Committee the notes of his meeting with Mr. Cohen on March 27, 1941, concluding:

> For your further information, I handled the entire negotiations with Savannah Shipyards myself and as you know I have kept your Committee informed of their progress. At no time did Mr. Corcoran ever participate in any meeting with the Savannah people at the Commission, nor did I learn of his association with Mr. Frank Cohen until several months after Mr. Cohen's first appearance.[23]

In the hearings on appropriations which were held that month and the next, Admiral Vickery and Admiral Land were called on to deny that the lobbying of Thomas Corcoran had any influence in the Commission either in connection with Savannah Shipyards or with Todd, by both of whom Corcoran had been employed. The report on the matter by the Truman Committee imputed no influence to Corcoran and no blame on the Maritime Commission. But some members of Congress voiced suspicions.[24]

In the case of Savannah Shipyards, Admiral Vickery notably made no attempt to use either the Commission or the Shipyard Site Planning Committee as a shield and buckler against political criticism. Indeed, the Committee was so much overruled and by-passed that it could hardly ever serve as a very effective defense, although sometimes it was called on in that connection. One case in which Commission and Committee action was carefully

---

[22] The copy in gf 507-4-15 is stamped "approved," but with "tentatively" written above in pencil. The conditions were set forth more fully in a letter, W. C. Peet, Jr., Secretary, to Savannah Shipyards, Inc., Oct. 11, 1941 in gf 507-4-15. The contract and record of its approval are in the same file.

[23] Vickery to Hugh A. Fulton, Dec. 16, 1941 in gf 507-4-15.

[24] House, *Independent Offices Appropriation Bill for 1943, Hearings*, Dec. 9, 1941, pp. 293-96; Truman Committee, *Additional Report* 480, pt. 5, Jan. 15, 1942, 77th Cong., 2d sess., pp. 84-90; House *Fifth Supplemental National Defense Appropriation Bill for 1942, Hearings*, Feb. 12, 1942, pp. 117-18.

recorded was in the first consideration and rejection of James Cromwell's proposal for a cooperative shipyard at Delair, New Jersey. It was a case which might become politically embarrassing since Mr. Cromwell was a man of wealth who had just been Minister to Canada and Democratic candidate for Senator from New Jersey. The shipyard was to be owned by the workers and was presented as an antidote to monopolistic tendencies. When Mr. Cromwell presented his plans in mid-January 1942, it was viewed favorably by some members of the Shipyard Site Planning Committee.[25] But Admiral Vickery frowned upon it. When Mr. Cromwell wrote Admiral Land protesting Admiral Vickery's rejection, Admiral Land wrote back, " Commissioner Vickery . . . will be the first to state that final action is not taken by an individual Commissioner but by the Commission as a whole." [26] The Shipyard Site Planning Committee drafted a letter rejecting the project for many reasons: first, because it was in the area of already heavily concentrated shipbuilding activities; second, because Cromwell's list of top personnel was only a plan, not a team of men accustomed to working together; and finally and at most length because the scheme for a cooperative and its financing did not seem practicable. Admiral Land softened the rejections as much as he could with a personal letter and the rejection was said to be the decision both of the Committee and of the Commission.[27]

After the President's call for expansion in February the Commission did not change its decision, although the Shipyard Site Planning Committee changed its attitude. Cromwell's staff had presented persuasive arguments that they could find labor and materials. Since Sun Ship was being expanded to 28 ways, it was hard to say that all expansion in that area was taboo. A report favorable to Cromwell's project was sent to the Commission in March but no action was taken.[28] Admiral Land wrote him again, after all the final wave of expansion had been completed, saying that no personal feeling against him was involved in the failure to take up his proposal, but that shipbuilding was one of the

[25] Minutes, exhibit to MC gf 507-2, Jan. 16 and 26.
[26] Land to James H. R. Cromwell, Jan. 30, 1942 in Land reading file.
[27] *Ibid.*, and MC Minutes, pp. 20962-64 (Feb. 3, 1942).
[28] Minutes of the committee, in MC gf 507-2, for Feb. 2, 11, 26, Mar. 5 and 16, 1942: MC Minutes, p. 21419 (Mar. 19, 1942).

industries least likely to succeed on a cooperative basis. Anyhow, there were to be no more new shipyards.[29]

In fact, as we have seen, the new yards needed to meet the Presidential directives were picked by Admiral Vickery, after consultation with Admiral Land, and Commission approval was close to being a mere formality. More than a year later, when testifying in a quite different connection, Admiral Vickery described part of the conference which took place at the White House during Mr. Churchill's visit at the end of December 1941. When the President asked if he could build the tonnage needed, Admiral Vickery replied, in part: " ' . . . can I select the managements and have these yards where I want them without interference from anybody? ' The President said, ' Did I ever interfere with you? ' I said, ' No you never did.' And he said, ' I have no intention of doing so.' "[30] Clearly Admiral Vickery felt that he personally had been given direct responsibility in the selection of shipyards, and the other Commissioners supported him in his assumption of that responsibility under the pressure of war. In February 1942 he became Vice Chairman of the Commission by election, succeeding Commissioner Woodward, who nominated Admiral Vickery for the post.[31]

Although the Shipyard Site Planning Committee did not have as weighty functions as its name implied, it performed those of a screening committee. It provided a formal and considered way of saying " No," or of keeping dubious cases under investigation. So many proposals were being put to the Commission that there was much to do in separating the ridiculous from the possible. Among the former, the most memorable is that of the man who wished government money to help him build ships in the eastern part of the state of Washington. When asked how he would send them to sea he said he planned a tunnel through the Rocky Mountains![32]

---

[29] Land to Cromwell, Apr. 25, 1942, in Land reading file.
[30] House, *Production in Shipbuilding Plants, Executive Hearings*, pt. 3, pp. 915-16, June 28, 1943.
[31] Minutes, p. 20965 (Feb. 3, 1942).
[32] Notes by F. C. Lane on interview May 6, 1947 with Carl Carroll Perry.

IV. THE COMMISSION UNDER THE WAR POWERS

As the story of the Shipyard Site Planning Committee illustrates, the strain of wartime duties decreased the importance of the Commission as a group and increased the importance of key Commissioners. There had been a tendency since the first creation of the Commission for each Commissioner to take one sphere of activity and be supreme within that field, but before the war both the law and the feeling of many Commissioners placed certain restraints on that tendency. During the war, the restraints were removed. In regard to carrying through the shipbuilding program, Admiral Vickery became the man directly responsible, the man running the program, while other Commissioners had other specialties.

This increased specialization within the Commission was partly a result of Presidential action and partly a matter of mutual agreement among the Commissioners. The increase came particularly in those fields in which Presidential intervention was real or potential. The First War Powers Act passed by Congress on December 18, 1941 enabled the President to take from the Commission and transfer to another agency many of its powers. Consequently, although he could not remove members of the Commission, he could determine what persons performed the functions hitherto held by the Commission. If the assignment of duties was not being arranged to his satisfaction within the Commission, he could at any time transfer some of its functions to the Office of Emergency Management where new agencies were rapidly being created on the eve of the war and were reshaped after Pearl Harbor to meet the many new difficulties which emerged as the nation moved into " the full tide of war." [1]

The Chairman of the Commission did not become entirely absorbed in any one specialty, for he remained the chief channel of contact with Congress and the President in regard to all the Commission's activities. But he was pulled away from shipbuilding by his appointment as Administrator of the War Shipping Administration on February 7, 1942. In creating the War Shipping Administration the President used his authority under the First War Powers Act to take away from the Maritime Com-

---

[1] *The United States at War*, pp. 104-6.

mission some of its former powers. Actually the same people kept on performing many of the same functions as before, since Admiral Land was at once made Administrator, and kept the WSA and the MC closely linked together. There were many joint services and much duplication of key personnel. Some new people with great influence, notably Lewis W. Douglas, took important administrative positions in the WSA, but it also required much of Admiral Land's attention.[2] Accordingly he had less time to handle the host of problems which were created by the acceleration of the shipbuilding program. They were left to Commissioner Vickery, whose rank in the Navy rose as his task increased in importance. A Commander when the emergency program started in 1941, Howard L. Vickery was named on February 4, 1942 to the rank of Rear Admiral.[3]

In organizing the WSA, Administrator Land made Commissioner Vickery the Deputy Administrator for Ship Construction within the WSA.[4] This served the purpose of tying the two organizations closer together. It also suggests that if the need of approval from the five-man Maritime Commission had at any time threatened to interfere with the production of ships, and if President Roosevelt had then used his war powers to transfer the function of ship construction to the WSA, Admiral Vickery could have continued in charge as WSA Deputy Administrator.

A third Commissioner clearly in charge of supervising a well-defined field was Captain Edward Macauley. When he became a Commissioner in April 1941, it was understood between President Roosevelt and Admiral Land that Captain Macauley would handle problems of maritime personnel,[5] and after the formation of the WSA, Captain Macauley became one of its Deputy Administrators. As such he supervised recruitment and manning, training, and labor relations of seamen without need of referring matters for even formal approval by the Commission. As Commissioner he supervised the Division of Shipyard Labor Relations until in October 1942 it came under Admiral Vickery.[6]

[2] See below, chap. 23, sect. II.
[3] Information from MC personnel files.
[4] WSA Administrative Orders Nos. 1 and 2.
[5] Land to the President, May 6, 1942, in Land file.
[6] *Ibid.*, WSA Admin. Order No. 2, supplements 3 and 3A and below chap. 9, sect. I, note 4.

THE COMMISSION UNDER THE WAR POWERS 163

Neither of the two civilian Commissioners had so large and well defined a sphere of action to call his own as did the naval officers. Commissioner Woodward had been especially interested in hearings on freight rate agreements, a type of work which markedly declined.[7] He was charged particularly with the handling of the many problems concerning railroads and their freight rates which arose in connection with the new shipyards, as well as with the screening work performed by his Shipyard Site Planning Committee.[8] The newest member of the Commission in 1942 was John M. Carmody who took office December 1941 to fill the vacancy caused by the resignation of John J. Dempsey. Almost immediately he took over the supervision of real estate transactions which had been done by Commissioner Dempsey.[9] He showed at once the interest in personnel management, public power, and a number of other topics with which he had had experience, but he had had no previous special contact with shipbuilding, shipping, or the special problems of the Commission. He became more influential in the Commission some years later.[10]

The tendency for the Commission to split up into a series of bureaus headed by the Commissioners was quite in accord with the general tendency to call for "czars" in the administration of war economy. But in the spring of 1942 there was talk of by-passing Admiral Land and Admiral Vickery and putting in someone else as a "czar" of shipbuilding. The committee created by the Senate under the chairmanship of Harry S. Truman to investigate the industrial aspects of mobilization for war was favorably impressed with the plans and progress of the Maritime Commission, but not everyone was equally satisfied. After the Truman Committee had released its report giving the Maritime Commission "a pat on the back," as Admiral Land said, "the Commission had been out in front to be shot at by everybody." "We are perfectly prepared to understand what is going to happen," he said, "and my head may be in the basket before the summer is over on account of it."[11]

---

[7] Notes by F. C. Lane on interview with Admiral Land, Sept. 30, 1946.
[8] MC Admin. Order No. 62, Jan. 22, 1942, Minutes, pp. 22345-47 (June 30, 1942).
[9] He countersigned memoranda to the Commission on this subject, see Minutes, p. 22336 et passim.
[10] See below, chap. 23.
[11] House, *Hearings on H. R. 6790 to Permit the Performance of Essential Labor*

One of the shots fired at the Commission in April 1942 was the criticism by a Senate subcommittee and by the columnist Walter Lippmann that no fair consideration had been given a strange new type of vessel called the *Sea Otter*. It was conceived as "a tin can with an outboard motor on it" while a retired naval officer on active duty and an automobile engineer were having luncheon. It was to use gasoline engines and so avoid one of the bottlenecks; it was to be small and of such light draught that torpedoes would miss it; and it was to be of the simplest construction so that it could be scrapped once it reached Britain (the pressure for shipping was all for outbound space). The idea reached and interested the President; and Roland Redmond, husband of Sara Delano, first cousin of the President, raised enough money to construct a model. Developing the project further was turned over to Joseph W. Powell. Mr. Powell was a most experienced shipbuilder, having been in charge of the shipbuilding division of the Bethlehem Steel Co. during World War I, and in World War II he was in the office of the Secretary of the Navy as a special assistant in charge of shipbuilding. He was admittedly skeptical about the *Sea Otter*, but while it was being built and tested he arranged with E. B. Germain to organize a private company, Ships, Inc., which was to be financed by the Reconstruction Finance Corporation in accordance with Mr. Roosevelt's wishes.[12] Admiral Land was one of the directors of Ships, Inc.[13] When the *Sea Otter* was subjected to trials, an adverse report on it was submitted to Admiral Land by R. T. Merrill, General Manager of the Merchants and Miners Transportation Co. Mr. Merrill concluded that even if the deficiencies apparent at the trial were rectified, and there were many, the Liberty ship was superior. The Liberty, he said, used "78% as much material to build; is built about two and one-half times quicker; costs 70% as much; transports over three and one-half as much cargo per member of the crew and two and one-half as

on *Naval Contracts Without Regard to Laws and Contracts Limiting Hours of Employment, to Limit the Profits on Naval Contracts, and for Other Purposes*, before the Committee on Naval Affairs, 77th Cong., 2d sess., pt. 2 (No. 207), Apr. 13, 1942, pp. 2976-77; Truman Committee, *Additional Report* 480, pt. 5, p. 90.

[12] "The Sea Otter, A Noble Experiment," a manuscript by Blanche Coll in the MC, Historian's file 111, which makes full use of the Senate Hearings, and papers in Land's file, Apr. 1942 and in MC gf 130-688.

[13] He answered the letters which he received about the *Sea Otter* by referring the authors to Mr. Germain. See Land reading file, Sept. 19, 1941 through Feb. 6, 1942, or gf 130-688.

much per gallon of fuel; and costs about one-ninth as much fuel to operate." [14] Mr. Powell, Admiral Land, and Secretary Knox were all unfavorable. Ships, Inc., decided not to build. But the original conceiver of the project, Comdr. Hamilton V. Bryan, felt he had not been given a fair deal, and the report released by the Senate, April 1, 1942, supported his view, citing particularly the way Commander Bryan, instead of being asked to help in making his idea work, had been forbidden to go near the shipyard where the *Sea Otter* was being constructed. Admittedly that was the responsibility of Mr. Powell and the Navy, not of the Maritime Commission, but question arose as to why development of a potential cargo ship had been turned over to the Navy. The Senate subcommittee and others who asserted that 'a promising new idea had been killed by the " brass hats " blamed the Maritime Commission also for lack of interest in a ship that was conceived as the solution of an admittedly difficult shipping problem.[15]

Criticism concerning the *Sea Otter* was not in itself to be taken very seriously, but about the first of April the shipbuilding program was not going well, and consequently the Commission was exposed to any attack on grounds that it had " missed the bus." On February 12, Admiral Land admitted " there might be some slight falling down in 1942," [16] and again on April 13 he said building was behind schedule.[17] At a meeting of the War Production Board on March 24, Mr. Harrison " stated that completion of more than 6,500,000 deadweight tons cannot now be forecast with certainty, and that remedial measures will be necessary in order to complete 7,500,000 deadweight tons." Its statistician Stacy May reported at the April 14 meeting that March delivery of ships was equal to February's but below schedule, below even the reduced schedule for March.[18] Unnamed members

[14] Land to Senator Josiah W. Bailey, Mar. 10, 1942, and Land to Secretary of the Navy, Mar. 25, 1942, with attachments, in gf 130-688 and Land reading file.
[15] Coll, " The Sea Otter," cited above.
[16] House, *Fifth Supplemental National Defense Appropriation Bill for 1942, Hearings*, Feb. 12, 1942, p. 108.
[17] One shipyard executive was reported as saying this was the first intimation he had had from official quarters that the schedule might not be met. *The Journal of Commerce*, Apr. 14, p. 3; Apr. 15, p. 24. Land's statement called for a 13 per cent speed-up. See House, *Hearings on H. R. 6790*, Apr. 13, 1943, p. 958.
[18] Civilian Production Administration, *Minutes of the War Production Board, January 20, 1942 to October 9, 1945* [Historical Reports on War Administration:

of the War Production Board were quoted in the press as being critical of the way the shipbuilding program was developing: there was not enough subcontracting, not enough pre-fabrication outside of the shipyards, insufficient use of the facilities of the structural steel industry which had been enlisted in shipbuilding in World War I. Some critics said the Commission was adhering too much to the building of standard cargo ships instead of concentrating on the emergency.[19] Even President Roosevelt, although not criticizing the Commission at all, in the survey of the war production program which he gave at a press conference on April 25 said that the only lag of any consequence was in shipbuilding.[20]

Among the reasons for this lag were factors which the Commission could not control. On April 23 Admiral Land explained the program and the relative position of the yards to the Truman Committee. When asked how the Committee could help speed the program, he mentioned first the need for more steel and second the importance of freezing labor-management relations.[21] Labor agitation, loafing, and lack of steel were difficulties for which the Commission did not feel to blame. But even if it had good alibis, the Commission was open to attack as long as it seemed unlikely to deliver the ships.

A move to appoint a shipbuilding czar, namely Joseph P. Kennedy, appeared in the newspapers.[22] It was mentioned as early as April 12, and commented on by Arthur Krock in the *New York Times* on the twenty-second, and in the *Journal of Commerce* on the twenty-third. The climax was the speech in favor of a " ship production administrator " made by Joseph W. Martin, Jr., Chairman of the Republican National Committee, at the Saturday luncheon of the Women's State Republican Club of New Jersey. " Some efficient capable man of proved experience like former ambassador Joseph P. Kennedy should be made head of such a division," said Mr. Martin, " and he should be given full and complete power to bring every shipyard, large and small, in the

WPB Documentary Publication No. 4] (Government Printing Office, 1946), pp. 33, 48.

[19] *Journal of Commerce,* Apr. 23, 1942, p. 20; Apr. 25, 1942, p. 1. These were partial anticipations of the Ezekiel Report which is discussed below in chap. 11, sect. V.

[20] *New York Times,* Apr. 25, 1942.

[21] Truman Committee, *Hearings,* Apr. 23, 1942, p. 5193. On labor difficulties at this time see chap. 9.

[22] MC Maritime News Digest, Vol. VI, No. 95.

United States into the picture to produce boats." [23] The *Journal of Commerce* said Mr. Kennedy was willing but had not been asked.[24]

A full month before this newspaper publicity, the connection of Joseph P. Kennedy with the shipbuilding program had been explored in an exchange among Mr. Kennedy, Admiral Land, and President Roosevelt. Mr. Kennedy had written the President from Palm Beach a general offer of his services and received a nice note in reply suggesting he talk to " Jerry " Land. In telephone calls the Chairman proposed to the Ex-Chairman of the Commission that he (a) organize, administer, and manage a new shipyard; (b) strengthen the organization and management of one of the weak yards, or (c) become the chairman of a labor board if one like the " Macy Board " of the last war was created. But no proposal was formulated which interested Mr. Kennedy. He then wrote the President again on March 12, 1942 saying that he had watched the shipbuilding program carefully and while it had done well to date, it might run into serious difficulty by the middle of the summer. The President's letter, said Mr. Kennedy, aroused in him the hope that they were contemplating divorcing the shipbuilding construction from all the rest of the shipping program and just making that a full-time job. Before this letter reached the President, he and Admiral Land had discussed the matter at a conference on March 14, and when Mr. Kennedy's letter arrived President Roosevelt sent it to Admiral Land, saying on the routing slip: " What do I do about this? My personal slant is that you offer him a specific, definite job: (a) To run a shipyard; (b) To head a small hurry-up inspecting organization under Vickery to iron out kinks and speed up production in all yards doing Maritime Commission work." Admiral Land's reply reviewed his previous proposals, said he had discussed the matter with Vickery, and continued:

> As time for developing new shipyards will not wait, we have gone ahead and have practically completed all new shipyard possibilities that have any real merit and any additional shipyards now in any locality whatsoever will, in our best judgment, not help the situation if for no other reason than we believe that steel is going to be the No. 1 bottleneck and propulsive machinery No. 2. In other words, the capacity of the country has been reached so far as shipbuilding is concerned unless we seriously encroach on the Navy program.

---

[23] *New York Times*, Apr. 26, 1942 (I), p. 35.
[24] *Journal of Commerce*, Apr. 23, 1942.

There are some weak shipyard organizations and managements. At this time the most outstanding example is the two yards at Portland, Maine. These are under the control of the Todd-Bath interests, and it is our understanding that Joe is a large stockholder in Todds.[25] The opportunity therefore exists to pull these yards out of the doldrums and put them on a basis of production comparable with leading yards on the West Coast and the Bethlehem yard at Fairfield, Maryland. . . .

With regard to divorcing the shipping program and the shipbuilding program, we believe that it is extremely difficult to separate the Siamese twins (Vickery-Land) as the shipping program is becoming more and more dependent upon the shipbuilding program.[26]

The public criticism of the Maritime Commission during the following month also quite failed to separate the " Siamese twins (Vickery-Land) ." Criticism from the War Production Board was chronic and was an expected part of the phase of war economy through which the country was passing in 1942; but talk of a shipbuilding czar disappeared as it became clearer that for all practical purposes Admiral Land had already developed such a " czar " for merchant shipbuilding in the person of Admiral Vickery, a man hitherto little in the public eye but becoming better and better known as shipbuilders made and broke new records under his goad and guidance.

V. Decentralization of Supervision

While far reaching decisions concerning new sites and new managements were being made at top levels, routine supervision was threatening to become unmanageable. The task of supervision would have swelled immoderately in the early months of 1942 even without the post-Pearl Harbor waves of expansion, for the new shipways planned earlier were now coming into production. When, in addition, new shipyards were planned and old ones enlarged as a part of the fourth and fifth waves of expansion, there was that much more need for hull inspectors, machinery inspectors, construction auditors, disbursement agents, expediters, schedulers, and so on. In the same months the staff concerned with ships at sea was assuming larger duties as part of the War

[25] Cf. *Congressional Record*, Mar. 22, 1937, pp. 2554-57. Kennedy's holding of 1000 shares of Todd was declared by a special resolution not to interfere with his appointment—*New York Times,* Mar. 18-31, 1938.

[26] Land to the President, Mar. 25, 1942, carbon signed also by Vickery, with attachments, in MC, Historian's Collection, Land file.

Shipping Administration. Very extensive administration reorganization had to be carried out in the midst of day-by-day decisions.

The main burden of decisions concerning construction, as well as part of the burden of supervision, fell on Admiral Vickery. He was working at a terrific pace which awed and inspired his subordinates but undermined his health and was to lead to a serious heart attack in September 1944. He carried the whole program in his head and had an intimate knowledge of all major yards as a result of his frequent tours of inspection. Under Admiral Vickery, detailed supervision of the construction of all yards and ships was gathered in the Construction Division, for the reorganization of July 1941 had placed it in charge of inspecting the older long-range yards as well as the new emergency yards. The Director of the Division, J. E. Schmeltzer, was also responsible for scheduling the production of all the yards and supplying them with materials.[1] His work piled up at a killing pace—more and more yard plans to be approved, more and more staff to be recruited and organized, more and more bottlenecks to be broken. In February 1942, Mr. Schmeltzer's health failed, and although he recovered enough to act as technical assistant to Admiral Vickery from August 4, 1942 to February 25, 1943, he was then stricken with a fatal heart attack while returning from inspection of the Providence shipyard.[2] Thus Mr. Schmeltzer's noteworthy career in modern merchant shipbuilding, which began in the drafting rooms of the shipyards at Chester, Pennsylvania, and included prominent participation first in the design of the standard types and then in the launching of the emergency program of multiple production, was closed by death in line of duty.

Shortly after Mr. Schmeltzer's illness in February 1942, the Construction Division was broken up and the supervision of the shipyards was decentralized into four Regional Offices of Construction, to which were assigned also important functions from the Technical Division. The burden of work at Washington had become insufferable; offices were overflowing with mail and the city itself was overflowing with persons employed by the many new wartime agencies. It became more and more difficult to handle from Washington alone the letters from the inspecting

---

[1] Admin. Order No. 37, Suppl. No. 42, July 25, 1941.
[2] Information from Personnel Division files.

staffs in all the yards. Moreover, in order to overcome their production difficulties, these yards began making requests for permission to depart in this or that particular from the working plans given them, and the key men of the Technical Division were soon spending most of their time on the road going to yards to pass on such changes on the spot. This kind of " spot " approval could not be given in the more distant yards although it was much in demand. Sometimes contractors called by telephone members of the Technical Division and secured authorization for changes of which the Construction Division's inspectors in the yards had no knowledge.[3] Additional expansion of the Washington staff to manage procurement and allocation was clearly in sight and would make the Construction Division even more unwieldy. An administrative center was needed near the yards with concentrated authority over inspection, plan approval, and minor changes.

Accordingly the Commission created by administrative orders of March 19, 1942 four Regional Directors of Construction, namely:[4] for the East coast, at Philadelphia, J. F. McInnis; for the Gulf coast, at New Orleans, L. R. Sanford; for the West coast, at San Francisco, C. W. Flesher; and for the Great Lakes, at Chicago, W. E. Spofford. See Plate XXV.

These four men, whose powers immediately made them second to none in importance among the officials under the Maritime Commissioners, were all " career men," promoted from within the Commission's own staff. All four had received professional education in naval architecture and marine engineering. The senior both in age and in years of service with the Maritime Commission and its predecessor, the U. S. Shipping Board, was Mr. Sanford. He had been the Section Chief in charge of inspectors within the Technical Division and then within the Construction Division. Mr. Spofford had been many years a civilian employee of the Navy before coming to the Maritime Commission in 1937. He was out of Commission employ from 1939 to 1942 working with the Consolidated Steel Co. in its new shipyard near Orange, Texas, and was called back by Admiral Vickery at the time of the decen-

---

[3] Director, Construction Division, to Director, Technical Division, Jan. 27, 1942, and Director, Technical Division to Acting Director, Construction Division, Feb. 12, 1942, both in gf 201-2-11.

[4] Minutes, pp. 21424-28 Admin. Order No. 37, Suppl. 58, named Baltimore as the location of the East coast office, but Philadelphia was decided on.

tralization. Mr. McInnis came to the Commission in 1938 and became chief of Hull Plan Approval Section in the Technical Division when Mr. Esmond left that position to work on the Liberty ship. The youngest of the new regional directors was Mr. Flesher, an Annapolis graduate who came to the Commission after about ten years with Westinghouse, and moved up to be chief of the Engineering Design and Specifications Section in the Technical Division and then, when Mr. Schmeltzer was taken ill, Acting Director of the Construction Division.[5] In general, the Regional Directors may be characterized as former section chiefs of the Technical Division.

Although procurement was kept centralized in Washington, nearly all the elements in the Commission concerned in any other way with ship production were represented on the staff of each regional office. The Finance Division was decentralized by creating Regional Construction Auditors, and Regional Disbursement Auditors.[6] A Regional Attorney and a Regional Industrial Adviser were also appointed.[7] The attorneys and the auditors were under the Regional Director "in administrative matters," but were "functionally" responsible to the Legal and Finance Divisions in Washington.

A tug of war between the Regional Offices and the Divisions located in Washington began almost from the moment that decentralization was ordered. Even in the two functions which were their basic reason for being, plan approval and the supervision of inspectors, the Regional Directors felt their authority threatened. To the Technical Division, in which they had but recently been section chiefs, was reserved authority over any changes in plans and specifications which would "affect design." How to draw the line and decide which changes did or did not affect design was a subject of controversy. To the Production Division, which was created out of all that had remained in Washington of the former Construction Division, was given authority over scheduling and regulating the flow of materials. It proposed to place representatives in the shipyards to send in

[5] Information from Personnel Division files. For biographical sketches of the Regional Directors and several other officials of the Commission, see "Key Men of the Maritime Commission," *Marine Engineering and Shipping Review*, Nov. 1942, pp. 121-27.
[6] Admin. Order No. 37, Suppl. 55, Mar. 19, 1942; Minutes, pp. 21428-33.
[7] Admin. Order No. 37, Suppl. 60, Apr. 3, 1942; Admin. Order No. 56, Mar. 19, 1942.

progress reports and check receipts. This moved Mr. Sanford to ask Admiral Vickery to reiterate that the Regional Offices were not parts of the Production Division, nor reporting to it. Admiral Vickery agreed that the Regional Offices should report only to him and should not be subordinate either to the Production Division or the Technical Division.[8]

Confusion over administrative jurisdictions was a natural result of the upheavals which affected the Commission's staff during the spring of 1942. At the same time that decentralization was being carried out in the divisions concerned with construction, the creation of the War Shipping Administration was shaking up drastically the rest of the staff and bringing in new people. The divisions which were kept most intact, Personnel, Information, Legal, Finance (in part), and the Secretary's Office, became known as the Joint Services and were subjected to the strain of serving two masters, the War Shipping Administration and the Maritime Commission.[9] It was the period when campaigns in both the Pacific and the Atlantic were going against the United States and all the war agencies were organizing or reorganizing for more intense efforts. There was patriotic energy and good will and also confusion. A suggestive touch is the plaintive letter sent out by the newly appointed Director of the Production Division to all Regional Offices saying that three bound volumes of very important ship cost data had disappeared and asking for a search to see if the books could be found anywhere among the boxes shipped out from Washington.[10]

By the summer of 1942 the turmoil of reorganization was subsiding. The separation of the WSA and the decentralization of the Maritime Commission staff had been carried through. The staff to supervise the new forms of shipbuilding contracts had been organized in such a way as to centralize procurement and decentralize the supervision of construction. This plan was adhered to throughout the war. All the emergency yards had been started and the corporations to manage them had all been selected. Shipbuilding programs had been expanded by the fourth and fifth waves to such a point that there were no big bursts of expansion thereafter.

[8] Sanford to Vickery, July 24, 1942, with longhand note by Vickery, in gf 201-3-3; and below, chap. 21.
[9] See below, chap. 23.
[10] W. F. Rockwell (by Barnes) to Flesher, June 10, 1942, in gf 506-1.

CHAPTER 6

EXCESS CAPACITY AND THE CANCELLATION OF THE HIGGINS CONTRACT

I. THE INCLUSION OF MILITARY TYPES AND THE SIGNS OF OVEREXPANSION

LOOKING BACK at the period when both the Navy and the Maritime Commission launched their emergency shipbuilding programs, Admiral Cochrane, Chief of the Navy Bureau of Ships, reminisced in regard to the expansion then contemplated: " It called for results which a great many individuals of long experience in the shipbuilding industry of this country characterized as utterly impossible of accomplishment. Even those of us who participated in the hurried planning and inceptive stages of the program shared something of this pessimism. . . ."[1] As late as April of 1942 the shipbuilders under contract to the Maritime Commission were on the whole behind schedule, and officials of the WPB calculated that the goal of 8 million tons in 1942 would be missed by a wide margin. Very little was said of the Presidential directive for 9 million tons in 1942, for even Admirals Land and Vickery did not hope to build that much. The best they hoped for as their goal in March and April 1942 was to come close to 8 million. The political attacks on the Commission at that time drew their force in considerable part from the fact that the performance of the new yards did not as yet give much hope that even the 8 million would be built.[2] Of the need there was no doubt. Submarine sinkings were alarmingly high. It was in this atmosphere of desperation that the Commission awarded contracts for the large number of new shipyards started during the fifth wave of expansion.

In May and June the picture changed. Emergency yards where

---

[1] Rear Admiral Edward L. Cochrane, " Shipbuilders Meet the War Challenge," *Marine Engineering and Shipping Review*, Dec. 1944, p. 149.

[2] See above, chap. 5, Sect. IV. All tonnage figures in this chapter are in deadweight tons.

CONSTRUCTION TIME OF LIBERTY SHIPS, AVERAGE OF ALL YARDS
Average Number Of Days From Keel Laying To Delivery For Vessels Delivered Each Month, January 1942 – June 1945

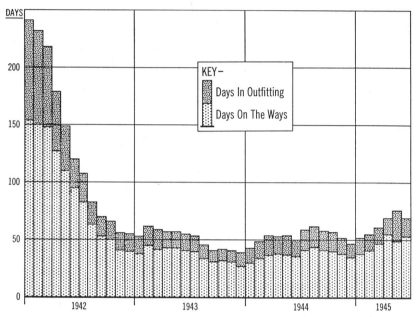

FIGURE 10. Average for each month for vessels delivered in that month. Source: Fischer, *Statistical Summary*, Table D-1, figures for Standard Liberty Ships.

the steam shovels had started work fourteen months before had now completed their facilities, built two or three rounds of ships off each way, and in the process of building learned how to do it faster. May 22, 1942 was Maritime Day and all the Commissioners were called on to make speeches. They spoke not of alibis but of achievements, of records broken, and of how much shipbuilding might exceed its schedule. Admiral Vickery cited one West coast yard that had on the previous day set a world's record by delivering three cargo ships: one built in 60 days, another in 65 days, and the third in 70 days—far ahead of the schedule of 105 days set by the Commission. " Assuming we get a break on the intangibles, by the end of 1943 American shipyards will be able

CONSTRUCTION TIME OF LIBERTY SHIPS IN SELECTED YARDS
Average Number Of Days From Keel Laying To Delivery For Vessels Delivered Each Month, December 1941 — March 1944

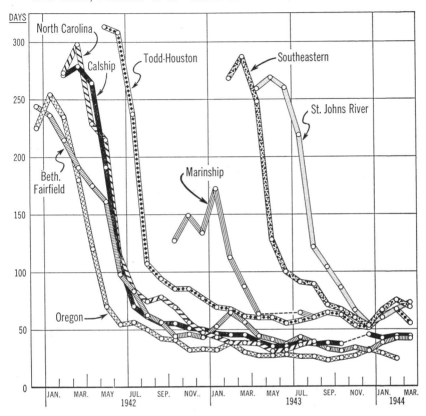

FIGURE 11. Source: MC, Division of Economics and Statistics, Chart 203, January 29, 1945, data from Production Division.

The first dot on each line shows the construction time of the first ship delivered from that yard. In every yard the construction time dropped rapidly after the yards had been delivering ships for about three months. The first group—Bethlehem Fairfield, Oregon, North Carolina, Calship, and Todd-Houston—all received contracts in January 1941. After April 1942, Oregon led the others in reducing construction time. Not shown on the chart but also in the lead was Richmond No. 1 (Todd-Cal), which was building ships similar to the Liberty ship for the British.

to turn out 28,000,000 tons," he said.[3] The two-year objective set by the President in the bedroom conference in February was 24 million tons. By May it was possible to think realistically of exceeding even this "terrific" objective.

Prospects continued to brighten during June. Whereas the total number of ships of all types delivered was 26 in March and 36 in April, it was 57 in May and 67 in June. The average building time of a Liberty ship, from keel laying to delivery for all yards, was 207 days in March, and 177 days in April; it was 151 days in May and 118 days in June. It was of course much faster in the yards which had been in operation for some time. At Richmond No. 1 the average building time of vessels delivered in June 1942 was only 69 days, at Oregon 54 days.[4] More important even than such figures was the ability of Admiral Vickery and the others in charge of the program to see from the way work was being organized in the fabrication shops and the assembly platens that the increase in speed would continue. They could foresee that the acceleration which had already occurred in the yards first built would soon be matched by a similar acceleration in yards more recently constructed. Calculation of this kind, based on the progress evident in June 1942, showed that the fifth wave of expansion provided facilities in excess of what was needed to reach the goal of 24 million tons of merchant ships for 1942-1943.

This excess capacity seemed like a godsend when it enabled the Maritime Commission to undertake the building of some military types which were suddenly found to be urgently needed and for which the Navy could find no room in the yards under contract to it. On May 28, 1942, Bethlehem-Fairfield and Kaiser Inc. were told to use for the building of landing ships (LST's) yards previously intended for Liberty ships. The plans being made for the invasion of Europe in 1943 gave priority to these tank-carrying landing ships, and it was fortunate that Maritime Commission yards could find room for them. Then at the beginning of June plans were made to devote the new Kaiser yard at Vancouver, near Portland, Oregon, to the construction

---

[3] MC, Maritime News Digest, Vol. VI, No. 121; *New York Times*, May 23, 1942, pp. 1, 3.

[4] Fischer, *Statistical Summary*, Tables B-2, and D-1 figures for "all Liberty Ships"; Reynolds' Way Charts.

ACTUAL SPEED OF PRODUCTION COMPARED WITH SCHEDULED SPEED
Time On The Ways Of Liberty Ships Launched Each Month Compared With Scheduled Time, September 1941 — April 1943

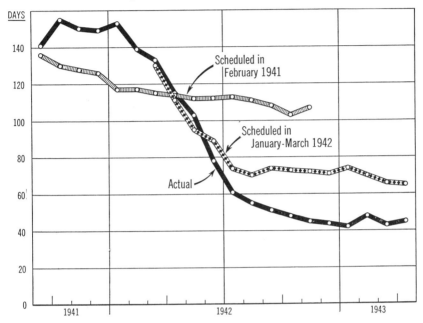

FIGURE 12. Source: MC, "Permanent Report of Completed Ship Construction Contracts," and the "work sheets" cited above p. 140, note 10.

of "baby flattops" as soon as it finished work on the LST's.[5] This further diversion of facilities from merchant ship construction was dictated by the obvious need of more escort vessels with which to combat the submarine menace. Finally, a third demand for the use of Maritime Commission facilities to build military types was voiced during the visit of Winston Churchill later in June. Admiral Ernest King, Chief of Naval Operations, was persuading the President and the Prime Minister of the necessity of building many small escort vessels similar to the Canadian corvettes. Since sinkings in June 1942 were at the new high level of close to 900,000 tons a month,[6] Admiral King's arguments had

---
[5] See below, chap. 18.
[6] *WSA Shipping Summary*, Vol. II, No. 1, p. 18 (January, 1945).

great weight. Admiral Land became concerned lest too many shipways be diverted from merchant shipbuilding to the building of escort vessels, and he requested President Roosevelt to call a conference of all concerned.[7]

This conference, which met on June 23, was the starting point of an intense conflict lasting about three weeks. The issue in the conflict was not, however, whether the Maritime Commission should build this or that military type; the issue was dividing up the nation's limited supply of steel. The battle for steel was a long fight which will be recounted as a whole in Chapter 10. The heated struggle between June 23 and July 10, 1942 was only an episode in this battle, but it had an immediate and sensational effect on the Maritime Commission's plans.

The way the issue arose gives importance to the conference held by President Roosevelt and Prime Minister Churchill on June 23, 1942. After Admiral King presented his arguments for giving priorities to escort vessels, Admiral Vickery stressed the point that steel was the limiting factor. It would not be too hard to find shipways on which to build the contemplated corvettes, but where was the steel to come from? Would it be taken from other Navy building, or would it be taken from merchant shipbuilding? The problem of allocating steel could hardly be gone into thoroughly at this meeting for it contained no representatives of the War Production Board, which as successor of the OPM was in general charge of industrial mobilization. There were five admirals present but the only civilians besides the Prime Minister, President Roosevelt and Harry Hopkins were men concerned with shipping—Sir Arthur Salter and Lewis Douglas. Consequently the President asked that a conference of those who were concerned with steel be held Thursday, June 25. According to Mr. Douglas' notes, the President said the assignment of the conference on the twenty-fifth was to report on " how escort vessels could be built and how the steel plate facilities would affect the merchant vessel construction program. He [the President] said he was sure that the bottom of the barrel could be scraped for steel plate."[8] The President thus lifted again the objective set for the shipbuilding of the Maritime Commission.

---

[7] Land to the President, June 19, 1942, in MC, Historian's Collection, Douglas files, " Construction."

[8] Douglas' notes on the conference and copies of his letters and those of Land concerning it are in his " Construction " folder.

How definitely President Roosevelt set a specific figure higher than 24 million tons as the objective of the Maritime Commission's shipbuilding at this time has been questioned. It appears that he did not do it in writing.[9] But there seems no doubt that he verbally urged Admiral Land to shoot at higher goals.[10] Even the figure of 24 million when it was set in February was viewed not as a limitation but as a spur to achieving the impossible. By June it was clear that "the impossible" and more too could be done if the steel were provided. Admiral Land felt that he had received either on June 23 or in earlier conferences the President's authorization to build up to the limit according to the new estimate of the capacity of the Maritime Commission's yards.

When the conference to scrape the bottom of the barrel for steel met on June 25 as the President had asked, Admiral Land proposed a plan based on his interpretation of the President's wishes. I have no evidence of anyone contesting his claim that the President had authorized maximum production of merchant ships.[11] But there were plenty of other questions. Were the escort

[9] The Commission's annual "Report to Congress" for fiscal 1943 (multigraphed), p. 3, says that at the start of the fiscal year covered by that report a further directive was given the Commission to add "another 2,890,000 deadweight tons. These expansions brought the total program to . . . 18,890,000 deadweight tons to be built in 1943." This is the basis for the statement in Bureau of the Budget's *The United States at War*, pp. 139-40: "and by June 30, 1942, a further directive has been given the Maritime Commission to add another 3,000,000 deadweight tons" (to the 24,000,000). On the other hand, there is no directive to that effect in the "Program Collection" compiled by the Maritime Commission's staff, and the Congressional committee which investigated the cancellation of the Higgins contract concluded: "Your committee is unable to find any formal directive for 29,000,000 tons of ships, but it is the opinion of your committee that there was considerable discussion of reaching the goal of 29,000,000 tons if the necessary steel could be provided." House, *Investigation of Cancellation of Higgins Contract, Interim Report*, from the Committee on Merchant Marine and Fisheries, Nov. 24, 1942, pursuant to H. Res. 281, 77th Cong., 1st sess., Report No. 2652, p. 3. The figure in the annual report is similar to the figure of 2,889,000 tons which the President did specifically authorize, but on Dec. 7, 1942. See chap. 10.

[10] Wayne Coy, Acting Director of the Bureau of the Budget, in his memorandum to the President on July 6, cited in note 19 below, says (p. 10): "I understand that you subsequently [subsequent to giving the directive for 24 million tons] instructed the Maritime Commission to build all possible merchant ships in 1942 and 1943."

[11] Douglas wrote of Admiral Land's first proposal of a statement to the conference: "I rather fear that your presentation of the case, however accurate a description of what occurred at the White House, will inspire immediate opposition on the part of the Army, the Navy, and every other interested party attending the conference."— Douglas' file, June 24. On June 25, 1942, after the conference, William L. Batt of the WPB wrote Admiral Land explaining how much an allotment to the Maritime Commission of more steel would upset all other programs, but did not question that the President had authorized such an allotment. Copy in Land reading file, June 26, 1942.

vessels to be built in yards and with steel previously intended for merchant ships, and so reduce the output of merchant tonnage? How much steel did the Maritime Commission have to have? Where was the steel to come from? Would the building of battleships or cruisers be postponed, or some other program? Putting it crudely, did the President know what he was doing when he called for a maximum output of merchant ships? Would he want to reconsider the decision when he realized its consequences? Such questions prevented any agreement that week on a report to the President.[12]

In one respect there was definite progress towards agreement during the later part of June, namely, in laying statistical bases for decisions of policy. Agreement was reached as to what ships were being counted when millions of tons were hurled back and forth in discussion—whether military types were included or excluded. While Admiral Land was conferring in the White House and the office of the Secretary of the Navy, his statisticians reached an understanding with their opposite numbers in the War Production Board (the WPB) and the Bureau of the Budget as to what ships their totals included.[13] If military types other than the corvettes were included, the total was close to 29 million tons. According to Admiral Vickery's estimates of probable speed

[12] Memoranda of the period June 24-July 10 in Land reading file and Douglas file " Construction."

[13] The WPB began issuing in April 1942 its monthly collection of official figures on the production of munitions, in accordance with the specifications laid down by the Advisory Committee on Official Production Statistics, but not until issue No. 3, that " as of June 1942," did it contain a complete picture of the Maritime shipbuilding program.—*Official Munitions Production of the United States . . . as of June 1, 1942*, pp. 107-11. This was based on an official schedule of the Commission dated June 19, 1942, prepared by Gerald J. Fischer, who had been on the staff of the WPB before he joined the staff of the Maritime Commission on May 8, 1942. It was the basis of a schedule of June 23 which provided for building aircraft carriers instead of Liberty ships at Vancouver. This June 23 schedule differed very slightly from a schedule dated June 25 prepared for Admiral Vickery by Norman Farmer directly on the basis of Admiral Vickery's estimates of the number of ships each yard would turn out. Mr. Fischer worked in consultation with Mr. Farmer and his schedule of the 23rd was the basis for reaching, by an adjustment, the figures in the memorandum of July 6, 1942 from Wayne Coy, Acting Director of the Bureau of the Budget, to the President, cited below in note 19.—Work sheets of Mr. Farmer and Mr. Fischer in MC, Historian's file, Statistics, and their oral testimony. Of this letter from Mr. Coy, Admiral Land said that it had been recommended in a conference with the Navy and Mr. Nelson, and that it was " considered that this was the best presentation of facts and figures that had ever been compiled with regard to the Maritime Commission and the shipbuilding programs."—Land to the President, July 11, 1942, Land reading file.

PROGRAMS FORMULATED IN JANUARY-JULY 1942
For Tonnage To Be Delivered In 1942-1943

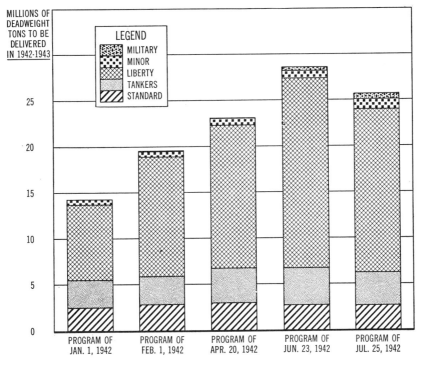

FIGURE 13. Source: Fischer, *Statistical Summary*, Table A-3.

of production, the yards working for the Maritime Commission could build 27,400,000 tons of shipping, exclusive of military and minor types, by the end of 1943.[14]

Admiral Land's goal then was to secure a sufficiently large allotment of steel for this tonnage of merchant ships. Building non-mercantile types for the Navy became a sort of side issue which was easily settled, the Navy building most, but not quite all, of the corvettes in its own yards.[15] The issue narrowed down to whether the War Production Board should allocate to the Maritime Commission additional steel to build 3,400,000 tons of

---

[14] This figure, which is in Land's letter of July 11 to the President, is the total for major types in the June 23, 1942 program.

[15] See below chap. 18.

merchant shipping in addition to the 24,000,000 tons ordered by the President in writing in February, so that the total for merchant ships would be 27,400,000 tons.

Unless the Maritime Commission obtained the steel for this additional 3,400,000 tons of ships it could not keep all its yards working at full capacity. It faced the extremely disagreeable possibility of having to cancel some of the new yards or of reversing its whole attitude towards the shipbuilders and telling them to slow down. To anyone who had his eyes fixed on the shipping situation, as did Admiral Land as War Shipping Administrator, or Lewis Douglas as Deputy Administrator, such a possibility was shocking. The shortage of cargo carriers was such that no one could question the desirability of launching as many as possible. In the *New York Times* of June 14, 1942, Admiral Land had called for the maximum possible production under a headline, " Ships, Ships, and More Ships," " We have too few; we cannot have too many." [16]

Donald Nelson, Chairman of the War Production Board, appreciated the needs of other users of steel, as was natural in view of his official position. Before the Truman Committee on June 25 he pointed out that with the speed of construction then being attained, it was necessary to review the schedules for shipbuilding to see whether they were not too high, since demands for steel plate were running 300,000 tons over production. The shipbuilders, those working for the Navy as well as those working for the Maritime Commission, were stepping up their output faster than were the steelmakers. Someone would have to get along with less steel than he hoped for.[17]

Donald Nelson had not been present at the White House conference of June 23 at which Admiral Land believed he received the Presidential directive to build additional merchant tonnage, but on July 4 Mr. Nelson discussed with the President the supply of steel plate, with particular relation to increasing the shipbuilding program.[18] Another presentation of the problem came to the President July 6 from Wayne Coy, Acting Director of the Bureau of the Budget. In an 18-page letter expounding at length the needs of the Navy, the need for cargo carriers, and

---

[16] *New York Times*, June 14, 1942, Magazine Section.
[17] Truman Committee, *Hearings*, part 12, pp. 5232-36.
[18] Nelson to Land, July 8, 1942, in MC, Program Collection; *Minutes of the War Production Board, January 20, 1942 to October 9, 1945*, pp. 88-89.

the amount of shipping which would be produced by proposed programs, he stated that the shortage of steel made imperative Presidential decisions: "If that portion of the over-all shipbuilding program . . . [as scheduled by the Maritime Commission] is to be accomplished as scheduled, the entire deficit of 560,000 tons of steel plate must be borne by programs other than shipbuilding. . . . It is imperative at this time for you to indicate your desire as to the carrying out of schedules set forth above or whether curtailments or modifications of such schedules should be made." [19] Conferences discussing steel were held in the office of the Secretary of the Navy both July 7 and after a cabinet meeting July 9. On leaving the former conference, while Mr. Nelson drove him back to the Commerce Building, the Chairman of the Maritime Commission urged on Mr. Nelson the need of more steel for shipbuilding. Later that night or the next night the Chairman called Mr. Nelson and told him that the largest of all the Commission's contracts, that with Andrew Higgins, might have to be cancelled if more steel was not allotted to Maritime.[20] On the morning of the ninth, Admiral Land and Mr. Douglas breakfasted with Harry Hopkins and set forth at length the urgent need for more ships in view of the high rate of sinkings. Mr. Hopkins was quite won and said he would try to have the President sign the necessary papers before nightfall.[21] But a little later that same morning, Admiral Land had a letter from Donald Nelson stating that the President had determined that "our total program of merchant ship completions for the years 1942 and 1943 should be 24 million deadweight tons, of which 8 million or slightly more, if possible, are to be completed in 1942." Next day, Admiral Vickery attended a meeting at the War Production Board at which Mr. Nelson expounded the steel shortage and the need of cutting the program of the Maritime Commission.[22]

[19] Memorandum for the President, from Wayne Coy, July 6, 1942, in Land file. A copy with "classification canceled" is in MC, Historian's Collection, Research file 206.1.1.
[20] Testimony of Donald Nelson cited in "Report of the American Federation of Labor Committee appointed by President William Green to investigate the cancellation of Higgins Corporation shipyard and shipbuilding contract," inserted in *Congressional Record*, Senate, Mar. 30, 1943, vol. 89, pt. 2, p. 2696.
[21] The memorandum prepared for this conference, demonstrating with colored charts the need of shipping, and notes on the conference are in Douglas file, "Construction," July 9, 1942.
[22] Nelson to Land, July 8, 1942, original in MC, gf 107-55, stamped "received 7/9,"

If the Presidential decision stated in Mr. Nelson's letter of July 8, 1942 proved final, then the Maritime Commission had to cut back its program. It moved to do so by a drastic step, one sure to stir a storm of criticism. On the afternoon of July 10 it voted unanimously to cancel the Higgins contract.[23]

II. Plans and Progress at the Higgins Yard

To appreciate the anger, the grave concern, and the confused outcries aroused by this decision, and to weigh the reasons for the decision, it is essential to consider the place which Mr. Higgins and his shipyard had won in the shipbuilding program and above all in New Orleans.

The choice of the Higgins company to receive a contract during the fifth wave of expansion was a natural one. Most of the contracts awarded at that time went to men known as good managers but without previous experience in shipbuilding. Andrew Jackson Higgins was a personality fabulous for his bounce, his unprintable eloquence, the faith which he aroused among his workers, and the gusto with which he and his four sons tackled whatever was presented to them as "impossible." The most striking of the stories of his achievement concerned his construction of lighters for the Navy early in 1941. Two days after a call from Washington asking for plans, Higgins had ready a model, and two weeks later the finished products were in Norfolk, having been painted enroute on flat cars so that delivery would be on time.[1] Although Mr. Higgins' fame as a manufacturer rested on his design and speedy production of small craft, his experience was closer to the shipbuilding industry than that of most of the new managerial talent enlisted by the Commission in the spring of 1942. Moreover he was on the Gulf, the area to which the Commission felt obliged to turn because of the pressure on the labor supply in both the Northeast and the West coast and because of President Roosevelt's interest in industrializing the South. As a leading Southern industrialist, a success in handling labor, a boat builder

and Land to the President, July 11, 1942, in MC, Program Collection. Nelson's letter says that its purport had already been telephoned over. It reads as if the decision it records was made by the President on July 4.

[23] MC Minutes, p. 22454.

[1] Herman B. Deutsch, "Shipyard Bunyan," *Saturday Evening Post*, July 11, 1942, pp. 22 ff.; "The Boss," *Fortune*, July 1943, pp. 101 ff.

and a go-getter, Andrew Jackson Higgins was an obvious man for the Maritime Commission to turn to when confronted with the President's "terrific" directive of February 1942.

In building small boats, Mr. Higgins had applied the principles of production on an assembly line in somewhat the same way in which they had been applied in the automobile industry. He dreamed of applying the same principles to the production of larger ships. On New Year's Eve of 1941 he called a consulting engineer, Walter B. Moses, about laying out a shipyard on an assembly-line basis.[2] The fundamental idea was to have moving platforms marching down to the sea carrying first the midship section, which would be assembled at the head of the line complete with the main engine and boilers, and finally all the hull as it was added section by section. There were to be four assembly lines, two on each side of the huge sheds in which plates and shapes would be fabricated and pre-assembled. The workmen in the fabricating and assembly shops would stay in one place doing one kind of work and their product would be put together so as to be added to the ship in large sections as it went by, moving down the assembly line to the launching basin.[3]

These plans were in embryonic form when, near the end of February 1942, Admiral Vickery looked over the Higgins boat factory, was very favorably impressed by it, and raised with Mr. Higgins the possibility of his building Liberty ships. Admiral Vickery later described the negotiations as he saw them. "At first Mr. Higgins was not very much interested in it, but at the end of the talks why I think I got a little under his hide because

[2] House, *Higgins Contracts, Public Hearings* before the Subcommittee of the Committee on the Merchant Marine and Fisheries (Unrevised Committee Print) 77th Cong., 1st sess., on H. Res. 281, part 2, Aug. 3, 1942 (cited hereafter as House, *Higgins Contracts, Hearings*), p. 215.

[3] A rough plan of Higgins' assembly line shipyard, dated Feb. 10, 1942, so rough there is no assurance it was designed for Liberty ships, is in Weber files, Designs, EC2's, Higgins Shipyard. C. D. Vassar has among his papers at the Maritime Commission a more detailed plan dated Mar. 16, 1942. The fullest explanatory description of the yard that I have seen is the testimony of A. J. Higgins and Frank Higgins, Truman Committee, *Hearings*, pt. 14, pp. 5691-94, 5704. The equipment planned for the yard is itemized in the appendix to the recommendation of the award to the Commission, which is in gf 507-3-29, Mar. 11, 1942. The largest cranes provided for were 30-ton whirleys. A general description of the Higgins yard is in gf 506-1, Vickery (dictated by Martinsky) to Walter P. Abel, Apr. 27, 1942. An aerial view of the yard and surrounding swamp June 30, 1942, showing the Vickery, Land, and Carmody canals or basins, as well as the canal connecting the yard to the sea, is in gf 503-54 under date of July 16, 1942.

I questioned his ability as to whether he thought he could do it or not, whereupon he took the bit in his teeth and found he could do it." [4]

The plan Mr. Higgins submitted formally to the Commisioners and which they approved March 13, 1942, called for the construction of 200 Liberty ships, 5 in 1942, the rest in 1943, in a yard which was then counted as the equivalent of 44 ways because it was designed to contain that many ships all at one time in various stages of completion before launching.[5] In arranging for an appropriate amount of working capital it was considered only a 28-way yard,[6] and the latter figure was more frequently used in comparing it with other yards, although it was admitted that the plan of the Higgins yard was so novel that its size could not be described in customary terms.[7] The only important basis for comparison was how fast it would turn out ships, and that no one could say for sure until it was in full operation. Admiral Land felt that the whole project was a fifty-fifty chance.[8]

One of the most attractive features of Higgins' proposals was his plan for training and getting maximum efforts from previously unused labor, especially Negroes. One pair of assembly lines would be all white laborers, the other pair all colored, with the huge assembly and fabrication sheds between. "This will result," Mr. Higgins wrote, "in a competitive spirit between each of the lines, white to white on adjoining lines and black to black on adjoining lines, and a healthy and patriotic competition will be insured between the two production lines separated by the fabricating shops." He was also operating, or planning, separate training schools, one for white, one for colored. "We have agreed with the principals of the organized crafts that there will be white foremen and supervision." [9]

Mr. Higgins' plans had enthusiastic support from the New Orleans leaders of the AFL. With the promise or prospect of large future employment for many skilled workers, they were

[4] House, *Higgins Contracts, Hearings*, p. 254.
[5] MC Minutes, pp. 21377-81 (March 13, 1942).
[6] House, *Higgins Contracts, Hearings*, testimony of R. E. Anderson, p. 307.
[7] *Ibid.*, Land's testimony, pp. 93-94; Truman Committee, *Hearings*, pt. 14, pp. 5724, 5729 ("I don't know any two men who could agree on whether it is a 28-way yard or a 44-way yard or a 4 assembly yard.")
[8] House, *Higgins Contracts, Hearings*, p. 82.
[9] A. J. Higgins to Vickery, Mar. 12, 1942, in MC gf 503-54.

THE HIGGINS YARD

ready to relax some of their restrictions on the number of beginners he could employ in relation to journeymen.[10] They submitted telegrams to be used by Mr. Higgins to support his bid for a contract. W. F. Donnels, AFL organizer, gave assurance: "There will be no stoppage of work for any reason. . . . Get this contract and we will show the world how to do the impossible in turning these ships out."[11] Mr. Higgins was holding out a vision of New Orleans as the future shipbuilding center of the world. His plans called for developing a new section and dredging new canals—one might almost say replanning the city. He boasted that he was opening vast new opportunities for Southern laborers who were suffering from lack of opportunities. His energy and the war, which was going very badly for the United States in March 1942, stirred intense emotional support for his imaginative proposal. When he came back from Washington successful, with the biggest of all contracts, New Orleans literally turned out the band and mobbed the railroad station. He was escorted home beneath banners proclaiming the pledge, "No strikes," "No stoppage," and listened to the following poem recited by its author Holt Ross, regional director of the International Hod Carriers, Building and Common Laborers' Union.

> Higgins Ships
> There'll be no loafing no delays;
> These ships must move on down the ways.
> Let's keep smiling. Speed up workers,
> We will tolerate no shirkers.
> The crucial need just now is ships.
> Let's build them fast and fill the slips.
> Then load them up with planes and tanks
> And also brave determined Yanks.
> Swing the gantries, hang the plate,
> For in your hands now rests the fate.
> Fire the forge now, light the torch,
> Uncle Sam is on the march.
> Papa Higgins has the plans
> The Federation has the men
> With Higgins ships we'll fill the ocean
> Through hard work and true devotion
> There's no time for speculation,

---

[10] A. J. Higgins to Vickery, Feb. 28, 1942, in MC gf 503-54.
[11] W. L. Donnels, Organizer, AFL, to A. J. Higgins, Mar. 10 and 11, attached to Higgins' letter of Mar. 12, 1942, in MC gf 503-54.

> Act right now and save the nation.
> Not two hundred ships but three,
> Anything to keep us free;
> For there'll be no joy or mirth
> If the Japs control the earth,
> And if Hitler rules the waves
> The workers here will all be slaves.
> So let's build the ways and turn out ships
> With no time lost and without slips
> Then Uncle Sam and our Allies afar
> Will know Higgins men help win this war.[12]

This warm labor support might be valuable to Mr. Higgins both in his shipyard and in a political way. He took pains to see that both President and Mrs. Roosevelt were aware of it. He suggested that his labor plans and the telegrams supporting them be submitted to the President with the suggestion that Presidential comment could be used to arouse enthusiasm in the Higgins plant.[13] Admiral Land accepted this suggestion. His letter to President Roosevelt enclosing the telegrams not only explained the plan to use white labor on one side of the fabricating shops and Negro labor on the other, each group having two assembly lines, to encourage a "healthy and patriotic competition," but explained the general layout of the yard, which was, the latter said, "somewhat of a departure from standard shipbuilding." In signing the letter, Admiral Land put a star opposite "somewhat," and wrote underneath, "This is where Vickery & I go overboard without a parachute!"[14] The President responded expressing his particular interest in the method of employment which had secured the approval of all groups.[15]

The social implications of Mr. Higgins' plans were stated by him in their broadest terms in connection with the Southern Conference for Human Welfare which was attended also by Mrs. Roosevelt. Mr. Higgins wrote its executive secretary, Mr. Dombrowski, a letter of which he sent a copy to Admiral Vickery saying that he had been "rooked into attending" but that the latter part of his letter would appeal to the President. Typical of its concluding paragraphs was the sentence: "Yes, we are

[12] Clark Salmon to Vickery, Mar. 16, 1942, in gf 503-54. Punctuation added.
[13] A. J. Higgins to Vickery, Mar. 12, 1942, in MC gf 503-54.
[14] Land to the President (dictated by Weber) Mar. 18, 1942, in Weber files, Designs, EC2's, Higgins Shipyard.
[15] Franklin D. Roosevelt to Land, Mar. 21, 1942 in MC gf 503-54.

going to build ships, by the strength of Americans, and we are going to tap that great reservoir of unskilled Negro labor, and by such educational training as we can give, equip him to do the job, with equal pay for the same work, and equal opportunity for advancement as Americans and for America." [16]

Mr. Higgins' energetic confidence in his ideas, as well as the size of his contract and the novelty of the method he proposed, gave his project national significance. He was claiming that once he got going he would produce more rapidly and more cheaply than all the other builders, and the Maritime Commission was helping him to get going. One of the very first ways in which the Commission helped him was to arrange for a tour by his son Frank Higgins, and his chief engineer Walter Moses, to inspect the Bethlehem and Kaiser yards building Liberty ships. The Higgins representatives were helped especially by the Richmond yard which furnished them with many drawings.[17] Another appeal to the Commission produced rapid action by Admiral Land with the Chief of Army Engineers as a result of which the Army dredges, the only ones quickly available, worked not only on the canal leading to the Higgins yard but on the yard itself.[18] Another successful appeal to the Commission produced action from the Department of Justice to break up what Mr. Higgins called an agreement among pile-driving contractors to quote high prices.[19]

In their Maritime Day speeches of 1942, members of the Maritime Commission pointed wih pride to the Higgins yard and spurred it on. Speaking at San Francisco, Captain Macauley while praising the accomplishment of the West coast yards pointed also to New Orleans where " four shipways are being laid down, each a mile long, from which would come 200 ships." [20] At New Orleans itself Commissioner Carmody " took Mr. Higgins up on the

[16] A. J. Higgins to Vickery, May 12, 1942 and attached letter of Higgins to James A. Dombrowski, in MC gf 503-54.

[17] Higgins to Weber, Mar. 2, 1942, and Weber to Higgins, Mar. 4, Higgins to Clay P. Bedford, Mar. 19, and Bedford to Higgins, Mar. 24, 1942, all in MC gf 503-54. House, *Higgins Contract, Hearings,* pp. 170-71.

[18] Dalton B. Shourds, resident plant engineer, to Land, Apr. 15, 1942, and Land to Shourds, Apr. 16, 1942, in gf 507-3-29.

[19] A. J. Higgins to Weber, May 14, 1942, in MC gf 503-54 and in Weber files, Designs, EC2's Higgins shipyard and, attached to this later copy the routing slip with message from Paul D. Page, Jr. to Admiral Vickery.

[20] Press releases cited in MC Maritime News Digest, vol. VI, No. 121a.

mountain and showed him the pearly gates." [21] Against the grim background of Allied defeats, Mr. Carmody said to Mr. Higgins " You have taken on a new job, a big job, the biggest shipbuilding job of all time in one yard. . . . The eyes of the shipbuilding world are upon you; the eyes of the fighting men of the United Nations are upon you. . . . What you do here may well mean the difference between success and failure. . . ." [22]

III. The Decision to Cancel

Nevertheless there were a number of reasons why the Higgins contract should be singled out for cancellation when it became clear that there was not going to be enough steel plate to keep all the shipyards working full speed. There was of course some skepticism concerning the ultimate success of his novel assembly line, but this skepticism was never endorsed by the Commission. Once having backed Andrew Higgins, the Commission officially maintained that his scheme was sound. What certainly did worry many of the Commissioners was the mounting cost. The first estimate approved by the Commission was for $25 million. In July the estimate was $59 million.[1] Moreover, it was discovered that an expensive housing project would be necessary in addition in order to attract workers to the yard. That came as a disagreeable surprise since the Commission had originally expected that a yard near a big city like New Orleans would not have that problem.[2] Another special difficulty was the need of constructing a new power plant to supply the yard.[3]

In the skyrocketing of his costs for the completion of his facilities, Mr. Higgins' yard was not so very different from many others,[4] and when Admiral Vickery told him that his fees would have to be cut in view of the higher cost of his facilities, " Mr. Higgins said he did not give a damn about the fee. It was perfectly all right with him for me to cut it to anything I saw fit." [5]

---

[21] The phrase is Congressman Herbert C. Bonner's, House, *Higgins Contracts, Hearings*, p. 239.

[22] Carmody, " Two-Fisted Shipbuilding," MC Press Release, Text 37.

[1] Vickery's testimony, House, *Higgins Contracts, Executive Hearings*, pt. 3, p. 258.

[2] Sanford's testimony, *ibid.*, pt. 2, pp. 143-45; Land's testimony, Truman Committee, *Hearings*, pt. 14, p. 5721.

[3] MC Minutes, p. 22361 (July 2, 1943) and below in chap. 23.

[4] See below, Table 9.

[5] Vickery's testimony, House, *Higgins Contracts, Executive Hearings*, pt. 3, p. 258.

Whether the Higgins yard when finished would have been the most expensive of all the yards can not really be determined because that would have depended on how many ships he could turn out per dollar invested in facilities.

But most of his cost had not been expended. There was $25 million to $30 million yet to be spent.[6] Moreover, Higgins would be the last "to really come into production." Original plans called for his delivering a ship in September, but by June it was clear his first deliveries would be delayed. It was doubted whether he would deliver any ships in 1942.[7]

A number of other yards, also in the South, which had been started in the last wave of expansion would similarly not come into production until 1943, but none of these other yards would have so voracious and insistent an appetite for steel as would Higgins' huge plant. As Mr. Higgins insisted when he first took the contract, his assembly-line methods required that all his material and components be on hand when wanted. Otherwise the whole assembly line would be stopped. In an ordinary yard the lack of particular kinds of steel plates or shapes might stop work on one ship; in the Higgins assembly line it would stop work on all the ships in the line.[8] Exploring the possibility of decreasing Mr. Higgins' need for steel, and also his cost for facilities, Admiral Vickery discussed with him in early July whether his yard could be replanned as two assembly lines instead of four, but no satisfactory half measures for reducing the Higgins costs and steel consumption were found.[9] Even if all the new 6-way yards in the South were cancelled there would probably not be enough steel to keep Higgins going at the rate he claimed he could build. At the same time that he was deciding to drop the Higgins contract, Admiral Vickery was planning to stabilize the other shipyards at 70 or 80 per cent of capacity, "so that without the Higgins yard, I had a 20 per cent margin in case anybody found materials to increase production. . . ."[10] If at any later date shortage either of materials or of manpower necessitated further contraction, one or two yards could be closed and

---

[6] Of the 29,000 tons of steel to be used in his facilities, only 16,000 were already shipped to him or ready to ship. Truman Committee, *Hearings*, pt. 14, p. 5699.
[7] House, *Higgins Contracts, Executive Hearings*, pt. 3, pp. 255, 261.
[8] A. J. Higgins to Vickery, Feb. 28, 1942, in MC gf 503-54.
[9] House, *Higgins Contracts, Executive Hearings*, pt. 3, p. 258.
[10] *Ibid.*, p. 259. On the attempt to stabilize see chap. 10, at the end of sec. II.

the others kept open.[11] Keeping a number of 6-way yards, instead of the Higgins "28-way" yard, would give more flexibility.

All these considerations entered into the decision as Admiral Vickery discussed the problem with his regional director for the Gulf and with Admiral Land on the morning after the arrival of Mr. Nelson's letter, and with the full Commission that afternoon, July 10, 1942.[12] Basically the Higgins yard was picked for cancellation because it involved the largest amount of unexpended steel and material, and it would be the last one to come into production. Admiral Vickery put it to the Commission on that basis. As he said soon after, "I told them of the political effects which would follow; that it was going to make a bad situation, but in my judgment for that amount of tonnage we would be expending Government funds without justification if we continued the project, because we did not need it with the amount of capacity we had and the production we were getting."[13] The Commission voted unanimously to authorize cancellation but directed the Chairman to ascertain the wishes of the President. According to the Minutes of the Maritime Commission: "The Chairman then called the Honorable Marvin McIntyre and discussed the matter over the telephone, and Mr. McIntyre asked that action be withheld until he had communicated with the President. Mr. McIntyre later called the Chairman and advised that the President was not prepared to authorize the suspension or cancellation of the Higgins contracts and requested that a memorandum of facts together with the Commission's recommendations be submitted to him at once."[14]

In the requested memorandum, Admiral Land reviewed for

[11] Statement by William A. Weber in interview, May 7, 1946.

[12] The morning conference was described by Mr. Sanford in his testimony at New Orleans Aug. 3, before the House investigating committee, *Higgins Contracts, Hearings*, pp. 111, 124.

The most complete statement of the factual considerations entering into the decision to cancel are in Admiral Vickery's testimony before the same committee on Aug. 20, 1942 (*ibid., Executive Hearings*, pt. 3, pp. 258-261). At one point Admiral Vickery said, "I must say nobody thought out the cancellation of the Higgins contract except myself. I discussed it with nobody until I went in to the chairman and discussed it with him."—*Ibid.*, p. 260. Carmody's testimony at the same hearing confirms the important role of Vickery in the decision (*ibid.*, pp. 234-37); but it seems, in view of the testimony of Nelson and of Admiral Land, that the Chairman was making up his own mind on the same solution.

[13] House, *Higgins Contracts, Hearings*, p. 259.

[14] MC Minutes, p. 22454 (July 10, 1942).

DECISION TO CANCEL 193

three pages some of the many conferences and memoranda which had culminated in " the reduction of 3,400,000 deadweight tons of shipping—from 27,400,000 to 24,000,000," and pointed out that it meant " scrapping of plans for delivery of 315 Liberty ships in 1943," although the saving in steel was a very small percentage of the steel-plate production. The Higgins plan for an assembly line producing ships was declared expensive but well engineered; " it is feasible and will produce ships at a rapid rate." He concluded: " The Commission will cancel the Higgins contract as soon as possible unless such action is contrary to your directives. The matter is most urgent as any delay involves financial losses to the Commission if the contract is to be canceled. Your instructions are respectfully requested at the earliest possible date." [15]

Admiral Land had not even at this late date stopped fighting for more steel. He followed his letter to the President with a letter to Donald Nelson and a copy of it to Harry Hopkins. While writing to Nelson mainly about specific details of the steel allocation, he frankly stated that he was " disposed to continue to press you and the President for a reconsideration of the directive contained in your letter of July 8." [16]

A week passed and no word came from the White House. On July 18, Land sent a briefer memorandum to the President saying again that the Commission was unanimously in favor of cancellation of the Higgins contract, that it was strongly recommended by Mr. Nelson, and that delay was very costly. " We have no idea of ' passing the buck ' to you," wrote Admiral Land to the President, " but assume full responsibility; however, in view of the seriousness of such action, I do not wish to have our actions reversed by you should you feel that our action was either incorrect or contrary to your desires." Later that same morning Mr. McIntyre telephoned a " go ahead." [17] The Commission met, voted unanimously to proceed with cancellation, and approved a press release saying that curtailment of shipbuilding was made necessary by lack of steel.[18]

Not till then was Mr. Higgins told. Although he had been in

---

[15] Land to the President, July 11, 1942, in MC, Program Collection.

[16] Land to Nelson, July 14, 1942, in Land file attached to letter of July 22. See also, Chaikin and Coleman, *Shipbuilding Policies*, pp. 43-44.

[17] According to Miss Eleanor H. Van Valey's note on the carbon copy in the general files, 107-1, July 18, 1942.

[18] MC Minutes, pp. 22522-24 (July 18, 1942).

Washington during the crucial days of decision, from July 9 to 11, no hint was given to him. On Thursday the ninth, the day that Admiral Land and Mr. Douglas explained to Mr. Hopkins at breakfast the need for more ships, Mr. Higgins spent most of the day with Admiral Vickery going over the possibility of reducing his costs by cutting his assembly lines from four to two. On the eleventh, Mr. Higgins, still in Washington, spoke with Admiral Land about another matter.[19] The Commission having voted and put the cancellation up to the President, nothing was said to Mr. Higgins about it. He went back to New Orleans and there was called at 11:30 on Saturday morning July 18 by Mr. Sanford and told that the Commission had cancelled his contract.[20]

IV. INQUEST POST-MORTEM

Mr. Higgins answered with a bellow from the bayous that reverberated for weeks through the corridors of Congress and was echoed in the headlines from coast to coast. He felt personally insulted and hurt, as he hotly protested, because the Commission had not called him in and told him of their decision and the reasons for it.[1] During the ensuing public uproar the Commission never gave reasons for not thus advising Mr. Higgins. The Commission took full responsibility and made no mention of referring the matter to the White House.[2] What might have happened if Mr. Higgins had been told that cancellation was being considered, instead of being confronted with an accomplished fact, is suggested

---

[19] House, *Higgins Contracts, Hearings*, pp. 112-13; Truman Committee, *Hearings*, pt. 14, p. 5710; *Congressional Record*, vol. 89, p. 2700 (Mar. 30, 1943).

[20] Truman Committee, *Hearings*, pt. 14, p. 5710. Mr. Sanford was informed by Admiral Land by telephone at 10:30 that same morning.—House, *Higgins Contracts, Hearings*, p. 124.

[1] House, *Higgins Contracts, Hearings*, pp. 34, 51, 138-40; Truman Committee, *Hearings*, pp. 5709-10.

[2] To be sure, at the start of the House hearings, Congressman F. Edward Hebert stated that he understood from the Commission's General Counsel, Wade Skinner, that it had been waiting for a week for the White House to tell it to go ahead and cancel (House, *Higgins Contracts, Hearings*, pp. 3-6), but Congressman Boggs said his understanding from the conversation with Mr. Skinner was that the question which had been submitted to the White House was "whether or not they had to abide by the ruling of the War Production Board" on steel, not the specific Higgins contract,—*Ibid.*, p. 5. During Admiral Land's testimony, at least in the published parts, he was not asked specifically about this point, and in his general statements he assumed full responsibility. Truman Committee, *Hearings*, pt. 14, p. 5715; House, *Higgins Contracts, Hearings*, p. 75.

by the amount of political pressure he brought to bear within a week after being told. President William Green of the AFL called at the White House next day.[3] Public interest was the greater because an article extolling Higgins as the Paul Bunyan of the shipyards had appeared just a week before, July 11, in the *Saturday Evening Post*. Resolutions were introduced into both houses of Congress to hold up the execution of the cancellation until after a Congressional committee had investigated. Although both resolutions were killed,[4] an investigation was at once launched by a subcommittee of the House Committee on Merchant Marine and Fisheries, with Representatives from Louisiana as the first witnesses. "The completion of this shipyard would have meant more to the city of New Orleans and to the South in general than almost any industrial development in modern times," said Congressman Thomas H. Boggs, expressing the way many people in New Orleans felt about it.[5]

Such a furor over the cancellation of the biggest and most advertised shipbuilding contract was certainly to be expected, especially since there was included in the very announcement of cancellation a re-endorsement of Higgins' ability to produce ships. The Maritime Commission was denounced for undermining the nation's morale by saying in its press release that the United States had not steel to build ships. Everyone seemed to agree that something was wrong, but opinions differed as to who was to blame. Some people pointed an accusing finger at the Maritime Commission and said there was more behind this than shortage of steel. Others turned on the War Production Board and demanded to know why steel for shipbuilding had been curtailed. Higgins hit out in both directions. He and his advisers claimed that the Maritime Commission had been forced, against its own will, to cancel his contract because the old-line shipbuilders of the Northeast, particularly Bethlehem, feared the competition which Higgins would offer them after the war both in building and repairing ships. Joseph W. Powell, who had been

[3] *New York Times*, July 19, 1942, p. 1.
[4] *Congressional Record*, vol. 88 (77th Cong., 2d sess.), p. 6421-22 (July 20, 1942), p. 6519 (July 23, 1942), p. 6574 (July 24, 1942), pp. 6617-18 (July 24, 1942), pp. 6635-40 (July 27, 1942). The resolution introduced by Senator Allen J. Ellender was attacked by Senator Charles L. McNary as undue legislative interference with administrative action.—*Ibid.*, pp. 6638-39.
[5] House, *Higgins Contracts, Hearings*, p. 125.

Vice President of Bethlehem Shipbuilding Corp. Ltd. in 1917-21 and who in 1942 was Chief of the Production Branch in the Navy's Office of Procurement and Material, was named as the instrument of this jealous policy of the old yards and also of the alleged resentment at Higgins' outspoken criticism of the Navy.[6] The War Production Board was also attacked, at least by implication, when Frank Higgins described the operations of the "black market" in steel.[7]

If the cancellation of the Higgins contract was a blow to all the boosters of New Orleans, it was a blow especially to the AFL labor leaders who had been counting on the rapid growth of their union as the number of Mr. Higgins' workers increased. The labor support which Mr. Higgins had carefully cultivated stood with him fully in his protest. Not only the New Orleans leaders but the AFL as a whole showed sympathy.[8] It was meeting strong competition from the CIO in many yards, especially on the East coast, but the AFL had union shop contracts with Mr. Higgins and on the West coast where Mr. Kaiser was the dominant figure.[9] The AFL encouraged talk that Mr. Kaiser and Mr. Higgins between them could build the whole of the President's directive of 24 million tons.[10] This vision helped the prestige of the AFL shipbuilders' unions.

All groups ready to find fault seized on the Higgins cancellation as a sign that matters were being seriously mismanaged. There was a flood of newspaper articles and letters to the President. Mr. Higgins was in Washington calling at the White House,[11] receiving, to no avail, a rehearing before the Maritime Commission,[12] and promising fireworks to the newspapers.[13] Within a week there were enough Congressional committees investigating

[6] House, *Higgins Contracts, Hearings*, testimony by Higgins, pp. 15, 35, 45-46 and by L. E. Detwiler, assistant to Mr. Higgins, pp. 69-71. On Mr. Powell's position, see Connery, *The Navy and the Industrial Mobilization*, pp. 147, 311-12.

[7] The discussion of the "black market" by Frank Higgins, son of A. J. Higgins, occurred during the hearings in New Orleans. House, *Higgins Contracts, Hearings*, pp. 158-61.

[8] MC, Maritime News Digest, Daily Summary, vol. VI, No. 171.

[9] See below, chap. 9.

[10] *Congressional Record*, vol. 89 (78th Cong., 1st sess.), pp. 2694-98 (Mar. 30, 1943).

[11] Maritime News Digest, vol. VI, No. 173, *New York Times*, July 23, 1942; House, *Higgins Contracts, Hearings*, p. 6.

[12] MC Minutes, p. 22544 (July 22, 1942), Executive session.

[13] M. B. Palmer, *We Fight with Merchant Ships* (Indianapolis, 1943), pp. 91-92.

so that one witness, at least, was not sure, and perhaps did not care, before what committee he was testifying.[14] The AFL, not satisfied with the findings of the House and Senate committees, appointed a committee of its own to take testimony.[15]

The Congressional investigating committees rejected Mr. Higgins' accusations that ulterior motives and outside influences controlled the decision of the Maritime Commission. Both the Truman Committee and the subcommittee, headed by J. Hardin Peterson, of the House Committee on Merchant Marine and Fisheries, accepted as a fact that shortage of steel was the crux of the matter [16] and, inquiring into that shortage, found much to criticize in the War Production Board's handling of steel.[17] Some of the personnel of the War Production Board, who apparently felt themselves publicly accused or unjustly put on the spot by the Maritime Commission's emphasis on the shortage of steel, took the position that the Maritime Commission had been given enough steel for the President's program, and the Commission should take the responsibility for deciding to close down Higgins since it allocated the steel among the yards.[18] Especially pointed was the testimony given in secret session by Mordecai Ezekiel, a statistician in WPB, who was just bringing to completion an elaborate survey of the use and need of steel in shipyards. Mr. Ezekiel said that many of the old established shipyards were receiving and holding more steel than they needed, and that the Maritime Commission could find steel enough for Higgins by a more careful control of the inventory of all yards. Mr. Ezekiel

---

[14] For the Senate there was the Truman Committee, and the Commerce Committee. For the House there were two subcommittees of the Committee on Merchant Marine and Fisheries, one with Mr. Boykin as chairman to investigate steel shortage, the other with Mr. Peterson as chairman especially for the Higgins contracts. Mordecai Ezekiel appeared before what he called a subcommittee of the House Rivers and Harbors Committee. Ezekiel to Robert Nathan, Chairman, Planning Committee, WPB, Aug. 3, 1942, copy in MC, Historian's Collection, Research file, 206.1.2. Actually it was the subcommittee under Mr. Boykin just referred to. *New York Times*, July 26, 1942, p. 19.

[15] Its report was placed by Senator Aiken in the *Congressional Record*, vol. 89, pp. 2692-2704 (Mar. 30, 1942).

[16] House, *Investigation of Cancellation of Higgins Contract, Interim Report*; pp. 5-6, 10-13. The attitude of Senator Truman was made clear in the Truman Committee, *Hearings* (pt. 14, pp. 5710-11, 5723-24) and in his reply to Senator Aiken (*Congressional Record*, vol. 89, p. 2705, Mar. 30, 1943).

[17] Truman Committee, *Additional Report*, 78th Cong., 1st sess., Feb. 4, 1943, No. 10, pt. 3, especially pp. 2, 4.

[18] *New York Times*, July 28, 1942, p. 11.

even went so far as to conclude that the other yards were incapable of increasing their output enough to produce the 200 ships for which Higgins had contracted.[19] Donald Nelson did not go out on that limb, but he did emphasize that the Maritime Commission was being given all the steel it had ever been promised—he meant enough for the 24 million tons in 1942 and 1943. He gave testimony before the AFL committee.[20] Using Donald Nelson's statements as its chief evidence, the AFL committee composed of labor leaders from Louisiana turned in a report which said shortage of steel was not the real reason for the cancellation, but blamed the influence of big business and the prejudice of the Navy against Mr. Higgins.[21] Their account of Mr. Nelson's account of what had happened was quite a contrast to the statement which the Chairman of the War Production Board sent to Chairman Truman of the Senate investigating committee on September 5, 1942. In his letter of September 5, Mr. Nelson said that there was indeed a shortage of steel and that the "fundamental reasoning underlying the cancellation of the contract with the Higgins Corporation was sound."[22]

The cancellation was political dynamite, but the Commission survived the blast without any substantial damage to its good relations with the White House and the Congress. There were weak points in the Commission's position, and Admirals Land and Vickery in their testimony admitted mistakes, without calling too much attention to them. The two points at which the Commission was most open to public criticism were those at which it was, in a certain sense, covering the President, as many people may have guessed. The fact was clearly brought out that the

---

[19] *New York Times*, July 26, 1942, p. 19. A copy of his report dated July 20, 1942, with a memorandum of transmittal to Robert Nathan is in MC, Historian's Collection, Research file, 206.1.4. A page entitled "Correction" says: "Just as this report was completed, the Maritime Commission announced it had canceled the Higgins contract for 200 EC-2 vessels. . . . For the reasons given on pages 18-26, it does not seem likely that other EC-2 yards can increase their production further to take the place of Higgins. Accordingly, all the estimates in this report for 1943 completions must be reduced by 2 million tons." Mr. Ezekiel's estimate for 1943 including Higgins was 19,362,200. As things worked out, actual deliveries without Higgins were 19,209,991 tons.

[20] *Congressional Record*, vol. 89, pp. 2696-97 (Mar. 30, 1943).

[21] *Ibid.*, vol. 89, pp. 2699-04 (Mar. 30, 1943). The AFL report is dated Nov. 9, 1942.

[22] Nelson to Truman, Sept. 5, 1942 in WPB file 323.22, cited and summarized in Chaikin and Coleman, *Shipbuilding Policies*, p. 56.

Commission voted in favor of cancellation on July 10 but made no move to stop expenditures till more than a week later. The House Committee chairmaned by Mr. Peterson severely condemned this waste of public funds. The second weak point was in the repeated references both by Admiral Land and by his critics to " the President's program," when they could not publicly agree on precisely what these Presidential directives were. In public hearings during a war it was not possible to go into detail as to how many tons, including or excluding escort vessels, aircraft carriers, and tank carriers, had been in the President's mind. In his slashing style, Admiral Land explained and defended the Commission's position on both these points, as far as it could be done without dragging in the President, when he told the committe:

Now questions have been asked, very properly, why did we not know this sooner; why did we delay our action. We have fought this steel battle since July 1941. The final decision was only made in July 1942.

Now we are just as much responsible as the rest of the Government for delay in this decision. I have no alibi except I never liked it. We kept fighting till we got it in the teeth. We still think somebody got more and we got less. That is natural and human, but the decision was made, and just like the tailor we had to cut the suit according to the cloth available. . . . [23]

Although shortage of steel was the decisive factor, it was not of course the only pertinent fact. The high cost was worrying many Commissioners. Mr. Higgins' personality aroused some dislike.[24] Mr. Carmody wrote to ex-Senator Norris that his distrust of how the power plant would be handled contributed to his decision to vote for cancellation.[25] Admiral Vickery was worried early in June by Mr. Higgins' quarrel with the Navy and by the way the yard was coming on.[26] Admiral Land publicly testified to his belief that it was a fifty-fifty chance [27]—but he may

---

[23] House, *Higgins Contracts, Hearings*, p. 86.

[24] Interviews with Commissioners Carmody, Land, and Woodward. A false rumor concerning the cancellation was Drew Pearson's statement that the site had proved a "bottomless pit." MC, Maritime News Digest, vol. VI, No. 190. The evidence to the contrary, which is quite convincing, is collected in *A Statement by the Higgins Corporation in regard to the site, construction and cost of the Micheaud Shipyards*, of which a copy is in gf 503-54, Sept. 18, 1942.

[25] Carmody to George W. Norris, Aug. 30, 1943, in Carmody reading file.

[26] Notes of telephone conversation between Weber and Sanford, MC, Historian's Collection, Weber files EC2's, Higgins Shipyard, June.

[27] House, *Higgins Contracts, Executive Hearings*, p. 82.

have thought some other contracts were gambles at even worse odds. The steel shortage was the cause of cancellation, however, in my judgment, in the sense that had there been no steel shortage there would have been no cancellation, whereas once the extent of the steel shortage was clear the contract would have been cancelled even if no other reason for cancellation had existed, even if there had been none of the above mentioned doubts.

"Steel shortage" as used in this analysis means the allotment to the Maritime Commission of less steel plate than it needed to operate all its yards. If less steel plate had been assigned to the Army and Navy, there would have been no shortage for the Maritime Commission and no need of cancelling the Higgins contract. The chief accusation of bad faith made by Higgins' supporters was that a man important in Navy administration, Joseph Powell, acting in private interests, was really responsible for the cancellation. When asked by the Congressional investigating committee concerning Mr. Powell, Admiral Land expressed high opinion of his competence and said, "I think it is very unfortunate that his name is introduced in this because there is absolutely no bearing whatsoever in any way, shape, or form." [28] Mr. Powell testified that his position as special assistant to the Secretary of the Navy and Deputy Director of the Office of Procurement and Material gave him occasion at times to fight for more steel for the Navy, but he denied that he had any private interest or anything to do with the Maritime Commission's decision to cancel the Higgins contract.[29] The relative distribution of steel as between the Maritime Commission and the Navy did have an influence, however, in forcing the Commission to cut somewhat in its program. Mr. Powell as part of the Navy Department was responsible for the cutback to the extent that his success in getting steel plate for the Navy reduced the steel plate available for merchant shipbuilding. But in accusing the Navy of being motivated simply by hatred for him and fear of his postwar competition with old-line yards, Mr. Higgins was exaggerating his own importance. The battle between procurement agencies for steel had a broader basis and will be more fully told in chapter 10.

Looking back at the expansion of American merchant shipbuilding after Pearl Harbor, it seems clear that the cancellation

---

[28] House, *Higgins Contracts, Executive Hearings*, p. 98, July 22, 1942.
[29] *Ibid.*, pt. 3, pp. 346, 348.

of the Higgins contract was a somewhat belated retrenchment in a program which had over-expanded shipbuilding facilities. Ironically, what put the Commission in so hot a spot in July 1942 was success in what had been its main concern—the development by shipyard managements of new and faster methods of construction and the utilization of unskilled labor in recruiting the necessary labor force. These measures increased sensationally the number of ships produced from each building berth or way. Speed on the ways was a cause of both congratulations and embarrassments.

CHAPTER 7

SPEED AND PRODUCTIVITY IN MULTIPLE PRODUCTION

I. Standardization and Subassembly

WHEN HITLER met with his admirals in September 1942 to survey the submarine war, he confidently asserted that American shipyards could not build ships faster than they were being sunk. The Presidential directive of 24 million tons for 1942 and 1943 was dismissed as mere propaganda.[1] In fact, the race between construction and sinkings in 1942 was nip and tuck, but the submarines were defeated whereas the shipbuilders achieved their goal. They not only fulfilled the Presidential directive of 24 million, they built 27 million tons in 1942 and 1943. In the post-Pearl Harbor programs for war production there was a good deal of overestimating, of "incentive scheduling." The total munitions production and war construction actually achieved in 1942 and 1943 was only 60 per cent of that planned on February 1, 1942.[2] Shipbuilders, however, attained their two-year objective.

Their success depended, first of all, at least from an engineering point of view, on having standardized the product. Speed in production came from building the same design over and over again with a continuity that made it possible both to learn from experience and to plan ahead. The second round of Liberty ships to leave the ways were built, on the average, in half the time from keel laying to the launching that was required for the first round. The tenth round required only one-fifth as much time

---

[1] The German figures are in British Registered Tons, but the total number of ships given shows the President's February objective was that being discussed. Navy Department, Office of Naval Intelligence, *Fuehrer Conference on Matters dealing with the German Navy*, 1942, lithoprinted, pp. 83, 118.

[2] War Production Board, *Wartime Production Achievements and the Reconversion Outlook: Report of the Chairman, War Production Board* [Historical Reports, Special Report] (Government Printing Office, 1945), pp. 11-12; also CPA, *Minutes of the War Production Board, January 20, 1942 to October 9, 1945.*

STANDARDIZATION AND SUBASSEMBLY

LOSSES vs CONSTRUCTION
Allied Merchant Shipping, December 1941 – June 1944

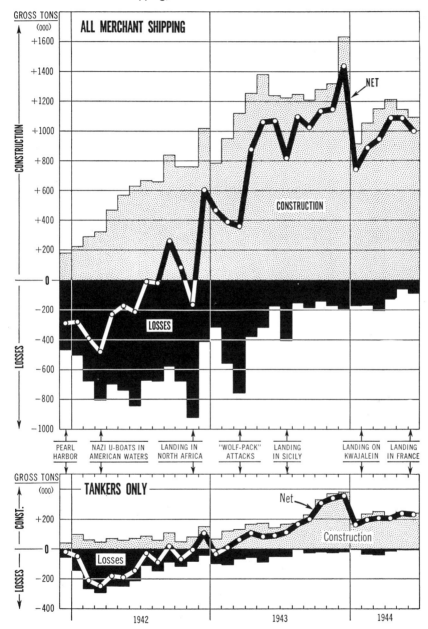

FIGURE 14. Source: W. K. Hancock and M. M. Gowing, *British War Economy* (London, 1949), p. 415 and *WSA Shipping Summary*, vol. II, no. 7 (July 1945), p. 22.

on the ways as the first. The advantages of continuity of production were realized in many other yards besides those that built Liberty ships. In the Kaiser-Swan Island yard, which built only standard tankers (T2-SE-A1), time on the ways was reduced from 149 days for the first round to 41 days for the tenth.[3]

Not only was production generally standardized in each yard at the beginning of the war, but a large measure of nation-wide standardization resulted also from adopting the Liberty design for all the first group of emergency yards. Later on the program became increasingly diversified by the inclusion of a variety of types, but during the crucial years 1942 and 1943, fourteen shipyards were all working on Liberty ships. Each of four yards delivered more than 150 Libertys in 1943, and the deliveries that year from all yards totaled 1,238 Libertys.[4]

Production of the same ship in fourteen yards made possible speed and economies not only in the yards themselves but also in procuring parts from manufacturers. With a thousand identical ships scheduled for delivery in a single year, a thousand rudders and sternposts had to be made all of the same design. Consequently the steel manufacturers could afford to set up special equipment. Once the equipment was prepared, individual frames could be turned out relatively rapidly and cheaply. Bethlehem Steel Co. installed special fabricating, welding, and assembly jigs that enabled it to produce rudders for Liberty ships at the rate of 120 a month.[5]

Another example of the advantages of standardization is provided by the joiner work. Much cutting and fitting of woodwork which had once been done by hand on the ship was now being done by machines in the shops of the contractors. Windows for the wheelhouse, for example, or parts of the refrigerator compartments, were cut in a shop to standardized measures and marked so that they could be rapidly and tightly fitted into place

---

[3] Computed from Reynolds' Way Charts. Round two averaged 110 days on the ways. On rounds of the ways of Liberty ships see Figures 15 and 16.

[4] If Richmond No. 1 and 2 are counted as two yards, New England as one, seventeen yards produced Liberty ships in those years, but Marinship, Vancouver, and Alabama were soon changed to other types.—Fischer, *Statistical Summary*, Table C-1.

[5] "Bethlehem's record of Rudder Production," *Marine Engineering and Shipping Review*, May, 1944, p. 184. Penn Steel Castings Co. of Chester received an order for 493 stern frames for Libertys and planned new facilities for their production. MC Minutes, p. 21212 (Feb. 26, 1942),

on the ship.[6] The preparation of working plans for all the emergency ships by Gibbs & Cox also saved money and time.

Working plans for all Libertys were made by one firm, but the manufacture of most components was dispersed among a number of companies—either because no single manufacturer could produce as many as were needed, or because public policy opposed monopoly, or because spreading the work gave better assurance of a steady supply. Construction of the reciprocating engine was spread among fourteen companies which were producing in October 1942 a total of just under one hundred engines a month.[7] By July 1943 orders for 2,387 boilers were distributed among eleven companies of which two were outstanding: Babcock and Wilcox building 673, and the Combustion Engineering Co. building 865.[8] Since the parts were interchangeable and the Commission was doing the buying for all the Liberty yards, the output of factories making boilers, for example, could be treated as a common pool from which boilers were sent to whichever shipyard had most immediate need for them. Consequently it was easier to keep the yards supplied when production was distributed.[9]

Because the production of components was spread among a number of manufacturers, the suppliers were turning out their products, not in thousands, but in hundreds, as were the individual shipyards. This being the scale of output, it is better to call the wartime shipbuilding an example of multiple production, not mass production. The basic principle of mass production is that if an identical operation is to be repeated tens of thousands of times it is worthwhile to prepare expensive apparatus to reduce the cost of each operation and the time consumed. With production at the rate of hundreds, instead of tens of thousands, it was not practical to use extensively in the shipbuilding industry the expensive dies and other such machinery used in the auto-

[6] Memorandum sent by Henry G. Laing, Laing Co., to U. S. Maritime Commission with covering letter of June 19, 1946, in W. G. Esmond's file on "Liberty Ships, Historical Data," now in Historian's Collection, Research file 111, Libertys.

[7] Allan D. MacLean, "Functions of the Maritime Commission Production Division," *Marine Engineering and Shipping Review*, Oct., 1942, p. 170.

[8] Senate, *Problems of American Small Business, Hearings*, before the Special Committee to Study and Survey Problems of Small Business Enterprises, 78th Cong., 1st sess., July 19, 1943, pt. 25 (Maritime Procurement: II), p. 3261.

[9] Rear Admiral Howard L. Vickery, "Shipbuilding in World War II," *Marine Engineering and Shipping Review*, Apr., 1943, pp. 184-85.

mobile and airframe industries. But producing hundreds of identical ships in one year from one yard was a big change from the practices of peacetime shipbuilding. Although not all the practices of mass production could be followed, some could.[10]

Another, closely connected contrast with peacetime shipbuilding was the extent to which shipyards became primarily assembly points. Admiral Vickery often said that the emergency yards were not engaged in shipbuilding in the old sense, they were in an assembly operation.[11] Work traditionally done on the ways was now performed outside the yard so that all that remained to be done was to put the parts together. Moreover, those items usually made outside the yard, such as engines, boilers, and winches, were arriving in such quantity that organizing their assembly, together with that of the material manufactured in the yard, became a most important problem.

The shipyards of World War II were less purely points of assembly than those of World War I, however, for steel fabrication was more largely done within the yard. In terms of mass, a ship consists mainly of pieces of steel, which have to be cut, shaped, and fastened together. The cutting, the shaping, and an uncertain amount of joining together is called steel fabrication. In World War I, the ships built at Hog Island were prefabricated in the sense that the steel was cut, bent, and drilled outside of the shipyard and even riveted together up to the size easily transportable. The prefabricated sections were then shipped from the structural steel plants to the shipyard in freight cars. There was relatively little fabrication of steel in the shipyard itself.[12] When the emergency shipbuilding program was launched in December 1940 and January 1941, Admirals Land and Vickery thought the new emergency ships could be prefabricated in the same sense in which the Hog Islanders had been. Admiral Land wrote of the new shipyards: " Such facilities must be backed up by proper structural fabricating facilities already in existence, such as bridge plants

---

[10] In the *Annual Report*, Apr. 1, 1943, of the National Council of American Shipbuilders, H. Gerrish Smith, President, said (p. 13) : ". . . the multiple production of ships has made possible the use of some of the factors involved in mass-production."

[11] House, *Hearings, Fifth Supplemental National Defense Appropriation Bill for 1942*, Feb. 12, 1942, p. 112.

[12] Edward N. Hurley, *The New Merchant Marine* (New York, 1920), pp. 63-66, 70-73.

and structural steel plants, as it will probably be impossible to get shipyard machine tools without serious priority changes." [13] But the idea of applying that method to the bulk of the program was abandoned within a few months; the steel plants could not be persuaded to undertake it. " I had tried to advocate that in our original shipbuilding contract. I didn't want to put any of these large fabricating facilities at these various shipyards that we finally had to put in," said Admiral Vickery, but " I couldn't get steel fabricators to do what we wanted." [14] In spite of the extremely tight bottleneck in machine tools, enough were finally provided to equip the shipyards so that they could fabricate for themselves. Libertys were not " prefabricated ships " in the sense in which that term is applied to the ships of World War I.

Fabrication within the shipyard was so organized, however, that less of the " building " of the ship took place on the ways. The phenomenal speed attained in shipbuilding during World War II consisted above all in reducing the length of time between keel layings and launchings. This was done by preassembly off the ways. If the work of putting together the steel plates and shapes which formed the hull was all performed on the ways or building berths, then each building berth was occupied by one ship for a relatively long time. If, in contrast, the pieces were joined into large sections elsewhere than on the shipway, the time on the way could be reduced.

Emphasis was put on reducing the length of time from keel laying to launching because that was the best measure of the speed of production of a yard. Outfitting the ship after it was launched took less time and could be arranged in such fashion that failure to complete the outfitting of one ship would not necessarily hold up the production of others. But laying a new keel depended on getting the space cleared by a launching. Over a period of continuous production such as a year or two, time on the ways was an index of a yard's speed in production.

The shortest time on the ways at which production was steadily maintained was about 17 days. The Oregon Shipbuilding Co. turned out Liberty ships at that clip throughout the middle of

---

[13] Land to Knudsen, Nov. 8, 1940, in MC gf 506-1.

[14] House, *East Coast Shipyards, Inc., Hearings*, pp. 380-81. See also Land to Knudsen, Apr. 11, 1941, in Production Division files, Administrative, No. 8.

TIME ON THE WAYS OF LIBERTY SHIPS, IN SELECTED YARDS
Average Number Of Days From Keel Laying To Launching For Each Successive Round Of The Ways

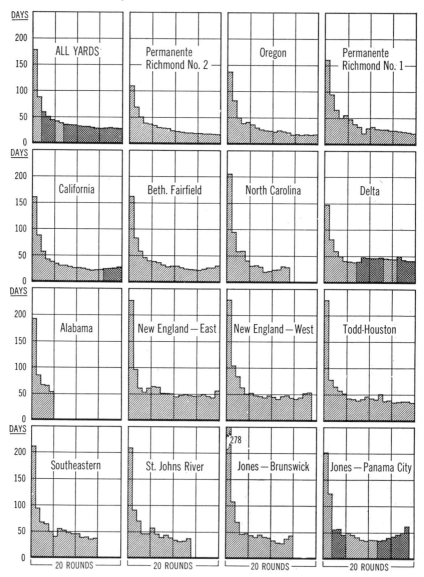

FIGURE 15. Darker bars indicate inclusion of Liberty tankers or other modified Liberty designs. Averages per round for first twenty rounds from Fischer, *Statistical Summary*, Table D-2.

STANDARDIZATION AND SUBASSEMBLY

MANHOURS PER SHIP FOR LIBERTY SHIPS, BY ROUNDS
Average For Each Successive Round Of The Ways In Selected Yards

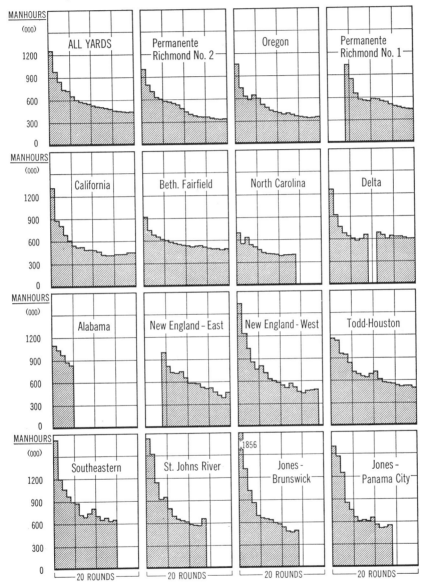

FIGURE 16. Omission of bars for certain rounds at Delta and Richmond No. 1 means that in those rounds all the ways were occupied in building modified Liberty types. Source: Fischer, *Statistical Summary*, Table H-3.

1943, and some other West coast yards were not far behind.[15] By additional preassembly off the ways and by special concentrations of manpower and of cranes, time on the ways could be cut to much less than 17 days. A Liberty was launched at Oregon Ship a mere 10 days after keel laying [16] and Richmond No. 2 responded to the challenge by assembling a ship in only 4 days on the ways.[17] These were called " stunt " ships because they were sensational achievements, aroused talk, focused attention on the new methods, and accordingly heightened morale. To maintain such a pace would have required a fantastic amount of space and of material in process. The yards made no attempt to keep up any such a pace as 4 days on the ways; 17 days on the ways seems to have been about the fastest rate at which construction could be maintained (Figure 15).

Of course the average for all Liberty yards shows a slower rate of construction. Until July 1943 the coming into production of new yards that were launching their first ships prevented the average for all Liberty yards from dropping below 40 days on the ways (see Figure 10). But Liberty ships delivered in December 1943 had been on the average only 28 days on the ways. Time in outfitting was reduced to 10 or 13 days. The total construction time from keel laying to delivery for all yards building Liberty ships was 41-42 days in September-December 1943. No comparable speed was attained in World War I. The speed record then for the entire Shipping Board program was set by the delivery of an 8,800 ton vessel by the Columbia River Shipbuilding Corp. at Portland, Oregon, in 52 days after keel laying.[18]

In a certain sense construction began in the fabricating shops and in the assembly areas long before the keel was laid. The timing of the work off the ways in relation to the erection after keel laying is shown in Figure 17.[19] For a few yards it plots the

---

[15] The 24th round off the ways at Oregon averaged 15 days on the ways. Fischer, *Statistical Summary*, Table D-3. See also Table D-4.

[16] Reynolds' Way Charts, Hull No. MCE 581, Sept. 13-23, 1942.

[17] *Ibid*. Hull No. MCE 440, Nov. 8-12, 1942. Outfitting was finished in 3 days.

[18] " Construction Period Analysis, U. S. Shipping Board Construction, 1917-1921," H. L. Deimel to Vickery, June 4, 1942, copy in MC Historian's Collection, Research file 117 (Statistics). Other records of speed in World War I are given in Hurley, *The New Merchant Marine*, pp. 92-93.

[19] For similar data on EC2's at Oregon, T2's at Sun, and C2's at Federal, see G. J. Fischer, " A Statistical Analysis of the Demand, Supply and Utilization of Steel Plate for Shipbuilding Under the U. S. Maritime Commission, 1941-1945," Chart

TIMING OF FABRICATION, ASSEMBLY, AND ERECTION OF STEEL

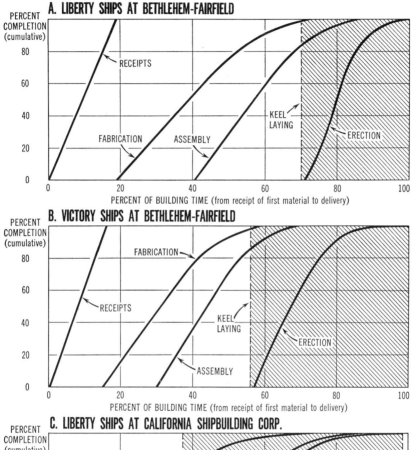

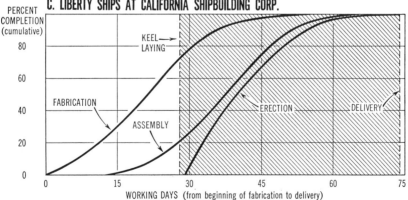

FIGURE 17. Note that chart C is different in its time scale from charts A and B and it does not include receipts.

chronological relation of the four essential stages in handling steel in a shipyard: receipt, fabrication, assembly, and erection. In view of the large extent to which the ship had been "assembled" even before the keel was laid, it is obviously somewhat misleading to refer only to the days between keel laying and delivery in stating how long it took to build a ship. If building time be considered the period between the date steel began to be fabricated for the ship and the date of the ship's delivery, the average building time for Liberty ships in 1943 would have to be stated, not as about 50 days as it appears on Figure 10, but as about 35 days in fabrication and subassembly, 40 days on the ways, and 10 days in outfitting, a total of 85 days.[20]

Preassembly enabled more men to work on the same ship at the same time, for the men "working on the ship" could be scattered over a wide assembly area. The yards that did most preassembly employed the largest number of men per way. In April and May of 1942, Oregon Ship had the lowest number of men employed *per ship delivered*, but the highest number *per way*. Using much preassembly, it employed 2,400 men per way in contrast to South Portland, which at that time used little preassembly and employed only 710 men per way. Part of the difference was due to fuller employment in the Oregon yard on the second and third shifts, but part was due to more use of preassembly.[21]

In addition to increasing the number of men who could be simultaneously engaged in work on the same ship, preassembly increased the output per man. Whereas the estimate made March 1941 of the number of manhours that would be required to build a Liberty ship was about 640 thousand, after multiple production was fully developed the average number of manhours actually expended, in 1943, was 414 thousand at North Carolina Ship and 352 thousand at Oregon Ship, and the average for all yards, including the least efficient, was 574 thousand.[22] Preassembly off the ways enabled more men to work under less crowded con-

---

11, in MC, Historian's Collection, Research file 206.1.4. Hereafter cited as Fischer, "Statistical Analysis, Steel."

[20] Cf. Fischer, *Statistical Summary*, Tables D-1 and F-5. The length of time steel was in process of fabrication was related to the problem of inventories. See below chap. 10.

[21] Ezekiel Report (cited above), pp. 29-32.

[22] See below, Table 10 in chap. 14; and Fischer, *Statistical Summary*, Table H-1. On manhours, see also sect. IV of this chapter.

ditions. It made it possible to plan their arrangement and to employ special kinds of framework, cradles or jigs, to make easier the joining of plates and shapes. Moreover, preassembly facilitated the use of a new technique which had become important in shipbuilding between World War I and World War II—welding.

Welding added to the advantages of preassembly because downhand welding could be done much more easily than either welding or riveting could be done in crowded or overhead positions in the hull. Accordingly, both time and manhours were saved by welding on the ground or with a jig that held the steel plates and shapes in the desired position. More and more, as the ship was put together in sections before going on the ways, methods were found to make the production not only faster but less costly.

Of course there were limits to how much could be assembled off the ways. The most immediate practical limitation was the lifting capacity of the cranes. There was no use assembling a section weighing 50 tons if there was no crane near the shipway that could lift more than 5 tons. Another limitation was the availability of space where preassemblies could be put together and kept until the ship was ready for them. Preassembly called for plenty of space and for powerful cranes.

Not only the increased use of preassembly but also the mere rise in the rate of production intensified the demand for spaciousness. It was found to be impossible to induce suppliers to send their products in well-timed deliveries so that they would move rapidly through the yards in a steady flow. Consequently more warehouses were needed for storing all kinds of materials and components. The need for storage space as well as of space for subassemblies swelled the size of the yards.

The final result was that the shipyards of World War II looked substantially different from those of World War I. The earlier yards had less powerful cranes, and those they did have were located near the outfitting docks, whereas in the new yards they were beside the building ways. At the head of the ways was far more space than in the yards of World War I. Yards such as Hog Island had relatively little space at the head of the ways and relatively small fabricating shops, for little fabrication or preassembly was done in the yard itself. They were " horizontal yards " needing a long frontage on the water. Those of World War II were mainly " vertical yards " in which a relatively small frontage launched many ships because of the work which went on a thousand feet inland. And in the sights and sounds within the yard, especially at night, there was the contrast between riveter and welder.[23]

In a typical shipyard of World War II, such as that shown in Figure 18A, there was a straight flow of steel. It went first to storage racks, then to the fabrication shop to be cut and bent. From the fabrication shop, it passed to an assembly building in which sections weighing 10 to 20 tons were welded together. These sections were then piled on the skids or platens at the heads of the ways or were there welded into larger sections weighing, normally, up to 45 tons. Thence they were put into their place in the hulls. See Plates V, VII, VIII, X, XIII.

Each part of this movement was handled by cranes. One group rolled back and forth through the storage yard picking up plates of the sizes desired, carrying them down to the end of the storage yard and handing them to the crane there which operated on a transverse track and so could deliver the steel to any desired bay of the fabrication shop. Other transverse cranes between the fabricating shop and the assembly building and between the assembly building and the pre-erection skids similarly delivered the steel or preassembly to the desired bay or skid. Within the buildings material was moved by overhead bridge cranes. On each side of the building ways were the high revolving gantry cranes. The tracks on which they operated ran back beside the skids so that they could pick up the preassemblies there and carry them to their proper places in the growing ship.[24]

II. The Plans of the New Shipyards

The advantages of preassembly and of welding were well known to shipbuilders. The New York Shipbuilding Corp. had pioneered since 1893 in methods of prefabrication and subassembly.[1] But

---

[23] The articles by Admiral Vickery and by H. Gerrish Smith, cited above, are the best brief summaries of the new methods. They are described in detail in many articles in the *Marine Engineering and Shipping Review* and in *Nautical Gazette*. I have been much helped also by discussion with H. Gerrish Smith, and P. J. Duff.

[24] G. V. Slottman, "Organizing Facilities to Promote Production Welding in Shipyards," *Marine Engineering and Shipping Review*, Dec. 1942, pp. 129-30.

[1] *Marine Age*, June 1949, p. 15; John F. Metten, "The New York Shipbuilding Corporation," in Society of Naval Architects and Marine Engineers, *Historical Transactions, 1893-1943* (1945), pp. 222-29.

only in time of war was there opportunity to apply these methods on a large scale. In normal times there was no point in equipping a yard to turn out hundreds of ships in a few years, for there was no market for them. The wartime demand for ships created a new unprecedented style in shipbuilding because it was an unprecedented demand, and because the Commission was paying for new yards fitted especially to the new purpose.

All the managements responded to the opportunities thus offered them, although not all with equal speed or in the same way. The Commission's policy of putting as much responsibility as possible on the contracting shipbuilders encouraged diversity. All plans of new merchant shipyards except of yards building for the British had to be approved by the Commission. During the first three waves of expansion, shipyard plans were sent to Washington and passed on by the Plant Engineering Section of the Emergency Construction Division. After decentralization they were passed on by the Plant Engineering Sections in the Regional Offices. The Commission's staff, more specifically Admiral Vickery, was thus the central clearing point of the ideas of various contractors. Moreover, the Commission encouraged and sponsored trips by builders from one district to inspect the yards of other builders. Since men who had never built ships before were being drawn into the Commission's program, the Commission was of assistance in helping them learn what others were doing. Experienced shipbuilders also were given a chance to learn new ideas. But the Commission did not try to impose any one type of yard. In dealing with the shipbuilders of established reputation, the Commission assumed that they knew their business and counted on their taking the lead.[2]

At the beginning of 1941, none of the plans showed a realization of how much space would be found necessary. One reason was that the Commission's program was not nearly so large then as it became later. Moreover, not everyone recognized how far preassembly would create the need for spaciousness. Even late in 1942 a writer in a technical trade journal praised a yard for its compactness, for the short flow from the railhead to the ways.[3] Some managements were more foresighted than others in getting

[2] See below, chap. 14; chap. 16, sect. II; chap. 20, sect. III.
[3] Charles F. A. Mann, article on Delta, in *Marine Engineering and Shipping Review*, Oct. 1941.

cranes with large lifting capacities. Each management faced a special problem the terms of which were set by the local terrain. There was not really any one ideal plan of a shipyard, since each site was a separate problem. To pick any particular shipbuilder or group of shipbuilders, therefore, and say that he or they introduced the techniques of mass production into shipbuilding would be misleading. It would ignore the extent to which each management went at solving the problem of speedy production by methods adapted to its special situation.

Notable among the yards which clung to the idea of compactness were those built under the management of William S. Newell at South Portland, Maine. Mr. Newell was accustomed to a tight yard, for the Bath Iron Works, which may be called the mother yard of the two built at South Portland, was more productive per square foot than almost any other yard in the country.[4] Its location in Bath left no room for expansion. To find the necessary space for a new fabrication shop, the company had built a plant at Harding, thirty miles from Bath. When Mr. Newell began work on merchant ships for the emergency, the Harding plant occupied a central position in the plan of operation.

At first, only the building of 30 ships for the British was involved. For the construction of these ships, Mr. Newell planned a yard at a site in South Portland on which he had had his eye for some time, on the east side of Cushing Point. Here subsoil conditions and the tide were favorable to the special type of yard which he had in mind, a basin type. The ships did not slide down the ways into the water, gates were opened to let the water into basins in which the ships were built. After the ship was towed to an outfitting dock, the basin was pumped dry and the keel laid for a new ship. This plan had been used to build the first " Dreadnaught," but was new for building merchant ships. It had the advantage for shipbuilding that materials moved along a level and moved down into position, not up. A yard so built would have utility as a repair yard also.

This yard later became the east-area yard of the New England Shipbuilding Co., but it was built by the Todd-Bath Iron Shipbuilding Corp. and was for a long time commonly called Todd-Bath. Its construction began when only the British building was

[4] Interview with P. J. Duff, June 12, 1947.

PLANS OF THE NEW SHIPYARDS 217

expected at South Portland.[5] Although the first wave of the Commission's program was being planned, it had not yet been announced and did not include South Portland anyhow. To build the 30 British ships, Mr. Newell planned to do most of the fabrication of the steel at the Harding plant and carry it by truck to the Todd-Bath yard. The plan of the yard called for no large fabricating building and no very large space for subassemblies. From the head of the ways there was only 100 feet to the assembly and fabricating sheds, each about 100 feet wide, on the other side of which plates were delivered by truck. There was a straight flow from the delivery platforms to the ways through little more than 300 feet.[6]

During the Maritime Commission's second wave of expansion a second yard was built at South Portland according to plans sketched by Mr. Newell. Its proportions were similar to those of the Todd-Bath yard, and it was badly hampered later, when the Maritime Commission's plans were further expanded, by the need for more space. Both the yards in South Portland were crowded into a settled community where expansion was difficult, and both were planned to take advantage of the existing steel-fabricating facilities at Harding.[7]

The Bethlehem-managed yard at Fairfield was larger, more spacious, and more productive, but it too was designed to put existing facilities to work and to fabricate at some distance from the shipways. Of the thirteen shipways in the original plans of the Bethlehem-Fairfield Co., two were existing ways which only needed restoration. To secure steel-fabricating machinery at a time when it was extremely scarce and when a speedy start in shipbuilding was most urgent, Bethlehem-Fairfield acquired some shops two and a half miles away. They had formerly been used to build Pullman railroad cars but could be put in shape to fabricate ship steel.[8] At the Pullman plant the steel was not only

---

[5] William S. Newell, "A Basin Type Shipyard," Society of Naval Architects and Marine Engineers, *Transactions*, vol. 51 (1943), pp. 25 ff.; M. P. Palmer, *We Fight with Merchant Ships*, pp. 165-68.

[6] Plans of the yards in MC, Historian's Collection, "Shipyard Facilities Data, U. S. Maritime Commission, East Coast Region, July 1, 1943," and in the collection made by C. D. Vassar in MC Bureau of Engineering. Plans of many yards are shown in Fassett, ed., *The Shipbuilding Business*, chap. IV.

[7] See below, chap. 15.

[8] Anderson to Vickery, Jan. 10, 1941 in Anderson reading file.

cut, rolled, furnaced, and drilled or punched, it was also welded or riveted into sections as large as could be handled by the freight cars which carried these subassemblies to the assembly areas at the head of the shipways. In the shipyard there was room to put together even larger sections. For example, innerbottom units, port and starboard, each weighing about 10 tons, which had been made up in the fabricating shop were joined with the vertical keel and rider plate to make a complete innerbottom unit measuring the full width of the vessel. These preassembled units weighing 22 tons were then stored in front of the ways. The largest subassemblies at Bethlehem-Fairfield were forepeak sections weighing 48.8 tons.[9]

Such methods of preassembly created the need for more room, as did the expansion which increased the number of ways from thirteen to sixteen. Many acres of adjacent land and land some distance away were acquired to gain more storage space of various kinds. In 1944 construction on two ways was discontinued to make room for an additional outfitting pier. Since the yard was sandwiched into an already developed part of the industrialized port of Baltimore, space had to be taken where it could be found.[10]

The third of the three emergency yards started on the East coast was North Carolina Ship. It depended on its parent company, Newport News, for the fabrication of the steel in its first ships[11] and was equipped initially with a very small fabricating shop.[12] Thus, for one reason or another none of the early East coast yards approached the layout which was later considered standard.

On the West coast the first two yards in operation were Richmond No. 1, then called Todd-Cal, which began by building British ships, and Oregon Ship. These yards were completely new, and their facilities accordingly were designed with a view to the specific purpose of producing one simple type of 10,000-ton ship rapidly. Richmond No. 1 was located on a swamp where

[9] These forepeak assemblies were for Victory ships.
[10] *Marine Engineering and Shipping Review*, Oct. 1942, p. 182, and personal inquiries.
[11] W. L. Marshall, Principal Construction Cost Auditor to J. A. Honsick, June 6, 1941, in gf 507-3-7.
[12] North Carolina Shipbuilding Co. Plant Layout, revised June 3, 1942, in Mr. Vassar's files shows that the only plate shop was a structure no bigger than a building way. It was directly behind shipway No. 5 and took space occupied behind other shipways by platens.

PLANS OF THE NEW SHIPYARDS

Kaiser's construction crews began dredging and filling on January 14, 1941. It was laid out according to a comparatively spacious plan.[13] There were about 330 feet from the head of the ways to the nearest building, which was a " plate shop and assembly bay " 260 feet by 540 feet; and behind the plate shop there was an extensive area provided with tracks and intended for the storage of the steel plates as they arrived.[14] This spaciousness was allowed also at Oregon which was begun at almost the same time on almost the same plan.[15] Later yards generally allowed at least as much at the head of the ways as Richmond and Oregon, over 300 feet, fully twice as much as there had been at Hog Island.

Allowing space between the ways and the fabricating shop became recognized as one of the basic principles in designing shipyards for multiple production.[16] Summing up in the fall of 1943 the experience in shipyard design, Admiral Vickery testified:

> There are two principles in the thing. One is the space between the ways and the fabricating shop, and the other is a straight flow of material. I think they are all laid out that way except possibly some of the earlier ones where there were some other ideas.

Even at Richmond and Oregon the managers very soon decided that they needed even more space in their yards. At Richmond it was found in assembly platforms built off to one side east of the yard.[17] At Oregon, the storage space behind the fabricating

---

[13] Maurice Nicholls recalls drawing the first plan of Richmond No. 1 on the back of an old envelope and figuring out the necessary equipment. Statements to F. C. Lane, Aug. 8, 1946.

[14] A plan of the yard dated Feb. 22, 1941 is in a booklet, " Richmond Shipbuilding Corporation, Richmond, California: Request for Funds for Additional Facilities at Todd-Cal Shipbuilding Corp. Yard. U. S. M. C., October 31, 1941," in exhibits 3-5 of gf 507-3-11.

[15] Francis H. Van Riper recalls going over the plans of Oregon Ship and Calship with Edgar Kaiser and J. A. McCone in Dec. 1940. Statements of Van Riper to F. C. Lane, June 6, 1946. I have found no plan of the yard earlier than that in Vasser's file revised to Sept. 8, 1942, which shows the assembly building and almost no storage space. The assembly building, mentioned below, was built in the summer of 1942 on what had been storage space, so that the original plan was much like Richmond's. MC Minutes, p. 21901-2; Charles F. A. Mann, " Emergency Shipyard of the Pacific Northwest," and " How They Break Records in the Oregon Shipyard," in the *Marine Engineering and Shipping Review*, Oct. 1941 and Oct. 1942, respectively, in Oct. 1941 at pp. 82 and 86, in Oct. 1942 at p. 202.

[16] House, *Investigation of Rheem, Hearings*, p. 65.

[17] " United States Maritime Commission, West Coast Yards, Shipyard Facilities Index, July 1, 1945," Richmond Shipyard No. 1, General Plant Layout. Copy in MC, Historian's Collection.

shop was used in the spring of 1942 to erect an assembly building, in which smaller assemblies were made before they were transported to the platens at the head of the shipways and there made into large assemblies. Locating the assembly building behind the fabricating shop had two disadvantages. It reduced the storage space for steel plates. In the spring of 1942 this difficulty was met by creating a whole new storage yard. The other difficulty was a traffic problem, since there was a loop back in the flow of steel. It went first to the fabricating shop, then back to the assembly building, and then forward again in subassemblies carried on cars around the end of the fabricating shop to the head of the ways. Being a traffic problem, this was solved by strict traffic control.[18] Although Oregon led the yards in developing speed of production, its layout was far from ideal. It violated one of Admiral Vickery's two basic principles in designing shipyards for multiple production, a straight flow, and yet it set the pace.

After Pearl Harbor the lessons learned were being applied in the building of new yards and the expansion of those which could find an adjacent space into which to expand. Under Edgar Kaiser's management two new yards were built near Portland, Oregon—Swan Island and Vancouver—which incorporated the lessons learned at Oregon Ship. They had large platens at the head of the ways for preassembly and the storage of the preassembled units. Immediately behind them was the assembly building from which came the subassemblies to be made into large units on the platens. Behind the assembly building was the plate fabricating shop, and behind that the racks for plate storage. Consequently in these yards there was straight-line flow from the plate storage to the shipways. They approached plan A in Figure 18, Swan Island coming the closest to it.

A third Kaiser-managed yard built after Pearl Harbor was Richmond No. 3. It was designed for the construction not of Liberty ships but of the larger transports, C4's, which were to be built in five dry docks or basins of the general character of those in the Todd-Bath yard at South Portland. The distance between

---

[18] Plant Layout in the "West Coast Yards, Shipyard Facilities Index," cited above; descriptions by Charles F. A. Mann in *Marine Engineering and Shipping Review*, Oct. 1941, pp. 82-90, 202; and in "Oregon Shipyard, a Record Breaking Yard," in *The Log*, vol. 37, no. 10, Yearbook and Review Number for 1942, pp. 61-62.

PLANS OF THE NEW SHIPYARDS 221

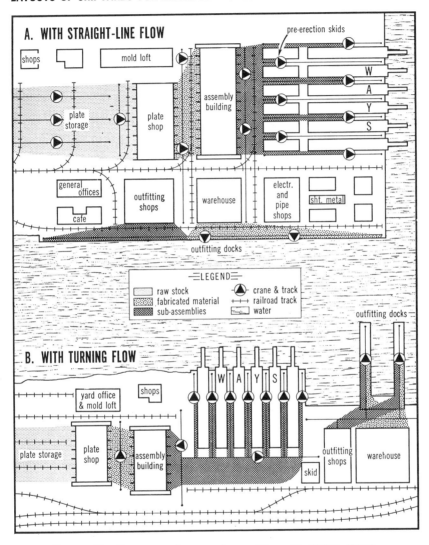

FIGURE 18. Compare photographs on Plates X, XXII, XXV.

the plate shop and the basins was about 500 feet and there were additional assembly areas on the sides.[19]

Plan A, shown in Figure 18 may be called that of a vertical yard with straight-line flow. Another layout, considered equally good is plan B in Figure 18. It might be called a horizontal yard with a turning flow. According to this layout the movement of steel was parallel to the shore line until it reached the head of the ways where it turned and flowed across the platens down to the shipways. Since there was no looping back, it may be called a straight flow although not in a straight line. This type of yard was adapted to a site where expansion inland was not practical, either because hills rose abruptly behind the shore or because a city was crowding the shipyard toward the water. When the Calship yard sought to expand, it had to grow up and down the shore by providing additional assembly areas at both ends of the yard.[20] When the Bechtels, who had been active in the management of Calship, built an entirely new yard, Marinship, at Sausalito, California, in 1942, they profited from the lessons learned at Calship. The result was like plan B in Figure 18. It had the added advantage that the steel and components for the hull flowed in at one end and all the material for outfitting flowed in at the other, so that two flows did not interfere with each other.[21]

On the East coast, the eight-way yard at Wilmington, North Carolina, was expanded after Pearl Harbor by the addition of a large fabricating shop and extensive storage area along the shore north of the shipways. After its expansion in the summer of 1942 North Carolina had a flow of steel much like that in plan B, and plenty of space, although it twisted to fit the terrain.[22]

Marinship was one of the five six-way yards started after Pearl Harbor in the belief that six ways was about the right size for

[19] Plant Layouts in the "West Coast Yards, Shipyard Facilities Index," Historian's Collection.

[20] Yard plans and description in *The Log*, March 1941, pp. 5-7; comment by Admiral Vickery in House, *Investigation of Rheem Manufacturing Co., Hearings*, p. 86.

[21] *Ibid.* The Marinship layout was mentioned and praised by Admiral Vickery in an address printed in the *Marine Engineering and Shipping Review*, Apr. 1943, pp. 182-90.

[22] Work was beginning early in April 1942. See letters and memoranda in gf 507-3-7 dated Apr. 7 and 11. Plant Layout dated Sept. 1, 1942, showing the additions is in "Shipyard Facilities Data, USMC, East Coast Region, July 1, 1943, No. 10."

efficient and flexible production. On the East coast, the plan used at Marinship was followed by the St. Johns River Shipbuilding Co. at Jacksonville.[23] The other new six-way yards were of the vertical type with straight-line flow more or less similar to the Kaiser yards on the West coast but some of them even deeper. In planning his yard at Providence, Mr. Rheem visited some Kaiser yards and then with Vickery's approval added 400 feet between the ways and the fabricating shop.[24] The climax in allowing plenty of space was at Brunswick, Georgia, where extra storage areas were provided between the fabricating shop and the ways so that the total distance from the ways to the shop was close to 1,500 feet.[25] See Plate X.

Nearly all the emergency yards built on the Gulf had some special feature because of the marshy character of the waterfronts or some other peculiarity of terrain, or because of the plans of their builders. Houston Shipbuilding Corp. adapted the plans generally used for a vertical yard to an extra small launching basin. The Delta yard at New Orleans was built for side launchings into a rather narrow channel. It had separate fabrication and assembly shops servicing each way and placed snugly beside them.[26] More spacious and equally distinctive was the layout of the Alabama Shipbuilding and Dry Dock Co. at Mobile. It was located on Pinto Island so that there was plenty of both space and waterfront. The shipways were spaced wide apart and the room between used for assembly platens.[27] Having the platens along the ways instead of at the head of the ways shortened the distance over which assemblies were carried by the cranes.[28]

This general survey of the actual development of shipyard layouts shows that the plans finally judged to be best were arrived at only by a process of trial and error. The success of Oregon

[23] St. Johns River Shipbuilding Company, Yard General Layout, in "Shipyard Descriptive Data, Gulf Coast Regional Construction Office." The plan was made by the Harry M. Hope Engineering Co. and approved July 30, 1942.
[24] House, *Investigation of Rheem, Hearings*, pp. 63, 65.
[25] Plan in the above cited booklet on facilities, No. 13, J. A. Jones Construction Co., Inc., Brunswick Shipyard, July 1, 1943; description in "Southern Shipyard Produces EC-2 Ships," in *Marine Engineering and Shipping Review*, Oct. 1942, pp. 235-36.
[26] Plan in "Shipyard Descriptive Data, Gulf Coast Regional Construction Office"; description by E. B. Williams, "The Delta Shipyard," in *Marine Engineering and Shipping Review*, Apr. 1943, pp. 192-202.
[27] Plan in "Shipyard Descriptive Data, Gulf Coast Regional Construction Office."
[28] Interview with P. J. Duff, June 12, 1947.

Ship is often said to have been due to its having been built expressly for the multiple production of Liberty ships, and it is true that the first yards to show realization of the importance of spaciousness were those built by the newcomers to shipbuilding who had the advantage both of a new approach and of new sites. But they did not allow, immediately, as much space as they later found desirable. The sensational records for speed made at Oregon Ship, at Richmond, and at Calship were made after these yards had been considerably expanded and remodeled; similarly, in the East coast yards with the best speed records, North Carolina and Bethlehem-Fairfield, original layouts were much modified later. By the spring of 1942, the new principles were generally recognized and by the end of that year yards had been remodeled accordingly.

A general survey also shows no simple correlation between the speed of output and the logical simplicity of the yard plan. Oregon made a high record for speed in spite of a loop in the flow. Not only the location of the main buildings and areas was important but also the exact placing of tracks, cranes, and other machines. Moreover, skillful application of other parts of the art of scientific management could overcome deficiencies in the layout.

III. Assembly-Line Methods

Comparison with the automobile industry was commonly used to describe the change brought to shipbuilding by the large and steady demands of war. " We are not only building ships, but, as Admiral Vickery says, we are assembling ships. We are more nearly approximating the automobile industry than anything else," said Admiral Land.[1]

Center and symbol of the production methods of the automobile industry is the assembly line with the car moving down it past one worker after another and rolling off the end a finished product. Could the same be done with ships? Could they be moved down an assembly line instead of being built in just two places, the shipway and the outfitting berth?

In the construction of small craft weighing less than 50 tons it was relatively easy to move the product down an assembly line

---

[1] House, *Fifth Supplemental National Defense Appropriation Bill for 1942, Hearing*, p. 112.

as in the automobile industry. The demands of the Navy for landing craft (25,171 landing craft of less than 50 tons were built for the Navy within the year 1944) occasioned a sufficient volume of production,[2] and landing craft were consequently built on assembly lines. These builders of small craft then engineered assembly lines for the construction of larger ships until vessels with launching weights of 1,300 to 1,400 tons (LST's) were being built in 1942 in series parallel to each other on even keels, moved horizontally to launching positions, and then launched sideways. The LST's constructed by this method were built for the Navy, and a similar method was applied in some Maritime Commission Class II yards: at the Ingalls yard at Decatur, Alabama, in building coastal cargo vessels (N3's) ; by East Coast Shipyards, Inc., at Bayonne, New Jersey, in building coastal tankers; and by yards producing some other minor types such as tugs.[3]

The idea of building large merchant vessels, even as large as 10,000 tons, on an assembly line, although it was not put into practice, was proposed to the Maritime Commission at the very start of its program. A Britisher, John Tutin sent in sketches of an "assembly-line shipyard" in January 1941.[4] The grounds then given by the Commission for rejecting the idea were that it called for overconcentration of men and materials in one yard, an argument which hardly applied fifteen months later when the enormously expanded program included the proposals of A. J. Higgins for building ships on assembly lines. At the time its contract was cancelled, the Higgins yard was being built to construct ships on four assembly lines each with eleven stations prior to the launching. Since the yard was never completed, it is difficult to say how all the engineering difficulties involved would have been solved. At the time of the cancellation there were still a number

---

[2] WPB, *Official Munitions Production of the United States*, No. 41, p. 62.

[3] H. Gerrish Smith, "American Shipbuilding Industry at War," *Marine Engineering and Shipping Review*, Nov. 1942, p. 118; George F. Wolfe, "Production Line Welding Plant Speeds the War Program," in Society of Naval Architects and Marine Engineers, *Transactions*, vol. 50 (1942), pp. 9-32; C. M. Taylor, "The LST, the Kingpin of Invasion Fleet," in *Marine Engineering and Shipping Review*, Feb. 1945, p. 138. Wolfe and Taylor both describe Dravo yards. See also "Mass Produced 'Landing Craft Infantry' at Todd Managed Shipyard," in *Marine Engineering and Shipping Review*, June 1944, pp. 164-68, and House, *East Coast Shipyards, Hearings*, pp. 385-96.

[4] MC gf 500-19.

of serious engineering problems for which no satisfactory solution had been worked out.[5]

Although merchant ships did not move down assembly lines on the way to their launching, there were assembly lines of a sort in the merchant shipyards of World War II. In the process of being fabricated and made into subassemblies, material moved down a series of lines towards the shipway. The part of the shipyard in which imitation of the automobile industry was easiest was in the fabrication shop. In the shipyards designed to use the new methods of multiple production, each part of the fabrication shop specialized in a particular operation, preparing plates or some other material for a particular part of the ship. Then the material moved on to be fitted into subassemblies.

Specialization in the fabrication shops made it worth while to buy machines designed for the particular operations of this or that bay. This is well illustrated by the way flame-cutting equipment was used. The use of torches in cutting steel was of prime importance. The torches were mounted on carriages of various kinds according to the work to be done. In a radiograph, several torches were mounted on a small self-propelling carriage that traveled on a track so as to cut to the desired pattern or template. Each bay in the fabricating shop had flame equipment designed for use on the kind of plates it was preparing, for example, flat keel plates in bay 1, shell plates in bay 2, tank top plates and others in bay 3, and so on. Some bays were equipped with furnaces and bending slabs, others were not.[6]

Whereas specialization in the work of the fabricating shop followed naturally from the use of various kinds of machines, the traditional association between assembly and the ship's way was harder to break down. At first, in most yards at least, the main subassemblies for each ship were prepared in the assembly area called a " skid " or " platen " at the head of its shipway. The crew of workers on each platen performed the same operations as did other crews on other platens, since each fitted and welded together all the subassemblies going into the ship on its shipway. Greater

[5] Statements by P. J. Duff in interviews in 1947 and 1948 with F. C. Lane. On endorsement of the plans, see above, chap. 6.

[6] Fully illustrated in J. R. Kiely and W. C. Ryan, *Construction Procedure used by California Shipbuilding Corporation in the Construction of USMC Type EC2-S-C1 Cargo Vessels,* July 1, 1942. Vickery's copy in MC Historian's Collection, Research file 112.

ASSEMBLY-LINE METHODS

application of the principles of mass production was made possible when a particular area was devoted to just one kind of subassembly, such as that of the double bottoms, or the prow and forepeak section. By repeating the same operation over and over again the workers became more expert at it. Moreover, special equipment for turning and holding the plates could be provided in the specialized area or assembly bay to facilitate the one assembly to which this area was devoted.[7]

These principles were applied generally in the new yards planned after Pearl Harbor and in the new facilities added to old yards in the summer of 1942. In the Kaiser-managed yards for example this specialization of assembly areas was worked out experimentally at Richmond No. 1 [8] and was made a feature of the assembly building added at Oregon Ship in April to August 1942.[9] It was applied in original plans for the Swan Island and Vancouver yards. Of the eleven bays in the assembly building at Swan Island, three were devoted to corrugated bulkheads, two were for side shell sections, one for tank top sections, and five for miscellaneous bulkhead and deck sections.[10] In some regions with mild climate, more assembly was done in the open. A large open area in the North Carolina yard at one end of the shipways specialized in the assembly of large sections of the side shell on a special framework which facilitated rapid and sound welding.[11] Nearly all yards came to have a special area for the subassembly of the prow and forepeak.

As soon as the various kinds of subassemblies were specialized in separate areas, it was logical to arrange the bays of the fabrication shop and those of the assembly areas so that the material could be moved directly from one to another. Then, as soon as a subassembly was finished, it had to be moved away so that the workers could start on another, and there had to be some place to put it. For that purpose was used the space at the head of the ways, once intended for welding the assemblies together but now

[7] G. V. Slottman, "Subassembly Welding and Erection," *Marine Engineering and Shipping Review*, May 1943, p. 177.
[8] Statements by Maurice Nicholls, Aug. 1946.
[9] *The Log*, vol. 37 (1942), no. 10, pp. 61-62, describes the work of each bay of the assembly building.
[10] "Swan Island Building Large Tankers," *The Log*, vol. 37 (1942), no. 10, pp. 75-76.
[11] North Carolina Shipbuilding Co., *Five Years of North Carolina Shipbuilding* (Wilmington, N. C.) [May 1, 1946].

freed by the building of separate assembly buildings. Thus a series of assembly lines flowing through different assembly areas converged on the shipway.

Another assembly line which terminated at the ship, but not necessarily at the shipway, was that for the construction of deckhouses. A separate assembly for this purpose was proposed in the fall of 1941 to serve both of the Richmond yards being managed by Clay Bedford. About halfway between the two yards, No. 1 and No. 2, was erected a building 400 by 200 feet in which were fabricated many parts of the upper structure of the ships. The idea was to build the whole upper structure in approximately three units.[12] These deckhouse sections were then carried on trailers to the heads of the shipways where they were picked up by the high gantry cranes which ran up and down beside the ways. Teaming together four cranes for one lift took skill, but was done.

The following spring, developing the idea which had been put to work at Richmond, the Kaiser managers proposed to make the prefabrication of deckhouses a special feature of its Swan Island Yard. It propounded an elaborate scheme for a separate slip where the ships would come to receive their deckhouses.[13] Although not put into effect at Swan Island, the idea was embodied in the building of the Vancouver yard. The "deckhouse slip" was between the building ways and the outfitting docks. It was thus planned that Liberty ships built at Vancouver be moved to three stations in being assembled.[14] The deckhouse in turn, before it was ready to be put on the ship, had passed through four stations receiving at each previously assembled features such as smokestacks.[15]

The arrangements at Vancouver for putting together previously fabricated subassemblies so that a 210-ton deckhouse would emerge complete, and for bringing the ship into a special slip to receive this part of its structure deserve attention because they

---

[12] R. J. Carroll statement in MC gf 503-55, Nov. 24, 1941. He says this type of assembly had been used at Hog Island.

[13] MC Minutes, pp. 22180-82 (June 11, 1942).

[14] "Kaiser Co. Inc., Vancouver Shipyard, Building Location Plan," June 1, 1945, in MC "West Coast Yards, Shipyard Facilities Index."

[15] "Proposed Superstructure Assembly Plant for Kaiser Company, Inc., Portland Yard, May 29, 1942," a profusely illustrated booklet, in MC, Historian's Collection, Weber's files.

so well illustrate the trend of development which resulted from attempting to apply to shipbuilding the assembly-line methods. In practice, however, the whole arrangement was not of great importance because it was designed for Liberty ships, and Vancouver, although originally intended for Libertys, was used mainly for landing ships, aircraft carriers and transports. Also, the arrangement received unfavorable publicity when on October 14, 1942, on the first attempt to use it, the cranes dropped the 210-ton steel deckhouse a distance of 20 feet onto the waiting hull.[16]

When the ship had become the terminus of several assembly lines, finishing a ship with the features that had been planned, even if those features were known to be wrong, was often easier than making any change after construction had begun. In some instances this may explain why a ship was finished with one kind of a door or bulkhead and then sent to a refitting yard where the door that had just been put in was ripped out and one of different dimension installed, even when it was known before the first door was put in that it would have to be changed. Why not put the right door in to begin with, or if it was not available, why not leave it off? If materials had started on their way through the yard, and if a particular lot of material, or a particular part of the structure was not put into its place in a subassembly and moved on down the line, it was in the way. It held up the assembly of other parts. Especially on the ways and the outfitting docks it was essential to finish the vessels in any way in which they could be finished quickly, which meant according to plan, and so get them off the ways and out of the outfitting berths so that the next keel could be laid and the next ship outfitted. In this sense the shipyard was like a production line and was so referred to by Admiral Vickery: " If you stop the leading ships in a production yard you not only stop the head ship, you stop the whole production line. Therefore it is better to take them out of the yard as quickly as you can and let the modifications go on some place else. . . ."[17]

The production lines of the shipbuilders, however much they resembled those of various factories in one way or another, were

---

[16] *The Log*, vol. 37 (1942), no. 2, pp. 30, 64.

[17] House, *Production in Shipbuilding Plants, Executive Hearings* before the Subcommittee on Production in Shipbuilding Plants of the Committee on Merchant Marine and Fisheries, 78th Cong., 2d sess., pt. 3, June 28, 1943, p. 899.

distinctive because of the massiveness of the final product and the bulkiness of components and subassemblies going into it. Each shop, each bay, each assembly area was equipped with cranes according to the weight of the material to be moved. In between wove railway cars and tracks. Shipyard management was becoming more and more a problem in moving things, especially heavy things, and in finding space for the immense amount of unused and half-finished material which had to be on hand when ships were being turned out in unheard-of numbers.

The savings of multiple production were applied most widely to the Liberty ships, of course, since they were built in largest numbers; but the Commission was supplying a sure market for all that could be built of other types also. Sun, even more than Alabama, developed the multiple production of tankers, and Swan Island was built for that purpose. North Carolina was changed over to C2's and built them with the same preassembly methods it had been using on Liberty ships. So many of the C-types were converted to special Navy uses that figures about speed of production, especially comparisons between yards, are largely misleading, but it is noteworthy that the same trends in shipbuilding practice which were carried so far in the construction of Liberty ships appeared also in the yards that were building standard types.

IV. The Productivity of Labor

The effect of multiple production was most evident in the speed of construction. Its effect on the productivity of labor was also substantial although not so sensational as the effect on speed. Speed (productivity per shipway) is most conveniently measured in days from keel laying to delivery. The productivity of labor is most conveniently measured by computing the manhours used in the construction of each vessel. Just as the decrease in building time expresses the increase in speed, so the fall in the number of manhours per ship expresses the rise in productivity.[1]

Yards building standard Liberty ships offer the most obvious example of multiple production. The rise both in speed and in

---

[1] This section is a brief summary of some of the conclusions reached by Gerald J. Fischer in his "Labor Productivity in Shipbuilding under the U. S. Maritime Commission during World War II," a typescript in Historian's Collection, Research file 210.11. See also, Fischer, *Statistical Summary*, Section H.

PRODUCTIVITY IN THE CONSTRUCTION OF LIBERTY SHIPS
Average Of All Yards

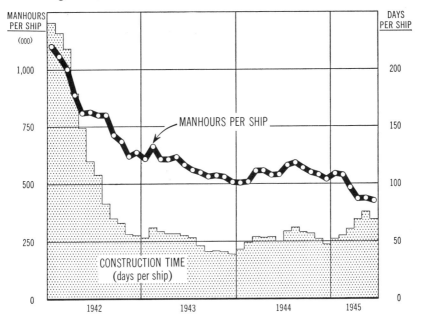

FIGURE 19. Averages for each month for vessels delivered that month, from Fischer, *Statistical Summary*, Tables D-1 and H-2, Standard Liberty Ships.

labor productivity in Liberty yards as a whole is shown on Figure 19. It emphasizes two points: first, the correlation between increase in productivity and increase in speed; and second, the effect of experience and continuity on the productivity of labor. Continuity in the operation of a shipyard is most easily measured in terms of "rounds of the ways." Launching one ship of a given design from each way completed one round of the ways for that type. When a second ship of the same design had been launched from each way in the yard the second round was complete. With each successive round, productivity and speed increased. This is demonstrated most effectively when an analysis is made yard by yard. Every yard shows a large reduction in manhours per ship on successive rounds (Figure 16), which is similar to the reduction in time on the ways (Figure 15).

If we disregard the first round of ships, which was often built while the yard was still under construction and was getting organized, we find that, on the average, manhours per ship were cut almost in half. That is, productivity increased by 93 per cent, to be more precise, between the second round and the thirteenth. By the twenty-sixth round in the four yards which produced that many, it had increased 6 per cent between the thirteenth round and the twenty-sixth. In general, continuity in the production of a single type in a yard made possible an increase in labor productivity of 100 per cent. Roughly speaking, that is the extent to which multiple production increased the productivity of labor, for continuity and multiple production are practically synonymous.

The importance for labor productivity of special features of the new techniques and new shipyard design may be sought by comparing performances of the yards when they were at about the same level of experience. For this purpose the productivity in the twelfth round launched off the ways of eight Liberty yards established in 1941 has been compared with the speed, the facilities, and the labor density.

Speed of construction and productivity of labor went together to the extent that the four yards having the greatest speed on the ways had also the highest productivity, and the yards standing fifth and sixth in both respects were the same. The same conclusion is supported by an analysis of the relation of speed and productivity within the same yard. Both at Oregon, the fastest, and in Delta, the slowest of the eight, increases in speed and in productivity went hand in hand, and show a very high correlation. The correlation is just as great in Oregon when it was going top speed. All the yards were working three shifts. If the "grave yard" shift had been eliminated it would have taken more days to build ships, but productivity per manhour would have increased. By working yards around the clock there was a sacrifice of productivity for the sake of speed. When that limitation is assumed, however, it may be said that the quest for speed was not pushed to a point at which more speed per way meant less productivity per manhour.

To estimate the effect of facilities on the productivity of labor requires a comparison of crane capacities, expenditures per way for machinery and equipment, and the space available for assembly.[2] The comparison shows that yards with more facilities had on the whole higher productivity of labor at the comparable level of experience; but there were very wide variations in area, or cranes, or other machinery from yard to yard for which there were no corresponding variations in productivity. The uncorrelated variations reinforce the conclusions that there was no single ideal plan of a shipyard in view of the diversity of sites and methods of operation.

To estimate the effect on productivity of the volume of labor employed per way, a calculation has been made to allow as far as possible for the differences both in facilities and in experience. Employment on the first shift only was considered since that is the best measure of how many men could efficiently be put to work in the yard at the same time. The result of the comparison was to show no correlation between variation in productivity ratio and variation in the number employed, at least within the range of 963 men per way to 1,314 men per way, which were the numbers actually employed in these yards when each was in its twelfth round. Similarly, when labor density is measured in terms of the number employed per square foot of yard or assembly area, comparison shows no significant correspondence between productivity and the degree of density. Accordingly the experience of World War II seems quite inconclusive in affording any precise indication of the point at which the addition of more labor led to decreasing productivity.

Of course a very large factor in the productivity of labor was the success of the management in arranging the work so that the men in the yard would always have work to do. Waste of labor might be due to failure of plans or of components to reach the yard at the right time or to failure by the management to plan correctly the use of the materials received. As a general rule it was important to have ships being worked on in various stages of construction. Labor in some yards was better trained or supervised than in others. Many such factors for which we have no statistics affected the productivity of labor.

Only Liberty yards provide a sufficient number of yards in continuous production of the same type to afford a basis for the

---

[2] The figures on these subjects, being compiled by different regional offices which used somewhat different terms of description, make accurate comparisons difficult.

MANHOURS PER SHIP FOR VESSELS BUILT IN NEW YARDS
Selected Types By Rounds Of the Ways

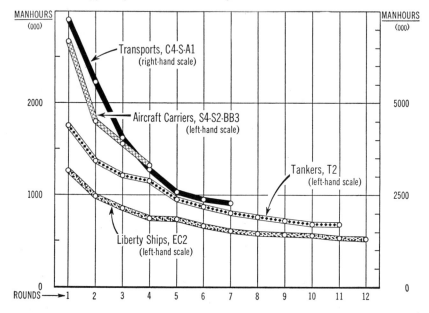

FIGURE 20. Source: Fischer, "Productivity," chart 7.

kind of comparisons just made concerning facilities or utilization of labor. But figures for yards building other types show that continuity of production was equally important in increasing productivity in these yards. Productivity of labor on T2 tankers increased 100 per cent between the second round and the eleventh, the last round built (Figure 20). In addition North Carolina, after making a record for low costs in building Libertys, was shifted to C2's and doubled labor productivity in work on these standard types. Consequently it is clear that the increase in productivity through multiple production was not confined to Liberty ships alone but was equally applicable to the ships of the Commission's long-range program.

The same conclusion results from analysis using a measure which allows not only for differences in displacement tonnage but also for differences in the complexity of construction in the various types. Such a measure is provided by the bogie or esti-

PLATE IX. Upper left, welding on a hull at Ingalls. Upper right, riveting on the *America*. Lower left, a "passer" with the white-hot rivet. Lower right, a frame emerging from the furnace and about to be shaped on the bending slab.

Plate X. The Jones-Brunswick yard (above) is an example of straight-line flow with maximum space between plate shop and shipways. A yard with turning flow such as Marinship (below) extended along the shore. Compare Figure 18.

mate prepared by the technical staff of the Maritime Commission stating the number of manhours considered reasonable for each type for the fourth round of the ways. When these estimates are compared with the actual number of manhours expended in the first, fourth, and seventh rounds of the ways, it is clear that increase in productivity by the seventh round was greater in the production of several other types than in the production of Liberty ships.

In summary it may be said that multiple production increased labor productivity in shipyards about 100 per cent both when it was applied to the construction of the relatively simple Liberty ship and to the building of standard cargo vessels and tankers. Yet in the average performance in the program as a whole the productivity of labor increased only 27 per cent from 1942 to 1945 (Table 23). Discontinuities through the introduction of the Victory ship and of military types or through other changes brought sharp decreases in productivity and affected the totals for later years (see Chapter 19), whereas starting new yards in 1942 kept down the average for that year. In short, a large portion of the building done for the Commission did not enjoy the full savings of multiple production.

The Commission's shipbuilding as a whole therefore showed far less gain in labor productivity than did some other war industries. Output per manhour in the airframe industry increased 98 per cent from the first quarter of 1942 to the fourth quarter of 1943. In that period the rise in productivity in the Commission's shipbuilding was 24 per cent. Because of the gradual bringing in of new yards and because of the changes in type, conditions in the shipbuilding industry were not so favorable as in many other industries for making large improvement in productivity to correspond to the increase in scale of operations. Shipbuilding made more phenomenal gains in speed than in the productivity of labor.

Chapter 8

BUILDING THE LABOR FORCE

I. Job Breakdown and Union Rules

AT THE beginning of 1941 labor was expected to be the bottleneck which would restrict production. In fact the increase in the output of the shipyards was accomplished largely by increased input of labor, since only one quarter was the result of increased productivity. Total employment in all shipyards—Navy, Maritime, and repair yards—was 168,000 in June 1940 and was about 1,500,000 when production reached its peak.[1] Where and how to find even a fraction of that number of shipbuilders seemed at the start an insoluble problem.

In 1940 and early 1941 reports of the Bureau of Employment and Security of the NDAC emphasized the shortage of craftsmen.[2] It was after a study of such reports and allowance for the construction which he knew had been planned that Admiral Land concluded as early as January 1941, "Any further dilution of shipbuilding brains (managerial, engineering, skilled, and unskilled labor) will only result in gross waste and inefficiency without accomplishing the ultimate results desired, namely the delivery of finished ships."[3] In concluding that more merchant shipbuilding, if it must be done, would interfere with the Navy's

---

[1] U. S. Dept. of Labor, Bureau of Labor Statistics, "Wartime Employment, Production, and Conditions of Work in Shipyards," [prepared by Edward M. Gordon, Eleanor V. Kennedy, and Albert A. Belman, under the direction of Herman B. Byer], *Bulletin No. 824* [May 10, 1945], pp. 1, 3.

[2] U. S. Social Security Board, Bureau of Employment Security, *Labor Supply Available at Public Employment Offices in Selected Defense Occupations, as of March 22, 1941* (mimeographed, April 1941), pp. 1, 3; H. O. Rogers and Jesse J. Friedman, Advisory Committee to the Committee on National Defense, Bureau of Research and Statistics, *Preliminary Survey of Shipbuilding Facilities in the United States* (mimeographed, Sept. 19, 1940), pp. 4, 13. The remarks on labor in Rogers and Friedman were based largely on J. Perlman, et al., U. S. Dept. of Labor, Bureau of Labor Statistics, "Earnings and Hours in Private Shipyards, 1936 and 1937," Serial No. 788.

[3] Memorandum for the Files, Jan. 27, 1941, in Land file.

program, which should then be modified, Admiral Land in this memorandum clearly based that conclusion not on shortages of facilities or components but on shortage of skill. In April a new Labor Department report concluded: " The shortage of available skilled workers in this industry is such that most of the new workers will require careful selection and extensive training before they can be utilized in the shipbuilding program." [4]

These predictions had in mind the kind of labor which had been used in the production of " tailor-made " ships as it was done before the emergency when more than 50 per cent of the workers in a yard were skilled craftsmen in the sense that they had served an apprenticeship of four years and learned a many-sided craft. The most famous of the apprentice schools training these craftsmen was that at the Newport News Shipbuilding Co. under the direction of G. Guy Via. His school was based on the principle that the type of work required of skilled workers in a shipyard was " of a comprehensive all-around nature." [5] Versatility was necessary as long as production was small. A craftsman might have to work one day in the fabrication shops, the next day on the ship. The school which Mr. Via had directed since 1919 taught 20 different crafts, but to learn even one of these crafts the apprentice had to master 40 to 70 skilled operations. Each craftsman could use a variety of tools and he had an understanding of ship construction which prepared him to work at any part of the ship where he was needed.[6]

Multiple production changed radically the amount of skill required of shipyard labor. A fixed schedule for months ahead made it possible to plan a specific task for a specific workman or type of workman. When preassembly off the ways permitted imitation of the automobile industry, and each kind of sub-assembly was placed in a separate area, workmen were assigned some to one area some to another. Then a workman specialized on just one kind of work. He could be useful as soon as he had learned to bolt two or three plates together, and since he concentrated on one operation he could learn that one quickly.

[4] U. S. Dept. of Labor, Bureau of Labor Statistics, *Estimated Labor Requirements for the Shipbuilding Industry under the National Defense Program* (mimeographed, Apr. 19, 1941), p. 35.

[5] U. S. Dept. of Labor, Division of Labor Standards, *Report on Apprenticeship System of Newport News Shipbuilding and Dry Dock Company* [prepared by Oswald L. Harvey] (mimeographed, Dec. 1940), pp. 1-2.

[6] *Ibid.*, pp. 1-2, 6, and Exhibit A.

This use of labor was the most important way in which the methods of mass production were applied to shipbuilding. In using machines explicitly designed for a particular operation the shipbuilders of World War II could go only a relatively short distance compared to the mass-production industries. In handling on schedule a large flow of materials they were facing problems familiar to men from the construction industry. It was in the utilization of labor that most was learned from the automobile industry. Knowing there was not enough skilled labor for the program they had contracted to fulfill, the managers were looking for ways of organizing the work so that they could use men who had never been in a shipyard before. They found the key to a solution in what they saw in automobile factories. Clay Bedford described as follows the impression made on him by visiting the Ford factory at the time he and the rest of the Kaiser organization were turning to the management of shipyards:

> One of the things that impressed me most was when I asked the personnel man how long it took to train a man to take the position of one of the men on the assembly line. He said that ordinarily it takes two days, after which the new man is watched for a day or two to see that he understands his problem.
>
> I went away from there quite amazed. I thought, these fellows really have something. If it were only possible to train the new people that we have in two days, wouldn't that be grand. So we attempted to set up a specialization program on the same basis, so that when making any certain section like a fore peak or welding in the pipes or doing any one of the single simple chores that are to be done on the ship, then that job was to be done by the same crew every day.[7]

When the work was thus planned so that the same crew had the same task every day, there was no need of teaching the new workmen the 40 to 70 operations which had formed part of the craft learned by an apprentice at Newport News. Instead a man who had only one skill could be kept busy doing that one thing. Detailed planning by management split apart the 40 to 70 skills which had gone to make a "craft." One job was "broken down" into many jobs, each of which required only a little training.

Breaking down the job is one of the commonplace principles of scientific management. But in a shipyard it was not practical to carry it as far as in the popular idea of an assembly line, where

---

[7] Clay Bedford's statement in the discussion of a paper by G. Guy Via, "The Wartime Training of Shipbuilders," Society of Naval Architects and Marine Engineers, *Transactions*, 1942, p. 325.

a worker gives just one nut a final turn. The common operations in cutting, joining, and shaping steel could hardly be learned in as little as two days. The most common of all shipyard workers in World War I and common even in World War II was the riveter. His was strenuous work and learning it required at least three weeks of training. Another example is the " shipfitter." He had the task, as his name implied, of seeing that the plates and shapes fitted together. By the breakdown of jobs during World War II, an individual shipfitter often knew nothing about any portion of the ship except the particular type of subassemblies on which he and his gang worked over and over again. In some yards he was not even required to be able to read blue prints; that was done by a supervisor. But he did have to know something about shipfitting markings, the language of shipbuilders, and the way parts were fitted together, so that most full-fledged shipfitters had served at least a year first as " handyman," or been through an intensive course of 8 or 12 weeks in a training school.[8] There was still a considerable range of skills in the shipyards even after scientific management had broken down the jobs.

A major change in the kind of skill needed by shipyard labor resulted from the extensive subsitution of welding for riveting. Not only was welding quicker than riveting, it also required less skill and muscle—at least when preassembly made it practical to do the welding in the easiest position, namely, as downhand welding of plates or shapes held in position by frames or jigs. Although women could not have been employed extensively as riveters, they could be employed as welders; and one manager reasoned optimistically by analogy that there would be as many good feminine welders as there were good sewers.[9] Welders were given the equivalent of three weeks in a training course. Their training continued while they were put to work as " tack welders," whose task, like basting, held pieces in place so that a good thorough welding job could be done later.[10] For all these reasons

---

[8] Federal Security Agency, Social Security Board, Bureau of Employment Security, U. S. Employment Service, Occupational Analysis Section, *Preliminary Job Study, Job Descriptions for the Ship and Boat Building and Repair Industry* (War Production Board, Shipbuilding Stabilization Branch, mimeographed, April 1943), p. 141; statement by Maurice Nicholls, Consulting Engineer, Kaiser shipyards, in interview with F. C. Lane, Aug. 8, 1947, in Historian's Collection, Research file 105.

[9] Rex R. Thompson of Brunswick to Vickery, Sept. 8, 1942, in MC, gf 508-1-1.

[10] James W. Wilson, Welding Inspection Coordinator, to Vickery, enclosing " Survey

welding helped solve the labor problem, especially for yards on the West coast and Gulf where the labor force grew most rapidly. Not welding itself, however, but the specialization of workers on a single operation which was repeated over and over again was what made possible the use of unskilled labor.

In introducing the dilution of skills which went with multiple production, shipyard managements met no opposition from labor unions. Indeed they received a large measure of active cooperation. Both the AFL and the CIO unions helped in recruitment, supported the new kinds of training programs required,[11] and agreed to some necessary modifications of working rules.

The craft unions, represented in shipbuilding by the Metal Trades Department of the American Federation of Labor which dominated all the yards on the West coast and many on the Gulf, had rules defining the distinction between skilled and unskilled labor and forbidding to each of these two classes many tasks reserved for the other. According to union tradition and experience, their purpose was to assure skilled labor that their wages and working conditions would not be lowered by forcing them to work in competition with lower paid, less skilled workers, as they would be forced to do if jobs traditionally part of their craft were assigned to the less skilled. Hence craft unions had frequently restricted membership to those who had served an apprenticeship. Rules were relaxed, however, during the war. A worker with very little training was allowed to do a job traditionally reserved for a " first-class mechanic," provided only that he was paid the wages of a first-class mechanic and was enrolled in the union.[12]

The readiness of even the craft unions to permit this lowering of the technical standard in the skill required of a first-class mechanic and to admit to their craft literally hundreds of thousands of new members with lower skills is explicable under the circumstances even before the declaration of war would have made objection seem unpatriotic. A basic purpose of craft rules

of Welding Conditions in Shipyards having U. S. M. C. Contracts," June 26, 1942, in gf 606-65.

[11] Metal Trades Dept. of the American Federation of Labor, *Proceedings of the Thirty-Second Annual Convention*, (Nov. 11, 1940), pp. 36-39.

[12] Blanche D. Coll, "Stabilization of Labor in Shipbuilding Under the U. S. Maritime Commission During World War II," typescript, May 12, 1948, in MC, Historian's Collection, Research file 205.8.

JOB BREAKDOWN AND UNION RULES 241

is to assure job security, and the boom in shipbuilding in 1941 made it clear that, as Admiral Land said, " no one need fear that he will work himself out of a job." [13] Moreover, leaders among the old-time shipbuilding craftsmen were moving up to become supervisors and foremen. The dues of the newcomers were swelling the union treasuries to the satisfaction of union officials, and since closed shop or union shop was guaranteed in most yards there was no immediate fear by the unions that managements would use the new situation to weaken the unions. Reduction of craft skill did not seem to threaten therefore either union status or the wage level.[14]

Craft rules also defined the jurisdiction between crafts. In the yards unionized in the AFL there were thirteen unions: Blacksmiths, Boilermakers, Electrical Workers, Operating Engineers, Laborers, Machinists, Metal Polishers, Pattern Makers, Molders, Plumbers and Steamfitters, Sheet Metal Workers, Carpenters, and Painters. Moreover the Metal Trades family had not necessarily stopped growing; other unions such as the Structural Iron Workers argued their legitimacy. Each of these crafts supposedly had well defined places in the operation of a yard, but as production methods changed, new kinds of jobs were created and questions arose as to which union had control of them. The normal way of settling such disputes was by conference between the national presidents of the craft unions concerned or through the long process of appeal to the Executive Council or national convention of the AFL. The Maritime Commission and the yard managements called for quicker methods of settlement. Arrangements were made to have jurisdictional disputes referred to the headquarters of the national unions immediately, but better regular methods of settlements were still being sought when the war ended.[15] Practically, under the pressure of war, " overlapping was accepted generally by common consent because of the scarcity of manpower," according to the AFL Metals Trades Department.[16]

[13] The Land Statement [made at a meeting at the Dept. of Labor], Jan. 17, 1942, attached to Memorandum for Files, Jan. 19, 1942, in Land file.

[14] Closed shops are discussed in the next chapter. On the general situation see Coll, " Stabilization," particularly chaps. I, II and III.

[15] Metal Trades Dept. of the American Federation of Labor, *Proceedings*, 34th Convention, Sept. 28, 1942, pp. 155-60; *Proceedings*, 36th Convention, Nov. 13, 1944, pp. 25-26; *Proceedings*, 37th Convention, Sept. 30, 1946, pp. 39, 122.

[16] Metal Trades Dept. of the American Federation of Labor, *Proceedings*, 37th Convention, Sept. 30, 1946, p. 48.

One manager testified that although craft distinctions were sometimes an impediment, they were being broken down and mentioned as an example that a phone call to the boilermakers' union saying 100 pipefitters were temporarily out of work resulted in immediate agreement to put the pipefitters on tasks usually done by boilermakers.[17] But there was much talk as late as June 1943 about the waste of labor resulting from union restrictions.[18] The House Committee on Merchant Marine and Fisheries directed many questions at this problem in the hearings it held on the West coast in the summer of 1943.[19] In return for increases in pay for certain classes of workers the AFL unions then agreed to abolish working rules like those which prevented a shipfitter from doing tack welding. The War Labor Board said that the restrictions lifted at that time would effect " economies which the parties estimate will save the procurement agencies from $50 to $75 million yearly." [20]

As manpower became tight in 1944 some yards complained that transfers from one craft union to another sometimes took so long they caused loss of manhours, unbalance between crafts, and layoffs.[21] But when the supply of electricians became desperately short the unions cooperated in allowing transfers, at least on a temporary basis.[22]

Jurisdictional disputes led to a few brief strikes involving a few thousand workers [23] and to one strike which threatened production seriously. The welders felt they had real grievances because they were not a separate craft union but were enrolled in various

---

[17] Testimony of Clay Bedford in House, *Production in Shipbuilding Plants, Hearings,* pt. 3, June 18, 1943, p. 616.

[18] Public attention was called to these practices by the article by John Patric in the *Reader's Digest,* June 1943, entitled " Remove Union Restrictions and Increase Shipyard Production by One-Third."

[19] House, *Production in Shipbuilding Plants, Hearings,* pts. 2-4.

[20] *War Labor Reports,* (Washington, D. C.: Bureau of National Affairs), vol. 12, p. 758; Paul R. Porter, " Labor in the Shipbuilding Industry," *Yearbook of American Labor, 1944* (New York: Philosophical Library, 1944), pp. 352-53.

[21] [Reports of] Manpower Survey Board, USMC, Richmond Shipyards, Jan. 1944; Kaiser Co.—Portland (Swan Island), Jan. 1944; Kaiser Co.—Vancouver, Jan. 1944, in MC, Historian's Collection, Research file 205.2.4a.

[22] [Reports of] Manpower Survey Board, USMC, Kaiser Co.—Vancouver, Mar. 1944; Oregon Shipbuilding Corp., May 1944, in file cited in note above.

[23] There were eleven strikes of this nature. The most serious of these lasted 4 days and involved 6,355 workers. Compiled from information in U. S. Dept. of Labor, Bureau of Labor Statistics, Division of Industrial Relations.

JOB BREAKDOWN AND UNION RULES 243

unions within the Metal Trades Department according to which sort of materials or part of the ship they worked on. Work-permit fees and the extra high initiation fees that had been paid by welders in some cases were abolished by the Metal Trades convention in September 1941, but the welders pointed to other injustices. The most important was the necessity of carrying membership cards in several unions in order to weld on various types of jobs, so that they had to pay numerous initiation fees and high dues. Welders began striking in the Puget Sound area in October 1941 and by November 1 it was reported that 5,500 were out in California. On November 4 their leaders announced secession from the AFL and the formation of an independent union. To meet the challenge the AFL unions agreed on November 18 that henceforth welders would not be required to carry more than one union card and that it would be accepted by any craft as a permit to work. Leaders of the welders still hoped to set up a separate organization and strikes continued sporadically until after Pearl Harbor, in fact until February 1942; but once the worst grievance had been removed and war was declared, public opinion, strongly voiced by Admiral Land as well as by other governmental leaders, was turned against the strikers, and the strike was broken.[24]

In addition to the rules defining jurisdiction between craft unions, there were rules for enforcing the specialization among members of the same union, particularly in the boilermakers' union to which belonged most AFL shipyard workers. This more intensive specialization of crafts within crafts—according to which tack welders could only tack, full-fledged welders make the final seam, and so on—seems to have been a result of the new ways of using semi-skilled labor in multiple production. It reflected the fact that workers were trained for one operation only. When objection was made to it, the boilermakers' union said that they favored permitting the interchangeability of such workmen, on condition, however, that their training had been sufficiently diversified.[25] Manpower surveys reported that interchangeability

[24] Metal Trades Dept. of the American Federation of Labor, *Proceedings*, 34th Convention, Sept. 28, 1942, pp. 21-23, 39-40; Coll, " Stabilization," pp. 83-85; and see below chap. 9.

[25] *Report of the International President and Executive Council to the Seventeenth Consolidated Convention of the International Brotherhood of Boilermakers, Iron Ship Builders, and Helpers of America*, Jan. 31, 1944, pp. 16-20; International

within the craft was practical whenever production methods warranted it.[26] This question arose in the later years of the program when the problem was no longer the expansion of the working force but the more thorough training and the more efficient utilization of a labor force whose total size was beginning to shrink under the general pressure of war upon the nation's manpower.

II. Wages and Recruitment

There was no general scarcity of labor when the new shipyards were begun, for the nation had not yet overcome the unemployment of the depression of 1929. In March of 1941 there were still millions registered as in search of work, although relatively few were qualified by craft training to be called shipbuilders. The problem was to attract and train the unskilled and to organize them into teams able to build ships. The Commission helped attract workers to shipyards by associating shipbuilding with national defense in the speeches and posters which publicized its program and (what was more important) by approving large wage increases. The " breaking down " of jobs was indispensable in order to draw on the general manpower of the nation; also indispensable was the attraction of high wages.

Traditionally, skilled shipyard labor was among the highest paid group of wage earners in the country. During 1940 shipyard labor worked a little less than 40 hours a week to take home $34.81; for all workers engaged in durable-goods industries the average earning was $28.44 in 39 hours; and in building construction, although labor received higher hourly wages than in shipbuilding, they did not get as much weekly " take-home." From the point of view of earning power, shipyard labor could boast itself an aristocracy.

Their earnings remained the highest of any group of industrial workers comparable in size even after the amount of skill required had been diluted by the breakdown of jobs. The average weekly take-home in July 1941 from the yards building merchant ships

---

Brotherhood of Boilermakers, Iron Ship Builders and Helpers of America, *Proceedings of the Seventeenth Consolidated Convention*, Jan. 31 to Feb. 9, 1944, pp. 259-62.

[26] Manpower Survey Board reports of Jan. 1944 for Oregon Ship, Richmond Shipyards, Consolidated Steel Corp., and others, in MC, Historian's Collection, Research file 205.2.4a.

WEEKLY EARNINGS IN SHIPBUILDING AND IN OTHER INDUSTRIES

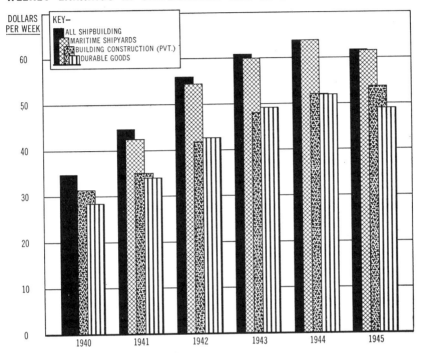

FIGURE 21. Annual averages from Fischer, "Wages," Tables I and II, based on data from Bureau of Labor Statistics.

was $42.38 for 41.8 hours worked. In 1944 it had gone up higher, to $63.90 for 45.2 hours. The standard straight-time pay of a first-class mechanic went from 70 cents or $1.00 to $1.20 per hour during the same period in which the amount of training and experience required to become a first-class mechanic was being drastically reduced—by as much as from four years to four months. The lead of shipbuilding over other industries which were competing with it for labor was maintained. Of course, to earn their average take-home of $63.90 in 1944 the shipbuilders, even if they were not versatile craftsmen, on the whole faced more strenuous outdoor conditions than did workers in, for example, aircraft factories, but a shipyard was the place to be to get big pay.[1]

[1] Fischer, *Statistical Summary*, Table G-10, p. 140; Gerald J. Fischer, "Wages in the Merchant Shipbuilding Industry," (typescript, Oct. 24, 1947), in MC, Historian's Collection, Research file 205.6.3.

High as shipyard wages rose in World War II, they rose less than in World War I. Although statistics are not available to permit exact comparison, roughly the increase in 1914-1918 was 150 per cent,[2] in 1940-1945 it was 57 per cent in gross hourly earnings, and even when measured in weekly take-home the increase was under 100 per cent. A national policy of freezing wages in all industries had the result that wages of all workers rose less during World War II than during World War I.[3]

These large increases in pay in 1941 and 1942 were attained by what were called "stabilization" agreements. How these agreements were reached, why they were called "stabilization," and how they increased the unionization of shipyard labor will be told in the next chaper. In connection with the building of the labor force the importance of these agreements lies primarily in the fact that they did raise the wage level high enough to attract workers, and secondarily in the fact that they made the labor leaders ready to help in the recruitment of new workers.

Finding workers was mainly the responsibility of shipyard managements, however. Each new yard contained a nucleus, even if very small, of experienced shipbuilders, and built around it until in most cases the nucleus was submerged. According to the policy which had governed the selection of sites, each yard was expected to make the most of the labor resources of its neighborhood. North Carolina Shipbuilding Co., for example, moved to Wilmington from the parent company at Newport News about 400 persons including at the top the president of the new company, Captain Roger Williams, and at the lower level 80 apprentices. They recruited at first from the small group of skilled mechanics in Wilmington—electricians, plumbers, and carpenters—who could turn their skills into shipbuilding, and then from the neighboring countryside, where there was little industry. A great many of the workers at North Carolina Ship came from the farms within a radius of 200 miles. Captain Williams was proud of the "character of the people in that part of our country" and the young men of high quality drawn into the shipyards by patriotic appeal. Having relatively little competition in its own

---

[2] Alexander M. Bing, *Wartime Strikes and Their Adjustment* (New York, 1921), p. 215.

[3] Fischer, "Wages in the Merchant Shipbuilding Industry"; Coll, "Stabilization," p. 203.

area, North Carolina had a comparatively stable labor force and was an instance in which the policy of locating yards in the South to tap its labor was eminently successful.[4]

North Carolina's nearest competitor for shipyard labor was Bethlehem-Fairfield. It too started from a strong nucleus of experienced shipbuilders, for the Bethlehem organization could supply them from the Key Highway repair yard and the Sparrows Point yard, across the Patapsco; and it also enlarged its force at first by using the unemployed skills in the city of Baltimore. Since there were many other defense industries in Baltimore the local labor market quickly became tight, however, and large numbers of workers were drawn to Baltimore from West Virginia and the Carolinas. These in-migrants came mostly from farms and had to learn how to live in a big city as well as how to work in a shipyard. Once in Baltimore they found other plants besides the shipyard ready to hire them. Whether because these in-migrants moved to other city jobs or because they went back to their Southern Appalachian farms, the turnover was relatively high at Bethlehem-Fairfield.[5]

Other examples show how small was the nucleus of experienced men even in the yards under the management of old-line shipbuilders. The yards at South Portland, Maine, like that at Wilmington, North Carolina, were expected to draw their labor from their own corner of the country and did so. A few hundred experienced men came from the affiliated yard at nearby Bath, but the bulk of the employees were new to shipbuilding and to the neighborhood of Portland.[6] At New Orleans the Delta yard was organized by about a dozen men sent by the American Shipbuilding Co. of the Great Lakes. They found in the city about 100 men who had previously worked in a shipyard, and building

[4] Williams to Land, Apr. 21, 1941, in MC gf 508-1-1; Truman Committee *Hearings*, pt. 17, Mar. 3, 1943, pp. 6993-96; North Carolina Shipbuilding Co., *Five Years of North Carolina Shipbuilding* (Wilmington, N. C., [May 1, 1946]), p. 15; Fischer, *Statistical Summary*, Table G-13, p. 144.

[5] William O. Weyforth, *Manpower Problems and Policies in the Baltimore Labor Market Area During World War II* (U. S. Dept. of Labor, U. S. Employment Service for Maryland, mimeographed, Mar. 27, 1946), pp. 23-72; Fischer, *Statistical Summary*, Table G-13, p. 144.

[6] Testimony of William S. Newell, Nov. 23, 1942, and Karl E. Klitgaard, Nov. 18, 1942, in Truman Committee, *Hearings*, pt. 15, pp. 6129, 6008, 6010-12, 6014-15; Wolff to Ring, Oct. 24, 1942, in Division of Shipyard Labor Relations file, Training in Shipyards.

on this nucleus they brought their payroll up to 13,300 in June 1942.[7]

The West coast yards started with even smaller numbers of experienced shipbuilders and grew until they were recruiting labor of all kinds from all over the country. The first shipyard of the Kaiser group was at Richmond building for the British. As hull superintendent they hired Edwin W. Hannay, who had been a trouble shooter in World War I, and had worked in many yards on the Pacific coast. In response to a telephone call from Edgar Kaiser he left retirement, and gathered about sixteen experienced shipbuilders to be top men under him; they brought along their friends; and so a nucleus was formed.[8] As new Kaiser yards were built, a few experienced shipbuilders were drawn from the old yards, although most of the organization, even the key men, were new to shipbuilding. Many had worked for Kaiser on his previous conquests of the " impossible," in dam building, and one of these who became impatient in the early days of Richmond No. 2 was heard to ask: " When do we pour the keel? " He was not fired; he stayed and became, according to Mr. Hannay, " quite a help to the shipbuilding." [9]

The Pacific Northwest had played a large part in shipbuilding during World War I and contained some skilled workmen, but it recruited labor from near and far. Dam-builders from Bonneville flocked especially to the Kaiser yards at Portland, Oregon. Lumberjacks, apple pickers, cherry pickers, and the like were hired until the local supply of labor was exhausted; and in August 1942, when the post-Pearl Harbor yards at Swan Island and Vancouver were getting ready, the Kaiser management began recruiting from other regions. Even before that, many in-migrants had come from midwestern and mountain states,[10] and by June

---

[7] The number of persons sent from the parent yard is estimated in E. B. Williams, " The Delta Shipyard," *Marine Engineering and Shipping Review*, Apr. 1943, p. 198, and J. M. Dalzell, General Superintendent, to Land, Apr. 13, 1942, in MC gf 508-1-6; Fischer, *Statistical Summary*, Table G-1, p. 126.

[8] Testimony of Hannay in House, *Production in Shipbuilding Plants, Hearings*, pt. 4, June 30, 1943, pp. 1000-01.

[9] *Ibid.*, p. 1015.

[10] Charles F. A. Mann, " Emergency Shipyard of the Pacific Northwest," *Marine Engineering and Shipping Review*, Oct. 1942, p. 199; statements by Maurice Nicholls, Aug. 8, 1947; " Shipbuilding Activity in Portland, Oregon," *The Log*, June 1942, p. 37; " Sources of Labor Supply in West Coast Shipyards and Aircraft Parts Plants " [prepared by Toivo P. Kanninen, under the direction of Louis

of 1943 they were coming from as far away as New York. Kaiser was advertising "help wanted" in eleven states.[11]

To the yards around San Francisco Bay also in-migrants came in large numbers. It has never taken much coaxing to get people to move to California. When carloads of "Oakies and Arkies" arrived at Richmond they found already working in the yards many folks from back home who had left in the migration born of drought and poverty in the mid-thirties.[12] Henry Kaiser had a large recruiting organization and under his agreement with the AFL unions, which will be described shortly, they also agreed to supply him with workers.[13]

The unions and the attractions of California were also among the chief means of recruitment at Calship at Los Angeles. It found on the lists of the employment offices 2,813 "usable" skilled applicants, which helped it get started. Later it drew on the lists for Arizona and New Mexico. The unions promised to assist by drawing on their membership in the oil fields.[14] After Calship was organized it supplied the nucleus which went to Sausalito to start Marinship on the northwest side of San Francisco Bay. The location of Marinship was designed to attract labor which could not so easily get to yards on the other side of the Bay, and it did draw several thousand from the agricultural and residential areas in the immediate vicinity. "Better than 90 per cent," said Kenneth K. Bechtel, President of Marinship, "had never worked in a shipyard before, and had never worked in the craft that they went to work in in our yard."[15]

M. Solomon], *Monthly Labor Review*, Nov. 1942, pp. 926-27; testimony of Edgar Kaiser in House, *Production in Shipbuilding Plants, Hearings*, pt. 4, June 30, 1943, p. 1100.

[11] Testimony of Clay Bedford in House, *Production in Shipbuilding Plants, Hearings*, pt. 3, June 18, 1943, pp. 622-23, 1099-1100.

[12] "Richmond Took a Beating," *Fortune*, Feb. 1945.

[13] Speech of Harry F. Morton in Metal Trades Dept. of the American Federation of Labor, *Proceedings*, 34th Convention, Sept. 27, 1943, p. 131; testimony of John P. Frey, in Truman Committee, *Hearings*, pt. 18, Mar. 26, 1943, p. 7384; Frank J. Taylor, "Builder No. 1," *Saturday Evening Post*, June 7, 1941, p. 11.

[14] California Shipbuilding Corp., Personnel Training Department, mimeographed report, May, 1941, in Gallagher's training file, pp. 17, 17a; W. E. Waste, Administrative Manager, to U. S. Maritime Commission, Apr. 26, 1941, in MC gf 508-1-1; W. E. Waste to McCone, Apr. 7, 1941, enclosed in McCone to Gallagher, Apr. 8, 1941, in Gallagher's training file.

[15] House, *Production in Shipbuilding Plants, Hearings*, pt. 3, June 22, 1943, pp. 691-92, 696, 706.

The same could probably have been said of the labor in the three new yards begun in the South after Pearl Harbor. They were all fairly close together, at Brunswick, Georgia, and at Panama City and Jacksonville, Florida. When the J. A. Jones Construction Co. took over two of these yards it brought in employees released from its construction projects, but the mass of workers was recruited from farms and small towns in Georgia and other southern states. Not all of them stayed. " Mistah," explained a departing welder, " ah'm fixin' to move back to Coffee County and raise me some more hawgs." Not all were farmers. There were several prizefighters who worked in the welding department next to a Ph. D. from the University of Pennsylvania; there were two professional golfers, and thirteen clergymen. A jockey and a former Wall Street investment banker were promoted to job instructors. And in the pipe-welding department there was an entire colored troupe, " The Original Silas Green New Orleans Shows." [16]

This variegated recruitment of a labor force may be numerically summarized in a few figures. The numbers working on contracts with the Maritime Commission were 47,300 in January 1941, and 650,900 at the peak in July 1943. Although rapid expansion began in 1941, it continued through to the end of 1942 at the rate of a net addition of 30,000 or 40,000 each month. For Maritime Commission building at least, the increase was proportionally largest on the West coast where the number at the peak was more than 30 times that in January 1941. On the Gulf it was about 20 times, in the Great Lakes 15 times, and on the East coast 6 times the figure of January 1941.[17] Not more than 3 per cent of the increase came through the re-employment of unemployed workers with shipbuilding skills. A somewhat larger portion was composed of workers who had skills that could be adapted to shipbuilding,[18] but the vast majority were from a shipbuilder's point of view completely unskilled.

[16] John A. Yancy, " Training a Democracy to Build Liberty Ships," *Marine Engineering and Shipping Review*, Nov. 1943, pp. 188, 193.

[17] Fischer, *Statistical Summary*, Table G-1, pp. 124-27. Figure of over 1 million given in a memorandum cited by Chaikin and Coleman, *Shipbuilding Policies*, p. 156, includes repair and Navy work in yards such as Federal which did work for both Maritime and Navy.

[18] " Characteristics of Recently Hired Shipbuilding Labor," *Monthly Labor Review*, May 1941, p. 1142; " Characteristics of Shipbuilding Labor Hired During First Six Months of 1941," *Monthly Labor Review*, Feb. 1942, p. 393.

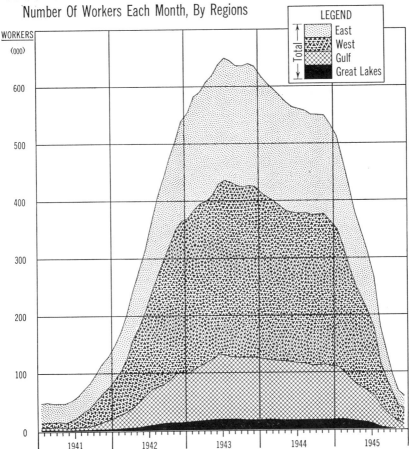

FIGURE 22. Source: Fischer, *Statistical Summary*, Table G-2.

Peak employment was reached in the summer of 1943; recruitment of new workers occurred mainly in 1942. West coast yards had substantially fewer employees than East coast yards at the beginning of 1941 but employed a larger number in 1942-1945. Employment dropped off in 1944 on the West and East coasts more suddenly than on the Gulf because of change-overs to new types and the scarcity of manpower in those regions.

III. NEGROES AND WOMEN IN SHIPYARDS

As the labor force approached its maximum, white male labor, even unskilled, could no longer be hired easily. Unemployment disappeared; the draft and mounting demands of the war industries created the prospects of a manpower shortage.[1] Consequently Negroes and women were hired, especially after 1942.

Negroes had been employed in shipyards as laborers but hardly ever for skilled work. Their importance if they could be trained and fitted into the new methods of multiple production was potentially greatest in the South. One of the attractive features of the plans of Andrew Higgins had been proposals for employing a large amount of Negro labor. Although he proposed to put them on separate "assembly lines," and to segregate them to that extent, his proposals were welcomed because they involved putting more Negroes into more skilled work.[2] In most yards there was no segregation, but the Negroes were restricted to lower paying, less skilled work; and the main problem was upgrading Negroes so that they could be employed more extensively. Accomplishing this required the cooperation of management, of unions, of communities, and of the government.

Discrimination on account of race was forbidden by a clause inserted in the Maritime Commission's contracts in accordance with the President's Executive Order to that effect,[3] and a special agency had been appointed to enforce the Presidential order against discrimination, the Fair Employment Practice Committee. When this committee called on the procurement agencies to enforce the Presidential order against discrimination, they said in reply that they were concerned also with policies "consistent with maximum production."[4] On that ground the Maritime

---

[1] *The United States at War*, pp. 429-32.

[2] Higgins to Vickery, Mar. 12, 1942; the President to Land, Mar. 21, 1942; Higgins to Vickery, May 12, 1942, and attached letter of Higgins to James A. Dombrowski, in MC gf 503-54.

[3] The Executive Order, No. 8802, was issued June 25, 1941. The clause in the Maritime Commission contracts read as follows: "*Fair Employment Practice*: The Contractor agrees that in the performance of the work under this contract, it will not discriminate against any worker because of race, creed, color or national origin." Skinner to Gibbs & Cox, Aug. 8, 1942, in MC gf 120-133.

[4] Malcolm S. MacLean, Chairman, Fair Employment Practice Committee, to Land, May 26, 1942; Ring to Chairman via Macauley, June 18, 1942; Stimson, Knox and Land to McLean, July 2, 1942, all in gf 120-133.

Commission objected to the issuance by the Fair Employment Practice Committee of orders to the shipyards without the Commission's being informed. Arrangements were made for prior consultation.[5]

Such consultation took place at length regarding the hiring and promoting policies of the Alabama Dry Dock and Shipbuilding Co. at Mobile when it was put under the spotlight as an instance in which local supplies of Negro labor were not fully used although white workers had been brought into the community in such large numbers as to create unsavory living conditions. After months of correspondence and conferences with the company,[6] Admiral Vickery reported in March 1943 to the Fair Employment Practice Committee that Alabama was employing over 6,000 Negroes (total employment in the yard was 18,500) and had recently upgraded Negroes to be chippers, riggers, painters, and even foremen.[7] Skilled Negroes were employed only on certain shipways, however. Two months later some additional upgrading or the assignment of Negro welders to shipways where all workers previously had been white touched off a race riot in which many Negroes were severely injured. A portion of the yard was shut down completely for almost a week.[8] After three weeks a "settlement" was reported by which skilled Negroes would be confined to 4 of the 12 ways.[9] To use with maximum productivity the labor of the region a change in the cultural pattern would have been necessary, and at Mobile there was no leadership which succeeded in bringing about much change.

The most definite successes in breaching the tradition of discrimination were in the areas in which Negroes were least

[5] Vickery to Cramer, Executive Secretary, FEPC, Nov. 2, 1942; Cramer to Vickery, Nov. 6, 1942, in gf 120-133.

[6] Testimony of Paul V. McNutt, Chairman, War Manpower Commission, in Senate, *Manpower, Hearings* before the Committee on Military Affairs, [Revised]. Oct. 21, 1942, p. 16; Hillman to Land, Mar. 6, 1942; Land to Dunlap, Mar. 17, 1942; Dunlap to Land, May 1, 1942; Cramer to Alabama Dry Dock, Nov. 19, 1942; Cramer to Vickery, Nov. 21, 1942; Cramer to Vickery, Dec. 9, 1942; Sanford to Vickery, Dec. 11, 1942; Dickerson to Vickery, Mar. 1, 1943, all in MC gf 120-133.

[7] Vickery (written by Tracy) to Dickerson, Mar. 23, 1943, in MC gf 508-1.

[8] *New York Times*, May 26, 1943, June 1, 1943; National Urban League for Social Service Among Negroes, *Summary of a Report on the Race Riots in the Alabama Dry Dock and Shipbuilding Company Yards in Mobile, Alabama*, (mineographed, June 25, 1943), in gf 508-1.

[9] *New York Times*, June 12, 1943.

numerous. At Moore Dry Dock on San Francisco Bay the President of the Company himself took an interest and wrote proudly to the Maritime Commission that a Negro had been accepted for the first time by the Metal Trades unions. Later, several thousand Negroes were employed in the yard, many fresh from the South, and were used for skilled work as well as unskilled.[10] When the Kaiser organization began hiring in New York for their Oregon yards, they were reminded that they must not discriminate, and in consequence filled some cars of their westbound trains with Negroes.[11] At Portland, Oregon, there was a mass meeting in December 1942 which was addressed by Daniel Ring for the Maritime Commission and by the leader of the AFL Metal Trades Department, John P. Frey, and by Edgar Kaiser. The aim of the shipbuilding program, the white workers were told, was production. Could production be maintained if there was segregation? Edgar Kaiser said "No," and the question was considered closed.[12]

It was reopened in a slightly different form in the summer of 1943 because of discrimination within the boilermakers' union. The boilermakers segregated Negroes into auxiliary locals, their grievances were taken up by the white officials of other locals, and they were not allowed to choose delegates to the union convention which elected national officers. Under these conditions several hundred Negroes at Portland refused to pay dues, and the shipyard fired them under the terms of the closed shop contract.[13] The FEPC ordered the Negroes reinstated. When the boilermakers refused to agree, the issue was fought out in the courts. In the spring of 1945 the Supreme Court of California held that the union having a monopoly of the labor supply could " no longer claim the same freedom from legal restraint enjoyed by golf clubs or fraternal associations " and therefore " Negroes must be admitted to membership under the same terms and conditions applicable to non-Negroes unless the union and

---

[10] Joseph A. Moore to Land, July 30, 1941 in MC gf 120-133; Katherine F. Archibald, *Wartime Shipyard: A Study in Social Disunity* (Los Angeles, 1947), pp. 59, 61.

[11] Statement of Maurice Nicholls, Aug. 8, 1947.

[12] Ring to Anna Rosenberg, n. d., but written sometime in Dec., 1942, in MC gf 120-133.

[13] Malcolm Ross, *All Manner of Men* (New York, 1948), pp. 142-46; Williams to Vickery, July 30, 1943, and Ring to Legal Division, June 3, 1944, in MC gf 120-133.

the employer refrain from enforcing the closed shop agreement against them." Letters of compliance were immediately sent in by the shipyards, but the boilermakers continued their protest, and the war ended without complete compliance with FEPC's orders.[14] Meanwhile the boilermakers led by their President, Charles J. MacGowan, had moved a long way toward eliminating discrimination when at their 1944 Convention they gave the Negro auxiliaries voting privileges at conventions, membership in local Metal Trades councils, and the choice of their own business agents.[15] Most Negroes were willing to pay dues to the union under these conditions.

A special effort to employ large numbers of Negroes was made at the Sun yards in Chester, Pennsylvania. In the fall of 1943 Sun was employing 18,000 Negroes in a total of 34,000. Out of this number, 7,500 were in yard No. 4, which was exclusively Negro except for 500 white supervisors. No Negro was a supervisor and the Negroes felt that this was due to discrimination, not to the Negroes' lack of experience.[16] Production in this No. 4 yard was very poor for many reasons. The ships being built there were transports for which the plans were changed many times, and these changes made it difficult to keep the workmen busy at all times. Idleness could be expected to lower morale. But the Maritime Commission's staff at Sun reported: "Labor is the root of the unsatisfactory conditions in that yard." They noted that supervisors had been subjected to bodily injury, and commented: "The pugnacious attitude of the workmen is a grave discouragement to the leaders."[17] The resident auditor characterized the employees as being until quite recently cotton pickers in South Carolina "whose mechanical skill was limited to repairing Model-T Fords with baling wire. The Maritime Commission," he philosophized sardonically, "has assumed the lead in transforming this group of people into the highest skilled class of workmen in the country. They are probably the least qualified

[14] Ross, pp. 146, 149-52.
[15] *Ibid.*, pp. 147-48; International Brotherhood of Boilermakers, Iron Ship Builders, and Helpers of America, *Proceedings*, 17th Convention, 1944, pp. 295-98; Ring to Legal Division, June 3, 1944, in MC gf 120-133.
[16] George M. Johnson, Assistant Chairman, FEPC to Vickery, Sept. 13, 1943 in MC gf 120-133.
[17] Hunt, Watt, Menzies, Wolnski, Limozaine to William H. Blakeman, Assistant Regional Director, East Coast, July 29, 1943, in Blakeman file, Yard Staff Committee, Sun SB and DD. Co.

for this over-night transformation. In embarking on this comprehensive training project the Commission should not depend on production from this Yard to materially assist in the war effort. This Yard should be regarded solely as a training and educational project, . . . The No. 4 Yard is nothing but a melting pot from which we expect much but the result depends on the admixture of patience, time, planning and much money put into it." [18] In the end it did build some complicated ships, but much behind schedule.[19]

A more conservative attitude towards the employment of Negroes was adopted by the Bethlehem management of the Fairfield yard in Baltimore. It employed in October 1942 only 8.5 per cent Negroes, 2,800 out of a total of 33,000 workmen, although it had just received a commendation for "the integration of Negroes." [20] Negroes composed 18 per cent of the population of Baltimore City and were traditionally engaged in heavy laboring work or personal service.[21] Vocational training was given in three Negro schools, however, and in the scholastic year 1941-1942 an additional vocational school was opened and developed courses for shipbuilding industries.[22]

For the country as a whole no figures are available to show how many Negroes were employed in the shipyards or how many attained the grade of skilled mechanic, even within the diluted wartime meaning of the word. Certainly some progress was made.[23] Some Negroes were employed on skilled jobs in merchant shipbuilding, but there is no means of measuring the extent to which racial barriers prevented others from exercising skills already mastered or from learning new ones.

[18] Hunt to Blakeman, July 30, 1943, in Blakeman file, cited above.
[19] Below, chap. 18, and Helen E. Knuth, "The C-4 Shipbuilding Program," (typescript, June 30, 1947), in Historian's Collection, Research file 111, C4's, pp. 21-22, 29-31.
[20] Berno, Industrial Relations Advisor, East Coast Regional Office, to Macauley, Oct. 13, 1942, enclosing copy of letter from Herbert R. O'Conor, Governor of Maryland, to John M. Willis of Bethlehem-Fairfield, Oct. 5, 1942, in MC gf 120-133.
[21] Weyforth, *Manpower Problems and Policies in the Baltimore Labor Market Area*, pp. 18-19.
[22] Sanford Griffith, *Where Can We Get War Workers* (Public Affairs Committee, 1942), reprinted in Senate, *Manpower (National War Service Bill)*, *Hearings*, 78th Cong., 1st sess., pt. 6, [Unrevised], p. 277, and Board of School Commissioners of Baltimore City, *Annual Report*, Scholastic Year Ending June 30, 1942, pp. 85, 87.
[23] Ring to Legal Division, attention W. Ney Evans, June 3, 1944, in MC gf 120-133.

NEGROES AND WOMEN IN SHIPYARDS 257

The employment of women in shipyards was more accurately recorded. It reached its maximum in 1944 and 1945 when female workers formed 10 to 20 per cent in most yards. The percentage of women was lowest in the yards of the North Atlantic and the Gulf, highest on the West coast. In the Richmond yards it was 20 to 23 per cent in 1944 and at Oregon Ship it rose as high as 31 per cent in May 1945.[24] This feminine influx was a far more striking change than any that took place in the employment of Negroes. There was never any question that Negroes should be admitted, at least in unskilled work. But the shipyard had been a man's world.

The female invasion began in the fall of 1942.[25] It was accompanied by much misgiving over the idea of women entering such heavy work full of physical and moral hazards, and it became a subject people wanted to read about, so that a pocketsized *Shipyard Diary of a Woman Welder* appeared in the nation's drugstores.[26]

Once the yards started hiring women they hired them to do practically everything. In the yards as at South Portland, Maine, which employed an average number, about half were welders; the rest were scattered through the yard as shipfitters' helpers, burners, crane operators, pipefitters, and so on.[27] The Kaiser management found that women could be trained quickly to do better than men in installing complicated electrical connections involving fine finger work.[28] There were plenty of women leadermen (leaderwomen) but only one woman, at Marinship, became a forelady; and it was generally said that men did not like to work under a woman.[29] The Women's Bureau of the Department

---

[24] Fischer, *Statistical Summary*, Table G-3, p. 129.

[25] Bureau of Labor Statistics, *Bulletin No. 824*, p. 6.

[26] Dorothy K. Newman, "Employing Women in Shipyards," *Bulletin of the Women's Bureau, No. 192-6* (Government Printing Office, 1944); Archibald, *Wartime Shipyard*, pp. 16-17, 19; *Shipyard Diary of a Woman Welder* was written by Augusta H. Clawson and published by Penguin Books (1944).

[27] Newman, "Employing Women in Shipyards," p. 21; Apprentice-Training Service, Bureau of Training, War Manpower Commission, "Report on Training at the South Portland Shipbuilding Corporation," *Technical Bulletin No. 107* (mimeographed, Feb. 1943), pp. 1, 8.

[28] Statement of Maurice Nicholls, Aug. 8, 1947.

[29] Report of Emma F. Ward, Shipyard Personnel Consultant, submitted July 6, 1945, in Division of Shipyard Labor Relations files; Archibald, *Wartime Shipyard*, p. 29.

of Labor warned that women had more difficulty than men in keeping their balance in high places, that they had one-half the muscular strength, and were more subject to fatigue. Consequently it recommended inside work in the fabricating shops.[30] The author of the *Shipard Diary* recounted the grim determination with which she conquered fear of heights, noise and fatigue, though conquer them she did.[31] On the whole women were employed most in the fabricating shops and in welding.

Although in some yards women formed 25 per cent of the employees, and Negroes were as many or even more in a few yards, and although in every yard there was some nucleus of experienced shipbuilders, taken as a whole the labor force of the shipyards was predominantly masculine, white, and inexperienced. The new methods of construction made it possible to draw on all parts of the nation's manpower, and womanpower, to carry on one of the most difficult of industries. Actual construction of the world's largest merchant marine was the handiwork of farmers, shopkeepers, housewives, and workers recruited from every walk of life.

IV. Training Programs

The task of training these inexperienced workers was divided between shipyard managements and the educational institutions outside. The original nucleus of skilled men in each yard naturally gave some instruction to newcomers. Thus the experienced men became supervisors and promoted their more energetic workers to be leadermen, that is, pace setters for other new workmen. The raw recruit who arrived without any acquired skill that could be applied to shipbuilding was generally called a learner or trainee, since the term apprentice was associated with a long and diverse course. The trainee advanced rapidly through various grades, such as handyman or helper, to attain as " mechanic " top ranking in what had once been an aristocracy of highly trained craftsmen. In June 1943, about 33 per cent were classified as first class mechanics. In the rush of expansion many a " mechanic " who had begun as trainee shortly before moved up higher to be a " snapper," " pusher," leaderman, or some other

[30] Newman, " Employing Women in Shipyards," pp. 8, 22-25.
[31] Clawson, *Shipyard Diary, passim.*

TRAINING PROGRAMS 259

kind of supervisor, for this supervisory staff formed from 7 to 10 per cent of the yard personnel.[1]

Although the supervisors were relatively inexperienced in shipbuilding, that was not the main source of their inadequacies. Many of them did not know how to instruct or how to handle the men under them. Consequently it was necessary to teach the more experienced men how to behave as supervisors at the same time that they were instructing the new trainees in the elements of their shipyard tasks. A report by the shipbuilders' professional association, the Society of Naval Architects and Marine Engineers, in April 1941 said that developing a corps of competent supervisors was the key to the success of the emergency program, but that little if any effort was being made to train such personnel.[2]

Efforts increased as the program grew. To help management train supervisors became in August 1941 the object of the Training Within Industry Service established in the Office of Production Management by C. R. Dooley, an executive experienced in industrial relations and training.[3] Its courses were brief, only five two-hour meetings, and it offered three " job " courses; one in how to teach a job, one in methods of organizing it, and one in the human relations of supervision.[4] Its services were used by the West coast shipyards especially.

Although much of what any new man learned was acquired while he was working at his job, separate schooling was also needed. Initial instruction in welding for example was everywhere to be gained in public vocational schools.[5] But on the North Atlantic coast many managements undertook to provide

---

[1] War Manpower Commission, Training Within Industry, *Basic Principles To Be Observed in Establishing Production Training Programs in Shipyards* (Government Printing Office [1942]); U. S. Dept. of Labor, Bureau of Labor Statistics, " Hourly Earnings in Private Shipyards, 1942," *Bulletin No. 727* (1943), pp. 20-23; Bureau of Labor Statistics, *Bulletin No. 824*, p. 29; U. S. Dept. of Labor, Bureau of Labor Statistics, " Basic Wage Rates in Private Shipyards, June 1943," [prepared by A. A. Belman], *Serial No. R. 1679* [1944], p. 5.

[2] H. E. Rossell, et al., *The Training of Shipyard Personnel* (The Society of Naval Architects and Marine Engineers, April 1941), p. 16.

[3] *Who's Who in America.*

[4] War Manpower Commission, Bureau of Training, Training Within Industry Service, *The Training Within Industry Report 1940-1945* (Government Printing Office, Sept. 1945), pp. 33-45.

[5] W. Daniel Musser, " Vocational Training for War Production Workers, Final Report," *Bulletin 1946 No. 10* (Federal Security Agency, U. S. Office of Education, Government Printing Office, 1946), pp. 59-61, 158-75.

this pre-employment training and to supplement it by offering courses to employees who wanted to move up. The company whose training system had the highest reputation of any in a Maritime Commission yard was that of Bethlehem at Baltimore which began in 1940.[6] The instructional booklet which it prepared was printed by the government for use in other yards. As this booklet shows, the Bethlehem system took off from the old ideal that the worker should understand what he was making, but it simplified instruction as much as possible and gradually shortened more and more the pre-employment training.[7] Approximately one hour a day was spent in the classroom; the remainder of the time on production work under the instruction of selected supervisors. In the first six months it trained about 1400 workers. Twelve weeks was at first considered about average time for in-school training for welders. Later, training time was reduced to eight, to five, to three.[8] Other old-line shipbuilders of the Northeast accustomed to training shipbuilding craftsmen also speeded up their programs somewhat.[9]

The Bethlehem training school and its booklet epitomize one aspect of the training which had to be given workers. Another aspect of what they needed to learn was epitomized in a booklet published for the Richmond yards as a manual for new workers.[10] It came to the " ABC of Shipbuilding " only on page 114. The previous pages were devoted to explaining the social enterprise in which the worker was becoming a part. The Richmond booklets opened with pictures of John Bull and Uncle Sam stretching out a friendly hand to each other, with the American flag, or with a statement of what the Maritime Commission was

[6] Wolff to Ring and Schmeltzer, Nov. 27, 1942, in Division of Shipyard Labor Relations file, Training in Shipyards, 1942; statement by W. B. Fairweather, Assistant Supervisor of Shipyard Training, attached to memorandum Wolff to Director, Division of Shipyard Labor Relations, May 31, 1943, in Division of Shipyard Labor Relations file, Training in Shipyards; Wolff to McInnis, att. W. H. Burns, Mar. 23, 1944, in MC gf 508-1-1.

[7] Bethlehem Steel Co., Shipbuilding Division, *An Introduction to Shipbuilding* (War Production Board, Labor Division, Training Within Industry, Government Printing Office, 1942).

[8] Testimony of A. B. Homer, Bethlehem Steel Co., in Truman Committee *Hearings*, pt. 6, July 16, 1941, pp. 1562-63.

[9] Rossell, *The Training of Shipyard Personnel*, pp. 12-13.

[10] [Kaiser Manuals] Todd-California, *Full Ahead* (1942); Richmond Shipyard Number Two, *Full Ahead* (1942); Richmond Shipyard Number Three-A, Kaiser Co., Inc. (1943).

and a welcome signed by C. P. Bedford, General Manager of the Richmond yards. It talked of pay days and the personnel office, of safety and first aid, and then, and only then, came to technical matters concerning shipbuilding. Even in discussing shipbuilding, the approach was from the social point of view, for the handbook dealt with the different kinds of jobs there were in the yard, and the illustrations at least reminded the reader that the jobs were being done by human beings. A good deal of technical knowledge about shipbuilding was included first and last, but the book's chief value was in explaining the human organization of which the beginner was becoming a part.

Next to the yard managements and supervisory staffs, the most important agencies of training were the public schools. On the Gulf and West coast the new yards depended at first almost entirely on the public schools to give basic technical training.[11] It was a major step in the general process of preparing the United States for war when the U. S. Office of Education began disbursing funds to the schools in June 1940 for pre-employment courses to provide specific and intensive training for definite jobs on the pay rolls of shipyards and other "defense industries."[12] After the growth of war industry had wiped out unemployment and industries were competing for workers, the shipyards found it desirable to hire workers and train them through "vestibule schools" after they were on the company pay roll. Also workers established in one kind of a job could take supplementary training, in order to be promoted to more skilled work. Courses were organized and instruction given according to job specifications and analyses arrived at jointly by industry and schools, and the length of the daily schedule was similarly determined. Most of this training was done in a portion of the plant, known as a training area, in which regular production procedure was gone through, and the usual standards of workmanship required, but the workers in training were not expected to meet production line schedules.[13] For shipbuilding, 436,930 were enrolled by the

---

[11] See letters to the Commission from Calship, Sun, North Carolina, Ingalls, Alabama, Louisiana, Todd-Bath and South Portland, Seattle-Tacoma and Richmond, in MC gf 508-1-1, Apr. 1941; Musser, "Vocational Training for War Production Workers," Table 9, pp. 156-57.

[12] Musser, "Vocational Training for War Production Workers," pp. 18-19.

[13] *Ibid.*, pp. 59-62.

public schools in pre-employment courses, 970,056 in supplementary courses.[14]

Supplementing the work of the public vocational schools were the activities of various agencies of the Federal government. The National Youth Administration, which was engaged in training unemployed young people to tide them over the period of slack employment, adapted itself to the defense program in 1941 and some of the youth it trained entered shipyards.[15] Others came from the Works Progress Administration.[16] The Apprentice Training Service of the Department of Labor not only continued to advise companies about apprenticeship programs with prewar standards, but assisted industry with all types of on-the-job training.[17] To coordinate all these agencies, and also the Training Within Industry Service mentioned earlier, was one of the tasks of Sidney Hillman when he was the head of the Labor Division of OPM. The President had directed him "to undertake full responsibility of getting the necessary workers into the industries claiming manpower shortages," not only shipyards, but also the other defense industries.[18]

With so many other government agencies active in regard to training labor, the Maritime Commission held back. It felt that training of labor, like recruiting it, was a responsibility of management—one of the functions to be performed by the companies in return for their fees.[19] Doubts did arise whether the shipbuilding companies were doing all that they could, but the Commission then acted only to bring the facilities offered by various government agencies to the attention of the yards. It drew back from anything which would seem to duplicate the efforts of Mr. Hillman's groups.[20] When proposals were first

[14] *Ibid.*, Tables 9 and 10, pp. 156-59.
[15] *The United States at War*, p. 180; "NYA Trains Shipyard Workers," *Marine Engineering and Shipping Review*, Feb. 1942, p. 104.
[16] *The United States at War*, p. 180.
[17] Fred W. Erhard, Principal Apprentice Field Representative, to Schmeltzer, Aug. 6, 1941, in Gallagher's training file.
[18] The President to Hillman, May 28, 1941, quoted in *The United States at War*, p. 180.
[19] Memorandum, Gallagher, King and Knight to Chairman, Mar. 14, 1941; Land (written by Scoll) to Alabama Shipbuilding Corp. and 16 other yards, Apr. 19, 1941, in MC gf 508-1-1.
[20] MC Minutes, May 1, 1941, p. 17152; Land (written by Scoll) to Hillman, May 13, 1941 in gf 508-1-1; testimony of Commissioner Dempsey in Truman Committee, *Hearings*, pt. 5, June 3, 1941, pp. 1467-68.

TRAINING PROGRAMS   263

made by members of the Commission's staff concerning training in shipyards, there was a question as to which Commissioner was chiefly concerned: Commissioner Macauley, who was inclined to act in the matter; or Commissioner Vickery, who was inclined to leave it to management.[21] In June 1942 the Commission seemed about ready to adopt a positive program, but cuts in steel allocations led Admirals Land and Vickery to caution the yards in August against overexpansion, whether in training or in anything else.[22] At least one observer of the general labor situation said in September 1942, " the Maritime yards are the weak spot in the training picture." [23] By that time there was much talk of a manpower shortage. Mr. Hillman had dropped out of the picture, and Mr. McNutt, the Chairman of the War Manpower Commission, was urging a labor draft.[24] It was feared that selective service would draw skilled labor from the shipyards so that there would be need of continually training more labor.[25] That September the Commission ceased to hold back. It approved a set of " Basic Principles in Establishing Production Training Programs " and appointed a Training Director.

If the Commission was resolved to use its power, it could exercise a more precise control of training than could the groups in the WPB or the War Manpower Commission—because it was paying the bill. Under the contracts between the Commission and the builders, the Commission was reimbursing the companies for what they spent in training their employees. By refusing to reimburse the builders whose training programs were inadequate the Commission could impose its policies. Moreover, if a trainee

[21] Deimel to Macauley, May 21, 1941 and Macauley to U. S. Maritime Commission, via Vickery, Nov. 26, 1941, encloses memorandum Tracy to Ring, in Division of Shipyard Labor Relations file, Training in Shipyards, 1942. The memorandum was not presented to the Commission. Tracy to Ring, Apr. 6, 1942, in Division of Shipyard Labor Relations file, Training in Shipyards, 1942.

[22] MC Minutes, pp. 22231-33 (June 18, 1942); pp. 22281-82 (June 23, 1942). A long hand note, Land to Vickery, n. d., attached to memorandum, Deimel to Land enclosing material on shipbuilding employment situation abstracted from Federal Security Agency confidential reports, July 3, 1942, cautioned against overexpansion. Land to Regional Directors of Construction, Aug. 24, 1942, in East Coast Regional Office file, Director, 1942.

[23] Memorandum, Aylwin Probert to C. R. Dooley, Sept. 3, 1942, in MC gf. 508-1-1.

[24] *The United States at War*, pp. 187-89.

[25] James W. Wilson, Welding Inspection Coordinator, to Vickery, enclosing Survey of Welding Conditions in Shipyards having U. S. M. C. Contracts, June 26, 1942, in MC gf 606-65.

was promoted to "first-class mechanic" without having gone through adequate training, the Commission might question such a classification on the pay roll and threaten not to give the companies full reimbursement of the wages they had paid out. Because of its power to approve or disapprove a builder's costs before reimbursement, the Maritime Commission could compel changes in a training program. In addition, it was more likely to be listened to than other government agencies just because industry recognized that a "customer" had rights to a say in how his money was spent.[26]

The principles for training which would be favored by the Commission were never in serious doubt. They were like those being developed in Bethlehem's training program. After being urged from time to time by members of the Commission's staff, they were fully and authoritatively set forth in June 1942 by a committee consisting of G. Guy Via of Newport News, Newell A. Hogan of Consolidated Steel's yard at Orange, Texas, and Jack Wolff of Kaiser-Richmond. Their work was initiated by the Maritime Commission, and, when modified and approved by representatives of the Navy, the Manpower Commission, and on behalf of the Maritime Commission by Commissioner Carmody, Mr. R. G. Hooker, Assistant to Commissioner Macauley, and Mr. Ring, was published and broadcast as "Basic Principles to be Observed in Establishing Production Training Programs in Shipyards."[27]

The "Basic Principles" endorsed fully "breaking down" jobs into tasks requiring less skill, teaching the trainee just enough to make him a productive worker, and then placing him on a job before giving him more training. Although the Director of the Technical Division early in the program had lamented this trend towards "the vertical organization of labor," he had recognized that it was sound procedure in the emergency.[28] The wartime expansion had caused more and more emphasis to be put on short, highly specialized training for a few weeks, or for as little even as a few days, followed by supplementary training a few hours

[26] Assistant General Counsel to Director, Construction Division, Oct. 24, 1941, in Gallagher's training file; and below chap. 13, sect. II.
[27] War Manpower Commission, Training Within Industry, Conference on Training for Shipyards, July 8-14, 1942, in Carmody file, Labor Relations; *Basic Principles To Be Observed in Establishing Production Training Programs in Shipyards*.
[28] Bates to Macauley, June 10, 1941, in MC gf 508-1-1.

a week after the trainee had become part of the productive labor force.

The importance of giving training to supervisors was also emphasized in the "Principles." The kind of training they needed was just the kind which Training Within Industry had been offering in its three courses: how to analyze a job, how to teach it, and how to handle men. These could be taught also in supervisors' conferences where men could bring up their problems and learn from each other's experiences.

Where the "Principles" entered furthest into management's field was in advising how the training program should be related to the rest of the yard organization. Yards employing a thousand or more workers should retain a Training Director, they were told, who should have the status of a superintendent and a voice in management. Training outside the yard, such as was being done in the public schools, was approved, provided companies were in control, recommending instructors and giving them guidance in determining the content of the courses so that they would prepare for specific shipyard jobs. Similarly, after a trainee was employed, he should be prepared for more difficult jobs by supplementary training planned by the yard management. All upgrading, it was recommended, should be systematized under the direction of a training supervisor. In addition to these emergency training programs, the yards were encouraged to provide long-term apprentice training. Monthly reports were requested from each yard to show the type and method of training, the hours involved, the numbers enrolled and the location of training centers.[29]

Some of the yards were already doing just what the "Principles" recommended, but many others were not. To get the "Principles" put into practice was the task of the training director whom the Commission appointed September 22, 1942, Jack Wolff. He had been Director of Personnel Training at Richmond Shipyard No. 1.[30] Mr. Wolff set to work to bring about a closer connection between the needs of production and the

---

[29] *Basic Principles To Be Observed in Establishing Production Training Programs in Shipyards.*

[30] Ring to U. S. Maritime Commission, via Macauley, Sept. 11, 1942, approved by Commission, Sept. 22, 1942, in MC gf 508-1-1.

teaching being done in the vocational schools. He objected that students were being kept on nonproductive work in welding booths for 300 to 400 hours when they could have been put to productive work as tackers much sooner. His criticism of shipyard managements, on the other hand, was that they were disinclined to look ahead and see the need for further training to replace the skilled men who would be taken by the draft. Some yards had good programs, others did not; and he felt there was a big task ahead of him.[31]

To carry it out, assistant training supervisors functionally responsible to Mr. Wolff were appointed in each of the four regional offices to make " periodic inspections of training activities," [32] and a flow of reports and instructions between Mr. Wolff and these assistants discussed the deficiencies of many yards. He was especially interested in organizing supervisory conferences in which top management and the supervisors most closely in touch with the workers went over problems together, for Mr. Wolff believed firmly that the lower levels of supervisory personnel could contribute to as well as learn from such conferences.[33] There is no written evidence, however, that his efforts were strongly backed by Admiral Vickery's personal prestige, for Admiral Vickery was not inclined to have the Commission take over a responsibility for training which was, he felt, properly the responsibility of management.[34] Neither was the training program enforced through the limiting of upgrading in connection with the checking over of wage rates.[35] Consequently, Mr. Wolff felt his efforts were successful in some yards, but training remained a function performed by management as it thought best, putting into effect in the main the policies in accord with those defined by the Commission.[36]

[31] Wolff to Ring, Nov. 6, 1942, Nov. 20, 1942, Jan. 29, 1943, in Division of Shipyard Labor Relations file, Training in Shipyards.
[32] Ring to USMC, Jan. 6, 1943, approved by Commission, Jan. 12, 1943, in MC gf 508-1-1.
[33] MC gf 508-1-1 and the Division of Shipyard Labor Relations file, Training in Shipyards, are full of Wolff's correspondence with the field.
[34] On Vickery's general attitude towards managerial responsibilities see chap. 14 below and note 21 above.
[35] See chaps. 13 and 20.
[36] Wolff to Director, Division of Shipyard Labor Relations, Feb. 2, 1944, in Division of Shipyard Labor Relations file, Training in Shipyards.

PLATE XI. Above, blacksmithing a deck fitting for a cargo ship. Below, apprentice welders learning their craft under the direction of a training supervisor.

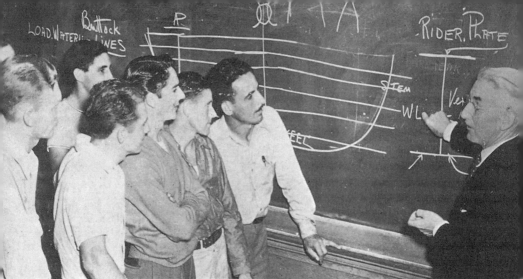

Plate XII.

Above, a class for shipfitters at the Richmond shipyards. On the left, changing shifts at the North Carolina Shipbuilding Co. Below, some of the women workers who were drawn into the shipyards in larger and larger numbers as manpower became scarcer during 1943. These are apprentices practicing burning. The use of torches for cutting steel and preparing it for welding was an important technique in the shipyards of World War II.

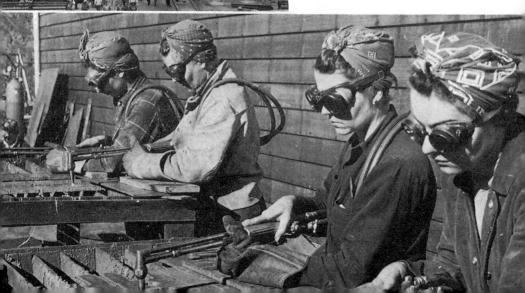

At the beginning of 1943 the labor force for shipbuilding had very nearly reached its maximum. High wages and the breakdown of craft skills had enabled managements to recruit a labor force as large as was needed. But the task of training labor was very far from finished. The need for skill in the labor force, although apparently overcome with surprising ease in the first years of the program, returned to haunt the Commission in its efforts at stabilizing wages, in its worries over cracks in welded ships, and in the new manpower and managerial crisis of 1944.

CHAPTER 9

COLLECTIVE BARGAINING

I. THE FORMATION OF THE SHIPBUILDING
STABILIZATION COMMITTEE

IN UTILIZING and recruiting labor, management played the leading role. In the fixing of wages and in the whole complex of problems relating to labor unions, management did not have an equally free hand. Collective bargaining between management and labor was guided or supervised by government agencies, and to some extent the procurement agencies became themselves parties to the process of collective bargaining.

To assume that collective bargaining was the way wages and similar issues would have been settled even if the government had not stepped in is, perhaps, to assume too much. In 1940 only a part of shipyard labor was organized into unions affiliated with the American Federation of Labor or the Congress of Industrial Organizations. Within the AFL, the Metal Trades Department was the coordinating body for thirteen craft unions covering the shipyard trades. They were strong in the shipyards of the Great Lakes, in most yards on the West coast, in some on the Gulf, and in a few on the East coast. Within the CIO, the Industrial Union of Marine and Shipbuilding Workers of America admitted workers of all shipbuilding trades to one union. It was strong in two big Eastern yards, New York Ship and Federal, and in some elsewhere. After its sensational success in organizing the previously unorganized workers in automobile and steel industries, the CIO was driving to organize all the big old-line shipbuilders. But local unions independent of both AFL and CIO dominated in some of the most important yards—Newport News, Sun, the big Bethlehem yards, Bath Iron Works, and the Todd repair yards. Leaders of the CIO accused these local "independent" organizations of being tools of management, and accused the big old-line shipbuilding corporations of wishing to determine

FORMATION OF STABILIZATION COMMITTEE

wages and working conditions by managerial decisions without real collective bargaining. The attitude of employers towards unions differed according to whether the employer was in the shipbuilding business temporarily, as were Kaiser, McCone, and Bechtel, or whether they had a past and future in shipbuilding as did old-line yards of the Northeast. The latter did not wish to establish conditions which would carry over into the postwar period and might affect the company's business unfavorably at a time when the government was not footing the bills. Certainly Bethlehem, Newport News, and Sun were opposed to the efforts of the CIO to organize their yards, but these companies and their subsidiaries dealt with local associations as representatives of their own employees, so that even they accepted, in form at least, the principle of collective bargaining which had been made the law of the land by the National Labor Relations Act. The relative role of "independent" unions and of national unions was one of the questions to come before the agencies through which the government exercised influence over collective bargaining.

The government agencies guiding collective bargaining in 1941-1945 were in part new, in part those by which the New Deal had already before the war been carrying through its policy of aiding the growth of national labor unions. The National Labor Relations Board (NLRB) was charged with the enforcement of the National Labor Relations Act. In the cases brought to it before and after 1941, it declared many local unions to be "company unions" dominated by the employer, and therefore illegal and ordered them disestablished. Such decisions opened the way for a national union to conduct an organizing drive and to secure recognition as the bargaining agent of the workers by winning the election subsequently held by the NLRB. If the union secured this recognition but failed to reach an agreement with the employer, another agency might enter the picture—namely, the National Defense Mediation Board which was replaced in 1942 by the War Labor Board.[1] The function of these Boards was to settle disputes between employers and recognized bar-

[1] The National War Labor Board was established on Jan. 12, 1942, after the National Defense Mediation Board was rendered impotent by the walkout of CIO representatives in Nov. 1941 following the refusal of the Board to grant the union shop in the "captive" mines case. Public membership in the persons of William H. Davis and Frank P. Graham was carried over from the National Defense Mediation Board to the National War Labor Board.

gaining agents if the dispute affected national defense. Quite often the closed shop or union shop was an issue. The decisions of the NDMB and WLB never converted a union shop into an open shop, and in many cases they provided for a union shop where there had previously been an open shop, and so strengthened the union's position.

In the agencies created expressly for the emergency in 1940 and 1941 organized labor was powerfully represented by Sidney Hillman, a leader of the CIO. He was in charge of the labor sections of the National Defense Advisory Commission and the OPM; and in OPM he was an associate, not a subordinate, of the Chairman. Himself the president of a national union, the Amalgamated Clothing Workers, Mr. Hillman believed labor was best served by national unions, and that leaders of such unions, if given power and responsibility, could be counted on to prevent strikes and slowdowns and to assure a high productive effort as well as to defend the recently won " social gains," such as the 40-hour week. He appointed a labor advisory committee of sixteen men all from national unions. Thus a common attitude dominated in all the agencies in direct charge of labor relations.[2]

The Maritime Commission—and the Navy, with which the policy of the Commission towards shipyard labor was coordinated—had a different point of view. As purchasers responsible for speeding production and keeping down costs, the procurement agencies had reasons to oppose drives for higher wages. Admirals Land and Vickery because of their official positions, industrial experiences, and personal associations were inclined to share the attitudes of employers. On the other hand, as part of the Roosevelt Administration the Maritime Commission was called on to play ball with Mr. Hillman. In fact the course of events pushed the Maritime Commission into a position where it seemed to many big employers to be fighting for the national labor unions and to many labor leaders to be fighting for the employers.

Within the Commission, Admiral Vickery expressed particularly the point of view that would be expected from the man in charge

[2] On the significance of Hillman's appointment see also Coll, " Stabilization," pp. 15-19, and Richard E. Purcell, *Labor Policies of the National Defense Advisory Commission and the Office of Production Management, May 1940 to Apr. 1942,* Civilian Production Administration, Bureau of Demobilization [Historical Reports on War Administration: War Production Board Special Study No. 23], Oct. 31, 1946, pp. 8-9.

FORMATION OF STABILIZATION COMMITTEE 271

of production, but he was not before 1942 so active in labor relations as Admiral Land or Captain Macauley. In appointing Captain Macauley a member of the Commission in April 1941, President Roosevelt discussed with Admiral Land the special qualifications of Captain Macauley for handling problems of "maritime personnel."[3] At that time the Commission's Division of Maritime Personnel was concerned with shipyard labor as well as seamen and stevedores, and it reported to the Commission through Captain Macauley.[4] The Director of the Division, Daniel S. Ring, had been a labor reporter, an attorney representing labor unions, and Assistant Administrator in charge of labor for the Works Progress Administration in New York. In June 1942 when separate divisions were set up to handle maritime and shipyard labor, Mr. Ring became Director of the Division of Shipyard Labor Relations. The Assistant Director was Edward J. Tracy who had for many years been employed by the American Federation of Labor.[5] In October 1942 the new division began reporting to the Commission through Admiral Vickery, instead of through Captain Macauley.[6] Throughout, however, Admiral Land personally took an active part in labor policy and, being as Chairman the spokesman of the Commission on all controversial subjects, had a great deal to say about it.

Admiral Land had a long record of taking public positions on labor questions. In 1937 John P. Frey, President of the Metal Trades Department, AFL, testified against his confirmation as a member of the Maritime Commission because Admiral Land had

---

[3] The President to Land, May 2, 1942, and Land to the President, May 6, 1942, in MC, Historian's Collection, Land file.

[4] Apparently there was no formal notification that the Division should do this, but memoranda from this Division were routed through Commissioner Macauley, e. g., Ring to U. S. Maritime Commission, Sept. 11, 1942, in MC gf 508-1-1. As early as July 24, 1941 the Commission placed interpretation of the zone standards agreements in the office of Commissioner Vickery (MC Admin. Order No. 59); but at the time of the national shipbuilding conference in Chicago in Apr. 1942 Mr. Ring was reporting to Commissioner Macauley. Telephone conversation between Commissioner Macauley and Mr. Ring, May 13, 1942, in Division of Shipyard Labor Relations file, Zone Standards—General.

[5] MC Admin. Order No. 37, Suppl. 70, June 2, 1942, and Personnel Division files.

[6] Memoranda from the Division of Shipyard Labor Relations began going through Commissioner Vickery's office in Oct. 1942, and Admiral Vickery was appointed to the Shipbuilding Stabilization Committee in place of Captain Macauley at this time. See Land to Paul R. Porter, Oct. 13, 1942, in MC gf 508-1.

opposed pro-labor bills in 1933 and 1936.[7] Admiral Land replied by saying that when he spoke against these measures he had been acting under orders from the Secretary of the Navy to oppose measures which would delay work on naval contracts. Denying any wish to attack labor, he told the committee what he repeated many times later: " I got my start in life by working from sunrise to sunset, with my hands, for $10 a month and keep."[8] In all his dealings with labor Admiral Land emphasized that he was no brass hat, that he understood labor's point of view, and that he was acting as was his duty as Chairman of the United States Maritime Commission.

One point on which the procurement agencies and the agencies specially charged with labor problems could agree, whatever their divergences on other issues, was the desirability of preventing a spiral of wage increases, one yard bidding against another and " scamping " each other's labor by offering higher and higher wages. Such a spiraling of wages would raise the cost of ships and bring wage disputes likely to lead to strikes. Such a situation had occurred during World War I when wages rose 150 per cent in 1915-1918, and there were 101 strikes in 1917.[9] Since Admiral Land recalled this experience, he asked Mr. Ring in the spring of 1940 to draw up some recommendations to prevent scamping and spiraling. Rejecting the use of coercion by the government to tie men to their jobs, Mr. Ring suggested that the disrupting movement of workers from yard to yard be prevented by standardizing wage rates as much as possible. Such standardization, he suggested, could be brought about by the cooperation of employers

[7] Senate, *Maritime Commission Nominations, Hearing*, before a Subcommittee of the Committee on Commerce, 75th Cong., 1st sess., Apr. 10, 1937, pp. 14-15.

[8] *Ibid.*, pp. 39-40.

[9] Bing, *Wartime Strikes and Their Adjustment*, pp. 215, 295. The author states that charts showing wages in shipbuilding and other industries " were prepared by the Bureau of Applied Economics, Washington, D. C., . . . based on data collected by the U. S. Bureau of Labor Statistics . . . Shipbuilding Labor Adjustment Board, and the U. S. Shipping Board, supplemented by material gathered from other sources." *Ibid.*, p. 211. Scattered data on wages, sometimes given as average earnings, sometimes as minimum rates established at selected dates, are given in Hugh S. Hanna and W. Jett Lauck, *Wages and the War, A Summary of Recent Wage Movements* [Bulletin of Applied Economics No. 1] (Cleveland, Ohio, 1918) and Bureau of Applied Economics, Inc., *Wages in Various Industries and Occupations, A Summary of Wage Movements 1914-1920*, Bulletin No. 8 (Washington, 1920). On the Shipbuilding Labor Adjustment [Macy] Board, see Bureau of Labor Statistics, *Bulletin 283*; Coll, " Stabilization," pp. 10-13.

FORMATION OF STABILIZATION COMMITTEE

and employees, and "setting up an efficient and impartial governmental group for the purpose of not giving dictatorial orders but coordinating all phases of the shipbuilding labor relations picture." During the summer of 1940, the memorandum was taken by Admiral Land to Mr. Hillman for discussion in the National Defense Advisory Commission (NDAC).[10] At the same time, President Roosevelt himself, who also had memories of experience with the problem in World War I when he was Assistant Secretary of the Navy, backed the formation of a committee of labor and management to stabilize wages.[11]

In September the NDAC formally authorized Mr. Hillman to organize a committee for this purpose, and on November 27, 1940 he announced the formation of the Shipbuilding Stabilization Committee.[12] Through this committee the government took part in and guided the process of collective bargaining.

Although wages, not union organization, was officially the subject to be considered by the Shipbuilding Stabilization Committee, its very nature and its membership raised the question who should represent shipyard labor in negotiation with management. Sidney Hillman appointed Morris L. Cooke, his aide, as Chairman; Admiral Land to represent the Maritime Commission; and Joseph W. Powell to represent the Navy. To speak for labor he named top officers of the national unions: for the AFL, John P. Frey, President of the Metal Trades Department, and Harvey Brown, President of the International Association of Machinists; for the CIO, John Green, President, and Philip H. Van Gelder, Secretary-Treasurer, of the Industrial Union of Marine and Shipbuilding Workers of America.

But the key executives of the big shipbuilding corporations were conspicuous by their absence. Executives of Bethlehem, Sun, and Newport News did not wish to be in the position of negotiating with the leaders of national unions in a meeting

---

[10] The memorandum with a longhand notation "around 6-2-40," is in Division of Shipyard Labor Relations file, Zone Standards General. Mr. Ring, in interview, Sept. 18, 1947, with Frederic C. Lane and Blanche D. Coll, explained he did not address it to anyone because it was in response to an informal request from Admiral Land.

[11] Testimony of John P. Frey, in Truman Committee, *Hearings*, pt. 18, Mar. 26, 1943, p. 7381.

[12] Minutes of the Advisory Committee to the Council of National Defense, pp. 91-92; Purcell, *Labor Policies*, pp. 217-18.

where the "independent" unions which they considered representative of their own workers were not included. Consequently the "industry" representatives were men familiar with the industry but not in important executive positions: Prof. H. L. Seward; Gregory Harrison and F. A. Liddell, attorneys; and the elder statesman of the industry, H. Gerrish Smith, President of the National Council of American Shipbuilders, who had been president of Bethlehem's Shipbuilding Corporation Ltd., in 1917-1921. Acting for the NDAC, Mr. Hillman announced that these men would represent employers of the whole industry in making plans for stabilization.[13]

The first act of the committee was to broadcast what was called a "no strike" pledge, and to that extent to fulfill what the procurement agencies considered one of the committee's two main purposes. It adopted on December 5, 1940 a resolution urging " that there should be no interruption of production on the part of shipyard employers and shipyard employees before all facilities . . . of the National Defense Advisory Commission for adjusting differences have been exhausted."[14] Events were to show, however, that the leaders of the national unions considered this pledge conditional, namely, a pledge not to strike so long as employers followed the recommendations of the NDAC.

In approaching the committee's next function, the standardization of wages, representatives of labor unions and of management consulted those they represented and formulated conflicting proposals. The consensus among managerial representatives from yards employing 90 per cent of the workers was that "existing methods of relationships between yards and employees should be maintained," that any board to deal with wages was of "doubtful value," but if such a board was to be established it should function by zones.[15] The CIO and AFL representatives, on the other hand,

[13] The membership is given in Purcell, *Labor Policies*, pp. 217-18. On their connections see *Who's Who in America,* and on J. W. Powell, see above pp. 195-96.

[14] Shipbuilding Stabilization Committee Meeting [Minutes], Dec. 5, 1940. A set of ditto copies of these " Minutes " is in the MC Historian's Collection, Research file 205.8.1. They are not verbatim and until July 1942 are simply reports of actions taken, without indication of the discussion preceding such action. Prior to March 1942 policies adopted were set forth as " the sense of the meeting." Another set in the National Archives contains rough notes of discussions.

[15] Digest of Remarks Made by H. Gerrish Smith before the Subcommittee of the Shipbuilding Stabilization Committee on Jan. 6, 1941, reporting meetings of industry on Dec. 27, 1940 and Jan. 3, 1941, attached to Minutes above cited.

saw in further activities by the Shipbuilding Stabilization Committee the possibilities of collective bargaining on a regional or industry-wide basis which would increase their prestige. They differed, however, in the amount of power they wished to lodge within the Shipbuilding Stabilization Committee itself, and how much to leave to regional conferences.

The AFL declared its desire to " endeavor to secure conferences with employers " in certain zones. Such conferences would work out collective bargaining agreements which would be a guarantee against strikes or lockouts.[16] The Industrial Union also visualized an era of peaceful production as a result of collective bargaining agreements, but it sought more initiative on the part of the Committee. Its representatives wished the Committee itself to recommend standard clauses relating to overtime, shifts, and grievance machinery. The goal for the industry, the Union said, was equalization of wages, and since this was impossible to achieve immediately, " a uniform, guaranteed hourly base rate for first-class mechanics " should be established because this was " the most significant single wage rate in the industry." The Industrial Union agreed with the AFL that the industry be divided into four zones for administrative purposes but wanted them to be " under the direct control of the Committee in Washington." [17] If the Industrial Union could get a raise for shipyard workers by its activity in Washington, it would have an easier time persuading the unorganized and the members of local independent unions to join the CIO.

Following this hearing from labor and management, the Shipbuilding Stabilization Committee formulated the following policy: standard wage rates would be established by zonal collective bargaining conferences.[18]

---

[16] Document signed by John P. Frey, Pres., and Jos. S. McDonagh, Sec.-Treas., Metal Trades Dept., AFL, Jan. 6, 1941 attached to Minutes, Stabilization Committee, Meeting of Subcommittee, Jan. 6, 1941.

[17] Proposals of the Industrial Union of Marine and Shipbuilding Workers of America, CIO, to the Shipbuilding Stabilization Committee, attached to Minutes cited in note 16 above.

[18] Shipbuilding Stabilization Committee, Meeting of Subcommittee [Minutes], Jan. 6, 1941.

II. THE ZONE CONFERENCES

The first zone conference to meet was that for the Pacific coast. Since a great many persons were anxious for the stabilization idea to "take," the choice of zone was important. There had to be reasonable assurance that this first conference would succeed or else the entire scheme would be endangered.

Mr. Frey, President of the Metal Trades Department, attributed the choice of the Pacific coast to "the much better degree of labor organization [there] than elsewhere in the United States, and the generation-long contract there had been between the shipyard workers and the management. . . ."[1] In the fall of 1940 AFL unions had agreements with the Albina Engine and Machine Works, Bethlehem's San Francisco yard, the Western Pipe and Steel Co. and Moore Dry Dock, located in the San Francisco Bay area, and the Todd-Kaiser yard, Seattle-Tacoma. The CIO Industrial Union of Marine and Shipbuilding Workers of America had made no headway in organizing in these areas, although it had agreements with two yards near Los Angeles, Craig's shipyard (later Consolidated Steel Co.) and Bethlehem-San Pedro.[2] But the CIO, through the Steel Workers Organizing Committee (an unusual arrangement), had agreements covering machinists with yards in the Oakland area—Moore's, and the General Engineering Works—in spite of the fact that the Boilermakers and other AFL craft unions also had agreements.[3] The Metal Trades Department of the AFL had showed willingness to cooperate with management on the Pacific coast in order to encourage the shipbuilding industry. Jurisdictional rivalry among the various crafts had caused strikes of small groups of workers in the early days of the Commission's program. One yard had been so disturbed it requested permission of the Maritime Commission to transfer its contracts to the Atlantic coast. To solve this problem, Mr. Frey in February 1940 sponsored a resolution outlawing jurisdictional strikes, and persuaded the Pacific Coast Metal Trades District Council to agree to ignore picket lines if a stoppage of

---

[1] Truman Committee, *Hearings*, pt. 18, Mar. 26, 1943, p. 7381.

[2] "Union Agreements in Shipbuilding," *Monthly Labor Review*, Sept. 1940, pp. 597-99.

[3] Testimony of James P. Smith, Business Agent, East Bay Union of Machinists, Truman Committee, *Hearings*, pt. 4, May 28, 1941, p. 1237.

this nature occurred.[4] On the whole the AFL dominated on the Pacific coast.[5]

There were reasons to think that the big newcomers to shipbuilding, the Kaiser group, could readily reach an agreement with the AFL. The Permanente Metals Corp., a Kaiser affiliate, had contracts with AFL unions which dated back to November 1939.[6] Henry Kaiser, according to his labor relations representative, Harry Morton, had not always been friendly to union organization; the Boulder Dam had been built with an open shop. But the closed shop was tested on the Grand Coulee Dam project, and it was found that the cost per yard of concrete was less than at the first project. Thereafter, according to Mr. Morton, the Kaiser group got labor " religion." Mr. Morton discussed certain " principles " with Mr. Frey late in the year 1940.[7] The Kaiser management saw a way of recruiting, if not skilled shipbuilders, skilled workers of allied crafts. The Metal Trades Department, since it had 99 per cent of the craftsmen in the area organized, could supply the yards with this kind of labor.[8]

The attitude of the new Kaiser companies was particularly important since it was reported that they were " willing to pay any rate necessary to secure the best men from the other yards." [9] Mr. Morton later pictured this competition as the force impelling the other shipbuilders to join in stabilization. He said: " My principles were to go into each port on the Pacific Coast in the shipyards. We did not have a labor force. We wanted some experienced shipbuilders, and the rest of them knew it well. We would have welcomed every one of them into our shipyards, and

[4] Metal Trades Dept. of the American Federation of Labor, *Proceedings*, 32nd Convention, Nov. 11, 1940, pp. 31-33.

[5] Clark Kerr, " Collective Bargaining on the Pacific Coast," *Monthly Labor Review*, Apr. 1947, p. 551.

[6] Metal Trades Dept. of the American Federation of Labor, *Proceedings*, 32d Convention, Nov. 11, 1940, pp. 35-36.

[7] Speech of Harry F. Morton in Metal Trades Dept. of the American Federation of Labor, *Proceedings*, 35th Convention, Sept. 27, 1943, p. 131. According to Frank J. Taylor (" Builder No. 1," *Saturday Evening Post*, June 7, 1941, p. 11), Kaiser became friendly to unions after the passage of the National Labor Relations (Wagner) Act. Probably both statements are correct.

[8] Speech of Morton, in Metal Trades Dept., AFL, *Proceedings*, 35th Convention, p. 131. See also John P. Frey's even more definite statement: " He [Kaiser] looked to the Metal Trades Department to assist him in securing these men, in recruiting them, and they did." Truman Committee, *Hearings*, pt. 18, Mar. 26, 1943, p. 7384.

[9] Sanford to Vickery, Feb. 6, 1941, in MC gf 506-1.

if they wanted to retain their forces they had to come into the stabilization picture." [10]

Because of the greater strength of the AFL on the West coast, it was assigned a leading role in the stabilization conference for that zone, although the Industrial Union was to " participate." [11] The important shipbuilders were all represented except Bethlehem, which in spite of urgings from the Maritime Commission, refused to attend because it did not want to be put in the position of discussing a general closed-shop agreement.[12] This was indeed one of the first questions raised when the conference convened on February 3, 1941,[13] and the first major achievement of the conference was to find a way to separate negotiations on this issue, at least formally, from the negotiations on wages. By February 12, Mr. Frey, chief negotiator for the AFL, and H. Gerrish Smith, the Chairman of the Conference, agreed with the recommendation of the government observers that " matters of local or plant significance, such as closed shop, particular working rules, etc., should be covered in supplementary agreements between particular plants and respective unions, and that only matters of general import applicable to all West coast employers whether CIO, AFL or open shop, should go into coastwise standard agreement." [14]

This separation of the formal negotiations over union status from negotiations over wages cleared the way for the representatives of the government agencies to take a more positive part

[10] Metal Trades Dept. of the American Federation of Labor, *Proceedings*, 35th Convention, Sept. 27, 1943, p. 132.

[11] Shipbuilding Stabilization Committee, Meeting of Subcommittee [Minutes], Jan. 22, 1941. The matter of CIO participation was brought up when the CIO representatives appeared at the Conference, and reference was made to those minutes for a ruling. It was then determined that the CIO representatives would sit as observers without voice or vote at the general conference " and without privileges of participating in the negotiations " going on in committees between the AFL and the employers with whom they had agreements. " Informal Joint Memorandum of Government Observers Covering Pacific Coast Shipbuilding Stabilization Conference," Feb. 3-13, 1941, in Division of Shipyard Labor Relations file, Zone Standards—General.

[12] Land to Bethlehem Shipbuilding Co., Feb. 7, 1941, in Land reading file; testimony of A. B. Homer, Truman Committee, *Hearings*, pt. 4, pp. 1143-47.

[13] Memorandum signed by George C. Castleman, Chairman [Subcommittee appointed by Conference of Labor and Shipbuilders of the Pacific Coast], Feb. 13, 1941, in Division of Shipyard Labor Relations file, Zone Standards—General.

[14] Telegram, Ring to Land, Feb. 12, 1941 in the same file; interview with Daniel S. Ring, Sept. 18, 1947.

in the conference. Capt. J. A. Furer from the Navy and Mr. Ring from the Maritime Commission were at first counted among the observers, as were T. L. Norton from the OPM (which had replaced the NDAC), and P. W. Chappell from the Labor Department. As long as closed shop was an issue at the zonal Shipbuilding Stabilization Conference itself, Mr. Ring did not feel that the government should put pressure on the parties negotiating or take part itself as a negotiator. When wages were clearly the issue, the government found it had to take a stand and become one of the parties to the negotiation, for under their contracts the shipbuilders expected to be reimbursed by the government for whatever wage they paid. Management and labor called on the government observers to say how high a rate the government would allow. Since the observers had no authority to answer that question, all agreed that it was necessary to recess and call on Washington for new instructions.[15]

During the period of the recess, from February 12 to March 10, 1941, the role of the Shipbuilding Stabilization Committee and its relation to collective bargaining was redefined in light of the trial made at the West coast conference. The fixing of regional standards of wages for all zones by negotiations in Washington was again urged by the CIO,[16] and definite rates to be used in such a settlement were proposed by the Navy and the Maritime Commission.[17] The Committee split over whether wages should be left to regional negotiation. The matter was referred to the OPM which decided in favor of negotiations in zone conferences, and the Shipbuilding Stabilization Committee then on March 3, 1941, outlined a definite agenda for these conferences as follows: (1) Basic wage rate for standard skilled mechanics; (2) overtime provisions; (3) shift premiums; (4) a no-strike and no-lockout clause; (5) a provision against limitation on production; (6) a clause outlining grievance and arbitration machinery; (7)

---

[15] Informal Joint Memorandum of Government Observers Covering Pacific Coast Shipbuilding Stabilization Conference . . . , signed by Ring, Furer and Norton, in Division of Shipyard Labor Relations file, Zone Standards—General; telegram, Ring to Land, Feb. 12, 1941, in the same file.

[16] *The Shipyard Worker*, Feb. 28, 1941, pp. 1, 4, reports a memo sent to Sidney Hillman urging this.

[17] Stabilization of Shipbuilding Outline, attached to Shipbuilding Stabilization Committee, Meeting of the Full Committee [Minutes], Feb. 21, 1941.

a two-year duration clause with provision for periodic wage adjustments.[18]

When the Pacific Coast Conference reconvened on March 10, there was a much better understanding of the place of everyone in the scheme and what each was supposed to do. The wage for skilled mechanics, the basic wage rate of the industry, became at once the main issue. Industry's offer had been the present scale (about $1.00) for the South Pacific yards and $1.05 for the North Pacific. Labor was asking $1.15 for the entire area.[19] The government's proposal had been in between the two—$1.02 for the South Pacific and $1.07 for the North Pacific. On March 21 Washington made a final offer of $1.10, adding that if this was "not obtainable conference should terminate."[20] The government observers on the spot were apparently alarmed at this stand. Labor, they telegraphed, had a "good basis" on which to bargain for $1.15 because wages in the locality were going that high. They feared the government would have to pay more without an agreement than if it agreed to $1.15 and was protected by the zone standards. Furthermore, the government would be blamed, they wrote, if the conference broke up. They suggested the government offer $1.12½ to get itself in a stronger position if the conference did fail.[21] Finally the parties agreed on the $1.12 rate, to become effective April 1, 1941.[22]

Labor also got higher shift premiums than those proposed by the government. Second-shift workers were given 10 per cent, and third-shift workers, 15 per cent. The government was also forced to modify its position on overtime. It had proposed penalty rates based on the 8-hour day and 40-hour week for weekdays including Saturdays, but labor was able to secure overtime rates for work on Saturdays regardless of the amount of time accumulated in the days preceding.[23]

[18] Shipbuilding Stabilization Committee, Meeting of the Full Committee [Minutes], Feb. 21, 1941, and March 3, 1941.

[19] "Informal Joint Memorandum of Government Observers," cited above.

[20] Telegram, Knox, Knudsen, Hillman and Land to Lubin, Fisher and Ring, Mar. 21, 1941, in Division of Shipyard Labor Relations file, Zone Standards—General.

[21] Telegram, Lubin, Fisher, Ring and Norton to Land, Mar. 22, 1941 in Division of Shipyard Labor Relations file, Zone Standards—General.

[22] Fisher and Ring to Land, Mar. 25, 1941, in file cited in note above.

[23] [Pacific Coast Zone Standards], submitted to the Office of Production Management by Pacific Coast Shipbuilding Stabilization Conference (mimeographed), Apr. 23, 1941 in MC, Historian's Collection, Research file 205.3.1.

THE ZONE CONFERENCES

In fixing these standards the procurement agencies had engaged in a sort of three-cornered collective bargaining. After the zone standards were agreed to, another type of collective bargaining, the usual type, remained to be done between management and labor to put the stabilization agreement into effect. The relation between these two types of collective bargaining was defined as follows:

> The zone standards arrived at between employers, representatives of organized employees, and representatives of Government shall be announced by O. P. M. as covering all shipyards doing construction work on the Pacific Coast.
> The shipyards and the organized employees shall undertake that these zone standards will be made a part of all collective bargaining agreements entered into between the shipyards and their organized employees.[24]

This formula left the way open for the management of a yard where the employees were not organized to set rates in accordance with the zonal agreement, but it assumed that union contracts would be made. In negotiating these contracts, the labor leaders asked for a closed shop and raised other questions of union status which had been excluded from the agenda of the stabilization conference. Other important questions, such as which workers should be classified as first-class mechanics, which as second-class mechanics or helpers, remained to be settled by these contracts. The zonal conference by no means superseded the usual collective bargaining between unions and management, although it decided basic issues about wages.

After the Pacific Coast Conference adjourned at the beginning of April, Mr. Frey undertook to " sell " to the locals of his unions not only the general wage settlement but also a master contract which he and other representatives of the Metal Trades Department had meanwhile negotiated with all Pacific coast shipyards except Bethlehem, Western Pipe and Steel, and the Craig-Consolidated yard.[25] This master agreement not only implemented the zone standards, it also provided for classifying all mechanics as first-class mechanics, and for a closed shop. The closed shop

---

[24] Shipbuilding Stabilization Committee, Meeting of Subcommittee [Minutes], Mar. 3, 1941.
[25] Memorandum of Observations and Recommendations—Shipbuilding Stabilization Conference, Apr. 4, 1941, in Division of Shipyard Labor Relations file, Zone Standards—General.

clause was a tight one: it stipulated that hiring was to be done through the AFL unions and made maintenance of union membership a condition of employment.[26]

The AFL unions appeared secure in their supremacy on the Pacific coast and could look forward to enjoying a great advantage in organizing the new yards beginning production in that area. The first such yard to sign up, on May 12, 1941, was Oregon Ship which had at the time only 66 employees, and by the summer of 1943 was employing 34,000.[27]

The success of the Pacific Coast Conference in overcoming obstacles and reaching an agreement must have encouraged its sponsors to hope for success in the other zones. They needed such encouragement, for difficulties were foreseen on the Atlantic coast where the CIO would have a dominant place in negotiating for labor, much as the AFL had spoken for it on the Pacific coast. Although the Industrial Union had contracts with only two of the big five—with Federal and New York Ship—it had organized a number of smaller yards, and was definitely stronger in the shipyards than the AFL.[28] Rather than accept a subordinate position, the AFL refused to participate,[29] and this gave the Industrial Union a chance to claim it was spokesman for all shipyard labor at the Atlantic Coast Conference which convened on April 28, 1941.[30]

Such a claim ignored, to be sure, the "independent" unions. The CIO union was in the midst of a vast organizing campaign to drive the independents out of Bethlehem, Newport News, and Sun, a campaign which had been going on in the courts since 1936. After appeals to the Supreme Court, independent unions or employee-representation plans at Bethlehem and Newport

[26] A copy of the Pacific Coast master agreement is printed in House, *Production in Shipbuilding Plants, Hearings*, pt. 1, May 11, 1943, pp. 89-96. Western Pipe and Steel signed the master agreement in May 1941. *New York Times*, June 1, 1941, p. 31.

[27] House, *Production in Shipbuilding Plants, Hearings*, pt. 1, May 11, 1943, p. 67; Fischer, *Statistical Summary*, Table G-1, p. 125.

[28] "Union Agreements in Shipbuilding," *Monthly Labor Review*, Sept. 1940, p. 599.

[29] It was agreed that AFL representatives would be recognized by the chair but since the AFL believed it should take part in the actual negotiations, it refused to participate. Shipbuilding Stabilization Committee, Special Meeting [Minutes], April 28, 1941; Metal Trades Dept. of the American Federation of Labor, *Proceedings*, 33rd Convention, Sept. 29, 1941, pp. 28-29.

[30] *Shipyard Worker*, May 2, 1941, p. 1.

News had been declared illegal company unions. New local unions, which claimed not to be company-dominated, sprang up, however, and were making it difficult for the CIO to organize these yards. At Sun the independent was in a solid legal position because it had been certified as a bona fide collective bargaining agent by the National Labor Relations Board in 1938. Employees of the Todd repair yards on the Atlantic coast, Bath Iron Works, Electric Boat Co. and some smaller yards were also represented by independent unions. Federal, New York Ship, Pusey and Jones, and some others of less importance were bargaining with the Industrial Union.[31]

Right before the Atlantic Coast Conference convened in April, the *Shipyard Worker* announced the triumph of the Industrial Union: "THE IUMSWA-CIO IS SPEAKING FOR THE WORKERS, THE WHOLE 100,000! Negotiating as it is with 75 per cent of the shipbuilding industry, the IUMSWA is achieving part of its greatest ambition—a nationwide agreement." The paper went on to document its claim by naming Bethlehem, Newport News, and so on among the companies who had sent representatives.[32] But when Bethlehem, Newport News, Sun, and Bath saw there was no possibility of getting the independents represented, they refused to participate.[33] They were determined not to encourage the CIO organizing drive by attending a conference where the Industrial Union alone represented labor. They demanded that the "independent" unions in their yards be represented, and the Navy and Maritime Commission supported them in this. Admiral Land argued that to deny representation to the independents was undemocratic and unjust.[34] But the OPM decided that the independents should not be admitted,[35] and Mr. Hillman

---

[31] For the legal battle over the independent unions see the cases before the NLRB and the courts cited in Coll, "Stabilization," pp. 52-56.

[32] *Shipyard Worker*, May 2, 1941, p. 1.

[33] See statements of H. Gerrish Smith and Mr. Woodward (of Newport News) on what took place in "Conference between Representatives of East Coast Independent Unions and Representatives of East Coast Shipbuilders, at Navy Department, May 20, 1941," in Division of Shipyard Labor Relations files, Minutes of Shipbuilding Stabilization Committee.

[34] Land to F. C. Lane, Mar. 15, 1948 in MC, Historian's Collection, Research file and below, p. 298.

[35] Statement of Joseph W. Powell, in "Conference," May 20, 1941, cited above. The Stabilization Committee did agree that the independents would be consulted by government representatives before approval of the agreement. "The Shipbuilding Stabilization Committee," [prepared by Rosalind S. Shulman in the office

was recognized as the President's representative in labor matters.[36] Consequently the CIO alone bargained for labor at the Atlantic Coast Conference, and the only employers attending were those having agreements with the Industrial Union.

Although only a fraction of the industry was represented at the Conference, it was supposed to set standards applicable to all yards. To secure their acceptance of these standards, Admiral Land and Joseph W. Powell, who was assistant in charge of ship construction to the Secretary of the Navy, met on May 20, 1941 with representatives of the independents and their employers. At this meeting Mr. Powell emphasized that the document which was expected from the zone conference was not an agreement at all; it was a zone standard. " When and if this committee [the zone conference] finally reaches something that they consider as a proper zone standard which they will sign," stated Mr. Powell, " those standards go to every shipbuilder and every union of the employees—the shipbuilders and the employees at each plant then sit down together to bargain out their final agreement which we hope will include these zone standards on the East Coast."

The independents, both labor and management, reluctantly accepted this interpretation, and Admiral Land and Mr. Powell promised that the local agreements negotiated would be subject to the approval of the government agencies—not the Shipbuilding Stabilization Committee.[37] This was a formal gesture because it was preposterous that the government would approve any local agreement which did not comply with the zone standards, but it was a face-saving device for the independents.

The wage offer of the government on the Atlantic coast was $1.06 to $1.08.[38] The Industrial Union first asked for $1.15 and later indicated it would in no wise consider accepting less than the $1.12 approved for the West coast yards.[39] The independents

of the Shipbuilding Stabilization Committee] (mimeographed, c. Nov. 1942), p. 10, in MC, Historian's Collection, Research file 205.3.1.

[36] Mr. Ring, in an interview Sept. 18, 1947, stated that the Maritime Commission recognized Hillman as the President's spokesman.

[37] " Conference between Representatives of East Coast Independent Unions and Representatives of East Coast Shipbuilders," May 20, 1941.

[38] Ring to Dempsey (Confidential), May 11, 1941, in MC gf 508-3; MC Minutes, June 3, 1941, p. 17530.

[39] " The Shipbuilding Stabilization Committee," cites Union Proposal on Atlantic Coast Zone Standards, submitted May 6, 1941; longhand notes on Atlantic Coast Conference Meeting of Subcommittee on May 9 [1941] in Shipbuilding Stabilization Committee file (in National Archives) 05/224: 41.

also asked for $1.12.⁴⁰ The Maritime Commission opposed this equalization vigorously. In a letter to Gerard M. Swope, Chairman of the Conference, Admiral Land set forth the Maritime Commission's opposition on several scores: first, shipyard labor was not adequately represented at the Conference; second, the Commission representatives were in reality only observers and could not make commitments; third, the West coast differentials should be preserved, both because the Merchant Marine Act of 1936 provided for a differential and because such a raise on the Atlantic coast would throw shipyards out of line with other industries and result in scamping; fourth, it would place the Commission in the position of paying premiums for advance deliveries and "the increased labor costs that enabled the yards to earn such premiums." ⁴¹

A deadlock occurred between the government agencies and the Industrial Union (IUMSWA) over the wage issue. According to Mr. Powell they " never did agree . . . but finally the matter was settled by higher authority . . . and the Office of Production Management determined that the scale for the Atlantic coast shipbuilders should be built around the base mechanic's rate of $1.12 an hour." ⁴² This was a large increase; the contracts IUMSWA had with New York Ship, Pusey and Jones, and Maryland Drydock were based at $1.00 an hour and were not due to expire for some time.⁴³ The independents met once more with Admiral Land and Mr. Powell on June 18 and approved these standards.⁴⁴

Organization on the Gulf was more evenly divided than in other areas, so at the Zone Conference which met there from May 13 to June 18, both the AFL and the CIO were represented. The AFL had agreements with Ingalls, Tampa, Pennsylvania,

---

⁴⁰ " Conference between Representatives of East Coast Independent Unions and Representatives of East Coast Shipbuilders," May 20, 1941.

⁴¹ Land (written by Carl F. Farbach of the Legal Division) to Swope, June 9, 1941, in MC gf 508-3.

⁴² Minutes of Meeting June 18, 1941 Held in the Office of M. L. Cooke . . . with Representatives of Independent Unions from Atlantic Coast Shipyards (verbatim, mimeographed), p. 1, in Division of Shipyard Labor Relations file, Minutes of the Shipbuilding Stabilization Committee Meetings.

⁴³ Ring to Dempsey (Confidential) May 11, 1941 in MC gf 508-3.

⁴⁴ Minutes of Meeting June 18, 1941 Held in the Office of M. L. Cooke; Atlantic Coast Zone Standards (mimeographed), June 4, 1941 in MC, Historian's Collection, Research file 205.8.

Todd-Galveston and Lykes Bros.; the CIO, with Alabama and Todd-Johnson.[45] The unions did not succeed in establishing a base rate as high as that for the other zones, but after the Conference learned that Atlantic coast workers had been allowed $1.12, they put it to $1.07 for the Gulf, which was a substantial increase for Southern labor.[46]

The AFL was the only representative for labor at the Great Lakes Conference which first met on April 23 and signed an agreement on July 11. All the agreements in existence in this area were with the AFL.[47] Here the wage rate matched that for the East and West coasts, $1.12.[48]

By the late summer of 1941 the seven basic points which the Shipbuilding Stabilization Committee had listed as essential had been embodied in four zonal agreements. In three zones the mechanic's rate was fixed at $1.12; on the Gulf it was set at $1.07. Time and a half was to be paid for all work over eight hours a day and forty hours a week and for work performed on Saturday; double time for all Sunday work. Shift premiums were highest on the West coast where men got 10 per cent for the second and 15 per cent for the third shift. Atlantic coast workers were to receive 7 per cent for both second- and third-shift work, and on the Gulf and Great Lakes workers were to get 40 cents extra on the second and third shifts. The agreements were broad in their provisions and left many details to local collective bargaining. They were to remain in effect for two years, but adjustments were to be made in wages if the cost of living rose.[49]

In the push and pull which took place in getting the zone

[45] "Union Agreements in Shipbuilding," *Monthly Labor Review*, Sept. 1940, pp. 598-99.

[46] "Proceedings Gulf Coast Shipbuilding Zone Conference beginning May 13, 1941" (dittoed "unofficial" personal notes of W. T. Crist, CIO representative) in Shipbuilding Stabilization Committee file (National Archives) 05/3234:41; telegram, Fisher, Ring, and Norton to Land, June 6, 1941 in MC gf 508-4; Report of Gulf Shipbuilding Labor (And Repair) Stabilization Conference (mimeographed), June 18, 1941 in Historian's Collection, Research file 205.3.1.

[47] "Union Agreements in Shipbuilding," *Monthly Labor Review*, Sept. 1940, pp. 598-99.

[48] Great Lakes Shipbuilding and Repair Zone Standards (mimeographed), July 11 [1941] in Historian's Collection, Research file 205.8.

[49] Pacific Coast Zone Standards, Apr. 23, 1941; Atlantic Coast Zone Standards, June 4, 1941; Gulf Shipbuilding and Repair Zone Standards Agreement, June 18, 1941; Great Lakes Shipbuilding and Repair Zone Standards, July 11, [1941], all in Historian's Collection, Research file 205.3.1.

standards agreed upon, some groups probably thought the trades had not been evenly balanced. Labor had been given a big raise, bigger than the Maritime Commission and the Navy expected, but as Admiral Land testified later, the Maritime Commission did not feel the price too high in return for agreements, binding for a period of two years, which would prevent spiraling and scamping. Moreover, he hoped that the no-strike, no-lockout clause was a promise of peaceful labor relations. The second stage of this scheme for collective bargaining, the embodiment of the standards in signed contracts between employers and employees, was still far from completed, however, and was not completed without strife.

III. Strikes and Loafing

The no-strike pledge given by the labor leaders as members of the Shipbuilding Stabilization Committee was a pledge subject to conditions. As the leaders of national unions interpreted the conditions, they retained the right to organize and were free to strike in case the employers did not accept the recommendations of agencies of the national goverment having authority in labor matters. Admiral Land thought the no-strike pledge meant no strikes and called it the " keystone " of the stabilization agreements. He wrote the President June 23, 1941, " We have been extremely liberal with shipyard labor and I believe that they should reciprocate in kind by producing to the limit without strikes." [1] The national labor leaders maintained that they could not be responsible for preventing strikes and speeding production unless they had the corresponding power, which depended, as they saw it, on having the workers organized under agreements akin to closed-shop contracts. The major strikes which occurred after the stabilization agreements were in part " wildcat " strikes opposed by national labor leaders, in part the results of their efforts to organize and to secure union recognition in the contracts which were being negotiated to put the zone standards into effect.

The ink was hardly dry on the Pacific Coast Agreement when, on May 10, a serious strike began in the San Francisco Bay area.

[1] Land to the President, June 23, 1941, in MC, Historian's Collection, Land file; and his statements in House, *Independent Offices Appropriation Bill for 1943, Hearings*, p. 270, and in Metal Trades Dept. of the American Federation of Labor, *Proceedings*, 35th Convention, Sept. 27, 1943, p. 63.

This strike was called by Local 68 of the International Association of Machinists (AFL) upon authorization of the national union headed by Harvey P. Brown, and by the East Bay Union of Machinists (CIO), affiliated with the Steel Workers Organizing Committee whose national president was Philip Murray. It was denounced as an outlaw strike by William Green, President of the AFL, and by Mr. Frey, head of the Metal Trades Department of which the International Association of Machinists was a member.

Two issues were involved, wages and recognition. Local 68 (machinists, AFL), which began the strike and maintained it longest, struck on the issue of wages, demanding $1.15 per hour instead of $1.12.[2] Whether or not Communist-controlled as charged,[3] this local was certainly in opposition to the national officers and unwilling to recognize the authority of the agreements they had negotiated. Its leader was not worried about union security, he said, because he had 98 per cent of the workers organized.[4] On the other hand, union status was the main reason why Harvey Brown, president of the machinists, for a time supported this strike. The Bay Cities Metal Trades Council at first refused to sanction the strike, but subsequently did approve the strike against Bethlehem because it wished to force Bethlehem to sign the master agreement for the Pacific coast. Harvey Brown told the Truman Committee that he sanctioned the strike on the basis of " no contract "; he did not consider the no-strike pledge given in the zone standards agreement valid until embodied in a local agreement. The national office, he said, would withdraw support when Bethlehem signed the master agreement.[5] The CIO was striking against the master agreement which they claimed had been " reached as a result of secret negotiations and collusion between shipyard employers, John P. Frey, and others," [6] and which they feared would result in their losing control. In spite

[2] Truman Committee, *Hearings,* pt. 4, pp. 1127-29.

[3] *Ibid.,* pp. 1141, 1184; *New York Times,* May 19, 1941, p. 10; Land to Matt Mehan, June 6, 1941 in Land reading file.

[4] Testimony of Harry S. Hook, Business Agent of Local 68, in Truman Committee, *Hearings,* pt. 4, pp. 1127-29, 1207; *New York Times,* May 11, 1941, p. 1. On a later strike by this local see *The Termination Report of the National War Labor Board: Industrial Disputes and Wage Stabilization in Wartime January 12, 1942–December 31, 1945* [Historical Reports on War Administration, National War Labor Board] (Government Printing Office, n. d.), I, 884 n.

[5] Truman Committee, *Hearings,* pt. 4, pp. 1251, 1173-75, 1186-89.

[6] *Ibid.,* pp. 1237-41; *New York Times,* May 11, 1941, pp. 1, 11.

of this tangle of conflicting issues, the local labor unions were not in conflict; for the AFL machinists struck yards on the San Francisco side of the Bay, CIO machinists those on the Oakland side.[7] The picket lines were to enforce the demands of only 1,700 men, but over 9,000 workers stayed away from their jobs. Labor solidarity was particularly evident in the way AFL members respected CIO picket lines in the Oakland yards.[8]

During the efforts to end the strike Admiral Land testified before the Truman Committee that the tie-up affected five yards having half a billion dollars' worth of contracts. He denounced it as an outlaw strike violating the principle of collective bargaining in which he believed. Accordingly he proposed that "a cessation of it be made at the earliest possible date, as far as I am concerned, without any limits whatsoever, including the necessary forces that the National Government has at its command to enforce the rights of the United States Government."[9] But the country was not yet at war and not apparently ready for extreme measures. The obvious step was to certify the disputes to the National Defense Mediation Board, but this was not done immediately, because, as Secretary of Labor Perkins explained in answer to criticisms, Mr. Hillman had requested it not be. Mr. Hillman sent Eli Oliver of his Labor Relations Division to investigate, and Mr. Frey, whose unions were so deeply involved, left for San Francisco to try to end the strike.[10]

Mr. Frey took the position that the no-strike pledge in the zone standards agreement was effective regardless of local agreements. He considered the strike one against the government, since the yards " operating under a fixed fee basis are simply the agents of the government and must follow zone agreements on wages, hours, and conditions. "The master agreement at issue here," he stated, "merely implements the zone agreement."[11] His plan was to lead workers through the picket lines,[12] and he announced that

---

[7] *New York Times*, May 11, 1941, p. 38.

[8] The *New York Times* (May 11, 1941, p. 38) stated that AFL and CIO machinists had always "cooperated."

[9] Truman Committee, *Hearings*, pt. 4, pp. 1121-27.

[10] *New York Times*, May 15, 1941, p. 19; May 16, 1941, p. 18; May 18, 1941, p. 38; May 24, 1941, p. 16.

[11] Quoted from San Francisco *Examiner*, May 16, 1941, p. 1, in George M. Keller, Jr., "Shipbuilding Stabilization History" (typescript, 1942), p. 178, in Historian's Collection, Research file 205.

[12] *New York Times*, May 19, 1941, p. 10. Land wired Frey voicing his support. Land reading file, May 14, 1941.

his followers would "go through every picket line except Bethlehem." He personally led the back-to-work drive through the CIO line at Moore Dry Dock, but his efforts failed. Only 500 to 1,500 returned to work under his leadership.

When it became clear that neither Mr. Frey nor Mr. Hillman—nor the Governor of California, who had also tried to bring the parties together—had met with success, the case was certified on June 1 to the National Defense Mediation Board. The following day the Bay Cities Metal Trades Council ordered its members to return to their jobs.[13] Local 68 voted to withdraw from the Council, but the other crafts "swarmed back to work."[14] The National Defense Mediation Board promptly recommended that Bethlehem sign the master agreement, which it said, "was the result of mature collective bargaining on an industry-wide basis." Bethlehem's opposition to the closed shop, the Board held, was outweighed by the decision of all the other shipyards on the Pacific coast. On June 23, Bethlehem announced its intention to comply with the Board's recommendation, and on June 26 the machinists of Local 68 returned.[15] The strike died out only after it had paralyzed the shipbuilding industry around San Francisco for about a month and threatened the entire plan of stabilization.[16]

The next serious strike was on a more clear-cut issue. In the yard of the Federal Shipbuilding and Dry Dock Co. at Kearny, New Jersey, the Industrial Union (CIO) began pushing for a closed shop when it opened negotiations for a new contract in the summer of 1941. Almost at once the case was certified to the National Defense Mediation Board. On July 26, 1941 the panel appointed to hear the case recommended that the following clause be included in the contract:

In view of the joint responsibilities of the parties to the National Defense, of their mutual obligations to maintain production during the present emergency and of their reciprocal guaranties that there shall be no strikes or lockouts for a period of two years from June 23, 1941, as set out in the "Atlantic Coast Zone Standards," incorporated herein and made a part hereof, the Company engages on its part that any employee who is now a member of

---

[13] Truman Committee, *Hearings*, pt. 4, p. 1191; *New York Times*, May 21, 1941, p. 45; May 22, 1941, p. 15; May 23, 1941, p. 14; May 24, 1941, pp. 1, 10; May 25, 1941, p. 1; May 27, 1941, p. 16; May 30, 1941, p. 10; June 1, 1941, p. 31; June 3, 1941, p. 14.
[14] *New York Times*, June 5, 1941, p. 14; June 10, 1941, p. 15.
[15] *Ibid.*, June 19, 1941, p. 15; June 24, 1941, p. 1; June 27, 1941, p. 12.
[16] Coll, "Stabilization," p. 73.

the Union, or who hereafter voluntarily becomes a member during the life of this agreement, shall, as a condition of continued employment, maintain membership in the Union in good standing.[17]

The union hailed this as a signal " that the government considers the Union Shop a desirable condition, which should prevail in shipyards where the Industrial Union has assumed the responsibility for the carrying out of the National Defense Program without interruption." [18] The company refused to accept the Board's decision on the ground that it was " contrary to the fundamental principle that the right to work is not dependent upon membership or non-membership in any organization, upon which principle is based the open shop policy of the company." [19] The union struck the yard just before midnight on August 6. Fifteen thousand workers were involved in the closing down of a yard which was building ships for both the Navy and the Maritime Commission. Two weeks of national tension followed. The country took sides and argued the issues.

On August 11, Mr. Korndorff offered his plant to the U. S. Navy.[20] On August 14 the Maritime Commission met in executive session and set forth its policy in a letter sent to the Navy, OPM, and the National Defense Mediation Board. The Commission opposed seizure by the government until " understandings " regarding the union security clause were put in writing or the " misunderstandings cleared up." On the basis of the information it had, the Commission " construed this clause as meaning a ' closed shop ' (at least a requirement akin thereto), prescribed by a governmental agency, hence the Commission believes that the decision . . . is so evolutionary if not revolutionary and may be so far-reaching in the precedent established that it may and probably will extend itself to the shipbuilding industry as a whole and may extend itself to all industry engaged in the national defense effort." The Commission recommended

[17] In re *Federal Shipbuilding Corporation and Industrial Union of Marine and Shipbuilding Workers* (CIO), July 26, 1941, in *Labor Relations Reference Manual*, vol. 8, p. 1267.
[18] Editorial, *Shipyard Worker*, Aug. 1, 1941, p. 8.
[19] L. H. Korndorff, Pres., Federal Shipbuilding and Dry Dock Co., to Ralph T. Seward, Exec. Secy., National Defense Mediation Board, July 30, 1941, Exhibit I, in " Chronological Statement of More Important Events Relating to Labor Situation at Federal Shipbuilding and Dry Dock Company Kearny, New Jersey," May 8, 1942, a booklet published by the company.
[20] *New York Times*, Aug. 12, 1941, p. 1.

that "final action be not taken except with the approval by and direction of the President of the United States." [21]

In reality there was nothing obscure about the maintenance of membership provision: it was *not* a closed shop, such as was in effect in the West coast yards, since it did not require anyone to join the union in order to be employed; it *was* a "requirement akin thereto," since it did force the worker to continue membership in the union for one year if he had already joined the union or if he subsequently became a member. The maintenance of membership clause was indeed revolutionary for Atlantic coast shipyards.

William H. Davis, Chairman of the National Defense Mediation Board, denied his organization had ordered a closed shop, or, as was maintained, that the union had violated its no-strike pledge.[22] The following day the Maritime Commission joined the Navy, the National Defense Mediation Board, and the OPM in an appeal to the President to seize the yard. On August 23 the President ordered the Navy to move in, whereupon the union voted to end the strike.[23]

When the plant was returned to the company in January 1942, the union still wanted a contract providing maintenance of membership. This was ordered by the National War Labor Board on April 27, 1942. Writing the majority opinion, Dr. Frank Graham said:

> The government of the United States is under a moral and equitable compulsion not to take advantage of the national agreement which has disarmed this union of its only weapon. It is under the same compulsion not to allow the most powerful corporation in the world to take advantage of this covenant of honor, and this covenant for maximum war production.

Continuing, he argued, "There is a basic relation between maintenance of membership, maintenance of the contract, and maintenance of production." [24] This time, although still protesting, Federal bowed to the decision of the Board.[25]

[21] MC Minutes, Aug. 14, 1941, p. 18499.
[22] *New York Times*, Aug. 17, 1941, p. 1.
[23] *Ibid.*, Aug. 18, 1941, p. 1; Aug. 24, 1941, p. 1; Aug. 25, 1941, p. 1.
[24] In re *Federal Shipbuilding and Dry Dock Company [Kearny, N. J.] and Industrial Union of Marine and Shipbuilding Workers of America, Local No. 16 (CIO)*, Apr. 25, 1942, in *War Labor Reports*, vol. 1, p. 143.
[25] Korndorff to Davis, May 8, 1942, Exhibit II in "Chronological Statement, . . . Relating to Labor Situation at Federal Shipbuilding and Dry Dock Company."

While winning this position at Federal, the Industrial Union was making progress towards its major objective, to establish collective bargaining contracts with the Bethlehem Steel Co. In May the company signed at Hoboken the first shipyard agreement with a national union.[26] A month later, on June 10, the Industrial Union and Bethlehem signed a " Memorandum of Understandings " in which the company agreed to negotiate a master contract for all yards on the Atlantic coast where the union was certified.[27] The Industrial Union then conducted a " whirlwind " campaign at the emergency Bethlehem-Fairfield yard in Baltimore which was crowned with success October 17, 1941 when the union won an NLRB election, 3,570 to 533 for " no union." At Sparrows Point, the other Baltimore yard building merchant ships, the Industrial Union got 2,567 votes to 1,386 for the Independent Shipbuilders Association of Sparrows Point, and 307 for " no union." [28] These victories for the Industrial Union brought to eight the number of Bethlehem yards to be negotiated for, because the union had previously won elections at the East Boston Repair Yards, two Brooklyn yards, Staten Island, and Baltimore Drydock (Key Highway).[29]

Negotiations dragged out for more than a year because of dispute over a union shop. The Industrial Union took the position it could not be " expected to extend its agreement for two years and waive the right to strike during that period without a protective clause that would insure the continued existence and strength of the organization." [30] The question was certified to the War Labor Board on February 12, 1942, and on September 1, the Board awarded maintenance of membership and the check-off to the union,[31] applying the formula previously used at Federal. The company and the union signed a master contract affecting some 70,000 workers on September 18, 1942.[32]

[26] Industrial Union of Marine and Shipbuilding Workers of America, CIO, *Officers' Report to the Eighth National Convention*, Sept. 22, 1942, p. 5; *Shipyard Worker*, May 9, 1941, p. 1.
[27] *Ibid.*, June 13, 1941, pp. 1-2.
[28] *Ibid.*, May 23, 1941, p. 6, and weekly reports thereafter.
[29] Industrial Union of Marine and Shipbuilding Workers, *Officers' Report*, 7th Convention, Sept. 23, 1941, p. 3.
[30] *Ibid.*, pp. 5-6.
[31] Industrial Union of Marine and Shipbuilding Workers, *Officers' Report*, 8th Convention, Sept. 22, 1942, p. 6.
[32] *Shipyard Worker*, Sept. 25, 1942, p. 1.

During this protracted struggle there was no large prolonged strike, but there were several short work stoppages, only one of which affected merchant shipbuilding. On Armistice Day 1941, the workers walked out for the afternoon at Bethlehem-Fairfield to demand overtime pay for the holiday.[33] This was not much time lost so far as the Maritime Commission was concerned, but the unsettled labor situation in the Bethlehem yards was a powder keg until a contract was signed.

After signing Bethlehem, the Industrial Union intensified its efforts to win other yards. It kept up its drive at Sun and was successful there in 1943. It tried and failed at North Carolina Ship, the emergency subsidiary of the Newport News Co.[34] The most serious strike connected with further growth of the CIO union was at South Portland, Maine. The situation there involved rivalry between CIO and AFL, bureaucratic tangles, and local mismanagement. The difficulties at South Portland, which will be described in connection with the story of the change in management at that yard, led to a short strike by the AFL on October 10, 1942, a longer, more complete strike beginning November 30, 1942, and a brief outlaw strike in May 1943.[35]

The Gulf and Great Lakes areas also had their share of labor troubles although strikes were mostly in the nature of brief protests at specific conditions. Groups of workers at Ingalls were restless throughout the year 1941. The Industrial Union had members in this AFL yard, and in January 1941 a hundred and thirty of them went out on a nine-day strike for a wage increase. There were strikes of from two to three days of various AFL trades in March and in October for wage increases and the closed shop.[36] On November 15, 1941 the National Defense Mediation Board denied maintenance of membership to the Metal Trades Council and recommended that the union agree to a clause promising company encouragement of union membership.[37]

At Alabama, where the Industrial Union was certified, labor

[33] *Ibid.*, Nov. 14, 1941, p. 8.

[34] Industrial Union of Marine and Shipbuilding Workers, *Officers' Report*, 9th Convention, Sept. 21, 1943, p. 3.

[35] See below, Chap. 15, and Coll, " Stabilization," pp. 172-83.

[36] Based on data collected by U. S. Dept. of Labor, Bureau of Labor Statistics, Division of Industrial Relations.

[37] In re *Ingalls Shipbuilding Corp. and Pascagoula Metal Trades Council (AFL)*, Nov. 15, 1941 in *Labor Relations Reference Manual*, vol. 9, p. 815.

troubles persisted over wages, working conditions, and the union shop.[38] On the latter issue the National Defense Mediation Board ruled that the question could not be opened until the expiration of the current contract in February 1942. The Industrial Union attempted to disturb established labor relations at Delta and at Southeastern, accusing the companies and the AFL of collusion, but it was not successful in upsetting the contracts.[39]

The Industrial Union also continued attempts to organize in yards on the Great Lakes, particularly at American and at the Great Lakes Engineering Co. Its members staged a protest walkout at American from September 4 to 16.[40] The AFL won consent elections in all the yards of the American Shipbuilding Co. in November 1941.[41] But the Industrial Union won at the River Rouge plant of Great Lakes Engineering.[42]

On the West coast, the first serious trouble after the settlement of the machinist strike in San Francisco Bay was the outlaw strike of the welders, October 1941 to February 1942, in rebellion against their status within the AFL craft unions. This was clearly an outlaw strike, but one in which public opinion placed the blame initially at least as much on the officers of the unions as on the outlaws. Mr. Hillman's first efforts to arrange a compromise failed to appease the impatient welders. Once they were on strike and had announced the formation of their own union, the concessions made by the AFL—permitting them to work in various crafts without having more than one union card—failed to halt the strike, even after Pearl Harbor.[43] But their chief grievance having been met, and the country being at war, Mr. Hillman denounced their strike as "a shocking act of disloyalty to the Nation." He and other national leaders successfully

---

[38] Based on data collected by U. S. Dept. of Labor, Bureau of Labor Statistics, Division of Industrial Relations.

[39] *Shipyard Worker*, Nov. 21, 1941, p. 1; Mar. 13, 1942, p. 6; Aug. 21, 1942, p. 1; Oct. 30, 1942, p. 5; Nov. 13, 1942, p. 5.

[40] Based on data collected by U. S. Dept. of Labor, Bureau of Labor Statistics, Division of Industrial Relations.

[41] In re *The American Shipbuilding Company and International Brotherhood of Boilermakers, Iron Shipbuilders and Helpers of America (AFL)*, Oct. 6, 1941, in *Labor Relations Reference Manual*, vol. 9, pp. 342-43.

[42] In re *Great Lakes Engineering Works [River Rouge, Mich.] and Local 2407, Steel Workers Organizing Committee (CIO)*, Aug. 26, 1941, in *Labor Relations Reference Manual*, vol. 9, pp. 75-76.

[43] See above chap. 8, sect. I.

appealed to the rank and file against the leadership of the new "outlaw" union.[44]

After the rebellion of the welders was quieted, the AFL reigned without serious challenge on the Pacific coast until in 1943 the CIO threatened to have the closed shop contracts between Kaiser and the AFL declared illegal. The Industrial Union, supported by Philip Murray, President of the CIO, protested that the existing agreements undemocratically denied the right of workers to select their own bargaining agent, for those agreements had been signed before the labor force was recruited, at a time when there were 66 workers at Oregon Ship, 191 at Vancouver, and none at Swan Island. There had been no elections. The CIO called attention to AFL initiation fees which ranged from $12.50 to $53.00 and dues from $1.25 to $4.50 monthly, and to the revolt going on in Portland against Tom Ray, the boilermakers' business agent.[45]

The Portland boilermakers' local was rich. Mr. Ray stated it had a turnover of 3,000 to 5,000 members monthly. Where did the money go? One place it had gone was into the construction of a $250,000 "labor temple" where the first floor was devoted to taking in cash from members and the bar above reserved for "special parties." The local would not admit women or Negroes to full membership, but they paid the full fees. In December 1942 a group of insurgents had held an election and voted to oust Mr. Ray, but he called this a rump election, and the insurgents had been unsuccessful in getting control.[46]

The National Labor Relations Board saw no alternative to ordering the closed shop dissolved, because it had a new policy of declaring a closed shop contract illegal unless 50 per cent of the potential number of workers had been hired at the time it was signed.[47] In rebuttal, Mr. Frey, speaking for the AFL, empha-

[44] Purcell, *Labor Policies*, p. 227. The *New York Times* gives extensive reports on this strike.

[45] Truman Committee, *Hearings*, pt. 18, pp. 7303-7330-31, 7350-51, 7353; House, *Production in Shipbuilding Plants, Hearings*, pp. 67, 75, 120-21, 134-35.

[46] Agnes E. Meyer in the Washington *Post*, and *Business Week*, Feb. 13, 1943, both partly reprinted in House, *Production in Shipbuilding Plants, Hearings*, pp. 73-75.

[47] Testimony of Gerard D. Reilly, member, NRLB, in House, *Department of Labor—Federal Security Agency Appropriation Bill for 1944, Hearings* before the Subcommittee of the Committee on Appropriations, 78th Cong., 1st sess., pt. 1, May 13, 1943, pp. 323-24. The previous policy of the NLRB had been to declare closed

sized the submission of the terms of the master contract to the local Metal Trades Councils which, he declared, represented the workers. He recalled the history of the stabilization agreement on the West coast and the active part taken by the government in getting it put through. Even the President, said Mr. Frey, had strongly urged its ratification, and the National Defense Mediation Board had ordered Bethlehem to sign the master contract. The closed shop was necessary to maintain discipline, he declared, and he cited his efforts to settle the machinists' and welders' strikes.[48] He called for a suspension of the National Labor Relations Act in this instance, and Congress responded by passing the "Frey amendment" as a rider to the appropriation bill of the National Labor Relations Board. The "Frey amendment" provided that no monies could be used to process complaints unless they were made within three months after a collective bargaining contract had been signed.[49] The boilermakers' national officers took steps to clean house in the Portland local. Tom Ray was removed from office and the labor temple was put to less private uses.[50]

The decision of Congress to prevent the National Labor Relations Board from upsetting union status in the Kaiser yards was based upon what is considered the interests of war production.[51] It was thus a victory for the position which Admiral Land, speaking as Chairman of the Maritime Commission, had been urging vigorously ever since Pearl Harbor. Before the nation was fully, legally at war, it was difficult to oppose publicly the right of labor to conduct organizing drives, but after December 7, 1941 there was a loud demand that labor stop fights over union status and devote all its energies to production.

In the chorus against agitation, strikes, and loafing, Admiral

shops illegal unless the union represented a majority of the workers, but this had been found unworkable when defense plants began to expand so quickly. *Ibid.*

[48] Truman Committee, *Hearings*, pt. 18, pp. 7381, 7388, and for a fuller discussion of the controversy, Coll, "Stabilization," pp. 183-90.

[49] *Congressional Record*, July 2, 1943, p. 7034.

[50] International Brotherhood of Boilermakers, Iron Shipbuilders and Helpers of America, *Proceedings*, 17th Convention, Jan. 31 to Feb. 9, 1943, pp. 154-56, 169, 172, 175.

[51] For the arguments of Congressmen on the basis of production see those of Senators Bridges and Brewster in *Congressional Record*, June 26, 1943, pp. 6567-68 and of Representatives Hare and Tarver in *Congressional Record*, July 1, 1943, pp. 6949-54.

Land's voice was one of the sharpest and, because of his official position, one of the most authoritative. During the desperate efforts being made after Pearl Harbor to get production rolling and meet "impossibly" high goals in shipbuilding, Admiral Land fired stinging barbs at shipyard labor. "Some signed agreements were noticeable in their breach rather than their fulfillments," he remarked caustically in December 1941,[52] and in February 1942 he testified that strikes during 1941 had "cost" the Maritime Commission between 7 and 12 ships.[53]

The formula which the Chairman of the Maritime Commission personally advocated for solving labor unrest was what he called "freezing relationships." If it had been applied it would have halted the organizing drive of the CIO and prevented the growth of the Industrial Union. By "freezing relationships" Admiral Land meant: "If there is a closed-shop agreement today, let it remain for the duration of the war. If there is an open-shop agreement, let it remain for the duration of the war. If there are any other agreements that are in existence today, or that were in existence yesterday, they are good enough for the duration of the war."[54] This step would do away with "the 1% of agitators, organizers and disturbing elements which should be eradicated."[55] Writing to the President in March 1942, he noted that in order to meet the projected program of 9 million tons in 1942 and 15 million in 1943 the productivity of labor would have to be increased 12 per cent in 1942 and 25 per cent in 1943. "Freezing relationships," he wrote, would accomplish this because "the minds of the men could be diverted from an unstabilized condition to a stabilized condition and their interest in production obtained."[56] But the War Labor Board, which replaced the National Defense Mediation Board on January 12, 1942 and was the chief spokesman thereafter of President Roosevelt's labor

[52] House, *Independent Offices Appropriation Bill for 1943, Hearings*, Dec. 9, 1941, pp. 261, 264, 269-70.

[53] House, *Fifth Supplemental National Defense Appropriation Bill for 1942, Hearings*, 77th Cong., 2d sess., Feb. 12, 1942, pp. 103, 108. For the basis of this statement and the extent to which it exaggerated see Coll, "Stabilization," pp. 93-94 and 100, n. 18.

[54] House, *Hearings on H. R. 6790*, 77th Cong., 2d sess., pt. 2, Apr. 13, 1942, p. 2952.

[55] "Memorandum for the Files: Subject: Ship Stabilization Conference, March 9th" (dated Mar. 12, 1942), in Land file.

[56] Land to the President, Mar. 13, 1942, in Land file.

policy, resisted all drives for a "national labor policy" on union recognition, and insisted on settling each case on its merits.

In mid-March 1942, the Administration became alarmed at the extent of anti-labor feeling, and a new note was sounded by President Roosevelt at his press conference on March 17. The President decried "'an amazing state of public misinformation' which he blamed in part upon the newspapers and irresponsible speeches in Congress." He stated that he was opposed to anti-strike bills before the Congress, because no strike problem existed.[57] Next day Admiral Land testified at committee hearings that "'our labor leaders are playing ball,' and drastic legislation might bring on a 'recalcitrant, bolshevik feeling'—'to hell with it,' which would seriously interfere with production." [58] Senator Thomas announced the agreement of his committee that a statement of labor policy should be forthcoming, but that the War Labor Board was the proper agency to enunciate this.[59]

Admiral Land now began a new line of attack. In a radio address on March 24 the Chairman said: "It is pleasant to report that the strike problem—formal strikes—is not a serious one at this time. Last year strikes cost us 8 to 12 ships. But what is serious—and serious right now—is loafing." [60] Again on April 13 he told a Congressional Committee there was "too damn much loafing going on in the shipyards right now." [61]

[57] *New York Times*, Mar. 18, 1942, p. 1. The President's statement followed a drive which began in mid-February to modify existing labor laws and proclaim a "national labor policy" for the duration. The National Association of Manufacturers published a report showing how strikes were "impeding production." (*New York Times*, Feb. 19, 1942, p. 13.) William L. Leiserson, a member of the NLRB, objected to policy being made by the War Labor Board. (*Ibid.*, Feb. 19, 1942, p. 1.) "Little Steel" announced its opposition to maintenance of membership. (*Ibid.*, Feb. 19, 1942, p. 1.) Arthur Krock noted that "only the public interest would get it in the neck" if maintenance of membership were awarded. (*Ibid.*, Feb. 22, 1942, sec. IV, p. 3.) On March 13, Gallup announced that 86 per cent answered "yes" to the question "Should Congress pass a law forbidding strikes in war industries until the war is over?" (*Ibid.*, Mar. 13, 1942, p. 12.) Congressmen responded to a flood of letters and telegrams by introducing "legislation which would suspend all laws and contracts calling for the forty . . . hour week, would outlaw the closed union shop and at the same time limit industrial profits on larger defense orders to 6 per cent . . . ." (*Ibid.*, March 17, 1942, p. 1.)

[58] *Ibid.*, Mar. 18, 1942, p. 18.

[59] *Ibid.* See Land's appeal to William H. Davis for a national labor policy in Land to Davis, Mar. 21, 1942, in MC gf 508-1.

[60] Radio Address by Rear Admiral Emory S. Land . . . Blue Network, Mar. 24, 1942, in MC gf 105-2.

[61] House, *Hearings on H. R. 6790*, pt. 2, Apr. 13, 1942, p. 2959.

Some labor leaders who were "playing ball" thought Admiral Land was pitching a fast game and was blaming labor unjustly for the fact that the merchant shipbuilding program was behind schedule, as it was in the early spring of 1942. In his efforts to increase production the Chairman was needling management privately in personal letters and begging for more production. For example, expressing to John C. Pew, President of the Sun Shipbuilding and Dry Dock Co., disappointment with the output of his yard, Admiral Land wrote him on March 27, 1942, " Here is the largest shipyard in the world whose record of deliveries shows that only one vessel this month is being delivered during this crucial period," a period in which deliveries were "vital to the security of the Nation." [62] While appealing personally to shipyard managers, Admiral Land spoke out publicly to stir the efforts of labor. The labor leaders, feeling that they were being attacked, hit back by blaming what loafing there was on inefficiency in management.[63] Admiral Land agreed that loafing was partly due to shortages of materials and lack of skilled supervision. The Chairman thought the shipbuilder "just as good as any other industrial worker, but it is a little easier to get away with it on a ship" where " you have watertight compartments and other places you can get away into and loaf." " They do not always do it deliberately. I am criticizing management as well as labor," he continued. " There is deliberate loafing on the part of some of them, and there may be some deliberate bad management, but I do not know of it." [64]

It should be recalled that in February and March of 1942 while this controversy was going on, the Nazi submarines were successfully pushing their attacks on shipping into American waters. The reality of war was being brought home by the light of flaming tankers off the Jersey coast. Yet it was not until the middle of April that the Army finally enforced an order to black out the neon lights of beach resorts, against which silhouetted ships made excellent targets.[65]

Loafing in shipyards was investigated by the Maritime Com-

[62] Land reading file, and in same file Land to William S. Newell, Feb. 23, 1942 and to Homer L. Ferguson, March 27, 1942.

[63] *Shipyard Worker*, Mar. 27, 1942, p. 1; and John Green to the President, Feb. 23, 1942, in MC gf 508-1.

[64] House, *Hearings on H. R. 6790*, Apr. 13, 1942, pt. 2, pp. 2970, 2977, 2979-80.

[65] Morison, *The Battle of the Atlantic*, p. 130.

mission's staff in March 1942. Eight yards were given a clean bill of health; seven rated fair; eleven, among them most of the emergency yards, were said to be " unsatisfactory " or " downright disgraceful." The Maritime Commission inspectors who made their reports gave inexperienced workmen under inexperienced supervisors as the cause in a majority of cases; the nature of shipbuilding work, such as waiting for crane lifts, and union objections to firing incompetents were given in one or two. Where conditions were " downright disgraceful " it was reported: " Men sit around in the holds of vessels, around the ships in the yard and all over. Men stand around in groups talking, while others wander around aimlessly." " I feel sure I could show you men sleeping on the vessels at any time of the day or night . . .," wrote one inspector. " Some lively crap games can be found going on during working hours." [66]

The Chairman was hitting at these disgraceful conditions in an attempt to eliminate at least some of them.[67] The *Shipyard Worker* accused the newspapers of playing up Admiral Land's loafing charge but not his qualifying phrases.[68] The White House and the Commission received a flood of letters affirming and denying loafing.[69] President Roosevelt commented that he thought Admiral Land's statement was " a good burr under the tail of both labor and management " but felt the shortage of steel plates was the chief cause of what Admiral Land termed " too damned much loafing." [70]

Although at his press conference President Roosevelt indicated that Admiral Land's criticisms were useful, he wrote him privately shortly thereafter suggesting in effect that he lay off. In a letter of May 2, 1942 the President said, ". . . a careful check-up reveals the fact that many of the so-called slow-downs in shipbuilding plants are due not to organized labor but to strict orders or suggestions from the foreman or their management that a slow-down would be advantageous because the non-delivery of shapes and plates will cause a lay-off if there is not a slow-down." He suggested that Admiral Land " turn over a large part of the

[66] Sanford to Vickery via Flesher, Mar. 17, 1942 in MC gf 508-1.
[67] See this clear statement of purpose in a letter to J. E. Otterson, May 4, 1942 in MC gf. 508-1.
[68] *Shipyard Worker*, Apr. 17, 1942, p. 4.
[69] MC gf 508-1 for this period is full of them.
[70] *New York Times*, Apr. 25, 1942, p. 1.

personnel problems . . . to Eddie Macauley . . . who seems to get on with them pretty well without yielding more than should be yielded in justice and fairness." [71] In replying to the President, the Chairman agreed that slowdowns existed "both on the part of management and labor." "With regard to management," he wrote, "inefficiency is one cause; lack of material is another cause; and fear of losing some of their good personnel is a third cause. . . . With regard to labor, there is some information to the effect that the amount of production should be geared to the rate of the slowest worker rather than to the rate of the average worker. Any form of incentive system tends to cure this. Some say the slowdowns are in accordance with Union rules. I don't know. My own ideas are that an improvement in morale from top to bottom and from bottom to top will go far to ameliorate such conditions." He had, he added, recently appointed Commissioner Macauley to the Shipbuilding Stabilization Committee.[72] Admiral Land gave up his campaign to "freeze relationships" and said no more publicly about loafing.[73]

This campaign by Admiral Land against strikes and loafing was at its peak in March 1942 when the Commission's shipyards were behind schedule. In May and June the effects of the new methods of multiple production began to show; prospects brightened. In July the Chairman of the Maritime Commission expressed wholehearted enthusiasm for labor's performance. He told the Senate Appropriations Committee that labor productivity had increased, that it had even surpassed the $12\frac{1}{2}$ per cent needed to meet the President's directive. Summing it up, the Chairman said, "The shipbuilding boys are going to town." [74]

There were two general reasons in May 1942 for easing off the campaign against "loafing." With the pick-up in production it was becoming clear that shortage of steel was the main difficulty.[75] Moreover, the cooperation of labor leaders was needed at that time in steps to stop inflation, which will be discussed shortly.

After April 1942, Admiral Land returned only occasionally to

---

[71] The President to Land, May 2, 1942, in Land file.

[72] Land to the President, May 20, 1942 in Land file.

[73] The word "publicly" was inserted by Admiral Land on a draft of Coll, "Stabilization."

[74] Senate, *First Supplemental National Defense Appropriation Bill for 1943, Hearings,* 77th Cong., 2d sess., July 8, 1942, p. 211.

[75] See below, chap. 10.

his advocacy of "freezing relationships" and his denunciation of agitators. On October 12, 1942, two days after the first AFL walkout at South Portland and while the CIO was trying to organize Sun and North Carolina, the Chairman of the Maritime Commission made an after-dinner speech to the Investment Bankers Association in New York. The *New York Times* reported him as declaring that "the principal obstructors of the shipbuilding program and the war effort generally were ' union organizers, profiteers, typewriter strategists and needle boys,' " and " ' as far as the organizers are concerned . . . they ought to be shot at sunrise.' " [76] Immediately truckloads of telegrams poured into the Chairman's office.[77] Admiral Land declared the newspapers had misquoted him; he had not, he said, modified the word " organizers " at all and there were various kinds of organizers.[78] Unlike his statements earlier in the year, these were impromptu after-dinner remarks, not part of a deliberate campaign. More soberly, Admiral Land had said only a few nights before: ". . . the support of labor has been so wholehearted that the shipbuilding program and the available supply of materials were threatened by an unbalance." [79]

Philip Murray and John Green immediately protested Admiral Land's proposal that organizers be shot. They asked the President to oust him from the Chairmanship of the Commission, but Admiral Land had lunch with Philip Murray the following week and the campaign was not followed up.[80] Sensation having been satisfied and the proper protests made, the parties retired to their customary positions.

The threat that the CIO would upset the AFL contracts on the Pacific coast gave the Chairman of the Maritime Commission a final chance to restate his position. In giving his support to the Frey amendment on that occasion, he said: " I differentiate between what I call stability and instability. Instability means unrest, uncertainty, agitation, and electioneering. That goes on

[76] *New York Times*, Oct. 20, 1942, p. 9.
[77] Statement of W. L. Nesbitt to Sarah C. Kieffer, Mar. 24, 1948.
[78] *New York Times*, Oct. 21, 1942, p. 17; Memorandum for Files, Oct. 21, 1942, in Land reading file.
[79] *New York Times*, Oct. 17, 1942, p. 5.
[80] *Shipyard Worker*, Oct. 23, 1942, p. 1; Oct. 30, 1942, p. 3. Arthur Krock, writing in the *New York Times* (Dec. 20, 1942, Sec. IV, p. 3), told of an alliance of union leaders and New Dealers who wished to kick Admiral Land upstairs, but apparently they never emerged from backstage.

when a jurisdictional dispute is in effect. It just cannot help but interfere with production." He attributed the West coast's outstanding production to stable labor conditions and repeated as his " final swan song "—" freeze relationships." [81]

The record as a whole shows that strikes in yards working for the Maritime Commission did not seriously impede production except in 1941, and even in that year they did not " cost " as many ships as Admiral Land said, at least not through man-days lost.[82] It is, to be sure, impossible to measure accurately the effects of the agitation connected with strikes and organizing campaigns, just as it is impossible to measure the effect of discontent in a yard which is entirely unorganized. As Table 3 shows, the percentage of working time lost by strikes was far larger in 1941 than in any subsequent year. The number of strikes was larger in 1943 and 1944, when the labor force was at its peak, but most of the strikes after 1941 were short and involved only a few thousand workers at a time (Table 4). Whereas the status of the union was the outstanding issue in the strikes of 1941, thereafter working conditions were the most important cause of strikes. Only in two cases were wages the sole issue in major strikes, and in 1941 wages were definitely less important than union status.[83] It seems that the " stabilization " agreements contributed markedly to labor peace in that year. If industry and the union had been left to bargain as usual, wages would have become a major issue in dispute, and coupled with the struggles between the AFL and CIO would have led to more serious conflicts.

After the nation was fully at war, the time lost in merchant shipbuilding because of strikes was very slight, less than one-tenth of one per cent of total working time in 1942-1944. But this good record was not peculiar to shipbuilding; the average for all industries in those years was almost as good.[84] Comparison with

[81] House, *Production in Shipbuilding Plants, Hearings*, May 11, 1943, pp. 28-33.

[82] Admiral Land based his statement on reports from the MC Division of Economics and Statistics, " Strikes in the Shipbuilding Industry, Jan. 1937 through Oct. 1941," Dec. 19, 1941, and " Strikes in the Shipbuilding Industry, Jan. 1937 through Mar. 19, 1942," May 5, 1942, in MC gf 508-1. These added up manhours lost due to strikes and divided by the number of manhours required to build a C1 or Liberty, but included manhours lost in repair yards and some working for the Navy, and used as devisor a smaller number of manhours than was actually being required at that time to build a Liberty. Coll, " Stabilization," pp. 93-94.

[83] Coll, " Stabilization," Tables I-III.

[84] *Ibid.*, Table IV and pp. 86-89.

TABLE 3

PERCENTAGE OF WORKING TIME LOST DUE TO STRIKES, 1941-1945

(In all industries and in selected industries)

Year	All Industries	Merchant Shipbuilding	Textiles	Iron & Steel	Transportation Equipment*	Mining
1941	.32	1.26	.36	.47	.49	4.52
1942	.05	.07	.14	.07	.03	.31
1943	.15	.07	.10	.14	.05	4.25
1944	.09	.09	.13	.22	.12	.56
1945	.47	.15†	.44	.81	.55	1.61

* Excludes automobile industry each year but 1941.
† First 9 months only.
Source: U. S. Dept. of Labor, Bureau of Labor Statistics, Division of Industrial Relations.

TABLE 4

STRIKES AND MAN-DAYS LOST IN STRIKES IN MERCHANT SHIPYARDS

JUNE 1940-SEPTEMBER 1945

Year	Number of Strikes	Man-Days Lost
1940 (last 7 mos.)	2	22,036
1941	23	246,453
1942	12	74,786
1943	45	129,990
1944	48	154,874
1945 (first 9 mos.)	18	115,633
Total	148	743,772

Compiled from data collected by U. S. Dept. of Labor, Bureau of Labor Statistics. Includes all Class 1 and Class 2 yards building for the U. S. Maritime Commission.

Britain in percentage figures is difficult, but more working days were lost through trade disputes in shipbuilding in the United Kingdom than in the United States in 1942, 1943, and 1944.[85] The small proportion of time lost by strikes in the United States during World War II must be attributed to such causes as the determination of management and labor generally to insure a continuous flow of production while taking their quarrels to the War Labor Board and abiding by its decisions. The collective bargaining in the shipbuilding industry, Admiral Land's barbs against striking and loafing, and labor's reactions are but manifestations in shipbuilding of the various forces and attitudes which, in the nation generally, formed the wartime effort at maximum production.

IV. INFLATION AND THE STABILIZATION COMMITTEE

The conferences creating the "stabilization agreement" obviously did not hold wages down to the level prevalent at the time the first zone standards were set, nor was this their purpose. On the contrary they raised wages about 25 per cent. Moreover each "stabilization" agreement contained an escalator clause providing wage increases to correspond to rises in the cost of living.[1] Although they served to reduce the movement of workers from yard to yard, from the point of view of the effort being made by the government to hold the economy as a whole at a stable level by preventing a general rise of wages and prices, these agreements did not deserve the name "stabilization." They were clearly a factor in the inflation which was occurring in 1941, and which threatened to grow to such an extent in 1942 as to interfere with the whole economic conduct of the war. Consequently the Shipbuilding Stabilization Committee was called into activity

[85] For the United Kingdom the figures are: 37,000 man-days in 1940; 110,000 man-days in 1941; 192,000 in 1942; 137,000 in 1943; and 370,000 in 1944.—*The Ministry of Labour Gazette* (London: H. M. Stationery Office), vol. XLIX, No. 1 (Jan. 1940), p. 5; vol. L (Jan. 1941), p. 5; vol. LI (Jan. 1942), p. 6, vol. LII (Jan. 1943), p. 10; LIII (Jan. 1945), p. 11.

[1] The Pacific Coast Zone Standards provided if the cost of living had increased by 5 per cent or more over the 12-month period beginning Apr. 1, 1941, wages would be "correspondingly adjusted." In the Atlantic and Great Lakes zones the average cost of living as of Dec. 15, 1941 was established as a base and if at the end of one year a 5 per cent or higher rise was reported wages would be raised accordingly. The Gulf Standards were to be amended on the basis of cost of living increases in ten shipbuilding centers in the area.

anew in the spring of 1942. This time it was really called on to contribute to general economic stability, for it was asked to correct some of the inflationary features of the agreements it had sponsored earlier.

The government also wished to modify another provision of the zone standards agreements, namely the payment of overtime rates for Saturday and Sunday work. The President's call for continuous operation in his January 1942 message to Congress led to considerable pressure to do away with overtime payments based on the 40-hour week.[2] Earlier Admiral Vickery had asked Congress to legislate the 48-hour week,[3] but Admiral Land did not directly support this movement. It was an economic problem, he said, and therefore up to Congress.[4] But all were agreed, and the Maritime Commission was most insistent on this point, that if overtime payments were made they should be only for the sixth and seventh day worked, not for working Saturday and Sunday specifically, and should thus encourage the arrangement of shifts which insured continuous production.[5] As the time drew near to make cost-of-living adjustments, the government became alarmed about the way these would increase the cost of ships and stimulate other workers to ask for raises.[6] The zonal agreements were valid for two years. Since the government did not want to be in the position of breaking its agreement, it had to find a way to induce labor leaders to relinquish voluntarily their contract rights. To meet this situation and to do away with Saturday and Sunday as premium days, the Shipbuilding Stabilization Committee resumed activities.

[2] The *New York Times* from mid-February until March 30, 1942 reports and supports these pressures.

[3] House, *Independent Offices Appropriation Bill for 1943, Hearings*, p. 271. Later Admiral Vickery stated he made this request of the President at the beginning of the program. House, *Production in Shipbuilding Plants, Hearings*, pt. 3, June 28, 1943, p. 915.

[4] House, *Hearings on H. R. 6790*, pp. 2953-54.

[5] See Land's first statement on this at conference at Dept. of Labor, Jan. 17, 1942, attached to Memorandum for Files, Jan. 19, 1942, in Land reading file, and others, for example, Memorandum for the Files, Mar. 12, 1942, in Land file. *New York Times*, Mar. 8, 1942, p. 39; Mar. 18, 1942, p. 11; Mar. 20, 1942, p. 1; Mar. 21, 1942, p. 1.

[6] Statement by Paul R. Porter . . . to . . . National Shipbuilding Conference, Apr. 27, 1942, in Division of Shipyard Labor Relations file, Minutes of National Shipbuilding Stabilization Conference on Wage Adjustments held in Chicago . . . 4/27 to 5/16, 1942.

It had not met since the Atlantic zone conference in April 1941. When the independent unions had been excluded from that conference, the leaders of big East coast shipbuilders had refused to be represented on the Committee, and although it had not been formally abolished it had been inactive. A new chairman, Paul R. Porter, was appointed on January 1, 1942 and he and Sidney Hillman then made efforts to induce the leaders of the big shipbuilding corporations to take a place on the Committee. The Maritime Commission vigorously supported them in this effort, and at the same time it again urged that the independent unions, which were still representing the workers at Newport News, North Carolina, Sun, Bethlehem's Fore River Yard, and Bath Iron Works, should also be represented. Speaking for the Industrial Union, John Green objected, saying that the President of the United States recognized only the two national federations in his councils. A meeting of the Shipbuilding Stabilization Committee on March 9 and 10, 1942, again adopted a resolution which excluded the independents from the bargaining table, but the question of representation and union shop were near settlement. Except for Newport News and its subsidiary, the big shipbuilders—even Bethlehem—sent representatives to the national conference which convened in Chicago on April 27, 1942.[7] Perhaps they were encouraged to come by the thought that this time labor was going to be pressed to give up something.

In opening the April conference Mr. Porter traced the problems which would arise to threaten the entire stabilization program, not only in shipbuilding but in the economy as a whole, if labor were given the raises to which it was entitled. It was generally expected, Mr. Porter pointed out, that the cost of living was to be stabilized at the level of March 15, 1942. If the escalator clause in the zone standards agreements was invoked, wages on the Pacific coast would rise 13 per cent and those on the Gulf 12 per cent. Wages on the Atlantic coast and on the Great Lakes would remain the same. Clearly it was the AFL which was called on to make the major sacrifice.[8]

On May 3 the Conference received a message from President Roosevelt:

[7] Coll, "Stabilization," pp. 128-32.
[8] Statement of Paul R. Porter cited in note 6 above.

The situation that now confronts you is that the full percentage wage increase for which your contracts call and to which, by the letter of the law, you are entitled is irreconcilable with the national policy to control the cost of living. Moreover I understand that the literal application of the contract would result in unjustifiable regional inequalities in wages in the shipbuilding industry.

Under these circumstances I suggest to the stabilization conference that you put your heads together, and try to work out a plan by which this may be resolved, so that the wage standards of the workers in the shipbuilding industry and in other industries, and the living standards of all persons of modest income may be preserved against an inflationary rise in the cost of living.[9]

A working committee of thirty put their heads together. The AFL asked for a 10 cent raise as against the 13 cents to which they were entitled. The Maritime Commission was offering 6 cents. Mr. Ring recommended to Commissioner Macauley from Chicago that the Commission offer 8 cents, which Mr. Davis of the War Labor Board felt "won't hurt" and to which the Navy had agreed.[10]

Agreement was reached. All shipyard workers were given at least an 8-cent-an-hour raise, which established the national base rate for skilled mechanics at $1.20 and increased the wages of other workers proportionately. In the Gulf yards, where rates had been lower than in the other zones, workers were given 9-cent to 13-cent increases. The unions agreed to accept the wage increase in War Bonds. These rates were to go into effect as of July 19, 1942 without the necessity of reopening collective bargaining contracts.[11]

Labor's agreement to the abolition of Saturdays and Sundays as premium days was anticipated. In March, just before President Roosevelt had warned Congress about hasty labor legislation, the AFL and CIO had accepted this modification,[12] and it had been agreed to for the Pacific coast at a zonal conference in January. The Chicago amendments called for time and a half for the sixth

[9] The President . . . to Paul R. Porter, May 2, 1942.

[10] Telephone conversation between Commissioner Macauley and Mr. Ring, May 13, 1942, in Division of Shipyard Labor Relations file, Zone Standards—General.

[11] Amendments to the Pacific Coast, Atlantic Coast, Great Lakes, and Gulf Coast Zone Standards of the Shipbuilding Industry, Approved by the National Shipbuilding Conference . . . (mimeographed) May 16, 1942 . . . , in Historian's Collection, Research file 205.3.1.

[12] *New York Times,* Mar. 25, 1942, p. 1. The Industrial Union had agreed to this at a national conference of the union on Apr. 10, 1942. See resolution No. 1 enclosed with Green to Porter, Apr. 13, 1942, in Division of Shipyard Labor Relations file, Zone Standards—General.

regular shift and double time for the seventh regular shift worked.[13]

The national agreement reached at Chicago in April 1942 was a compromise between an Administration trying to fight inflation and unions which had succeeded in getting very advantageous conditions written into the first stabilization agreements. Shipyard workers received less than their contracts called for but were given another substantial raise, a raise which enabled them to remain the highest paid workers in any major industry in the country.

In this account of the three-cornered negotiations between labor unions, managements, and government, conflicts and recriminations have perhaps bulked larger than cooperation. The end result, however, was clearly satisfactory in one essential respect: labor was not the limiting factor in the production of ships. Although it had been expected at the beginning of the program that the lack of skilled labor would prove the toughest bottleneck, the breakdown of skilled jobs into simpler tasks and the offering of high wages to the relatively unskilled labor used for these tasks had broken that bottleneck. Once recruited, the labor force was kept at work with relatively few interruptions arising from discontent among the workers. Controversies over wages and union status kept recurring, but after 1941 they rarely led to strikes. If labor had been the only problem in 1942-1943, production would have been even higher than it was. The extent to which shipbuilding expanded to meet the needs of war was limited, not by the supply or productivity of shipyard labor, but by the supply of steel.

[13] Amendments to Zone Standards, May 16, 1942.

CHAPTER 10

THE BATTLE FOR STEEL

I. Opening Engagements, 1941

THAT STEEL should prove the limiting factor in shipbuilding was quite contrary to expectation at the time the emergency program began but was recognized as soon as the methods of using labor in multiple production had been worked out. Part of the battle for steel has been described already in connection with the cancellation of the Higgins contract, but it is desirable here to go back and survey the problem as a whole.

Before the war shipbuilding was not particularly important among consumers of steel, for Navy and merchant yards together took only sightly over 2 per cent of the total output. During the war, shipbuilding came to take nearly 20 per cent. Moreover, the shipbuilders' demand for steel was primarily a demand for plate (of a ship's steel hull 77 per cent is steel plate), so that at the peak of wartime shipbuilding the shipyards were using 60 per cent of all the steel plate rolled.[1] The steel mills increased production of plate from 4 million tons in 1940 to 13 million in 1943, but the shipyards increased their consumption in even higher ratio—from half a million tons in 1940 to 7.5 million tons in 1943. At the peak of consumption in 1943-44, the Maritime Commission program alone was using steel plate at the rate of about 5,700,000 tons a year, taking about 70 per cent of the total going into shipbuilding of all kinds.

The shortage of steel plate was attributed sometimes to the limited capacity of the plate rolling mills, sometimes to the limited capacity to produce the steel ingots which were basic to all forms of steel production. Shortage from either of these causes affected

---

[1] A breakdown by agencies of the distribution in the peak period, 1943-44, shows 63 per cent going to the Navy and Maritime, but that includes not only ship repair but also docks, and some forms of armament. All the figures on steel here given are in short tons (2000 lbs.) and are based on those in G. J. Fischer's "Statistical Analysis, Steel." Compare Fassett, ed., *The Shipbuilding Business*, I, 7.

the Commission's shipbuilding by cutting down the delivery of plate to the yards. There were severe shortages in some special steel items also, such as steel valves, because of excessive demand on the manufacturers of those products. But since shipbuilding consumed steel in the form of plate far more than in any other form, its pressure on the steel industry was mainly expressed by demands for plate.

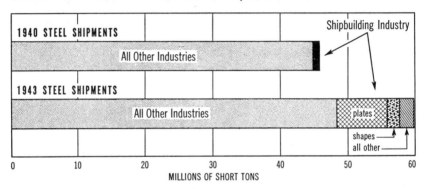

FIGURE 23. Source: Fischer, "Statistical Analysis, Steel," Table 8.

This demand did not appear all at once, of course; it was presented in the piling of one estimate on another as the Commission's program went through wave after wave of expansion. Only at the end of 1943 or in 1944 was the situation eased. Meanwhile the various defeats suffered by the Maritime Commission during 1942 in its battle for steel had left their impress on the Commission's wartime program.

As in labor matters, the activities of the Maritime Commission in providing steel for the yards were closely related to those of the National Defense Advisory Commission, the Office of Production Management, and the War Production Board. But matters affecting the supply of materials were not centralized in these bodies to the same extent to which matters affecting labor were centralized around Sidney Hillman. In the NDAC and the OPM there was a "splintering" of authority in regard to steel, as in regard to most materials, among the different Divisions and Branches;[2] but they all consulted the same staff of experts on

[2] *Industrial Mobilization for War*, pp. 30, 93-113.

steel.³ Walter S. Tower, previously President of the American Iron and Steel Institute, was in the Iron and Steel Branch of OPM until public agitation led to the adoption by OPM of a general policy prohibiting altogether the employment of paid trade-association executives.⁴ In the WPB, the Iron and Steel Branch (later a Division) worked with an Iron and Steel Industry Advisory Committee.⁵ The experts on steel were moved about organizationally so as to be now under one superior, now under another, first in OPM and later in WPB under the general authority of Donald Nelson, its Chairman—or more immediately under the Program Vice Chairman, in November 1942 Ferdinand Eberstadt; but the fact that they were experts gave them a measure of autonomy. Finally H. G. Batcheller, president of the Allegheny Ludlum Steel Corporation joined WPB to enable it to "get things under control." ⁶ During all these moves, the experts from the steel industry were the group in OPM or WPB on which the Maritime Commission depended when trying to get more steel for its shipyards.⁷

In securing efficient distribution of steel plates, there was a large gap between laying down general policies and implementing them effectively. The decisive mechanism in implementation, the key point, was the schedule of a steel mill. The rolling schedules drawn up by the mills determined what steel was produced and to whom it was shipped. From the point of view of the Maritime Commission, the function of the OPM was to see to it that these rolling schedules included the steel plates needed to build the merchant ships called for by the President's programs. The Maritime Commission showered the OPM with complaints and requests in efforts to have the OPM force the steel mills to increase their production, change their methods, and revise their schedules so that there would be steel in the shipyards.

Among the methods used by the OPM was the establishment

³ *Ibid.*, p. 174. Arthur D. Whiteside, President of Dun and Bradstreet, was the head of the committee on steel in the Priorities Division during the first half of 1941 and there used the staff experts of the Production Division.
⁴ *Ibid.*, p. 109; Truman Committee, *Additional Report*, 77th Cong., 2d sess., Jan. 15, 1942, No. 480, pt. 5, pp. 7-10; Truman Committee, *Additional Report*, 78th Cong., 1st sess., Feb. 4, 1943, No. 10, pt. 3, p. 2.
⁵ Chaikin and Coleman, *Shipbuilding Policies*, p. 35.
⁶ Nelson, *Arsenal of Democracy*, p. 356.
⁷ *New York Times*, Mar. 4, 1942, p. 12, for men from the industry who came in at that time to head "units."

of a priority schedule. Such a schedule was a list of products ranked in order according to the urgency with which they were needed. By agreement between the OPM and the Army and Navy Munitions Board (ANMB), the preparation of the priority schedule for military orders and items of such military importance as ships was done by the ANMB. To enforce the priority schedule, priority certificates were issued to contractors, giving a rating to their contracts and authorizing some to obtain materials for their execution ahead of producers whose contracts had a lower rating. These priority certificates, signed in blank by the OPM, were issued by the procurement officers of the Army and Navy and, somewhat later, by the Maritime Commission. The shipbuilders who received these priority certificates could then place on their orders to the steel mills the same priority ratings which they had received. When arranging these orders into a rolling schedule the managers of the steel mills were supposed to roll first the steel plates for those orders which bore the highest priority.[8]

A system of priorities had governed orders for steel plates ever since March 15, 1941 when "ship plates" were added to the Priorities Critical List,[9] but the way the system of priorities worked was found unsatisfactory by the Maritime Commission, for it gave the Army and Navy many advantages. Behind them was the authority of the statute authorizing priority in production for the armed forces. Until August of 1941 there was no such statutory basis for giving priorities for materials going into merchant ships; although such priorities were given, they could not have been enforced in the courts.[10] Moreover, if priorities were issued for more steel than could be immediately supplied, and if the producer had to choose which orders he would fill, he had reasons to fill those of the Army or Navy, for the Army and Navy were old customers and had representatives at the mills to push the steel makers.[11]

The shortcomings of priorities from the point of view of the

[8] *Industrial Mobilization for War*, I, pp. 171-73, and below chap. 11.

[9] Coleman, *Shipbuilding Activities*, p. 67. This list was compiled by the ANMB and issued through the Priorities Division of OPM.

[10] Although the President was given statutory authority to set up priorities for other than military uses by law of May 31, 1941, he did not delegate the authority so that it could be used until he established SPAB on Aug. 28, 1941. *Industrial Mobilization for War*, I, pp. 28, 96, 110, 192.

[11] Coleman, *Shipbuilding Activities*, pp. 113-14. Schmeltzer to Vickery (written by Hammond), Jan. 23, 1942, in MC gf 507-4-1-3.

Maritime Commission became apparent as soon as there was a shortage of steel plate, namely, in the summer of 1941. Earlier the Commission had been assured by the steel men in OPM that there was plenty of capacity to roll steel; if the steel appeared scarce it was only because of the large consumption by the automobile manufacturers.[12] On May 3, 1941, Mr. Knudsen called on the automobile industry to reduce production 20 per cent beginning August 1, and on May 22 the steel expert, Gano Dunn, reversing his earlier estimate, predicted a 6 million ton deficit of steel for 1942.[13]

In the middle of growing recognition of a coming shortage, a few shipyards began screaming in June 1941 that they were already short. Henry J. Kaiser particularly complained to the Commission that the shops in one of his Richmond yards ran out of steel on June 12, while the other yard was being delayed a month or more on the steel work for its first ships.[14] Admiral Vickery passed Kaiser's complaints on to the White House to Mr. Hopkins with the comment that the Commission was " having a great deal of trouble with steel." [15]

Although the violence of the outcries seemed to indicate that the steel mills were not meeting their schedules for the shipyards, that was not yet the difficulty in the overall picture. The main cause of difficulty was that some shipyards were actually getting under way faster than had been anticipated. After the President's proclamation of an " unlimited Emergency " the Commission had prepared new and faster schedules for the delivery of ships from the emergency yards. When totaled up, they showed 27 Liberty ships to be delivered in 1941, instead of the 3 shown in the original schedules.[16] Steel with which to meet the expedited

[12] Memo, Chief, Machinery and Scientific Specifications Section, to H. L. Vickery, Commissioner, Apr. 18, 1941, in files on steel of L. C. Hoffman of the MC, Technical Div.

[13] *Industrial Mobilization for War*, I, pp. 137, 190-91.

[14] Kaiser for Richmond Shipbuilding Corp. (of California) to Vickery, June 26, 1941, in gf 507-4-1-3. See also Kaiser for Todd California Shipbuilding Corp., to Vickery, June 26, 1941, and letters, Kaiser to Benjamin F. Fairless and Eugene G. Grace, June 30, 1941 (attached to letter of Kaiser to Land, June 30, 1941, all in gf 507-4-1-3), giving particulars of delays because of their failure to deliver.

[15] Coleman, *Shipbuilding Activities*, p. 108.

[16] Report of Progress as of May 30, 1941 (No. 41), June 1, 1941, in Statistical files of Inspection and Performance Division; USMC, " Estimated Delivery Schedule of Cargo and Combination Vessels for the Period May 31, 1941 to July 1, 1942," June 24, 1941, in MC, Historian's Collection, Research file 117, Programs 1941.

schedules was needed at once, because the steel that went into a ship had to be at the shipyards many months before the ship was launched. The steel mills were called on to increase their shipments to the yards correspondingly above the earlier schedules, but said they could not do it immediately.[17] In August, Kaiser reported that his yards at Portland, Oregon and Richmond could not meet the expedited schedules because steel was lacking. " It is beyond the scope of human ability " even he confessed, " to construct steel boats without steel." [18]

In the same month, the Maritime Commission made its first general criticism of the way steel was being handled in the OPM. A " Proposed Report to the President " of August 18, 1941, related the Commission's efforts to comply with the President's instructions to expedite ship construction, the generally favorable results with respect to the supply of materials, equipment, and machinery, and the bottleneck presented by steel. It mentioned specific cases of idle shipways, construction delays, and attempts at speeding up construction frustrated by actual or prospective deficiencies in the supply of steel. It concluded that too much time had already been lost to permit the delivery of the entire group of ships promised in 1941, although everything was being done to secure as many ships as possible.[19]

Armed with this memorandum Admiral Land began needling OPM to correct the situation. He reminded Mr. Knudsen and Edward R. Stettinius, Jr. of OPM that in early July he had given the President his opinion that the steel difficulties could be ironed out satisfactorily with the assistance of OPM, and that he had made a " stump speech " about the shortage to the producers, but found that " the results at the shipyards are not satisfactory." He said he knew three things about steel and steel manufacturers: " (1) They can do this job if they have the will to do it; (2) the customer is always right; (3) kissing goes by favor—and the Maritime Commission wants to be kissed." In concluding, he asked

---

[17] W. F. Gibbs to MC, July 19, 1941, in gf 507-4-1-3. Data collected from 9 of the 11 emergency shipyards Aug. 19-23, 1941 showed that the tonnage scheduled for the period of $2\frac{2}{3}$ months was approximately 175,000 tons, while receipts in the same period totaled about 125,000 tons. Telegrams to J. E. Schmeltzer, in gf 507-4-1-3. The steel deliveries scheduled to meet the requirements of the original contract dates amounted to approximately 125,000 tons for the 9 yards in the same period. Historian's Collection, Research file 117, Plates and Shapes—Requirements.

[18] Kaiser to Vickery, Aug. 13, 1941, in gf 507-4-1-3.

[19] " Proposed Report to the President," Aug. 18, 1941, in gf 107-1.

for advice before taking the matter up with the President in writing.[20]

On the same day that Admiral Land was calling for kisses from the steel industry, the ANMB approved and issued priority ratings which put merchant vessels on a par with naval vessels, except for a few Navy items having the emergency AA rating. It gave the highest regular rating, A-1-a, to merchant vessels to be completed in 1941,[21] but the next day, August 21, Admiral Land lodged with the Priorities Division of OPM a complaint about the difficulty of obtaining steel even with an A-1-a priority. Deploring the lack of action in the face of the disrespect for priorities shown by the steel companies, he called for aggressive action to implement priorities. Material found its way into nondefense items or less important defense equipment when it should have gone directly to critical defense needs. It was not yet decided, however, who was to determine the relative importance of material and equipment to national defense and what was to be the relation between the Army and Navy Munitions Board and the Priorities Division of OPM. He said that the basic priorities policies should be more clearly defined, and added, " we believe you should have more authority and execute it; also ' execute ' those who don't ' come across.' " [22]

When Admiral Land was notified that Knudsen agreed with the proposed report to the President, he decided to send it along somewhat simplified on September 2.[23]

By that time OPM had been reorganized and placed under the Supply Priorities and Allocations Board, for many other pressures in addition to that from the Maritime Commission showed the need of change. In SPAB, as the new board was called, was lodged the control over civilian priorities for which a contest had been going on for months; and SPAB reaffirmed the ameliorative steps initiated by OPM just before the creation of the new board.[24]

---

[20] Memorandum for the Office of Production Management, Attention Mr. Knudsen and Mr. Stettinius, Aug. 20, 1941, in gf 506-1-1-21.

[21] Coleman, *Shipbuilding Activities*, pp. 76-77. The OPM Director of Priorities by Preference Order P-7 assigned this A-1-a rating on June 12, 1941; *ibid.*, p. 72.

[22] Memorandum for Blackwell Smith, Priorities Division, OPM, from Land, Aug. 21, 1941, in Land reading file. On the transmittal slip, Land wrote: " You asked for comment. I trust these may be constructive. We are up against it for steel and I have today written to Knudsen and Stettinius. We need help badly."

[23] W. H. Harrison to Land, Aug. 28, 1941, in gf 506-1-1-21; Land to the President, Sept. 2, 1941, in gf 506-1-1-21.

[24] *Industrial Mobilization for War*, I, pp. 109-11; Gerald J. Fischer, " Problems

Attack on the steel problem now went forward on four fronts: (1) building additional steelmaking capacity, (2) developing machinery for allocation, (3) curtailing the automobile industry, and (4) changing the specifications of the steel plate used in shipbuilding so that it could be produced in the strip mills which had hitherto been supplying the automobile industry.

A report on the possibility of expanding the steel industry by 10 to 15 million tons of annual capacity had been requested by the President after the second Dunn report of May 1941. Four months later, on September 30, SPAB approved a 10-million-ton program,[25] but it had to be approved over again the following May.[26] Production from the new mills eased the situation in 1944.[27]

Meanwhile the effects of the shortage were alleviated by developing a system of allocations. By allocations a specific amount of steel was assigned to Maritime Commission shipbuilding and to other uses. Priority schedules, periodically revised, continued in effect and gave a needed guidance to the officials in OPM or WPB who made the allocations. Priority certificates continued to be issued but became less important, for allocations introduced a new method of carrying priority ratings into effect through controlling the rolling schedules of the steel mills. When production was regulated only by priority certificates, each steel company made up its own rolling schedules. Enforcement depended on the steel companies' finding it possible to fill the orders with higher priority ratings first and yet make what the mill managers regarded as a practical workable rolling schedule for each of their mills. Under a fully developed system of allocations, the officials of OPM or WPB went over the rolling schedules of the mills and rearranged them if necessary to provide that the amount allocated for various purposes would actually be manufactured.

Allocation of steel plate was begun during the fall of 1941.[28]

---

of Steel Production and Distribution in relation to Shipbuilding under the U. S. Maritime Commission during World War II," typescript in MC, Historian's Collection, Research file 206.1.3. Nearly all of this section is based on Fischer's research.

[25] *Industrial Mobilization for War*, I, 154.
[26] Nelson, *Arsenal of Democracy*, p. 143.
[27] See below, section IV and Figure 24.
[28] General allocation was ordered by SPAB on Nov. 29, 1941, thus reaffirming Preference Order M-21 of Aug. 9, 1941, which supplied the authority for the

Shipyards were having much trouble getting any steel mill to accept their orders,[29] and the Iron and Steel Branch of OPM went over the rolling schedules of the steel mills in order to see to it that orders from the shipyards were included. In November a Chief of the Iron and Steel Branch wrote, "we simply took the schedules apart and while we may not make a world's over-all record, the Maritime Commission will get plates."[30] Results were not so satisfactory as promised, however; for until the spring of 1942 the Iron and Steel Branch lacked adequate information to do a complete month-by-month job of allocating steel from specific mills.[31]

Supplemental to the direction of material to desired places by allocation was its diversion from undesired uses by curtailment orders. Orders to curtail the consumption of steel by automobile producers were drawn in August and issued in mid-September, but they did not require substantial reduction in output until nearly the end of 1941.[32] By that time the attack on Pearl Harbor brought full conversion of the automobile industry to the production of trucks, airplanes, and other munitions.

Once automobile production was curtailed, there was a chance of rolling ship plates in mills which normally rolled thin strip steel for the auto-builders. Since the equipment of these strip mills was different from that of the plate mills, the shipbuilding industry and the Maritime Commission were asked to change their specifications regarding steel plates.

The shipyards were asked to accept plates unsheared, since many strip mills lacked the equipment for thus trimming the plates to stipulated sizes. Also the number of different sizes and shapes were to be reduced (there were 750-800) so that more steel could be rolled without readjusting rollers, and plates for at least

---

allocation. Coleman, *Shipbuilding Activities*, pp. 112-14, and letter of Blackwell Smith, signing as Assistant Director of Priorities, to Land, Sept. 5, 1941, in MC gf 107-55, and the President to Land, Sept. 20, 1941, in MC gf 506-1-1-21. On Oct. 29, 1941, R. B. Paterson declared that allocation had so much improved the situation in regard to steel plate and pig iron that he proposed it be applied to all steel products.—CPA, *Minutes of the Supply Priorities and Allocations Board*, p. 22.

[29] Director, Construction Division to Vickery, Nov. 6, 1941, in gf 507-4-1-3; Land to Arthur D. Whiteside, Nov. 10, 1941, in gf 506-1-1-21.

[30] Whiteside to Land, Nov. 14, 1941, in MC gf 507-4-1-3.

[31] Memorandum Prepared by the Iron and Steel Branch . . . dated Apr. 29, 1942, attached to Land to Vickery, Apr. 30, 1942, in gf 506-1-1-21.

[32] *Industrial Mobilization for War*, I, 196.

12 ships were to be ordered for rolling at one time.[33] Both these changes meant more work for the shipyards in the fabrication of the plates, and the use of a higher tonnage of steel plate per ship. For Liberty ships the increase was from 2,313 to 2,406 tons.[34] But the Maritime Commission agreed to both. Between November 6, and December 23, 1941, its experts, working with the steel men of the WPB and with the American Bureau of Shipping, arranged to reduce the number of thicknesses of plates used in the hull of the Liberty ship from 75 to 27.[35] The number of different lengths and widths was also reduced. At the same time, the long-range yards, which bought their own steel, were directed to confer with their suppliers with a view to making similar modifications in their demands.[36]

In the period from December 1941 through January 1942 many changes were being made in the plate models and in the erection sequences used in various yards because the new yards were then working out new and faster methods of multiple production. Changes to permit the use of strip mills for ship plate and changes to increase subassembly and to speed shipyard fabrication were all going forward together. Accepting unsheared plates and more uniform sizes increased somewhat the amount of cutting to be done in the shipyards in preparing the plates for assembly. But this cutting was frequently done by burning and was combined with the normal preparation of the plate for welding. Under the industrial conditions of the moment, the operation could be performed in the shipyards with fewer delays than if done in the steel mills. Since receiving steel in a steady flow was more important to the yards than the slight addition in manhours, they were

[33] Director, Construction Division to Vickery, Nov. 6, 1941, in gf. 507-4-1-3; S. B. Adams, Asst. Chief, Iron and Steel Branch, OPM to J. E. Schmeltzer, Dec. 10, 1941, and Adams to W. E. Hammond, Dec. 18, 1941, both in MC, Prod. Div. Adm. files, No. 8 Misc.

[34] Cf. J. E. Schmeltzer to Glenn E. McLaughlin of the National Resources Planning Board, Nov. 24, 1941 and C. E. Adams, Chief, Iron and Steel Branch, WPB to Vickery, May 7, 1942, both in MC gf 507-4-1-3.

[35] W. G. Esmond to J. E. Schmeltzer, Dec. 23, 1941 in Esmond reading file, MC, Technical Division; historical notes by W. G. Esmond in Historian's Collection, Research file.

[36] Schmeltzer to all shipyards, Nov. 15, 1941 in gf 506-1-1-21, and same to same Jan. 3, 1942 in " Steel Plate " file of L. C. Hoffman, Technical Division; Schmeltzer to Weirton Steel Co., et al., Jan. 30, 1942 in gf 507-4-1-3:

willing to accept the products of the rolling mills in the less finished form.[37]

A third change asked for by WPB in the specifications for steel plates was a redesign of the plate model so that no plates would be wider than 72 inches. This would involve extensive readjustment, for more than one half of the hull plate actually in use at the beginning of 1942 was wider than 72 inches.[38] Moreover, the change would increase the amount of welding by 40 per cent.[39] But since the Commission was told that strip mills would have to be used and that many of them could not roll plates wider than 72 inches, it proposed to meet the request of the steel men by ordering the narrower plates for the new emergency yards authorized after Pearl Harbor. A changeover in yards already in operation would be more difficult because it would mean preparing new templates; but the long-range yards were urged to use the narrower plate where practicable.[40] In one way or another the strip mills were finding that they could roll what the shipyards wanted.[41] When in May the Commission presented its new bill of material for the Liberty ship based on a 72-inch maximum width of steel plate, strip mills limited to that width were already carrying as heavy a load as the wider mills. The WPB discovered also with surprise that the new bill of material using narrower width required 246 tons more per ship than the former list, an increase of 10 per cent. Considering the relation then existing between different kinds of demands for rolled steel from all sources and the kinds of mills, WPB concluded that the change " would actually be disadvantageous." The Commission cancelled its instructions to the new yards accordingly.[42]

Thus in the end the Commission continued using a bill of material which retained the wider plates very much as provided in the original material list prepared by Gibbs & Cox. Production

---

[37] Statements of Maurice Nicholls on Aug. 8, 1947, of Harvey H. Hile on Dec. 9, 1947 and of J. T. Gallagher Dec. 10, 1947.

[38] Esmond to C. W. Flesher, Feb. 23, 1942, and notes, March 9, 1942 in Esmond reading file, MC Technical Division.

[39] W. F. Rockwell to H. Leroy Whitney, July 14, 1942 in gf 507-4-1-3.

[40] Fischer, " Problems of Steel Production," chap. II, sect. A.

[41] *Ibid.*, and the statements of Admiral Vickery in Truman Committee *Hearings*, 77th Cong., 1st sess., pt. 12, p. 5198.

[42] C. E. Adams to C. W. Flesher, March 17, 1942; Adams to Vickery, March 19, 1942 and May 7, 1942; Vickery to Adams, May 11, 1942; all in MC gf 507-4-1-3.

at the strip mills was facilitated by combining sizes so that rollings could be in large quantity. This increased about 4 per cent the amount of steel required, but the Commission was saved the cost of any wholesale revision of design and the sizable increase in manhours and welding materials involved in the change to narrower plates. To be sure some extra cost was involved in paying for plates from a strip mill the same prices paid for plates from a plate mill—extra freight charges, two or three times the normal amount of scrap, cutting costs, and marking costs. R. H. Mohler, in charge of the Emergency Construction Audit, suggested early in 1942 that these extra costs be compensated for by a differential between the price paid for plate from strip mills and plate from other mills. No such differential was set up. A few days after Pearl Harbor it had been said that some form of price relief would be necessary to get the mills to roll plate in preference to other steel products which brought a higher return. As it worked out, there was no rise in the basic price of plates, but insofar as the more liberal tolerances of the new bills of material did reduce the work required at the steel mill on each plate the steel companies were granted a hidden price increase.[43]

The adjustments made were sufficient so that plate production increased very rapidly. Between October 1941 and April 1942 it rose over 50 per cent and most of the gain came from strip mills, which more than doubled their output of plate in that six month period and more than tripled it by July 1942.[44]

II. "Taking it in the Teeth"

After Pearl Harbor there was no longer any doubt that industry would concentrate on producing what was needed for war, but the huge Army supply program and the Navy's post-Pearl Harbor plans competed fiercely for steel with the Maritime Commission's program, swollen by the fourth and fifth waves of expansion. In 1942 the problem was how to divide the total steel production of the nation among various uses, each of which seemed essential to the successful prosecution of the war. In this division the Maritime Commission received far less than its leaders claimed was necessary to fulfill their directives.[1]

[43] Fischer, "Problems of Steel Production," chap. II, sec. A.; Mohler to Honsick, Mar. 20, 1942 in gf 507-4-1-3.
[44] *Wartime Production Achievements and the Reconversion Outlook*, p. 47.
[1] See Table 4 in Fischer's "Statistical Analysis, Steel."

One source of dispute in 1942 between the WPB and the Maritime Commission was what the latter called the "deficit" resulting from small deliveries of steel in the previous year and in the opening months of the new year. Admiral Land and his staff referred to this deficit as something which would have to be made up if the President's objectives of 8 or 9 million tons in 1942 were to be met.[2] They were stating the matter in terms of an annual total or of a global allotment for the whole program. In January, Carl E. Adams, Chief of the Iron and Steel Branch of WPB, was also discussing the problem in those terms.[3] But later the discussion centered on the current situation: how much steel could be used by the yards and how much could be supplied by the mills in any given month?

The total output to be planned for, whether by year or month, was a more basic cause of confusion. The President had named an impossibly high objective in February 1942 when he called for 9 million tons. Admirals Vickery and Land said in February that production could come to between 7.5 and 8 million. Some of the staff of WPB thought the completion of more than 6.5 million uncertain. They were at the same time trying to bring down the whole program of war munitions and construction reasonably near to their estimate of the nation's capacity. In March the war munitions and construction program asked by the Armed Forces was at $62.5 billion and WPB's estimate of the possible was about $20 billion less. The way in which these two figures would be reconciled might affect the Maritime Commission's shipbuilding.

To complicate the picture further, the President from time to time added special objectives such as landing ships and escort vessels to be built by the Maritime Commission as well as the Navy.[4]

A subject of dispute which was still of major importance at the beginning of 1942 but which decreased in importance as 1942 wore on, was the priority schedule. Immediately after Pearl Harbor, merchant shipbuilding lost the degree of parity with Navy building which it had gained in August 1941. On December

[2] Land to Nelson, Apr. 18, 1942, and Land to the President, May 16, 1942, both in gf 506-1-1-21.

[3] *Minutes of the War Production Board*, pp. 17-18.

[4] *Industrial Mobilization for War*, I, pp. 273-81; *Minutes of the War Production Board*, p. 33. Above, chap. 6, sect. 1.

17, 1941 new rulings on priorities were issued favoring Navy vessels. The Maritime Commission soon protested both to the ANMB, which gave the ratings, and to the WPB, which had authority to review them, but to no immediate avail. Maritime was denied even the special rating to complete conversions of C3's to Navy carriers and so clear the shipways for further construction, although this request was supported by the Navy's Bureau of Ships. Of course, many other components and materials aside from steel were involved in these priority schedules and in the delays resulting from low ratings. WPB called on the ANMB to present a comprehensive new system of priorities. When it did so on May 20, 1942, it gave a new top rating, then called AA-1, to half, only, of the Commission's " 9,000,000 deadweight ton program for 1942." The rest of that program was given the relatively low rating AA-3 along with naval vessels to be completed in 1943 or 1944. Since materials for nearly 50 per cent of the 1942 program had already been procured by the end of May 1942, the AA-1 meant very little, so said the Planning Committee of the WPB. This Committee criticized the ratings given the Maritime Commission even more severely than did Admiral Vickery.[5] Half of the 1942 merchant ship program seemed likely to be lost.[6]

But the Commission kept up its efforts to have the priority ratings changed,[7] and the revised draft dated June 8, 1942 saved practically all the 1942 program by giving it the rating AA-2. At the beginning of the specifications under the rating AA-2 was the following sentence: " *a.* An additional amount equal to that provided for under AA-1 above, not to exceed however in combined aggregate one hundred per cent, for any program listed under AA-1 above and therein limited to a state[d] percentage." Since an AA-1 rating was given the Maritime Commission for 50 per cent of " that part of the 9,000,000 Deadweight Ton program of 1942 consisting of transoceanic shipping, tankers, and

---

[5] Chaikin and Coleman, *Shipbuilding Policies*, pp. 23-26; Vickery to Eberstadt, May 20, 1942, in gf 120-124-2.

[6] *Industrial Mobilization for War*, I, p. 296.

[7] A. J. Goldenthal, Planning Committee, WPB, to Weber, May 27, 1942, and Land to Goldenthal (written by Weber), May 29, 1942.—Bland to Vickery, Mar. 31, 1942, and Vickery to Bland, June 1, 1942, all in gf 120-124-2. On the revised draft of June 5 see Douglas to the Administrator, June 8, 1942, attaching comparison of May 20 and June 5 drafts, in gf 120-124-2.

oil barges with their complementary tugs," the clause added in the draft dated June 8 had the effect of giving an AA-2 rating to the other 50 per cent of that 9,000,000 ton program. The part of the Maritime Commission's 1942 program left in the AA-3 rating consisted of ore carriers, coastal vessels and barges, and construction for 1943.[8]

The approval of the ANMB rating, whether in the form of May 20 or in the revised version of June 8, was bitterly opposed by most of the staff of WPB on the grounds that it gave the armed forces control of the nation's economy to the detriment of civilian needs.[9] But with a reservation in favor of civilian requirements the revised directive was approved by the President.[10] Mr. Nelson then acquiesced and accepted the new rulings as effective July 1, 1942, although his staff continued to criticize it and some resigned in protest.[11]

As soon as nearly all of its 1942 program had been lifted to the category of AA-2, the immediate concern of the Maritime Commission was with special items which would need to be lifted to AA-1 or to the new super-rating AAA. For example, the ANMB directive had made no special provision for the military types being built by Maritime [12] such as the aircraft escorts being built by Kaiser.[13] Relief in such cases was sought through the appointment of a "Special Priorities Committee in connection with AAA priorities."

Although Admiral Land named Admiral Vickery to that special committee, in the spring of 1942 he was shifting as much as

[8] The ANMB directive "Priorities" of June 8 from the files of the Statistical Analysis Branch of Production Division, MC. With it are the estimates of its meaning in terms of specific ships made June 18, 30, and Aug. 1, 1942. See also G. J. Fischer to Vickery, Aug. 27, 1942, in Historian's Collection, Research file 117, filed under "Steel Priorities."

[9] *Industrial Mobilization for War*, I, pp. 296-98.

[10] *Minutes of the Planning Committee of the War Production Board*, p. 62.

[11] *Industrial Mobilization for War*, I, p. 298; Nelson, *Arsenal of Democracy*, pp. 363-64; C. H. Mathessen, Jr., Chief of Bureau of Priorities, to James S. Knowlson, Director, Div. of Industry Operations, in WPB Documentation file 142.36. Perhaps the revision which had lifted nearly all the Maritime Commission program into the class AA-2 was not known, or perhaps not so interpreted, by some at least of the staff of WPB. Bertrand Fox to Stacy May, July 3, 1942, in WPB Policy Documentation file 201 R. The estimate there given of the need of steel plate is irreconcilable with ratings in the Maritime Commission copy of the June 8 directive.

[12] Memorandum titled "Priorities" of that date, cited above in n. 8, and Fox to Stacy May, July 3, 1942, in WPB, PD file 201 R.

[13] Vickery to Eberstadt, June 18, 1942, in gf 120-124-2.

possible of the fight over priorities on to the shoulders of the practical businessman who was added to the Commission's staff as Director of the Production Division, Col. Willard F. Rockwell,[14] who had had wide experience in companies in which production scheduling was important.[15] Admiral Land began urging as early as April 23, 1942 that Colonel Rockwell be made a co-chairman of the ANMB.[16] Finally in August 1942, Rockwell was appointed to the ANMB's Executive Committee but to act only "when questions concerning coordination between the Army, and Navy, and Maritime Commission in procurement matters are under consideration."[17]

Priorities were rapidly becoming of diminishing importance, however, between June and August in regard to steel. One reason was the inflation of high ratings. In the revision of June 8, the same clause which lifted 50 per cent of the Maritime Commission program from AA-3 to AA-2 was phrased so it could apply to many items in the Army program. Anyhow, the year was running out. Two months later, effective August 11, 1942, all AA-2 ratings were made equal to AA-1 ratings.[18] Materials wanted in 1942 for merchant ships to be delivered in 1943 still had only AA-3 ratings, and in July and August the Maritime Commission was declaring that such low ratings on its orders to the mills made difficult the securing of steel plate.[19] But as Rockwell had begun to find out some months before, allocation by the WPB was becoming both more important and relatively independent of the ratings on the priority schedule prepared by the ANMB.[20]

Ever since allocation of steel plate began in the fall of 1941,

[14] Land to the Chairman of the ANMB, June 8, 1942, and Rockwell to Vickery, June 20, 1942, in gf 120-124-2.

[15] Information from Personnel Division files. He was a Colonel in the USA Reserve Corps for whom Admiral Land wrote in vain to secure the rank of Brigadier General.

[16] Land to Nelson, Apr. 23, 1942, in gf 107-55; Land to the President, May 16, 1942, in gf 506-1-1-21. On June 1, 1942, Vickery wrote of the matter to Judge Bland, saying it was believed this would ease the difficulties in regard to "steel, machinery, and machine tools." In gf 120-124-2.

[17] Land to Brig. Gen. W. B. Smith, Secretary of the Joint Board, Aug. 24, 1942, in MC Personnel Division files.

[18] C. E. Walsh, Jr., Chief Procurement Section, to All Vendors, Aug. 15, 1942, in MC Statistical Analysis Branch files, Priorities.

[19] Chaikin and Coleman, *Shipbuilding Policies*, p. 27. On ratings see G. J. Fischer to Vickery, Aug. 27, 1942, in gf 120-124-2.

[20] Rockwell to Vickery, May 23, 1942, in gf 107-55. He had learned that Iron and Steel Branch of WPB might make their own.

the Iron and Steel Branch of OPM and WPB had felt that they were not bound to act strictly in accordance with priority ratings. These ratings were in disrepute both because of overissue and because of the feeling that they were not accurate indications of the importance of particular uses of materials in the war effort. Putting high-rating orders into rolling schedules without regard to the production runs of the steel mills would have disrupted production and diminished the final total output.[21] Consequently, although the intensity of the struggle over priorities in May and June of 1942 indicated that they were still of general importance, allocation became increasingly the key issue in regard to steel plate. With the creation of WPB, a Requirements Committee was established therein which doled out the available steel among the various agencies and programs asking for it. Appeals from the Requirements Committee and efforts to affect its future decisions went all the way to the President, and the enforcement of its decisions depended on the machinery for allocation in the Iron and Steel Branch. The battle of the Maritime Commission to improve its position in preference ratings was only one part in the complicated battle for steel. As late as August 13, 1942, Admiral Land expounded the obscurity which he found prevailing in the relation of priorities to allocations as follows in a letter to Mr. Nelson:

> We understand that you have endorsed, or at least approved, the new ANMB priority ratings. Under these it would appear that the total steel allocation to the Maritime Commission, including plates and shapes under AA-1 and AA-2 preference ratings, would have an ingot equivalent of 866,000 tons from July to December of the current year. If extension of these ratings should make it difficult to obtain plate under AA-3 ratings—as appears very probable at present, may we count, nevertheless, on the monthly delivery of 368,800 tons of plate or shall we be subject to further curtailment? We understand from Mr. Knowlson that the allocation of 368,800 tons was determined by the Combined Chiefs of Staff. If their allocation is at variance with the amount obtainable under the ANMB ratings, which controls?[22]

On the same day Rockwell characterized the way priorities were still used by steel mills, as follows: " Although both the WPB and ANMB have advised us that the mills can and will run AA-1 and AA-2 requirements, the steel mills advise us they cannot run the AA-2 requirements on their September schedules."[23] Obviously

---

[21] *Industrial Mobilization for War*, I, 455-56.
[22] Land to Nelson, Aug. 13, 1942, in Land reading file.
[23] Rockwell to J. W. Fesler, Acting Secretary, Planning Committee, WPB, in gf 506-1-1-21, and 507-4-1-3.

the top officials of the Maritime Commission felt they had to fight two battles for steel, one over priorities, the other over allocations, and they did not know just what connection the one had with the other.

Discussions over allocations had been going on concurrently with those over priorities throughout 1942, with Admirals Land and Vickery expressing their discontent often both to the WPB itself and to the President. When shipments for January 1942 were only 174,400 tons where 200,000 to 220,000 were expected, Admiral Vickery at once wrote of it to Mr. Hopkins at the White House, noting that the shortage was equal to about 10 ships.[24] When the Commission's allotment on the May rollings of the mills was 20 per cent less than the Commission had asked, it rushed to WPB a protest that this would curtail production by an average amount of 5 to 10 days in each yard.[25] When for July only 370,000 tons of plate were allotted the Commission in spite of a stated requirement of 468,000 tons, Admiral Land called the President's attention to the seriousness of the situation and the need of a conference of all those concerned.[26] The conferences which followed were linked to the problems of building escort vessels and led to the cancellation of the Higgins contract, as has been described in Chapter 6.

These protests over the lack of steel for shipbuilding were reinforced by public statements both from the Commission and from shipyards. On April 23, 1942, shortly after protesting to the WPB against their small allotment for May, Admirals Land and Vickery appeared before the Truman Committee and called attention to the failure of steel deliveries to come up to needs during the past 8 months, so that a deficit of about 169,000 tons had accumulated. They named specific delays due to lack of steel.[27] In May Admiral Land summed up the situation in a report to the President by saying that there was ample capacity in the shipyards " to convert into shipping all the steel that can be allocated by the War Production Board for Commission contracts," [28] and in June several shipyards emphasized this dra-

[24] Vickery to Hopkins, Feb. 7, 1942, in gf 507-4-1-3.
[25] W. F. Rockwell to Vickery, Apr. 19, 1942, in gf 506-1-1-21.
[26] Land to the President, June 19, 1942, in Douglas file, Construction. A memo on steel prepared at that time for use at the White House is referred to in Vickery to Hopkins, June 16, 1942, in gf 506-1-1-21.
[27] Truman Committee, *Hearings*, pt. 12, pp. 5188, 5197.
[28] Land to the President, May 16, 1942, in gf 506-1-1-21.

matically. One yard, Calship, was making a determined drive in June to set a new record by delivering 15 ships in a single month. Steel was the bottleneck and by the middle of the month the management was threatening to lay off 15,000 to 20,000 men because of the lack of steel in the yard. The Los Angeles Metal Trades Council let out a howl, letters flowed in to Congressmen, and another Congressional committee was holding hearings on steel.[29]

Thus public interest had already been turned towards the steel shortage when it was placed under the spotlight by the Commission's announcement on July 18 that the Higgins contract was being cancelled for lack of steel. The contest preceding the decision to cancel and the furor created by cancellation have already been described. The Commission's efforts in that case to get more steel were but part of a long contest. The President's decision to support Donald Nelson and hold the merchant shipbuilding program down to 24 million tons meant the loss of one round, an important round, but it did not end the fight.

A new round opened immediately. On receiving Mr. Nelson's letter of July 8 stating the President's decision, the Maritime Commission not only moved to reduce its program, but began restating its needs. On July 14 Admiral Land wrote Mr. Nelson asking clarification and assurance that minor and military types were not being counted in making the total of 24 million tons. He also sought to straighten Mr. Nelson out as to the number of aircraft carriers the Maritime Commission was to build, 50 not 44. These 50 being included, the Maritime Commission had scheduled minor or military types totalling 814,000 deadweight tons for delivery in 1943, and about 435,000 deadweight for delivery in 1942. As to cargo ships in 1943, Admiral Land said that the continued rate of sinkings and the increased demand for shipping for military operations led him to believe that construction should be accelerated to the utmost until next June 30 at least. Accordingly he said he was " disposed to continue to press you and the President for a reconsideration of the directive contained in your letter of July 8th." [30] He sent a copy to Harry

---

[29] Rockwell to MacLean and Gallagher, June 16, 1942, in gf 507-4-1-3, and House, *Shortage of Steel in Shipbuilding, Executive Hearings* before the Committee on the Merchant Marine and Fisheries, 77th Cong., 2d sess., June 25, 1942, pp. 1, 4, 15.
[30] Land to Nelson, July 14, 1942, in Land file.

Hopkins with an added word of protest against the recent action of the WPB in cutting by 5 per cent the Maritime Commission's allotment of plate for August. When the President called on Nelson to reply, Nelson assured the President that the 5 per cent cut was necessary because if it were not made, although there would be enough steel plate, there would not be enough structural shapes for the Maritime Commission or enough other "steel products to go with the plate so that it would be fully employed in producing tanks, ships, and synthetic rubber and 100-octane plants."[31] According to Colonel Rockwell, the shortage at this time was in basic ingot capacity, which was attributable to shortage of scrap.[32] To Admiral Land also, Mr. Nelson reiterated his previous figures and the necessity for the cut.[33]

When allocations for September came up for discussion, there was new occasion for protest, although now the practical effects would be felt only in the program for 1943. The Maritime Commission had stated that it would require 427,000 tons, but it was allocated only 368,800, compared to 395,000 for the previous month. On August 13 Admiral Land wrote to Mr. Nelson the most scathing of his protests at the way allocation was being done.

> I must point out to you that of even more present embarrassment to us than this reduced allocation of plate is the seeming uncertainty as to future allocations. In default of final and definitive instructions from you I am instructing our Production Division to assume that the cut to 368,800 tons for September will also apply to October, November and December. On that basis there will be a reduction in production of EC-2 ships of 70 in the first and second quarters of 1943 or equivalent reduction in naval and auxiliary vessels. We will also be required to cancel further plans for expansion of facilities and since such cancellations have been shown to have difficult and somewhat time-consuming repercussions, you will understand that we should be definitely advised on the highest authority as to what our program is going to be. Since we accepted your letter of July 22d as an authoritative definition of the President's directive, we believe that you should now re-define it. Any process of endeavoring to seriously modify it every month in accordance with variable hand-to-mouth monthly allocations can lead only to further confusion, criticism and loss of efficiency.
>
> I believe I mentioned to you briefly the other day that the situation is made

---

[31] The quotation is from Nelson to the President, July 24, 1942, copy in the Land file. See also Chaikin and Coleman, *Shipbuilding Policies*, pp. 44-45.

[32] This explanation is given in a letter, Vickery to Eberstadt (written by Rockwell), Dec. 16, 1942, in gf 506-1-1-21. Contrast Nelson, *Arsenal of Democracy*, p. 173.

[33] Nelson to Land, July 22, 1942, original in Land file.

PLATE XIII. Preassemblies. Above, sections behind the ways at Marinship waiting to become part of tankers like that shown below. Lower right, at Oregon Ship a deckhouse assembly is lifted onto a Liberty hull.

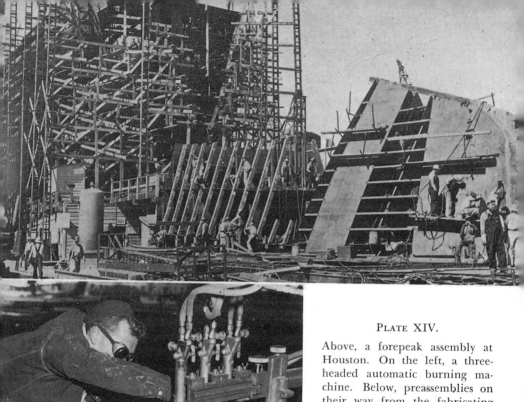

PLATE XIV.

Above, a forepeak assembly at Houston. On the left, a three-headed automatic burning machine. Below, preassemblies on their way from the fabricating shop to the shipways of the Bethlehem Fairfield yard in Baltimore. By using equipment in an idle Pullman plant two and a half miles from the ways, Bethlehem Fairfield gained a head start on the other yards and was the first to launch a Liberty ship.

the more obscure for us by the lack of conformity in the allocations of plate by the several agencies assuming the authority to allocate it.

Admiral Land then described, as quoted earlier, the obscurity over the importance of priorities compared to allocations, and continued:

> I think I mentioned to you our disappointment with the procedure leading to the determination of such questions as those under consideration by the Combined Chiefs of Staff. Our first knowledge of the decision as to this present cut was communicated to representatives invited to attend a meeting on August 6th of the "Working Committee of the Combined Steel Plate Committee of the Combined Chiefs of Staff." Mr. Nathan was present at the meeting of the Working Committee and he and Mr. Knowlson were present at the meeting of the Combined Steel Plate Committee. The report of the Working Committee was fully written and mimeographed when the Working Committee was called to order. Discussion was permitted but given no serious consideration and the mimeographed report was adopted without change in rapid succession by the Working Committee, the Combined Steel Plate Committee and (apparently) by the Combined Chiefs of Staff.
>
> Thus in essence the allocation of steel plate was made not on any high plane of strategic consideration by the Combined Chiefs of Staff but on some basis of compromise and relative cuts by a military committee under Captain Sweet and, as far as we are advised, no review of comparative importance or even breakdown of military needs was made in any gathering that had representation from the Maritime Commission or the War Production Board. Under such procedure attendance by our representatives or yours at meetings of the higher committees is purely informative. The procedure would appear to put into effect the proposal of General Somervell's letter of May 15th which was rejected in your letter of May 21st.[34] Unless the Maritime Commission and the War Production Board have effective representation on the actual working level of these proceedings, it would appear to us that any procurement agency, other than the Army and Navy, given your directive under authority of the President, is being put in an impossible position.
>
> Pending further instructions, we shall schedule ships on a basis of 368,800 tons of plate per month but we await and expect a modification of your letter of July 22d and would greatly appreciate information as to where the allocating authority now lies and to whom we should present our conception of the injury that is being done to the cause by the dwindling fleet of merchant vessels already insufficient to meet the requirement of the armed forces.[35]

The next day Admiral Land addressed a plea to the President, whom he had already approached indirectly by sending to Messrs. Hopkins and McIntyre on August 13 copies of a personal note addressed to Admiral J. M. Reeves in the Secretary's office of the

---
[34] See *Industrial Mobilization for War*, I, 258.
[35] Land to Nelson, Aug. 13, 1942, in Land reading file.

Navy Department.[36] In this personal note Admiral Land mentioned some large recent requests from the Army and Navy for ships for military use and also the cut which had been made in his steel allocations. He concluded that it looked to him very much like "'playing both ends against the middle.' The various Planning Committees, up to the Combined Chiefs of Staff, can't have their cake and eat it too." In his formal presentation of his case to the President, the Chairman of the Maritime Commission itemized how much the Commission's allotments had been cut month by month beginning in July and explained how big a reduction this would mean in deliveries in 1943. Many yards had been instructed not to produce to the limit and were therefore laying off men, while others which were yet to come into production would have to be told to stabilize below capacity. He concluded: " The Commission's chief aim at the present is to have a stabilized schedule of steel deliveries which will meet the stabilized schedule mentioned in the first part of this letter. . . . If, however, it is impossible to supply the required materials, then a new directive should be issued in accordance with the capacity of the material suppliers as set up by the War Production Board." [37] No reply came from the President, so far as I can find. The Commission's allotment was only 368,000 for September and then fell to 360,000 for October, 352,000 for November, and 361,000 for December.[38] Steel was going into army supplies and into the needed new construction, such as synthetic rubber plants.[39]

The decisions of July and August 1942 regarding both immediate allocations and future expectations influenced many changes in the Commission's program in addition to the cancellation of the Higgins contract. Changes were made to decrease the amount of steel which would be consumed by the Commission's other yards in 1943. Some of these changes were desirable anyhow regardless of the steel supply. More tankers were needed because in the summer of 1942 enemy attacks were being concentrated on tankers with devastating effect (Figure 14). Accordingly the

[36] August 10 in Land reading file, under dates given in text.
[37] Land to the President, Aug. 14, 1942, and " Summary, Maritime Commission's letter of August 14th to the President," both in Land reading file.
[38] Figure 24.
[39] *Industrial Mobilization for War*, I, 367, 407, 652.

contract with Marinship for Liberty ships was cancelled and it was given a contract to start delivery of tankers at the end of 1942.[40] Of the six-way yards started after Pearl Harbor, Marinship had been most successful in getting under way rapidly. Its conversion served the positive purpose of meeting the urgent need for tankers, and also caused the yard to consume less steel. Another conversion slowing down a yard which was rapidly turning steel into Liberty ships, was the award to North Carolina of a contract for C2's, on which it began laying keels in midsummer of 1943.[41] Both these changes were shown on tentative schedules drawn up at the end of August 1942.[42]

In the same month a circular letter was sent to all Regional Directors over the signature of Admiral Land telling them that because of the steel situation they should survey the programs for housing, transportation, and facilities of all kinds to cut off any excesses, and make sure that trainees were not taken on so fast that it would be necessary to lay off men with consequent loss of morale. Simultaneously, Admiral Vickery instructed the Regional Directors to stabilize production at stipulated monthly quotas.[43] Of course some of the quotas were either never attained, as that of Walsh-Kaiser, or not attained until many months later, as that of South Portland. But in other cases the quotas set were relatively low. For example, the quota of Oregon Ship was the rate of deliveries already attained in April 1942, namely 10 ships, and actually Oregon exceeded that quota constantly in 1943, averaging 16 ships a month.[44] Similarly, Calship's quota of 12 had already been exceeded in June and was substantially exceeded in every month of 1943. Spurred by the terms of their contracts, the shipbuilders wished to turn out ships as rapidly as possible in spite of the Commission's efforts temporarily to restrain them in view of the steel situation.

By the end of the fall of 1942 the Maritime Commission had sustained its worst defeats in the battle for steel. The effects of these defeats were felt throughout the rest of 1942. Allocations

[40] Minutes, Sept. 10, 1942; Reynolds' Way Charts.
[41] Minutes, Dec. 17, 1942; Reynolds' Way Charts.
[42] Schedule in MC, Historian's Collection, Research file 117 (Statistics).
[43] Land to Regional Directors of Construction, Aug. 24, 1942; Vickery to Regional Directors, Aug. 20, 1942; General memorandum by J. F. McInnis, Aug. 7, 1942, in East Coast Regional Files, " Director, 1942 " in Historian's Collection.
[44] Reynolds' Way Charts.

STEEL-PLATE REQUIREMENTS, ALLOCATIONS, AND SHIPMENTS
For The Maritime Commission, September 1941 — December 1944

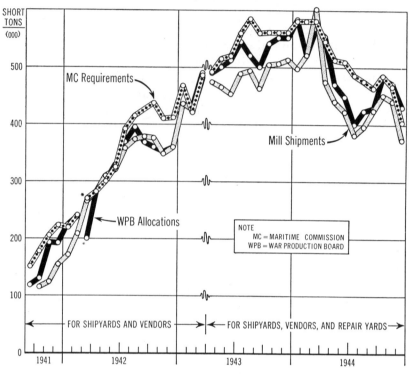

FIGURE 24. Source: Fischer, "Statistical Analysis, Steel," Table 4. Asterisk indicates that requirements and allocations for vendors were not included prior to March 1942.

by the WPB and shipments of steel plates to the yards in each of the last six months of 1942 were about 10 per cent less than the amounts stated by the Maritime Commission as necessary (Figure 24). Moreover, the schedules for ship construction, on which were based the Maritime Commission's estimate of its requirements, had themselves been lowered because of the lack of steel. The Higgins contract had been cancelled in July; in August a "hold-back" order had been issued to all yards, and two of the fastest builders of Liberty ships had been told to shift and construct other types which would consume less steel per month.

Late in the fall of 1942, however, the Maritime Commission launched a new drive to increase its steel allotments. It regained enough lost ground so as to obtain larger allocations and schedule a larger ship-construction program in 1943.

III. Regaining Some Lost Ground

The story of this counteroffensive can be understood only by grasping at the same time the role which the Joint Chiefs of Staff assumed later in 1942 in the economic direction of the war. Earlier, in June and July, the contest for steel had been decided in the White House since it involved the political potentialities of the cancellation of the Higgins contract. In August, Admiral Land had again appealed to the White House—in vain. A new drive to obtain the steel to increase the program of merchant shipbuilding began in the latter half of October. Again it went through White House conferences without winning a definite order in favor of the Commission's proposal. But this time the President permitted the Joint Chiefs to have the final word and they decided in November to increase the Maritime Commission's program for 1943 from 16,000,000 tons to 18,889,000 tons.

Putting this question up to the Joint Chiefs, of whom Admiral W. D. Leahy had late in July 1942 become the head, was a result of the procedure worked out for reconciling military programs with the country's capacity to produce as estimated by the WPB. There had been a contest between Mr. Nelson in the WPB on the one hand and the Armed Services on the other. The Armed Services felt that Mr. Nelson was not bearing down hard enough on the civilians, that he was not ready to do as much as should be done to meet military needs. Mr. Nelson felt that the Army and Navy were making unrealistic demands. He objected also to control of the economy by the military. But he found no way of forcing the military to schedule the production of their orders within the limits which the staff of WPB declared possible.

In October and November of 1942 a way of solving the problem was found, a way which was followed in essentials during the rest of the war. It was General Somervell's suggestion. Mr. Nelson wrote to the Joint Chiefs telling them that only $75 of a $97 billion total objective for 1943 could be achieved.[1] The Joint

---

[1] On Oct. 19, 1942, *Industrial Mobilization for War*, I, 289; Nelson, *Arsenal of Democracy*, pp. 380-81, 358.

THE BATTLE FOR STEEL

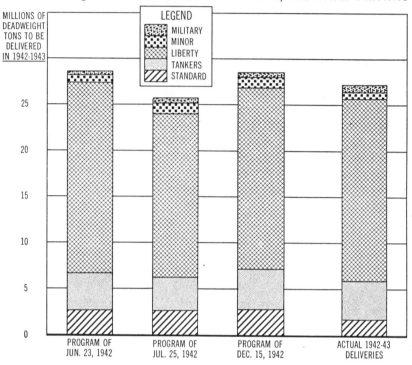

FIGURE 25. Source: Fischer, *Statistical Summary*, Table A-3.

Chiefs then revised the program, cutting it down to $80 billion, and determining which elements were to be dropped out in order to attain this objective.[2] Although the figure of $80 billion was $5 billion more than WPB's estimate of what was possible, it was accepted as capable of accomplishment. At least it was a considerable reduction in the military programs, and to that extent Mr. Nelson's insistence that the military take into account the civilian needs and the total possibilities triumphed. But within the limits set, the military won the power to decide which among the various contestants for steel should receive increases or decreases in their allotments. They decided not only the

[2] By "Objectives" dated Nov. 26, 1943, transmitted Leahy to Nelson, Nov. 24, 1942.—*Industrial Mobilization for War*, I, 289.

contests between claimants within the Army and Navy but also the contest waged by the Maritime Commission for a large share. By allowing this contest to be decided by the Joint Chiefs, President Roosevelt acted as if he now regarded the question as one that could be settled primarily on the basis of the strategic considerations on which the Joint Chiefs were the authority. It is noteworthy that about this time also the ship construction programs of the Maritime Commission were regularly submitted to the Joint Chiefs in order that they might decide on strategic grounds how many ships of various types should be included. Although appeals were still made to the White House, after November 1942 decisions on the points at issue between the Maritime Commission and other agencies were received by the Maritime Commission from the Joint Chiefs.

Before the position of the Joint Chiefs in the chain of command had been thus clearly established there had been a period in the spring and summer of 1942 when it seemed as if a central position in the direction of the war would be taken by the British-American Combined Chiefs of Staff and the other "Combined" boards which contained both British and American members. In November it became clear that the purely American Joint Chiefs were to have the last word concerning the military use of American resources, but the Maritime Commission's successful drive for more steel began in the "Combined" boards. Its initial impetus came chiefly from those in touch with the need for shipping. A prominent part in it was taken by officials of WSA, by Lewis Douglas, Deputy Administrator, and by Fred Searls, Jr., who represented the WSA on the Combined Production and Resources Board. The need for shipping was set forth by the Combined Shipping Adjustment Board to the Combined Production and Resources Board and by both the Combined Shipping Adjustment Board and the Combined Production and Resources Board to the Combined Chiefs of Staff.[3] When these channels proved not to be the best for obtaining a decision at the highest level, Mr. Douglas pushed the issue directly at the White House through

---

[3] R. M. Bissell to Vickery, Sept. 23, 1942 with attachments and Sir Robert Sinclair to Sir Arthur Salter and Douglas, Oct. 3, 1942 with attachments, in Douglas file, "Construction." On Searls' position, see Nelson to Douglas, July 9, 1942 in gf 106-2-5 and Land to Aurand, Aug. 27, 1942 in gf 107-80.

Harry Hopkins.[4] A memorandum of October 16, 1942 signed by Admiral Land and Mr. Douglas, but apparently drafted by Douglas, began by pointing to the great deficiency of maritime transportation.[5] It stressed that additional merchant ships and escort vessels were more important than almost anything else, stated the capacity of the shipyards to build more, and recommended to the President that he direct the Chairman of the WPB to allocate enough steel to use existing facilities fully.[6] Mr. Douglas went over this memorandum first with Harry Hopkins, and then he, Hopkins, and Admiral Land went over it with the President on October 21.[7] Mr. Higgins and Mr. Kaiser were also at the White House that day, and on leaving, Mr. Kaiser told reporters of his needs for steel and his plans for expanding his Fontana, California steel plant,[8] plans which the WPB promptly approved.[9] In the evening the President discussed the steel shortage with Mr. Searls.

To follow up the conference of October 21,[10] another conference was held at the White House on October 23 attended not only by Admiral Land, Admiral Vickery, Mr. Douglas, and Mr. Hopkins, but also by Mr. Nelson, the Secretaries of War and Navy, General Marshall, and Admirals King and Leahy. According to Mr. Douglas' notes the President set forth the proposals in the Land-Douglas memorandum and the need of curtailing civilian use of steel. Mr. Nelson questioned the figures and Mr. Douglas defended them. After further discussion of where the steel was to come from, the conclusion was that " the President asked for a report Monday as to how the steel was to be made available." [11] To the press next day, the President said he talked to this special conference, as to his Cabinet the day before, on the necessity of further limitation of nonessential, semi-luxury goods to get more steel and more everything for war production.[12]

---

[4] Note from LWD Diary, Oct. 9, 1942, tells of going over the joint CPRB and CSAB memoranda with Hopkins. Douglas file, Construction.

[5] Authorship is inferred from the fact that there is no carbon in Land's reading file whereas there is a copy in Douglas.'

[6] Land file, Oct. 16, 1942.

[7] Notes of the conference of October 21, in Douglas file, Construction.

[8] *New York Times*, Oct. 22, 1942.

[9] *Ibid.*, Oct. 23, 1942.

[10] Notes on the conference of Oct. 21, in Douglas file, " Construction."

[11] " Notes on Conference at the White House " Oct. 23, 1942, in Douglas file, " Construction "; Chaikin and Coleman, *Shipbuilding Policies*, p. 45.

[12] *New York Times*, Oct. 24, 1942.

These conferences were just a few days after Mr. Nelson had put it up to the Joint Chiefs to shape a program according to the capacities of the country. On October 29 Mr. Douglas heard from Robert Nathan that the Joint Chiefs were inclined to postpone the whole matter.[13] On November 18 the Joint Chiefs of Staff, in making suggestions as to how the Commission might adapt its program more directly to military needs, spoke as if the total size of the program was still being decided: "The Joint Chiefs of Staff are considering your recommendation that 1,200,000 tons of steel be allocated to the Maritime Commission for additional shipbuilding in 1943, sufficient to enable you to make full use of existing facilities." [14]

Meanwhile the availability of steel for merchant shipbuilding in a balanced program became clearer. On the whole there was less demand for structural steel shapes for new plants in 1943 than there had been in 1942, relative to the total output, and therefore more steel ingots could be devoted to rolling plates.[15] The British proposed in view of the shipping situation "a reduction in the allocation to them by 150,000 tons in the last quarter of 1942 and by 150,000 tons in the first quarter of 1943, on the understanding that this steel will be used to augment the shipbuilding program here." [16] In the United States many programs were being delayed by lack of other scarce materials—copper, alloy steel, and aluminum—so that they would not use all the carbon steel which had been expected. More was found to be available for merchant shipbuilding.[17] When on November 24, 1942, the Joint Chiefs sent WPB their decision concerning the production objectives for 1943, they included 18,800,000 tons of ocean-going merchant shipping, 50 to 70 escort vessels, and 50 escort aircraft carriers. Whereas the Navy, Air Force, and Army Supply programs were reduced 10 to 23 per cent, the Maritime Commission program was increased from $3.6 billion to $4.4 billion.[18]

The enlarged program totalling 18,889,000 tons was given Presi-

---

[13] Notes dated Oct. 31, 1942, in Douglas file, "Construction."
[14] Leahy to Land, Nov. 18, 1942, in Program Collection.
[15] Fischer, "Statistical Analysis, Steel," Table 8.
[16] L. W. Douglas to Isador Lubin, Nov. 2, 1942, in Douglas file, "Construction."
[17] Robert R. Nathan to Nelson, Nov. 16, 1942; copy in Douglas file, "Construction"; cf. Chaikin and Coleman, *Shipbuilding Policies*, pp. 46-47.
[18] *Industrial Mobilization for War*, I, 289-90; Chaikin and Coleman, *Shipbuilding Policies*, pp. 12-13.

dential approval in a memorandum to Admiral Land, but the President also sent this memorandum first to Admiral Leahy with the note, " I am going to send this to Land unless the Joint Board protests. Let me know." Admiral Leahy wrote on it " Joint Chiefs of Staff approve. W. D. L." [19]

Even after the Commission's program for 1943 had been restored to the height which had been planned before the cancellation of the Higgins contract, the Commission did not believe that full capacity was being provided for. The productive capacity of the other yards had been so increased that they could do far more than just make up for the disappearance of Higgins from the program, it was hoped. This capacity was due to the increase in speed resulting from experience with methods of preassembly. To a considerable extent, it was the result of improving the equipment of existing yards through additions of shops and machinery authorized in May, June, and July before the cancellation of the Higgins contract (Figure 26). After this cancellation also large additions to facilities were authorized, but these were almost entirely to enable yards like Marinship, designed originally for building a relatively simple ship, the Liberty, to build more complicated and larger types, such as tankers.[20] The big improvements in the Liberty yards, such as the new plate shop at Oregon,[21] and the new cranes at Calship,[22] were authorized in May and June, and were the main reason why it looked by the end of 1942 as if the existing yards could produce 20 million tons of merchant shipping in 1943, provided they had the steel. The possibility of enough steel was suggested by Robert Nathan's memorandum to Nelson on November 16 in which he spoke of adding to the 16 million-ton program. He said there would be sufficient carbon steel " to cover the demands of the four million ton accelerated merchant ship program." [23] A

---

[19] MC, Program Collection, placed between Dec. 2, and Dec. 7, 1942. Although there are no original dates on this memorandum, the attendant circumstances described above indicate that it was written between Nov. 18, when the Joint Chiefs said the matter was under consideration, and Nov. 24, when decision was reported to Nelson.

[20] Fischer, *Statistical Summary*, Table E-7; MC, Division of Finance, Facilities Contracts and Awards as of Sept. 30, 1942. In Historian's Collection, Research file 117.

[21] MC Minutes, pp. 21901-02 (May 12, 1942).

[22] See Minutes, pp. 22211-13 (June 16, 1942).

[23] Nathan to Nelson, Nov. 16, 1942, in Douglas file, " Construction."

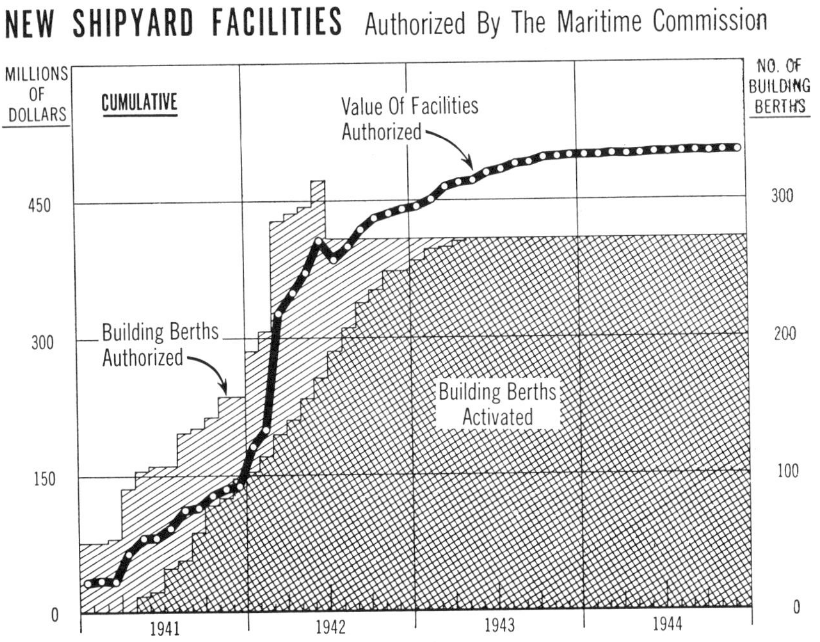

NEW SHIPYARD FACILITIES Authorized By The Maritime Commission

FIGURE 26. Source: Fischer, *Statistical Summary*, Tables E-6 and E-7.

memorandum of Admiral Land to the President dated November 17 informed him that an output of 20 million tons in 1943 was possible, and asked his approval to make fiscal plans accordingly.[24] On November 25 Admiral Land informed Admiral Leahy that the shipyards could build 20 million tons of merchant ships and asked the Joint Chiefs to consider using fully this capacity.[25] On December 7, 1942, President Roosevelt wrote the Bureau of the Budget that he approved the Commission's making fiscal plans for 20 million tons, but closed by saying: "The Maritime Commission, of course, will not proceed beyond the 18,889,000 without the approval of Mr. Nelson."[26]

In fact the 20 million-ton directive never became important except for planning Congressional appropriations.[27] Obtaining the steel and other essentials for the 18,889,000 remained the

[24] Land to the President, Nov. 17, 1942, in Program Collection.
[25] In Program Collection.
[26] *Ibid.*
[27] That is for merchant shipping. The total actually scheduled was over 20 million tons, for it included minor and military types.

important practical problem even after it had been approved explicitly by both the President and the Joint Chiefs. The next step in line logically in execution was the action of the Requirements Committee in WPB. After September 20, 1942 this Committee was headed by Mr. Eberstadt, who had previously been the head of the Army and Navy Munitions Board (and so on the side of the Armed Services in their struggle with WPB) and whom Mr. Nelson made a Vice Chairman in WPB in an effort to end the friction with the Armed Forces. Mr. Eberstadt as WPB Vice Chairman on Program Determination was the chairman of the Requirements Committee and dominated its action.[28] He did not consider an increase from 16,000,000 to 18,800,000 " definitely approved," [29] and questioned its priority.[30] He faced the fact that Navy shipbuilding had not been rescheduled to accord with the directive issued November 26, 1942 by the Joint Chiefs.[31] In answer to his questioning Admiral Land finally wrote him on February 8, 1943:

> There are few things in my experience that have been booted up and down the alley as this 2,800,000 tons have been. To my personal knowledge it has been approved a half a dozen times all the way from the President to the char-lady. On every occasion, however, one high ranking Naval Officer * (* not Admiral Leahy), and possibly two, have been dissatisfied and kept entering protests one way or the other. . . .

Admiral Land then cited the high authority back of the 18,800,000 ton objective, including a letter from F. D. Roosevelt to Winston Churchill, and action by the full ANMB.[32]

So the battle for steel was carried over into 1943 and seemed likely to go on unendingly at the levels concerned with translating into practicalities the topside decisions. Gaps between what the Maritime Commission asked for and what it was allocated kept recurring, for the Commission was persistently passing on to the WPB the pressure exerted on it by the yards. As a sort of Christmas present Edgar Kaiser wired Admiral Vickery an offer to produce 17 ships a month in 1943 if the materials were supplied, adding: " If Santa Claus comes your way

---

[28] *Industrial Mobilization for War*, I, 257-62.
[29] Chaikin and Coleman, *Shipbuilding Policies*, p. 47.
[30] Eberstadt to Land, Jan. 21, 1943, in Land file.
[31] Chaikin and Coleman, *Shipbuilding Policies*, pp. 13-14. Note that 89,000 tons are passed over.
[32] Land to F. Eberstadt (personal and confidential) Feb. 8, 1943, in Land file.

tell him the boys at Oregon want no less than 16 ships of steel in January and we will be listening for the sleigh bells." [33] Actually Oregon Ship produced just about 17 ships a month in 1943. The total deliveries from all Maritime Commission yards were 19,209,991 tons, including 632,293 tons of military types.[34]

IV. THE OUTCOME: PRODUCTION AND INVENTORY

Although the battle for steel was in a sense continued in 1943 and even in 1944, the details of later struggles are of less interest. Steel continued to be a limiting factor on the Commission's plans for construction throughout 1943. Production was slowed down at the end of the year in the three yards which had been producing Liberty ships most rapidly because they were changed over to the construction of a new type, the Victory ship, one reason for the change being the shortage of steel. In 1944 the expansion of steel facilities, which had been begun in 1942 came fully into effect; and after the record delivery in March 1944 of over 600,000 tons of steel plate to Maritime Commission yards in a single month (Figure 24), steel was not so scarce as to cause a really hot fight. Even in 1943, the issues at stake were not so vital as in 1942, for by the end of 1942 the battle for steel had a decisive effect: the full capacity of the country to build merchant ships was not used.

To be sure, any analysis of what limited the building of merchant ships is subject to chronological qualification. Other limiting factors—facilities, trained labor and management, propulsion machinery and similar scarce components—were important at certain times. Before 1942, in spite of some temporary local steel shortages, shipbuilding was limited mainly by shipyard facilities. But by June or July of 1942 facilities got ahead of steel. At the rates of production then reached, existing facilities could fabricate and erect more than they were receiving. Also, the facilities planned would be able to consume more steel than was provided for in the plans for future allotments of steel. Consequently the planned-for facilities were substantially reduced by cancelling the Higgins contract. Even after that reduction, the facilities were adequate so that deliveries would have been higher

---

[33] Chaikin and Coleman, *Shipbuilding Policies*, p. 49.
[34] MC, *Official Construction Record*.

in the last quarter of 1942 and throughout 1943 if more steel had been allocated to the yards. This is indicated not only by the plans the Commission made, and gave up, for scheduling larger production, but also by a statistical analysis of the relations between actual inventory and the minimum necessary inventory (See Figure 27.) Then in the last quarter of 1943 the situation began to change. During 1944 faster cargo carriers and various military types occupied a larger and larger number of building berths. Once these changes were made in the types of ships being built, propulsion machinery and skill in labor or management became more important as limiting factors than steel, as was evidenced by the rise in inventories in Liberty and Victory yards.[1] The general principle used by the Commission in allocating steel among the yards was to set aside what was needed for the yards building tankers, standard types, and military types, and then give as much as was left to the Liberty yards. Accordingly a sharp rise in the inventories of Liberty-ship yards in 1944 indicated that the Commission was no longer pinched for steel (Figure 29)

This chronological distinction is essential. When it is said that steel set the limits to the Maritime's ship production the statement applies directly only to the fifteen or eighteen months from mid-1942 to October or December of 1943. In that period more steel would have resulted in more ships. The shortage in that period affected action at two levels. In the shipyards work was slowed because of empty racks. In Washington plans were cancelled because there was no prospect of steel for their execution. These changes of plan reduced the whole wartime program below the total tonnage that might have been planned and built had their been no steel shortage.

While it is fairly clear that more ships would have been built in Maritime Commission yards if they had received more steel, it can not be said with equal definiteness that the capacity of the steel industry was what limited ship construction. The question naturally arises whether more steel than necessary was allotted to activities other than merchant shipbuilding. Whether the Navy was not being oversupplied was a question raised frequently, and especially during the hubbub over the cancellation of the Higgins contract. How completely the Navy succeeded

[1] Fischer, " Statistical Analysis, Steel."

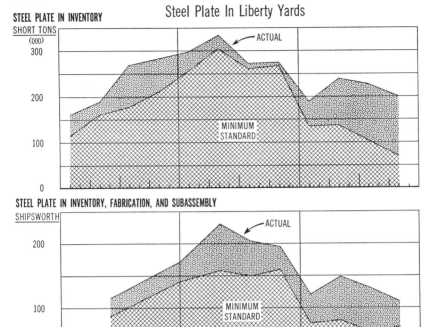

FIGURE 27. Source: Fischer, "Statistical Analysis, Steel."

in securing full allocation of its estimated requirements could be learned only from a close study of Navy shipbuilding. From the records of the Maritime Commission and the WPB it seems that the Navy did do better, for its protests over allocation were less frequent or urgent.[2] Some of the staff of the WPB felt that the Maritime Commission (headed by Admirals) was not aggressive enough in attacking the Navy's requests and in urging that cargo carriers were more important than some combat vessels.[3] And

[2] Chaikin and Coleman, *Shipbuilding Policies*, pp. 35-54. To be sure there was some cancellation of facilities for Navy building also, but only small changes in the Navy's program are attributed to the steel shortage by R. H. Connery in his *The Navy and the Industrial Mobilization*, p. 303.

[3] *Ibid.*, p. 26.

once the steel was allocated, the Navy had more inspectors at the steel mills, and inspectors of rank and prestige, to see that its orders were rolled and shipped.[4] Only a comparative study of Navy records would enable one to say whether the Navy's requests for steel were more fully met than those of the Maritime Commission all the way down the line from decisions in the White House to rollings in the steel mill. Whether they should have been is a question quite outside the scope of this history. But it should be made clear that the limit to merchant ship production was set not by the nation's steel capacity alone, but by it in conjunction with the allocation of steel to Army, Navy, and other users.

Another question which was raised during the battle for steel was whether the Maritime Commission could not have produced more ships by reducing the inventories of its yards. The charge was made frequently by the WPB in its counterattacks against the Commission's criticisms. It was urged with a mass of statistical documentation in the report of Mordecai Ezekiel, partially made public at the time of the Higgins cancellation. He said then that there had been no steel shortage up to that time which the Maritime Commission could not have corrected by better distribution of the amount allotted to it.[5] A half year later, Donald Nelson answered the Maritime Commission's complaints over inadequate allocations in exactly the same vein, saying in a letter of January 5, 1943, that the plate inventory of Maritime Commission yards was equal to something over 600 ships, and concluding, " active and continuous efforts on your part to effect a more equitable distribution of plate among all the yards could have done and would still do much to relieve the shortage now faced by some." [6]

Actually, inventories were only a part of a larger problem which Nelson may have had in mind in his reference to about " 600 ships," and which may be called the problem of the " pipe line." Steel plate which had been shipped by the steel mill but which had not yet become part of a completed vessel delivered

[4] Schmeltzer to Vickery, Jan. 23, 1942, in gf 507-4-1-3; *Industrial Mobilization for War*, I, 634.

[5] A copy of the Ezekiel Report is in MC World War II Historian's Collection, Research file 206.1.2. See also *Minutes of the Planning Committee of the War Production Board*, pp. 55-56, 57, 60-61, 81-82.

[6] Nelson to Vickery, in MC gf 606.96.

PRODUCTION AND INVENTORY

to the Commission was in one of the following forms: (1) in transit to the shipyards, (2) in the raw stock inventory of the yards, (3) being fabricated within the yards, (4) being erected on the ways, and (5) in vessels launched but in process of being outfitted. The disposition of steel in the pipe line, which can be imagined as flowing from the point of shipment by the steel mill to the point of delivery of a completed vessel by the shipyard, is shown on Figure 28.[7] The total amount of steel filling the pipe line is impressive; at the peak of production in late 1943 the steel in the pipe line was about 2,500,000 tons, enough to build a thousand Libertys. The volume of steel in the pipe line grew in correspondence with the increasing number of building berths activated until the number of berths reached its peak in May 1943. Compare Figure 26 and 28. Subsequent swelling of the pipe line is to be attributed either to the fact that as the war progressed heavier ships were put into construction or to the growth of inventories and delays in outfitting.[8]

Reducing the tons of steel in the pipe line was, to be sure, one way in which more ships might have been produced in a given month or quarter without any increase in shipment from the mills. Reduction of the pipe line could not have gone on very long without diminishing the total production; in the long run what comes out of a pipe line depends on what flows in. But if at any given period the steel in the pipe line could be reduced by passing it more rapidly through production, the steel shortage would be relieved to that extent, in that period. If in the 18 months from July 1942 to December 1943 all yards had utilized steel as efficiently as did the three yards with least inventory, more ships would have been produced without drawing steel from the many other urgent demands for it.

How much steel might have been squeezed from the pipe line by maximum efficiency in utilization? Although no answer can be given with certainty, it seems to me for reasons shortly to be explained that perhaps a figure as high as 300,000 tons would be reached if all the elements involved were taken into considera-

---

[7] Although it is difficult to determine to what extent these figures include the "emergency purchases" made by West coast yards.—See below, chap. 21.

[8] The average plate required per vessel under construction increased from about 2,800 tons in mid-1942 to slightly over 3,600 tons in mid-1944. Fischer, "Statistical Analysis, Steel." Cf. growth in military types shown on Figure 39.

STEEL PLATE BEING USED IN SHIP CONSTRUCTION
Of The Maritime Commission

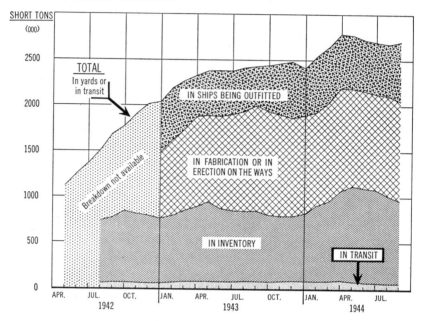

FIGURE 28. Source: Fischer, "Statistical Analysis, Steel."

tion and if all these elements were considered malleable.[9] Three hundred thousand tons of steel would build 120 Libertys. Admiral Vickery claimed that the size of the pipe line was reduced in 1942, relative to the scale of production, and that this reduction enabled the Maritime Commission to meet its 1942 objective in spite of a deficit of 250,000 tons in allocations.[10] Certainly squeezing the pipe line was a matter of no small consequence.

After 1942 the part of the pipe line which varied least in size—apart from the small percentage in transit (which is estimated)—was the tonnage being erected on the ways. This cannot be measured directly for all yards (it cannot always be distinguished

---

[9] For more cautious conclusions and a fuller explanation of the whole problem, see Fischer, "Statistical Analysis, Steel," on which my ensuing discussion is based.
[10] Vickery to Eberstadt, Dec. 16, 1942, in MC gf 506-1-1-21.

from the steel fabricated in subassemblies off the ways); but an acceptable substitute figure is the tonnage of plates intended for use in ships of which the keels had been laid but not launched. This figure varies relatively little, much less, than the tonnage of steel embodied in ships which were launched but not yet delivered to the Commission. The latter, the tonnage in outfitting, increased notably after the middle of 1943 when considerable difficulty was encountered in outfitting military and special types. A high tonnage in outfitting was a sign of trouble, but not the kind of trouble which could be cured by steel distribution.

The most variable factor from month to month was the inventory of raw stock. Most of the controversy which arose during the war with respect to the utilization of steel centered on the question of inventories. Unfortunately, no specific standard was accepted then to measure the adequacy of inventories, nor is one available now. Prewar standards were hardly applicable because under the conditions of irregular production in many prewar yards the shipbuilder expected to have all his steel in the yard before he laid his keel, and might be 6 months or more building according to the type of ship; yet it was obviously ridiculous under the wartime conditions of rapid multiple production to have 6 months' supply. The chief of Maritime's Production Engineering Section developed the rule of thumb that the yards should have a 2 months' supply.[11]

An effort has been made in preparing this study to review the statistical record. In a search for conclusions which would be of interest in showing the effects on inventories of multiple production and of the system of distribution used, the best that could be done was to record what the inventories actually were on the average in yards producing different types of ships and also to note what were the smallest inventories with which yards building various types succeeded on occasion in operating. The latter figures may be roughly averaged to give a minimal figure which may be used as a standard in the sense that it states a minimum to which inventory might fall in relation to production, at least if all other factors affecting production were favorable. For an overall picture, the minimum standard may be compared with the actual average.

Consider first the actual inventories. From the beginning of

[11] Interview with J. T. Gallagher.

1942 to the fall of 1943 the average inventory in Liberty yards declined from about 2 months' supply to 1.04 months' supply (Figure 29). Thereafter it rose substantially, one main indication that the third quarter of 1943 was the period when steel ceased to be the limiting factor in the program. Tanker yards on the other hand had about 3.5 months' supply on hand at the beginning of 1942 and their average inventories declined until 1944, and remained then relatively stable at about 2.3 months' supply. The ability of the tanker yards to operate with that small an inventory in 1944 suggests that they could have operated earlier with less inventory. The inventory of "all other yards," which were mainly, like the tanker yards, long-range yards, the group of which Mr. Eziekiel had particularly complained, rose from 2.5 months' supply to 3.5 between mid-1942 and early 1943 in spite of the fact that during that period the Commission assumed more responsibility for scheduling their steel plate. It rose again sharply in September 1943 and in 1944 came down almost, but not quite, to the level of the other yards (Figure 29). As this behavior of inventories suggests, there were special difficulties in scheduling accurately the future production of these yards.

One sort of minimum standard for comparison with actual inventories may be obtained by averaging the record of the three yards which produced most rapidly—Oregon, Calship, and the Permanente-Richmond yards No. 1 and 2 considered as one yard. These three most rapid producers were of course all in the Liberty programs, and comparison between their inventories and those of yards producing different types is meaningful only when account is taken of the fact that they were consuming steel at different rates. Analysis in these terms also shows very wide deviations from the minimum standard by most yards most of the time, but approach to it by many yards on some occasions. For the Liberty yards alone, comparison of the average actual inventories with the minimum standard set by the fastest yards, in terms of shipsworth of plate per way, shows that the average actual inventories were about 30 per cent above the minimal standard during most of 1942, but were brought down very close to the minimum during 1943, and diverged much above the standard in 1944. See Figure 27.[12]

[12] These figures may be inaccurate because of the extent to which the yards used in establishing the minimum standard actually received, through emergency pur-

INVENTORY RATIOS OF STEEL PLATE In Maritime Commission Shipyards

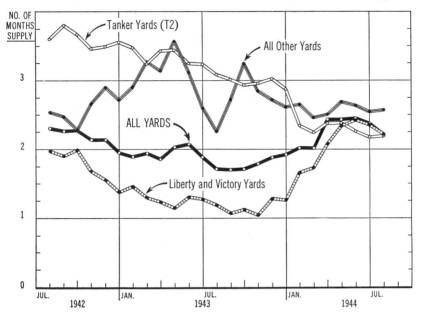

FIGURE 29. Source: Fischer, *Statistical Summary*, Table F-2.

Closely connected with the raw-stock inventory in the yards was the amount in process of fabrication and assembly but not yet erected on the ways. These two together may be called "the steel plate behind the ways." The combined figure of steel behind the ways proved better than the inventory figure alone as a measure of whether yards were oversupplied with steel. Some yards withdrew steel plate from inventory, fabricated it, and stored it in more or less subassembled form even when the erection was held up for one reason or another. They thus kept employment steady in the fabrication shops. In other cases a yard might deliberately reduce a swollen inventory of raw stock by stepping up fabrication considerably beyond the amount needed to meet scheduled erection, with a consequent stockpiling of material in process. A general tendency to stockpile fabricated plate made it

chases and reallocation within the West coast region, more steel than was reported to the Statistical Analysis Branch, on whose figures these graphs are based. See chap. 21 below.

possible for yards to build up reserves of steel even when their inventories were falling.

Statistical analysis shows that stockpiling of this kind did occur. Using again as a minimum standard the performance of the yards which produced Liberty ships fastest, a comparison can be made between the actual amount of steel plate behind the ways for the whole Liberty program and that in the yards which had a minimal amount. During the period when steel was short there was about 30 per cent more steel plate behind the ways on the average in Liberty-ship yards than would have been needed if all the yards had been able to operate with as little plate behind the ways as did those which had least in relation to output. See Figure 27.

The reasons why there was this 30 per cent difference in inventories of steel behind the ways were many. It is not to be called a 30 per cent excess. The record indicates that as a practical matter it was not an excess at all, but was a necessity in the sense that total production could not have been the same without it, yard plans, yard managements, plan changes, inventory controls, and other conditions being what they were. That question will be re-examined after describing the administrative organization through which the Commission attempted to regulate the flow of steel and other materials and the difficulties which were encountered in the attempt.

This broad study of inventory must close with a conclusion that is vague as well as tentative: Some additional ships, just how many it is impossible to say, could indeed have been built during a crucial period even without larger allocations if methods had been found of reducing substantially the 2,500,000 tons of steel in the shipbuilders' pipe line. Steel plate was the limiting factor in merchant shipbuilding during World War II, but the limit was set by the way inventories were handled and by the allocations to Navy and other military uses, as well as by the capacity of the nation's steel industry.

CHAPTER 11

GUIDING THE FLOW OF MATERIALS

I. PRIORITIES

CONFLICTS at the top of the administrative hierarchy have been described in recounting the battle for steel. Such conflicts were part of the process of planning the use of the nation's resources. The execution of the plans required much detailed scheduling and regulation which was carried out by the specialization of functions among a large corps of workers occupied with details—in short, by a bureaucracy. I use bureaucracy not in the popular invidious sense in which it means loafing at public expense, but in the sociological sense to mean salaried specialists who work together as a hierarchy in bureaus, divisions, sections, and branches. Since the distribution of goods was not left to the market where it would be governed as in ordinary times by willingness to pay the price, a bureaucracy was necessary to guide the materials to the producers who needed them.

Steel was only a part of the problem. Cranes, rollers and presses, welding machines and welding rods, forgings and castings of many kinds, valves, gears, turbines, and all sorts of machine tools were as fiercely fought over in detail as was steel. Careful organization was required to deliver even items usually in comparatively easy supply, such as reciprocating engines, at the times and places they were needed. Many elements in the organization for guiding the flow of materials have been introduced in describing the battle for steel, but in a systematic survey of the whole process attention must be given to many other items as well.

In the full regulation of the flow of materials, it is useful to distinguish five parts: (1) Scheduling ship production. Logically this was the basic activity governing all the others. It involved setting down as precisely as possible what ships would be constructed at future dates and what would be their stages of completion in all the yards having contracts with the Maritime

Commission. Practically it began with compiling the record of prospective keel layings, launchings, and delivery dates. (2) Scheduling the arrival at the shipyards of the materials and components, and scheduling correspondingly the production of these materials and components in the mills and factories of the suppliers. (3) Placing orders with the suppliers for the materials and components. (4) Securing priorities or allocations which would permit or compel the producers to deliver materials and components according to schedule. (5) Expediting, that is, following up with the individuals concerned, by telephone or interview, to see that specific items which were urgently needed were actually delivered under the priorities and allotments to specified yards on scheduled time.

At the beginning of the emergency program, most of these functions were being performed more or less by shipbuilding corporations or their naval architects, for the yards in the long-range program were accustomed under their lump-sum contracts to secure their own supply of materials, and Gibbs & Cox, as we have seen, was expected to schedule the delivery of materials for the first 312 Liberty ships in the process of placing the orders. But the issuance of priorities could not be delegated to a private contractor. While Admiral Land presented at the White House or the OPM the need for high preference ratings on merchant shipbuilding, and maneuvered to secure effective representation of the Maritime Commission on the Army and Navy Munitions Board which assigned the ratings to classes of commodities or groups of ships, members of the Commission's staff issued priority certificates to specific contractors. The chief need for priorities at the beginning of 1941 was not for steel, but for machine tools, cranes, valves, and forgings. Only in certain periods did the priority certificates issued by the Maritime Commission for these materials carry ratings high enough to enable them to compete with the Navy, but they took precedence over civilian competition. At the beginning of 1941 the main purpose of priorities was to keep critical materials out of civilian uses not essential to the defense effort.[1] For this purpose, blanket priority orders were issued requiring manufacturers of stipulated articles to deliver only to holders of priority certificates. The first such

---

[1] Quarterly Report, Technical Division, June 30, 1941; Flesher to Vickery, Oct. 3, 1941, in Production Division file L 4-3 Administration.

PRIORITIES

order was issued on March 12, 1941, for traveling cranes, lack of which was holding back the new shipyards. Machine tools also were placed under priorities. The existence of priorities assisted substantially in the rapid equipping of the shipyards by enabling them to get their orders filled ahead of civilian users who had no priorities.[2]

The first unit set up within the staff of the Commission to handle priorities was under Mr. Flesher at the time when he was Chief of the Engineering Design and Specifications Section of the Technical Division. He was named in December 1940 the Commission's representative on the priorities committee of the Army and Navy Munitions Board,[3] and the following year added "Priorities" to his title as Section Chief.[4] That section was concerned with turbines, gears, and valves—the components for which priorities were most urgently needed in order to keep going the long-range program of C-type ships. When the Construction Division was created in July 1941 it was placed in charge of inspecting these long-range yards as well as the emergency yards, and its Production Engineering Section, continuing the work of the section of that name in the Division of Emergency Ship Construction, was in charge of scheduling, allocation, and expediting. But the handling of priorities remained in Mr. Flesher's section of the Technical Division and a conflict in jurisdiction promptly developed. To Mr. Schmeltzer, the Director of the Construction Division, Mr. Flesher seemed to be reaching out into the field of expediting, and by attempting to make production and its associated problems a corollary of priorities was trying to make "the tail wag the dog," as Mr. Schmeltzer put it.[5] After Mr. Schmeltzer's illness and the decentralization of March 1942, priorities were taken over by the newly formed Production Division.

Priorities involved such a mountain of paper work that efforts of many kinds were made to simplify it. In the quarterly report of September 30, 1941, the priority unit said they had in that quarter handled 1,000 incoming long distance telephone calls, 129,100 pieces of paper (letters, telegrams, or forms) and 672

---

[2] Coleman, *Shipbuilding Activities*, pp. 52-72, 103.

[3] Land to Hines, Secretary of ANMB, Dec. 4, 1940, in MC gf 120-124-2.

[4] Chief, Classification Section to Personnel Files, May 28, 1941, in MC Personnel files; Quarterly Reports of Technical Division, Sept. and Dec. 1941.

[5] Schmeltzer to Vickery, Dec. 13, 1941, MC gf 120-124-2; and Flesher to Vickery, Oct. 3, 1941, in Production Division file, L-4-3 Administration.

visitors inquiring about priorities, as well as sending out 4,500 pads of preference rating certificates.[6] One method adopted to reduce the paper work was to issue a general preference order for certain ships or shipbuilding facilities to the prime contractor empowering him to pass on his priority ratings to subcontractors and the vendors from whom he purchased. This system (using form P-7) was adopted by OPM about the time when the Commission's staff was taking over from Gibbs & Cox procurement for Liberty ships.[7] Consequently the Commission itself was treated as a prime contractor with reference to its direct purchases and in this connection it took on a new mountain of paper work.

Priorities were of use in excluding from competition those users who were not connected with defense or war and who could get no priorities at all, but over-issue of priority certificates made them little better than " hunting licenses " when supplies became really short.[8] Even after a quantitative control had been set up by stipulating how many ships were to rate as A-1-a, how many as A-1-b, etc., and after priorities were supposed to be issued only for the quantities of material needed for the tonnage of ships stipulated by the ANMB, there was no effective control over the actual issue of the priority certificates. The Maritime Commission official in charge was well supplied with blank forms and there was no regular system of outside check which restricted him to issuing certificates for only the amounts of material needed to supply the number of ships stated in the directive from the ANMB. While he felt that he was being painfully honest in resisting temptation and never issuing more certificates than he was supposed to, or at least hardly ever, he felt sure that other procurement officers similarly placed in other agencies with pads of priority certificates at their elbow were less scrupulous.[9]

Indirectly, to be sure, there was a check against over-issue of priority certificates by the Commission's officials. Procurement officials of the Army or Navy who found these certificates getting

[6] Quarterly Report, Technical Division.

[7] Chief, Engineering Design Specifications and Priorities Section, to Clerical Section, June 18, 1941, with attached Memorandum to all Shipyards, signed by Vickery, telling them how to apply for these ratings, in gf 120-124-2; telegram, Schmeltzer to all yards June 16, 1941, in gf 507-3-1.

[8] *The United States at War*, p. 114.

[9] Statements by John G. Conkey in interview with F. C. Lane, Mar. 10, 1948.

PRIORITIES

in their way would complain to their superiors until the matter reached the ANMB. The Maritime Commission itself complained on occasion of the issuance of priorities by other agencies in ways that crippled the program for merchant shipbuilding. Colonel Rockwell, Director of the Production Division, wrote that other agencies had admitted using AA priorities issued for large projects to cover items not essential to the original project but warned against the dangers of the Maritime Commission doing the same:

> We have received AA priorities for the construction of certain vessels and we find that these AA priorities are being extended to cover the purchase of special items which disrupt the A-1-a priority of other agencies. When this happens often enough there will be investigations resulting in the issuing of AA priorities to many other agencies or a very decided restriction against our further use of the AA priority. . . . We shall continue our fight against any other agency which abuses the priority privilege but we cannot expect much consideration if other agencies prove that we are indiscreet in the use of high priorities.[10]

The abuse of priority certificates was similarly possible under a blanket preference rating extended to a shipbuilder as prime contractor. Reprisals were also the only effective check if the shipbuilder who received a high rating for a particular group of ships and facilities used those ratings to secure materials and equipment for other ships or facilities. Colonel Rockwell recounted the details of one such use of priorities in a letter of July 31, 1942: "The Kaiser Company was given the right to use AA priority rating in May to secure equipment for the ATL [landing ship] shipbuilding program. You will note that they requisitioned 3216 welding machines, 144 cranes and 6 locomotives as listed below. . . ." The list showed that a very large percentage of this equipment had gone to Kaiser yards other than those building the landing ships.[11] When the mechanism for administrating priorities was so set up that such things could happen, and when the ANMB itself was authorizing so many priorities that A-1 soon had to be superseded by A-1-a, then by AA-1, and finally by AAA, it is easy to see why a priority was considered only a hunting license.

Even with the best of licenses, the hunting was not always easy. Certain companies were accustomed to taking orders from one

---

[10] Rockwell to Berlage, June 27, 1942, in gf 120-124-2.
[11] Rockwell to Berlage, July 31, 1942, in gf 120-124-2.

agency and were indebted to that agency for previous awards of contracts, for aid in the expansion of plant and equipment, and for assistance in securing the patents under which they operated. Such companies might be reluctant to accept for prior delivery orders from any other agency. The Maritime Commission profited from having established such a relation with the companies producing anchor chain by other than the Di-lok process and with some companies making steering gear, windlasses, and winches.[12] On the other hand, they found their priorities almost useless if they tried to get their orders into plants thoroughly committed to Navy contracts and watched over by a Navy inspector. Complete illustration is found in the report of the expediter or inspector, who visited a plant in Titusville, Pennsylvania, to see why the crankshaft forgings on order there were not forthcoming. An OPM official who accompanied him asked to see the schedules of what had been shipped the previous month and against what priorities. " This report was brought forth and it indicated that Navy A-1-b's and A-1-c's had been shipped ahead of Maritime Commission's A-1-a's. When the management was questioned on this point they stated that they had no alternative but to ship in accordance with the wishes of the Naval Inspectors at the plant." The Navy had expended $2,250,000 in expanding the Titusville plant under a contract which specified that those facilities were to be used exclusively for the Navy and that in other shops of the company Navy orders would be given priority. When the Maritime Commission's representative accused the Navy inspectors present, asking who was running the plant, the Navy inspectors openly avowed that they were, and the company's general manager agreed. The only satisfaction obtained was the promise that when products began to come from the new press which was being installed, the Maritime Commission orders would be filled, many weeks late.[13]

These weaknesses of the priority system caused it to be supplemented not only in steel distribution but also in handling other materials and components by direct government allocation and scheduling. The Priority Section in the Production Division began to shrink and split up. Some of its staff were then assigned to work with the ANMB to get AAA ratings for special emer-

[12] Statements by John Conkey.
[13] F. F. Noonan to C. W. Flesher, Dec. 2, 1941, in Technical Division file A-9.

gency items, some worked with WPB on the new forms of controlling materials being developed there, and many went into the Procurement Division when that new division was formed in January 1943. There they filled out priority forms to go with the purchase orders.[14] Thus priorities became subordinated either to allocation or to procurement.

II. Placing Orders

In contrast to the administration of priorities, which was an activity entirely alien to peacetime shipbuilding, procurement and the attendant scheduling were normal activities of shipbuilders and their naval architects. In the long-range program before 1941 the only buying done by the Commission was of items needed in the outfitting—such as radio apparatus, galley equipment, hand tools, textiles, furniture and so on, which were traditionally provided not by the builder but by the purchaser. These supplies and equipment were known as the allowance list, and since under the long-range program the Commission was frequently in the position of purchaser, it bought the " allowance list " items through a Division of Purchase and Supply.[1] Materials and components for ship construction were a different sphere of activity, and established firms in the shipbuilding fraternity were prepared to procure these materials for themselves. The Commission accordingly took over procurement only gradually and never altogether.

The Commission began its broad entry into procurement by ordering for Liberty ships when the contract with Gibbs & Cox was terminated. A Purchasing Section for this purpose was then established in July 1941 in the Construction Division. Charles E. Walsh, Jr., who was brought in from the purchasing organization of Bethlehem Steel, headed this section and its continuation, under the name of Procurement Section, in the Production Division.[2] To assure coordination between this new unit and the group purchasing the allowance list, the Director of the Division

---

[14] Quarterly reports of Production Division, Sept. 30 and Dec. 31, 1942 and March 31, 1943.

[1] Admin. Order No. 37, Suppl. 23, Jan. 7, 1941.

[2] Admin. Order No. 37, Suppl. 42, July 25, 1941; Admin. Order No. 37, Suppl. 45, Aug. 19, 1941; Personnel files.

of Purchase and Supply was ordered to report to the Director of the Production Division in so far as concerned supplies for new vessels.[3] When Procurement was detached from the Production Division and made a separate division under Mr. Walsh on January 1, 1943, it absorbed the Division of Purchase and Supply. The new Procurement Division was made responsible for "all procurement operations in connection with the Maritime Commission's activities." [4]

By 1943 the Procurement Division was placing the orders for components and materials for other types of ships as well as for 1,274 Liberty ships. It was buying for many tanker yards, for Liberty ships of modified design, and for parts of the tug, barge, and long-range programs. When the smaller cargo vessels, combat loaders, coastal tankers, and the Victory ships were added to the Commission's programs, the Procurement Division ordered materials and components for these types also.[5] On the other hand, ordering for the C-types of items other than those on the allowance list continued to be done by the yards producing these types or by the naval architects they hired. Not only was this true of yards which had been established before the emergency; North Carolina after it started building C2's and the Kaiser yards in their work on troopships (C4's) and escort carriers (S4-S2-BB3) also arranged their own procurement. When the corvette program was undertaken, Kaiser Cargo Inc. did all procurement for the 90 corvettes except that steel, boilers, and engines were procured by the Commission.[6] For many types the Commission bought the turbines and turbo-generators. The size and distribution of the task of procurement is indicated by the figures for December 1, 1941-April 1, 1943. In ships delivered during that period materials procured by the Maritime Commission totaled $957,554,750, materials procured by the shipyards for themselves totaled $168,980,025.[7] The Procurement Division made purchases for the War Shipping Administration as well as for the shipbuilding

---

[3] Admin. Order No. 66, July 14, 1942.

[4] Admin. Order No. 37, Suppl. 83, and Vickery to the Commission, Dec. 31, 1942, in gf 201-2-43. John G. Conkey was made Assistant Director, F. E. Hickey, Special Assistant to the Director.

[5] Quarterly reports of Procurement Section, Production Division; quarterly reports of Procurement Division, 1944: also the compilation dated Sept. 15, 1945 in material requirement schedules in MC Historian's Collection, Research file 117.

[6] Senate, *Problems of American Small Business, Hearings*, pt. 21, pp. 2959, 2986.

[7] *Ibid.*, p. 2959.

activities of the Maritime Commission. Its activities thus as a joint service swelled the total volume of its work,[8] and the supplies for ship construction alone cost $2,789,967,770.[9]

These figures represent a very great variety of materials and manufactured goods. The function of the Procurement Division was to take the specifications prepared by the naval architects and send out to possible suppliers the information they needed in order to submit bids. At the same time, Procurement obtained from the Production Division an estimate of approximately when the material would be needed and incorporated such information in requests for bids. In awarding contracts after the bids were received they had to take into account not only the price but such factors as assurance of delivery, competition in priorities, the policy of spreading work among small plants, and the areas of labor shortage.[10]

The starting point of the work within the division was the arrival of a "requisition," or bill of materials which stated the specifications to be met, the number of items, and ideally, the time of delivery, although this often had to be added later. The requisition was supplied to the Procurement Division by whoever was doing the work of naval architect on the ship in question. It might be the yard itself, as in the case of the requisitions for T2 tankers, which were provided by Alabama Shipbuilding and by Kaiser-Swan Island, or it might be a naval architect engaged by the Commission. In any case the requisition had to be approved by a representative of the Technical Division before it reached the Procurement Division.[11] There it was sent to a Records Section which checked it for completeness and then distributed copies to those who were to act on it. One of the buying sections (of which there were ten) then sent out letters to ask quotations. It kept all the bids with the requisition in an active file with a "follow-up date" seven days later when it was to be reviewed to see if it could be filled. Each buying section was responsible for maintaining a list of prospective bidders for the various types of material which it handled. When

---

[8] Although WSA did most of its purchasing through its general agents. Statement by Conkey to Lane, Apr. 13, 1948.
[9] House, *Independent Offices Appropriation Bill for 1948, Hearings*, pt. 2, p. 904.
[10] On the profits of suppliers see chap. 20.
[11] Statements by Conkey, Apr. 13, 1948.

sufficient time had been allowed, the bids of various manufacturers were tabulated and the award made, and the actual purchase order drawn up in the Purchasing Section and processed and recorded in the Clerical Section. Other sections in the Division had the function of advising the buying sections in regard to their specialties—namely: Small Business; Liaison with the OPA, WPB, etc.; Rates and Routing; Surplus Material; and Priorities.[12]

III. SCHEDULING SHIP CONSTRUCTION

Although Procurement placed the orders, the final timing of deliveries was a responsibility of the Production Division, which was in charge of scheduling.

Scheduling of ships was logically the first and basic function since it governed the scheduling of materials, and the Production Engineering Section of the Maritime Commission set up schedules of ship production from the very beginning of the emergency program. Three kinds of schedules need to be distinguished, however: (1) a schedule based simply on dates in the contracts, (2) a schedule based on the fast building times which contractors were being urged to achieve, and (3) a schedule presenting an estimate of what the yards were really most likely to do in view of all the facts available concerning the degree of completion of the yards, their previous performance, recent keel layings, and so on. Since some yards got into production much faster than others, a schedule compiled in June 1942 on the basis of contract dates alone would show ships about to be launched from ways on which no keels had yet been laid, and in other yards would show as future events the keel laying of ships already launched.

Schedules had to be constantly revised to make them realistic. During the battle for steel just before the Higgins crisis, new schedules were carefully compiled on the basis of Admiral Vickery's own estimates of what rate of production could be obtained from the yards. A statistical group was organized by David Scoll, Special Assistant to Admiral Vickery and representative of the Maritime Commission on the Advisory Committee on Official Production Statistics; and it was amalgamated into Mr. Gallagher's Production Engineering Section which had pre-

[12] Procurement Division Purchasing Instruction No. 88, July 28, 1943, and organization charts, Jan. 1, 1944.

PLATE XV. Above, an escort aircraft carrier, built at the Kaiser-Vancouver yard, hitting the waters of the Columbia River. To the right, James L. Bates, Director of the Technical Division of the Maritime Commission. Below, left to right, R. Earle Anderson, Director of the Finance Division, Willard F. Rockwell, the part-time Director of the Production Division, and Allen D. MacLean, his Assistant Director and later the Director of the Division.

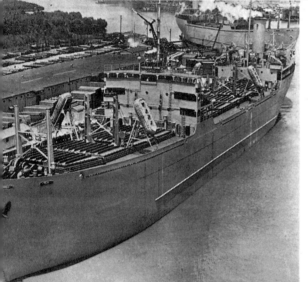

PLATE XVI.

The heavy loss of tankers by enemy action caused some yards originally planned for Libertys (Alabama Ship, Marinship, and Swan Island) to build tankers, the T2-SE-A1 shown above. Calship and Delta also built tankers, using the modified Liberty ship design shown below (Z-ET1-S-C3). These tankers took heavy deck loads, carrying planes to Europe as well as fuel. The Delta yard at New Orleans also built another special type of Liberty ship, the Liberty collier (left).

viously assisted Admiral Vickery in charting and statistical work.[1] On the basis of past experience and current conditions, both as reported way for way by the inspectors and as known generally to Admiral Vickery, they calculated the length of time each shipway would take to be ready for a keel laying, how long it would be between keel laying and launching, and when delivery might be expected.[2] This group became the Statistical Analysis Branch and began in July 1942 to recalculate ship-production schedules anew every month.

For the Statistical Analysis Branch to be able to make up accurately these schedules of the expected production of ships, it needed frequent reports of conditions in the yards. Partly because of this need the Production Division sent out to the yards representatives called " Material Coordinators," but the Regional Directors objected to reports on construction by officials not under their jurisidiction. Consequently the material coordinators were placed under the regional offices.

The Statistical Analysis Branch was therefore forced to base its estimates on the schedules which were sent it by the Regional Directors, and often found these schedules very far from perfect. Being absorbed in getting the yards into operation. or in launching ships already behind schedule, or in completing those tied up at the outfitting docks, the staffs in the regional offices and the yards had little patience with requests for judicious guesses of what ships would be built six months or a year later. Being in a sense on the firing line, they were not particularly sensitive to the need of logistic planning.

The schedules sent in from the Regions were naturally based on those made out by the companies. If they had always been the latest schedules developed by the yard managements, there would not have been so many difficulties, but many times a schedule was sent to Washington which reflected mainly a company's obligations and was therefore simply a compilation of dates given in the contract, even though conditions in the yard had already made clear that actual construction would be ahead of or behind the contract. Merely by examining these schedules, the Statistical Analysis Branch could in some cases see their defects. For example, the schedules submitted as of October 31, 1942

---

[1] Renamed on some organization charts the "Scheduling and Planning Section."
[2] See above, pp. 140n, 180n.

showed for several of the newer yards no speed-up in the construction time of successive rounds of ships, although the experience with the older yards indicated there was almost sure to be such a speed-up. The schedule for the Fairfield yard showed two ways unoccupied at the end of one month, and, at the end of another month five ways with two keels on each way—a confusion created by the fact that Bethlehem-Fairfield was building both landing craft and Liberty ships and that an old, out-of-date schedule for the Liberty ships had been used. The schedule for Calship showed 114 days on which two ships would have to occupy one way simultaneously. At Houston an idle period of more than a month per way was shown between launchings of the hulls of one contract and the laying of the keels for those of the next. These inconsistencies and improbabilities were called to the attention of the Regional Directors by a letter over the signature of Admiral Vickery.[3]

Almost a year later the head of the Statistical Analysis Branch while on a field trip personally pointed out similar difficulties to the staffs of the Regional Offices. Only in the Gulf did he find scheduling of construction being given "the attention it merits." That office already had a production section preparing schedules, although they were scheduling on the assumption that facilities, management, and labor would all improve considerably. The Great Lakes and the West coast prepared also to edit the schedules sent in from their yards in the same way that the Statistical Analysis Branch had been doing.[4] The East coast office responded to the suggestion of the Statistical Analysis Branch by calling on the shipbuilding companies to submit revised schedules which would take a "realistic view of actual possibilities."[5]

Another tendency which appeared in the work of the Regional Offices was overscheduling. It was done in order to be sure of getting delivery of components and materials. Since this was the kind of exaggeration apparent in the schedules which expediters of

---

[3] Vickery to Sanford and Vickery to McInnis, Nov. 30, 1942, with attachments dated Nov. 27 and Nov. 30, 1942, in MC, Historian's Collection, Research file, Statistics 117, "Notes on Regional Progress Reports."

[4] G. J. Fischer to Vickery, Oct. 4, 1943, in Fischer reading file.

[5] W. H. Blakeman, Assistant Director, Field, East Coast, to Chester L. Churchill, President, New England Shipbuilding Corp., Aug. 17, 1943, and similar letters to the Brunswick, Savannah, and Sparrows Point yards, MC, Historian's Collection, Research file, Statistics 117, "Notes on Regional Progress Reports."

some shipbuilding companies presented at Washington when clamoring for materials, they might be called "expediters' schedules." Each region wanted also to get its full share of the steel plate, valves, and innumerable other scarce items, so they too had a tendency towards this kind of overscheduling. Indeed, the Statistical Analysis Branch, when computing the whole program for submisssion to the WPB as a basis for claiming steel, was under a similar pressure to err on the side of overscheduling in order that the Commission might get a larger allotment. But in making the schedules which would govern distribution between regions and yards after a steel allocation was received from WPB, the Statistical Analysis Branch was equally interested in correcting "contract schedules," "incentive schedules," or "expediters' schedules," and arriving at a scheduler's schedule, namely, as accurate an estimate as possible of future keel layings, launchings, and deliveries.

IV. Scheduling Components

The Maritime Commission was the only government agency charged with scheduling the production of merchant ships and other ships built in its yards, but in scheduling the production of the materials and components which entered into these ships other agencies, especially the WPB, was at least equally active. In this scheduling of materials the Maritime Commission acted in a double capacity. On the one hand it was a regulatory body assisting in industry-wide scheduling by which the government told industry what and when to produce in wartime. On the other hand, as a procurement agency it was under contractual obligation to certain shipyards to supply them with the steel, engines, boilers, and so on. The "manhour" contracts relieved many shipyard managements from responsibility for providing themselves with most of their materials and components. When that type of contract was adopted it was understood that Gibbs & Cox, acting as agent for the government, would schedule the delivery of materials to the yards as well as arrange their purchase. Almost at once, however, the Commission's Production Engineering Section began revising the schedules made by Gibbs & Cox and in the middle of 1941 took over entirely the scheduling of materials. Those yards which had fixed-price or price-minus contracts continued to take responsibility for their own supplies.

To schedule the delivery of the materials which the Commission purchased, two branches were immediately formed in 1941 within the Production Engineering Section: one for items connected with hull and deck activities; the other for engines, valves, piping, and related items. The Engineering Scheduling and Planning Branch, for example, prepared charts and card files showing the allocation of machinery to various hulls according to the expected delivery dates. Piping proved so extensive a problem that in August 1944 a separate branch was created to assure its proper delivery. Although these branches began in connection with the Commission's purchase of materials for the yards, they later fitted their work into the methods of industry-wide allocation developed under the WPB. They kept records, for example, of when the reciprocating engines used in the Liberty ships would be finished by the engine makers, and they allotted these engines to the various Liberty-ship yards according to when hulls would be ready. Similiarly they kept tab on the stern frames, anchor chain, life boats, and dozens of other items. The volume of Commission-procured material of this kind reached it peak in 1943, although attention was then focused rather upon procedures for general allocation.[1]

General allocation meant that the whole product of an industry was scheduled under national control. In that case the scheduling had to be done by WPB, which could consider the needs of all claimants. By nation-wide scheduling the government first regulated production of a few items which were especially scarce—turbines, gears, valves, and steel forgings, for example—and then spread out to impose comprehensive controls on industries in 1943.[2]

The scarce components which were the particular concern of shipbuilders began to come under nation-wide scheduling as early as 1941 on a voluntary basis. These arrangements were worked out in the Shipbuilding Section of the OPM, which became later the Shipbuilding Branch, and then Division, in WPB. Admiral Land had headed the shipbuilding units in the NDAC and OPM. His assistant, Captain J. O. Gawne, gradually took over more and

---

[1] Quarterly reports of the Emergency Ship Construction Division, the Construction Division, and the Production Division; Admin. Order No. 37, Suppl. 42, July 25, 1941.

[2] *Industrial Mobilization for War*, I, 683, 687.

SCHEDULING COMPONENTS

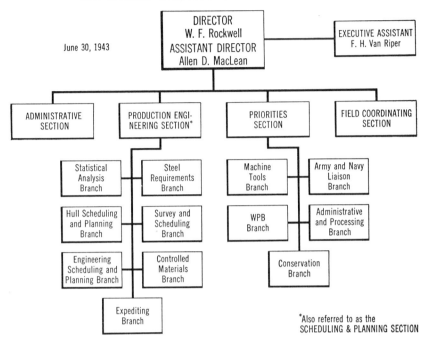

FIGURE 30. Source: Quarterly reports of Production Division, April-June, 1943, and Rockwell to Vickery, February 20, 1943 in MC gf 201-2-31.

more responsibility, and in the WPB, Captain Gawne became Director of the Shipbuilding Division.[3] The later, more comprehensive controls of WPB did not abrogate those worked out in the shipbuilding section, but included them in a more general system.

First came a nation-wide scheduling of all turbines and gears, those being made for the Navy as well as those for merchant ships and other purposes. Their assignment, turbine by turbine, to specific hulls was indicated on charts showing the monthly output of each producer. These schedules were worked out in the spring of 1941 in connection with the plans for expanding the industry and in consultation with the Turbine and Gears Defense Industry

---

[3] Coleman, *Shipbuilding Activities*, pp. 2-3; Chaikin and Coleman, *Shipbuilding Policies*, p. 31.

Advisory Committee consisting of representatives of seven manufacturers, the Navy, the Maritime Commission, and OPM. In November 1941 the plan was put into effect for the entire industry and remained operative throughout the war, for it fitted into the method which WPB later applied more generally.[4]

Steel valves became a scarcity attracting the attention of OPM as early as gears and turbines. The Steel Valve Industry Advisory Committee was the first to be legally authorized within OPM, on July 20, 1941.[5] But scheduling the production of steel valves was much more complicated than scheduling turbines. The producers were more numerous and the number of different sizes and kinds of valves ran into the thousands. Not only shipbuilding but other vital industries, especially the synthetic rubber industry, had good claims to the products of the valve makers. Effective scheduling of valve manufacture was therefore dependent on the development of WPB's whole program to control scheduling and was not fully and effectively developed for the whole industry until 1943.[6]

Electric motors was another component which was not only in urgent demand on ships—in deck machinery, ventilation, etc.—but was also wanted by many other users. In the main the Maritime and the Navy were left to act as procurement agencies in increasing the new sources of supplies and arranging with manufacturers the schedules for deliveries, but the Shipbuilding Division cooperated with their efforts, and the WPB's power was used before the end of 1942 to freeze the production schedule of at least one producer so that he could operate with assurance that his plans would not be disrupted by sudden changes in the priorities of his orders.[7]

Many components entering into shipbuilding were put under industry-wide schedules when General Scheduling Order M-293 was issued February 26, 1943 by order of C. E. Wilson as Production Vice-Chairman of WPB.[8] The part of the Maritime Commission most affected was the Survey Branch in the Pro-

---

[4] Coleman, *Shipbuilding Activities*, pp. 118-19; Chaikin and Coleman, *Shipbuilding Policies*, p. 113.

[5] Coleman, *Shipbuilding Activities*, pp. 122-23.

[6] Chaikin and Coleman, *Shipbuilding Policies*, pp. 71-83, 115; Nelson, *Arsenal of Democracy*, p. 383.

[7] Chaikin and Coleman, *Shipbuilding Policies*, pp. 68-69.

[8] *Industrial Mobilization for War*, I, 687.

duction Division. It had been formed in the fall of 1942 to work with WPB in making the schedules for turbines. Renamed the Survey and Scheduling Branch, it worked with the WPB in establishing production schedules under form M-293. This Branch also took the responsibility of submitting forms PD 903 which set forth scheduled needs on all critical items, both those purchased by the Maritime Commission and those for the long-range yards which did their own buying.[9]

Before the Survey Branch was developed to handle the Maritime Commission's function in connection with industry-wide schedules for components, a Materials Requirement Branch was formed to work with WPB in regard to industry-wide scheduling of materials. Since steel was the critical material, this Branch became a Steel Requirements Branch, and it took over from the Hull Scheduling and Planning Branch the recording of requirements and deliveries of steel plates and shapes in yards for which the Maritime Commission was procuring as well as in the long-range yards. In a sense it did for steel what the Survey Branch did for gears and turbines.

V. Scheduling Materials, PRP and CMP

Scheduling a material like steel was a more complicated problem administratively than scheduling the production of a component such as turbines because the number of claimants was so much greater. In fact the distribution of steel plate raised major issues concerning the administration of the war economy, as described in the last chapter. The development within the staff of the Maritime Commission of the means of coping with the allocation of steel was affected by these conflicts. Only gradually did the staffs of the WPB and the Maritime Commission work out the distribution of functions between them and finally mesh together like two gears in a single mechanism which distributed steel in accordance with the program of the Joint Chiefs. One important consequence of the battle for steel was the improvement of this mechanism; indeed, the mechanism developed out of the conflict.

At the beginning of 1942 neither the WPB nor the Maritime Commission had adequate staffs working on allocation, nor were

[9] Quarterly reports of the Production Division for Mar. 31 and June 30, 1943. *Industrial Mobilization for War*, I, 688, explains PD 903.

reports from the shipyards and steel mills adequate for efficient scheduling. Moreover, there was not then any directive from the President which was clearly and generally understood as indicating which agency was primarily responsible for obtaining the reports and developing the staff to handle them. By the executive orders creating it, WPB had been authorized to exercise "general direction" over procurement and scheduling, but WPB deliberately left procurement to the Army and Navy and the other agenices which like the Maritime Commission were already placing orders and awarding contracts. The Armed Forces went beyond procurement in its narrow sense. They developed their own staffs for scheduling deliveries. With Presidential approval they backed up their schedules with priorities assigned by the Army and Navy Munitions Board, as has been illustrated. The result was much overlapping between the Services and the WPB and a stiff contest between them for power. In the middle of 1942 the WPB was struggling to put into effect a plan which placed directly with the WPB the responsibility for dividing up among industrial plants the materials they needed.[1] Consequently it was a possibility that the WPB would develop the staff to calculate the steel needs of the shipyards, asking the Maritime Commission only for its program of ship construction.[2]

One essential of good scheduling, whoever was to do it, was adequate control of inventories. Looking back at the whole wartime experience the section chief who was most constantly at the center of the problem of allocations and scheduling in the Maritime Commission has emphasized repeatedly the need of a uniform way of "keeping store."[3] "In a future emergency," wrote another member of the staff, "it is absolutely essential that a standard steel report be established with enforcement of specific instructions to insure that results will be applicable to the problem and comparable."[4] Both WPB and the Maritime Commission began

[1] *Industrial Mobilization for War*, I, 208, 212-22, 453 ff.; Nelson, *Arsenal of Democracy*, p. 202. Compare the interpretation in Lincoln Gordon, "An Official Appraisal of the War Economy and its Administration," in *The Review of Economic Statistics*, XXIX, No. 3 (Aug. 1947), p. 185.

[2] W. H. Harrison to Vickery, June 9, 1942 (in MC gf 506-1-1-21) for example, emphasizes that for the WPB to do its allocating it must collect directly from the yards data on inventories, consumption rates, and expected future requirements.

[3] Statements by J. T. Gallagher.

[4] Gerald J. Fischer, Head of Statistical Analysis Branch, in his "Statistical Analysis, Steel," p. 3.

in 1941 developing instruments for control of inventory but before the middle of 1942 neither had adequate information. In September 1941 the Maritime Commission instructed its inspectors in the shipyards to report weekly the amounts of hull steel received, fabricated and erected, yet it was done in such form that the amount of raw stock in inventory as distinct from that being fabricated could not be learned from these reports. Inspectors were told that in making the report " any convenient tabular or narrative form may be used," and in October a different kind of report was asked for.[5] Reports from different yards were often not comparable. A uniform way of keeping steel records might have been imposed by the Maritime Commission on the emergency yards at their inception, if the Commission had had the will to do so and known what system to impose, although it might have been hard to get the established yards to change their systems. But diversity was freely permitted to all yards, in systems of inventory control as in erection schedules.[6] This was in accordance with the general policy of encouraging initiative among yard managements. Brilliant engineering achievements justified part of that policy, but the time came when uniform inventory records were fervently desired.

The first step toward placing a limit on inventories was taken by the OPM in May 1941 as part of a general plan for limiting the use of sixteen basic metals. It left enforcement up to the producers, and set the limit vaguely to accord with " the Customer's usual method and rate of operation."[7] After Pearl Harbor WPB moved to gain adequate information both from shipyards and from steel mills, and the forms sent out in February 1942 were the basis for the system of steel allocation used thereafter. To all consumers of steel plate, which of course included the shipyards, went form PD 299. It requested not only information on consumption, actual and anticipated, but also statements of the inventory at the end of the reported month, receipts during the reported month, and receipts anticipated in the following

[5] L. R. Sanford to all Principal Hull Inspectors, Sept. 16, 1941, in gf 507-3-1, and J. T. Gallagher to Ingalls Shipbuilding Corporation and others, Oct. 29, 1941, both in gf 507-4-1-3.

[6] Ezekiel Report, p. 32, illustrates the wide differences from yard to yard in the way operating statistics were kept.

[7] *Industrial Mobilization for War*, I, 181. The same principle was embodied in " Priorities Regulation No. 1 as amended June 26, 1942." *Ibid.*, pp. 501-02.

month. Another form, PD 298, requested the consumers of steel plate to list the orders that they had placed on each steel producer and needed to have shipped in a given month. Beside the order numbers, the consumer filled in his priority ratings, allocation numbers, the tonnage, and so on. The shipyard was to send this form PD 298 to the steel company which, after filling in the names of the mills on which each order was to be rolled, was to send copies to the WPB, to the Maritime Commission, and to the shipyard.[8] By another form, PD 169, addressed directly to the steel mills, WPB required them to show all orders scheduled for rolling in a given month. But some mills failed to show on their forms PD 169 all the orders which had been listed by the shipyards on the forms PD 298. The Maritime Commission asked WPB to instruct the mills to list all orders received, including those carried over from the previous month, whether or not these orders were placed in the rolling schedule of the month they were reporting.[9] That was necessary in order to check on what the mills were actually doing and what orders were not being taken care of. The use of this set of forms was intended, the WPB announced, to enable it "to secure accurate information as to monthly requirements for steel plates and to correlate this data between producers, consumers, and government agencies," and "to promote more accurate and practical control over steel plate schedules and allocations."[10]

While this announcement made it appear that the staff of WPB was about to take responsibility for allocating steel, the WPB was soon criticizing the Maritime Commission for not doing enough on its side. There was no doubt of the responsibility of the Commission for placing its orders for ship steel in the hands of the mills in plenty of time.[11] But whose responsibility was it to distribute the steel as between yards building merchant ships? At the end of April, the Iron and Steel Branch of WPB spoke as if it were ready to allocate directly to the yards provided the Maritime Commission would "specify the yards and the orders

---

[8] Forms PD 298 and PD 299 attached to letter of Gibbs & Cox to Schmeltzer, Feb. 21, 1942, in gf 507-4-1-3.

[9] C. W. Flesher to C. E. Adams, Apr. 2, 1942 in gf 507-4-1-3.

[10] *New York Times*, Feb. 17, 1942, p. 31.

[11] The need of placing orders on time was stressed by Flesher to all shipyards, Feb. 24, 1942 (in gf 507-4-1-3).

most critically needed." [12] But in the long run the Maritime Commission was held responsible for distributing steel among all shipyards under contract to it, and already in April 1942 it was being criticized by WPB for not doing so efficiently.

The most elaborate criticism from the WPB concerning the distribution of steel among the Commission's yards was in the special report by Mordecai Ezekiel, a statistician on loan to the WPB from the Department of Agriculture where he had been Chief Economist of the Federal Farm Bureau. Preliminary versions of his report were discussed in the WPB Planning Committee in April and May, and the final report submitted July 20, 1942. Mr. Ezekiel criticized the Commission for permitting too much steel to accumulate in some yards while others, mainly new yards on the West coast, did not have enough to operate at full capacity. As he pointed out, the yards building standard-type ships ordered their own steel, usually months ahead, and the Commission made no effort to schedule its delivery. These yards were on the whole oversupplied with steel; the limiting factor in their deliveries was the propulsion machinery. Another weakness was the unevenness of inventories. As an analysis of Calship showed, a yard might have many shipsworth of one kind of steel plate but lack altogether some other of the 700-odd kinds of plates needed in a ship. The Maritime Commission, he said, should develop a staff and act vigorously to even out these inequalities and plan future needs more accurately yard by yard. The whole tenor of his criticism implied that the Maritime Commission was still responsible for the distribution of steel both to the yards for which it purchased and to the long-range yards.

Although Admiral Land objected to the way Mr. Ezekiel permitted his report to be used against the Commission during the hubbub raised by Mr. Higgins, the truth of some of his criticisms was admitted.[13] Colonel Rockwell, who was in the process of

---

[12] Memorandum prepared by the Iron and Steel Branch, Apr. 29, 1942, in gf 506-1-1-21. This was in answer to Land's testimony before a Congressional Committee. Admiral Land's comment at that time indicated that he thought that this "specifying of . . . orders most critically needed" was a matter to be worked out through the not yet entirely discredited method of priorities.

[13] With the copy of the Ezekiel Report in the Historian's Collection, Research file 206.1.2, is a copy of Admiral Land's Memorandum to the Files concerning his telephone conversation with Ezekiel on July 30, 1942, and his Memorandum to Admiral Vickery, June 25, 1942, commenting both on the report and on a memorandum from Rockwell of June 19, which I have not been able to find. The main

reorganizing the Production Division, admitted that the long-range yards had placed orders as they wished and secured favorable treatment from the steel companies, but said a scheduling unit was being formed to exercise control. He accepted Mr. Ezekiel's suggestion that the Maritime Commission must expand considerably the part of its staff devoted to inventory control and scheduling.[14]

Under this goading from the WPB the planning and scheduling staff of the Maritime Commission was enlarged. We have already seen how this led to the more thorough and systematic scheduling of ship construction by the formation of the Statistical Analysis Branch. That Branch also developed estimates of steel requirements. At the same time was formed the Materials Requirements Branch, later known as the Steel Requirements Branch, which worked directly with the Iron and Steel Branch of WPB.

A new attack on the problem of inventories was then made by the Maritime Commission, and its Statistical Analysis Branch developed two ways of checking on the amount of steel in the yards. One check on inventory was through the weekly steel reports sent in from each yard. These reports gave the cumulative totals of plates and shapes received, fabricated, and erected prior to the reporting period (at least the totals for that type of ship), the amounts received, fabricated, and erected during the period covered by the report, the amounts withdrawn for other than ship construction purposes, and the total on hand at the end of the period. Although only the latter might be considered "inventory," the other figures were necessary to make checks which would permit the Statistical Analysis Branch to detect errors in the reporting and were desired to measure progress in the yard. On these reports, steel was considered "fabricated" as soon as it was moved from the storage yard to be taken to the fabricating shop. The most significant figure was the difference between tons received and tons erected, since various yards differed in the extent to which they accumulated fabricated steel prior to erection, and might be accumulating excess either in raw stock or in fabricated form. Consequently

---

reply to Ezekiel from the Commission is Rockwell's letter to Fesler, Aug. 13, 1942, MC gf 506-1-1-21 and 507-4-1-3.

[14] *Minutes of the Planning Committee of the War Production Board*, p. 57 (May 14, 1942); Rockwell to Fesler, Aug. 13, 1942, in gf 506-1-1-21.

SCHEDULING MATERIALS, PRP AND CMP 375

the "total of steel behind the ways" (i. e. steel in stock plus steel in fabrication), as well as the total allocated compared to the total turned into completed ships, helped determine the amount needed by a yard to meet future schedules.

Since the officials who sent in these steel reports were the material coordinators who were under the regional directors, their reports and queries about the reports had to be sent through the regional offices, at least until October 1943, when they were permitted by administrative order to correspond directly with officials in the Production and Procurement Divisions. The cumbersomeness of the earlier system is illustrated by the roundabout channels through which any explanations concerning the steel report were routed. It was necessary for the head of the Statistical Analysis Branch to write a memo to the chief of the Field Coordinating Section of the Production Division, who then wrote a memo to the Regional Director, attention Principal Material Coordinator, who in turn wrote to the material coordinator in the yard involved. Then the reply followed the same course in reverse. With this cumbersome way of securing explanations, errors sometimes accumulated faster than corrections.

A second method of checking on the amount of steel in a yard depended only on the records which the Statistical Analysis Branch could itself keep in Washington. These showed the cumulative total of allocations which had been made to each yard. By deducting from that figure the total amount of ships delivered from the yard, it estimated how much was in the "pipeline" of that particular yard. Some of the steel previously allocated might not have been delivered by the steel mills, but since past allocations were carried over from month to month, the allocations were still good.[15] The weakness in this method was that a yard sometimes obtained from a regional director permission to make "emergency purchases" in advance of its allocations. Moreover, steel was switched from one yard to another to meet urgent needs by the authority of a regional director. These emergency purchases and transfers from one yard to another were reported

---

[15] Gerald J. Fischer, "The Organization and Procedures Utilized by the Maritime Commission in the Procurement and Distribution of Materials and Equipment Required for the Merchant Shipbuilding Program During World War II," pp. 16-18. MS in Historian's Collection, Research file 206.5 and statements to F. C. Lane, Apr. 7, 1948. On the Field Coordinating Section see Production Division Quarterly Report of Dec. 31, 1942.

to Washington by yard expediters and regional offices; but when reports did not arrive promptly enough or did not get compiled promptly enough in the Production Division, steel might be allocated to a yard which had already been supplied by emergency purchases or transfers. That yard would then have surplus, although some other yard might be short.[16]

Imperfect though it was, the work of the Statistical Analysis Branch and the Steel Requirements Branch showed that the Maritime Commission came to take seriously the problem of allocating among the yards. It did so in response to pressure from the WPB. By their criticism of each other the WPB and the Maritime Commission were impelled to tighten their controls—by the one over the steel industry, by the other over the shipyards. Thus by their pressure on each other they played their roles in getting the war economy organized.[17]

The organization developed by the Maritime Commission in this conflict fitted it to act as a claimant for steel on behalf of its shipyards. Ironically, the WPB was pushing the Maritime Commission into building this organization at the very time when the upper levels of the WPB hierarchy were disputing whether allocations should be made through claimant agencies or should be made directly by the WPB to industrial units. The latter process was called a horizontal system of distribution; allocation through claimant agencies was called a vertical system. In the summer of 1942 WPB was officially committed to a horizontal system known as the PRP (Production Requirements Plan), but the difficulties of putting the plan into operation proved stupefying, especially since it was begun hurriedly and amid much opposition both within WPB and in the ANMB. When Ferdinand Eberstadt was brought to WPB as Program Vice Chairman on September 20, 1942, he began preparations for putting into effect a vertical system of allocation which became known as the Controlled Materials Plan. Under it the WPB made allocation of all the most critical materials to seven claimant agencies, one of which was the Maritime Commission.[18] Thus the same sort of

[16] Conflicts between the Production Division and the Regional Directors over surpluses are described in chap. 21.
[17] Much detail of this give and take is recounted in the MS by Gerald J. Fischer, " Problems of Steel Production . . .", Historian's Collection, Research file 206.1.3.
[18] *Industrial Mobilization for War*, I, 259-62, 457-89, and in the relation of PRP to control of inventory see *ibid.*, p. 504.

system that had developed for distributing steel plate was applied as a general scheme for control of the economy.

Reconsideration of the strong and weak points in the Ezekiel report throws some light on why vertical allocation was the form found workable. Mr. Ezekiel was right in many of his facts but wrong in many of his conclusions. His statistics revealed the need of better statistical organization in the Maritime Commission, and showed which yards were oversupplied and which undersupplied. But he was wrong on the most important practical issue of the moment; for he concluded that the cancellation of the Higgins contract was unwise, arguing that the other yards would not be able to increase their production enough to make up the difference and meet the President's objective for 1943. Events proved that Admiral Vickery's judgment on that matter was better—and no wonder—since Mr. Ezekiel was no shipbuilder and Admiral Vickery had been following almost every move in each of those shipyards from the beginning. Decisions of detail in allocating yard by yard could also be done best by members of the Commission's staff, for by experience in the activities of the Commission as a procurement agency they had acquired knowledge about the ships, the equipment of the yards, and the possibilities of the shipbuilding companies as working, human organizations. To the extent that other procurement agencies were in a similar position, vertical allocation took advantage of their specialized competence. Other industries presented other problems, but from the point of view of shipbuilding it seems in retrospect inevitable that the procurement agencies should have become also the channels for allocation.

The general plans for the control of the economy added to the paper work of the Maritime Commission's staff, but PRP added only a little, relatively. The Priorities Section of the Maritime Commission attended to sending out the necessary instructions to the yards concerning how to fill out new forms, and it compiled reports received; [19] but most of the paper work under PRP passed between the yards and WPB and between sections of WPB. On the other hand, the Controlled Materials Plan was expected to call for considerable additional paper work and led to the forma-

[19] J. G. Conkey to All Buyers, July 15, 1942, in Historian's Collection, Research file 117 (Statistical files), " Priorities," and quarterly reports of Production Division for Sept. 30, 1942.

tion during the last quarter of 1942 of a Controlled Materials Branch in the Production Division.

One appealing characteristic of the Controlled Materials Plan was that it seized on just three materials: copper, aluminum, and steel—carbon steel and alloy steel being allocated separately—and regulated the whole economy through a close control of those three metals. For airplanes aluminum was decisive; for some munitions, copper; but for shipbuilding carbon steel was the determining factor in the sense that the consumption of the other two metals was so slight, relatively, that if enough steel could be found for a big program there would be little difficulty in finding enough aluminum and copper. The task of the Controlled Materials Branch was to make estimates for each quarter of how much of each of these three metals would be needed. The Branch was subdivided into five units, each working on one or more types of ships. Starting with the bill of materials and the information supplied them by the Technical Division on the amount of the three metals in the components and in the auxiliary equipment of various kinds, they compiled an estimate of the total weight of each metal in each type of ship. This provided a basis for checking and correcting the statements of requirements which the shipyards were sending in. These statements were edited and totaled up according to the number of ships expected per quarter. These estimates were made in rough form for a year and a half ahead, but revised quarter by quarter, since the allotments by the WPB were made quarterly.[20]

After a full set of branches had been formed to cooperate with the WPB, most of the personnel of the Production Division, 414 out of 544, was in its Production Engineering Section. The organization of the section is shown on Figure 30.[21] Its chief, J. T. Gallagher was named Assistant Director in January 1944 after Allen D. MacLean succeeded Willard Rockwell as Director of the Division.[22] Mr. MacLean had been acting as director a good deal of the time ever since he with Colonel Rockwell joined the Commission in April of 1942, for Colonel Rockwell worked for

[20] Quarterly reports of the Production Division for Dec. 31, 1942 and Mar. 30, 1943; *Industrial Mobilization for War*, I, 485-93.

[21] An earlier general description by Allen D. MacLean and organization chart is in the *Marine Engineering and Shipping Review*, Oct. 1942.

[22] MC Minutes, p. 27560 (Feb. 15, 1944).

THE SYSTEM IN OPERATION 379

the Commission on a per diem basis, kept his business connections, and was at his office at the Commission only a few days a week or less.[23] Mr. MacLean, who was an executive under Colonel Rockwell in the Pittsburg Equitable Meter Co., worked full time for the Maritime Commission both before and after he received the title of Director.[24]

VI. THE COMPLETED SYSTEM IN OPERATION

The system for regulating the flow of materials operated by the fully developed organization of the Maritime Comision and the WPB under the Controlled Materials Plan was as follows.[1]

A " Ship Construction Program of the United States Maritime Commission " by shipyards was compiled in the Statistical Analysis Branch on the fifteenth of each month, taking into account the latest decisions of the Joint Chiefs as to what was desired and the latest information from the Regions as to what was possible in the yards. When it had been approved by Admiral Vickery, the Ship Construction Program was circulated among the Commission's staff and in the WPB. In a condensed form of a summary by types, the Programs were systematically reviewed every quarter by the Joint Chiefs of Staff, but of course changes were called for also by the Joint Chiefs whenever they thought changes strategically desirable. If ships were scheduled for which no contracts had been awarded, the Commission's Committee on Awards arranged to negotiate the contracts. If the " Program " called for types of ships requiring turbines, the Survey Branch looked for means of supplying those turbines. Actually, both contracts and turbines were normally taken into account even before the " Ship Construction Program " was prepared. They were among the data considered by Admiral Vickery in giving the Statistical Analysis Branch the instructions which enabled them to draw up the programs.

Once the schedules for ships were worked out, the two essentials in scheduling materials were of course to know just what materials to order and at what stage in construction they would

[23] Statements by J. T. Gallagher, June 4, 1948.
[24] Files of Personnel Division.
[1] The following section was worked up by studying first the quarterly reports and administrative orders, by interviews, and by submitting a preliminary draft to persons who had personal knowledge of the process described.

be needed. Requisitions for those materials and components which were purchased by the Commission, having been approved by representatives of the Technical Division, were sent by the yards or the naval architects to the Production Division. There materials and components, other than steel plates and shapes, were broken down by the Hull Scheduling Branch into three groups according to their "lead times." Group I was to be delivered well before keel laying, Group II at about the time of keel laying, Group III between keel laying and launching. The tonnage of steel plates and shapes which would be required was calculated by the Statistical Analysis Branch. All this steel had to arrive before keel laying, in view of the large amount of pre-assembly, but how long in advance was one of the most debated problems of "lead times."

Having itemized the bill of materials, and classified it according to lead times, the Hull Planning Branch sent to the Procurement Division a "materials requirement schedule." This gave, for a whole group of contracts concerning the same type of vessel, the delivery time, month by month, for each group of material. The Procurement Division then wrote purchase orders, fixing the delivery dates roughly according to this schedule and issuing also appropriate priorities. Copies of the purchase orders were sent to the Hull Planning Branch, and for propulsion machinery to the Engineering Scheduling and Planning Branch. These Branches immediately posted the orders in allocation books and then wrote to the supplier "allocation letters" telling specifically which yard to deliver to and at what dates.

The Hull and Engineering Branches kept the allocation books in order to have a record of orders and deliveries. These books aimed to show whether for any reason Procurement had failed to place an order for any of the innumerable components, and whether the suppliers were filling the orders as instructed. But there was difficulty in keeping posted in these books the actual deliveries to the yards. Events moved too fast. Suppliers were ordered not to switch orders or ship to other yards than those indicated by the letters of allocation, although many of the shipbuilders sent out expediters to persuade suppliers that they needed the goods more than anyone else. Moreover Regional Directors could authorize transfers or "emergency purchases" by the shipyards. Out in the Regions, components were transferred from one yard to another, or were secured by the shipyards through "emergency purchases" authorized by the Regional Directors. When these transfers and purchases were not reported promptly to Washington or not recorded rapidly enough in the allocation books, components and materials were shipped as originally allocated to yards which had already been supplied by borrowing from other yards or by emergency purchases. Therefore "surplus" or "excess materials" accumulated in some yards while the same materials were urgently needed in other yards. Avoiding this situation, or remedying it where it arose, was made more difficult by the contest between the Regional Directors and the Production Division for authority in the matter, and for control of the material coordinators who were stationed in the shipyards to report on such matters. If all the components of a certain sort, for example all the lifeboats for a given type of ship, were used within the same Region, the Regional Director was allowed to do the allocating among "his" yards, and he was also held responsible for seeing that "borrowings" by one yard from another within his region were paid back. But when several Regions were involved there was no way to bring any measure of reason into the distribution except to have the allocation controlled from Washington, even if these allocations did have to be changed around many times at the last moment.

The scheduling based on the "Ship Construction Program" was mainly long range and was subject to much short-run revision. Short-run planning, for ships to be delivered within the next six months, was based directly on the "Monthly Report of Progress" sent in by the Regional Offices and edited and compiled by the Statistical Analysis Branch. The Regional Offices also sent in weekly reports on vessels on which delivery was expected in 45 days or 60 days. But the progress reports which came in on the first of each month were more complete, showing dates of keel layings, launchings, and deliveries for all vessels under contract, and were widely circulated in the Maritime Commission and WPB. Although they were supplemented by many telephone calls to the Regions and to yards, and by the activity of the expediters which the yards maintained in Washington, these reports from the Regional Offices were the general basis for short-range scheduling.

As the time of scheduled delivery drew near, the Hull

Scheduling Branch or the Engineering Scheduling Branch might find that the items they watched were more badly needed at some other yard than those to which they had been allocated. New letters of allocation were then sent.

Many goods were so scarce that no attempt was made to allocate them far in advance. They were handled by "spot" allocation, that is, the manufacturer wired or called up when he had products ready, and the Hull or Engineering Branch immediately or "on the spot," assigned the items to the shipyards needing them, and told the manufacturer, item by item, just where to ship.

Since it is impossible to deal fully with all the materials and components allocated, we will, at the risk of seeming to overemphasize its importance, give detail only about the monthly allocation of steel plate. This had two stages: the allocation by the WPB to Navy, Maritime, and other claimant agencies; and the allocation by the Maritime Commission among its yards. Action to secure allocations from the WPB began in the Statistical Analysis Branch which revised its long-range estimates of steel requirements on the basis of the latest reports, and compiled an estimate of what would need to be rolled by steel mills for the Maritime Commission in the month after next. For example, it submitted to WPB at the second half of June its estimate of the tonnage required from the August rollings of the steel mills.

If a quarter was coming to an end, as it would be in June, the Controlled Materials Branch was also making ready a new estimate of requirements. It made an estimate for the next three months based on the monthly estimates of steel-plate requirements, adding the carbon steel needed in other forms and the amounts of other metals in the components and equipment for a corresponding volume of production.

These detailed estimates of steel needs, together with rougher indications of plans for the next year or so, were then considered in WPB by a subcommittee, the Steel Division Requirements Committee, to which came also the requests for steel from the other six claimant agencies.[2] Since most of the Commission's requirement was for steel plate and the Commission consumed more plate than any other claimant, its affairs were at the center

[2] *Industrial Mobilization for War*, I, 446-48, 630.

when the distribution of plate was discussed. Usually its claims were presented by Mr. MacLean or William Floyd, his assistant, who was often accompanied by men from the Statistical Analysis Branch. The discussion in the Steel Division Requirements Committee sometimes reviewed the whole statistical process by which the estimates were arrived at, examining the amount of plate calculated for different types of ships, the lead times or the level of inventory considered justified, the actual size of inventories, and the allowances for contingencies. As a result the figures submitted by Maritime were sometimes reduced by what were called "statistical adjustment." The program was also reviewed in terms of the end product to see if more ships were provided for than had been approved by the Joint Chiefs, or whether plate was being demanded for ships for which other components such as turbines would not be available.[3] Since the total amount of steel plate being asked for was more than the supply, at least until 1944, the requirements stated by the claimant agencies had to be pared down somewhat, somewhere, and the subcommittee squeezed out as much padding as it could and made a recommendation to a higher committee, the Program Adjustment Committee of WPB.

The authoritative decision was usually formulated in the Program Adjustment Committee, which concerned itself primarily with the relative urgency for war needs of the programs of the various agencies. The Requirements Committee itself gave formal decision on the first of the month, in the example we have taken, July 1, but it had become "for many purposes simply an appellate body."[4] Debate did not entirely cease when the Requirements Committee made its decision, but was continued within the divisions of WPB while the decision was being implemented.[5]

The overall quarterly allotment of carbon steel came back to the Controlled Materials Branch and was then allotted by it to the yards, using the estimates they had sent in but screening them and cutting them down a good deal. When in doubt the shipbuilders would naturally take the chance of asking for too much rather than too little. The Maritime Commission could

---

[3] Bertrand Fox to J. A. Krug, June 9, 1943, and other copies of memoranda, in Historian's Collection, Research file 206.1.4.
[4] *Industrial Mobilization for War*, I, 439.
[5] *Ibid.*, I, 434.

prune off a good deal in making allotment for the quarter and see whether the yards came back for a supplementary allotment, as they could be expected to do if they found they really needed it. After the third quarter of 1943 the Maritime Commission never allotted to its yards the full amount which was allotted to it by WPB under the Controlled Materials Plan, for the Commission could not obtain assignment of specific rollings from specific mills up to the amount allotted under the Controlled Materials plan. Those in WPB operating the plan distributed among the claimant agencies about 10 per cent more steel than was being produced.[6] Consequently the Iron and Steel Division of WPB said they could not find room in the schedules of steel mills for all the allotment, especially not in the form of plate. So under the CMP, as before, the vital question for the Maritime Commission was not the overall allotment for carbon steel of all kinds but the monthly allotment of steel plate.[7]

When this vital monthly allotment of plate came back to the Commission from the Requirements Committee of WPB, the Statistical Analysis Branch of Maritime prepared a plan for distributing it among the yards. At this stage, since a new month had just begun, a new progress report from the Regions was being compiled and edited. The division of steel among the yards was made on the basis of this latest progress report, of the past records of the yards, and of the latest steel reports received from the material coordinators. When the Statistical Analysis Branch had worked out an allocation of plate yard by yard for the next month, it was sent to the Chief of the Production Engineering Section for approval.

At this stage there was a good deal of last minute negotiation between the yards and Mr. Gallagher, Mr. MacLean, and Admiral Vickery. Could the yards actually use as much steel as was indicated by the schedules they had sent in and which had been more or less discounted in the figures prepared by the Statistical Analysis Branch? Sometimes Admiral Vickery would decide to allot to a yard all the steel its managers claimed they had to have even though he thought they were overestimating their capacity, because he could then use the fact as a club over their heads, or as a point with which to needle them to greater efforts.

[6] See Fischer, "Statistical Analysis, Steel."
[7] Statement by Gallagher, June 11, 1948.

THE SYSTEM IN OPERATION 385

By such methods for getting under their skins he drew out of various managements the extra efforts on which the success of the whole program depended.

Once the allotments to yards had been approved, it went to the Steel Requirements Branch. Their job was to see that the steel plates and shapes allocated to the yards were actually included in the rolling schedules of the steel mills in the thousand different sizes and shapes required for the many different types of ships in the program. To do this they worked very closely with the Iron and Steel Division of WPB, a representative of which was given an office in the Maritime Commission while a representative of the Maritime Commission spent all his time in the offices of the Iron and Steel Division at WPB. The function of the Iron and Steel Division was to arrange the rolling schedules of the steel mills to accord with the decisions of the Requirements Committee. It sent to the Maritime Commission lists of steel mills and showed mill by mill the amount of rollings allotted to the Commission. Within these allotments appeared the orders which had already been placed with the steel mills either by the shipyards or by the Maritime Commission's Procurement Division. The Steel Requirements Branch was the unit in the Maritime Commission which received these schedules from the Iron and Steel Division. Steel Requirements took the mill schedules on the one hand and the allotments to shipyards on the other and assigned steel from particular mills to particular shipyards.

The assignments made by the Steel Requirements Branch were then sent to the Procurement Division. If the steel was being bought by the Commission, as so much of it was, the Procurement Division itself placed the orders for all the 700-odd different sizes and shapes of steel plate needed for the specific shipworths which the Steel Requirements Branch had ordered stipulated mills to ship to stipulated shipyards. Since these orders had all to be made out in a few days, mill orders were filled out ahead of time with everything except the name of the shipyard and of the steel mill.

At this stage the work had to move fast, not only in the Procurement Division but also in the pertinent Branches of the Production Division and of WPB. The decision of the Requirements Committee of WPB having been made on July 1, according to our example, the allocation to yards by the Statistical Analysis Branch had to be finished about the tenth. At that date also the Iron and Steel Division was supposed to send over the allocations on the mills. That left about seven days for Steel Requirements to match up yards and steel mills and Procurement to write the orders, for orders were supposed to go out on the seventeenth. Then they would surely reach the steel mills by July 25, and give the mills a few days in which to finish arranging their rolling schedules for August. If the orders did not get to the mills on time, WPB was provided with an alibi when the Maritime Commission complained of not getting its steel. To answer that the WPB had not sent its allocations on the steel mills over early enough, as was sometimes the case, did not make up for shipments lost.

While the Procurement Division was rushing in the orders to fill spaces reserved for the Maritime Commission in the schedules of the steel mills, the yards which ordered their own steel were negotiating with the Steel Requirements Branch and with the steel mills to make sure that their most needed orders were included in the mill schedule. If an order they had placed with one mill was squeezed out of its schedule, the best they could do was to arrange to have it substituted at some other mill. The yards knew better than anyone else which of their orders were most urgent.

A letter from the head of the Steel Requirements Branch (then Materials Requirements) to the head of a shipbuilding company which ordered its own steel is worth quoting at length because it describes so many of the problems handled by that Branch:

> As suggested in our telephone conversation yesterday we wish to outline a procedure for scheduling plates which will be of assistance in obtaining for you the orders which you require to meet your monthly requirements.
> First of all we wish to outline the mechanics by which the schedules are brought to their final approved form. Each steel mill makes its own preliminary schedule based on the PD298 forms which it receives. In doing so they schedule not only Maritime orders but also those of other agencies such as Army, Navy, Defense Plants, Lend Lease, etc. The completed schedule as prepared by the mill accounts for the mill's rolling capacity and the Maritime classification is only a part of it. After the preliminary schedule is completed according to the mill's convenience, it is sent to the War Production Board who break the schedule down according to the various classifications and send each separate part to the agency involved. We, of course, receive only the Maritime portion and can have no idea of what may be scheduled for other

agenices. The preliminary schedule is subject to revision by the War Production Board. This office cooperates with the Board with respect to Maritime orders only.

In revising the schedule we are first of all limited by the overall tonnage ceiling placed on the Commission, a fact which has forced us in past months to defer one-third or more of the request for plate for Maritime purposes. It is, therefore, important for us to know your rolling preferences because it is very possible that all orders cannot be included in schedule.

Since our schedule must be integrated with those of other agencies in order to balance the rolling capacities of the mills, we are sometimes limited in the amount of tonnage which can be included on any one mill. For this reason it is necessary for us to know either the dimensions of the plates, the mill order numbers, or the producing mills on which they are to be rolled in order that we may make some alternative arrangements for rolling, if this is possible.

The final revised schedule is completed under the above mentioned conditions. After approval this schedule is official and supersedes the preliminary schedule originally submitted by the mill. The mill is obliged to roll orders approved by the War Production Board without regard to their preliminary schedule.

* * * * *

After the schedules are revised you may find that the orders to be produced do not meet your needs exactly. Relief from this situation can be obtained by substituting orders in place of others currently scheduled, providing that these changes occur on the same mill of the same producer and do not interfere with mill conditions. It is our suggestion that you make arrangements for such substitutions directly with the mill and then notify this office regarding changes that can be arranged.[8]

Not only the long-range yards but all the shipbuilders were expected to keep track themselves of the hundreds of sizes and shapes of steel they needed, place orders so as to balance their stock, and check up at the mills to see that vital orders were not skipped over. A reason one yard operated with much lower inventories than others was that it had a good steel man who spotted deficiencies as they developed and filled them.

VII. Expediting

After all the allocating and ordering was finished there remained, for the thousand and one components as well as steel, the very important task called expediting. From the point of view of ordinary peacetime operations, expediting may be said to begin with charting promised deliveries, keeping up to date

[8] A. W. Larsen to Consolidated Steel Corp., Sept. 9, 1942 in MC gf 507-4-1-3.

on probable changes in delivery dates, and checking on actual deliveries.[1] During the war also, each yard did this for itself, more or less; but in addition the Maritime Commission's staff did it for Commission-furnished components and materials, and for items which were under general allocation. Charting and checking deliveries on hull items was done by the Hull Branch, on reciprocating engines by the Machinery Branch, and so on. If a needed component failed to arrive at a shipyard, or if a particular supplier seemed to be falling behind his schedule, someone in the appropriate Branch would get on the telephone to find out what the trouble was, or the Chief of the Production Engineering Section would take a hand, or a man would be sent to the plant to look for the cause of trouble and clear it up. This latter group of traveling trouble shooters formed the Expediting Branch.

Nearly all the men in this branch were away from Washington scattered over the country near the sources of supply. They were selected as persons experienced in factory operations with a special knowledge of one or another of the many products which went into a ship. As soon as Procurement had placed orders, the expediters were told to visit the plants to see whether the manufacturer was equipped to do the job, whether he could meet his schedules, and whether he had placed his orders and secured the priorities he might need to obtain his materials. The expediter's job was to break any bottlenecks he discovered, by visiting subcontractors, if necessary, or by securing changes in priorities or allocations. Thus the expediters on their own initiative furnished warning of where schedules might not be met, and at the same time they were kept busy responding to the appeals made to them by other branches of the Production Division to find out whether this or that firm could live up to its promises or why not. But the Expediting Branch was quite small compared to the volume of orders being handled; it contained 19 in December 1941 and about 120 at its peak in 1943.[2] Indeed, the Maritime Commission staff prided itself on the fact that its whole expediting organization, including those in various branches checking on deliveries, was relatively small compared to the number of Navy expediters.

As the various allocating branches and the Expediting Branch

---

[1] Charles R. Holton, "Procurement and Storekeeping," in F. G. Fassett, ed., *The Shipbuilding Business*, II, 134-35.

[2] Quarterly reports in Production Division Administrative file A 9-3.

checked week by week or day by day on the suppliers, there was much last-minute shifting around of orders, especially for the innumerable small items. If a manufacturer fell behind in filling an order for castings or refrigerator motors, for example, the Hull Planning Branch threatened to cut down the size of his order next time, and, when he promised big production, called on the nearest expediter to visit the plant and see whether the promise could be kept. Should the expediter's report be unfavorable, the head of the Hull Planning Branch would call up the Procurement Division and tell them to switch the orders from that manufacturer to someone else. In both the Procurement and Production Divisions there were men keeping track of all the suppliers of the product which was their specialty, so that if one seemed likely to fail to produce, his order could be placed elsewhere.

One way of speeding up deliveries was to obtain an emergency priority rating (AA, or later AAA) or a special scheduling order. This required appeal to the Special Rating Branch of WPB and that contact was handled for the Maritime Commission by the Priorities Section, but expediters often supplied the drive behind the claim for a particular AAA rating.[3]

The same tendency prevailed in handling steel, but in expediting steel there was also a brief appeal to private enterprise. After Colonel Rockwell became Director of the Production Division he proposed that the Commission employ the Robert W. Hunt Co. of Chicago to expedite. The steel crisis was then just about to come to a head with the cancellation the following month of the Higgins contract. As Colonel Rockwell saw the situation in June, the Commission's representatives only visited the offices of major producers each month to check on the schedules of deliveries. The Robert W. Hunt Co. already had experienced men in all steel companies, and by working part of their time for the Commission, they could be in every mill as often as once a week to check on rollings.[4] This arrangement was confirmed August 27, 1942,[5] but by October it was allowed to lapse.[6]

[3] *Ibid.*, report of Expediting Branch, Jan. 20, 1943.
[4] Rockwell to Vickery, June 15, 1942 in gf 507-4-1-3.
[5] Land to Robert W. Hunt Co., Aug. 27, 1942, initialed by Rockwell, in gf 507-4-1-3.
[6] Larsen to Robert W. Hunt, Oct. 5, 1942, saying he would inform them of any expediting desired thereafter. In gf 507-4-1-3.

Instead of employing a private firm, the Commission began organizing in September 1942 special regional offices for steel expediting. They were set up at Detroit, Oakland, Chicago, and Pittsburgh under A. W. Larsen, the Head of the Steel Requirements Branch. Being in constant contact with the mills they could see that the agreed-on schedules were adhered to, or report any failure by the mills to do so.[7]

Although administrative orders and administration charts emphasized functional divisions and separated scheduling, expediting, and priorities, there was a tendency for anyone who handled a particular commodity or component to follow through in regard to his specialty, checking orders, deliveries, and priorities, as need might be. The Survey Branch followed through on some turbines, but in regard to materials for copper bearings, the Expediting Branch kept track of the whole supply and formed a special unit to deal with WPB concerning this material. As electronic equipment increased in importance, the Priorities Section organized an Electronics Branch which was concerned at first with helping WPB set up schedules, and later undertook allocation to the yards and pushing deliveries. The experts on machine tools had for a long time formed another branch in the Priorities Section. Nearly all other branches called on the Expediting Branch for reports on conditions in the factories of suppliers, but even the Survey and Scheduling and Electronics Branches had a few men in the field.[8] Thus specialization on particular materials or components overrode functional divisions.

VIII. The Causes of Large Inventories

In spite of all the efforts of the schedulers and the expediters, the actual flow of materials was far from being perfectly adjusted to the needs of production. The statistical analysis reported in the last chapter shows that inventories in most yards were on the average well above the level at which some yards were able to operate effectively. Eight reasons for this difference deserve consideration: (1) differences in yard layout and management, (2)

---

[7] Rockwell to Vickery, Sept. 18, 1942, in gf 507-4-1-3; Quarterly reports for 1942, in Prod. Div. Adm. file A 9-3; Rockwell to Vickery Feb. 20, 1943 in gf 201-2-31.

[8] Quarterly reports, especially of Priorities Section in 1944, and interviews.

changes in the Commission's program, (3) changes made during construction in the specifications of ships being built, (4) strikes or other labor difficulties, (5) unforeseen deficiencies of management, (6) unusual weather, (7) failure of the steel mills to deliver as anticipated, (8) mistakes in scheduling arising from inadequate reports or from other administrative defects.

There was much difference from yard to yard in their facilities for moving inventory quickly, but an analysis of the inventories of steel plate at individual yards shows very wide fluctuations in *the same yard* at different times, and these fluctuations are not merely those which accorded with higher or lower rates of production. Clearly therefore the differential in inventory between the average for all yards and the average minimum standard was not due simply to differences in layout and management between the average and the yards setting the standard.

Changes in the Commission's program on a number of occasions had the result of making materials accumulated at one yard become suddenly excess inventory. For example, the Kaiser Vancouver yard was given a big contract for Libertys and then was changed over to the construction of military types. Many shipsworth of steel plates of the sizes and shapes required for Libertys had been received before the changeover and were transferred to nearby Oregon Ship, thus swelling its inventory temporarily.[1] Another example, complicated by the number of agencies involved, was the steel for the LST's built at Bethlehem-Fairfield. In June 1942 the yard was given a contract for 45 of these landing ships; in October it was cut to 30, according to a request made by the Navy in September when the program was in full swing. Steel for all 45, having been allocated before the cancellation was decided on, continued to arrive. Letters by company officials addressed to the Maritime Commission, the Navy (which handled procurement for the LST's) and the WPB failed to stop the flood of steel piling up on the siding of the shipyard—until finally an official of WPB called up very irately and denounced the company for thus hoarding steel.[2]

[1] "Semi-annual report of plate steel forms PD-299," by Dougherty, Material Requirements Branch, Jan. 18, 1943, in Production Division Statistical files, Inventory (Plates and Shapes).

[2] Helen E. Knuth, " The Building of LST's by the Maritime Commission," a typescript in the Historian's Collection, Research file 111-8-3, pp. 9, 13. Cf. Reynolds' Way Charts for building dates.

Another reason for excesses was change in design which occurred after construction had begun and forced the builder to wait for new designs. This occurred mainly in vessels which were being built for the Army or Navy, or in vessels which were taken half-finished by the Navy and converted to Navy auxiliaries. In such ships, changes to meet military needs interfered with standardized or prearranged procedures. A striking example is the C4 type of transports. These design changes account in part for the huge inventory accumulation at Richmond No. 3 although over-optimistic scheduling also contributed. In general this difficulty affected the long-range yards most, for they built most of the special and military types, at least until 1944. It is one main reason why inventories were higher in these yards than in any other.[3] See Figure 29.

There are no clear cases in which strikes alone caused a big piling up of inventory, but the combination of labor trouble and deficiencies of management had that effect at South Portland in late 1942.[4] In that year a good deal of steel was taken from Eastern and Gulf yards and shipped to the West coast—four shipsworth were sent from Portland, Maine to Portland, Oregon—for the production in a number of East coast and Gulf yards was not up to expectations, whereas West coast yards were ahead of schedule.[5] More experience with different yard managements gradually provided a better basis for estimating correctly the needs of the yards.

Quite unpredictable, on the other hand, was the hurricane in the Gulf which inspired the Regional Director to wire Vickery that they had missed their schedule by five ships because his staff had not succeeded in controlling acts of God,[6] or the solid week of rain at Portland, Oregon, in January 1943,[7] or the extreme cold at South Portland, Maine, in January and February 1943.[8]

[3] See chap. 16 and Fischer, " Statistical Analysis, Steel."

[4] Blanche D. Coll, "Labor in South Portland," typescript, Oct. 1, 1946, in Historian's Collection, Research file 112.3.3. Cf. below, chap. 15.

[5] Statement of H. H. Hile, Nov. 17, 1948; Rockwell to Fesler, Aug. 13, 1942, and Vickery to Eberstadt, Dec. 16, 1942, in gf 506-1-1-21; Vickery to Flesher, Oct. 1, 1942, in gf 201-3-4.

[6] Gulf Coast Regional Construction Office, Minutes of the Staff Meeting Held Oct. 2, 1943.

[7] Weekly Steel Report, Plates, shows erection for the week of Jan. 23, 1943 as 5,000 tons whereas the preceding week it had been 12,000 and the two weeks following was 7,000 and 10,000 respectively. In Production Division Statistical file, Oregon, Plates and Shapes.

[8] Weekly Steel Report, Plates, shows erection of less than 5,000 tons in January

Even if the needs of the yards were successfully guessed, inventories could accumulate if the steel mills delivered some of the orders included in allocations, but not others. Sometimes the mistake was made of placing an allocation on a mill which could not handle that particular kind of plate or shape.[9] To a certain extent the mills were sure to omit some orders because WPB followed the policy of allocating from each rolling mill 5 or 10 per cent more than they would probably produce.[10] In part this was done to spur the mills to maximum effort and to make sure that they would not lack work even if they happened to have a very good month and produced more than anticipated. In part it was done for the sake of flexibility.

A mill which had this degree of flexibility could combine its orders in a way which would use the mill's equipment more efficiently. There were 43 items, for example, which the mills at one time said were in quantities so small that they had to either run a 6 months' supply all at once, or roll none. Actually they rolled some and not others. Then the yards made up the shortages in such small items, especially in small plates, by cutting up large plates of which they had a supply. " Such discrepancies are necessary to get out ships but they do create discrepancies in inventories," wrote Rockwell in his reply to the Ezekiel report. And he complained that the Maritime Commission was often unaware of these deficiencies and so could not take steps to correct them.[11]

Since the mills had more orders each month than they could fill, they had some power of decision as to which ones to omit. Whatever orders were carried over from one month to the next were supposed to be rolled first thing the next month, but this

and February, whereas during November and December it was 7,000. In Production Division Statistical file, New England, Plates and Shapes.

[9] For an example affecting Sun Ship see MC Minutes, p. 23526.

[10] *Industrial Mobilization for War*, I, 494, 633-34. In June 1944, it appeared that the carryover of second quarter orders for steel plate into the third quarter would be approximately 15 days rolling and would remain about the same at the end of that quarter.—Bertrand Fox to S. W. Anderson, June 27, 1944, in Fischer's file, now in MC Historian's Collection, Research file 206-1-4. On Feb. 8, 1943, Rockwell wrote Eberstadt that he found the average carry-over in the last 7 months was 3 per cent and therefore he was increasing his estimate of requirements to allow for this factor.—In gf 506-96. In the ensuing months the carry-over was far larger than that, as is shown on Chart 5 in Fischer, " Statistical Analysis, Steel," by the difference between the " Mill Shipments " and " Total Orders."

[11] Rockwell to Fesler, Aug. 13, 1942, in gf 506-1-1-21.

was not always done.¹² Priorities and allocations were sometimes in conflict. Under these conditions the steel mills retained considerable discretion concerning the distribution of their output and could use it in ways not anticipated by the schedulers. It was believed that they made better deliveries for the standard yards which they considered regular customers than they did for Liberty yards.¹³

One policy of the Maritime Commission had a similar influence. In allocating among the yards, the Commission's practice was to meet the needs of the yards building standard and military types and then divide what was left among the yards producing Libertys.¹⁴ Consequently, restriction on steel meant more reduction in deadweight tonnage than would have occurred if a different policy had been followed. If the only objective had been to obtain a maximum deadweight tonnage in ships per ton of steel plate consumed, another policy would have served better. But the demand for tankers, for military types, and for the C-ships was sufficient so that the Commission felt justified in supplying first the yards producing them. They were in direct demand by the Armed Forces as auxiliaries, and as merchant ships they were more valuable than Libertys because they were faster.

All of these reasons why the flow of materials was not perfect might be considered mistakes in scheduling in the broadest sense. They were all matters which an absolutely omniscient scheduler would take into account. If he knew everything, he would allow in his allocations for the differences in yard layout, for changes in the program, for alterations in specifications, for labor troubles and management troubles, for tornadoes and for the failures of the steel mills to deliver all orders. Practically, that was of course impossible, even if the individuals doing the scheduling had been the smartest human beings conceivable. But in addition to the inevitable failure to foresee everything, there were some failures inherent in the administrative setup, especially the long dispute over supervision of the material coordinators. Sometimes a company was allocated more steel than it was really expected to use

---

¹² Rockwell to Fesler, Aug. 13, 1942, in gf 506-1-1-21.
¹³ Nelson, *Arsenal of Democrcay*, p. 249.
¹⁴ Statements by G. J. Fischer and the letter of Rockwell to Land (written by Fischer), July 1, 1943.

CAUSES OF LARGE INVENTORIES

because Admiral Vickery decided that was the best way to handle a particular team of managers. And at the statistical level there were failures to transmit swiftly among the decision-making groups the most accurate possible reports. Looking back at the problem, those who were in the thick of it say that if they had to do it again they would certainly try to set up at the start more uniform and more adequate methods of keeping track of inventory.

CHAPTER 12

INCREASING THE SUPPLIES OF COMPONENTS

I. Propulsion Machinery

ALTHOUGH shipbuilding is often thought of as a seaboard industry, of interest only to the nation's coasts, actually it demands products manufactured all over the country. The goods flowing to the shipyards were about as varied as the equipment needed to supply any small community. A Liberty ship needed about 7,000 items, which included curtains, medical supplies, kitchen equipment, bolts, screws, and so on. Since only a fraction of the national output of such items was bought for the shipyards, there were few special problems in obtaining supplies of this kind. Manufacturers' representatives came to the shipyards and to the Maritime Commission seeking orders, especially when priorities made it more and more difficult to produce goods for ordinary civilian use. At the same time, the buyers for the Commission and the shipbuilders learned of new possible suppliers through their many industrial contacts, through chambers of commerce, and through their expediters in the field. Thus business pressures arising from the usual efforts of private enterprise produced adequate supplies of such items.

Other and very important components, such as marine engines, anchor chain, booms, and compasses, came from specialized producers whose only market was in the shipbuilding industry. Expanding the plants of these producers was just as essential as extending facilities of the shipyards. Consequently the Navy and the Maritime Commission together stimulated an investment of $406 million in additional plants for the production of components (Table 5).

The Maritime Commission was particularly active in stimulating the production of propulsion machinery. The first step in this field was to revive the manufacture of triple expansion marine reciprocating engines. Production of these engines for the Liberty ships was arranged early in 1941 without building new plants, because this type of engine could be built with the equipment available. After the acceleration of the program, efforts were made to enlist Ford and some others in this work, and the eleven firms already under contract expanded their output.[1] Since the industrial capacity of the nation was fully

TABLE 5

NEW FACILITIES AUTHORIZED FOR SHIPYARDS AND FOR PRODUCTION OF SHIP COMPONENTS

PUBLICLY AND PRIVATELY FINANCED, JULY 1940-JUNE 1945
(Projects of $100,000 and over)

Type of Facilities	Cost (in millions)		
	Public	Private	Total
Grand Total	$2,157	$243	$2,400
Shipyards			
Total	1,877	117	1,994
Navy	1,376	91	1,467
Maritime Commission	499	17	516
War Department	2	9	11
Producers of Components			
Total	280	126	406
Turbines and Gears	127	56	183
Marine Diesels and Parts	78	25	103
Valves and Fittings	20	12	32
Instruments	14	2	16
Reciprocating Units	9	7	16
Deck Equipment	5	6	11
Generator Sets and Bearings	5	1	6
Anchor Chain	3	3	6
Marine Boilers	.2	4	4
Shafting for Ships	4	1	5
Propellers	1	2	3
Miscellaneous Ship Components	14	7	21

Source: Chaikin and Coleman, *Shipbuilding Policies*, p. 58.

employed in 1942, new plants were called for. The Maritime Commission gave contracts enabling General Engineering Co., the pioneer in the manufacture of this engine, to raise its production from 18 a month to 25, and enabling the Joshua Hendy Iron Works to increase its facilities and raise its output from 10 a month to 30, thus becoming the largest producer.[2] Before

[1] H. L. Vickery, "Shipbuilding in World War II," *Marine Engineering and Shipping Review*, Apr. 1943, p. 184, refers to 11 firms but A. D. MacLean, "Functions of the Maritime Commission Production Division" in *Marine Engineering and Shipping Review*, Oct. 1942, p. 170, says 14 companies.

[2] MC Minutes, pp. 21691-93, 21136, 21211, 21950-51, 22159, 22415-17.

the emergency program there had been no demand in the United States for this type of engine so that the industry may be said to have grown from producing nothing to producing at a rate which supplied the thirteen hundred engines of this type in ships delivered in 1943.

Since these reciprocating engines could not be used in fast merchant ships, provision was made as early as June 1941 to expand production of turbines and gears. The machine tools necessary, especially the gear hobbers, were so scarce that new ones had to be built. Consequently contracts were made providing for large new plants to be constructed by the General Electric Co. at West Lynn, Masachusetts, and the Westinghouse Electric Co. at Lester, Pennsylvania.[3] Smaller contracts made later with six other firms [4] and gear hobbers brought over from England added to the capacity of the industry.[5] There were many irritating delays—from strikes,[6] from the contest with the Navy for turbines and especially for the machine tools for use in the new plants,[7] and from difficulties with the British gear hobbers. By the fall of 1943, however, it could be said that no yard was being held up because of lack of turbines and gears.[8] Peak production was reached in 1944 when the industry produced altogether 1,097 main propulsion turbines, about half for the Navy, half for Maritime. This contrasts with an output in 1940 of only 68 for both.[9] In addition, since gears to go with the turbines formed the tightest part of the bottleneck, turbo-electric units were used.[10] The heated controversy which developed over standardizing turbines will be described in chapter 17.

With the steam turbine and engine production strained to

[3] MC Minutes, pp. 17702, 17805, 17938, 18182, and 18264. The last action, approving financial arrangements with the Defense Plant Corporation, was on July 29, 1941.

[4] *Ibid.*, p. 18449 (Aug. 12, 1941—Allis Chalmers); p. 19102 (Sept. 30, 1941—DeLaval); p. 19593-94 (Nov. 4, 1941—Hendy); and Table 6 below.

[5] E. R. Wisner to W. H. Sullivan, Feb. 6, 1943 in MC gf 500-3-1.

[6] House, *Independent Offices Appropriation Bill for 1943, Hearings*, pt. 1, p. 264.

[7] On the contest with the Navy for turbines, see Land to the President, Oct. 17, 1941, and Land to Admiral Stark, Dec. 30, 1941, both in the Land file. On the contest for machine tools, see W. F. Rockwell to Nelson, Aug. 22, 1942, in gf 507-4-1-3.

[8] Quarterly reports of Survey Branch, in Production Division Administrative file A 9-3.

[9] Chaikin and Coleman, *Shipbuilding Policies*, p. 62.

[10] Table 6.

the utmost, Diesel engines were substituted for main propulsion wherever possible in new designs. Producers of that type had begun expanding their facilities before Pearl Harbor and afterward accelerated their expansion;[11] but Diesel-engine production for the Navy was twenty-four times that for the Maritime Commission in the spring of 1944, although the Maritime had tripled its requirement during the last year.[12] Among manufacturers of

TABLE 6

MAIN PROPULSION TURBINE PRODUCTION FOR MARITIME COMMISSION VESSELS, 1944

(Figures denote delivered units)

PLANT	UNITS		
	Turbines	Gears	Turbo Electric
All plants.....................	513	585	184
G. E. Co. Erie Defense Plant.........	183	...	...
G. E. Co. Schenectady Works.........	...	...	78
G. E. Co. Lynn Defense Plant.........	...	137	...
G. E. Co. Lynn River Works.........	33	42	82
Westinghouse, Lester, Pa............	198	165	...
Joshua Hendy.....................	53	53	...
Allis-Chalmers.....................	31	...	...
Falk Corp.........................	...	103	...
Elliott Co.........................	...	...	24
DeLaval...........................	15	15	...
Koppers Co., Baltimore.............	...	70	...

Source: Compilation of C. D. Vassar for the Maritime Commission in 1948 from the records of the Production Division.

Diesels, the Nordberg Manufacturing Co. of Milwaukee was outstanding in supplying main drives for the Commission's shipyards, and also for producing the uniflow engines for the aircraft carriers built at Vancouver.[13]

II. Financing Suppliers

This expansion in the production of main engines was to a very large extent financed by the Federal Government, as was the expansion in the plants of many other suppliers of components.

[11] Chaikin and Coleman, *Shipbuilding Policies*, pp. 64-67.
[12] Quarterly report of the Survey and Scheduling Branch, Apr. 28, 1944, in Production Division Administration file A 9-3.
[13] MC Minutes, pp. 22371-72; Anderson to Robert E. Friend, President of Nordberg Manufacturing Co., July 1, 1942, in Anderson reading file concerning how the uniflow engines would be paid for.

Some corporations built for themselves the new plants they needed and asked only for a "certificate of necessity" which, by declaring the investment necessary to the defense or war effort, enabled the manufacture to amortize it in five years in his tax returns.[1] The Maritime Commission through its Production and Procurement Divisions recommended to WPB the grant of such certificates and so encouraged suppliers to build for their own account the needed additions to their facilities.[2]

The largest new plants were built through the Defense Plant Corporation, however, under a plan by which the government kept title to the plant and leased it for use during the emergency to the company that had built it. Leases generally provided a clause giving the lessee an option to purchase the plant at cost plus interest and minus depreciation, or at a negotiated price. Under this system of governmental financing, the government took all the risk of putting money into factories which might not be wanted after the war was over. By leasing the plants at low rates it prevented the cost of the plant, which might be very high in view of the short period of its use, from raising the wartime price of the products.[3] Giving to the same company both the contract to build the plant and the lease on it for the length of the emergency, as well as an option to purchase it later, made the operating company feel that it owned the plant. That was the general attitude; Maritime Commission records commonly refer to the leasing arrangements made by suppliers with the Defense Plant Corporation as loans or advances.[4]

The activity of the Defense Plant Corporation (DPC) in providing fixed capital for industrial expansion was initially limited to activities associated with the manufacture of "arms, ammunition and implements of war,"[5] but a law of June 10, 1941 permitted its application for all goods "necessary to the national

---

[1] Ethan P. Allen, *Policies Governing Private Financing*, pp. 23 and 89.
[2] Statements by J. G. Conkey, Dec. 6, 1948.
[3] White, "Financing," pp. 175-77.
[4] For example, "loan" is used in the "Explanatory Note" to the list compiled by C. D. Vassar, "Defense Plant Corporation Projects (Plancors) sponsored by the U. S. Maritime Commission," although it is explained that they are lease agreements. Office of Director of Construction, Feb. 21, 1946. Photostat in Historian's Collection, Research file 207.6.2. In House, *Independent Offices Appropriation Bill for 1946, Hearings*, 79th Cong., 1st sess., pp. 584-85, they are called "Advances."
[5] 54 Stat. 573-74.

defense."[6] Two days later, June 12, the Commission voted to approve the plans of General Electric which intended to use this method to finance construction of a new plant for making a hundred turbines a year.[7] Altogether the Commission sponsored 62 such projects to the DPC. Their total value was over $103 million. About $69 million of this went to the six turbine manufacturers, including Hendy; about $7 million to shipyards (for ten of the smaller yards used this form of financing); and the rest, $27 million, went to forty-six manufacturers of various sorts.

When the Maritime Commission sponsored a plant to the DPC it agreed that on the termination of the lease agreement and at the demand of the Defense Plant Corporation the Commission would request from Congress appropriations to reimburse the DPC for the unpaid balance, plus $1\frac{1}{2}$ per cent interest.[8]

In addition to capital for new plants, the Maritime Commission was also involved in providing working capital. Companies were undertaking to do business on so large a scale that they asked for advance payments. Gibbs & Cox as procurement agent had provided for such advance payments in many cases.[9] Prime contractors made advances to their subcontractors and then looked to the Commission for help.[10] Some companies received advance payments from the Army and Navy as well as from the Maritime Commission, each advance being subject to restrictions connected with the purchase which was the basis for it. From the contractor's point of view this led to awkward restrictions on his use of funds; from the Commission's point of view, it was increasingly difficult to determine the necessity for cash advances or the conditions which should be attached to them. Some way of coordinating the activities of the different procurement agencies in providing working capital was needed, as Mr. Anderson, the Director of Finance, expounded in a letter which Land sent to the Chairman of WPB on March 7, 1942.[11] After various conferences a method was provided March 26, 1942, by Executive Order No. 9112 and the regulations issued thereunder by the Board of Governors

---

[6] 55 Stat. 249-50; White, "Financing," p. 180.
[7] MC Minutes, p. 17702.
[8] See lists cited above in note 4, and e. g., MC Minutes, pp. 22371-72.
[9] Anderson to Director of Construction Division, Sept. 18, 1941, in gf 201-2-31.
[10] Anderson to C. E. Walsh, Feb. 20, 1942, and Anderson to File, May 23, 1942, both in Anderson reading file.
[11] In Anderson reading file.

of the Federal Reserve System. Administration of short-term financing of war contracts was placed as far as possible in the hands of the ordinary commercial bankers. Thereafter bankers made the loans, but only after the manufacturer had obtained from the Maritime Commission or some other procurement agency a Production Certificate stating that his contract was essential, that he could produce, and that the procurement agency would guarantee a percentage of the loan. The loan would be repaid, as advances had been, out of the progress payments which the Commission was to make to the manufacturer as his work progressed. This was the system called V-loans. The Commission took the risks but the administration of the loan was left to the banks.[12]

Mr. Anderson praised the system of V-loans because it brought in the local banks in such a way as to make them " interested guardians of the whole situation." [13] The banks were not strict enough, however, to suit all the Maritime Commissioners. Within three months the Commission had guaranteed thirty loans totaling $14 million, and on the average the Commission gave its guarantee to a very high percentage of the loan, as Commissioner Woodward noted in a memorandum in which he suggested that all these guarantees be approved by Commission action. The Director of Finance proposed a compromise, which was adopted, according to which all loans in excess of $250,000 and all where the guarantee exceeded 90 per cent were referred to the Commission.[14]

In passing on loans the Commission soon began laying down conditions. On September 28, 1942, it voted the general policy that firms for which V-loans were guaranteed should pay no dividends and pay no salaries over $25,000 a year. If a company was so short of working capital as to need a V-loan, how could it afford such payments? But this policy was adopted at a meeting

[12] American Bankers Association, *Special Bulletin* 85. Executive Order 9112 was a result of the letter of March 7, from the Chairman of the Maritime Commission, according to Commissioner Woodward's memorandum in MC Minutes, pp. 22811-12. The specific plan was worked out by Col. Paul Cleveland.—Anderson to Lane, May 31, 1950.

[13] Anderson to the Commission, May 9, 1942 in Anderson reading file.

[14] MC Minutes, pp. 22811-12 (Aug. 25, 1942). Approval of guarantees of $100,000 or less up to 90 per cent could be given by the Regional Offices.—Anderson to Regional Directors of Construction and Regional Construction Cost Auditors, May 27, 1942 in Anderson reading file.

of the Commission attended by only three of the Commissioners, Mr. Woodward, Captain Macauley, and Mr. Carmody.[15] Its wisdom was disputed by the Director of Finance who declared it contrary to the fundamental principle of V-loans, namely, " the credit facilities and the business judgment of the banking institutions of the country are made available to the War and Navy Departments and the Maritime Commission, the banks being interested parties because of their participation in the credit to the extent of the unguaranteed portion thereof." Only where the Commission guaranteed 100 per cent of the loan should it pass on the salary and dividend policies of the companies, he contended; in other cases these matters should be left to the judgment of the local banks. If the policy voted was rigidly followed, he said, firms would do business only with agencies having " a more realistic policy." On October 25, 1942, at a meeting attended by all five Commissioners, a move was made toward putting the policing of the loans back in the hands of the local banks, for the Commission approved a letter to the War Loans Administrator of the Board of Governors of the Federal Reserve Banks asking that the limitations on dividends and salaries be made part of his policy.[16] Banks did make such restrictions, yet the Commission in a number of instances later added " boiler-plate " clauses limiting dividends and salaries.[17]

Even V-loans did not solve all the needs for working capital; advances were still necessary occasionally. The Commission was willing to go a long way to prevent a firm which was building badly needed engines from closing down because of lack of cash. Accordingly, the Commissioners held a special meeting March 11, 1943 to authorize payment of a voucher to the Harrisburg Machinery Co. in an amount not to exceed $40,000 to meet payroll for another week.[18]

All the Maritime Commission's methods of financing vendors

[15] MC Minutes, p. 26343.
[16] MC Minutes, pp. 26592-95 (Oct. 25, 1942), where Anderson's memorandum of Oct. 9 is recorded. For letters of Anderson to the War Loans Administrator, see also Anderson reading file, Sept. 22, 1942 and Oct. 5, 1942.
[17] MC Minutes, pp. 27257-58 (Jan. 11, 1943); pp. 24936-40 (Apr. 22, 1943); pp. 26221-22 (Sept. 9, 1943); and Anderson to Land, Oct. 20, 1943; Anderson to Commission, Oct. 19, 1942 and Dec. 1, 1942; Anderson to Schell, Dec. 2, 1942; all in Anderson reading file.
[18] MC Minutes, p. 24564 (March 11, 1943). Cf. on the same company MC Minutes, p. 25092.

were involved in the expansion of the biggest West-coast manufacturer of engines, the Joshua Hendy Iron Works. In addition to some V-loans and to financing through the Defense Plant Corporation, and in spite of the fact that some Commissioners thought Hendy was receiving too high a price per engine, a direct loan of $10 million was made to that company, July 26, 1943, to keep it going while a study was made with a view to its financial reorganization.[19] In one form or another the Maritime Commission was then responsible for $35 million which had been invested there by the Federal Government. The expert who analyzed the company's finances for the Commission recommended changes both in its financial policy and in its top financial personnel.[20] At a special two-hour meeting devoted solely to this subject, the Commissioners discussed his report and then called in the chief executives of the Hendy company. Admiral Land then told them in no uncertain terms on what conditions the Commission would renew its loan to the extent required, namely $7 million, and the company officials agreed.[21]

There were quite a number of other cases also in which the Maritime Commission became deeply involved in the affairs of the manufacturing companies. Being urgently in need of their products, the Commission loaned them money, and having loaned the money it assumed more and more control over the management.

III. The Case for Small Business

For the manufacture of main engines the Maritime Commission necessarily turned to the large corporations experienced in that kind of work. Components less difficult to manufacture could be ordered either from a few large firms or from small businessmen. There were some forces pushing towards one method, and some forces pushing the other way. In favor of concentrating the orders in the hands of a few firms was the saving of time and personnel involved. It might not take much more time and labor

---

[19] MC Minutes, pp. 22324-25 (June 26, 1942); pp. 23614-15 (Nov. 27, 1942); pp. 23883-85 (Dec. 29, 1942); pp. 25598 (July 1, 1943); p. 25709 (July 15, 1943); pp. 25730-32 (July 20, 1943); and Anderson to Quarg, July 26, 1943 in Anderson reading file.
[20] MC Minutes, pp. 26917-20 (Nov. 25, 1943).
[21] Ibid., pp. 27032-34 (Dec. 8, 1943).

to send out 600 invitations to bid instead of 20, but it would take much more time to analyze the bids, especially if a serious effort was made in each case to determine whether the manufacturer could actually both meet the specifications and deliver on schedule.[1] From the point of view of the officials charged with procurement it was easier to do business with established firms whose products were well known. The pressure of time made them inclined to give repeated orders to the same firm or small list of firms, provided the work was satisfactory, and so did the pressures which might be called goodwill, corporate or personal.

On the other hand, pressure to spread orders among many suppliers arose out of the need for having more output. For example, the American Hoist and Derrick Co. of St. Paul developed the manufacture of certain kinds of winches but was unable to supply as many as were needed. It suggested two other Minnesota firms, the Appleton Iron Works and Beloit Iron Works, a manufacturer of paper-making machinery, and supplied them with engineering services.[2] Having a good number of suppliers had the advantage also of flexibility, especially in the production of interchangeable components. If one supplier fell behind on deliveries, another might be able to make up for it, and thus the flow to the shipyards kept up.[3]

These pressures in favor of dispersion did not go nearly far enough, however, to satisfy the champions of small business. In its defense the Commission said in May 1943 that 75 per cent of its contracts for procurement of material was with firms having fewer than 500 employees; [4] but its critics said that the better measure was the value of the contracts. By value, only 36 per cent of procurement contracts placed in January-June of 1943 were with firms having 500 employees or less. Counting all contracts, including those with shipyards, only 16 per cent of the total value was placed with these small applicants.[5]

[1] Senate, *Problems of American Small Business, Hearings,* 78th Cong., 1st sess., pt. 21, p. 2977.
[2] Statements by J. G. Conkey, Nov. 22, 1948; and Historical Notes by W. G. Esmond, in Historian's Collection, Research file 111, Liberty Ship.
[3] On flexibility and interchangeability see H. L. Vickery, " Shipbuilding in World War II," *Marine Engineering and Shipping Review,* Apr. 1943, p. 185.
[4] Senate, *Problems of American Small Business, Hearings,* pt. 21, p. 2959.
[5] *Ibid.,* p. 2936; and " Monthly Report on Procurement Distribution to Smaller War Plants Corporation," Report No. 1, Photostat in Historian's Collection, Research file 207.5.1.

Those who feared that the bigness of "big business" would be increased by wartime procurement had enacted their policy into law in June 1942. Its enforcement was watched over by a Smaller War Plants Division of WPB, which was linked to a Smaller War Plants Corporation, and by a Senate special committee under the chairmanship of Senator James E. Murray.[6] The Maritime Commission was presented by the Smaller War Plants Corporation with a list of 6,500 companies which might be able to fill some of its orders, but Maritime placed with the companies thus designated only one tenth of 1 per cent of its total expenditures (for October 1, 1942-January 31, 1943). Of the total business awarded by all services through Smaller War Plants Corporation before May 1943, the Maritime Commission furnished only seven-tenths of 1 per cent. Desiring more co-operation, the Smaller War Plants Corporation in March 1943 asked the Commission to create within its Procurement Division special units to see that orders were placed with small businesses, presumably by culling over the list of 6,500 firms and investigating prospects.[7] Mr. Walsh, Director of the Procurement Division, advised against such a step and it was rejected at a Commission meeting of April 13, 1943. Instead, one man was to be appointed in each Regional Office to act as liaison.[8] This did not satisfy the Smaller War Plants Corporation, which made public its complaints against the Commission at hearings held by Senator Murray's Special Committee to Study and Survey Problems of Small Business Enterprises. Admiral Land was chided for not having fulfilled the mandate of the law " to mobilize aggressively the productive capacity of all small business concerns. . . ." His defense was that he had to consider not only helping small business but also the need for economy and speed—above all speed —with a limited staff.[9]

The hearings in May 1943 also aired two other grievances which the spokesmen of small business had against the procurement policy of the Maritime Commission. They complained that, whereas the long-range yards managed their own procurement

---

[6] *Industrial Mobilization for War*, I, 527-32, 573-74.
[7] Senate, *Problems of American Small Business, Hearings*, pt. 21, pp. 2936-42.
[8] MC Minutes, p. 24841-42.
[9] Senate, *Problems of American Small Business, Hearings*, pt. 21, pp. 2936-42, 2954-89, especially pp. 2953, 2981.

for most materials, the Commission did nothing to make these yards spread their work among small businesses.[10] On the other hand, even louder complaints were made against instances in which the Commission had taken procurement out of the hands of the yards and centralized it in Washington or in the hands of one agent who bought all the materials for one type of ship. These complaints in May came from Duluth, Minnesota, from firms there which had been supplying the local shipyards—Walter Bulter, Barnes Duluth, Globe—and found their business gone when all procurement for the frigates, which those yards began building in 1943, was placed in the hands of Kaiser Cargo, Inc., of Oakland, California. Kaiser made arrangements with West coast firms so that life boats, for example, made from steel sent from the East, were manufactured in California and hauled over the transcontinental lines from California to Duluth. Procurement for the new contracts for coastal vessels also was taken out of the hands of the local shipbuilders and was handled by the Procurement Division in Washington. Duluth manufacturers complained that although the Navy had an office in Duluth, the Maritime Commission practices made it necessary for them to come to Washington, often unsuccessfully.[11] Admiral Land voiced suspicion that people in Superior had had a middleman organization to get a rake-off, but confessed he was " in a mystic maze " after a personal investigation of that angle.[12] Better control of prices and greater speed were the two reasons given for centralizing procurement.[13]

Admiral Land's defense of the Commission's procurement practices at these hearings in May 1943 included a promise to try to do more for the smaller concerns. At meetings later in May and in June, the Commission voted to set up Small Business Sections in the Procurement Division and the Regional Offices.[14] The *Report to Congress* for the fiscal year 1944 said that during the first three quarters of the fiscal year contracts with small war plants represented more than 40 per cent of the total number of purchases. Thereafter, such contracts were handled through a

---

[10] *Ibid.*, p. 2937.
[11] *Ibid.*, pp. 2865-2952, testimony of many witnesses, on May 4, especially.
[12] *Ibid.*, pp. 2982-83.
[13] *Ibid.*, p. 2886 in testimony of H. W. Clark, and *passim*.
[14] MC Minutes, p. 25272 (May 27, 1943); p. 25355 (June 4, 1943); pp. 25988-89.

representative of the Smaller War Plants Corporation who worked in the Commission's Procurement Office.[15]

Attempts to place orders with small business men were at least sufficient to add to the headaches of the Production and Procurement Divisions. Some new suppliers found in this way proved very satisfactory to be sure—for example, one stove manufacturer who turned to making life boats—but others did not meet schedules.[16]

IV. Inspection and Conservation

Many other headaches also arose from the attempts to increase supplies suddenly and enormously by placing orders with unknown firms. Inspection became a troublesome problem. The general policy of the Maritime Commission throughout its activity was to follow standard commercial practice. In sharp contrast to the Navy, which set standards adapted to its special needs and installed its own inspectors to make sure the standards were met, the Maritime Commission accepted the standards of the American Bureau of Shipping for steel, engines, shafting, and all structural members, the standards set by the Coast Guard for life boats, firefighting equipment, and everything concerning safety at sea, and the standards of various trade and professional associations such as the American Institute for Electrical Engineers. Inspection also was the responsibility of the American Bureau of Shipping and of the Coast Guard, through the Bureau of Marine Inspection and Navigation, within their spheres. Many other items, such as electric motors, were checked on by the shipbuilding companies and by the Commission's inspectors in the yards. But there remained in addition some materials—notably paints and textiles—for which the Commission had to work out its own specifications and enforce them by a special corps of inspectors.

These tasks were entrusted to the Materials Section in the Technical Division. It made tests, of various kinds of paints for example, to see how the specifications should be written in purchase orders, and it arranged for tests to be run by the Bureau of Standards. Inspection to make sure the product delivered was up to specification was gradually shifted as much as possible from the point of delivery to the point of manufacture where sampling

---

[15] Report to Congress for the period ending June, 1944, p. 9.
[16] Statements by J. T. Gallagher, June 4, 1948.

methods could be more effectively employed. Medical supplies could be bought on the basis of established reputation, but a large number of products not checked by anyone else had to be checked on by the Commission—pillows full of chicken feathers instead of duck feathers, mattresses full of sweepings, and so on.[1]

With the Technical Division setting the standards and the Procurement Division doing the ordering, conflicts sometimes arose between them. The procurement officers were in a hurry to place orders; and, according to a protest on at least one occasion from the Technical Division, they sometimes placed orders without requiring adherence to the standards set by the Technical Division. Then, of course, the material inspector had no basis for action.[2] To the Procurement Division the engineering experts seemed slow to approve the products of unknown manufacturers which could, after all, serve just as well.[3]

Another method of increasing supplies was to substitute less scarce materials for those most in demand. A conservation committee was created in the Technical Division in September 1942 for the purpose of pointing out portions of the specifications which called for materials especially hard to get, for which something could be substituted. The appropriate Section, Hull or Engineering, then tried to arrange a change.[4] One such change was that in condenser tubes. The Navy continued the use of the best kinds, those of cupro-nickel; the Maritime Commission used for some ships those, easier to get, of aluminum brass and admiralty metal. After agreeing to the use of condensers of admiralty metal on the standard tankers (T2) the Commission found that they did not last.[5]

In 1943 the WPB increasingly emphasized conservation and the Maritime Commission created in the Production Division a special Conservation Section to work with WPB.[6] The rules for

[1] Statements by James L. Bates, and J. G. Conkey, Nov. 22, 1948. C. R. Brown, Chief, Materials Section to Director, Technical Division via A. MacPhedran, Asst. Dir., May 2, 1944, in gf 201-2-11, lists the functions of the section. On tests by the Bureau of Standards see MC Minutes, p. 20040 (Aug. 24, 1943).
[2] Bates to Walsh, Dec. 4, 1944.
[3] Statement by J. G. Conkey, Nov. 22, 1948.
[4] Esmond to Wanless, Sept. 15, 1942, in Production Division files, QM8 A3-0-1. The Committee voted to disband, Oct. 13, 1944. MacPhedran to Bates, Dec. 4, 1944 in gf 201-2-11.
[5] Statements by James L. Bates, Nov. 22, 1948.
[6] Admin. Order No. 37, Suppl. 77-B, July 20, 1943.

substitution kept changing as now one, now another, material became particularly scarce. In 1942, wood was considered a good substitute for steel in some products, but when, during 1943 lumber became very scarce and steel supplies increased, this substitution was reversed.[7] The function of the Conservation Section was to review on the one hand the "Conservation and Limitation Orders" issued by WPB and, on the other, the ships' specifications and bills of material, and to recommend to the Technical Division changes which would conserve critical materials.[8]

The whole business of conserving materials and developing new suppliers was closely tied together, as is illustrated by the problem of supplying booms for Liberty ships. Whereas the original plans specified 5-ton booms only, the increasing size of tanks made it necessary to put 50-ton booms at the forward end of the No. 2 hatch, and 15-ton booms at the No. 4 hatch. Demand for steel booms was increased also by the failure of the wooden booms which had been supplied to the first 122 Liberty ships built on the West coast. A new, recently developed process of cold rolling tapered tubes of heavy gauge steel was applied to boom manufacture by the Union Metal Manufacturing Co. To experts from the Maritime Commission and the American Bureau of Shipping, it demonstrated that one of its new-type "Monotube" cargo booms weighed 20 per cent less than a standard pipe boom of the same strength. Altogether 16,931 booms of this type representing a saving of 2,337 short tons of steel were ordered by the Commission between July 1942 and V-J day.[9]

Compared to the big shipyards, the suppliers of the Maritime Commission received little publicity. Since they were so numerous and so varied it is difficult here to give an adequate picture of their activities and importance, or of the many problems they created for the Maritime Commission. In the process of building up the productive capacity of widely scattered plants of extremely varied output, the Commission had to take new responsibilities for inspection, for conservation, for spreading work to small businesses, and for the profits and salaries of companies which it supplied with funds.

[7] *Industrial Mobilization for War*, I, 659-60.

[8] It also represented the Commission on a joint board dealing with the problem. Quarterly report of the Conservation section, for last quarter of 1943, in Production Division Administrative file A 9-3.

[9] Historical Notes of W. G. Esmond in Historians Collection, Research file 111, Liberty Ship.

Chapter 13

STABILIZATION AND MORALE IN THE LABOR FORCE

I. The Problem

THE MOTIVES that induced the government to enter the field of labor relations were in part the same as those that led it to seek control over the flow of materials. Competitive bidding for materials was limited because it would have drawn the short supply into channels other than where it was most needed in the national interest and would have raised enormously the prices paid by the procurement agencies. Similarly, wages and other issues between employers were not left to collective bargaining free of governmental interference. To do so in an industry expanding as fast as shipbuilding would have led to strikes and a costly spiraling of wages. Because of its social objectives, the Roosevelt administration was already before 1940 encouraging collective bargaining; and with the formation of the Shipbuilding Stabilization Committee at the end of 1940 the Navy and Maritime Commission took part in the negotiations, described in chapter 9, which fixed the basic wage rate of the shipbuilding industry.

Wages and union status constituted only one aspect of stabilization, however. Broadly speaking, stabilizing labor may be defined as establishing conditions which make workers contented with their jobs and ready to make an effort to increase production. If the shipyard worker was to remain satisfied in his job, he had to be convinced that he was being paid a wage which was fair compared to that of other workers in the same yard and of comparable workers in other industries. It was also important that the community in which he lived be such that he could obtain with his wages what he regarded as a decent living. High morale involved more. If the worker was to do his best work and use his ingenuity, he had to feel that his ideas would be welcomed and rewarded and that the rate of production was important to him, whether because he worked on a bonus system or because he wanted to do his share in winning the war. Equitable wage levels, adequate facilities in the community, and the stimulation of ideas and effort in the labor force were all essential to maximum production.

The various measures affecting the morale of labor do not fit together into a closely knit narrative. If the various policies and problems had a common root, that does not appear clearly in a history dealing with a single industry. To some extent this is indicative of a lack of coherence or comprehensiveness in the national policies affecting labor morale.

Strikes, loafing, absenteeism, and turnover were the most talked about indications of low morale. The stabilization agreements together with the general atmosphere of cooperation after Pearl Harbor reduced strikes to a negligible factor in 1942-1945. Loafing, which has already been discussed in Chapter 9, will be considered again in Chapter 20 in connection with the charges that the shipyards hoarded labor. Deliberate loafing undoubtedly occurred, although its extent was often exaggerated, but it cannot be measured and thus evaluated as a factor in slowing down production, as can the two other manifestations of low morale: absenteeism and turnover.

Absenteeism ran as high as 7 to 8 per cent in April to October of 1942.[1] In March 1943 Admiral Land estimated that 200 Libertys could have been built with the manhours lost through absenteeism.[2] The biggest loss was on the East coast according to figures collected monthly by the Maritime Comimssion in 1943 and after. The national average for 1943 and 1944 of manhours lost due to absenteeism was about 10 per cent, and the regional averages were: East coast, 12 per cent; Gulf, 10 per cent; West coast, 10 per cent; and Great Lakes, 7 per cent.[3]

A public outcry treated the term "absentee" almost as the equivalent of the term "slacker" in World War I.[4] Particularly

---

[1] Eleanor V. Kennedy, "Abstenteeism in Commercial Shipyards," *Monthly Labor Review*, Feb. 1943, p. 211.

[2] House, *Hearing before the Committee on Naval Affairs on H. R. 1876, a Bill to Require All Department of the Navy Contractors and Subcontractors to File Information with Respect to Absenteeism*, 78th Cong., 1st sess., Mar. 3, 1943, pp. 327-28.

[3] Fischer, *Statistical Summary*, Table G-11, p. 140.

[4] Labor Research Section, Division of Economics and Statistics, "Report on

to those whose close relatives were in the armed forces, the idea of workers taking time off to amuse themselves or to sleep off the effects of a night out seemed damnable. And to the man in the street an absentee was thought of as a person who did just that, for little was done to make the public understand the fact that the figures published on absentees included sick persons and others with valid reasons for staying away from work. The natural reaction of many persons was that absentees should be jailed or drafted. A bill was introduced in Congress requiring naval contractors to file information with Selective Service about persons absent without leave.

The complex nature of absenteeism was explained in the Congressional hearings held on this and similar proposals and in the special studies made by the Bureau of Labor Statistics and by Maritime Commission representatives at Beth-Fairfield, South Portland, and North Carolina Ship in the spring of 1943. About 10 per cent of the time lost attributed to absenteeism was the result of counting unannounced quits as absentees.[5] Moreover, the figures did not distinguish between justified and unjustified absences. Separating excused absences or those due to illness was impractical because some yards recorded them only if reported on the day they occurred. Many men drove long distances from rural communities to work in shipyards.[6] Every once in a while and in certain seasons particularly such men would stay home to help on the farm. At least one shipyard manager felt proud his employees were doing double duty for the war effort. Captain Roger Williams of North Carolina Ship felt that the rate of absenteeism among his workers was no sign they were slackers, that it merely reflected the semi-rural character of his labor force.[7] Henry Kaiser emphasized that the men and women living in the half-built, overcrowded cities near new shipyards had difficulties

---

Absenteeism," March 1943; also various memoranda filed under Apr. 23, 1943 in MC, Historian's Collection, Land file.

[5] "Effects of Unannounced Quits on Absenteeism in Shipbuilding," in *Monthly Labor Review*, June 1943, p. 1047.

[6] Kennedy, "Absenteeism," in *Monthly Labor Review*, Feb. 1943, p. 21, and "Report on Absenteeism," of MC Division of Economics and Statistics, cited above.

[7] Williams to F. C. Lane, May 5 and 26, 1947, in MC, Historian's Collection, Research file 112, North Carolina; and Truman Committee, *Hearings*, pt. 17, pp. 6985-87, 6994.

in making purchases to provide for their families when store hours did not "click with shifts."[8]

Some absentees were loafers and drunks, but many others were persons working to their full capacity who through fatigue were forced to be absent now and then because of the extra efforts they were making. The absenteeism of the loafer might be reduced by appeals to his patriotism or by penalties; the absenteeism of those who were working to the limit of their strength was not altogether an evil that it was desirable to eliminate. Insofar as it was desirable to eliminate it, the way to do so was to improve the conditions promoting health and safety in the shipyard and the living conditions in the shipyard communities.

Probably the most serious indication of dissatisfaction in the labor force was the high rate of turnover. The best measure of turnover is the number of workers being separated from employment in shipyards each month compared to the total number employed. The number of separations in March 1943 was 12.6 per cent of the total number employed. During the last six months of 1943 the separation rate was between 8.2 and 12.0 per cent each month. It dropped somewhat during 1944, when the highest rate for any one month was 10.8 per cent, but was still relatively high.[9] More persons left their jobs in merchant shipyards than in any other key war industry: in ordnance plants the separation rate was between 6.4 and 9.0 per cent; in aircraft between 4.5 and 7.9 per cent during 1943 and 1944. Other industries had lower rates.[10] The largest element in the separations were those classified as "quits." At their peak in July-September 1943, quits were 8.1 to 8.3 per cent per month of the total number of workers. They fell to 5.5 in November 1943 but rose again and were about 7 per cent throughout 1944.[11]

Among the workers quitting were many who moved from one yard to another to get better positions or higher pay from longer

---

[8] Truman Committee, *Hearings*, pt. 15, p. 7018.
[9] Fischer, *Statistical Summary*, Table G-15, p. 147.
[10] Comparisons of the following industries were made from tables on file at U. S. Dept. of Labor, Bureau of Labor Statistics, Division of Employment Statistics: aircraft, aircraft parts; radios, radio equipment, and phonographs; blast furnaces, steel works, and rolling mills; bituminous-coal mining; anthracite mining; engines and turbines; motor vehicles, bodies, and trailers; ordnance.
[11] Fischer, *Statistical Summary*, Table G-15, p. 147.

hours and bonuses.[12] Those who moved about in search of the largest possible "take-home" pay were mainly men with youth and strength or long familiarity with rough, outdoor labor. Women and older men attracted to unfamiliar work by the high wages or by patriotism were more likely to quit for other reasons. Some found at once that the strenuous, dirty labor of a noisy shipyard was too much for them; others stuck it out for a while and then gave in to fatigue or illness.[13] The separation rate for women was higher than that for men, since to sickness and exhaustion were added the difficulties of keeping up with home duties.[14] In general, workers were more apt to stay on the job if fortunate in finding a place to live near the shipyards.[15] As time went on observers accounted for the high rate of separations by pointing to the inadequacy of housing, transportation, and other community facilities.

From this brief examination of the symptoms of instability it appears that there are three main topics which require further examination: (1) wage rates, (2) living conditions of shipyard workers, and (3) working conditions affecting health. Pertinent in a somewhat different way is a fourth factor, namely the various incentive systems aiming to make the worker identify himself more fully with high production.

II. Policing Wages

A standard of uniform wages for each of the four zones—$1.12 for the Atlantic coast, Great Lakes, and Pacific coast; and $1.07 for the Gulf—had been established by the zone conferences arranged in the opening months of 1941 by the Shipbuilding

---

[12] House, *Production in Shipbuilding Plants, Hearings*, pp. 567-68, 595, 607, 614-15, 619, 1096-97.

[13] The few available studies of reasons for quits show that health factors were predominant. California Shipbuilding Corp., Monthly Report by the Industrial Relations Division, Jan. and Feb. 1944; Swan Island, Reasons for Termination, Clearance Office Report, Dec. 12-Dec. 18, 1943, Jan. 17-Jan. 22, 1944, Feb. 14-Feb. 19, 1944; Feb. 28-Mar. 4, 1944; Oregon Shipbuilding Corp., Reasons for Termination, Clearance Office Report, Dec. 12-Dec. 18, 1943; Jan. 17 to 20 (?), 1944; Feb. 14-21, 1944; Mar. 13-20, 1944, in Production Division files; MC Manpower Survey Board [Report] Kaiser Co.-Vancouver, January 1944.

[14] MC, Manpower Survey Board [Reports] Oregon Shipbuilding Corp., Jan. 1944; Kaiser Co.-Vancouver, Jan. 1944; Kaiser Co.-Portland, Jan. 1944.

[15] Data Respecting Manpower, Housing, Community and Commercial Facilities at Richmond Shipyards, Apr. 8, 1943, Exhibit 1 to MC gf 115-1-1-3.

Stabilization Committee. Thereby the weekly take-home wages of shipyard workers were raised so that they continued to be the highest in any industry of comparable size. This raise was agreed to by the government when it seemed necessary to attract workers to an industry which was expanding its labor force tenfold. Without it, competitive bidding by yards for workers would probably have raised wages even higher than did the stabilization agreements.[1]

These agreements fixed the wages of first-class mechanics. But what percentage of the workers in a yard would receive that wage? How many would be classified as first-class mechanics? That was a question which had not been settled by the stabilization agreements. The answer depended on how the various jobs in the shipyard were classified: whether, for example, all welders were classified as first-class mechanics, or as second-class mechanics, or whether tack-welders and others not fully trained were classified as helpers or trainees. The classification of jobs was one of the subjects dealt with in the second phase of collective bargaining, the negotiations between management and labor leading to a contract between them, and was a subject of renegotiation even after the basic wage rate, that of first-class mechanics, had been frozen.

Although the procurement agencies did not take part in the negotiations which established the classification of jobs, they undertook to pass on these agreements to see whether or not they conformed to the spirit and letter of the stabilization agreements. After all, the zonal agreements were only a set of principles to guide the negotiation of contracts between labor and management. To see that the principles were carried out, there was need of some machinery for enforcement. At the time the contracts were being negotiated between employers and workers, the Shipbuilding Stabilization Committee lacked the support of many of the big industrialists and made no attempt to develop any supervisory staff. Accordingly the Navy and Maritime Commission undertook the necessary enforcement. To coordinate their activities they agreed that "all interpretations and decisions with respect to the administration of zone standards will be issued jointly by the office of the Assistant Secretary of the Navy and

---

[1] See above, chaps. 8 and 9.

the office of Commissioner Vickery." [2] All the yards were called on to send in lists showing the trades employed, the new and old hourly rates of pay, and a summary list of premium men.[3]

Premium men were those paid more than first-class mechanics because they had outstanding skill or some supervisory duties. If a large number of workers were classified as premium men, such a practice might lead to a break through the ceiling set for the basic mechanic's rate. Accordingly on July 25, 1941 an Administrative Instruction was issued over the signature of Ralph A. Bard and Admiral Vickery which ordered the yards to submit for approval all changes in rates for such men or the addition of any premium men to the payrolls. Such additions were to be approved by representatives of the procurement agencies " provided, that the resultant total number . . . in any one trade does not exceed 3% of the total mechanics in that trade, and provided further that such representative shall, in no case, authorize payments of premiums exceeding 5¢ per hour above the standard scale." [4] Unless the designated officials of the Maritime Commission approved the addition of premium men, the Commission's resident auditors in the shipyards would refuse to approve reimbursement of their extra wage as part of the shipbuilders' costs. Through the resident auditors who checked the payrolls of the shipbuilders, the Maritime Commission had a very effective mechanism for enforcing their supervision of wage rates to see that they were in accord with the zone standards.[5]

The same method could be applied to enforce Commission disapproval if yards gave too high a classification to jobs that required little or no skill. The task of checking wage rates sub-

---

[2] Within the Maritime Commission a Shipbuilding Stabilization Zone Standards Committee was appointed under the chairmanship of J. W. Slacks, financial assistant to Admiral Land. The Committee was composed of the Directors of the Divisions of Finance, Maritime Personnel, Emergency Ship Construction, Maintenance and Repairs, and the Chairman of the Legislative Committee. MC Admin. Order No. 59, July 24, 1941.

[3] Administrative Instruction No. 2, Aug. 9, 1941. These Administrative Instructions were the formal orders issued jointly by the Maritime Commission and the Navy over the signatures of Admiral Vickery and Ralph A. Bard, Asst. Secretary of the Navy. Later the Army joined and General Somervell's signature was added. A set of these Administrative Instructions Nos. 1 to 19 is in MC Historian's Collection, Research file 205.4, and also in the files of the Division of Shipyard Labor Relations.

[4] Admin. Instruction No. 1, July 25, 1941.

[5] On the work of these auditors, see chap. 14 sect. II, and 22, sect. III.

mitted by the yards against the zone standards and those of other yards, in order to recommend disapproval of any that were out of line, fell to the Division of Maritime Personnel (later the Division of Shipyard Labor Relations) under the supervision of Mr. Ring and Mr. Tracy. Since the main worry in 1941 was "scamping" (the enticing of workers from one yard to another by offers of higher pay) the chief objective in checking the wage rates was to see that comparable jobs in yards near each other were paid for at about the same rate.[6] On the whole this was achieved; a yard-by-yard analysis of wages shows that the average earnings in yards within competing distances were very nearly the same.[7]

There was no attempt in 1941, however, to establish a system of classifying jobs which would be uniform for the whole industry. As a result the average hourly earnings of shipyard workers differed considerably from region to region. By October 1941 when rates based on the zone standards had been put into effect in all yards, West coast workers averaged $1.01 for straight time and $1.17 counting overtime and shift bonuses. By working 45.5 hours a week they were able to make $53.24. Although workers in North Atlantic yards worked more than an hour longer they earned almost $7.00 less a week. Their average straight-time hourly earnings were only 86 cents. Overtime, shift bonuses and incentive payments brought the gross average to 99 cents. (See Table 7.)

These differences reflected the higher classification of jobs on the Pacific coast. In West coast yards all "mechanics" were first-class mechanics and received at least the $1.12 per hour. Below these were handymen, helpers, laborers, apprentices, and trainees. In those Gulf yards where the AFL had agreements, job classifications were like those on the Pacific coast, whereas CIO yards followed the Atlantic coast pattern.[8] Just what jobs would be

[6] Statements by Edward J. Tracy, Oct. 31, 1946, noted in MC, Historian's Collection, Research file 105.

[7] Fischer, *Statistical Summary*, Tables G-8, G-9, G-10. Coll, "Stabilization" Tables VI, VII, IX, X, and pp. 114-17, 164-65.

[8] Master Agreement Covering New Ship Construction Between the Pacific Coast Shipbuilders and the Metal Trades Department, A. F. of L., the Pacific Coast District Metal Trades Council, the Local Metal Trades Councils, and Affiliated International Unions (Apr. 23, 1941, Seattle, Wash.), reproduced in House, *Production in Shipbuilding Plants, Hearings*, pt. 1, p. 95; Dept. of Labor, Bureau of Labor Statistics, *Bulletin No. 824*, p. 23.

TABLE 7

AVERAGE HOURS AND EARNINGS FOR SELECTED MARITIME
COMMISSION SHIPYARDS, BY AREAS 1941-1945

(Annual Averages of Monthly Data)

	1941	1942	1943	1944	1945
(A) Average Hourly Earnings (Gross)					
North Atlantic...	$ .9715	$ 1.1530	$ 1.3468	$ 1.4482	$ 1.5261
South Atlantic...	.8585	.9072	1.1927	1.3047	1.3358
Gulf............	.8227	1.0319	1.2013	1.2625	1.2309
West Coast......	1.0880	1.3002	1.4002	1.4706	1.4382
Great Lakes.....	.7853*	1.0868	1.2382	1.3156	1.2973
All Areas......	$ .9931	$ 1.1789	$ 1.3326	$ 1.4138	$ 1.4150
(B) Average Hourly Earnings (Straight Time)					
North Atlantic...	$ .8474	$ .9298	$ 1.0532	$ 1.1237	$ 1.1501
South Atlantic...	.7477	.7561	.9623	1.0465	1.0846
Gulf............	.7683	.8669	.9916	1.0765	1.1057
West Coast......	.9704	1.0862	1.1407	1.1785	1.1992
Great Lakes.....	.7211*	.8953	1.0178	1.0818	1.1349
All Areas......	$ .8863	$ .9784	$ 1.0772	$ 1.1349	$ 1.1606
(C) Average Weekly Hours					
North Atlantic...	44.1†	48.6	48.1	45.7	44.2
South Atlantic...	42.9†	50.3	46.9	45.9	44.3
Gulf............	35.8†	47.9	45.2	46.8	45.2
West Coast......	42.0†	43.2	42.5	43.9	41.2
Great Lakes.....	....	50.1	54.7	54.8	49.6
All Areas......	41.8†	46.2	45.0	45.2	43.5
(D) Average Weekly Earnings§					
North Atlantic...	$44.581†	$56.036	$64.781	$66.183	$67.454
South Atlantic...	37.036†	45.632	55.937	59.886	59.176
Gulf............	25.336†	49.428	54.299	59.085	55.637
West Coast......	48.161†	56.169	59.509	64.559	59.254
Great Lakes.....	......	54.449	67.730	72.095	64.346
All Areas......	$42.378†	$54.465	$59.967	$63.904	$61.553

* Average for 5 months, August-December.
† July 1941.
§ D = A x C.

Sources: For parts (A) and (B), compilations based on monthly data reported by Maritime Commission auditors. Data and methods used described in Fischer, "Wages." For part (C), Bureau of Labor Statistics data for midweek of each month, July 1941 on.

placed in what classification was settled on the basis of local tradition and through local negotiation in the three zones where there was no master agreement.

In Atlantic coast yards in 1941 only 15.2 per cent of the labor force was classified as first-class mechanics. The remaining mechanics, 38.5 per cent of the labor force, fell into the second and third class, and thus received a wage less than $1.12. Compare this with the Pacific coast where 52.1 per cent of its 57.6 per cent of mechanics was in the first class! Even on the Gulf more workers, 24.6 per cent, got the first-class rate than did those on the Atlantic coast. And on the Great Lakes 27.1 per cent received $1.12. The West coast had more supervisors also—7.3 per cent of the labor force, while Atlantic coast yards had 5.3 per cent engaged in supervision; the Gulf, 4.2 per cent; and the Great Lakes, 4.7.[9] The differential between the Pacific coast and other regions would have become yet wider if raises had been made during 1942 in accordance with the rise in the cost of living and the provisions of the original zonal agreements, but the modification of the zone standards agreed to at the meeting of the revived Shipbuilding Stabilization Committee in April 1942 prevented this. It set $1.20 as the basic rate and brought up the rate in Southern yards nearer to that of the North Atlantic coast, the Great Lakes, and the Pacific coast.[10] But it left the different schemes of classifying jobs unchanged, so that in October 1942 the straight time hourly wage of shipyard workers on the Pacific coast averaged 13 or 14 cents higher than in the North Atlantic coast area and about 20 to 30 cents higher than in the South Atlantic area.[11] See Figure 31.

These differentials were attacked by some of the labor leaders, especially those of the Industrial Union (CIO). They wished to bring about more uniform classifications on the Atlantic coast. The yards on the West coast, where the AFL had succeeded in getting jobs classified to so high a level, were mostly managed by industrialists who were new to shipbuilding and not likely to stay in the business after the war. On the East coast, however, most of the yards were managed by companies with a future in

[9] Dept. of Labor, Bureau of Labor Statistics, *Bulletin No. 727*.
[10] See above, chap. 9.
[11] Fischer, *Statistical Summary*, Table G-9; Coll, "Stabilization," Table VIII, and pp. 132-35.

shipbuilding. They resisted attempts to alter the wage pattern of the industry in a way that might raise labor costs in the long run, and the Navy and Maritime Commission put no pressure on them to yield to labor on this matter. The procurement agencies saw no need to raise the cost of ships by setting up standard classifications. Moreover, during 1942 the government adopted the general policy of freezing all wages and prices.

The agency charged with primary responsibility for the general stabilization of wages was the National War Labor Board. Previously the War Labor Board had had authority to determine wage rates only when they were the subject of dispute between management and labor. On October 3, 1942, by Executive Order 9250, the Board was given power to pass on all increases, even those arrived at voluntarily. This same order froze wages at levels attained by September 15, 1942, and the Board was to allow increases only " to correct maladjustments or inequalities, to eliminate substandards of living, to correct gross inequities, or to aid in the effective prosecution of the war." [12] The Board's interpretation of these exceptions followed the policy laid down in its decision in July 1942 regarding workers in the " little " steel companies. The " Little Steel " formula allowed general wage increases in an industry only if straight-time hourly earnings had not already risen the 15 per cent deemed necessary to meet the rise in the cost of living which took place between January 1, 1941 and May 1, 1942. To mollify labor, which objected to this " freeze," the Board allowed increases in take-home pay through generous action on " fringe " issues such as premiums for night shifts.[13] In the shipbuilding industry the most important " fringe " item was the reclassification of jobs.

When the War Labor Board was given supreme authority over all wage increases, the Shipbuilding Stabilization Committee had at last become representative of the industry. In August 1942 even Bethlehem was on management's side of the table.[14] The suggestion was therefore made that this Committee act as the agent of the War Labor Board for the shipbuilding industry, but industry representatives objected, and consequently the War Labor Board appointed a new body, the Shipbuilding Commis-

[12] *Federal Register*, Oct. 6, 1942.
[13] *The United States at War*, pp. 197-99.
[14] Shipbuilding Stabilization Committee Minutes, Aug. 3-5 and 10-12, 1942.

sion, containing much the same membership as the Shipbuilding Stabilization Committee, to pass on matters coming within the national wage stabilization order.

The Shipbuilding Commission as originally constituted did not work out to the satisfaction of either the War Labor Board, the labor unions, or the Maritime Commission. The unions, particularly the CIO, accused the Maritime Commission and the Navy of failure to view cases according to equity and the broad policies of the War Labor Board, and of acting instead only to keep down costs. The representatives of the Maritime Commission felt they were in a difficult position, torn between the specific interests of their agency in costs and their responsibilities as general respresentatives of "the public." Difficulties came to a head in a dispute between Bethlehem and the Industrial Union over the reclassification upward of ten jobs. When management and the procurement agencies voted down the union request, the War Labor Board first told the Commission to reconsider, and then, on its refusal to reverse itself, overruled it and decided the case in favor of the union. Shortly thereafter the War Labor Board created an entirely new Shipbuilding Commission on which the procurement agencies were not represented.[15]

The procurement agencies were also unhappy about their position in the Shipbuilding Stabilization Committee, which still survived. Although it had no final authority, there was nothing to prevent it from making recommendations to the War Labor Board if it wished; and the unions, supported by the Committee's Chairman, Paul Porter, wanted the Committee to recommend increases, arguing that they were needed according to the original agreements to keep wages in line with the cost of living. This put the procurement agencies in an awkward position; for on the one hand they were parties to the agreements which provided for such changes, and on the other hand they felt responsible not only for policing costs but also for supporting the general stabilization policy ordered by the President in October 1942. They were accused of "ganging up" with management. But withdrawal from the Committee seemed likely to be viewed even more as an act of bad faith. So the Navy and Maritime Com-

---

[15] Coll, "Stabilization," pp. 137-45; *War Labor Reports*, vol. 8, pp. 709-11, 725-26, vol. 9, pp. 589-92.

mission cooperated in sponsoring a few additional zone conferences under the auspices of the Committee.[16]

An annual review of wages had been provided for in the 1942 amendments to the zone standards. It was conducted in 1943 by the War Labor Board. The Chairman of the Shipbuilding Stabilization Committee argued for a wage increase but the procurement agencies opposed it. The War Labor Board denied any general increase, finding that shipyard labor had already received more than was allowable under the "Little Steel" formula; but the Board tempered its ruling by ordering a review of wage-rate structures and job classifications. Such adjustments, said the Board, could be initiated on the Pacific coast through a zone conference, which was arranged by the Shipbuilding Stabilization Committee, and elsewhere by recourse to the War Labor Board's Shipbuilding Commission.[17] West coast workers quickly got some reclassifications, but adjustments in other zones were not so readily made. Employers would not agree to them in zone conference, and although the Shipbuilding Commission in 1944 finally announced a standard system of job classification for the Atlantic coast, and others for the Gulf and Great Lakes, they were announced as "Tentative Conclusions," were opposed by the Maritime Commission and the Navy, and went into effect slowly.[18]

In spite of complaints on the part of labor leaders that they had been caught in the wage freeze, the earnings of shipyard workers continued to rise during the period from October 1942 until V-J day.[19] Part of the rise was due to reclassification of jobs, but a large part also was the result of upgrading. Whereas reclassification raised the wage for a particular job, such as welding, upgrading raised the pay of a particular worker by promoting him to a job with higher classification. A yard might establish all the approved types of jobs, duly classified with approved rates of pay, but then fill more of the higher paid jobs and few of the others. As the program became more complicated in 1944 and there was more need of skilled workers, the percentage of workers in jobs of higher classification increased. After the labor force

[16] Coll, "Stabilization," pp. 145-54.
[17] *War Labor Reports,* vol. 10, pp. 237-44.
[18] Coll, "Stabilization," pp. 150-53, 156-60.
[19] Fischer, *Statistical Summary,* Tables G-8, G-9, G-10.

was fully recruited and better trained, it was natural to promote workers. Thus the Bureau of Labor Statistics reported that 36 per cent of the production force on the Atlantic coast was made up of first-class mechanics in June 1944 as compared with 19.3 per cent a year earlier.[20]

In the rise in shipyard wages between October 1942 and V-J day (August 1945), the differentials between regions was strikingly reduced, and a marked degree of uniformity resulted in the whole industry (Figure 31). In July 1943 the differential between straight time hourly earnings in the Pacific and North Atlantic zones was 9 cents; between the Gulf and North Atlantic, 5 cents. One year later workers in North Atlantic yards were earning only 5 cents less than those on the Pacific coast, and while the differential was the same between the North Atlantic and the Gulf, Gulf workers had gained in relation to those on the Pacific coast. South Atlantic workers had also gained slightly in relation to those on the North Atlantic and Gulf.

In gross hourly earnings the differential between North Atlantic and West coast yards was even less (Figure 32). In April 1944 workers in North Atlantic yards got a penny more than those on the West coast and they continued for the remainder of the program to get about the same or slightly more. Workers on the South Atlantic, Gulf and Great Lakes never made as much in gross earnings as did the other two areas and there was a considerable differential maintained between North and South Atlantic yards. So far as gross earnings were concerned, workers in South Atlantic yards did slightly better than those on the Gulf although in straight time earnings the positions were reversed.

Because of the long hours they worked the take-home pay of employees in Great Lakes yards was the highest in the country. In July 1944 they were earning $71.26 weekly. The average take-home of workers in South Atlantic and Gulf yards was lower than elsewhere. In July 1944 those in South Atlantic yards earned $59.61 and those in Gulf yards, $61.21. Workers in North Atlantic yards took home more than those on the West coast, even though their hours were not much longer. In January 1943 North

---

[20] Coll, "Stabilization," pp. 158-67; Shipbuilding Commission, Atlantic Coast Wage Review, Tentative Statement of Policy Governing 1944 Contracts . . . July 7, 1945, in *War Labor Reports*, vol. 25, pp. xxv-xxviii.

STRAIGHT-TIME HOURLY EARNINGS In Maritime Commission Shipyards
Actual Earnings Compared With Standards Fixed By Stabilization Agreements

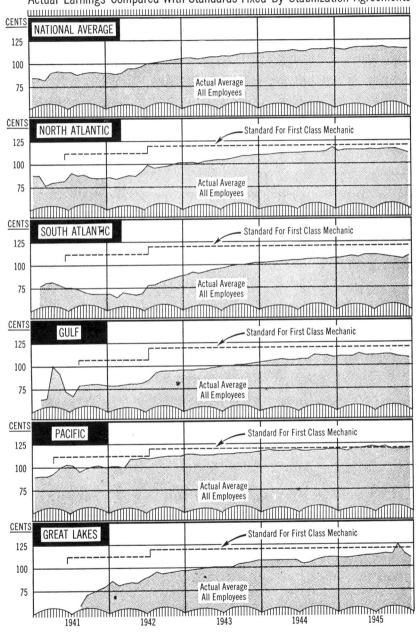

FIGURE 31. Actual earnings are averages of the sample of the yards in Fischer, *Statistical Summary*, Table G-9.

GROSS HOURLY EARNINGS In Maritime Commission Shipyards
Averages At Six-Month Intervals, By Zones

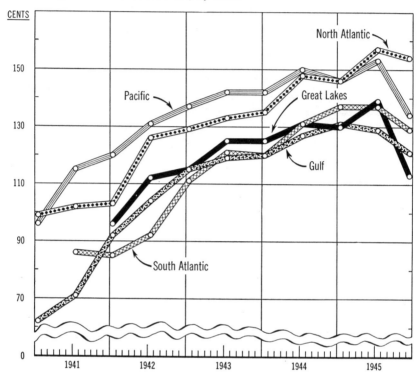

FIGURE 32. Source: Fischer, *Statistical Summary*, Table G-8.

Atlantic workers averaged $60.76 while those on the West coast earned $56.86; in July 1943 North Atlantic workers took home $64.90 to the West coast worker's $60.07; in July 1944, $66.75 compared with $64.50.[21]

This review of earnings in shipyards makes clear that wage rates as such were not an important cause of the high turnover. Shipbuilding was well able to offer as much and more than other industries. In yards located to compete in the same labor market, average hourly earnings were about the same, showing that the classification of workers and jobs was similar. Regional differences

[21] Fischer, " Wages."

were hardly so large as to provoke in themselves transcontinental migration, and it is significant that the region which had the highest hourly rate, the Pacific coast, had also the highest turnover, a fact indicating the importance of factors other than wage rates.

So far as the search for higher pay was the reason for "quits," it was not because of differences in wage rates but because of opportunities to earn bonuses for overtime or to win individual promotion. Nearly all repair yards and a good many Navy yards were using nine- or ten-hour shifts and paying time and a half— or in repair yards, double time—for all time over eight hours. Take-home pay in repair yards in 1944 averaged $66 to $75 a week as compared to $60 to $68 in yards used for new construction.[22] In general, workers who had acquired some skill and experience frequently found they could get employment at a higher level in a new yard more easily than they could get themselves upgraded in the yard where they were already employed. To some extent the regulations handed down by the Navy and Maritime Commission to prevent too rapid upgrading gave workers occasion to change yards in order to get promoted.[23] But the policing of wage rates at least prevented wages from becoming a means of scamping or a major cause of high turnover.

III. Federal Aid in Transport and Housing

Whereas the role of wages in stabilization was recognized from the beginning, the importance of governmental action to provide transportation, housing, and other necessities in shipyard communities was not immediately realized. The first emergency programs, in January and April of 1941, were large enough to push up wages, but were small compared to what came later. After the final waves of expansion, in January-March of 1942, the shipyards were attracting tens of thousands of inhabitants to localities

---

[22] House, *Production in Shipbuilding Plants, Hearings,* pp. 1096-97. See also testimony of Clay Bedford in *ibid.,* pp. 595, 607, 614-15, 619; and Flesher, *ibid.,* pp. 567-68; Dept. of Labor, Bureau of Labor Statistics, *Bulletin No. 824,* pp. 21-22.

[23] Admin. Instruction No. 17, Oct. 23, 1943, and No. 17, Suppl. No. 1, Apr. 1, 1944, in MC Historian's Collection, Research file 205.4; Edgar Kaiser to Vickery, Nov. 14, 1943; Wolff to Director, Division of Shipyard Labor Relations, Nov. 19 and Dec. 8, 1943, all in MC gf 508-1-1; Wolff to Tracy, Feb. 14, 1944, in Division of Shipyard Labor Relations file, Training in Shipyards; and compare on this point chaps. 8 and 20.

which had no facilities for handling such numbers. It became apparent that the Federal government would have to act on the problems of these communities just at the time when the nation was threatened with a "rubber famine" because of Japan's conquests, and the Northeast was threatened also with a "gasoline famine" because of the sinking of tankers off the Atlantic coast.[1] Since shipyard workers drove to work in their automobiles as did many other Americans, the shortages of rubber and gasoline made it seem necessary to provide other means of transportation to the shipyards and also to provide more housing close to the yards. These problems were hastily tackled by many government agencies in search of the quickly needed solutions.

To be sure, at the beginning of the emergency programs, Mr. Hillman's office in the National Defense Advisory Commission made plans to meet community problems through the housing agencies sponsored by the New Deal and through state and local governments.[2] When Henry Kaiser asked the Maritime Commission for help with housing problems in August 1941 he was told to negotiate with the housing authorities.[3] After Pearl Harbor the President created two new agencies to deal with transportation and housing. The first was the Office of Defense Transportation (ODT) which was given broad responsibilities for seeing that the means of transportation already in operation were used to best advantage and for helping local groups get necessary equipment through the Supply Priorities and Allocations Board. The ODT had no money to spend itself and not until January 1943 was it given a veto over the plans of other agencies.[4] The second was the National Housing Agency (NHA) whose function was to coordinate various housing groups already in existence such as the Home Owners Loan Corporation and the Federal Public Housing Authority (FPHA). As its administration was worked out, the NHA formulated housing programs on the basis of

---

[1] The rationing of tires and the elimination of unnecessary driving was ordered right after Christmas, 1941. Office of Defense Transportation, *Civilian War Transport, A Record of the Control of Domestic Traffic Operations, 1941-1946* (Government Printing Office, 1948), pp. 157-59.

[2] Purcell, *Labor Policies,* pp. 10-12.

[3] Kaiser to Land, Aug. 13, 1941 and Land to Kaiser, Aug. 28, 1941, in MC gf 115-1-1.

[4] Executive Order No. 8989, Dec. 18, 1942 and Executive Order No. 9294, Jan. 4, 1943.

information received from the procurement agencies, from war plants, and community leaders. The FPHA let the contracts, supervised construction and managed the projects after they were finished.[5] These agencies being in general charge, the Maritime Commission acted only in particular situations which urgently threatened to interfere with the production of ships.

Within the Maritime Commission, problems of community development were not routed through the Division of Shipyard Labor Relations, where other aspects of stabilization were handled, but came before the Commission through several divisions and Commissioners. In the early months of 1942, when the rubber shortage loomed as a special problem, much initiative was also taken by W. Lalley, a special assistant to the Chairman. By the summer of 1942, Commissioner Woodward was paying particular attention to problems of transportation. He was the member of the Commission who had had most experience with the Interstate Commerce Commission, and was designated for liaison with its allied Office of Defense Transportation.[6] In the Commission he had been in closest touch with the work of the Chief Examiner, D. E. Lawrence. This official, normally concerned with conference agreements about shipping rates and having less of that work after the formation of the War Shipping Administration, was placed in general charge of housing and transportation, and delegated much of the work to T. D. Geoghegan, adviser to the Commission regarding rail transportation. They represented the Commission in dealing with the ODT and NHA and advised the Commission on policy. The execution of remedial measures, on the other hand, if carried through by the Maritime Commission, depended largely on the Regional Offices of Construction which reported to Admiral Vickery and brought up to him directly many proposals.[7]

In March 1942 the Commission warned the yards that desperate measures would be necessary to overcome the effect of the rubber

[5] *U. S. Government Manual,* Summer 1944, p. 129; National Housing Agency, *Third Annual Report,* 1945, pp. 6-8.

[6] MC Admin. Order No. 62, Jan. 22, 1942.

[7] Woodward to Eastman, Director of ODT, Feb. 10, 1942 in MC gf 107-70; Peet to C. F. Palmer, Office of Defense Housing Coordinator, Aug. 29, 1940 in MC gf 115-1-1; MC Minutes, p. 21502 (Mar. 27, 1942), and *passim,* the record of the many Commission actions on these problems cited below.

shortage on transportation.[8] On May 7, 1942 it began taking action directly when it requisitioned two ferries to ply between downtown Portland and the Swan Island and Oregon yards.[9] On the nineteenth it approved the purchase of eight school buses for the new Higgins yard and on June 2, fifty tractor-trailer buses for the same plant. After the Higgins contract was cancelled the Commission distributed them to several yards on the Gulf. These, along with some others bought later, served nearly all the yards in the South Atlantic and Gulf areas. Alabama was the only yard in the south that used ferries; and Houston, the only yard that had a train. The yards at Baltimore and Bayonne had a ferry service, but the boats were owned, not by the Commission, but by the private companies that operated them. At South Portland, however, the Commission initiated a ferry service. On the Pacific coast, ferries were requisitioned to serve Richmond and Marinship; and four yards—Vancouver, Calship, Richmond, and Moore—were furnished with trains. Altogether the Commission expended money for transportation to twenty of its yards.[10]

The transportation systems provided by the Maritime Commission were not used as fully as anticipated because the shortage of tires never became as acute as expected. As long as it was possible, workers preferred to use their own cars or to get a ride with a fellow worker. In March 1943 after Lawrence and Geoghegan returned from an inspection trip to the Gulf they protested to ODT and OPA " that almost without exception the efforts of the Commission to provide quick, cheap and efficient transportation . . . are nullified in a large part by the supplemental gas ration rules at the various localities and the issuance of war certificates to independent owners of nondescript buses and cars." [11] Nevertheless, Lawrence and Geoghegan were convinced that the rubber situation would get worse and recommended that the Commission authorize them to order fifty buses for distribution as need arose among the South Atlantic and

---

[8] Lalley to California Shipbuilding Corp., March 7, 1942, and Lalley to Guy A. Richardson, ODT, March 2, 1942, in MC, Lalley file, 18-2.

[9] MC Minutes, pp. 21859-60.

[10] Director, Construction Division to MC, May 2, 1946, in MC Historian's Collection, Research file 205.7.2.

[11] Lawrence and Geoghegan to Richardson, Butkiewicz, and Marston, March 26, 1943 in MC gf 509-1.

Gulf yards.[12] But this time the Commission moved slowly and refused to allow the plan to go forward without assurance from ODT that it was necessary. The bus pool was never acquired.[13] By August 1943 Lawrence couldn't imagine where he could use more than five surplus buses in the Gulf area.[14]

The Gulf area was not the only place where workers continued to use their own cars—the story was the same nearly everywhere in the summer of 1943. " Peculiarly enough," Lawrence reported, concerning the situation in Oregon, " the average individual would prefer fighting traffic all the way from Portland to the Vancouver Yard (and it is very heavy and slow . . .) than to ride comfortably on the train into the yard." In the San Francisco area the trains could carry twice as many passengers, but this area appeared for the moment to be " ferry minded " so it was expected the Portland ferries would be transferred since continuance at Portland did not appear justified. On the other hand, trains at Los Angeles and Houston were doing reasonably well. Bus operations in some of the Southern yards had been cut down, but the ten or fifteen buses running at most spots were performing necessary service.[15]

Although none of the transportation services provided by the Commission boomed to the extent expected in 1942, about half were continued into 1945.[16]

All the arrangements about transportation were worked out in consultation with the ODT, but the Commission rarely called upon this agency for help in getting priorities for equipment. The Commission found it easier to get priorities directly from SPAB than to work through ODT. This was even more true when it came to getting houses built.

In March 1942 Admiral Vickery protested to WPB about the

[12] Lawrence and Geoghegan to USMC, Mar. 29, 1943 in MC gf 509-1, MC Minutes, pp. 24747-48 (Apr. 2, 1943).
[13] Peet to Geoghegan and Lawrence, MC Minutes, pp. 24747-48 (Apr. 2, 1943); Lawrence to Murray, Aug. 18, 1943, in gf 509-1; Schnapp to Lawrence, Aug. 27, 1943 in gf 509-1.
[14] Lawrence to Duff, Aug. 25, 1943 in gf 509-1.
[15] Lawrence to MC, July 1, 1943 in gf 509-1. On the ferry at San Francisco Bay see MC Minutes, pp. 30077-79 (Oct. 31, 1944). At the peak of ferry travel in March 1943 it carried 573,981 passengers.
[16] Director, Construction Division to MC, May 2, 1946 in MC Historian's Collection, Research file 205.7.2; Lawrence to MC, Nov. 22, 1943 in MC gf 115-1-1 and Nov. 25, 1943 in gf 509-1.

low priority given NHA for Maritime Commission housing projects. He pointed out that Army and Navy housing was favored with A-1 ratings, while that slated for shipyard workers had been given only A-2.[17] At the same time the Commission directed Mr. Lawrence to investigate and try to speed up construction.[18] For three months he made efforts to better the NHA's position and that agency was optimistic.[19] In April, WPB made a general agreement to see it got necessary material,[20] and early in June, after Admiral Land strongly urged the highest priorities for projects at Vancouver, Portland, Richmond, Sausalito, Los Angeles, and Mobile, NHA promised to finish construction there by fall. Mr. Blandford, NHA administrator, also assured the Commission his agency would go to work immediately on housing at Pascagoula; Panama City; Wilmington, North Carolina; South Portland; Savannah; New Orleans; and Sturgeon Bay.[21] But Mr. Lawrence did not believe NHA could make good its promises, for he did not believe that it would obtain sufficiently high priority ratings from WPB. Accordingly, the Commission directed him to suggest to NHA that the Commission itself build housing and turn it over to FPHA to manage.[22] The Commission was assured of getting priorities because it planned to build houses under its shipyard facilities contracts, which had high priorities already assigned them. After the NHA assured the Commission it was agreeable to this plan, Lawrence was authorized on July 17, 1942, to go ahead.[23]

About this time NHA succeeded in getting A-1-a priorities on the Commission's most essential projects, but priorities were now inflated to AA's. The NHA had started in on the West coast,

---

[17] Vickery to William Harrison, WPB, March 25, 1942 in MC gf 115-1-1.

[18] MC, Minutes, p. 21502 (March 27, 1942).

[19] Lawrence to USMC in Minutes, pp. 22747-50 (Aug. 18, 1942); Lawrence to Seaver, April 16, 1942; Blandford to Land, April 21, 1942 in MC gf 115-1-1-2; Blandford to Land, May 21, 1942 in MC gf 115-1-1-2.

[20] War Production Board and National Housing Agency Policy for War Housing for the Period Apr. to Dec. 1942, Apr. 15, 1942, reproduced in House, *War Housing, Hearings* before the Committee on Public Buildings and Grounds, 77th Cong., 2d sess., June 1942, pp. 11-13.

[21] Blandford to Nelson, May 27, 1942, countersigned by Land (not stamped); Land to Kahler, WPB, June 9, 1942; Schedule of War Homes to Become Available for Occupancy, attached to Blandford to Land, June 15, 1942 in MC gf 115-1-1.

[22] MC Minutes, pp. 22241-42 (June 18, 1942).

[23] Blandford to Land, June 22, 1942 in MC gf 115-1-1; MC Minutes, pp. 22517-18 (July 17, 1942).

but progress was slow according to Mr. Lawrence, and more housing than had been planned for was needed. Nothing at all had been done on the Gulf or at South Portland or Sturgeon Bay. On August 18, therefore, the Commission authorized the construction of 6,000 apartment units and a school for the Portland-Vancouver area. The three yards there were employing 46,000 at the time and planned to expand to 97,000. The management estimated an immediate need for 20,400 units and FPHA was building about 15,000. But the Commission did not have to carry through this project, for the NHA was able to take over.[24] During the next few months the Commission authorized the construction of apartments and dormitories at Richmond, apartments at Oakland, Pascagoula, Mobile, Panama City, Tampa, and Beaumont, and dormitories at South Portland.[25]

These projects were almost finished by the time the Vice Chairman of WPB warned Admiral Vickery in February 1943 that "drastic action" was about to be taken against the Maritime Commission because in constructing houses it had used ratings which had been assigned with the expectation that they would be used to build shipyards. Admiral Vickery defended the Commission's action by saying the housing was directly related to the production of ships, called attention to the agreement with NHA, and pointed out that WPB's Facilities Clearance Board had given its approval.[26] The matter seemed to indicate that one arm of WPB did not know what the other arm was doing, for the Director of the Program Bureau admitted approval had been given by the Clearance Board but that this Board had nothing to do with priorities.[27] Since the houses had been built, the question was highly academic by this time and Admiral Vickery promised to forbid the regional directors to issue any more priorities for housing without specific approval of WPB.[28]

In the midst of this exchange Mr. Lawrence reported, " The overall situation is becoming steadily more favorable." The only

[24] *Industrial Mobilization for War*, p. 295; Lawrence to USMC in Minutes, pp. 22747-50 (Aug. 18, 1942); p. 24805 (Apr. 8, 1943).
[25] MC Minutes, Aug.-Nov. 1942, indexed under " Housing Projects."
[26] Eberstadt to Vickery with attachments, Feb. 2, 1943 and Vickery to Eberstadt, Feb. 11, 1943 with attachments in MC gf 115-1-1.
[27] Donald D. Davis to Vickery, Feb. 7, 1943 in MC gf 115-1-1.
[28] Vickery to Sanford and other Regional Directors, Feb. 22, 1943 in MC gf 115-1-1.

requests for more housing were from the Kaiser yards.[29] Meanwhile, the NHA, once well started late in 1942, moved ahead quickly. Its progress was extremely important, for this agency constructed a great many more homes for shipyard workers than did the Commission. It projected over a hundred thousand FPHA units, and by April 1943 it had over 67,000 available for occupancy. The Commission had finished about 9,000. In all the areas where it was active, the FPHA built also. In addition to this it carried the entire load at Sausalito; Los Angeles; Savannah; Wilmington, North Carolina; Baltimore; and Sturgeon Bay.[30]

As family units on the West coast, both those built by Maritime and by the FPHA, were completed, the FPHA took over their management, but the dormitories serving the Kaiser yards were managed by the company itself which was reimbursed by the Commission.[31] On the Gulf, the Regional Office went into the housing business for a while. When Mr. Sanford, the Regional Director, was notified that some units were ready for occupancy at Panama City and that workers were clamoring to move into them, he authorized the construction company to rent them and then proceeded to follow the same policy at the other yards. Then, without consulting the Commission, he hired personnel and paid their salaries out of the rents they collected. Out of these receipts also the Regional Office supplied the projects and their occupants with certain necessities—maintenance tools, garbage cans, pick-up trucks, and furniture, the furniture to be paid for in instalments.[32] The chief accountant discovered this several months afterward and called for immediate suspension of such activities, since there was no authorization to spend rent monies for expenditures which were for construction and therefore should be charged to the facilities contract.[33] Mr. Sanford emphasized

---

[29] Lawrence to MC, Feb. 15, 1943, in MC gf 115-1-1.

[30] Lawrence to MC, Apr. 19, 1943, mimeographed, in Carmody file.

[31] MC Minutes, pp. 23627-29 (Dec. 1, 1942); Flesher to MC via Vickery, June 5, 1943 in MC gf 115-1-1-2; Minutes, pp. 25405-08 (June 10, 1943); Williams to MC, Aug. 31, 1943 in MC gf 115-1-1-2; Land (by Lawrence) to Blandford, Sept. 7, 1943; National Housing Agency, *Third Annual Report*, 1945, p. 31.

[32] " A Record of Facts, Causes and Events Leading Up to the Management and Operation of Maritime Housing Projects Gulf Coast Area United States Maritime Commission " and attachments, May 29, 1943, attached to Sanford to USMC, May 29, 1943, in MC gf 115-1-1.

[33] Telegrams, Slattery to Marshall, May 19, 1943, and May 20, 1943; Sanford (by Duff) to Walston Brown, May 23, 1943, in MC gf 115-1-1.

the great emergency with which he was faced, and he was granting the necessary interim authority to manage the projects until they were turned over to FPHA on July 1, 1943.[34] A year later, on July 1, 1944, the Commission transferred to NHA the land and buildings it had acquired for housing, excepting only the dormitories on the West coast.[35]

In general, dormitories were not popular; they were less used and brought in less revenue than was expected.[36] Those at South Portland contained many double rooms but workers did not want to double up, so that rentals brought in about $100,000 less a year than was estimated. As soon as the FPHA family housing was finished, toward the end of 1943, it was recommended the dormitories be closed, although union officials protested they should be kept going for the use of single men and women but with rents reduced and a better management.[37] At Vancouver the Hudson House had rooms for over 5,000; in April 1944 only half were occupied. Another dormitory equipped to take care of 7,000 was started but construction stopped after 5,000 units had been built. There were only 850 residents in this new dormitory in April 1944.[38] For the most part West coast dormitories were used as receiving centers for in-migrants to stay while they were being cleared with the U. S. Employment Service and the unions and given indoctrination courses and lectures about yard procedure.[39] Facilities of this kind were needed but only on a much smaller scale.

The pressure to do something fast, without delaying to investigate thoroughly and without any past experience to indicate how it would work, explains much of the wasted motion involved in the attempts of Federal agencies to help with the problems of

---

[34] " A Record of Facts," attached to Sanford to MC, May 29, 1943; MC Minutes, pp. 25250-52 (May 25, 1943); Peet to Vickery, June 8, 1943; Land to Blandford, June 1, 1943; Blandford to Land, June 12, 1943 attached to Marshall to Lynch, July 5, 1943 in MC gf 115-1-1.

[35] MC Minutes, pp. 28960-63 (July 4, 1944), p. 29128 (July 20, 1944), pp. 30040-41 (Oct. 26, 1944); Page to MC, June 28, 1944 and Blandford to Land, around this date, in MC gf 115-1-1.

[36] Lawrence to MC, Nov. 22, 1943 in MC gf 115-1-1.

[37] McInnis to Vickery, Dec. 2, 1943; McInnis to MC via Vickery, Dec. 7, 1943 approved by MC, Dec. 14, 1943, Simon A. Stopel, IUMSWA, to Sen. Brewster, Jan. 7, 1944 in MC gf 115-1-1-6.

[38] Lawrence to D. W. Fernhout, Oct. 20, 1943; Quarg to Slattery, Apr. 29, 1943, enclosing clipping from a Vancouver newspaper in MC gf 115-1-1-2.

[39] Page to MC, June 28, 1944 (approved by MC June 29, 1944) in MC gf 115-1-1. See also Lawrence to MC, June 19, 1943 in gf 115-1-1-3.

the shipyard communities. The men in the Maritime Commission struggling with these problems acted on the recommendations of other agencies in Washington or of yard managements, but were facing unfamiliar situations. They lacked means of consulting fully the persons most affected by their plans. In the communities which were being turned upside down, local leaders were sometimes cooperative, sometimes hostile to the sudden influx of outsiders being attracted by shipyard wages. The Maritime Commission was sometimes regarded as an outsider also, even as a particularly stupid, malicious, and dangerous outsider since it was able to use its powers of condemnation to take real estate and pull up a community by its roots. This appears clearly in the story, told below, of South Portland, Maine, and there are hints of it in the events at Mobile, Alabama.[40] To replan a community in a hurry, requiring all the sacrifices necessary to speed production, without inflicting undue hardship or permitting private parties to reap personal profits, was a most difficult task. It called for honest and intelligent cooperation by local business leaders, local politicians, labor leaders, shipyard management, and several government agencies.

An effort to provide this kind of cooperation was made when in April 1943 the President appointed a Committee for Congested Production Areas.[41] Representatives of the Committee acted as trouble shooters in 18 communities, in 9 of which the Commission had contracts—Portland, Oregon; Vancouver, Washington; San Francisco; Los Angeles; San Diego; Portland-South Portland, Maine; Brunswick, Georgia; Mobile, Alabama; Pascagoula, Mississippi; and Beaumont-Orange, Texas. The Committee's activities in these areas embraced everything from securing priorities for water tanks to releasing a flock of turkeys that had been "frozen," and the Maritime Commission acknowledged its helpfulness in straightening out many difficulties.[42] But the Committee on Congested Areas went to work relatively late; basic decisions about the acute problems had to be made in the spring and summer of 1942. An examination of some typical shipyard com-

[40] See chap. 8, sect. III and chap. 15. sects. II and III.
[41] Executive Order No. 9327.
[42] President's Committee for Congested Production Areas, *Final Report*, Dec. 1944 (Government Printing Office, 1944); Committee on Congested Production Areas, Status Report, Portland-Vancouver Area, Dec. 4, 1943, in MC gf 120-159; Schell to Gill, Nov. 8, 1943 in Division of Shipyard Labor Relations file, Housing.

munities will show both how much was achieved and how much remained to be done.

IV. SHIPYARD COMMUNITIES

From the point of view of the living conditions offered the workers, shipyards may be divided into two groups: those in or near metropolitan centers equipped to handle a large population, and those placed in what had been mere towns or very small cities. Extreme examples of the latter type were Panama City in Florida and Pascagoula in Mississippi.

When the Ingalls Iron Works decided to build a shipyard at Pascagoula, that town had a population of 4,000; in 1944 it had 30,000. The growth began before the war, and in 1940 the Commission arranged with the Navy for the building of 700 new houses there. It then found it had to take over for a while the operation of the housing and solve such problems as water mains, sewers, and schools. By 1943 those houses were reported the most sought-after homes in Pascagoula because, being built before material became short, they were much better than those built later.[1] During the war the NHA and the Maritime Commission put up 1,000 prefabricated " efficiency apartments " that had two beds, two chairs, and a hot plate. They formed a community without adequate sewerage, stores, or pavements. The tenants' association complained: " We have nothing but sharp shell walks and frequently these are under water over shoe tops. The shells are so sharp they cut cheap soles all to pieces. It is nearly impossible to get shoe repair work, and with rationed shoes the problem is acute, considering the fact that you can't get enough gasoline to do your shopping and the women often walk two miles a day for supplies." [2] One mother told a Congressional committee she ran " all over " trying to get milk for her children and then decided to get a job in the shipyard so that she could put them in the nursery school where they would be served milk. Garbage stood in the streets and the town was overrun with rats. Workers preferred to live quite a distance away and commute to the ship-

---

[1] Testimony of Harrison G. Otis, FPHA representative, in Senate, *Wartime Health and Education, Hearings* on S. Res. 74, before the Subcommittee of the Committee on Education and Labor, 78th Cong., 1st sess., pt. 2, Dec. 16, 1943, p. 785. Extensive correspondence in MC gf 115-1-1-1, Sept. 19, 1940 to Sept. 1941.

[2] Senate, *Wartime Health and Education, Hearings*, p. 764.

yard; for they "would rather go to the trouble of riding that distance . . . so they can leave their children in a school, get a doctor once in a while, get their groceries, get their checks cashed, and have a few chickens to help them out in this high cost of living."[3]

Panama City grew from 20,000 in 1940 to 60,000 in 1943.[4] The Maritime Commission arranged for the company building the shipyard to provide at the same time the essentials to attract workers. The J. A. Jones companies built not only the 6-way shipyard but also houses and cafeterias. It even delivered milk and ice to the workers' families, and was allowed to set up a consolidated service account so that losses encountered in delivering ice, for example, since they were not separately reimbursable as costs of shipbuilding, could be charged off against profits made in other activities.[5]

The most difficult situations were encountered in cities somewhat larger to begin with but overwhelmed by the increase occasioned by a large shipyard. Such was the fate of Mobile, Alabama, the metropolis called on to service not only the workers at the 10-way yard of the Alabama Shipbuilding and Dry Dock Co. in Mobile harbor, but also many of those at nearby Pascagoula, and at the Gulf Shipbuilding Co. at Chickasaw a few miles up the river. The Mobile Metropolitan District contained 114,906 persons in 1940; in 1944 it had 201,369.[6] In spite of assistance from Federal agencies, Mobile could not keep up with the demands made upon it.

On September 19, 1942, Alabama Dry Dock was shut down because its employees protested paying 5 cents to ride the ferry. A union official summed up the situation, " The toll ferry . . . is a symbol in the minds of the workers of all the foul hell they have had to endure in Mobile, not only at work but throughout their daily lives." In 1941 there had been two ways of getting from the mainland to Pinto Island where the yard was located— through a tunnel where the toll was 25 cents each way or by

---

[3] *Ibid.*, p. 619, and other testimony, pp. 610-1030.
[4] *Fortune,* July 1943, pp. 100, 218.
[5] MC Minutes, pp. 29804-07 (Sept. 26, 1944).
[6] Committee for Congested Production Areas, Status Report, Mobile, Feb. 19, 1944, in MC gf 120-159; " Housing for War," *Fortune,* Oct. 1942, pp. 190, 193; U. S. Dept. of Commerce, Bureau of the Census, *Population, Congested Production Areas* (Series CA 1944) CA-1, No. 1, June 15, 1944, and CA-3, No. 1, July 20, 1944.

riding free on a ferry operated by the company. In the fall of 1941 the tunnel toll was reduced to 10 cents and at the same time a city ordinance was passed which prevented the shipbuilding company from operating the ferry. Another private company ran it for awhile and then the Maritime Commission arranged that the city of Mobile operate the ferry.[7] On September 17, 1942, the City began charging fares on the ferry, and the strike resulted. Following this outburst by the workers and charges that one ferry was unsafe, the Commission authorized purchasing the boats from the city at a cost of approximately $100,000 and the further expenditure of $99,000 for repairs and improvements in landing facilities. Three more boats were chartered, and a contract for operating them was given to the same company that had been running the ferry before the city took over.[8]

The other complaints of the employees of Alabama Dry Dock in the fall of 1942 centered around housing and prices. Single men had no trouble getting a room provided they did not object to paying high rents. Alternatives were a jalopy, or a park bench, or " hot beds " with as many as seven in a room. Some couples located places with cooking facilities, but if they had children they usually ended up in a tent, a trailer, or a shack. The city was so short of water that housewives kept their taps open all day to capture the trickle. Sewers were desperately needed. There was only one doctor for every 3,000 persons, and there was a shortage of hospital beds. It was reported that the price of bed sheets had gone from 98 cents to $1.98.[9]

NHA scheduled a huge program of about 9,000 units in Mobile. A few had been finished in October 1942 and most of the rest were under construction, but Mr. Lawrence considered the immediate emergency warranted the Commission's building 1,500 apartment units for Alabama employees.[10] The site chosen was Blakely Island which was connected to Pinto Island by a bridge and was within walking distance of the shipyard.[11]

[7] MC Minutes, pp. 23464-67, 23469 (Nov. 12, 1942).
[8] MC Minutes, pp. 24398-400, 23464-67, 23469 (Feb. 19, 1943); Sanford to MC, May 2, 1946 in MC Historian's Collection, Research file 205.7; statement of T. E. Shaw, Aug. 26, 1948.
[9] MC Minutes, pp. 23467-68 (Nov. 12, 1942); H. Hill, Gulf Shipbuilding Corp. to Lawrence, May 18, 1942, enclosing resolution adopted by State Council of Administrators in MC gf 115-1-1; " Housing for War," *Fortune*, pp. 190, 193.
[10] MC Minutes, pp. 23134-35 (Oct. 6, 1942); p. 24381 (Feb. 18, 1943).
[11] Statements of T. E. Shaw, Aug. 26, 1948.

By February 1943 the NHA had almost half its units ready for occupancy.[12] In July 1943, 1,230 of the Commission's apartments were finished and Mr. Sanford recommended that no more be built for the time being. Instead he asked that the rest of the money be spent to make the Blakely Island community "self-contained." There was urgent need for topsoil, sprigging, fences, sidewalks, parking spaces, and street signs. Lest anyone question the need for fences, Mr. Sanford explained they were to keep the children off the railroad tracks and from falling into the river. The community needed a school, a day nursery, police station, fire house, barber shop, beauty parlor, and a "town hall" for meetings and recreation. The Commission approved this change in plan but reduced the proposed expenditure from $350,000 to $200,000.[13]

In spite of these efforts the housing shortage continued, and there was even more of a lag in supplying community services. In February 1944 the opening of the filtration plant was being held up for want of a single valve.[14] The war was almost over before Mobile caught up with itself, and in the meantime the discontent of workers with their living conditions had been expressed in "quits," in absenteeism, and even in strikes.[15]

Relatively favorable, in contrast, was the situation surrounding the yard in the small city of Wilmington, North Carolina (population 33,407 in 1940). Most of the employees were recruited from within a radius of about 200 miles and a great many traveled back and forth daily from their farms.[16] Nevertheless the population of the county in which Wilmington was located increased from 48,000 to 60,000 in two years. Contributing to this increase were several military installations which made the towns around Wilmington more congested than the shipbuilding center itself. Temporary housing trailers were moved in, and the workers were so enthusiastic about them that the trailer camp not only became permanent but was enlarged. Twenty-three hundred persons were

---

[12] MC Minutes, p. 24381.

[13] MC Minutes, pp. 25540-41 (July 24, 1943).

[14] Committee for Congested Production Areas, Status Report, Mobile, Feb. 19, 1944.

[15] On the race riots at Mobile see chap. 8, sect. III.

[16] North Carolina Shipbuilding Co., *Five Years of North Carolina Shipbuilding*, pp. 15-17.

living in 730 trailers at the beginning of 1944.[17] The Maritime Commission scheduled some housing for Wilmington but abandoned the project when NHA proved it was able to do the job instead.[18] This agency provided 6,039 family units which housed 25,350 persons, or about 4 to a unit. Private groups built an additional 1,423 units. All this housing was within walking distance of the yard. To serve those who drove to work new roads were built and parking areas provided. The Maritime Commission helped out by sending eleven 100-passenger buses to carry those dependent on public transportation.[19]

The residents of Wilmington faced other difficulties. They had trouble buying food, and the presence of so many military personnel made eating in restaurants almost impossible. Hospital beds had always been few in this community, but satisfactory progress was made in increasing them. The sewerage system was also improved. There were enough schools. By the beginning of 1944 the greatest need in Wilmington was for more stores and for recreational facilities.[20]

Among large cities, Baltimore is an example of one relatively well able to take care of the addition to its population. In the first wave of expansion, housing, health, education, and recreation facilities were sufficient. City officials were confident they could take care of shipyard workers as well as those in aircraft and many other prime defense jobs.[21] The Commission built no housing here and supplied no transportation. The NHA built 2,100 units for shipyard workers.[22] Access to the big Fairfield yard was difficult because it was across the river from the main part of the city and there was only one narrow bridge. Many workers used a ferry and the transit company was able to arrange for enough streetcars to service the yard. Children went to school a bit later in the morning in order to ease the transportation problem, but two shifts were never necessary. The several large

[17] Committee on Congested Production Areas, Report on the Adequacy of Services and Facilities in the Wilmington Area, North Carolina, n. d., about Jan. 1944, in MC gf 120-159.
[18] MC Minutes, p. 23168 (Oct. 8, 1942).
[19] *Five Years of North Carolina Shipbuilding.*
[20] Committee on Congested Production Areas, Report on Wilmington Area.
[21] House, *National Defense Migration, Hearings,* before the Select Committee Investigating National Defense Migration, 77th Cong., 1st sess., pt. 15.
[22] MC Minutes, p. 24381 (Feb. 18, 1943).

department stores were open one night a week, and although natives told with solemn faces stories of women in bare feet purchasing mink coats and were indignant at the sight of others strolling down the most fashionable street in slacks, the majority of Baltimore's citizens were proud of the shipyards. The city was crowded and " hot beds " were by no means unknown, but it was not bursting at the seams. Nevertheless " quits " at Bethlehem-Fairfield ran very high because Baltimore, unlike Wilmington, had many other defense industries and a good many workers preferred working indoors even if they earned a bit less money.[23] So far as the community was concerned it was equipped to absorb whatever number of employees were needed because it started with facilities for almost a million, and when the population tipped the scale on the other side it did not have the relative impact experienced in smaller centers.

Although cities on the Gulf and the East coast had their difficulties, the biggest concentration of new shipyards was on the West coast. The shipyards there were all in or adjacent to large cities, but operations were so large and the in-migration so great that cries for assistance from this area began earlier and lasted longer than in any other. Around San Francisco Bay alone the Commission's shipbuilding program cost over $2\frac{1}{2}$ billion dollars; around Portland, Oregon, over 2 billion dollars; compared to about $1\frac{1}{2}$ billion in the entire Gulf area. Topping all communities in amount of expenditures and accounting for over a million and a half of San Francisco Bay's share was Richmond, California, where four of the Kaiser yards were located.[24]

Richmond is not far from Oakland, the home and headquarters of Henry Kaiser. It had space, a deep water port, and, in 1940, 23,000 inhabitants. By 1943 the little town had mushroomed to 100,000. The Negro population rose from 400 to about 14,000. Workers came from all over the Middlewest and Southwest—the Oakies, Arkies and Texies were most evident. Thousands of workers lived in Richmond and depended on it for various services while other thousands lived in outlying districts, adjacent towns, or San Francisco across the bay.[25]

---

[23] Statement of W. O. Weyforth, Dec. 1946.
[24] Fischer, *Statistical Summary*, Table H-7, p. 164.
[25] " Richmond Took a Beating," *Fortune*, Feb. 1945, p. 264; " An Avalanche Hits

Transporting workers to and from the Richmond yards was first approached by the management in terms of access roads, not public transportation. Construction of new roads dragged on and was still being discussed in February 1942. Meanwhile plans had changed; instead of one 7-way yard, provision had to be made for four yards with a total of 27 ways. Then attention was diverted from roads to public transportation because of the rubber shortage.[26] The Commission authorized the requisitioning of ferries to ply between San Francisco and Richmond and the expenditure of over 7 million dollars for construction and equipment of a railroad line to serve the Richmond-Oakland area. Its coaches were old New York El cars.[27] But the vast majority of workers continued to use automobiles to come to work. In January 1944 7 per cent were using the train; 6 per cent, the ferry; 6 per cent, buses, and 66 per cent, automobiles.[28] Faced with this fact the Commission agreed to the expenditures of money for improving access roads.[29]

The FPHA scheduled 3,000 dormitory units, 2,000 apartments, and 1,000 houses, and in the fall of 1942 construction was under way, but by that time it was clearly foreseen they would not be sufficient.[30] To ease the situation the Commission authorized the lease of a Montgomery-Ward warehouse which would house 500 single men, and the construction of 6,000 apartment units and a school building.[31] Big as this seemed in prospect it was not enough. In December 1942 the company asked the Commission for more housing. They wanted 4,000 apartment units plus stores, schools, and a hospital. The company reported it had on hand 10,000 applications for housing for persons with families.[32]

Richmond," A Report by the City Manager, City of Richmond, Cal., July 1944 (mimeographed), pp. 1, 9, 13, in Labor Department Library.

[26] Minutes of a Conference enclosed in Carroll to Hope, Oct. 4, 1941; Henry Kaiser, Jr. to J. A. McVittie, City Manager, May 27, 1941; Bedford to Schmeltzer, June 3, 1941; Schmeltzer to Bedford, June 11, 1941; Lalley to Bedford, Jan. 8, 1942 in MC gf 507-3-11; Jago to Lorne, Feb. 11, 1942 in MC gf 503-55.

[27] MC Minutes, pp. 22046-47 (May 26, 1942); *Fortune,* Feb. 1945, pp. 264-65.

[28] MC, Manpower Survey Board [Report], Richmond Shipyards, Jan. 1944.

[29] MC Minutes, p. 24755 (Apr. 2, 1943).

[30] MC Minutes, pp. 22747-50 (Aug. 18, 1942) and Lawrence to MC, Aug. 31, 1942 in MC gf 115-1-1-3.

[31] Lawrence to MC, Aug. 31, 1942 [approved by the MC, Sept. 10, 1942] and Sept. 8, 1942 [approved by MC, Sept. 10, 1942] in MC gf 115-1-13, referred to in memo as Chevrolet warehouse but later corrected.

[32] Bedford to Vickery, Jan. 14, 1943 in MC gf 115-1-1-3, and Status of Housing in

On March 25, 1943 the Commission authorized the construction of 4,004 apartment units and 71 utility buildings.[33] Altogether the FPHA and Farm Security Administration built 14,000 units; the Maritime Commission 10,000; and private industry, 6,000.[34] All but a fraction were " temporary " with all the deficiencies of construction which that word implies.[35]

While waiting for a decent place to live, many set themselves up in shacks and trailers, some of which were unfit for human habitation. In one, originally a chicken house, there was a man, wife, and seven children. About half of the trailer camps were deemed " good " by inspectors in April 1943 and about 30 per cent with about 600 trailers were " fair." But about 500 trailers were located in " bad " parks where conditions were felt to be beyond correction.[36]

Although housing had not begun to keep pace with needs, community facilities had lagged even more. No new markets had been built since the boom. Fire protection was woefully inadequate. There were not sufficient hospital beds nor doctors to care for the sick. The water supply was adequate, but the sewerage system was not. Although garbage collection was regular, such a shortage of garbage cans had developed that much refuse was scattered about. Schools were overcrowded; one in the heart of the shipyard workers' district with a normal capacity of 245 had an enrollment of 518; another supposed to take care of 800, had 2,540 pupils. As a result the children attended in shifts, and because no one knew when they were supposed to be in school, juvenile delinquency rose to alarming proportions.[37]

Unless both husband and wife worked, weekly wages for persons with families were not lush. The average worker who made $61.00 a week spent $18.50 for food, $8.50 for rent, $6.75 for

Richmond, Data Respecting Manpower, Housing, Community and Commercial Facilities at Richmond Shipyards, Apr. 8, 1943, exhibit 1, MC gf 115-1-1-3.

[33] Flesher to MC, Mar. 24, 1943 [approved by MC, Apr. 25, 1943] in MC gf 115-1-1-3.

[34] *Fortune*, Feb. 1945, p. 264; MC Minutes, pp. 22933-35 (Sept. 10, 1942), p. 23815 (Dec. 17, 1942); pp. 24188-89 (Feb. 2, 1943).

[35] " An Avalanche Hits Richmond," pp. 33-34.

[36] Federal Security Agency, Office of Defense Health and Welfare Services, Visual Inspection of Trailer Camps and Shack Developments in the Unincorporated Area Surrounding Richmond, California, in Status of Housing in Richmond, in MC gf 115-1-1-3, exhibits.

[37] *Ibid.*, and " An Avalanche Hits Richmond."

recreation and household expenses, $6.10 for war bonds, $6.70 for taxes, $5.50 for clothing, $2.45 for transportation, 50 cents for hospitalization insurance, and had $6.00 left for savings and incidentals. All this to live in a town the editors of *Fortune* summed up this way:

> Huge barracks-like public housing projects cover the mud flats between the harbor and the town. The sidewalks are blocked by gaping strangers in cowboy boots, blue jeans and sombreros. Women in slacks and leather jackets and shiny scalers' helmets wait in long lines to buy food. Pale-haired children and mangy hound-dogs wander the treeless streets. Nobody knows anybody.[38]

Conditions in many ways like those at Richmond could be found elsewhere on the Pacific coast in housing projects, such as that at Vanport near Portland which was finally swept away by floods in 1948. Yet for many they were an improvement over a sharecropper's farm; there was excitement; there was knowledge of helping Uncle Sam win the war; it was, if they stopped to think about it, far better than a foxhole.

The main reason living in these and other shipyard communities became so difficult was the suddenness with which the shipbuilding program was expanded. Any person intrepid enough to recommend in 1940 that vast communities be built to take care of thousands of shipyard workers would have been considered an alarmist.

In retrospect, after experience with the problem of maintaining a stable labor force, the Director of the Commission's Division of Shipyard Labor Relations summed up the importance of community planning as follows:

> Once the lure of the higher "take-home" is out of the picture in the geographic area which is determined to be a common labor market . . . other stabilizing forces begin operation . . . [namely] domestic and social satisfactions. Workers who have families desire education for the children, a place in community life for themselves, friendships and associations, religious opportunities by some, fraternal connections by others—in other words, some sort of domestic and social roots. Workers who do not have families rarely overlook the opportunity, if it presents itself, of setting up a family life where they observe that domestic and social satisfactions tend toward happy existences. All these matters flow from stability of labor and are in turn contributing factors to such stability and its continuance. In contrast, where the

---

[38] *Fortune*, Feb. 1945, pp. 262-64.

labor market is constantly churning, so that roots have no chance to take any hold, the "restlessness" primarily produces irresponsible attitudes and behaviors, and the irresponsibility in turn produces readiness to listen to advocates of strikes, slow-downs, production interruptions of various types.[39]

Although shipyard workers showed little readiness to listen to those who might have advocated strikes, many who were dissatisfied showed it by quitting or absenteeism, which slowed production and increased costs.

V. Health and Safety

Another aspect of the struggle to reduce turnover and absenteeism consisted of efforts to improve health and safety within the shipyards themselves. Ship construction is an industry with many hazards, for extremely heavy materials are lifted and placed; much moving and climbing about on heights and in narrow spaces is necessary; the processing of steel sends off fumes and sparks and splinters of metal, and the tools themselves are either heavy or hot. The danger of accidents and of respiratory diseases was increased many times when the shipyard labor force was swelled by the inexperienced. Yard managements of course had the initial responsibility for meeting these dangers, and from the beginning of the emergency program the Maritime Commission assigned the task of checking on safety measures to the Resident Plant Engineer. Nevertheless some reports caused alarm. In the summer of 1942 therefore the Commission, following the recommendations of Commissioner Carmody and Mr. Ring, approved the appointment of Dr. Philip Drinker of the Department of Industrial Hygiene of the Harvard School of Public Health and John M. Roche of the Safety Engineering Staff of the National Safety Council to make a study of health and safety in the yards.[1] As a result of their survey, it was decided to enlist the Navy's help in promulgating some minimum standards and to send consultants into the field to enforce them. Standards were agreed upon by representatives of the shipyards in December 1942.[2]

---

[39] Comments by Daniel S. Ring, Dec. 18, 1947 in MC Historian's collection, Research file 105. See also: Ring to Vickery, May 23, 1945 in MC gf 508-1.

[1] Interview with Daniel S. Ring, Dec. 15, 1947 in MC, Historian's collection, Research file, 105; MC Minutes, pp. 22246-47 (June 18, 1942); p. 22651 (Aug. 4, 1942).

[2] Ring, and Fisher of the Navy, to Vickery and Bard, Nov. 9, 1942; Ring to MC,

Actually the requirements were "minimum," and there were not many yards which did not already meet them.[3] Yet they were extremely important because they focused attention on health and safety and thereby prevented the relaxation of effort which so often occurs once a " system " has been installed.

"The Minimum Requirements for Safety and Industrial Health in Contract Shipyards," which were approved by the Navy on January 20 and by the Commission on February 9, 1943, was divided into *shall* and *must* items and *should* and *may* items. Pre-placement physicals had to be given all employees and periodic check-ups given those working in dangerous occupations. It was recommended but not required that yards having up to 5,000 men have two full-time doctors and one additional doctor for each additional 5,000 employees. Inspection of food and its preparation and of the water supply and disposal system had to be made at frequent intervals by qualified personnel.

The manual listed occupational diseases most likely to occur among shipyard workers and enumerated methods for their prevention. These diseases were of two types: injuries to the eye as a result of exposure to light from a welding arc or its penetration by flying objects; and injuries to the respiratory tract from inhaling various types of fumes. Two preventive measures were insisted upon—protective equipment and ventilation.[4] When the welder got down to work he might look like " a man from Mars," but he had for that reason a better chance of remaining a healthy earthling.

The safety program depended on constant vigilance by a safety director and on inspection committees which were to be set up for each department. Dr. Drinker and Mr. Roche were pleased with the progress made. A year after the inauguration of their program they could write to the regional consultants that " while compliance with suggestions and with Minimum Standards is sometimes slow we feel that compliance, by and large, has been gratifyingly good." [5]

Nov. 24, 1942 [approved by MC, Nov. 26, 1942] and Ring to MC, Jan. 4, 1943, [approved by MC Feb. 9, 1943], in MC gf 503-1-6.

[3] Roche to C. G. Ewerts of Manitowoc yard, Feb. 24, 1943 in MC gf 503-1-6.

[4] U. S. Navy Department, U. S. Maritime Commission, *Minimum Requirements for Safety and Industrial Health in Contract Shipyards* (Government Printing Office, 1943).

[5] Drinker and Roche to All Health and Safety Consultants, Dec. 3, 1943 in MC gf 503-1-6.

During 1942 there were 9.2 injuries for every 100 employees; in 1943 there were 7.3; in 1944, 5.5; and in 1945, 4.9.[6] The number of deaths occurring as a result of accidents was small in relation to total number of disabling injuries and to the large number of persons employed. Nevertheless the total was large: there were 700 fatal accidents in the shipyards—that is in both Maritime Commission and Navy yards—during 1943. Those who studied the causes of these accidents and compared them with other accidents which did not result in fatalities concluded it was only chance which kept the figure down to this level.[7]

Besides stressing the importance of protective devices against welding fumes and flash burns, and keeping tab on sanitation, the Commission's health consultants went into such common shipyard complaints as welder's cough and "shipyard eye." Welder's cough was accompanied by symptoms such as spitting blood, and its victims naturally feared tuberculosis.[8] The conclusion reached by the U. S. Public Health Service after a study of a selected group of shipyard workers was that inhalation of welding fumes brought on mild respiratory infections and the more serious "arc-welder's siderosis," but did not cause or predispose to tuberculosis. All infections caused only temporary disability.[9] However, these findings were not available until after the war was over, and although efforts were made to convince welders their symptoms were not serious, many refused to continue the occupation, either because they had serious fears or because they wished to rid themselves of the discomfort they were suffering.

As to shipyard eye, it was just plain pink eye and not peculiar to workers in shipyards. It spread through the West coast and

[6] Computed from figures on employment in MC, Historian's collection, and from the Final Report, Joint Navy Department-Maritime Commission Safety Program in Contract Shipyards 1943-44-45, Prepared by the Division of Shipyard Labor Relations, United States Maritime Commission, Based on Information Submitted to the Bureau of Labor Statistics, Department of Labor (mimeographed, Washington, D. C., Dec. 1, 1947), pp. 5a-6; "Shipyard Injuries During 1943," prepared by Frank S. McElroy and Arthur L. Svenson, *Monthly Labor Review*, May 1944, p. 1005; "Work Injuries in the United States During 1943," *ibid.*, Nov. 1944, p. 905; "Work Injuries in the United States During 1944," prepared by Max D. Kossoris, *ibid.*, Oct. 1945, p. 638.

[7] "Fatal Work Injuries in Shipyards, 1943 and 1944," *ibid.*, July 1945, p. 75.

[8] Drinker to MacPhedran, Aug. 27, 1943, in MC gf 503-1-6.

[9] Waldemar C. Dreesen, *et al.*, "Health of Arc Welders in Steel Ship Construction," *Public Health Bulletin No. 298* (Government Printing Office, 1947), p. 167.

HEALTH AND SAFETY

Gulf yards in the winter of 1942-1943.[10] Shipyard eye gave the yards a bad name and it was difficult to convince persons it did not come from welding-arc flashes or fumes.

The work of the Commission's Health and Safety Consultants was of value in two ways—by allaying fears that working in a shipyard was more dangerous to life and limb than working somewhere else, and by making this true through insistence on a high standard of protection and precaution.[11] The Maritime Commission cooperated with them by allowing funds for special studies, providing the yards with clinical equipment, and by paying salaries to doctors and nurses.[12] It followed through on the recommendations of the health consultants when they pointed out the need in some yards for better sanitary facilities.[13]

When, in 1943, women began to be hired in substantial numbers, the President wrote Admiral Land urging the necessity for more cafeterias, rest rooms, toilets and locker facilities. Admiral Land assured the President the Commission had authorized and would continue to expand such facilities.[14] During 1943 and 1944 the Commission voted funds for increased facilities in many yards to take care of at least the minimum needs of the women employed there, and to improve facilties in general. But its attitude was cautious. As Admiral Land told a Senate Committee early in 1944, " Apparently the sentiment of the country and the policy of the country . . . requires us to do more than we have ever done before [in the way of sanitation, health, safety, in-plant feeding, housing, transportation . . .hospitalization, child care for women workers]. I do not say that it is 100 per cent justified, but I am not going to buck the tide on in-plant feeding or hospitalization or sanitation, because it is just futile." [15] The Commission moved with particular caution in the matter of in-plant feeding.

---

[10] Philip Drinker, " Health and Safety in Shipyards," American Merchant Marine Conference, *Proceedings,* 1944, p. 111.

[11] " Health of Arc Welders in Steel Ship Construction."

[12] MC Minutes, pp. 23972-73 (Jan. 5, 1943) ; pp. 24011-12 (Jan. 7, 1943) ; pp. 24313-14 (Feb. 12, 1943) ; pp. 24827-28 (Apr. 9, 1943).

[13] MC Minutes, pp. 25535-36 (June 24, 1943) for example; Land to Rep. Hugh Peterson, Mar. 20, 1943 in MC gf 503-1-6.

[14] The President to Land, Dec. 22, 1943 and Land to the President, Dec. 22, 1943 in Division of Shipyard Labor Relations file, Feeding.

[15] Senate, *Independent Offices Appropriation Bill for 1945, Hearings* before the Committee on Appropriations, 78th Cong., 2d sess., Feb. 17, 1944, p. 189.

The Commission had, to be sure, supplied funds for building cafeterias, but except in some yards on the East coast they were for executives and office employees only. In nearly all the yards workers either carried their lunches or bought cold sandwiches which they ate sitting down somewhere on the platens or ways.[16] There were increasing complaints about this way of doing things or about the quality of the food where there were cafeterias. During the summer of 1943 the Committee on the Merchant Marine and Fisheries inquired about the problem of in-plant feeding everywhere it went on the Pacific coast, and in October adopted a resolution recommending that such facilities be provided.[17]

In July 1943 the Maritime Commission appointed Horace D. Willis, who was connected with an industrial cafeteria concern, to make a survey.[18] He found what was being done—that is, the cleanliness, the quality, the price of the food—to be on the whole good. He recommended equipping the yards with mobile canteens which would serve hot meals, the expansion of kitchens where necessary, and the provision of mobile shelters in which to eat.[19]

His recommendations were vigorously opposed by Mr. Flesher, the West coast Regional Director, who charged that the whole matter had been stirred up for political reasons, and who foresaw that the Commission would be " criticized upon the concessioners we select, the price of meals, the quality and quantity of food, and in case of ptomaine poisoning, the workers will insist that they be paid for the time lost and for hospitalization and medical services." [20] The unions pushed for it, particularly through Congressman Welch of California, a member of the Merchant Marine and Fisheries Committee, and Judge Bland, its Chairman, announced his intention to explore the entire matter thoroughly.[21]

---

[16] Horace D. Willis, MC Chief Feeding Consultant, to Ring, Sept 27, 1943, in Division of Shipyard Labor Relations file, Feeding; testimony of Ring in House, *Production in Shipbuilding Plants, Hearings,* pt. 4, pp. 1192, 1195.

[17] House, *Production in Shipbuilding Plants, Hearings,* pts. 2-4 *passim.* The resolution of the Committee was Oct. 26, 1943. *Ibid.,* pp. 1199-1200.

[18] MC Minutes, pp. 20572-03 (July 15, 1943).

[19] Willis to Ring, Sept. 27, 1943, in Division of Shipyard Labor Relations file, Feeding.

[20] Telegram, Flesher to Land, Sept. 23, 1943 in Division of Shipyard Labor Relations file, Feeding.

[21] San Francisco Labor Council to Hon. Richard J. Welch, Sept. 10, 1943, enclosing

On October 28, 1943 the Commission approved compromise recommendations on in-plant feeding made by Mr. Ring. He suggested a start be made in the San Francisco Bay Area and gradually extended in yards where the need was "urgent and pressing." [22] The Commission finally sponsored large-scale feeding facilities in only four yards—Calship, Consolidated, Western Pipe, and Moore.[23]

In testifying before the Merchant Marine and Fisheries Committee in January 1944 Mr. Ring confessed that in-plant feeding was something that "should have come to my attention far sooner and that I should have done a lot more yelling about it right at the outset." He continued in explanation, "But when we were trying to get a program started, when all of the urge was on production, production, production, we sometimes lost sight . . . of what would achieve the greatest production until other things were brought to my attention." [24]

VI. Worker Participation

Incentive systems of various sorts commonly used in industry were put into effect by shipyard managements individually in order to secure greater efforts from their workers and to utilize their suggestions. General, industry-wide efforts to stimulate incentive took the form of (1) labor-management committees, (2) appeals to patriotism through posters, broadcasting, and so on, (3) honorific awards to yards of pennants and other insignia on a competitive basis, and (4) awards in war bonds to individuals who made suggestions to increase efficiency.

A proposal that joint committees of labor and management be set up in all yards to discuss how to break bottlenecks which were holding up production was made on March 3, 1942 by Donald Nelson, Chairman of the War Production Board, after consultation with President Roosevelt.[1] The suggestion came at the

---

Resolution adopted Sept. 10; Carmody to Chairman, Oct. 11, 1943; Bland to Ring, Sept. 27, 1943 in Division of Shipyard Labor Relations file, Feeding.

[22] MC Minutes, pp. 26651-53.

[23] MC Minutes, pp. 27298-301 (Jan. 11, 1944); pp. 28186-88 (Apr. 18, 1944); pp. 29432-35 (Aug. 15, 1944); pp. 30294-95 (Nov. 21, 1944); pp. 30415-17 (Dec. 5, 1944).

[24] House, *Production in Shipbuilding Plants, Hearings*, pt. 4, p. 1195.

[1] *New York Times*, Mar. 4, 1942, p. 10; War Production Board, *Official Plan Book, War Production Drive* (Government Printing Office, n. d.).

time when ship production was behind schedule, in the midst of controversy between capital and labor, and during Admiral Land's barrage of attacks on strikes and loafing. Because of this atmosphere, Admiral Land had some reservations, although the Maritime Commission formally endorsed the plan. "I am 100 per cent in agreement with his [Nelson's] statement as he promulgated it," Admiral Land told a Congressional committee, "but it can be misinterpreted both by capital and labor." [2] There was talk to the effect that such committees might "sovietize" American industry.[3]

In spite of Mr. Nelson's assurance that his proposal would "not put management in labor or labor in management," [4] the Maritime Commission did not immediately throw its weight back of his suggestion.[5] Some yards acted directly in response to the appeal of the WPB Chairman, however, without waiting for urging from the Commission; [6] and within the Commission, Mr. Carmody took a lively interest in this development. In June he wrote several companies inquiring about their experience. The tenor of the replies varied from substantial enthusiasm at Houston to the feeling at Federal that the committee served no useful purpose. Although most of the managers reported growing pains due to misunderstandings about the purpose of the committees, since unions looked to them as forums where they could air grievances, most saw in them something useful.[7] The Commission voted on June 30, 1942 that Commissioner Macauley should write all shipyards urging them to set up labor-management com-

---

[2] House, *Hearings on H.R. 6790*, p. 2969.

[3] Bruce Catton, *The War Lords of Washington* (New York, 1948), pp. 145-46.

[4] WPB, *Official Plan Book, War Production Drive*, p. 5.

[5] A letter drafted by Paul Porter to be sent over the signatures of Rear Admiral Fisher of the Navy and Commissioner Macauley asking the yards to set up such committees was approved by Macauley. Porter to Fisher and Macauley, Apr. 13, 1942, and Macauley to Porter, Apr. 15, 1942 in MC, gf 508-1. But later correspondence indicates that the Navy sent a separate letter, and on July 3, Mark O'Dea wrote as if no letter had been sent by the MC previously. O'Dea to Macauley, July 3, 1942 in MC gf 508-1.

[6] On June 17, 1942 the following had such committees: Ingalls, Calship, Consolidated, Seattle-Tacoma, Moore, Houston, Concrete Ship, Pacific Bridge, Sun and Federal. Mark O'Dea to Vice Chairman, June 17, 1942 in MC gf 508-1.

[7] Carmody to Korndorff, Pew, Lanier, Stout, Bechtel, etc., June 22, 1942; Stout to Carmody, enclosing announcements and minutes, June 27, 1942; J. H. Love, Federal, to Carmody, July 7, 1942; and others around this date in Carmody file, Labor-Management Committees.

mittees which would operate along the lines the WPB had laid down in its *Official Plan Book* for the War Production Drive.[8] He received replies indicating willingness to cooperate and by January 1943 such committees had been formed in all yards except Rheem, MacEvoy, and North Carolina.[9] Indeed the shipbuilding industry (including Navy yards) was second only to the ordnance industry in the relative number of workers covered by labor-management committees (15 per cent).[10]

On other features of the drive proposed by WPB the Maritime Commission acted more vigorously. As part of a Victory Fleet Drive efforts were made to appeal directly to the individual worker. The Division of Public Relations was placed in charge to work with the WPB in spreading information on the War Production Drive and in developing an Award of Merit; a new set of posters was approved;[11] and a speakers' bureau was organized for an experimental three months' period under Col. C. M. Paul, who proposed to enlist the services of workers who had received merit awards, survivors of torpedoed ships, and others to arouse the workers to greater productive effort.[12]

Propaganda directed towards urging shipyard workers to greater efforts was of course but a part of the general campaign being carried on by newspapers, radios, newsreels, and every other organ of publicity to arouse patriotic spirit throughout the nation. Shipyard workers were influenced by these general appeals as well as by those directed at them specifically.

The Maritime Commission's system of merit awards centered about the right of a yard to fly an " M " pennant. The Army and Navy were making a practice of awarding to their suppliers " E " pennants for excellence; and in March 1942 the Navy was proposing to give its " E " to Oregon Ship. Thereupon the Maritime

[8] MC Minutes, June 30, 1942, pp. 22353-54 (June 30, 1942), pp. 22281-82 (June 23, 1942).

[9] H. G. Tallerday to Macauley, Aug. 5, 1942, mentioning Macauley's of Aug. 1; Korndorff to Macauley, n. d., received Aug. 4, 1942; Pew to Macauley, Aug. 7, 1942; Rheem to Macauley, Aug. 7, 1942, and Lalley to Vickery, Jan. 9, 1943, all in MC gf 508-1.

[10] Carol Riegelman, *Labour Management Co-operation in United States War Production: A Study of Methods and Procedure* (Washington, D. C., International Labour Office, 1948), p. 184.

[11] MC Minutes, pp. 21621-22 (Apr. 10, 1942); pp. 21753-55 (Apr. 24, 1942); pp. 21958-59 (May 15, 1942).

[12] *Ibid.*, pp. 21796-801 (May 1, 1942).

Commission established its own pennant, or rather a whole series of pennants, and gave them to yards which met or exceeded their schedules. Lapel pins bearing the Maritime Commission insignia and " Award for Merit " were distributed to the employees of the winning yards. Later, " M " pennants were given very generally to yards, taking into account their workmanship and their success in overcoming difficulties. Admiral Vickery kept the Division of Public Relations on its toes designing new elaborations on flags to reward the extra merit of the outstanding yards. The actual conferring of the awards was a ceremonial occasion usually combined with a launching and attended by a Commissioner and local celebrities.

Awards were made to suppliers of components as well as to shipyards and went therefore to every part of the nation. Since the criterion was good work delivered on schedule, a pennant could be won by small plants as well as large. A one-man plant making spokes for the steering wheels of Victory ships flew the " M " pennant on a flagpole erected in his front yard, showing that although he was only a sub-subcontractor, he had met all schedules and been so certified. A total of 175 manufacturing plants and 35 shipyards were awarded " M " pennants.[13]

In addition to conferring honors on the whole staff of a shipyard, the Maritime Commission offered pecuniary rewards to individuals for suggestions which improved efficiency. This project began modestly August 11, 1942 with the allotment by the Commission of $187.50 monthly in war bonds to be given as prizes for such suggestions.[14] The suggestions were handled in the yards by labor-management committees, and twelve yards reported in March 1944 that they had received and applied suggestions which saved $11 million at cost to the Commission for bonds and cash awards of only $22 thousand. Valuable suggestions were passed on to other yards.[15] At the end of 1944,

---

[13] MC Minutes, pp. 21507-08 and 21603-04; " Maritime ' M ' Award Record," in *Marine Engineering and Shipping Review,* June 1945, pp. 160-62; statements by J. G. Kendrick of the Division of Public Relations on Aug. 9, 1948. See also below, chap. 14, sect. I.

[14] MC Minutes, pp. 22693-94 (Aug. 7, 1942).

[15] Ring, Johnson, and Lalley, Shipyard Efficiency Awards Committee, to MC, Mar. 7, 1944 in MC, Historian's Collection, Research file 205.15 and MC Minutes, p. 27757. By October 1944 it was stated more than $25 million had been saved. Ninety percent of the suggestions had been distributed to other yards. Permanente

3,029 suggestions had been reported by the yards through the regional offices and 90 per cent of them were judged worthy of awards. The savings had amounted to 45 million dollars and 31 million manhours, without including the savings through improved health and safety.[16] These results are evidence of the ingenuity and interest of shipyard labor.

Metals Corp., St. Johns River, and Calship ranked first, second, and third in total savings. MC Press Release, Oct. 25, 1944, in MC, Historian's Collection, Research file 205.15.

[16] Report from Ring and A. L. Jordan, Feb. 22, 1945 in MC Minutes, pp. 31275-76.

CHAPTER 14

MANAGING MANAGEMENTS

I. Varied Talents and Planned Competition

MANAGERIAL initiative and ability were no less needed for the success of the shipbuilding program than steel, components, and labor. Indeed, it was managerial initiative and ability that developed the techniques of multiple production, overcame the lack of skilled labor, and thus precipitated the crisis over steel.

The managements selected at the beginning of the program worked out well on the whole. Although some proved inadequate, most of them were successful, some brilliantly. Whether the managers were proving themselves good or bad selections, the Commission's role in its relation to them was very far from passive. Much of the activity of the Maritime Commission, especially of Commissioners Land and Vickery and the Regional Directors, was devoted to managing management. Other agencies such as the WPB shared in responsibility for labor and materials; management was a factor in production for which the Maritime Commission alone was fully responsible.

In spurring on the executives of the shipbuilding companies, appeals to patriotism and appeals to what is commonly called the profit motive were both vital. The prospects of profit were defined in the contracts for ship construction, described in chapter 4. The relative importance of the three main types—lump-sum, price-minus, and manhour or cost-plus-variable fee—can be indicated by a few figures rounded off in billions. Out of a total of about 13 billion dollars' worth of ship construction contracts awarded in 1941-1945, about 5 billion was for Liberty ships which were all built under manhour contracts; 1 million was for Victory ships, a type added in 1943 and built about half under manhour, half under some form of lump-sum contracts; 3 billion was for tankers and C-types, of which .5 billion was under lump-sum contracts and the rest under price-minus contracts; and about 4 billion was for minor and military types which were built partly under price-minus contracts, partly under fixed-fee contracts or contracts that were converted into fixed-fee contracts as a result of the changes in design that occurred so often in military types during construction. These figures are only rough approximations. An exact calculation is impractical because some contracts are difficult to classify, and there are many cases in which the form of contract was changed while the ships were being built, or even after the completion of construction. Many lump-sum contracts (including the type called selective price, which is explained below) were conversions of earlier contracts. But from 1941 until 1945 something like nine tenths by value of all shipbuilding awards were in price-minus contracts or manhour contracts, and were about equally divided between these two types.[1] They both provided for the reimbursement of costs, and they fixed maximum and minimum fees which were intended to give incentives to speed and efficiency. The manhour contract was so called because it measured efficiency and inefficiency with reference to the manhours of labor used in building instead of with reference to a cost in dollars, as did the price-minus contract.

In an effort to keep pressure on the shipbuilders, the length of building time allowed by the dates in their contracts was reduced again and again. When contracts for Libertys were first awarded, construction time was figured as 210 days for the first round, 150 days for the third or fourth round off the ways. For each day beyond that time there was a penalty; for each day ahead, a bonus, up to a certain maximum. Immediately after Pearl Harbor, the standard building time for Liberty ships was cut to 105 days, 60 days on the ways and 45 for outfitting. As described in chapter 5, the schedules called for by the contracts of January 1942 seemed fast when signed, but all the nine Liberty yards then operating except the two slowest, New England and Houston, made the maximum allowable fees on these contracts. On May 14, 1942, new contracts allowing only 65-76 days from keel laying to delivery, about 40-45 days on the ways, were awarded to Richmond No. 1 and No. 2.[2] In December 1942,

---

[1] Estimates from the figures for the total costs of different types given in Fischer, *Statistical Summary*, Table H-6, and from card files of contracts in Historian's Collection showing types of contracts.

[2] MC Minutes, pp. 21936-37 and gf 507-4-11.

when the next large group of new contracts for Libertys was awarded, most yards were allowed only 27-30 days on the ways. The way time allowed for in awarding these contracts and the actual time on the ways of these ships, which were built in the second half of 1943 when production had reached about its maximum speed, are shown in Table 8. It does not include Delta and Calship because they were building a modified type, the Liberty tankers, nor does it include New England Ship, which received no new contract in December 1942.

TABLE 8

ACTUAL WAY TIME AND CONTRACT ESTIMATES

LIBERTY SHIP AWARDS OF DECEMBER 1942

Yard	Days on the Ways per Ship	
	Actual	Contract Estimates
Bethlehem-Fairfield	24.7	30
Houston	44.1	45
North Carolina	26.9	27
Oregon	18.5	27
Richmond No. 1	23.1	30
Richmond No. 2	19.4	30

Source: The "way time" allowed in contract award is taken from the memorandum approved by the Maritime Commission in making the awards. Minutes, pp. 23866-68. Actual way time was calculated from Vickery's bar charts, in Historian's Collection, Statistical files from Production Division.

Actual building time was much faster than required by the contracts. Fees depended on manhours as well as building time, but the fast yards were also earning large bonuses for savings in manhours. All these yards except Houston were building Liberty ships much faster in 1943 than was required to earn the maximum fees.[3] Their situation was much the same as it would have been if they had been operating on contracts guaranteeing the payment of their costs plus a fixed fee, namely $140,000 per ship, the maximum actually specified in their contracts. The fact that their contracts provided variable fees, with the maximum depending partly on speed, may have contributed to the initial speeding up

---

[3] North Carolina, although not earning large bonuses for speed, was earning them for savings in manhours. See Table 10. Its lower speed may be explained by the fact that it changed to C2's in mid-1943. On the fees of these companies see chap. 24, sect. I.

PLANNED COMPETITION 459

of construction, but some incentive other than the variable fee must be sought to explain why the speed continued to increase in yards where further increases in speed did not mean any increase in fee. To be sure, the sooner one contract was finished, the sooner a yard could start earning fees on a new contract; there was that connection between speed and profit, whatever the form of the contract. Yet the profit motive was not, I believe, the only driving force behind the extraordinary acceleration of construction.

Incentives beyond those implict in the contracts were created by the fact that the nation was at war and were intensified by the personal efforts of the members of the Maritime Commission, especially by Admirals Land and Vickery. The fact that these Commissioners were also officers of high rank of the U. S. Navy affected the role they were able to perform during the war both in dealing with other branches of the government and in dealing with industrial executives. Although Admiral Land was on the retired list, he was raised by a special act of Congress in July 1944 to the rank of Vice Admiral.[4] Howard L. Vickery, who was a younger man, being 50 in 1942 when Admiral Land was 63, was raised to Rear Admiral in April 1942 and to Vice Admiral in October 1944. A special act of Congress had enabled Admiral Vickery to remain on the active list in spite of his civilian appointment.[5] These promotions and their approval by Congress were evidence that the shipbuilding program for which these two men were primarily responsible was generally considered a success. Being on the active list, Admiral Vickery wore his Navy uniform, and that may have been of some psychological importance in the man-to-man dealings with officers of shipbuilding corporations. Both Admirals Land and Vickery were esteemed also for their technical competence as naval constructors. Their personal relations with the leaders of the shipbuilding industry was a very important part of the Commission's management of management.

Before he paid the penalty for overwork and suffered a serious heart attack late in 1944, Admiral Vickery was the driver who

[4] House, *Hearings No. 283 on H. R. 7576* before the Committee on Naval Affairs, and *Report No. 2487*, 77th Cong., 2d sess.; *Report No. 1204 on H. R. 634*, 78th Cong., 2d sess., Mar. 1, 1944; Senate, *Report No. 904* of the Committee on Naval Affairs, 78th Cong., 2d sess.; *Congressional Record,* Index.

[5] Senate, *Report No. 2076* of the Committee on Commerce, 76th Cong., 3rd sess., Aug. 5, 1940.

by daily conferences and telegrams directed, advised, and goaded the managers of the shipyards. He was the sort of a man who preferred seeing and doing things to reading and writing about them. He wrote almost no letters or memoranda, but scrawled, "See me, V" on the flood of papers that flowed across his desk and then routed them to those who wrote letters for him—his aide, Lt. Comdr. William A. Weber, an employee of the Commission since 1937, or Miss Irene Long, his secretary and Confidential Assistant, or some Division Director.[6] He had photographs sent regularly to Washington from every yard showing the state of completion of the facilities and of each hull, and took time to study them.[7] Man-high charts on which pins and bars showed each keel-laying, launching, and delivery in every Maritime yard in the country were kept in a special, locked room where he spent much time, so that he had a picture of the progress of the whole program day by day, could compare the records of the various yards, and forecast what they could do in the future.[8]

Restless when tied to a desk, Admiral Vickery spent about a third of his time visiting shipyards. Flying from Washington with his aide, Commander Weber, he would often circle a yard a few times before landing so that he already knew before he talked to the management how far along ships were and whether material was piled up in the stock yard or on the assembly platens. Then he went all over the ships and through every corner of the yard to point out mistakes, or tell how much faster or more cheaply some fabrication or assembly was being managed elsewhere. Caustic comments and heated conferences did not prevent evening entertainment which often went far into the night, but breakfast was always at 7:30 to start a new day of climbing around shipyards.[9] Powerfully built and seemingly tireless, Admiral Vickery won the personal admiration of the shipbuilders by his own drive and thorough knowledge.

Besides Admiral Vickery's nation-wide trips, there were also inspections of shipyards quite frequently by the Regional Direc-

---

[6] Interviews with many of the Commission's staff, especially William Weber, and Irene Long.

[7] Vickery to Flesher, May 8, 1942, in gf 201-3-4.

[8] Many persons who worked with Admiral Vickery emphasize how much he used these charts.

[9] Statements by W. Weber May 7, 1946, and other interviews.

ADMIRAL VICKERY'S CHART ROOM IN 1943

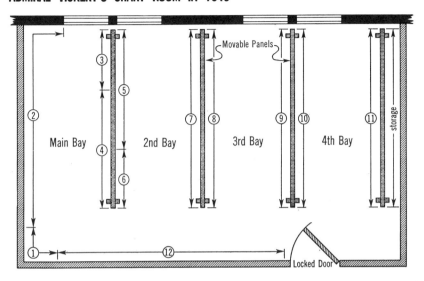

FIGURE 33. The charts, as recalled by Joseph T. Reynolds, who prepared most of them, were:

(1) MONTHLY SCORE BOARDS—Number and deadweight tonnage of deliveries that month compared with Admiral Vickery's quotas and other schedules, cumulative, by types, by yards.
(2) WAY CHARTS FOR MAJOR TYPES (Other than military)—A graphic record of production in each yard showing for each vessel the length of time that it was on the shipway and in outfitting.
(3) TOTAL DEADWEIGHT TONNAGE DELIVERED, cumulative, monthly, by types.
(4) ANALYSES OF LIBERTY SHIP PROGRAM INCLUDING SHIPYARD PROFILES—The analyses showed average time on the ways by rounds of the ways for all yards; average time on the ways by rounds of the ways for each yard, and monthly trends of deliveries and average construction time for all yards; the "profile" was a cumulative visual summary, corrected each day, which showed the vessels delivered, their hull numbers, and the status of incomplete vessels for each shipway in each yard.
(5) MANHOURS PER SHIP, LIBERTY SHIPS AND T2 TANKERS, by yards.
(6) SHIPYARD PROFILES FOR VICTORY SHIP PROGRAM.
(7) WAY CHARTS FOR MILITARY AND SOME MINOR TYPES.
(8) SHIPYARD PRODUCTIVITY OF MILITARY TYPES AND TANKERS—Ships produced per way per month, by yards.
(9) SHIPYARD PRODUCTIVITY OF STANDARD- AND LIBERTY-CARGO TYPES.
(10) PLANS AND PROCUREMENT FOR VICTORY SHIPS—Semi-monthly reports of progress in completion of working plans and ordering of major components for the Victory ship program. This space was subsequently used to show employment by regions and by yards.
(11) THE STEEL SITUATION—Receipts, fabrication, and erection of steel in selected yards.
(12) SHIPYARD PROFILES FOR ALL YARDS—A cumulative visual summary, corrected each day, showing for each shipway in each yard the hull number and type of each vessel delivered, under construction, or under contract.

tors, in whose judgment Admiral Vickery placed much confidence,[10] and by his technical assistant, Mr. Schmeltzer. When someone had to be sent for an extended stay in a yard either to find out what was wrong or to take over the management temporarily, Admiral Vickery called on Henry E. Frick, who had been Chief of the Construction Section of the Technical Division before 1938 and then Vice President of the American President Lines. Mr. Frick was engaged on loan from his company in March of 1942 to help out at Calship, and he became known as Vickery's "trouble-shooter."[11] Many visits to shipyards were made also by Commissioner Carmody [12] and by Mr. Anderson, the Director of Finance.[13] Admiral Land went occasionally, but the nature of his duties, especially those as Administrator of the War Shipping Administration, required him to stay most of the time in Washington. He always arranged to be in Washington when the Commission's Vice Chairman, Admiral Vickery, was away.[14]

Direct personal inspection of the shipyards by an expert such as Admiral Vickery was all the more important because many of the Commission's yards were being managed by men who had never built ships before. An essential feature of the Commission's management of management was the utilization of several kinds of industrial executives. One group consisted of the executives of such firms as Sun Ship, Bethlehem, the men from Newport News who were operating North Carolina Ship, and the men from the American Shipbuilding Co. who were at the head of the Delta yard. Members of the "shipbuilding fraternity," they felt that building ships was different from other industries, and were inclined to be skeptical of the extent to which the methods of industrial efficiency developed in other kinds of plants could be applied in shipyards. Of course, they recognized that the methods used when ships were tailor-made, each an individualized product to suit an individual customer, needed to be changed more or less now that the demand was for multiple production. But they were mostly trained in naval architecture or marine engineering

---

[10] House, *Walsh-Kaiser Co., Inc., Hearings,* p. 267.

[11] House, *Investigation of Rheem Manufacturing Co., Executive Hearings,* p. 92; MC, Personnel files; Minutes, p. 34322 (Jan. 8, 1946).

[12] Cohen, "Carmody."

[13] Anderson to Schell, July 4, 1942; Anderson to Honsick, June 24, 1943; Anderson to Quarg, Nov. 3, 1944, all in Anderson reading file.

[14] Statements of Eleanor Van Valey, Admiral Land's secretary, Aug. 18, 1948.

rather than in the general efficiency methods of industrial engineering. Their experience was in the traditional ways of doing things.

Another group consisted of construction engineers with the companies of H. J. Kaiser, W. A. and S. D. Bechtel, John A. McCone, and J. A. Jones who had little or no experience in shipbuilding before 1941. For their importance, see Figure 34. They came into shipbuilding disposed to try new methods. They were skeptical of the notion that shipbuilding was necessarily different from other industries and that "shipbuilding brains" were something special and mysterious. They thought the same methods which had increased industrial efficiency in other industries could increase efficiency in shipyards. Their experience in heavy construction work made them think in terms of the importance of space and of arranging and moving materials.

A third group should perhaps be distinguished, the executives from ship repair companies. Their experience made them thoroughly familiar with ships, but in some cases they had no appreciation of the problems of large industrial operations and construction. Of the three groups, this last proved on the whole the least successful in multiple ship production.

Since the ideal was a combined knowledge of the special problems of shipbuilding, of how to move materials in mass, of the efficiency methods of more standardized industries, and of personnel management in a big plant, the problem arose within each company of harnessing together in one team men of these various abilities. At Wilmington, North Carolina, the top man, Roger Williams, was Vice President of Newport News Shipbuilding & Dry Dock Co., but employed many executives under him who "knew nothing in the world about shipbuilding" when they came to the shipyard.[15] The Kaiser management contained many former dam builders who were impatient to "pour the keel," but it was mightily helped at the start by a few old-line shipbuilders.

The very first Kaiser-managed yard, that at which both Edgar Kaiser and Clay Bedford made their plunge into shipbuilding, was the yard originally named Todd-California, later Richmond, No. 1. It began work for the British before the American Liberty program was underway. The British inspectors, who were old

[15] House, *Investigation of Rheem Manufacturing Co., Executive Hearings*, p. 99.

hands at shipbuilding, helped the Kaiser engineers out of difficulties in the early days.[16] Outstanding also in the Richmond organization was Edwin W. Hannay who had been hull superintendent at Moore Dry Dock as early as 1917 and during the period between wars had been superintendent for a number of years in Bethlehem yards. Henry Kaiser called him from retirement to become general superintendent of the Richmond No. 1 and No. 2, and hugged him when those yards beat their competitors.[17] He was shifted to Richmond No. 3 when that yard began a difficult assignment of building transports. But Mr. Hannay seemed too conservative to some of the younger men who wanted to try new ways of doing things. Among the innovators was at least one trained marine engineer, Maurice Nicholls, who had been in Kaiser's employ since 1939. On one occasion Mr. Hannay was sent to Washington on a mission while a preassembly of the bow for Liberty ships was tried out. When he came back and found it done, he said it was fine, " Go ahead." [18] As the engineers of the Kaiser organization learned shipbuilding, they were moved up, and Mr. Hannay was made a consultant and finally retired. He was quite philosophical about it, but among the foremen and various superintendents there was much dissatisfaction when called on to work under men who had never built ships before and who said it was just like any other construction job. " Family quarrels " Mr. Hannay called them.[19] Through a Congressional investigation the troubles at Richmond were thoroughly aired in public. Many other cases of friction in fitting old-timers and newcomers together were probably worked out " in the family."

At the same time that sandhogs and seadogs were being harnessed together and learning from each other within the shipbuilding companies, a similar exchange between companies on a nation-wide scale was going on under the sponsorship of the Maritime Commission. When the Liberty ship program was launched there were big conferences in Washington, a general exchange of information, and some loaning of personnel. From

[16] *The Log*, vol. 37, Oct. 1942, p. 25, and Thompson and Hunter, " The British Merchant Shipbuilding Programme," discussion in North East Coast Institution of Engineers and Shipbuilders, *Transactions*, vol. 59, pp. D 47-64.
[17] House, *Production in Shipbuilding Plants, Executive Hearings*, pt. 4, p. 1001.
[18] Statement by Maurice Nicholls in interview, Aug. 8, 1947.
[19] House, *Production in Shipbuilding Plants, Hearings*, pt. 4, pp. 1000-07, and the testimony of Alonzo B. Bryant and others at those same hearings.

information supplied by the Newport News Shipbuilding Co., for example, the Technical Division of the Commission prepared a list of the pneumatic tools needed for the outfitting of the new shipyards.[20] Later, visits between yards were encouraged. For example, Mr. Newell, the old-line builder from Bath, Maine, went to the Pacific coast to see how things were done in the Kaiser yards, and Henry Kaiser paid tribute to the help he received from Mr. Newell.[21] The new industrialists brought into the program in the spring of 1942 were given letters of introduction for a tour of the yards which had started in 1941.[22] And Admiral Vickery carried knowledge of new methods on his trips from yard to yard.

In addition to these informal methods of exchanging knowledge of new methods, two more formal arrangements deserve notice. At the initiation of any new type of ship there were conferences regarding the drawing of the working plans. These conferences were attended by the naval architects preparing the plans, by the shipbuilders who were to construct the new type, and by members of the Maritime Commission staff. They were the occasion for a systematic cooperative survey of many of the technical processes of construction.[23] Another formalized arrangement for exchanging knowledge, more concerned with methods of using labor, was through the Shipyard Efficiency Awards Committee which supervised the award of prizes by labor-management committees. Suggestions deemed worthy of awards were distributed to all shipyards through the regional offices.[24]

The natural rivalry between the diverse groups of managers was accentuated by Admiral Vickery's needling. When the Oregon yard set a new record for speedy construction, Admiral Vickery called up the head of the Liberty yard run by the world's largest shipbuilding organization to say: "You haven't matched that. And you're supposed to be a shipbuilder!" Another manager, whose old-line yard was more renowned for quality than speed,

[20] C. W. Flesher to P. J. Duff, Mar. 18, 1941, in gf 507-1-1.

[21] House, *Investigation of South Portland Shipbuilding Corporation* before the Committee on Merchant Marine and Fisheries, Interim Report, 77th Cong., 2d sess. [title page gives 1st sess., which is a misprint], Nov. 24, pp. 5, 11-12; *Portland Press Herald,* Nov. 18, 1942.

[22] See above in chap. 6 and House, *Investigation of Rheem Manufacturing Co., Executive Hearings,* p. 63.

[23] Examples of such conferences are given in the chapters on the Liberty ship, on the Victory ship, and on military and minor types.

[24] MC Minutes, pp. 31275-76 (Mar. 6, 1945), and above in chap. 13.

tells the story that Vickery " never put the needle in me as he did some people," but that one day he remarked that a Kaiser yard had built a tanker in 100 days. Having work on some ships delayed for lack of material, the old-timer concentrated workers on one of his tankers to finish it in 99 days.[25] Of course Admiral Vickery was equally willing to pit one old-line yard against another or play up rivalry among the many new yards. When the new 6-way yards were started in 1942, Admiral Vickery told them " You have 6-way yards, and I consider you are all in competition," and then he kept them informed of where they stood in the race.[26] Marinship was soon so far ahead of the others that it was changed over to tankers, since some yard had to be changed to meet the need for that type.[27] But in spite of Marinship's relatively good record, Admiral Vickery kept needling them to do better. In the fall of 1943 the yard was having difficulties completing both standard T2's and tankers of a special military type which were being built at the same time. Admiral Vickery wired the president of the company, K. K. Bechtel, on November 16: " Where is tanker you were going to deliver on November second? Regards." [28] The West coast Regional Director reported that such needling wires upset Mr. Bechtel very much.[29] When the long delayed tanker was finally delivered Mr. Bechtel wired exuberantly: " It's a girl, *Purissima*, delivered at 6 p. m. November 23rd. Child very healthy. All parents weak but will recover, Regards." [30] Admiral Vickery wired back. " The delivery was stimulating news. I hope the next period of gestation will not be that of an elephant." [31]

Admiral Vickery's needling soon acquired something of the aura of a legend. Adorning the wall of his outer office was a glass-covered case lined with black velvet containing a series of small medicine bottles labelled: " Insidious Comparison," " Sarcasm," " Mickey Finn for Visiting Firemen—Dosage by Miss Long " (on the largest bottle), " Heckling," " Jeering," " Belittling," " Sneering," " Irony," " Fury," " Nagging," " Profanity,"

---

[25] Interviews.
[26] House, *Investigation of Rheem Manufacturing Co., Executive Hearings*, p. 96.
[27] House, *Production in Shipbuilding Plants, Hearings*, pt. 3, p. 916.
[28] In MC gf 506-8-10.
[29] Flesher to Vickery, Nov. 20, 1943, in gf 506-8-10.
[30] Nov. 23, 1943, in gf 503-86.
[31] Nov. 24, 1943, in gf 503-86.

and on the tiniest bottle "Compliments." In the center was a poem ending:

> So here's to our Admiral Vickery
> Who's never been known to wheedle
> But prescribes the Victory Vitamins
> "Quick, Weber, the needle."

Dealing with industrialists of large affairs proud of their "know how," their organizations, and their capacity to do the impossible, Admiral Vickery put it up to them to show they were as good as they thought they were.

Supplementing Vickery's jibes, and serving as a basis for the general system of encouraging rivalry between yards, were the statistical methods for rating yards that were building Liberty ships. These ratings were developed by Mr. Anderson and the Special Studies Branch in his Division of Finance. The first method was in terms of percentage of completion of contract. It depended largely on the amount of expenditure,[32] and a company could therefore rate high by having ordered needed materials rapidly even if it had done relatively little with them. Although imperfect, it was the only system of rating available at the start while the yards were still being built, and it served to inject from the beginning the feeling of competition. It was soon replaced by ratings in terms of costs, number of ships delivered per way, and manhours used per ship. The comparative status of each yard was reported to it monthly with appropriate comments. That of June 30, 1941, one of the first, sent when yards were being built, concluded: "Congratulations Oregon—I hope you will do as well on building ships."[33] When the Senatorial Committee Investigating the National Defense Program under Senator Truman began hearings, comparative ratings were reported to it in April 1942 and regularly thereafter, and to other Congressional committees. The competition became one for public reputation.[34]

---

[32] Anderson to Land, Apr. 21, 1942, in Anderson reading file. Use was also made of actual keel laying and progress reports of inspectors, and a standard was set by delivery dates in the contracts and by comparison with previous experience with C2's. The statistical conception was Mr. Anderson's. Statements by Wesley Clark and Frank R. Hunter, Jr., June 7, 1947.

[33] Land to South Portland Shipbuilding Co., June 30, 1941 in gf 507-4-14.

[34] Truman Committee, *Hearings*, 77th Cong., 1st sess., Apr. 23, 1942, pt. 12, pp. 5177-79; House, *Independent Offices Appropriation Bill for 1943, Hearings*, pt. 1, pp. 264-65; Anderson to Bland, May 14, 1942 and July 14, 1942, in Anderson reading

In its fullest development, after most of the Liberty ships had been built, the system of rating used by the Maritime Commission considered four factors: dollar cost, cost in manhours, speed of construction, and economical use of facilities. Ratings based on consideration of these four factors were published by the Truman Committee, which collected figures not only from the Maritime Commission but also directly from the shipbuilders, and which gave praise especially to yards operating at low cost.[35]

In both the Truman Committee's ratings and those of the Maritime Commission, the same five yards made outstandingly good records on Liberty ships, namely: Oregon, Richmond No. 1 and No. 2, North Carolina, California, and Bethlehem-Fairfield.

Of these five, the Western yards, led by Oregon had the highest rating for speed (number of ships delivered per way). Oregon led also in having the lowest manhours per ship delivered, but North Carolina led in having the lowest dollar cost per ship, partly because wages were lower in the South Atlantic states.

Although often caustic to the shipyard managements in private communications, in public hearings Admirals Land and Vickery, stressed especially the achievements of the leaders and explained the dangers that the figures would be misinterpreted. There was, and is, much such danger, for in spite of efforts to put yards on a comparable basis so that they could be raced against each other, actually there were innumerable differences in equipment or other conditions. Particularly noted was the effect of interjecting the program for landing craft into one Liberty-ship yard, Bethlehem-Fairfield. It was one of the leaders, the first to launch a Liberty ship. When the Commission was ordered to build landing craft in a hurry, half were assigned to Bethlehem-Fairfield, half to Kaiser yards. The Kaisers successfully objected to interrupting production in any of their yards in which the construction of Liberty ships was fully underway, and a small new yard, called Richmond No. 3A, later called No. 4, was built so that continuity of production at Oregon Ship and Richmond No. 1 and No. 2 was preserved.[36] But twelve of Bethlehem-Fairfield's sixteen ways

file; *The Journal of Commerce*, Apr. 28, 1942, p. 20, headline: " Pacific Yards Lead U. S. Shipbuilders."

[35] Truman Committee, *Report 10*, 78th Cong., 1st sess., Apr. 22, 1943, pt. 8, pp. 11-15, 32-33; *Report 10*, 78th Cong., 2d sess., June 23, 1944, pt. 18, pp. 8-11, 30-32; *Marine Engineering and Shipping Review*, June 1943.

[36] Helen E. Knuth, " The Building of LST's by the Maritime Commission."

were full of landing ships from August to December 1942. Even when the yard went back to building Liberty ships it did not attain a rate of production as high as the West coast yards or North Carolina Ship.[37] The more extensive use of riveting at Bethlehem-Fairfield was one factor, but another was the interruption in order to build landing craft. As Mr. Anderson wrote to the chief counsel of the Truman Committee, " We have endeavored to build the greatest rivalry between yards, between coasts, and between shipbuilders. Very naturally those who were compelled to divert their production from one type to another felt that such diversion was seriously interfering with their tonnage production." [38] This is but one illustration of the very many difficulties in making public comparative ratings of yards and doing it fairly. Other yards had other obstacles. But publication of the ratings intensified the competitive spirit.

Another form of public recognition intended to provide incentive to management as well as to labor was the award of " M " pennants. Stars were added to these pennants for each three-month period in which schedules continued to be met. Eleven stars earned the Maritime Merit Eagle; twenty-two stars earned the Gold-Wreathed Merit Maritime Eagle, which was added to the pennants of only five yards (to June 1945). Another flag was the " 250 club " for yards completing that many ships. To spur on the tanker-building yards there was a pennant inscribed " Tanker Champs " for which the competition between Swan Island and Marinship was so close that it changed hands between them five or six times.[39]

On the whole the palm of victory in the race set by the Maritime Commission was awarded by popular acclaim and by Admiral Vickery to the Kaiser yards. Although Henry Kaiser was at the top of their management, he did not as Admiral Vickery put it, " understand shipbuilding "; [40] the shipyards were managed by Clay Bedford at Richmond and Edgar Kaiser in Oregon, young men in their thirties. " I would say of all the yards and of the management in all the yards that the outstanding people who have

[37] Reynolds' Way Charts.
[38] Anderson to Hugh A. Fulton, Chief Counsel of the committee, Feb. 27, 1943, in Anderson reading file.
[39] Statements by J. G. Kendrick, Aug. 9, 1948; " Maritime ' M ' Award Record," in *Marine Engineering and Shipping Review,* June 1945, pp. 160-62.
[40] House, *Walsh-Kaiser Co., Inc., Hearings,* Apr. 26, 1944, p. 271.

470 *MANAGING MANAGEMENTS*

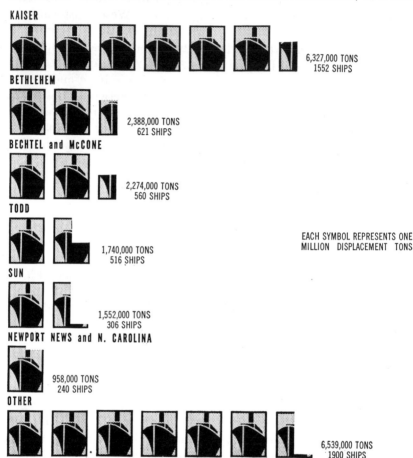

FIGURE 34. Source: Fischer, *Statistical Summary*, Table C-1.

done an outstanding job are Edgar Kaiser and Clay Bedford," said Admiral Vickery in June 1943; [41] and in July 1944 he said of the management by Edgar Kaiser of the three yards near Portland, Oregon, "It is the finest operation in the United States." [42]

[41] House, *Production in Shipbuilding Plants, Executive Hearings*, pt. 3, p. 905.
[42] House, *East Coast Shipyards, Inc., Hearings*, p. 407.

The praise for the Kaiser yards, and praise given also to Marinship and Calship, was recognition by Admiral Vickery that building ships did not necessarily require experienced shipbuilding organizations, as had been supposed in early 1941 when Admiral Land was cautioning against excessive expansion. "In this type of shipbuilding, which is ship manufacturing," declared Admiral Vickery, "shipbuilding experience has not been necessary, and those people who have not had shipbuilding experience have done a better job than the people who have had it, with the exception of the North Carolina yard, which has done the best job of the old shipbuilders." [43] To the heavy construction industry there seemed nothing remarkable about this success. "This was a 'natural,'" says their historian, "for contractors were used to assembling huge quantities of heavy materials, hiring and training crews of skilled and unskilled workers, coordinating operations for maximum speed, and improvising new methods when old ones were inadequate. To them, building a ship was simply a matter of building another kind of structure." [44] But not all the contractors from the construction industry made distinguished records in shipbuilding; much is to be attributed to the individual ability of men in both groups, as Edgar Kaiser at Portland, Oregon, and Captain Roger Williams at Wilmington, North Carolina. The outstanding record of the yards in those two cities is also to be credited in part to the quality of the labor force.

II. Freedom of Management vs. Bureaucratic Controls

The managed competition between managements which was created by the terms of the contracts, by the published systems of ratings, and by Admiral Vickery's personal relations with the executives was highly successful in encouraging speed. It was not equally successful in cutting down the costs. Admiral Vickery was not indifferent to cost. "Simply because I drive these people to get high production and do the job quickly does not mean I expect them to run wild with funds," he said. "I want speed, but at the same time I want it economically." [1] How to get

---

[43] House, *Investigation of Rheem Manufacturing Co., Executive Hearings*, p. 94.
[44] Van Rensselaer Sill, *American Miracle* (New York, 1947), p. 161.
[1] House, *Investigation of Rheem Manufacturing Co., Executive Hearings*, p. 76.

economy without interfering with high speed was a problem not solved to the Commission's satisfaction.

At the beginning in the construction of the yards, the government was reimbursing for all costs and there were no penalties on the contractor if he ran over the estimate. A limit was stated in the contracts but the contractors came back for permission to spend

TABLE 9

ANTICIPATED COST AND ACTUAL COST OF SELECTED SHIPYARDS

Yard	Cost (in thousands)	
	Anticipated in June 1941	Actual estimated in December 1944
Total, 5 yards....................	$42,327	$122,591
Beth-Fairfield, 16 ways..............	9,889	35,149
Calship, 14 ways...................	10,004	26,928
Houston, 9 ways...................	7,606	14,907
North Carolina, 9 ways.............	7,632	20,436
Oregon Ship, 11 ways..............	7,196	25,171

Source: House, *First Supplementary National Defense Appropriation Bill for 1942, Hearings*, p. 423; House, *Independent Offices Appropriation Bill for 1946, Hearings*, pp. 546-47.

The yards compared were selected because of the availability for those five yards of estimates of June 30, 1941 which planned for as many ways as were in the completed yard.

more and received it. Actual expenditure for five large Liberty yards was $122 million, whereas in June 1941 it had been estimated that those yards would cost $42 million (Table 9). A good part of the increase was for more equipment to do added fabrication and use new methods of subassembly, or for housing and transportation, but costs higher than the estimates were also due in part to mistakes by contractors and to high cost in early operations.[2] In the general confusion when bulldozers, steam shovels, and pile drivers arrived to build a yard, and men were being hired almost before any management was organized, it was even difficult to keep track of who was spending what. Resident auditors had a hard time getting the company's books set up, or keeping track of the primary documents such as subcontracts and receipts for

[2] MC Minutes, pp. 17182-83 (May 2, 1941) (Louisiana Shipyard); pp. 20489-91 (Jan. 10, 1942) (Houston); p. 22133 (June 5, 1942) (Calship).

materials.³ Since work had to be started before the shipyards were fully planned, it is hard to see how else they could have been built except by contracts calling for reimbursement of cost; but these contracts, in which there were no incentives for economy, launched the program in an atmosphere of extravagance.

After the yards were fully operating, they kept on asking for more and more equipment to get things done more speedily or with fewer manhours. To some extent it was a matter of " everybody thinking of something new they wanted . . . ," as Admiral Vickery said, " like the youngster with candy who wants more," ⁴ but the Commission generally favored thus increasing the facilities; indeed the additions were often suggested by Admiral Vickery himself as he went from yard to yard, telling each of what was being done better elsewhere. Of course, any shipbuilding company was free to pay for new machines out of its own pocket, but when it asked to be allowed to buy them under the facilities contract according to which it was reimbursed for costs, then the question of fees was raised. Why should the Commission pay out money indefinitely to enable the shipbuilding companies to earn the higher fees that they received as the result of the speed or saving in manhours effected by the additional equipment? It seemed fair to take some of the cost of the added equipment out of fees. For example, on March 9, 1943 the expenditure for Richmond No. 1 and No. 2 of $89,000 for welding machines was approved with the proviso that " one half the cost of the welding machines be credited to the Commission on account of fees otherwise payable." Similarly in many other yards and often on a much larger scale, improvements in equipment were balanced by reduction in the scale of fees.⁵

The ship construction contracts also did not prove so conducive to economy as had been hoped. After the program had been underway almost two years, the Director of Finance confessed that the cost-plus-variable-fee contracts put " insufficient pressure on the contractor to reduce over-all costs," ⁶ in spite of their clauses designed to create incentives to efficiency. The contracts specified

---

³ For example, Hartline's report from South Portland, E. E. Hartline to Mohler, Aug. 8, 1946, in gf 507-3-14.
⁴ House, *Production in Shipbuilding Plants, Hearings*, pt. 3, p. 912. See also Vickery's statements, House, *Investigation of Rheem, Hearings*, pp. 75-78.
⁵ MC Minutes, p. 24516. See also Minutes, p. 26110 (Aug. 30, 1943).
⁶ Anderson to Land and Vickery, Nov. 25, 1942, in Anderson reading file.

an "estimated average number of manhours" required for the construction of a ship. If the actual manhours, for which the contractor was reimbursed as cost of the ship, exceeded the estimate, the fee was decreased by $33\frac{1}{3}$ cents times the difference between the actual and estimated manhours, figuring for the contract as a whole. If the actual manhours were less, the fee was increased by 50 cents times the difference. But the total increases and decreases were limited by the maximum and minimum fees fixed in the contracts. The estimated average manhours, called the bogie, was set in the first contracts as 510,000 hours of direct labor plus about 22 to 28 per cent more of indirect manhours for supervision and overhead. The larger the shipyard the less was the amount of indirect manhours allowed per ship.[7] At the end of 1942, the most efficient yards, which were also the faster-building yards, were well under their bogies, but the less efficient yards, and those more recently built, had not yet brought their cost in manhours per ship down to the estimated average fixed in their contracts. In the new contracts which the Commission awarded in December 1942 the manhour bogies were still at the level of the first contracts. Then, in the contracts awarded the following spring there was a general reduction in the manhour bogies. A comparison of these bogies with the actual manhour cost incurred in the execution of these contracts is shown in Table 10.

Clearly, in manhour costs, as in speed, the variation from yard to yard was so great that some yards could easily earn bonuses that would have put their fees above the maximum allowed by the contracts, while other yards would have made far less than the minimum fees if penalties for high cost or for delay had been deductible from the minimum fees. The profits reported by the shipbuilders show few instances in which a company earned a figure halfway between the maximum and minimum; nearly all earned, before renegotiation, either the full maximum fees or the bare minimum.[8] In many cases the management could foresee

[7] Vickery (by Weber) to J. V. Hayes, Counsel of House Committee on Merchant Marine and Fisheries, Aug. 20, 1942, appendix B, in MC, gf 507-3-14 states the general principles. Illustrative contracts which I examined include Nos. MCc 7797, 2181, 13096, 15734 (with Beth-Fairfield) and MCc ESP 12, MCc 213, 13099, 15923 (with Houston).

[8] House, *Investigation of Shipyard Profits, Hearings* (1946), pp. 290, 437, 438, 442, 476, 480, 486, 598, 500 and compare pp. 237-42 517, 518, 548, 611.

before much work had been done on a contract that its fees on that contract were, practically speaking, fixed, either by the guarantee of a minimum or by the maximum limit. To that extent the incentive to efficiency disappeared.

TABLE 10

ACTUAL MANHOURS AND CONTRACT ESTIMATES, LIBERTY SHIPS

Yard	Manhours Per Ship (in thousands)	
	Actual	Contract Estimate
Awards of December 1942		
Beth-Fairfield	451.9	622.6
Houston	678.8	640.7
North Carolina	403.4	640.7
Oregon Ship	345.5	634.0
Richmond No. 1	406.4	622.3
Richmond No. 2	347.5	622.3
Awards of April-June 1943		
Beth-Fairfield	499.0	548.8
Houston	531.9	564.7
Panama City	600.4	573.7
Brunswick	584.3	573.7
New England	477.1	553.9
Oregon Ship	385.8	558.9
St. Johns	671.5	573.7
Southeastern	706.6	573.7

Source: Compilation by Blanche Coll using the contracts to find the estimated manhours, and using the sources described in Fischer, *Statistical Summary*, p. 148, to find the actual manhours for the hull numbers of the contract. Delta and Calship are omitted because they built Liberty tankers. Difficulties in separating figures for Richmond No. 1 and No. 2 are the reason for omitting these yards from the 1943 list. New England received no new contract in December 1942.

A conceivable way of avoiding this would have been to abandon any general standard of average estimated manhours and vary the manhour bogie of each yard, setting lower bogies for the more efficient. Such action would have been open to the objection that it penalized the more efficient yards. As has been explained, the time schedules set in the contracts did impose harder terms on the more efficient and faster building yards, but did so in ways somewhat obscured by the complexity of schedule dates, and only to a degree which fell far short of preventing those yards from making maximum fees. Such a yard as Oregon Ship was so far ahead that the bonus clauses of the contract were not important as incentives either to speed or to efficiency.

At the other extreme, the contract's clauses concerning bonuses and penalties failed to provide adequate incentive to economy for contractors whose costs were running so high they could hope for no more than minimum fees.

So far I have been analyzing primarily the position of the contractors who were building Liberty ships and were under the cost-plus-variable-fee contracts. The same situation existed to some extent among the contractors who were building C-types, tankers, or military types with price-minus contracts. Under these contracts also there were many instances in which the shipbuilder, either through advance delivery of ships, or through dollar savings in cost, " earned " more than was allowed him under the maximum fee fixed in the contract, and many other instances in which the shipbuilder made only minimum fees. For example, on one price-minus contract after another in 1942-1944 Sun Ship made all the bonuses allowed by its contracts for tanker construction, while in contrast Alabama Ship on its price-minus contracts for tanker construction made minimum fees only, until August 1944.[9] But the terms and the timing of the price-minus contracts varied so much from company to company, even in the few cases where several companies were building the same form of ship under this contract, that comparisons are difficult. Whereas Alabama Ship contracted for 4 vessels per contract, Kaiser-Swan Island contracted for 56 tankers on its first contract and 47 on its second. On the first contract, costs at Swan Island went over the estimate by $40 million and it earned only minimum fees. On the second contract, the builder's costs were much below the estimate, the Commission " saved " about $16 million in what it paid for the ships, and the contractor earned the maximum fee.[10] There were instances in which a company's earnings under price-minus contracts fell in between the maximum and minimum fixed in the contract,[11] but

---

[9] MC, " Accounting Sheets, Ship Construction Costs, Arranged alphabetically by builders," prepared in the Special Studies Section of the MC Finance Division under Wesley Clark, in Claims Division in Sept. 1949. See also on Sun, MC Minutes, pp. 34497-98.

[10] *Ibid.*, and House, *Investigation of Shipyard Profits, Hearings* (1946), p. 388.

[11] For example, fees of Consolidated Steel Corp. were more often than not in between the maximum and minimum, as reported in House, *Investigation of Shipyard Profits, Hearings* (1946), pp. 543-49.

under price-minus as under cost-plus-variable-fee contracts, the contractor earned in a large number of cases either the maximum or the minimum.[12]

On the whole the price-minus contracts seemed to Mr. Anderson, the Director of the Finance Division, to be more effective in reducing costs than were the cost-plus-variable-fee contracts. In his view, price-minus contracts had at least the advantage of definitely throwing the burden of cost reductions on the contractor.[13] It seems also that since they were not tied to a particular generalized formula, such as the manhour bogies in the cost-plus-variable-fee contracts, they permitted more flexibility in negotiation.

A disadvantage under the price-minus contract, as under the cost-plus-variable-fee, was that it failed to provide adequate incentive for saving when a contractor fell so far behind that he had no hope of making more than the minimum fee, since this fee and all costs were guaranteed him. In regard to at least one such company, Mr. Anderson suggested that the next contract include a heavier penalty for late delivery and that the penalty be deducted from the minimum fee until that had been cut in half.[14] Admiral Vickery did not desire to make contracts so harsh, however, as to leave contractors working without any prospect of fee. " Very frankly, I do not like to have a contractor working for nothing, . . . because there is no incentive or drive in it, outside of one particular man's patriotism. It has been my experience that you have got to depend upon a certain amount of profit to drive a job through." [15] Even a contractor doing so badly that he knew he could get only minimum fees at best still had something to work for, and if he did too badly he could be declared to have defaulted on the contract, in which case he would lose both all fees and future contracts. So every price-minus and cost-plus-variable-fee contract provided for a minimum fee not impinged on by the penalties, and the contractor had almost no chance of losing on the operation unless he defaulted for incompetence or spent money for items not reimbursed under the terms of the contract.[16]

[12] *Ibid.*, pp. 389, 390, 524; and Anderson to B. J. Zincke, Feb. 28, 1944 in Anderson reading file; in addition to the instances above cited.
[13] Anderson to Land and Vickery, Nov. 25, 1942, in Anderson reading file.
[14] Anderson to Vickery, Dec. 2, 1942, in Anderson reading file.
[15] House, *Investigation of Rheem Mfg. Co., Executive Hearings*, p. 78.
[16] Suggestive of the kind of spending which might cause a contractor financial

In view of the weaknesses of the price-minus and cost-plus-variable-fee contracts, the lump-sum contract was held up as the ideal. There is no denying that the mental attitude of the managers was affected by being under a lump-sum contract instead of under a contract where the government was obligated to pay all the costs. Different habits seem to have been associated with the two forms of contract. But the prices in the lump-sum contracts were high, these contracts contained a profit limitation, and any profit above the maximum was recaptured by the Commission.[17]

Although none of the contracts set terms which forced the contractors to cut to the bone in order to profit, the cost-plus-variable-fee contracts were the least effective in inducing managers to economize. Consequently the Commission undertook to supervise the costs under these contracts with increasing care.

One of the items the Commission regulated increasingly was the size and number of managerial salaries. The initial contracts provided that any amount over $25,000 a year would not be reimbursed. In March 1942 a survey was made of all salaries over $5,000; and Admiral Land commented caustically on the resulting compilation: "Efficiency of yards (production) about in inverse order to number of high salaries."[18] Mr. Anderson, after consulting with Navy officers who had to deal with similar problems,[19] worked out a formula for deciding what was a reasonable total reimbursable by the Commission for salaries of $5,000 and over. The formula took into account the size of the yard, the character of the product, and any special complexity to the managerial problems.[20] The yards which were paying disproportionately high salaries were then told by him or Admiral Vickery or the Regional Directors that they should cut.[21] The policy was

difficulty is the arrangement of the St. Johns River Shipbuilding Co. to pay a fee of $300,000 to the subcontractor who built the yard. The Commission decided to reimburse on account of this fee only $100,000. See House, *Investigation of Shipyard Profits, Hearings* (1946), pp. 311, 320.

[17] "Accounting Sheets, Ship Construction Costs, Arranged alphabetically by builders," prepared in the MC Special Studies Section, in Claims Division in Sept., 1949.

[18] The report was "noted" in the Minutes of Mar. 26, 1942, p. 21484. Admiral Land's comment is on a copy in the files of the Director, Technical Division, Budget.

[19] Anderson to Files, Oct. 10, 1941, in Anderson reading file.

[20] The memorandum is the bottom first item in his reading file for Jan.-May 1942.

[21] Anderson to James C. Merrill (Pres. of St. Johns River Shipbuilding Co.), May 25, 1942; Flesher to Concrete Ship Constructors, Feb. 28, 1942 filed under Mar. 4, 1942; and Anderson to Vickery, July 24, 1942—all in Anderson reading file.

not to say what should be the salaries of particular individuals, but to propose cuts in the totals. For example, a total of $140,000 was decided on as appropriate for certain large yards. Actually one East coast yard was paying $182,000 in salaries of $5,000 or over, and a West coast yard producing about the same was paying $371,00.[22] After August 1942 the Regional Directors were given authority to approve salaries as items of cost, but they were asked to submit the lists of salaries to the Director of Finance in Washington who noted any that were out of line.[23] Of course the Commission had no authority to prevent the paying of higher salaries, but it could refuse to reimburse for them.[24]

A special aspect of salaries was presented by the system of bonuses adopted in some yards. The Commission refused to reimburse for bonuses unless it found they were truly part of an incentive system.[25]

In new companies the finances also were subjected to even more scrutiny by the Commission. The problem was not fixed capital, except in the case of those Class II yards which were sponsored by the Commission for help from the Defense Plant Corporation.[26] The operators of big new yards did not need fixed capital, since they were directly reimbursed for the cost of facilities, but they did need working capital to meet current bills and payrolls. By supplying the companies with their steel and many components, the Commission reduced the amount of working capital needed for building Liberty ships. Moreover the contracts provided that the Commission reimburse the yards for payrolls within ten days and other costs semi-monthly or sooner upon presentation of a certified public voucher.[27] This assurance enabled the companies to borrow money on their contracts, since the Assignment of Claims Act, October 9, 1940, had abrogated a long-standing statutory prohibition against assignment of claims against the government.[28] Interest on bank loans was counted as part of the

---

[22] Anderson to Vickery, July 24, 1942, in Anderson reading file.
[23] Vickery (by Anderson) to Spofford, Aug. 28, 1942, and many letters in September, in Anderson reading file.
[24] Anderson to J. A. Farrington, Jan. 9, 1943, in Anderson reading file.
[25] Anderson to David R. Dunlap, Dec. 27, 1941, in Anderson reading file.
[26] See list in House, *Independent Offices Appropriation Bill for 1946, Hearings,* pp. 584-85.
[27] Contract clauses 10 and 22. See above, chap. 4.
[28] Miller, *Pricing of Military Procurement,* p. 120.

reimbursable costs of construction. The highest rate that the Commission would reimburse was $2\frac{1}{2}$ per cent, according to the general policy adopted by the Commission as early as 1941.[29] Bank loans of this type, or the plowing back of profits on early contracts, was the source of much of the working capital of the shipbuilding companies.[30]

A certain proportion of the working capital, however, 50 per cent according to the early contracts, was required to be in " non-interest bearing funds supplied by the stockholders or the parent corporation of the Contractor in such form as to be fully subordinated to all obligations of the contractor under the provision of this contract or share capital, fully paid in cash." [31] Oregon Shipbuilding Corp., for example, was required by one contract to have $1,100,000 of this kind of working capital.[32] Total working capital, including the half in bank loans on which interest was reimbursed by the Commission, was estimated in 1941 as $2,400,000 for a 13-way yard.[33] Admiral Land told a Senatorial committee that the contractors would need that much, adding: " You cannot bite into this cherry on a shoestring." [34]

But some contractors did " bite in . . . on a shoestring," during later waves of expansion.[35] Moreover, operators of the emergency yards needed larger amounts of working capital as they increased their volume of production per way. The yards with price-minus contracts had to meet many bills from their suppliers. Shipbuilders found it necessary to make advances to the manufacturers from whom they bought, and this led, as explained in chapter 12, to the institution of V-loans. Soon, shipyards also began applying for V-loans. Since their contracts contained a general provision (Article 22) requiring the shipbuilding companies to have adequate working funds, the Commission would have been within its legal rights if it had refused aid to contractors who were short

---

[29] Anderson to Honsick, Oct. 14, 1941; Anderson to J. Stephenson, Mar. 25, 1942; Anderson to Farrington, Feb. 19, 1943; all in Anderson reading file.

[30] House, *Investigation of Shipyard Profits* (1946), *Hearings*, pp. 313-15 and *passim*.

[31] These words were added at the Commission meeting, March Minutes, p. 16545.

[32] Contract MCc 15751, Apr. 1943.

[33] Anderson to various companies, Mar. 11, 1941, in Anderson reading file.

[34] Senate, *Emergency Cargo Ship Construction, Hearings*, Jan. 29, 1941, p. 12.

[35] Compare Land's statement in Mar. 1944, House, *Investigation of Shipyard Profits* (1946), *Hearings*, p. 9; and the financing of East Coast Shipyards, Inc., described below, chap. 17.

of money and declared them in default. But unless the contractor was failing to produce ships and the Commission wished to put a new management in his plant, to hold him thus to his contractual obligation would interfere with ship production—it would defeat the main purpose of the contract.

Consequently the Commission guaranteed V-loans for shipbuilders and made them special advances. Shipbuilders who received V-loans were placed under the same restrictions regarding dividends and salaries as were applied when these loans were made to suppliers. Since the only reason for such loans was lack of working capital, it was obviously undesirable to have the company pay out money in higher salaries or in dividends; and therefore it was provided that the company would not pay dividends or increase salaries without prior approval of the Commission.[36]

Another way to ease the need of shipbuilders for working capital was to speed up the payment of fees. Minimum fees were payable partly at the launching, or within fifteen days thereafter, partly at delivery, but a company might earn twice the minimum fees on scores of ships and have payment of the amount above the minimum delayed a long time because the audit, although practically complete, was held up on account of a few items in dispute. In such cases applications were made for partial payment in advance of settlement. Millions were voted in this fashion to Permanente Metals and Oregon Ship on February 15, 1944 to enable them to meet a tax payment due that day.[37] In other cases companies had performed much of the construction but were unable to complete delivery through lack of components or changes in plans which were not their fault. In some such cases also advance payments on the fees were made.[38]

In one way or another, a number of shipyards as well as a number of manufacturers of components, became dependent for their financial health on Commission action, and the Commission therefore had occasion to look into their whole financial structure.

---

[36] Anderson to Edgar Kaiser, Oct. 20, 1942; Memorandum for Mr. Griffith, Feb. 24, 1943, both in Anderson reading file. Minutes, p. 24548 (Mar. 9, 1943); p. 24560 (Mar. 11, 1943), pp. 24821-23 (Apr. 9, 1943); p. 29746 (Sept. 19, 1944).

[37] Minutes, pp. 27538-44 (Feb. 15, 1944), and buck slip, Anderson to Schell, Feb. 11, 1944, in Anderson reading file.

[38] Minutes, pp. 23265-66 (Oct. 22, 1942). This policy of advance payments on fees was reversed when the membership of the Commission changed. See action concerning Sun Ship, Jan. 29, 1946, Minutes, pp. 34497-98.

To be supervising the finances of private companies was not a new experience either for the Commission or for its Director of Finance. Before the war Mr. Anderson had spent much of his time on the finances of the ship operating companies which were financially dependent on the Commission. Now he was doing the same for many construction companies. The scattered location of the shipyards made him feel the need of assistants familiar with local situations. Accordingly the Commission in May and June 1942 approved the appointment of Regional Assistants to the Director of Finance who were attached to the regional offices. According to the usual formula, they were under the Regional Director administratively and under the Director of Finance functionally. In function they differed from the Regional Auditors as a company treasurer differs from a comptroller. They were to advise concerning the credit standing of the firms in the region which were doing business with the Commission and particularly on V-loans.[39] Although Mr. Anderson in 1942 expected much of these assistants, by the end of 1943 they were dropping out of the picture, because there was not much for them to do once credit standings were known and banks were policing the V-loans; and in the more important questions of managing management, even those with financial angles, the Regional Directors acted personally on matters not handled in Washington.

In addition to the supervision of salaries and finances exercised by Washington and the regional offices, there was the supervision exercised in each yard by the resident staff of auditors. In general their functions were the same that had been given them at the beginning of the program as described in chapter 4, but the volume of work and the kinds of problems had multiplied many times. Policing the application of zone standards is only one example. As the yards were enlarged to two or three times the size first planned and as they made provisions for recruiting labor, housing it, feeding it, providing for its safety, general welfare, and high morale, the number of questions to be ruled on by the Commission's engineers and auditors increased correspondingly.

While the technical staff approved plans and specifications, the auditor had to decide concerning the host of vouchers which

---

[39] Anderson to J. A. Farrington, May 22, 1942; Anderson to Schell, May 30, 1942; Anderson to Commission, June 10, 1942; Anderson to Flesher, Aug. 11, 1942; and Anderson to S. R. Kirby, Nov. 16, 1942, in Anderson reading file.

flowed over his desk whether they all represented costs of ship construction under the meaning of the contract. Especially difficult were those which might be called part of the general cost of doing business and be charged to overhead. For example, was a contribution to the local community fund to be allowed? Or a club house or some other eating place for the executives? Or publication of a plant newspaper by the management? If the cost of a plant organ was counted as part of the cost of building ships, should it be restricted to shipyard subjects or could it boom the whole " Kaiser organization " or " Bethlehem organization "? Should the auditor therefore be censor of the publication? If the management hired consulting engineers, should it be reimbursed for what they charged, or was that the same kind of managerial service for which it was receiving fees so that such consultants should be paid by the company out of its fees? If new supervisory officials were created in the yards, did that violate the restrictions in the zone standards? These are only a few of the thousands of questions confronting the auditors.

Out of all this checking there naturally arose a large number of conflicts between government inspectors or auditors on the one hand and the representatives of the companies on the other. Settling these conflicts in ways which would maintain the morale of the Commission's staff and also that of the companies, so that the latter would not feel that they were being hamstrung with regulations and the former would not feel that they were being let down by their superiors, was one of the main problems of the Regional and Division Directors. The auditors' point of view and the value of their efforts in saving millions for the government will be considered more fully in chapter 22. That some such auditing was essential under contracts providing for reimbursement of costs is certainly obvious. From the point of view of managing managements, the difficulty was that it tended to curb managerial action so much that no one had the power essential to responsibility. Management, the Commission, and the labor unions could each point its finger at the other when things went wrong and say, " My hands are tied, it's up to the other fellow."

An episode will show the problem. Commissioner Carmody made a speech at Portland, Oregon, January 12, 1943 explaining certain aspects of the shipbuilding program and lauding the contribution of Edgar Kaiser.[40] Mr. Carmody's speech moved one of the employees at the Swan Island yard to write two letters. One concerned very specific suggestions which had been made to the company for increasing efficiency of operation—such as providing an aide to each foreman to make time checks and do work for the cost department. It was answered, said the writer, as were all other suggestions for change, by saying, " The Maritime Commission wants it like that, just as it is."[41] The other letter was more general and expressed sentiments that were certainly widely felt about the mushrooming shipyards. " As you made clear in your Portland speech the Kaiser Co. has no capital invested in Swan Island and exercises only supervised administrative functions. But employees and the public perceive that the contracts are quite evidently such that the Company does not suffer financial loss from any inefficiency, laziness or wastefulness of management or workers. These evils go unchecked or with only feeble and ineffective pretense of checking, as minor officials admit." Mr. Carmody was told that he had not heard the real opinion of Oregon people at the " First Citizen banquet," for the real-estate persons there were " naturally feeling as satisfied as the cat just after eating the canary." " Your Commission," wrote Mr. Carmody's correspondent, " is forfeiting public confidence by allowing the appalling waste of taxpayers' money. . . . When patriotic workers, anxious to do their best for the war effort see drunken incompetents (frequent absentees) working alongside them at the same hourly rate . . . they become disgusted with the hypocrisy of pep publicity. . . . The men often have to stand around idle for hours and even days because ordered to do so by leadmen and foremen who have no work laid out for them."[42]

The letter which contained specific suggestions was sent by Commissioner Carmody to Mr. Ring, who in turn sent it to Edgar Kaiser, who took the opportunity to set forth at length management's view of the situation. Perhaps, replied Edgar Kaiser, some suggestions had been answered, by the statement, " The Maritime Commission wants it like that, just as it is," for

[40] Copy in Historian's collection, Research file, 118.3.
[41] A. B. Harrison to Carmody, Feb. 8, 1943, and A. B. Harrison to Suggestion Supervisor, Kaiser Co., Inc., Dec. 31, 1942, which are both attachments to Anderson to Carmody, June 8, 1943, in Carmody files, Kaiser folder.
[42] A. B. Harrison to Carmody, Mar. 28, 1943, in Carmody file.

government auditors were exercising so much control that management felt unable to manage. "We are constantly being told, 'We can't do this, We can't do that'—because the Audit section will not reimburse." In regard to the specific suggestion for engineering aides for foremen, he said that even if qualified persons could be hired, which he thought unlikely, "there isn't any question in my mind that the Audit Section would immediately say, 'Too much supervision. We are sorry, but we will not reimburse you for this.'"

Enlarging on this general theme, he continued:

> Unknowingly and perhaps unavoidably, as time passes, more governmental procedures and controls are creeping in—building up—and it all centers back on this question of what the General Accounting Office will say or what the Audit Section will say, and management is now reaching a point where its very thinking is controlled and limited by these factors. I'll give you a good example. No one has effectively stopped, over a continuous period of time, quitting early, including ourselves. We make drives on it and it improves. The fundamental reason for quitting early is that it is a practice over the nation as a whole to permit a workman to put away tools on company time, because there is a long-established union custom originating from the time when the majority of the workmen had their tools. Now they are, in the main, company tools and must be checked in so that they won't be lost. Power curves will show that if the quitting time is 4:30, the power will begin to slack off somewhere between 3:45 and 4:00, down to almost nothing by 4:15. It is of utmost importance to efficiency per manhour that we secure continuous production. Therefore, we have suggested that it be a requirement, and the unions have agreed, that all personnel work right up to the quitting whistle at 4:30 and that they be paid for ten minutes after 4:30, which rate must be time and a half, or payment for fifteen minutes. There isn't any question in my mind that if this were permitted, we could really control quitting early. This is just one example.

> There are literally hundreds of other things that can be done. They must be done within a reasonably short time after the organization has developed the idea. Management and labor have talked the problem over and have decided that such and such a system merits trial. They are enthused and inspired, but then it actually may be weeks or months if you wait for approvals before you can try the system, and by that time the inspiration and enthusiasm are gone. Letters have been written about "Why doesn't the company do this"—or—"It's a cost-plus job and the contractor is merely trying to build up more cost for more profit." In my judgment, if the Maritime Commission is to continue to lead the field, it must take account of this factor. It must decide that it has or hasn't confidence in its contractors, and it must tell them that they have freedom of management. . . .

> In all sincerity, I think I understand what has happened—a gradual building up of regulations so that the yards will operate identically to govern-

mental agencies. If you accomplish this, you will lose incentive, initiative and the very things for which you hired the contractors to do this job. . . .

Again I say; I think we are at a turning point where the Maritime Commission must make its decision. We either operate as a government agency or we go back to freedom of management. . . .[43]

Edgar Kaiser's plea was not entirely in vain. A special committee consisting of Henry E. Frick, the trouble shooter on construction, W. L. Slattery, General Auditor of Construction, and Walston Brown of the Legal Division was sent West to study the " definition of duties and clarification of responsibilities and relationships between the representatives of the Commission and of the contractor." [44] They reported that the Pacific Coast shipbuilders had really nothing to complain of regarding disallowances, and only a little in regard to delays; but the Committee did make recommendations for better coordination among the officials of the regional office so as to prevent the issuance of conflicting instructions.[45]

Admiral Vickery's general attitude towards the proper role of the inspectors and auditors was clearly stated by him to Congressional committees. Their jobs were to audit and inspect but not to try to run the yard or " run the management." [46] In a sense, running the managements was Admiral Vickery's own job. But he did not propose to do it by having the Commission's staff say how a yard should be run and then impose their recommendation on the management.

Neither did he have much use for " efficiency engineers." Said he at one Congressional hearing, " . . . my office has been loaded with them coming in and wanting to go into these plants to correct every shipyard in the country. I get reams of correspondence on the damned things." [47] Commissioner Carmody believed strongly that industrial engineers could make improvements in the shipyards by showing how efficiency methods developed in other kinds of manufacturing could be applied in shipbuilding. He thought that the men directly in charge of the program and

[43] Edgar F. Kaiser to Ring, May 31, 1943, in Carmody files, Kaiser folder.

[44] Anderson to Flesher, June 21, 1943 and Land (by Anderson) to many shipbuilding companies, in Anderson reading file.

[45] Frick, Slattery and Brown to the Chairman, Aug. 21, 1943, in Carmody files, folder " Br-Bz," quoted more at length in chap. 21, sect. III, at note 14.

[46] House, *Walsh Kaiser Co., Inc., Hearings*, pp. 267-68.

[47] *Ibid.*, p. 267.

the yards, being primarily naval architects, marine engineers, and constructors, did not know much of modern efficiency methods in industry nor realize how much their application would increase efficiency in shipbuilding. But consulting firms of industrial engineers were employed in only a few shipyards.[48] Of course the bigger and better shipbuilding organizations had efficiency engineers of their own, and the principle on which Admiral Vickery operated was to leave it to management as far as possible.

Even in his own personal "running" of the managements, Admiral Vickery sometimes allowed them to do things which were against his own judgment in order to make them feel free and responsible. "You have got to allow something for the enthusiasm of people when they are competing, and very often I have allowed launchings and keel layings ahead of time when it is not the most economical or the most productive way to do it, because I believe you gain something on the morale in a new shipyard starting up." "We talk to management, but we do not go in and try to run their plant. You have got to give management freedom to a certain extent." [49]

III. The Selective-Price Contract

Under such a policy, the main hope of cutting costs was to reshape the contracts for ship construction so that the companies would have compelling incentives to save. Attempts were made to do exactly that, but the difficulties were manifold. "As has been pointed out by the reports of various Congressional committees," said a memorandum approved by the Commission on June 22, 1943, "and as shown by certain cost studies made by the Finance Division of the Commission, the cost-plus-a-fixed-fee type of contract tends to discourage economy on the part of the contractor and to result in an ultimately higher cost to the government than would be the case under a lump-sum contract, which provides for the payment of a reasonable price." [1] But how could the shipbuilding companies be persuaded to go over to lump-sum contracts at reasonable prices? It meant taking more risks: first, the risk of underestimating their costs; and second, the risk of running out of working capital because they would not be reim-

---

[48] Interviews with Mr. Carmody, and see below in chap. 19, sect. III.
[49] House, *Investigation of Rheem Manufacturing Co., Hearings*, p. 96.
[1] MC Minutes, p. 25485.

bursed for their costs semi-monthly as the work progressed. Although progress payments could be provided for, as they usually had been, in lump-sum contracts, these contracts might increase the contractors' investment in work in progress and his requirements for working capital.[2] A third kind of risk was less a matter of finances than of achievement in the war effort. Suppose they cut down their labor force for the sake of economy, and then were called on to build a new type quickly because of military need. Would they be able to meet these sudden military needs which might, and did in fact, arise? Or would they be caught short of workers and staff, unable to handle bigger contracts, and to fulfill their boasts of doing " the impossible " in order to win the war?

To balance these risks, especially the financial risks, prospects of profit had to be offered. By 1943 profits were subject to recapture at three different levels. First there were the provisions in the contracts themselves which specified that all profit above a stated sum should be recaptured by the Commission. Since the sum was fixed as a percentage of the estimated cost per ship it might mean a very high yield per year on the capital and managerial ability.[3] But, second, there was a chance for additional recapture under the renegotiation statute passed in 1942. Finally there was the excess profits tax taking as much as 80 per cent.[4] With these limitations on the profits, there was inadequate incentive to take the risks of an ordinary lump-sum contract.

In spite of these difficulties the Commission took a step toward a return to lump-sum contracts in June 1943. It began using a type of contract in which the first risk at least, that of underestimating the cost, was almost entirely eliminated. The new type was called the selective-price contract, because the builder could select at any time prior to keel laying the lump sum for which he would build the ship. The contract stated a list of prices from which the contractor could make his choice and stated the amount of retainable profit for each price. If he chose a high price, all

---

[2] Anderson to Chairman and Vice Chairman, Nov. 25, 1942 in Anderson reading file.

[3] Land testified Apr. 13, 1942: " Our contracts do not permit of bonuses in excess of 8.2 per cent to 8.3 per cent at the maximum." House, *Hearings on H. R. 6790*, pt. 2, p. 2967. On legal limitation see below chap. 24, sect. I, notes 2 and 38.

[4] Miller, *Pricing of Military Procurement*, pp. 170-85. Below, chap. 24. Fear that renegotiation might operate to remove incentive is expressed in Anderson to Land, Oct. 11, 1943, in Anderson reading file.

profit except a relatively low amount would be recaptured by the Commission; if he chose a low price, he was allowed to retain a relatively high profit. In the first cases proposed, the maximum retainable profit allowed was 5.7 per cent of the Commission's estimate of the normal cost of that type of ship. The incentive to economy lay in the fact that the shipbuilder would know under this contract that if he built successive ships at lower and lower costs he would get higher and higher fees. If the contractor spent more than the lump sum he selected, he would not be reimbursed for it, but since he could postpone his selection until just before keel laying he was in little danger on that account.[5] By that time, when much of the fabrication and assembly had already been done, he took little risk of underestimating his costs, especially when he had already completed some ships of the same design and was reaping the benefits of continuity of operation (see Figure 17).

Whether the selective-price contract was more like a cost-plus contract than a true fixed-price contract is debatable. T. H. Reavis, then specialist on the Maritime Commission in the General Accounting Office, declared, " The so-called selective-price contract is nothing more than a cost-plus contract with provisions for increasing profits through the reduction of the admittedly high cost." He cited in support of this view the escalator clauses for labor and material and the clause stipulating that the Commission-furnished material would be reckoned at prices fixed in the contract, regardless of their actual cost.[6] Moreover, practically speaking, the selective price contract could be considered really a form of cost-plus because the builder was allowed to delay setting his price until he knew his costs, at least with a high degree of accuracy. On the other hand, the shipbuilding executives argued that they could not be sure of all costs at the time they "selected" their price, and that these contracts did involve risk.[7] Close as they are in effect to a kind of cost-plus, I shall refer to them as a type of lump-sum contract since they are of that legal form.

[5] Minutes, pp. 25485-86 (June 22, 1943). A memorandum by Mr. Anderson explaining the selective price contract, Mar. 18, 1944, is in the Weber file folder " Selective Price."

[6] House, *Investigation of Shipyard Profits, Hearings* (1946), pp. 691-92; also pp. 685-92.

[7] *Ibid.*, pp. 643-45, 651-52.

The profits made under selective-price contracts, although limited to sums stated in the contract, were not subject to renegotiation. These contracts were excluded from renegotiation on the grounds that the profits to be derived under the contract could be determined with reasonable certainty when the contract price was established.[8] In some respects the Maritime Commission's selective-price contracts were similar to the " incentive contracts " which the Navy Department was urging on its contractors in the fall of 1943. The War Department was seeking the same objective by adding an array of price-adjustment clauses to its fixed-price contracts. The aim of all these moves was to increase incentives to efficiency by close pricing under lump-sum contracts and to allow for the different costs of production among contractors.[9]

The opportunities which the selective-price contracts offered the shipbuilding companies to increase their profits by reducing costs and to have these profits exempt from renegotiation were not, however, sufficiently attractive to persuade the builders to conclude their new contracts under this form of lump-sum. Only one company, North Carolina Ship, which was already the low-cost yard, accepted the new form of contract. While shifting to the construction of C2's, it concluded in June 1943 a selective price contract for additional vessels of this type.[10] Not until a year and a half later did any other big yards accept even this especially modified form of return to lump-sum.[11]

Companies could not be forced to take lump-sum contracts as long as the demand for ships was practically unlimited, but Admiral Vickery tried to scare them into greater efficiency by a picture of what would happen when the demand for ships became less intense. On August 2, 1943 he wrote to all shipyards and to all Regional Directors that the Commission " is looking ahead to a period of curtailment of its present building program," and

---

[8] MC Minutes, p. 25486.

[9] Connery, *The Navy and the Industrial Mobilization,* pp. 217-8; Miller, *Pricing of Military Procurement,* pp. 137-40, 186; Anderson, " United States Maritime Commission Procedure," in Fassett, ed., *Shipbuilding Business,* II, 27-31.

[10] The contract was dated Apr. 4, 1943, but the Commission action approving it was on June 22, 1943. MC Minutes, pp. 25485-86. The award made to North Carolina Ship Apr. 20, 1943 approved the same conditions as the previous contract for C2's, namely, cost-plus-variable-fee. MC Minutes, pp. 24913-14. The first keel of a C2 (first contract) was laid at North Carolina, June 19, 1943. Reynolds' Way Charts.

[11] See below, chap. 20.

PLATE XVII.

Above, a standard cargo type, the C2-S1-A1, converted for use as a Navy auxiliary. While attention was focused on the records for speed being made in emergency yards such as Oregon Ship (right) by multiple production, American shipyards were also building so many standard cargo types that by the end of the war their number exceeded that envisaged in the Maritime Commission's prewar long-range program. At North Carolina Ship methods of multiple production were used to build the C2-S-AJ1 shown below.

PLATE XVIII. Upper left, Howard L. Vickery, Vice Admiral (USN), Vice Chairman of the Maritime Commission. Upper right, Transfer of a double-bottom section at Richmond. Below, the Kaiser-Swan Island yard after a snow storm.

concluded: "The least efficient yards will be tapered off first in order that the most economical use can be made of man-power, dollar value, and facilities."[12] But this threat could not be carried out; although the Liberty program was curtailed, the demand for special and military types kept the shipyards running full tilt throughout 1944.

As long as maximum production was being authorized by the White House and the Joint Chiefs of Staff, the Commission's efforts to push through any new system of contracts was under an impossible handicap. The Commission could not allow a shipyard to be idle or even allow a company operating a major shipyard to go bankrupt and be liquidated to the benefit of its creditors in a way which would hold up production. Inducing management to save depended entirely on offering rewards, therefore, not on financial penalties. This is brought out most clearly in the letter, written by the Director of Finance, in which the Chairman of the Maritime Commission gave advice concerning excess profits taxes. After pointing out that too high a tax took away incentive and made contractors so wasteful that the U. S. Treasury might really lose in the end from too high a tax rate, he continued: "As a matter of fact, we have experienced just this kind of a condition in a number of situations where it has been necessary, in order to save a contractor from default under a fixed-price contract, to convert it into some form of cost-plus contract, and we have found it difficult, and almost impossible, to avoid a state of mind on the part of such a contractor whereby he proceeds regardless of cost."[13]

Return to lump-sum contracts was hampered not only by the factors cited, especially the difficulties over working capital, but also by the introduction of new types of ships into the Maritime Commission's program. When the Director of Finance explored the possibilities in November 1942 he emphasized to what extent costs were becoming standardized and stabilized in the Liberty yards. Once costs were predictable, there was hope of getting away from contracts guaranteeing reimbursement of costs. The introduction of the Victory ship and of many new military types in 1943 and 1944 re-infected the program with uncertainties about cost. What steps in the direction of lump-sum

---

[12] Historian's Collection, East Coast files, Director, 1942, Aug. 1943.
[13] Land to Colin F. Stem, Anderson reading file, Aug. 19, 1943.

contracts were taken under those conditions will be discussed in chapter 20.

But there was one step the Commission could and did take if a management was proceeding regardless of cost and failing to produce—change the management. The threat to oust a contractor was made in many instances, and the threat was actually carried out in a number of big yards. Those cases illustrate further the many aspects of managing management: use of the government's financial power; appeals to the profit motive, to patriotism, and to vanity; the invoking of public opinion; and the application in private of needling and of praise.

CHAPTER 15

CHANGING MANAGEMENTS

I. Tampa and Savannah

"IF YOUR SHIPS are not coming out and the costs are going up, I change the management," said Admiral Vickery.[1] First to last the Commission forced at least ten major changes: six in Class I yards and four in Class II yards. Congressional committees investigated at least half of these cases. By the hearings they held on many aspects of ship construction, the House Committee on Merchant Marine and Fisheries and the Senate Committee Investigating the National Defense Program were an ever-present factor in the management of management, as much by the possibility of what they might do as by what they did. In nearly every case of a change in contractors, the Commission was subject to severe public criticism from someone, on the ground that it should have changed sooner, or that it should not have changed at all, or that it put in the wrong people, or that it paid too much or paid too little—usually on several conflicting grounds by different critics. Changing managements was politically the most difficult part of the process of managing managements.

When a management was judged so bad that a change was decided on, there were four ways in which the Commission might proceed. If it was a government-owned yard, for the building of which the Commission had reimbursed the contractor, then it could, without needing to give any reason, cancel the contract for ship construction and take possession of the yard. The clauses permitting the Commission to cancel the ship construction contract whenever it desired were framed with a view to a sudden ending of the war, however, and they provided for substantial payment of fees for half-built ships.[2] In cases of bad management, there was no desire to pay these fees; it seemed like rewarding failure.

[1] House, *Walsh-Kaiser Co., Inc., Hearings*, p. 271.
[2] Article 25 of the cost-plus-variable fee contract. House, *First Supplemental National Defense Appropriation Bill for 1942, Hearings*, pp. 439-41 (July 17, 1941).

A second possible method of change in government-owned yards was to declare the contract terminated for default. In case of default the contract gave the Commission full right to take possession not only of the facilities but also of the partly completed vessels, equipment, supplies, and so on. These provisions were inserted to protect the Commission against cases of excessively bad management; but proceeding under these clauses did not prove practical in most instances, as the story of particular cases will show. One weakness was that the Commission was required to prove the contractor's failure to " use due diligence " to perform his covenants under the contract; and another was that in addition the Commission had to give notice to the contractor as to such failure, and the contractor after being so notified was then allowed thirty days in which to remedy the failure.[3] The need of thirty days' notice came into conflict with the need for speedy action. If a contractor was once so discredited as to be declared in default, it was not desirable to leave him in undisturbed possession for another thirty days. As matters worked out, the main value of the clause concerning default was in its use as a threat.

A third method was to institute condemnation proceedings. This power had to be invoked in order to oust the management from a yard which was privately owned. The power of the government to requisition property in time of war on the ground that it was needed for national defense was sufficiently sweeping so that this method was effective as a way of obtaining possession, but it involved so abrupt a change that operation was likely to be disrupted.

The fourth and most often used method was negotiation under pressure. The Commission preferred that changes be made by negotiation, for that was the method most likely to maintain and increase production. It had powerful weapons it could use to persuade an existing management to step aside in favor of a new group. Many shipbuilding companies were dependent on the Commission for sponsorship of loans, or for advances, or for new contracts to maintain their credit. If financial pressure was not effective, the Commission could threaten to use the more extreme methods already mentioned—condemnation on grounds of national necessity or termination for default—and by threats negotiate a change.

[3] Articles 23 and 24 in the contracts. MC Historian's Collection, Research file 209.11.

A number of the most important of the changes in management were made in the winter of 1942-1943. By that time the yards had been operating long enough to show the glaring contrast between the efficient and inefficient yards. Military leaders were demanding ships; laggards must be replaced by " go-getters " wherever possible. But two earlier instances in which the management of a yard was changed by action of the Commission deserve attention: Tampa and Savannah. They are instructive in themselves and they help explain the policies followed in making later changes.

The Tampa yard had had a unique relation with the Commission from the beginning of the long-range program. Its bid on the first C-types was accepted in 1938 in order to beat down the bids of the other shipbuilding companies, and the company arranged at that time with the Reconstruction Finance Corporation for a loan which was guaranteed by the Maritime Commission. In 1939 the Tampa Shipbuilding and Engineering Co. was awarded a contract for four more Diesel-powered C2's. The Commission assumed certain responsibilities regarding the development of the yard and took a hand in various ways as soon as its head, Ernest Kreher, began to run into difficulties, as he promptly did.[4]

The first difficulty was with labor. A representative of the AFL complained to Mr. Ring that the Tampa yard was violating the National Labor Relations Act by signing up its workers in a company union. Mr. Kreher was told he should not do that,[5] and subsequently he signed a closed shop agreement with the AFL. That did not end his troubles; there was Ku Klux Klan activity, some skilled Negro and Spanish laborers were forced out by the union, and at the same time there were strikes and objections by the union to Mr. Kreher's actions.[6]

These and other difficulties in the operation of the yard made the Maritime Commission feel that changes in management were

[4] See above chap. 4, sect. I. Unless otherwise indicated this account of the difficulties with Tampa is based on House, Committee on Merchant Marine and Fisheries, *Investigation of Certain Transactions of the Tampa Shipbuilding Co., Report No. 938,* 78th Cong., 1st sess., Dec. 6, 1943. (Hereafter cited as *House Report No. 938.*)

[5] Statement by Mr. Ring in interview, Sept. 18, 1947, in MC, Historian's Collection, Research file 105.

[6] House, *Tampa Hearings,* pp. 104, 109-10. See also *Shipyard Worker,* Aug. 25, 1939, and Metal Trades Dept. of the American Federation of Labor, *Proceedings,* 32d Convention, Nov. 11, 1940, pp. 61-62, 89-90.

necessary. The Commission objected also to the amount Mr. Kreher was spending on enlarging his facilities. Consequently they required him to appoint as general manager and as superintendents men in whom the Commission had confidence, former members of the Commission's staff. Mr. Kreher was told that these men were to act under him, after receiving certain delegations of authority specified in letters from the Commission; but conflicts arose, as might be expected, when the supervisory force was thus subject to two masters.

The right of the Commission to interfere thus in Mr. Kreher's choice of assistants was not stated in the original contract but was established as a result of his financial difficulties. His costs were increased by the contract with the AFL and subsequent labor troubles. To obtain working capital to keep the yard going, he entered into an agreement with the Commission by which it was empowered to require changes in the management. The Commission enforced these changes by threatening to withhold the progress payments with which the yard was meeting operating expenses.[7]

In spite of the changes in staff, higher labor costs prevented the company from completing the ships at the prices stipulated in the contracts. The company owed to the vendors who had sold it supplies far more than it could pay. Attempts to sell the yard to some stronger shipbuilding organization failed. In this situation the Commission had the legal right to declare that the company had defaulted on its contract, to take possession of the uncompleted ships, finish them in any way it could, and leave the creditors of the company to suffer the losses which might arise from the inability of the company to pay its debts.[8] To have followed this course of action would have ruined the shipyard as a going concern. Since there was a shipbuilding boom developing, its staff could easily have obtained jobs elsewhere as soon as they heard that the company was bankrupt. The disruption was prevented when the Commission made arrangements which put

---

[7] House, *Tampa Hearings*, pp. 99-100, 104; Anderson to Kreher, Jan. 12, 1940; Land to Kreher, Feb. 8, 1940; Vickery to Kreher, Apr. 19, 1940, all in Anderson reading file.

[8] House, *Tampa Hearings*, p. 208. S. B. Tulloss for the Comptroller General declared: "Therefore, there was no obligation on the part of the United States to asume the outsanding debts of the contractor." Mr. Anderson, on the other hand, said the government would have had to pay these claims.—*Ibid.*, p. 32.

in charge of the yard a new management headed by George B. Howell, a banker of the company, who had been in close touch with its affairs throughout and whose banking interests in Tampa made him wish to have that important local industry kept going. A new company, Tampa Shipbuilding Co., Inc., was formed under his control in November 1940 and took over the plant and most of the officials from the old company. Although after April 1941 it worked on Navy contracts, not for the Maritime, it was one of the shipbuilding organizations which was called into being by the Commission's long-range program and which did valuable work during the war.

In keeping the yard operative as a going concern, the Maritime Commission followed in Tampa the basic principle it was to follow in a number of other later cases. But the method by which Mr. Howell was placed in control and the yard kept going had unique features, as indeed did nearly every case in which management was changed by Commission action. In this case the Navy helped provide a solution. At the time that the old company seemed about to fold up, the Navy asked to have for conversion to naval auxiliaries some ships of just the type that Tampa was building. Arrangements were then worked out by which the Navy bought the ships from the new company at prices which enabled it to pay all the debts of the old company. By arranging the sale on these terms, the Commission created a situation in which Mr. Howell was willing to take over and Mr. Kreher acquiesced in the change.

This method of changing management was later denounced by the Comptroller General as illegal because it involved paying for the ships $1,926,568.52 more than the value set on them under the contracts already entered into for their delivery to the Commission. In a special report to Congress on June 10, 1942, the Comptroller General referred to " the donation of approximately $2,000,000 appropriated moneys, and stock ownership and control for which only $500 was paid by George B. Howell." [9] The extensive hearings held on the subject clearly show, in my judgment, that in return for the " donation " in question Mr. Howell's new company assumed debts of the old company to the value of nearly $2,000,000, and undertook the task of transforming an ineffectively managed company into an effectively managed one

---

[9] House, *Report No. 938*, p. 6.

at a time when managerial ability, especially in shipbuilding, was able to command high prices.

The ships cost the Navy less than it would have cost to have had comparable ships built elsewhere. Judge Bland, chairman of the investigating committee, probed whether the legality of the proceeding could not be justified by the clause in the Merchant Marine Act which vested in the Commission " authority to exercise business discretion and business judgment in all matters coming under their jurisdiction," [10] a contention later supported by the Attorney General.[11] The report of Judge Bland's committee found some practical justification for the Commission's acts because they had kept the yard going, but said that this could also have been done by the unquestionably legal method of simply transferring to the Navy all the rights of the Maritime Commission under its contracts and letting the Navy take the lead in arranging to finance Mr. Howell's company. It suggested that this could have been done through loans without the government being put in the position of buying ships through one agency, the Navy, for $2,000,000 more than the sum for which a shipbuilder was under obligation to deliver them to another agency, the Maritime Commission.

Changing the management at Tampa reveals the embarrassments which could arise from bailing out a weak company in order to insure continuity of production. Changing the management at Savannah illustrates the complications in pushing a weak company aside by insisting on the Commission's right under its contract to take possession of facilities and in the acquisition of a privately owned yard by invoking the right of eminent domain.

Savannah was the first instance of changing management which arose after the declaration of war. The contract which the Commission had made with Savannah Shipyards, Inc. was loaded with special provisions, explained in chapter 5, because the management and capital structure was not one in which the Commission had much confidence. The contract had been awarded because Frank Cohen and his associates in Empire Ordnance had made appreciable progress in building a 3-way yard without use of any

---

[10] House, *Tampa Hearings*, p. 218. Compare Section 207 of the Merchant Marine Act.

[11] *Correspondence Between Comptroller General and Attorney General Relative to Tampa Shipbuilding Co.* [House, Committee on the Merchant Marine and Fisheries, Document No. 45], pp. 654-55.

money from the Maritime Commission; and the terms required the company to recruit an adequate staff, to show $750,000 of working capital within 30 days, and to complete the facilities within 60 days. On December 26, 1941 the 30-day term expired. At the time the contract was entered into, the Commission contemplated the possibility in case of such a failure of simply terminating the contract and abandoning any plans for building cargo ships in Savannah. "The Commission would have spent no money and been under no obligation...." But after Pearl Harbor it was imperative to utilize every shipbuilding facility, even a half-finished or "incipient" shipyard. Admiral Vickery and Mr. Anderson advised the Commission that Savannah Shipyards had not supplied adequate staff nor evidence of the required working capital.[12] The facilities were far from completed.[13] Accordingly the Commission on December 30 authorized its agents to take possession (as the contract authorized) of the half-finished facilities,[14] and at the same time to take by condemnation proceeding the land and the other property in the shipyard.[15] The plan was to finish on the site a 6-way yard and award to some new company, under management which the Commission considered adequate, a contract to operate it.[16] A maze of legal and administrative complications resulted.

Until a new company took over, administration fell upon members of the Commission's staff. The Resident Auditor was authorized to take necessary steps to pay workmen.[17] The Chief of the Plant Engineering Section and other officials went at once to Savannah, closed the yard, and engaged a firm to take inventory, while the Commission's Solicitor, Paul D. Page, Jr., was securing the necessary court actions.[18] All construction stopped for a week.[19] To prevent the labor from scattering, a press release was given out stating the intention of the Commission to have work begin

[12] MC Minutes, pp. 20341-44.
[13] *Ibid.*; J. Kirkpatrick to Sanford, Feb. 9, 1942, in MC gf 503-40; Long report of Dec. 24, 1941 of W. L. Marshall, Principal Construction Cost Auditor, in gf 507-4-15; F. D. Graves, Hull Inspector, to W. H. Lalley, Dec. 30, 1941, in gf 507-4-15.
[14] MC Minutes, pp. 20341-44.
[15] Land to Attorney General, Jan. 1, 1942, in gf 507-4-15.
[16] Lalley to Vickery, Jan. 3, 1942, an outline of the general plan followed, including mention of Rieber and Rentschler as possibilities.
[17] MC Minutes, pp. 20344-45.
[18] *Ibid.*, p. 22503. Various papers under dates Jan. 1-5, 1942, in gf 507-4-15.
[19] MC Minutes, p. 22089.

again shortly and to expand the working force.[20] Letters began to come in from companies that had shipped materials to Savannah Shipyards, Inc., or had them ready to ship, asking how they would be paid and whether to continue shipments and to whom. In replies the Commission declined responsibility for the debts of Savannah Shipyards, but said that if Savannah Shipyards cancelled its orders the Commission would be interested in acquiring the materials in many cases.[21] Particularly complicated, for example, was the question of who now owned and who would pay for some cranes of which some parts were delivered and some parts were on the way.[22] The Hull Inspector Coordinator, Mr. Kirkpatrick, visited the yard on February 6 to see how much of a beginning had been made on ship construction. He reported that the mold loft had been finished although no templates had been made. Of a skeleton force of approximately 20 shipbuilders collected by Savannah Shipyards, about half had left Savannah and the other half, like the foreman shipfitter and others of the supervisory staff, were working on the construction of the yard.[23] The total force which Savannah Shipyards had employed was only about 300 so that the quantitative importance of the interruption was not large,[24] but the incident shows the nature of the disruption involved in changing management by such drastic methods.

Completing the facilities was a more pressing problem than installing a new shipbuilding company, and accordingly the Commission made a contract directly with a construction company, although its usual practice was to have the operating company subcontract the work of constructing the yard. The Daniel Construction Co. of Savannah, Georgia, a firm highly recommended by the Navy, was engaged to finish the project as a 6-way yard. Under an oral understanding, work on the yard was resumed January 12, and on January 30, 1942 the Daniel Construction Co. was sent a letter stating the intent to give them a contract for

[20] Press release 1118, under Jan. 5, 1942, in gf 507-4-15.
[21] Schmeltzer to Vendors, Jan. 20, 1942, in gf 507-4-15; Schmeltzer to Vendors, Jan. 14, 1942, in gf 503-40, and letters from vendors and subcontractors in both these series of files. See also William C. Cross to Page, Jan. 30, 1943, in gf 503-40.
[22] Many letters in the gf 503-40 and 507-4-15, for example telegram from Brooklyn Contractors Machinery Exchange to Paul Page, Jan. 6, 1942, and the contract MCc-2207 under Feb. 10, 1942, in gf 507-4-15.
[23] Kirkpatrick to Sanford, Feb. 9, 1942, in gf 503-40.
[24] PR 1118 of Jan. 5, 1942, in gf 507-4-15.

the work at cost plus a fixed fee of $70,000.[25] Another firm, Sirrine and Co., which had made the inventory, was also engaged at cost-plus-a-fixed-fee to finish the design of the 6-way yard and supervise its construction.[26] The Commission's representatives supervising these contractors, the Resident Auditor and the Resident Plant Engineer, had difficulties between themselves in defining their spheres of activity under the unusual arrangements.[27] After a new company was formed to operate the yard, the contracts of Daniel Construction and of Sirrine and Co. were assigned to the shipbuilding company so that the contractual arrangements were finally fitted into the pattern for emergency, government-owned shipyards.[28]

The new management to operate the yard was determined on January 20, 1942, after negotiations had been carried on with ten different men or managerial teams. The team selected was headed by men having much experience with ships and their components —G. A. Rentschler, President of General Machinery Co., the leading manufacturer of the reciprocating engines used in the Liberty ships; William H. Smith, a former executive of Todd Shipbuilding Corp.; and Captain T. Rieber, who had directed the tanker fleet of the Texas Company.[29] Captain Rieber's energies were then available for shipbuilding because he had recently resigned his position with the Texas Company after agitation over his close connections with the Nazis.[30] In 1942 he was supervising the Charleston Shipbuilding and Dry Dock Co. of Charleston, S. C., and Admiral Vickery believed this affiliation would assist the shipbuilding operation at Savannah. These men organized the Southeastern Shipbuilding Corp., showed evidence of strong financial backing, and were awarded by a 3 to 2 vote of the Commission the contract to operate the Savannah yard, Commissioners Macauley and Woodward voting nay.[31] Since the yard had been

[25] Page to Files, Jan. 16, 1942, and Schmeltzer to Daniel Construction Co., Jan. 30, 1942, in gf 507-4-15; MC Minutes, p. 22089-91.
[26] MC Minutes, pp. 22089-91.
[27] Albert G. Keen to J. A. Honsick, Feb. 23, 1942 and Mar. 21, 1942, in gf 503-40.
[28] MC Minutes, pp. 22089-91.
[29] MC Minutes, pp. 20649; Anderson to File, Feb. 21, 1942, in Anderson reading file.
[30] Herbert Feis, *The Spanish Story*, who cites an article in *Life*, July 1, 1940, by Joseph L. Thorndike.
[31] MC Minutes, pp. 20649. Concerning Woodward's attitude see "Minutes of Shipyard Site Planning Committee," Jan. 19, 1942.

partly built before they took over responsibility for it, the fees were less than the standard.³² To spur them on, Southeastern was considered as starting about even with other entirely new 6-way yards, even though contracts for the others were awarded only in February and March 1942. Compared to the three other 6-way yards which built Libertys, it delivered substantially more ships than any other in 1943, because of its head start, but fewer in 1944,³³ and its showing in manhours was relatively poor.³⁴

The legal complications growing out of the Commission's seizure of the Savannah yard are of interest here chiefly because of the heavy drain they made on the time of Maritime Commission officials in a period when time was worth more than money. The legal points involved are of less general interest because the pertinent clauses in the contract between the Commission and Savannah Shipyards were not the standard form used in most contracts, but a unique arrangement to meet the unique situation which Frank Cohen had created.³⁵ At one time the Commissioner's Solicitor believed he could negotiate a settlement out of court of all claims arising from the whole matter of condemnation and termination for $1,000,000, and the Commission authorized settlement at that sum,³⁶ or at $1,076,000, which the shipyard interests were at one time willing to accept.³⁷ But the decision in matters of condemnation lay with the Department of Justice, which would not concur in these proposals for settlement. The case went to trial in Savannah in July 1942 while Admiral Vickery was on the West coast looking into the steel shortages there, and the Gulf Regional Office was busy with clearing up after the cancellation of the Higgins contract. Yet many key men from the Washington office and the Gulf Regional Office were required to give technical testimony concerning the condition of Savannah Shipyards, Inc.³⁸ As Admiral Vickery said later in explaining why some things did not get done properly in July 1942: " While I was terribly short of people, my people were held down there for a month. They

---

³² Anderson to Vickery, Jan. 12, 1942, in Anderson reading file.
³³ *Official Construction Record.*
³⁴ Fischer, *Statistical Summary,* Table H-3.
³⁵ MC Minutes, p. 22385 (July 4, 1942).
³⁶ Page to Chairman, July 11, 1942, in gf 507-4-15.
³⁷ Land (written by Page) to Attorney General, July 17, 1942, concerning settlement at $1,076,000, in gf 507-4-15, and memoranda. Statements by Page.
³⁸ Telegrams calling for their attendance at the trial, dated July 3, 1942, in gf 507-4-15.

were yelling to high heaven that they had other important work to do, . . ." [39] Admiral Vickery would have rushed back to testify except that the judge refused to delay and the attorney for the Department of Justice optimistically decided that his presence, while desirable, was not necessary.[40] Decision went against the U. S. Government and on appeal also Savannah Shipyards was awarded substantially more than the $1,076,000 which the Commission was willing to pay as just compensation and which the Cohen interests had been willing to accept as such.[41]

II. SOUTH PORTLAND, THE LOCAL SCENE

The backfire from the forceful ejection of Savannah Shipyards Inc. was still being felt in the Commission when it became clear that something drastic was going to have to be done about one of the largest emergency yards, that at South Portland, Maine. The change in management at South Portland overshadowed in importance the others made by the Commission, for it involved one of the largest shipyards (fifth largest in number of ways), it touched important personalities and powerful companies, and it engaged the attention of two Congressional committees and of the White House. The circumstances furnish concrete illustration both of the community problems created by the shipyards and of the difficulties of managing a management that proved to be a disappointment.

South Portland was assigned a prominent part in the emergency shipbuilding program from its very inception, because, as Admiral Land said in his memorandum for President Roosevelt, November 29, 1940, " It is backed by the shipbuilding brains of Mr. Newell of the Bath Iron Works and the design brains of Mr. Gibbs' design firm in New York." [1] In Maine, shipbuilding was a memory of which the state was proud and it was proud also of William S. Newell. " Pete " Newell, as he was called by the shipbuilding fraternity, had received honorary degrees, or the corresponding accolade of membership among the trustees and directors, from

[39] House, *Investigation of Rheem, Hearings*, p. 83.
[40] Statements to me by Page. Also Land to Attorney General, July 17, 1942; Land to Joseph F. McPherson, Office of U. S. Attorney, Savannah, July 21, 1942; McPherson to Page, July 24, 1942, in gf 507-4-15.
[41] Page to the Commission, Aug. 15, 1942, in gf 507-4-15; Page to the Commission, Jan. 20, 1944, in gf 503-40.
[1] Gf 500-3.

many a Maine institution—from Bowdoin College, Colby College, the University of Maine, the Maine Central Railroad Co., the Bath Trust Co. and from other solid institutions inside and outside the state. Forty years before, just after his graduation from Massachusetts Institute of Technology, he had gone to work as a ship draftsman at the Bath Iron Works. In 1940 he was president of that company and was celebrated for the business brains shown in keeping it alive during the doldrums of the shipbuilding industry, and for the high quality of his work in building destroyers for the Navy. On two ways at his Bath yard he was also building cargo ships for the Maritime Commission.[2] As head of the largest shipbuilding enterprise of northern New England he had on his payroll a large part of its skilled shipbuilding craftsmen.

To meet the British demand for emergency ships, Mr. Newell built a basin type of shipyard (already described in chapter 7) which was particularly adapted to the tides and the condition of the subsoil, a hard rock ledge 20 feet below natural grade, at the site selected. The site was the east side of Cushing Point, South Portland, on the edge of a residential community having many of the features of a pleasant summer colony. Since it was built by the Todd-Bath Iron Shipbuilding Corp., the basin yard acquired the name " Todd-Bath." It became known as the east-area yard later when it was operated by the New England Shipbuilding Corp., and I will call it the east-area yard (Figure 35).

Work began on this yard January 3, 1941, and long before it was ready to go into operation work began also on another yard on the west side of Cushing Point where ships were to be launched in the usual fashion. The west-area yard was often referred to as the South Portland yard because it was built by the South Portland Shipbuilding Corp. Of course, in terms of location they were both South Portland yards. They were both managed by the same group of men, by Mr. Newell and the vice presidents from his Bath organizations.[3] A rigid legal separation existed until 1942 because one yard held contracts from the British government, the other had its contracts with the Maritime Commission, but immedi-

[2] Reynolds' Way Charts; Society of Naval Architects and Marine Engineers, *Transactions,* Vol. 51 (1943), p. 25n; Palmer, *We Fight With Merchant Ships,* p. 165.

[3] Those receiving salaries from the three corporations, according to the files of the Resident Auditor, O. H. Schulze, were W. S. Newell, A. M. Main, G. V. Pach, T. R. Allen, and R. F. Hill.

ately after Pearl Harbor plans were made for operating the two yards under one corporation.[4] In 1942, the United States bought the east-area yard;[5] thereafter both yards became the property of the government, and were operated for it first by the South Portland Shipbuilding Corp. and later by the New England Shipbuilding Corp.[6]

A main cause of later difficulties was the lack of space and the need for enlarging the facilities for fabrication and transportation when the number of ways was increased and the speeding up of production raised the volume of the operation. The west yard was much too crowded. Mr. Newell made the basic plan for it as a four-way yard in March or April, working on an aerial photograph and marking the location of machine shops, piers, and ways.[7] When the decision was made to expand it to six ways, the design was not changed correspondingly.[8] As Karl E. Klitgaard, later manager of the plant testified, " The arrangement of the yard, Senator, for a four-way shipyard building ships in about 140 days, was perfectly all right. That was a nice compact little yard where the general manager could sit up in his office and see all over the yard. But when you . . . put six ways in instead of four, and on top of that, you again reduce the number of days . . . you are shoving your yard in, whereas your yard should expand with it. You need three times the amount of room and you haven't got it, and there is no way to get it." [9]

One reason the west area was thus faultily designed was the delay in deciding whether the four ways should be expanded to six ways. Estimates had been made, the matter discussed, and the site inspected by Admiral Vickery and Mr. Schmeltzer during June

[4] The contract with the South Portland Corp. to build ships in the east-area yard for the Maritime Commission was dated Jan. 17, 1942, and was approved by the Commission Apr. 7, 1942. MC Minutes, p. 21586.

[5] The facilities were bought from the British by contract No. MCc-8456, dated Sept. 30, 1942; the land from the Todd-Bath company by contract No. 8462, dated Oct. 19, 1942.

[6] Work was being done in the east area in some basins and docks for the British and in other basins for the Maritime Commission from Sept. 7, 1942 when keel was laid on MC Hull 768 until Nov. 17, 1942 when Hull B 30 was delivered. MC, " Permanent Report of Completed Ship Construction Contracts," p. 24A and Reynolds' Way Charts.

[7] Palmer, *We Fight With Merchant Ships,* p. 169.

[8] See below, and Vickery's testimony in House, *Investigation of Rheem, Hearings,* pp. 85-86.

[9] Truman Committee, *Hearings,* pt. 15, pp. 6017-19.

1941. On July 21, 1941, Frank E. Wall, Resident Plant Engineer, wrote, "... let us know as soon as possible. It will help the general yard layout to do it now rather than later."[10] Yet the contract for the additional two ways was not finally executed until the following October 6. Although the whole cove was filled during the summer of 1941, pile driving for the additional ways did not start until that October.[11] See Plate XIX.

This delay was all the more important because the necessity of moving materials across the end of the other four ways to build the two new ones interfered with shipbuilding. The delay occurred partly in the Commission, partly in the South Portland Shipbuilding Corp. Not until August 19, 1941 did the Commission approve the plan and estimate for the additional facilities.[12] Then Mr. Newell wrote back that he would have to have the approval of his Board of Directors.[13] Only after the Commission and the Corporation had reached agreement on the fees and conditions in a contract for building ships on these two ways did the South Portland Shipbuilding Corp. on October 6, execute the contract.[14]

Transportation was another aspect in which plans that looked good for a relatively small operation proved bad for the much larger one that developed. The Commission's Resident Plant Engineer recommended from the beginning that a railroad spur be built to the yard, and the Plant Engineering Section agreed,[15] but the final decision did not lie in their hands. Mr. Newell, when planning the east-area yard to build for the British, had investigated the cost of bringing a railroad to Cushing Point. The Maine Central Railroad, whose tracks were about three miles away, estimated at $350,000 to $400,000 the cost of building a spur to the shipyard. They proposed to charge the shipbuilding company also for the cost of operating and maintaining it. Mr. Newell

---

[10] Wall to Harry M. Hope, July 21, 1941, in gf 507-3-14; cf. John D. Reilly to Vickery, June 10, 1941, in gf 507-3-14.

[11] Dated photographs in MC, Production Division files.

[12] MC Minutes, p. 18555.

[13] Newell to MC, Nov. 29, 1941, in gf 507-3-14 Exhibit (photostat).

[14] Letters in general files 507-4-14, Sept. 11, 22, 1941 and in gf 507-3-14, Oct. 4, 6, 8, 1941.

[15] Report of Bonner to Gallagher, July 28, 1941; Wall to Hope, July 29, 1941, in Production Division Adm. file E201-S3-1; Hope to Wall, Aug. 8, 1941, in gf 507-3-14. The engineering surveys referred to in these letters are described more fully in the Portland *Press Herald,* July 3, 12, 18, 28, Aug. 2, 4, 1941 and in with a map in the Portland *Evening Express,* July 29, 1941.

RAILROADS AND FACILITIES AT SOUTH PORTLAND, MAINE

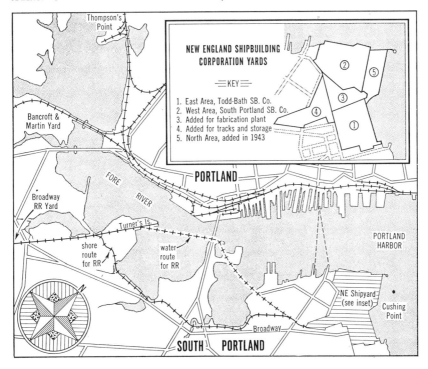

FIGURE 35.

investigated the cost of trucking materials from sidings only about three and a half miles away, the extensive Broadway yard. Kenneth T. Burr, a local businessman experienced in handling steel, said it could be trucked across at the rate of 75 cents a ton. When Mr. Newell quoted this figure to an official of the Maine Central, the latter called it " very reasonable " and showed no interest in naming a competing figure. So arrangements were made for trucking,[16] first for trucking to the east area and then for trucking to the west area.[17]

A third aspect of the plan for facilities, fabrication, also envisaged a relatively small operation. The original plan for the west area included no fabrication building, for it was expected

[16] Truman Committee, *Hearings*, pt. 15, p. 6135.
[17] *Ibid.*, pp. 6026-27.

that the fabrication would be done either in the east area, or, mainly, at the Bath Iron Works plant at Harding, a suburb of Bath about thirty miles from Portland.[18] Transportation from Harding was also by truck.[19] It should be recalled that all these arrangements were worked out before the declaration of war, and before the possibilities and principles of multiple production had been demonstrated.

By the spring of 1942 when war called for all-out production, these arrangements proved quite unsatisfactory. Building of the British Libertys in the east area was speeding up and their steel was piling into the Broadway yard. To handle the steel for the west area, the South Portland Shipbuilding Corp. first leased the Bancroft and Martin yard, a little farther away,[20] and then leased a much larger area $7\frac{1}{2}$ miles away, Thompson's Point. "Steel was diverted from the Bancroft and Martin yard, racks were built, but the steel came in so fast that the facilities at Thompson's Point were simply buried. The track layout was not suitable for handling steel in such quantities. There were no parallel tracks and no roadways for trucks, and there was no time to build them. Steel was arriving at 25 or more cars per day, and it was unloaded from cars to trucks, moved to every available space, and unloaded with crawler and truck cranes. The result was considerable confusion."[21] When trucks came from the yards to get plates, they had to go to several places and often stand around several hours to collect a load.[22] Even if this inefficiency in handling the steel was overcome, no form of trucking could, in the spring of 1942 be considered satisfactory for a shipyard. The Japanese were conquering the rubber plantations and German submarines were cutting into the gasoline supply, especially that of New England.

In the fabrication also, the original plans, made before Pearl Harbor, were proving inadequate to meet the conditions existing after Pearl Harbor. Plans for using the plant at Harding were

---

[18] W. K. Carter to Sanford, Nov. 18, 1941, in gf 507-4-14.

[19] Wall to Herman Lame, Dec. 6, 1941, in gf 507-3-14. Wall was writing in support of an access road.

[20] Schmeltzer to Wall, Feb. 2, 1942, in gf 507-3-14 approves extension of the lease but protests its high cost.

[21] Extract from statement submitted by Mr. Burr, who did the trucking, in Truman Committee, *Hearings*, pt. 15, p. 6241.

[22] *Ibid.*, on Thompson's Point, see also Stephan to Sanford, Mar. 12, 1942, in MC gf 503-53.

upset when the Navy's program was speeded up so that the Harding plant was busy preparing material for naval building at the Bath Iron Works. In this Bath shipyard Mr. Newell was making an enviable reputation in the construction of destroyers for which he received the Navy " E " in February 1942. To keep up the work on Libertys, Mr. Newell proposed that the Thompson's Point property be used for fabrication, as well as for storage, and that old railroad shops there be rebuilt for the purpose.[23] This recommendation was rejected, however, by the Maritime Commission in March 1942 and from that time on the Commission began to take a more active part in planning for and providing facilities.[24]

There were several reasons why the Commission should intervene more actively. The relative influence of these reasons at various times is difficult to determine, but one reason closely connected with Mr. Newell's proposals regarding Thompson's Point was the suspicion of improper expenditures. The rawness of the unsavory practices was not proved until later in 1942, but rumors thereof may have circulated earlier and affected the attitude of the Commission towards the South Portland Shipbuilding Corp. They certainly affected the attitude of the people of South Portland towards the shipyards and towards the Commission which was held responsible for them. While rumors generally exaggerate, there were many grounds for them.

The worker in the yard met the situation when he went to lunch. At the recommendation of the Commission's Resident Plant Engineer, Frank E. Wall, a separate cafeteria was built for the west area. The contract to operate it was taken by Mr. Wall's son-in-law.[25] The citizens of South Portland had occasion to think of it every time a truck rumbled by, for the trucking was being done by a special company organized for the purpose by Mr. Burr, " a kind of bleeding corporation " it was later called by the then Senator Harry S. Truman.[26] Construction work in the west area was increasingly passed over from the original subcontractor to the A. C. Stanley Co., a newly incorporated company

[23] W. S. Newell to MC, Jan. 30, 1942, in gf 507-3-14.
[24] The memorandum of Flesher and Vickery to the Commission with the personal notes of the Commissioners is filed under Mar. 10, 1942 in gf 507-3-14, and under the same date, in the Exhibits to the same file.
[25] See MC gf 503-52-1.
[26] Truman Committee, *Hearings,* pt. 15, p. 6036.

of which Mortier D. Harris, a director of the South Portland Co., became treasurer.[27]

Since Mortier Harris was a director of Todd-Bath and of the South Portland Shipbuilding Corp., he will serve to illustrate the style of business which many people of South Portland associated with the shipyards. He had indirect dealings with the corporation of which he was director not only in the construction but also in the trucking business. When Todd-Bath was considering how to bring in its steel, he went with the traffic manager to Kenneth T. Burr, Treasurer and General Manager of the Bancroft and Martin Rolling Mills Co., for advice. As a result, Messrs. Burr and Harris planned to form a new company to take the trucking contract. Thus was conceived the Materials Handling Corp. But it did not come to delivery as planned, for Wadleigh Drummond, Mr. Newell's lawyer, called in Mr. Burr and Mr. Harris and told them Mr. Harris should not be in the company because he was a director of Todd-Bath. Mr. Drummond suggested that Burr could go ahead, form the corporation alone, borrow money and take the contract. Mr. Burr did so, delivered the steel, and piled up profits.[28] But Mr. Harris kept contact with the trucking business, for his three brothers, as The Gilbert Partnership, although they had never rented trucks to anyone else, rented trucks to Mr. Burr's concern, buying them on the installment plan out of rentals and making $74,000 on it in less than two years.[29]

This picture is in sharp contrast to the ideal of competitive bidding producing maximum service at lowest possible cost. At higher levels, conduct was also far from that ideal. To secure Thompson's Point, Mr. Newell as President of the South Portland Shipbuilding Corp. entered into a contract with the Portland Terminal Co., a subsidiary of the Maine Central Railroad, of which he was a director. The counsel to South Portland Ship, Wadleigh Drummond (who, because of his position, can be

---

[27] A letter from the Acting Secretary of State of the State of Maine to Miss Jean Maxwell, June 30, 1942 (in Historian's Collection, Research file, 112.3.9) shows Mortier D. Harris as the Treasurer on the 1942 return.

[28] Truman Committee, *Hearings*, pt. 15, pp. 6025-32, 6230-40; House, Committee on Merchant Marine and Fisheries, *Report No. 2653*, 77th Cong., 2d sess., pp. 8-9. (Hereafter cited as Bland Committee, *Report No. 2653*.)

[29] Bland Committee, *Report No. 2653*, pp. 9, 10; Truman Committe, *Hearings*, pt. 15, pp. 6042-43.

supposed to have known the whys and wherefores of the lease) was a director of "one of the underlying railroad corporations involved." By this lease South Portland Ship proposed that the Maritime Commission pay a yearly rental of $36,000 for a property to which title was obtained later through condemnation proceedings for $183,000. The lease contained some other provisions which Mr. Newell found it difficult to explain, and he admitted to a Congressional committee that either "the railway company of which he is director had driven a sharp bargain" or somebody was asleep in his organization.[30] Actually, although the lease was dated January 20, 1942 it was signed after the Commission made known its objection to it in April.[31]

As evident in the spring of 1942 as the inadequacy of the facilities, and more evident than were, at that time, the scandalous contracts concerning them, were the poor methods of ship construction being used in the west area. Ships were being "tailor-made," with very few men available who knew how to "cut the cloth to fit." Very little if any preassembly was attempted. Instead, deck houses and other portions were constructed piece by piece and there was much overhead welding.[32] These methods were objected to from the very start by the Commission's Hull Inspector, Herman Stephan. He tried to insist that the welding sequences which had been prepared by the Commission's staff should be used, but the South Portland Corp. had him overruled by telephoning to New York to Mr. Esmond, Chief of the Hull Section of the Construction Division, and securing his approval.[33] Mr. Esmond felt that it was up to the shipbuilders to decide on the fastest methods of building according to the facilities they had available and other local conditions. As long as the builder's proposals met the safety standards of the American Bureau of Shipping, neither Mr. Esmond nor Mr. Stephan's other superiors in the Construction Division were disposed to object initially.[34]

[30] Bland Committee, *Report No. 2653*, pp. 7-8. Congratulating Admiral Land on the killing of the lease, Judge Bland wrote that it was "conceived in sin, born in iniquity, and prepared in complete disregard of national interests." Bland to Land Aug. 24, 1942, in gf 509-10.

[31] Minutes, p. 23545 (Nov. 20, 1943).

[32] Klitgaard to Newell, July 6, 1942, in East Coast Regional Office files, Director, 1942.

[33] Herman R. Stephan, Principal Hull Inspector, to L. R. Sanford, Sept. 18, Oct. 2 and 28, 1941, in gf 507-4-14; A. M. Main, Jr. to Stephan, Oct. 27, 1941 in Prod. Div. Adm. file E201-L6-3.

[34] E. G. Esmond, Chief, Hull Section, Construction Division to South Portland

Mr. Stephan's opposition to the piecemeal methods being used was supported by visiting inspectors. James W. Wilson, Welding Inspection Co-ordinator, wrote back, February 27, 1942, " It is my opinion that Mr. Stephan and his staff are working zealously under trying conditions to secure better work. . . ." [35] But from the office of the Inspection Section of the Construction Division, Mr. Stephan received quite unsympathetic replies to his complaints.[36] He seemed better at finding things wrong than at putting them right. He complained about many things, about a plate which passed through the fabricating shop without being rolled; [37] about the lack of realism in the delivery schedule proposed by South Portland Ship,[38] about how much welding wire and other material was being " loaned " by South Portland Ship to Todd-Bath,[39] and about removal of scrap metal without his consent.[40] Mr. Stephan was not on speaking terms with the chief company executive actually in the yard.[41] Early in 1942, he was transferred to North Carolina. The South Portland company was complaining of him and their complaints were felt to be justified,[42] but after the Construction Division had been through the upheaval of decentralization in March 1942 and the Regional Offices were established, the East Coast Office also became highly critical of the methods of hull construction being used at South Portland.[43]

Shipbuilding Corp., Nov. 5, 1941, in Prod. Div. Adm. file E201-L6-3 and interviews with Mr. Esmond.

[35] Report in MC gf 507-3-14. See also reports of W. K. Carter in MC gf 507-4-14, Nov. 18, 1941, Jan. 9, 1942, and Apr. 15, 1942.

[36] *Ibid.,* Oct. 31, 1941, Nov. 13, 1941, Feb. 24, 1942.

[37] *Ibid.,* Nov. 13, 1941, in gf 507-4-14.

[38] Stephan to Sanford, Dec. 2, 1941, in gf 507-3-14.

[39] Stephan to Sanford, Feb. 19, 1942, in gf 507-4-14.

[40] Stephan to Sanford, Feb. 9, 1942, in gf 503-52.

[41] *Ibid.,* and interview with J. W. Wilson, July 19, 1946.

[42] " As far as shipbuilding goes, they had some complaints about the shipbuilding inspector up there, which were found to be justified, six weeks or a month ago, and we removed the inspector." Testimony by Admiral Vickery, Tuman Committee, *Hearings,* pt. 12, p. 5183. Stephan was transferred Apr. 13, 1942 to North Carolina.—Land to Bland, May 28, 1942, in gf 503-52. Later he was at Panama City.—Vickery to Sanford, Apr. 20, 1943, in gf 201-3-3.

[43] MC gf 503-52.

III. SOUTH PORTLAND, THE INTERREGNUM

The general fact that all was not well at South Portland, whatever the cause, was clear from Mr. Anderson's system of comparative ratings. For some time Admiral Vickery had been needling Mr. Newell by telling him how well other yards were doing,[1] and when the Truman Committee in April 1942 inquired how the shipbuilding program was coming along, Admirals Land and Vickery presented and publicly expounded charts which emphasized the contrast between Henry Kaiser's yards in the Portland area on the Pacific and William S. Newell's yards near old Portland on the Atlantic.[2] Senator Brewster of Maine showed his chagrin, if not incredulity, and Senator Truman, as Chairman of the Committee, remarked: " It is a Maine child, Senator, and I thought you ought to spank it if it needed to be spanked." " That is right," replied Senator Brewster, and questioned Admirals Land and Vickery at length about the reason.[3] Admiral Land reiterated the formulas he had been emphasizing all along: " spreading shipbuilding brains too thin " and " morale."[4] Admiral Vickery explained more specifically the contrast between Mr. Newell's brilliant record in building destroyers at Bath and his disappointing showing at South Portland. " It is not a real shipbuilding job," he explained. " It is a mass-production job of erecting materials."[5] The Liberty ship had a heavier, less complicated structure than a destroyer, and was to be built more rapidly. The Liberty ship required " a great deal more weight to move into it in a much shorter period of time. . . ." The leaders in West coast shipbuilding were " used to moving a mass of material." " When they jumped into this building of ships it was just like any other structure. . . ." They saw it as another " production construction job where they can move a lot of mass material and move it quickly."[6]

At South Portland the mass was not moving fast enough, and in

---

[1] Vickery to Newell, July 15, 1941; Land to South Portland Ship, June 30, 1941 in MC gf 507-4-14. In regard to the east area, Sir Arthur Salter, of the British Merchant Shipping Mission, wrote to Land, Jan. 9, 1942 (in gf 507-3-14) " I assume, however, that as progress in Maine has been so much less rapid than in California, you are not contemplating placing a follow-on order in the immediate future."
[2] Truman Committee, *Hearings*, pt. 12, pp. 5177-79.
[3] *Ibid.*, pp. 5182-83.
[4] *Ibid.*, p. 5185.
[5] *Ibid.*, pp. 5183-84.
[6] *Ibid.*, p. 5185.

April 1942 the urgency was pressing. Moreover, a wholly new situation had arisen since Pearl Harbor, and new provisions for facilities were necessary. Mr. Newell was very busy making his brilliant record in building destroyers at Bath.[7] Although Admiral Land, who was an old personal friend of Mr. Newell, tried to move in ways which would not hurt his pride,[8] the Maritime Commission felt obliged to interfere increasingly in the South Portland yards and assumed much responsibility for the top management.

Early in April, Admiral Vickery, William H. Harrison of the War Production Board, and John Reilly of Todd Shipyards visited South Portland.[9] Shortly thereafter a new executive was put in charge as Deputy President. This "trouble shooter" was Karl E. Klitgaard who for years had supervised the building of the ships of the Standard Oil Co. of New Jersey. Whether Mr. Klitgaard's presence in the yard was "due entirely to the intervention of the Maritime Commission" is a disputed point, but both Mr. Reilly and Mr. Newell thanked Admiral Vickery warmly for his aid in getting Klitgaard.[10]

The new manager faced three main problems: to introduce a maximum amount of subassemblage with the best welding sequence, to improve the supervision of work in the yard, and to complete the facilities. In regard to the first, he had full and adequate authority. He changed the work over from building ships tailor-made, as he said had been done, to introducing all the prefabrication and subassembly there was space for or the cranes could lift.[11]

---

[7] The Navy "E" pennant was raised over that yard Feb. 17, 1942. Program of ceremonies in gf 503-28.

[8] Land to Newell, Aug. 26, 1942, disclaiming intention of butting in to his affairs.—gf 503-52. On June 25, 1945 Admiral Land sent the following telegram to be read at a banquet in Mr. Newell's honor. "Heartiest Congratulations to Pete Newell from all hands and the ship's cook on the U. S. Maritime Commission and W. S. A. Pete is a top-flight shipbuilder, an A-1 industrial leader, an outstanding citizen and a loyal friend. Nufced! "—gf 503-28.

[9] Truman Committee, *Hearings*, pt. 12, pp. 5183, pt. 15, p. 6143.

[10] The House report says so in the language quoted above (Bland Committee, *Report No. 2653*, p. 13). Mr. Klitgaard dodged the question (Truman Committee, *Hearings*, pt. 15, pp. 6007-08), and Mr. Reilly, while not saying who proposed Klitgaard, claimed some credit for Todd Shipyards, and said that he (Reilly) asked Klitgaard's release by Standard and asked Vickery to urge it (*ibid.*, pp. 6143, 6144). The letters exchanged by Newell and Reilly with Vickery (MC gf 503-52, Apr. 18, 1942, gf 507-4-14, Apr. 20, 1942; Apr. 22, 1942) imply mutual agreement.

[11] Truman Committee, *Hearings*, pt. 15, pp. 6009-10; Bland Committee, *Report No. 2653*, pp. 11-12.

SOUTH PORTLAND, THE INTERREGNUM

In regard to the supervisory force, the east-area yard, still called Todd-Bath and being operated by the company of that name, had a much better reputation than the west area at the time Klitgaard took charge. The main difference, it was generally agreed, was in the foremen or superintendents.[12] The top officers of the two companies were identical, but the supervisors and the foremen represented in the one case Mr. Newell's first choices from among his workmen at Bath, in the other case second choices. The supply to pick from had been quite small, only the 3,200 men employed at the Bath plant at the beginning of 1941, many of whom had relatively little experience. A few hundred were taken out to start the building for the British, and four or five months later a few hundred more were picked as a nucleus for the building in the west area. Many men who were made foremen proved unfit and had to be replaced. There had been much favoritism to friends and relatives, and when Mr. Klitgaard came to South Portland, he began weeding out incompetents chosen because of their family connections.[13] But the task of supervision was indeed overwhelming, as the labor force in the west area increased from 4,000 to 11,000 between April and August 1942.[14]

The third problem was facilities. On that subject Mr. Klitgaard did not have full authority; indeed, it is doubtful if anyone did. Before Mr. Klitgaard became Deputy President, the Maritime Commission had ceased to follow Mr. Newell's judgment in the planning of facilities. At times he and the Commission seemed to be working at cross purposes. The plans of the Commission ran violently contrary to local sentiment. Many inhabitants of the Cushing Point section of South Portland had always resented the intrusion of the shipyards into their residential village and summer colony and rumors of the unsavory contracts increased their indignation. The Commission, with its eyes on the war and the need for ships, bit more and more deeply into the community, while Mr. Newell's company disclaimed responsibility. Without

---

[12] Truman Committee, *Hearings*, pt. 15, p. 6014; Bland Committee, *Report No. 2653*, p. 13.

[13] Mr. Newell's testimony, Truman Committee, *Hearings*, pt. 15, pp. 6127, 6129. Bland Committee, *Report No. 2653*, p. 12. Although this report gives the inaccurate impression that some of the Vice Presidents were removed, Mr. Schulze's records show them still on the pay roll. Memorandum for Land by W. H. Lalley, Special Assistant to the Chairman, Dec. 9, 1942, Lalley's file 2-8.

[14] Fischer, *Statistical Summary*, Table G-1.

the full cooperation of the management the Commission could not get the yard operating efficiently; the Commission was more effective in violently disrupting the community by large condemnations of property for fabricating shops, housing, and a railroad.

Instead of using Thompson's Point for fabrication, both Mr. Wall and Mr. Stephan recommended building a fabrication plant as part of the yard. To get space it was necessary to condemn the strip of land still known as Cushing Point and occupied by cottages, which lay between the east area and the west area. Since the Maritime Commission was at this time taking steps to acquire the east area from the British and had already given the South Portland Shipbuilding Corp. a contract for the operation of the east area, taking the land on Cushing Point would consolidate physically what was to be combined into one enterprise.[15] Late in February Mr. Newell was still urging that Thompson's Point be preferred because it could be made ready more quickly,[16] but at conferences in Washington on March 6 it was agreed that Thompson's Point should be used for warehousing purposes only and that "all fabricating operation" would ultimately be accomplished at Cushing Point.[17] Money for these purposes was voted by the Commission at the same time it voted to take Thompson's Point, which it took, not by the expensive lease Mr. Newell's company had arranged, but by condemnation proceedings.[18]

The Commission decided also to build a railroad track to the yard. Fears regarding supplies of rubber and gasoline were sufficiently intense from March to June 1942 so that it seemed unwise to plan a shipyard scattered in parts separated by as much as eight miles and connected by trucks, as Mr. Newell seemed to be doing. The expense of the existing trucking and storage arrangements was also an argument against them.[19] Spokesmen for the company

---

[15] Wall to Lame, Jan. 30, 1942, and Feb. 18, 1942, in MC gf 507-3-14; Stephan to Sanford, Mar. 12, 1942, in gf 503-52.

[16] Newell to MC, Feb. 21, 1942, in gf 507-3-14.

[17] Wall to Lame, Mar. 6, and 8, 1942; Newell to USMC, Mar. 8, 1942, in gf 507-3-14 Exhibits.

[18] Flesher and Vickery to the Commission, Mar. 10, 1942, in gf 507-3-14 and exhibits; Flesher to South Portland Shipbuilding Corp., Mar. 11, 1942, in gf 503-52. Award of contract for the plant to Brown Construction Co. was approved by the Commission May 12, 1942.—Minutes, pp. 21928-29.

[19] Schmeltzer to Wall, Feb. 2, 1942, in gf 507-3-14.

SOUTH PORTLAND, THE INTERREGNUM 517

wrote as if they had no need of a railroad to Cushing Point,[20] but the Maritime Commission for its part decided that a shipyard without a railroad was preposterous and went ahead on its own to decide where and how it should be built.

On April Fool's day 1942, Mr. Wall announced to the people of South Portland that the railroad would probably go down Broadway (Figure 35).[21] Such a proposal had produced protests the previous summer and did so again. It was made just at the time that old residents on Cushing Point were being evicted from their homes to make room for the fabricating plant and were being treated, many of them thought, quite unjustly.[22] James C. Oliver, who represented the district in Congress, received protests from hundreds of constituents.[23] Mr. Wall continued to urge a railroad and persuaded T. D. Geoghegan, Transportation Advisor to the Commission, who visited South Portland on April 28, that his ideas were sound. Mr. Geoghegan called up Commissioner Woodward, who had been designated by the Commission as liaison officer with Office of Defense Transportation,[24] and urged the need of prompt action to get a road built. But Mr. Geoghegan called the three estimates of cost which had been prepared by the Portland Terminal Co. absolutely fantastic, and suggested to Mr. Woodward that he have the Interstate Commerce Commission send a construction engineer to Portland to make an unbiased report on the cost. Mr. Woodward did so the same day.[25] John E. Hansbury was loaned by the Interstate Commerce Commission and within a week he had visited South Portland and prepared a report.[26] On May 6 this was discussed at a meeting in Mr. Woodward's office attended by T. G. Sughrue, Chief Engineer of the Maine Central, by E. W. Wheeler, General Counsel of the railroad, by W. Lalley, who was acting in this matter as Admiral Land's special assistant, and by the Commission's technical advisers, Messrs. Hans-

[20] See below n. 34; and Arthur Sewall II, Assistant to President, Todd-Bath Iron Shipbuilding Corp., to Robert G. Albion, May 23, 1942 (in Mr. Albion's possession).
[21] Portland *Press Herald,* Apr. 1, 1942.
[22] *Ibid.,* Apr. 5, 9, and 15, 1942, and the memoranda concerning these condemnations in MC Minutes, pp. 2459-63 (Mar. 9, 1943).
[23] Oliver to Lalley, Apr. 27, 1942, in gf 509-10.
[24] See chap. 13.
[25] Woodward to Charles D. Mahaffie, Apr. 28, 1942; Geoghegan to Woodward, May 1, 1942, in gf 509-10.
[26] Lalley to Land, May 4, 1942, in gf 509-10.

bury and Geoghegan. The conference decided in favor of a route along the shore.[27]

To meet the clamor arising in South Portland against this decision, and to look into alternatives, Mr. Lalley went down to Maine on May 8 and conferred there with Congressman Oliver and with Mr. Klitgaard, who, he found, also urged a railroad.[28] Next morning he wired Admiral Land that there was no merit in any alternatives proposed to the shore route. Land wrote on the telegram " O. K. Go ahead." [29] But in the combat between outside experts and local residents this was only the first round.

Accepting the inevitability of a railroad, many citizens of South Portland at this point concentrated their efforts on persuading the Commission to build by a route which would cut across the water instead of twisting along the shore. The over-the-water route, they argued, would avoid grade crossings, would save demolishing property, and would be more direct (Figure 35).

On May 14, Representative Oliver and other spokesmen for South Portland were in Washington laying the case for this route before Commissioners Woodward and Carmody.[30] The Commission was induced to reconsider, and Mr. Hansbury, the expert from the Interstate Commerce Commission, prepared a new report comparing more specifically the possibilities of the over-water route.[31] This was debated for fifteen minutes, with Mr. Hansbury present, at the regular meeting of the Commission on May 19. Mr. Hansbury's reindorsement of the shore route was approved,[32] and Mr. Woodward wired that afternoon to the Boston & Maine engineers to get work started as soon as possible.[33]

Democracy permits plenty of protests, however, and the local citizens concerned had a third and more adequate chance to

---

[27] Woodward to Sughrue, May 13, 1942. The participation of the Boston & Maine was secured by a telephone call from Land to E. S. French. See French to Land, May 4, 1942, in MC gf 509-1.

[28] Lalley to Land, May 11, 1942, in gf 509-10.

[29] Lalley to Land, May 9, 1942, in gf 509-10.

[30] MC, Maritime News Digest, VI, 126; Oliver to Woodward, May 14, 1942, in gf 509-10-1.

[31] John E. Hansbury, Cost Engineer of the Bureau of Evaluation, Interstate Commerce Commission, to Woodward, May 18, 1942. The Boston & Maine engineer, Mr. Sughrue also recommended the shore route. Sughrue to Lalley, May 12, 1942, in gf 509-10.

[32] MC Minutes, p. 21973.

[33] Woodward to Sughrue, May 19, 1942, in gf 509-10.

challenge the decisions reached by the technicians from Washington. A South Portland Citizens Club met 300 strong on Sunday afternoon, May 30, and listened to a presentation of the case against the shore route by a local lawyer, Clinton T. Goudy, and Charles H. Prout, engineer. The two Senators from Maine and the Governor attended and encouraged the Citizens Club to marshall their facts and figures so that they could present their case yet again to the Commission; but these office holders did not offer to be spokesmen themselves. Senator Brewster read Hansbury's report but was recorded by the local newspaper as saying that " he understood that William S. Newell, President of the yards, did not feel ' particularly concerned ' about a railroad to the yards and apparently felt he could handle the situation on the present basis." Taking the cue, Senator Wallace H. White, Jr., drew loud applause by saying, " If the shipyards don't need the railroad and are not anxious for it I don't see why the Maritime Commission insists on building it." [34]

When the pros and cons were seriously threshed out, however, in a long session in Mr. Woodward's office on June 8, no one questioned the need of building a railroad and at once. The stability of the foundations on the shore route or the over-water route was the central point of controversy. It was Hansbury's judgment against Prout's, Hansbury arguing effectively that there were almost sure to be some soft spots on either route. " If the track settles on the shore route, a few carloads of gravel will take care of the situation without delaying operation, but if the trestle settles, operation must be discontinued until the trestle is repaired." [35] A trestle, he said, would take longer to build. Pleas that decision be delayed so that soundings could be taken along the over-water route were overridden by Commissioner Woodward.[36] Fast action was wanted. On June 9, 1942 the Commission reaffirmed its approval of the shore route and on July 13 awarded

[34] Portland *Press Herald,* June 1, 1942.
[35] Memorandum by Hansbury, June 8, 1942, in gf 509-10.
[36] Four pages of typescript dated June 8, 1942 in gf 509-10 appear to be summary minutes of the meeting that morning. In an interview with Mr. Prout in South Portland on Aug. 20, 1946, I read the quoted extract from these minutes to him and he said they were substantially correct. In the same interview Mr. Prout said he had assurances from the Bennet Contracting Co., an experienced Portland firm, that there was ledge foundation.

a contract for its construction to A. C. Stanley Co., who agreed to finish in 60 days.[37]

The local citizenry had had several chances to state their case and in the end at least had voiced their arguments even into the ears of one or two of the Commissioners. Many, many letters and telegrams had been addressed to President Roosevelt, Mrs. Roosevelt, and Admiral Land, and many answered over Admiral Land's signature, although the replies do not show whether Admiral Land read any letters and reports on the situation except those of Mr. Lalley and Mr. Hansbury. Mr. Woodward was the Commissioner handling the problem; he listened to the pleas from South Portland and reheard the case twice without wavering in his complete confidence in the engineer furnished him by the Interstate Commerce Commission.[38] To the Senators from Maine, Mr. Woodward wrote that the shore route was necessary because it could be constructed in half the time.[39]

When the sixty days were up, in September 1942, Commissioner Woodward, although then busy with other matters, wired to Mr. Wall, " Let me know status of spur." [40] There was no answer.[41] Two months later its completion was reported,[42] but in January 1943 Mr. Woodward was trying vainly to find out why steel was still coming into the yard by truck instead of coming over the railroad,[43] and the following May, almost a year after the hearing, the rail facilities were not being " fully utilized." [44] Part of the village of South Portland had been pulled up by the roots in hope of gaining two or three months. Yet those months were lost.

The delay occurred in matters of detail which Mr. Woodward and the other Commissioners could hardly personally attend to, as they had personally pushed for a decision on the route. A. C. Stanley Co. protested that they did not receive until too late proper

[37] MC Minutes, p. 22201; contract in gf 509-10.

[38] The protests and replies fill two fairly thick files, gf 509-10-1 pt. 1 and pt. 2, and a part of gf 509-10, pts. 1-3. The carbons of Admiral Land's letters show that they were dictated by P. D. Page and D. E. Lawrence, or in a few cases by Mr. Woodward. Mr. Woodward dictated his own replies.

[39] June 9, 1942, in gf 509-10-1.

[40] Sept. 24, 1942, in gf 509-10.

[41] Woodward wired on Sept. 24, " Answer my wire of 14."—gf 509-10.

[42] Woodward to Wall, Nov. 13, 1942, and McInnis to Vickery, Oct. 22, 1942, in gf 509-10.

[43] Transcript of hearing entitled " Before the Maritime Commission," Jan. 12, 1943, pp. 17-19, in gf 503-52.

[44] MC Minutes, p. 25175, May 18, 1943.

SOUTH PORTLAND, THE INTERREGNUM 521

priority numbers for the purchase of materials, especially of bolts and spikes; that some of the houses in the way were not removed soon enough; and that the amount of fill proved to be twice the amount which had been estimated by Mr. Wall. Accordingly the contractor applied for and obtained an extension of time without penalty,[45] and, in fact, took twice as long as estimated to bring the line to the shipyard. Then the problem became one of trackage within the shipyard, and again the Commission was presented with an issue easier to decide in principle than to resolve in all its details.

The decision in principle was made in the middle of the fight over the railroad, when speed seemed so important. On May 28, 1942, the Commissioners approved the recommendation, which had been under discussion for some time, that a large triangle of land adjoining the shipyard and containing about eighty homes be taken by condemnation proceedings to be used for tracks and a storage yard [46] (Figure 35). Consequently, while the railroad was being built, homeowners were being ousted and their land taken at prices which many of them thought grossly unfair. Evaluations were not made by the Maritime Commission, they were made by local real estate men, and yet the Maritime Commission and especially Mr. Wall were blamed. It was all felt to be the work of a misguided far-away power.[47]

While during the summer of 1942 eviction and demolition proceeded rapidly, there was little progress towards an agreement between the Maritime Commission and the shipbuilding company about who should bear expense of tracks and other new facilities. Mr. Klitgaard submitted in June 1942 estimates of the cost of various facilities.[48] In accord with its general policy the Commission proposed that fees be reduced since the company was receiving more facilities paid for by the Maritime Commission.[49] Since

---

[45] A. C. Stanley Co., Inc. to Wall, Aug. 4, 1942, in the files of MC Resident Plant Engineer, South Portland, Maine, copy in the Historian's Collection, Research file 112.3.9.

[46] MC Minutes, pp. 22061-62 (May 28, 1942). It was part of the plan recommended by Lalley May 11, 1942, Lalley to Land, in gf 509-10. On failure of South Portland Ship to build tracks, see Minutes, pp. 27215-17 (Jan. 4, 1944).

[47] Interviews with a number of persons in South Portland; *Portland Sunday Telegram*, Aug. 16, 1942, Sec. D; MC Minutes, pp. 24529, 24533.

[48] Klitgaard to Wall, June 27, 1942, and Wall to McInnis, July 16, 1942, in gf 507-3-14 exhibits.

[49] Vickery to Newell, Sept. 2, 1942, in gf 507-3-14.

this was a question of finance it was quite outside Mr. Klitgaard's sphere; it was handled by Mr. Newell, who refused to accept the reduction. He claimed that when his company consented to join in the accelerated program in mid-1941, it was agreed that the Maritime Commission would pay for some additional facilities to enable his company to meet the earlier delivery dates.[50] A conference in Washington in late September failed to resolve this financial dispute, and it became absorbed into the larger issue, cancellation of the contract.[51]

The dispute over facilities reveals a vital weakness in the Klitgaard regime at South Portland. There was no unified command and therefore no full responsibility. This is one way of explaining the poor showing which the New England yards were making in the fall of 1942.

In addition to its troubles over methods of construction, over supervision, and over facilities, South Portland was afflicted with labor troubles; and since strikes in wartime make good copy, its labor troubles were probably the most advertised. Again, the establishment of the east and west areas as separate yards originally and division of authority later contributed to the difficulties.

At his Bath Iron Works Mr. Newell had an independent union. The east area in South Portland was organized by the CIO but the AFL won the election held in the west area in March 1942. But the AFL did not win a clear-cut victory even in the west area, for the National Labor Relations Board indicated it would entertain a new petition for election when it could be shown that there had been a " substantial increase " in employment and that the petition represented a " substantial number " of workers.[52] This decision encouraged the CIO to continue organizing.

[50] Newell to Vickery, Sept. 18, 1942, filed under Nov. 18, in gf 507-3-14.

[51] Vickery to Newell, Sept. 26, 1942, and Vickery to the Commission, Oct. 24, approved Oct. 29, 1942, in gf 507-3-14. The House investigating committee declared categorically that this quarrel held up the construction of the additional facilities. Bland Committee, *Report No. 2653*, p. 14. On the other hand, when the dispute was finally settled in January, 1943 out of the $4,151,000 then approved for " new " facilities, $2,457,764.14 had already been spent by the South Portland Shipbuilding Co. Schulze to Kell, Mar. 9, 1944, in the file at South Portland of Schulze, Resident Auditor, " Disallowances and suspensions." Also Kell to Schulze, Feb. 25, 1943.

[52] *Shipyard Worker*, Mar. 20, 1942; National Labor Relations Board, *Decisions and Orders*, vol. 39, pp. 485-88; vol. 40, p. 271. The contracts between the companies and the unions came some months after the elections: that of the Industrial Union with Todd-Bath on May 14, 1942; that of the AFL with South Portland Shipbuilding Corp., on July 2, 1942. Each left to subsequent negotiations specific wage rates and rules for grading. See Agreement between Todd-Bath Iron Shipbuilding

PLATE XIX. Above, Cushing Point, South Portland, Maine, July 1941. Summer cottages being crowded out by Todd-Bath (east-area), in background, and the west-area yard, in foreground. Below, the crowded west area in April, 1942.

PLATE XX.

The most startling fractures of welded ships occurred on tankers, namely: the *Esso Manhattan* outside New York harbor (above) and the *Schenectady* at its outfitting dock (below). Fractures on Liberty ships were most serious in the cold stormy waters of the North Pacific, where the *Valery Chkalov*, one of the Libertys transferred to the Soviet Union under lend-lease, split in two. The aft section (left) remained afloat and was salvaged.

Rivalry between the AFL and CIO by no means explains all the labor troubles at South Portland, however. The rank and file in the west yard were conscious of working for a management that had not succeeded in organizing for the job it was to do. They complained of poor supervision, of nepotism, and of not having enough work to keep them busy. They did not have faith in the company. Although a labor-management committee existed on paper, no system for acting on suggestions had been worked out.[53] Wages were another reason for discontent. In the spring of 1942, the average hourly earnings at South Portland were considerably lower than in any other North Atlantic yard.[54] Both unions were negotiating with the management for contracts that would classify jobs in such a way as to raise wages. Delay in obtaining higher wage rates was the immediate cause of the strikes.

By July 2, 1942, the AFL had reached agreement in principle on basic provisions, but negotiations for putting the principles fully into effect by classifying jobs, department by department, dragged on for months and were not completed until September 23, 1942. Before the rates went into effect they had to be approved by the Maritime Commission which had taken the position that they would make no commitments on a departmental basis, that wage rates would have to be approved " as an entirety." [55] Just at the time when the full system of classifications was sent to the Maritime Commission for approval, accompanied by a warning that unless they were acted on speedily there would be a strike, authority to approve wage scales was transferred from the Maritime Commission to the War Labor Board as a result of the national stabilization act.[56] This transfer delayed definite action.

Corp. and Industrial Union of Marine and Shipbuilding Workers of America— C. I. O., Local No. 50 [May 14, 1942], in gf 503-52, and Agreement [between the Metal Trades Department, American Federation of Labor and Its Affiliates and the South Portland Shipbuilding Corp., July 2, 1942], in Division of Shipyard Labor Relations files.

[53] Office of Government Reports, Excerpts from Special Report of Acting State Director for Maine on Shipyard Situation, South Portland, Maine, Apr. 30, 1942, attached to Lowell Mellett to Land, May 7, 1942, in gf 508-3; Memorandum, Field Division to R. Keith Kane, Bureau of Intelligence, Dec. 3, 1942, in gf 508-3; Bland Committee, *Report No. 2653*, pp. 12-13.

[54] Fischer, *Statistical Summary*, Tables G-8 and G-9.

[55] Tracy to McClellan, Sept. 19, 1942, and other letters of about that date in MC, Division of Shipyard Labor Relations file, Wage Stabilization, South Portland, and in East Coast Regional Office file, Labor Relations, South Portland, and in gf 503-52.

[56] See above p. 421. For warning of the strike, Buckley to Tracy, Sept. 22, 1942 in gf 503-52.

Unrest increased and on October 10 welders and fitters began a walkout which continued a little more than three days. The number staying off the job was estimated at 400 by the company and at 50 by the union.[57]

Although AFL leaders pointed to delays in approving the wage rates as the major cause of dissatisfaction, they blamed the CIO for increasing unrest by seizing this opportunity to organize.[58] The CIO admitted it was organizing—at the request of AFL members, its representatives said.[59] The secretary of the Industrial Union called the walkout " another AFL disgrace " and blamed " AFL bureaucrats " who had attempted to placate the rank and file with " empty promises." [60] The Industrial Advisor of the East Coast Regional Office, who visited the yard during the strike, recommended that the promised new election be held immediately in order to eliminate " persistent union sniping by both the CIO and the American Federation of Labor." [61] The Industrial Union filed a petition for an election with the National Labor Relations Board the following week.[62]

Meanwhile the Maritime Commission had sent a recommendation to the National War Labor Board on the proposed rates. Its proposals were a blow to the AFL for they called for reductions in 20 out of 71 classifications. Moreover, the Maritime Commission now objected to two provisions of the agreement in principle, provisions which had been known to the Division of Shipyard Labor Relations since July 1942 without any objection having been registered. The provisions rejected were indeed novelties. One provided that all mechanics be paid the rate of first-class mechanics, as was done on the Pacific coast but not in any yard in the Atlantic area. The second provided that workers on the first shift be paid during a half-hour lunch period. The company had been paying for this luncheon period since the beginning of its

[57] Baltimore *Sun,* Oct. 11, 1942, in MC Press Digest, Oct. 12, 1942; New York *Herald-Tribune,* Oct. 13, 1942, in MC Press Digest, Oct. 13, 1942; Memorandum, Young to Buckley (taken over the telephone), Oct. 12, 1942, in Division of Shipyard Labor Relations file, New England Shipbuilding Corp.
[58] Memorandum, Young to Buckley, Oct. 12, 1942, above; New York *Herald-Tribune,* cited in note above.
[59] Portland *Press-Herald* clipping, n. d., in Historian's Collection, Research file 112.
[60] *Shipyard Worker,* Oct. 16, 1942, p. 4.
[61] Berno to McInnis, Oct. 10, 1942, in East Coast Regional Office file, Labor Relations, South Portland.
[62] *Shipyard Worker,* Oct. 23, 1942, p. 1.

operations without any protest from the Maritime Commission. But in October the Commission announced that it would not reimburse the company for these payments, and at the same time refused to approve the rating of all mechanics as "first class." [63] The AFL labor leaders protested and predicted trouble, but on November 11 the War Labor Board issued an interim order which followed the recommendations of the Maritime Commission and rejected these provisions.[64]

The result was a second and more serious outlaw strike which began on Monday night, November 30, among the welders, tackers, and drillers. Next day a total of 4,838 or 72.5 per cent of the labor force were out. The number refusing to work continued to grow during the first shift Wednesday, but by Thursday, perhaps in response to appeals from the union, the Commission, and the War Labor Board, about 60 per cent had returned to work. By December 8 production was practically normal.[65]

AFL-CIO rivalry was evident in this as well as the October walkout.[66] Whether union leaders of either group were directly responsible is anybody's guess, but the winds of the coming union election were undoubtedly blowing strong. In another sense responsibility for the strike, like responsibility for the confusion over facilities, lay in the division and shifts of authority, in this case the shift of authority among government bodies as well as the division between the government and the management, and within the management.

No one held Mr. Klitgaard primarily responsible either for the labor troubles or for the contest over facilities. Both Admiral Land and Congressman Bland, whose committee was investigating the shipyard, exerted pressure during the summer to see that he

---

[63] Vickery to McClellan, Oct. 22, 1942; L. J. Riso to Tracy, Sept. 23, 1942, in MC gf 503-32; Memorandum, Dec. 22, 1942 by W. H. Lalley, in Lalley file.

[64] Bard and Land to Porter, n. d.; MacGowan, Calvin and Buckley to Porter, Nov. 6, 1942; in re *South Portland Shipbuilding Corporation* [South Portland, Maine] *and Metal Trades Department (AFL)*, No. BWA-359, Nov. 11, 1942, all in *War Labor Reports*, vol. 4, 318-28.

[65] Portland *Press-Herald*, Dec. 1 and Dec. 4, 1942; Riso to MC, Report on Strike of Dec. 1, 1942, in Division of Shipyard Labor Relations file, New England Shipbuilding Corp.

[66] Telegrams, Herrick to Land, Nov. 30, 1942 and Buckley to Ring, Nov. 30, 1942, cited in n. 82; Portland *Press-Herald*, Dec. 3, 1942; *Shipyard Worker*, Dec. 11, 1942, p. 3; *ibid.*, Dec. 25, 1942, p. 1; Portland *Sunday Telegram*, Feb. 7, 1943.

put aside thought of leaving in September as he had planned.[67] But by Thanksgiving, it was clear that further change was necessary. The west area was still far behind in its deliveries, although the east area was, by September, consistently earning bonuses for delivering ships to the British ahead of time. Mr. Klitgaard was finding his position difficult and resolved to leave.[68] His friends could hardly urge him to keep up the killing pace at which he was working, " down in the yard at 7 o'clock in the morning " and " back after dinner at night until 10 or 11 o'clock." [69] To have the management slip back then into the hands of the team from Bath was intolerable in view of the strikes and the stories set forth in November and December 1942 by Congressional committees.

IV. South Portland, A New Team

The Congressional Committee with which the Maritime Commission was customarily most closely in contact was the House Committee on Merchant Marine and Fisheries. Since its Chairman was Judge Schuyler Otis Bland, Congressman from the district around Newport News, it was commonly called the Bland Committee. One member of this committee was the young Congressman representing the Portland district, James C. Oliver, who alone among the Maine politicians had definitely opposed the Commission in regard to the spur railroad. As a member of the Bland Committee, Mr. Oliver could appropriately spearhead an investigation of the poor record of these yards. The committee also employed special counsel and investigators who visited Maine for some months, and it held extensive hearings. At the time the committee reported, in October 1942, Mr. Oliver had been defeated in the fall campaign, but Mr. Bland fully endorsed the findings, which were a blistering attack on the management Mr. Newell had headed. It was accused of not laying out the yard properly to begin with and of not keeping track of materials delivered. Connivance in profiteering was imputed by making public the details of Mr. Burr's trucking contract and the terms

---

[67] Land to Klitgaard, Aug. 20, 1942 and Bland to Klitgaard, Aug. 24, 1942, in gf 509-10; Klitgaard to Land, Aug. 24, 1942; Land to Reilly, Aug. 26; Reilly to Land, Sept. 1, 1942, and Newell to Land, Sept. 2, 1942, in gf 503-52.

[68] Memorandum by Lalley, Dec. 9, 1942, in Lalley file 2-8.

[69] Mr. Reilly's testimony in " Before the Maritime Commission," Jan. 12, 1943, p. 24, in gf 503-52.

of the lease of Thompson's Point. It was revealed, for example, that Mr. Burr, by investing $3,000 in a new company, Materials Handling, had made a profit of $87,294, in addition to paying such salaries as $9,875 to his twenty-two year old daughter, formerly a $25-dollar-a-week stenographer. Mr. Newell was reproached both for permitting the terms of Mr. Burr's contract and of the Thompson's Point lease, for refusing to reduce the fees asked by his company, and for not paying enough attention to the yard. Mr. Newell's capacity as a shipbuilder was not denied, but it was asserted that the South Portland Shipbuilding Corp. had not to any great extent had the benefit of his attention. Deliveries were far behind schedule, yet, according to the contracts, the fees would at a minimum be $5,040,000. The report concluded that since the South Portland Shipbuilding Corp. had contributed practically no capital, only $250,000, and no management except bad management, it should not receive these fees. The contract should be terminated immediately and a competent manager appointed by the Maritime Commission.[1]

Congressman Bland followed up this attack by sending a copy to Admiral Land personally,[2] by presenting his case orally to the Commission,[3] and finally by writing to President Roosevelt a letter which attacked both the company and the Commission's attitude. " South Portland," he wrote to the President, " is receiving a fee (of $5,040,000) for the trouble of incorporating a company, choosing a name for that company, holding an occasional directors' meeting and delegating the performance of its duties." Claiming an absolutely convincing case, " founded upon facts ... which are admitted by the Maritime Commission," Mr. Bland blamed the Commission severely for removing Mr. Stephan, the inspector who protested the company's methods, while retaining Mr. Wall, who approved Mr. Burr's succulent contract. He complained that the Commission had given little time personally to the matter and should have terminated the contract in October when his Committee revealed the true situation.[4]

[1] Bland Committee, *Report No. 2653*, and its *Confidential Committee Print*, Oct. 6, 1942, a report by James V. Hayes, General Counsel, and Bernard J. Zincke, Associate General Counsel, containing full details concerning the trucking contract.
[2] Bland to Land, Nov. 25, and Dec. 4, 1942, in MC gf 503-52; Bland to Land, Dec. 28, 1942, in gf 507-4-14. Cf. Bland to Macauley, Dec. 4, 1942, in Macauley file.
[3] MC Minutes, p. 23683 (Dec. 5, 1942). At this meeting Senators Brewster and Kilgore and Congressman Oliver also presented their points of view.
[4] Bland to the President, Dec. 8, 1942, attached to note of Dec. 10, 1942, the President to Land, reading " To prepare reply for my signature." In gf 503-52.

Against this blast the first line of defense was the Truman Committee, which, being charged with investigating war production in general and having South Portland in its general hearings in April 1942, prepared in the fall to go into the matter more thoroughly. Two members of the Committee were particularly interested, Senator Brewster of Maine who had faith in Mr. Newell as an outstanding shipbuilder,[5] and Mr. Kilgore of West Virginia who had visited Bath in June 1942 and been greatly impressed by Mr. Newell and his famous yard.[6] In August Mrs. Brewster and Mrs. Kilgore christened two of the ships launched for the British by the Todd-Bath Co. in the east area.[7] Together with Senator Truman, their Chairman, Senators Brewster and Kilgore formed the subcommittee on shipping. Just at the time when the findings of the Bland Committee were being reported to the House they held hearings and formulated their own report. Mr. Newell and Mr. Reilly were given a chance to state their case and to plead that they had earned their fees. Mr. Newell expounded his difficulties on obtaining materials and skilled workmen. He was asked embarrassing questions about the trucking contract, about labor, and about the number of manhours he was using per ship, but he was not asked about Thompson's Point.[8] Mr. Klitgaard testified at length without blaming anyone for anything. Mr. Burr, although repentent and offering restitution, was raked over the coals not only for the size of his profits but also for having formed to collect these profits a corporation separate from that of which he had been and continued to be president.[9] An even worse impression was made by Mr. Wall. He claimed that Mr. Burr's contract had been let by competitive bidding and that he knew nothing of Mr. Burr's large profits until he "read it in the newspapers." But he volunteered the information that since the termini for the trucking were very indefinite, it was orally agreed with Mr. Burr that if the contract price did not prove profitable it would be revised upward. When Chairman Truman emphasized the favoritism involved, Mr. Wall protested that the matter "had been threshed out here in Washington by everybody connected

[5] Truman Committee, *Hearings*, pt. 12, p. 5183.
[6] Portland *Press Herald*, June 2, 1942.
[7] *Marine Engineering and Shipping Review*, Sept. 1942, pp. 138-39.
[8] Truman Committee, *Hearings*, pt. 15, pp. 6007-25, 6123-44.
[9] *Ibid.*, pp. 6025-44 and 6239-46.

with the Commission." His attempt to duck responsibility did not carry conviction.[10] He was discredited publicly and in the eyes of the Commission. The report of the Truman Committee, submitted by Mr. Kilgore, unreservedly condemned him and Mr. Burr. On the other hand it excused the South Portland Shipbuilding Corp. from blame for failure to provide adequate facilities and expressed respect for Mr. Newell's confidence that he could make good on his contracts. It favored renegotiation to reduce fees but opposed cancellation. " The difficulty with making a change in management," it concluded, " is that it would take months for a new management to become familiar with the yard, and the disruption and dislocation involved in discharging the old management and installing a new management would cause an immediate loss which could not be made up for many months no matter how competent the new management might be." [11]

So long as the Truman Committee was investigating, Admiral Land answered the rising clamor for cancellation by saying that he was waiting for its report.[12] Once it was prepared, he balanced the findings of the Senators against the findings of the Bland Committee. President Roosevelt had referred to him Judge Bland's letter of December 8. In replying, Admiral Land invited the President's attention " to the fact that the Commission finds itself in the unhappy position of being between the fire of the Bland Committee and the fire of the Truman Committee." [13] Senators Kilgore and Brewster reinforced this remark by writing themselves to the President on the same day in defense of Mr. Newell and Mr. Reilly. " With the reputation and pride of these two men and their organizations at stake we feel there is every assurance of success," said their appeal.[14] Admiral Land placed before the President a collection of pertinent facts and proposed that the President reply to Judge Bland by assuring him that both the

[10] *Ibid.,* pp. 6148-55.
[11] Truman Committee, *Interim Report on Shipbuilding at the South Portland Shipbuilding Corporation* in *Senate Reports, Miscellaneous,* Vol. I, 77th Cong., 2d sess., Report No. 480, pt. 12, pp. 1-4.
[12] As in his answer to Congressman Bland, Dec. 1, 1942 in gf 503-52.
[13] Land to the President, Dec. 17, 1942, in gf 503-52.
[14] H. M. Kilgore and R. O. Brewster to the President, Dec. 17, 1942, in gf 503-52. It is noteworthy that both Mr. Newell and Mr. Reilly were at just this time in Washington negotiating with the Commission.

House Report and the Senate Report were being given "serious consideration" by the Maritime Commission.[15]

Cancellation or termination for default was actually being discussed among the Commissioners. After hearing arguments by Judge Bland, Mr. Oliver, Senator Kilgore, and Senator Brewster on Saturday, December 5, the Commission discussed the matter on the following Tuesday without coming to any conclusion.[16] Next day, the ninth, Admiral Land wrote a Memorandum for the Files summing up.[17] Much of this summary was based on previous memoranda prepared for the Chairman by Mr. Lalley placing the record of South Portland in a relatively favorable light.[18] Included was a comparison with other yards, a comparison of which the misleading character was clearly demonstrated by Admiral Vickery when the company attempted to use it in a private hearing before the Commission.[19] Legal complications and expense of cancellation were emphasized. "We should have learned by bitter experience in Savannah," Admiral Land wrote, "that cancellation is not child's play but extremely serious and may be frightfully expensive." Land's conclusion was that renegotiation should proceed and that additional supervisory personnel be put in at South Portland. "The questions are who are these personnel, how can they be obtained."[20]

But some other members of the Commission were ready for the more radical change which Judge Bland was urging.[21] On December 10 Commissioners Woodward and Macauley wrote memoranda to the Commission recommending "cancellation" for "default," and Captain Macauley presented his on December 29, 1942.[22] Some change was clearly essential, in view of Mr. Klitgaard's

---

[15] Copies of two letters for the President to choose to use in answering Bland are under Dec. 17 in gf 503-52.

[16] Minutes, pp. 23683, 23705.

[17] The original with his signature and with changes which were incorporated in the copy sent to President Roosevelt Dec. 17, 1942 is in gf 503-52.

[18] Memorandum of Nov. 25, 1942 and attachments to MC, Lalley's file 2-8.

[19] Transcript of hearing "Before the Maritime Commission," Jan. 12, 1943, in gf 503-52, p. 6.

[20] Memo cited in note 19 above.

[21] Having received from Admiral Land a copy of his memorandum for the files, Judge Bland wrote an extended reply. He concluded that the "outside chance that some lawsuit might be instituted" should not prevent "strong and immediate action" to get a "new capable and efficient management." Bland to Land, Dec. 28, 1942 in gf 507-4-14.

[22] Dec. 10, 1942 in gf 503-52, Minutes, p. 23890.

retirement, and Mr. Anderson and Mr. Lalley prepared for Admiral Land memoranda outlining various proposals.[23] Cancellation was not practical under the contract, for it would have required paying all cost plus about $1,800,000 in fees, and the payment of any such fees was exactly what Congressman Bland considered monstrous. Termination for default would require giving thirty days' notice, and if the company used due diligence during those thirty days, its failure to have done so at some past time might not be adequate legal grounds for termination.[24] That course also was not likely to satisfy the opposition. Under the circumstances, to appeal simply to the Commission's strict legal rights under the contract did not offer a politically desirable issue. Neither would it advance the Commission's main concern: production of ships. Resort had to be had therefore to negotiations.

On December 16, 17, and 19, Mr. Newell and Mr. Reilly were in Washington negotiating but reached no agreement with the Commission. To be sure these conferences did outline a new managerial team to be installed at South Portland by Todd Shipyards,[25] but they revealed no inclination by the businessmen to give up their hopes of profit and their " hedges " against possible loss.[26] In the Commission the main concern was speeding construction, but it was held to be " intolerable and in the public interest wholly undesirable " that the companies should make money out of the Commissioners' fears that cancellation would delay the construction of ships. " I feel therefore that the last drop of profit possible should be squeezed out of the contract," wrote Captain Macauley on January 5, 1943, in regard to Mr. Reilly's latest offer.[27]

Essentially there were five points at issue: (1) Removing the possibility that any profit be made on the sixteen, very much delayed ships of the first contract; (2) reducing fees on future ships in accordance with the provision by the Maritime Commis-

[23] These are most complete in MC, Lalley's file 2-8.
[24] Memorandum of Dec. 5, 1942 in gf 503-52.
[25] Land to Newell, Dec. 10, 1942, Reilly and Newell to Land, Dec. 19, 1942, in gf 503-52; memorandum of Dec. 15, 1942 by Lalley in Lalley's file, 2-8.
[26] The letter from Reilly and Newell cited above. The importance of fees in the deadlock is indicated also by a memorandum from Lalley to Land, Dec. 31, 1942 in gf 507-4-14, in which Lalley states the reduction in fees which he hopes will be volunteered by the company, and on which in the margin Admiral Land has written, " No soap I fear. L."
[27] Macauley to Land, Jan. 5, 1943, in gf 503-52.

sion of increased amounts for facilities, (3) claims by the company for reimbursement denied by the Commission for the wages which had been paid for the lunch period: (4) the exclusion in the future of Mr. Newell's Bath Iron Works from ownership and profit; and (5) provision of a unified, capable management. The question of reimbursement for lunch-hour pay was removed when the Comptroller General ruled that it was allowable.[28] Agreement was forecast at a hearing before the Maritime Commission on January 12, 1943.[29] When it went into the history of the South Portland yards, that meeting somewhat resembled Hamlet with Hamlet left out, for none of the men from Bath who formed the original management was present. Mr. Reilly, however, made an effective presentation and his proposals appealed to the Commission as reasonable.[30] Some kind of an agreement must have been reached during the previous month between Bath Iron Works on the one hand and Todd Shipyards on the other by which Mr. Reilly and the other executives of Todd Shipyards took over both the management in South Portland and the negotiations in Washington. The following items were agreed to, January 28, 1943: [31] (1) On the first 16 ships the company would claim no fees, and in claiming expenses would see that Mr. Burr's corporation made only 6 per cent of their net sales before taxes; and that the rental paid on Thompson's Point would be based on the value given the land in condemnation proceedings. (2) Facilities costing $4,151,000 would be paid for by the Commission, and the scale of fees on the ships to be constructed would be cut to $32,735, $82,735, and $112,735. The company protested that this was unjust, saying these reductions would deprive them of $750,000 in guaranteed minimum fees already " earned," and nearly $3,000,-000 in guaranteed minimum fees anticipated. (3) A new man-

---

[28] Anderson to Land, Jan. 7, 1943 in Anderson's reading file tells of the ruling, and asks how Reilly is to be told. Under Jan. 8 is a memorandum by Anderson setting forth why he considered this ruling unreasonable.

[29] A typescript record of the discussion, full of passages in which the stenographer obviously did not understand what was said, is under Jan. 12, 1943 in gf 503-52.

[30] Captain Macauley's change of attitude is clear in his letter of Jan. 16, 1943, Macauley to Land in gf 503-52.

[31] These terms are embodied in the exchange of letters formally recorded in the Minutes, Jan. 19, 1943 (pp. 24096-02) and Jan. 28, 1943 (pp. 24161-64). Originals, photostats, and related memoranda are at about those dates in gf 507-4-14 or 507-3-14. Settlement of contract DA-MCc-19 for 16 ships at cost was approved Nov. 2, 1944. Minutes, pp. 30110-11.

agement would be put on trial for a period of 60 days in which it was expected to deliver 12 vessels, fabricate 36,000 tons of steel, and erect a like amount; otherwise the Commission could immediately terminate the contracts. Actually the new management had been installed January 4, 1943 and discussed man by man, together with the organization chart, at the hearings before the Commission on January 12. At the head were C. L. Churchill, as president, and W. L. Green, vice president and general manager, both brought by Todd from their subsidiary, the Seattle-Tacoma Co. Two other key figures were Mrs. Elinore Herrick, the Todd labor adviser, and A. B. Sides, who had been Washington representative for Todd for many years. (4) Bath Iron Works would keep nearly 50 per cent ownership in South Portland. Suggestions that it sell out at a " price which would allow no profit past or future to Bath Iron Works . . ." were rejected indignantly " as a matter of principal [sic], equity and fair dealing " [32] and with the comforting belief that these suggestions came from sources outside the Commission.

Todd Shipyards did not retreat from the position they had formally placed on record before the Commission:

> It has hitherto been the consistent policy of the Todd Company, which operates in many localities on both the Atlantic and Pacific seaboards, to join in every venture, whenever possible, with a local partner, with local capital and local knowledge of conditions, to make sure that the Todd policy is not at variance with the locality in which it is operating.
>
> For a Maine venture, Todd would ask no more suitable local partner than the Bath Iron Works Corporation which has been a Maine institution for many years, the principal modern embodiment of the Maine shipbuilding tradition and one of the chief reliances of the Navy in another part of the Government's shipbuilding program.
>
> We should not think the Commission would be inclined to go along with any suggestion from outside, no matter what its political source, which would say, in effect, that the most important shipbuilding people in Maine should be forced out of a Maine shipbuilding business.[33]

So ended the long argument over fees, facilities, and management which had begun, unknown to the general public, during the spring or summer of 1942 and had been under the glare of publicity since October. Near the end of March 1943 the new management was judged to have performed satisfactorily during

---

[32] " Memorandum: Re South Portland . . .", Jan. 5, 1943, in gf 503-52.
[33] Minutes, pp. 24051-54.

the test period.[34] The facilities were brought to conclusion and consolidated when an area known as the north area, lying out towards the sea from the west area, was obtained from the Navy Department and provided enough more storage and trackage space so that the distant yard at Thompson's Point was entirely dispensed with.[35] Although New England Ship, as it was now called, never became one of the fastest yards and was slow in reducing its costs,[36] it built a total of 274 ships. Moreover it should not be overlooked amid all the hot argument in Washington in the winter of 1942-1943 that some people were doing hard work in the icy Maine winter, handling steel in frost and snow, three shifts around the clock. "During the past week we have had Rain—Snow—Subzero weather—Temperature as low as 12 below—high winds—loss of power due to ice jam at the . . . power house," reported the Hull Inspector on December 19, 1942, in explaining from his point of view some of the difficulties of stepping up production.[37]

As an example of the difficulties which a wartime shipbuilding program is likely to encounter, the story of South Portland is instructive from several points of view. Much of it must be accepted simply as verification of Admiral Land's prophecy that the enormous expansion demanded in shipbuilding would inevitably lead to inefficiency, for managerial brains would be spread too thin. When the need for changing managements arose, the inadequacy of the provisions made in the contracts was more clearly revealed by the case of South Portland than any other. For almost a year, from March 1942 to the end of January 1943, there was no clear line of authority. More drastic action in March 1942 might have remedied the situation sooner. Once the situation had reached the point arrived at in November and December 1942, the course of action pursued by the Maritime Commission seems the only wise one. If they had followed the suggestion of Congressman Bland's committee and cancelled or terminated the contract, it seems unlikely that they could thereby have obtained

[34] Vickery to the Commission, Mar. 22, 1943, in gf 503-52.
[35] Minutes, p. 25175, May 18, 1943.
[36] Anderson wrote to Land, Aug. 27, 1943, "When you pin the rose bud on South Portland next Tuesday you may wish to have in mind that while they have done a good job in getting production up, they are still running high in their costs per ship." In Anderson reading file.
[37] Soren Willeson to J. Kirkpatrick, in East Coast office files New England, L-5.

a management which would have speeded up production as promptly as the management which was installed by Todd in January 1943. The Congressional investigations were not, however, entirely in vain. They created a pressure on Todd Shipyards to do something to save their contract and reputation. The Maritime Commission in its December negotiations was, in effect, letting that pressure beat on Todd so that Todd would search its organization and from it bring into South Portland the kind of managerial talent needed. In terms of how the issue worked out, that was the answer. I am tempted to say it was the answer Land had in mind to the question in his memorandum for the files: "... who are these personnel, how can they be obtained?" To answer that question for South Portland was all the more difficult because in January 1943 the Commission was also seeking new management for two or three other yards.

V. Minor Instances and Conclusions

The elements entering into the story of the shipyards at South Portland enter also in other cases of changing management, which may be treated more briefly. The Commission decided in January 1943 that the Brunswick Marine Construction Corp., although it had substantially completed its yard at Brunswick, Georgia, was not making satisfactory progress in building ships. The Brunswick company agreed to turn over the yard and its contracts to the J. A. Jones Co. which agreed to compensate it for building the plant.[1] Later the ousted company claimed that it accepted the settlement under "duress," namely statements by Mr. Anderson threatening to terminate the contract for default. The Commission rejected the claim.[2] The performance of the new management, J. A. Jones Co., both at that yard and at Panama City was not always up to Admiral Vickery's demands, but it was a large company engaged in many construction projects, and could call in men from other parts of the organization. Under the prodding of Admiral Vickery and of the Gulf regional director it brought in men to strengthen the managements of its yards at both Panama City, Florida, and Brunswick, Georgia.[3]

[1] MC Minutes, pp. 24118-9, Jan. 21, 1943.
[2] Minutes, pp. 25368-69, June 8, 1943.
[3] House, *Investigation of Rheem, Hearings*, p. 94; also statements by Sanford to F. C. Lane.

At Providence, Rhode Island, the change was made because the contractor, the Rheem Manufacturing Co., was unable to finish the facilities within even the very high costs allowed. The site there proved to be unexpectedly difficult. Instead of the $6 million originally estimated, Rheem spent $18 million.[4] Even so the plant was far from complete in February 1943. Consequently the Commission arranged to install a new management formed by combining the Kaiser organization with the Walsh Construction Co., a firm with which Kaiser had been allied on some occasions before the war in various sorts of construction. There was an awkward interim between the decision " in principle " on February 5, 1943 to terminate the contract with Rheem,[5] and the putting into effect of the new contract with Walsh-Kaiser Co., Inc. Rheem was promised reimbursement for carrying on in the meantime, and the Commission authorized its Director of Finance to advance money to meet the payroll. In general the attitude of Mr. Rheem was that he was willing to give up his contract but wished to lose as little as possible, although he had spent $2 to $3 million above the very top limit set by the Commission.[6] After Walsh-Kaiser was in charge, Rheem's lawyers claimed the Commission had had no right to terminate Rheem's contract and threatened suit.[7] The Commission settled on December 28, 1943 by paying $1,500,000. It was severely criticized for this by the Bland Committee which said of the Rheem episode as a whole: " The Maritime Commission did not properly safeguard Government expenditures through adequate supervision and control of an inexperienced contractor's activities and expenditures." [8] That committee was also highly critical of the record made in Providence by the new management, Walsh-Kaiser.[9] Providence

---

[4] On whether Admiral Vickery was responsible for the site, or Mr. Rheem, or both, see House, *Investigation of Rheem, Hearings,* pp. 68-70, 80, 89, and House, *Interim Report* No. 2057, of the Committee on the Merchant Marine and Fisheries, Dec. 12, 1944; 78th Cong., 2d sess., p. 2. This *Report* seems to me an excellent summary of the whole episode.

[5] MC Minutes, p. 24212. The final decision to terminate was made Feb. 11. *Ibid.,* p. 24295.

[6] MC Minutes, pp. 24268-72 (Feb. 10, 1943) (Rheem states his position); pp. 24387-91 (Feb. 18); p. 24356 (Feb. 16); p. 24405 (Feb. 19); pp. 24423-26 (Feb. 23); pp. 24461-62 (Feb. 25, 1943).

[7] House, *Interim Report,* No. 2057, pp. 9-10, 15.

[8] *Ibid.,* p. 13.

[9] House, *Walsh-Kaiser Co., Inc., Hearings,* and House, *Report of Activities of*

provided headaches both before and after the change of management.

Reorganization of top management without any change in name occurred at Mobile, Alabama. In August 1943 the management of the Alabama Dry Dock and Shipbuilding Co. acknowledged that "the magnitude of the operation had gotten beyond the capacity of their executive staff." They were willing to restrict themselves to repair work. Various solutions were discussed at a meeting in Admiral Land's office at which the only other Commissioner was Admiral Vickery, but which was attended by regional directors of the Gulf and the East coast, by Commander Weber, W. S. Brown, and Mr. Anderson. The possibility of Todd Shipyards taking over was discussed, but the conclusion reached placed the management in the hands of two engineering firms, Turner Construction Co. and Spencer, White, and Prentice, acting in conjunction. Since the incumbent executives of the Alabama Dry Dock and Shipbuilding Co. did not wish the formation of a separate corporation, they formed instead a Shipbuilding Division and agreed that its head should be an executive vice president chosen by the above-mentioned engineering firms. The repair work was to be entirely separate from the shipbuilding division in services, personnel, accounts, etc., and the Maritime Commission would reimburse only for the costs of the Shipbuilding Division.[10] Thus, although there was no change in ownership, Alabama Ship is really a case of complete change of management. Together with the yards at Tampa, Savannah, South Portland, Brunswick, and Providence, it brings to six the number of such changes in Class I yards.

The changes in Class II yards also include some borderline cases. There were both drastic changes and reshufflings in the management of the yards building concrete ships,[11] and radical changes were made when Walter Butler Shipbuilders, Inc. took charge of the Barnes Duluth yard,[12] and when Todd-Galveston Dry Docks, took charge at Gray's Iron Works.[13] Only one remain-

Investigating Committee for the Year 1945, of the Committee on Merchant Marine and Fisheries, Committee Doc. No. 82, p. 4.

[10] Memorandum for Files, Aug. 7, 1943 in Anderson reading file. Cf. on Mr. Sanford's part especially, House, *Investigation of Rheem, Hearings,* p. 59.

[11] See chap. 18.

[12] MC, Historian's Collection, Research file 112, Barnes Duluth, now out of file.

[13] MC Minutes, pp. 25438-40, June 15, 1943; Anderson to Walston S. Brown, May 27, 1943, and Vickery to Sanford, Aug. 25, 1943 in Anderson reading file.

ing instance seems to merit close attention, however, that at Bayonne, New Jersey. The yard itself, which was first under the Marine Maintenance Corp. and later under East Coast Shipyards, Inc., was not of first-class importance in ship construction; although it was a valuable repair yard, it had only two building berths. But two difficulties likely to arise in any governmental management of management in an all-out shipbuilding effort are illustrated especially vividly in the story of this yard. It shows how the Commission might, when it dictated a change of management, get dragged into supplying working capital for a private company, and how it and its staff were likely to be accused of favoritism towards old friends.

Contracts for the construction of coastal tankers had been given to the Marine Maintenance Corp., in spite of its relatively inadequate facilities, because even among Class II shipyards there were no other plants left. In presenting to the Commission the builders of the small tankers Admiral Vickery said, " Now these are the cats and dogs of shipbuilders. We are going to have trouble with every company that we have brought in here, and we are just taking a chance, but we have to get the ships. . . . That's the best there is, there isn't anything else." [14]

Marine Maintenance Corp. did not make such a bad showing comparatively,[15] although the regional office was far from satisfied.[16] But in December 1942 the district auditor of the War Shipping Administration charged Marine Maintenance with intentionally defrauding the government in its repair work by forging labor sheets and other documents, and on January 26, 1943 indictments were returned against the company, its treasurer, and its president.[17] Mr. Anderson, while acting as if he thought them guilty, took the ground that the indictment would so impair their credit even if they were ultimately acquitted that they could not complete their contracts. Accordingly the Commission took possession of the yard by condemnation proceedings, while Mr. Frick, the trouble shooter, was placed in temporary charge.[18]

[14] House, *East Coast Shipyards, Inc., Hearings,* p. 378. The quotation is from his testimony in July 1944.
[15] *Ibid.,* p. 415.
[16] Blakeman to McInnis, July 4, 1944 in Historian's Collection, East Coast Regional Office, Director's Files, East Coast Investigation.
[17] House, *Investigation of East Coast Shipyards, Inc., Interim Report,* No. 2075, from the Committee on Merchant Marine and Fisheries, 78th Cong., 1st sess., pp. 2-3.
[18] *Ibid.,* pp. 3-4; House, *East Coast Shipyards, Inc., Hearings,* pp. 232-34, 352;

Thus a new management had to be found for the yard at Bayonne just at the time that the South Portland imbroglio was being wound up and new managers were being sought for the yards at Providence, Rhode Island, and Brunswick, Georgia. Admiral Vickery had to leave by air for Portland, Oregon, on the twentieth of January because of the splitting of the tanker *Schenectady*, an event creating the new crisis in the Commission's program which will be discussed in the next chapter. But before leaving, Admiral Vickery talked over plans for completing the four coastal tankers which were unfinished at Marine Maintenance and twelve more which were wanted by the Joint Chiefs of Staff. The central figure in these plans was John E. Otterson, who had been Coordinator of Ship Repair and then had become president of the New Jersey Shipbuilding Corp. This company was partially owned by Todd interests and was building for the Navy under a plan whereby large sections of the hull were fabricated and assembled for it by the Harris Structural Steel Co. at nearby Plainfield, N. J. Admiral Vickery thought the same plan could be applied to the coastal tankers, and left for the West expecting the Director of Finance to negotiate a contract accordingly with the New Jersey Shipbuilding Corp.[19] But the controlling interests in that company decided to stick to Navy contracts. Mr. Otterson then resigned as president and proposed to form a new company, East Coast Shipyards, Inc., to take over the Marine Maintenance yard and apply there the same methods, utilizing facilities at Harris Structural Steel Co. for prefabrication.[20] These matters were discussed in meetings of the Commission January 26, 28, 29 and February 4; and on February 10, after Vickery had returned, the Commission unanimously voted to award contracts to East Coast Shipyards, Inc.[21] Court proceedings were pushed through by the Department of Justice, and on February 17 a notice of condemnation was ready to be served, so that East Coast shipyards could take charge of operations.[22]

New York Times, Sept. 16, 1943. After the war Marine Maintenance regained possession of the yard and was paid $845,000 for its use by the Maritime Commission. MC Minutes, p. 33948 (Nov. 29, 1945). See also Minutes, pp. 29883-85, 30540-42, 31044, 31083, 31194-98, 31231-32, 31290.

[19] House, *East Coast Shipyards, Inc., Hearings,* pp. 381, 385, 401-02. Vickery's testimony.

[20] *Ibid.,* pp. 497, 564-65, Otterson's testimony; pp. 228, 339, Anderson's testimony.

[21] MC Minutes, pp. 24142, 24164-65, 24201-03, 24264-68.

[22] House, *East Coast Shipyards, Inc., Hearings,* pp. 163-68. Testimony of R. I.

Embarrassingly enough, something went wrong that same day with Mr. Otterson's plans for financing his new corporation. As presented to the Commission and as explained by Mr. Anderson, who had communicated with Mr. Otterson's backers, Mr. Otterson's arrangements for working capital were obviously adequate to the operation he was undertaking, but at the last moment the lawyers for the insurance company to which he planned to sell debentures and the lawyers for the Bankers Trust at which he expected his line of credit developed a "long legal battle." [23] Mr. Otterson made other financial arrangements that same night, but arrangements which gave him much less working capital.[24]

When it developed that Mr. Otterson's financial backing was not so good as promised, the Commission had really only two alternatives, it seems to me, if it was to get the yard at Bayonne operating effectively. It could drop Mr. Otterson out of the picture, and while keeping Mr. Frick there temporarily, look for another contractor who would take over the yard; or it could put East Coast Shipyards in possession and help it build up its credit, as a number of other contractors had been helped. It did the latter. A small contract only, providing for the completion of the four half-finished vessels, was executed on February 18 on the grounds that the small amount of working capital which East Coast Shipyards had secured was adequate for a contract of that size. Thus East Coast Shipyards had a contractual basis for taking over the management. In helping the new corporation to meet its bills, the Commission and its staff seem to have done even more than was done to help out other companies that it tried to save from default in order to keep up production. Certainly the House Committee on Merchant Marine and Fisheries felt that it went much too far.[25]

That Congressional committee in its interim report criticized also, although only by implication, the fact that Mr. Anderson had taken a dominant part in the negotiations with Mr. Otterson in spite of their having been close business associates for almost thirty years before Mr. Anderson became the Commission's

Cameron describes the situation at the yard when the notice was ready to be served but East Coast Shipyards was not yet in shape to take over.

[23] So Anderson called it, *ibid.*, p. 276.

[24] *Ibid.* The financial plans and difficulties are recounted at length in the testimony of Messrs. Anderson, Otterson and C. Y. Palitz. See especially, p. 342.

[25] House, *Investigation of East Coast Shipyards, Interim Report,* No. 2075, pp. 5-6.

Director of Finance. They had been together in the Winchester Repeating Arms Co. in 1915 and again in Paramount Pictures.[26] Admiral Land was also an old friend of Mr. Otterson, both having been at the Naval Academy at the same time and Naval Constructors in the Bureau of Ships. Admiral Land's high opinion of Mr. Otterson is evidenced by his having recommended him in 1941 for appointment as a member of the Maritime Commission.[27] Because of their friendship, Admiral Land kept out of the negotiations with East Coast Shipyards. Admiral Vickery, who had no personal relations with Mr. Otterson, insisted that he himself decided to award the contract. As soon as Admiral Vickery returned from the Pacific coast, Mr. Anderson suggested on February 5 withdrawing himself from negotiations with East Coast Shipyards, but Admiral Vickery said he did not think it necessary.[28] At that time Mr. Otterson's financial difficulties had not yet appeared and the approval of his arrangements for working capital may have looked like a routine matter. It was the collapse of Mr. Otterson's plans on February 17 which made the situation delicate. Later, after East Coast Shipyards was operating the yard but before the final execution of the contract for 12 more tankers, Admiral Land suggested to Mr. Anderson that there had been some feeling that he might possibly be influenced by his long association and acquaintance with Otterson.[29] At Mr. Anderson's request, the Commission appointed a committee to act in his place in making recommendations concerning East Coast Shipyards,[30] but this withdrawal was too late to prevent a situation in which a Congressional committee paraded testimony implying favoritism to old friends.[31]

Additional embarrassment for the Commission resulted from the unsatisfactory way in which Mr. Otterson's company operated the yard after it was placed in charge. The conditions there in March 1944 moved Commissioner Carmody to write to the Commission

[26] *Ibid.*, p. 8.
[27] Land to the President, July 1941, in Land reading file.
[28] House, *East Coast Shipyards, Inc., Hearings,* p. 400.
[29] *Ibid.*, pp. 277, 367.
[30] Minutes, pp. 24706-07, Mar. 30, 1943 and Peet to Anderson, Mar. 30, 1943 in Anderson reading file.
[31] House, *East Coast Shipyards, Inc., Hearings,* p. 78 and *passim*; *Report of Activities of Investigating Committee for the Year 1945,* Committee Document No. 82 cited above in note 9.

after a visit of inspection: "The idleness, clearly observable, was almost unbelievable. . . . I have never observed so little work being done by so many men, nor so many men on a job who had absolutely nothing to do." Commissioner Carmody recognized that it was not all the fault of the management. Extenuating circumstances were the late arrivals of necessary cranes and of revised Navy plans. Commissioner Carmody concluded, "None of these, however, seems to me to justify the clear overloading of the payroll, yard and office."[32]

In view of how much all men experienced in ship construction were thrown together, and how many decisions the Commission and its staff had to make for financing private companies, charges of favoritism are not surprising. The surprising thing is that there were not more such open charges. Changing managements required decisions of the kind particularly likely to be charged with favoritism—namely, decisions based on judgments about personalities.

Indeed the whole art of managing management was so largely dependent on personal appraisals and personal confidence that it is difficult to reconcile with the ideal of government by law, and not by men. It depended so much on individual, personal qualities that they have been given a large place in these chapters. Any generalization about it is difficult. But insofar as it can be reduced to a system and judged as such, one conclusion stands out: its rewards were much more effective than its penalties. With the contracts offering substantial rewards for outstanding performance, there undoubtedly were some cases of brilliant management; but the relatively slow and high-cost producer received substantial fees also, unless he could be ousted from control in an atmosphere of public condemnation. There were three main reasons for changing management: failure to produce, the prospective bankruptcy of the contractor, and the suspicion of fraud. If all three reasons occurred together—and in fact there usually were more reasons than just one—the contractor was indeed in a weak position. But in most cases of ejection, the ousted managements made comparatively good terms, either by negotiation with the Commission, as in the case at South Portland, or by appeal to the courts, as in the case at Savannah. So the main incentive which

[32] Carmody's letter in MC Minutes, p. 27740 (Mar. 7, 1944). Cf. Vickery's testimony in House, *East Coast Shipyards, Inc., Hearings,* pp. 385-96.

the government could use was not fear of punishment but hope of reward.

A superficial view might regard the terms of the ship construction contracts as the reason for the government's inability to command more efficiency from the laggards. Both the manhour contract and the price-minus contracts gave assurance of minimum fees in addition to reimbursement of costs, and the clauses concerning termination required that thirty days' notice be given. From a legal point of view, these provisions protected the companies that failed in their managerial functions from being severely penalized for their failures.

Looking deeper one can see that the legal situation was mainly a reflection of the facts about the nature of management and the way it was organized in the shipbuilding industry. There were a certain number of companies with which experienced engineers and executives were associated. Outside of these companies it would have been very hard to find the needed managerial ability or to put it to work effectively. Consequently contracts were made in terms agreeable to the companies, minimizing penalties and emphasizing rewards. Management as well as labor was "organized," although the form of organization was different; and neither could effectively be coerced.

CHAPTER 16

CRACKS IN WELDED SHIPS

I. Crises, Criticisms, and Correctives

AT THE BEGINNING of 1943, just when the shipyards were proving they could reach the " impossible " goals set by the President, the basic methods used to achieve this speed were suddenly called into question by a series of instances of ships breaking in two. The cracking of steel plates under the pounding of heavy seas was a commonplace in shipping history. With the introduction of the welded ship this form of casualty assumed a new importance, for a crack in one plate, instead of stopping at the edge of the plate as it usually did in a riveted ship, could readily extend into adjacent plates. The overloading of ships during the war—especially the carrying of heavy deck cargo—increased the strains. But cracks induced by the stresses and strains of a ship at sea were only a part of the crisis confronting the Commission in 1943. More alarming was the splitting of ships before they even got to sea. On the night of Sunday, January 16, 1943, at 11 p. m., the brand new tanker, *Schenectady*, while it was calmly tied to the outfitting dock, split across the deck and down both sides with a loud bang. She was the first vessel built at Kaiser's Swan Island Yard and had just completed her sea trials.[1] At midday of February 12, 1943, the ore ship *Belle Isle*, while she also was at her outfitting dock, that of the American Ship Building Co. at Cleveland, Ohio, split across the deck and part way down the sides. Since the *Belle Isle* had riveted seams, the crack stopped after crossing the deck and severing the sheer strakes on both sides.[2] Five days later the Liberty ship *Henry Wynkoop*, which had made only one or two voyages, fractured her deck and upper side shell, port side, while being loaded in New York.[3] About the

[1] David B. Tyler, " A Study of the Commission's Experience with Welding during World War II," a typescript in MC, Historian's Collection, Research file 204.1.1., cited hereafter as Tyler, " Welding," pp. 8 ff.
[2] *Ibid.*, pp. 46-52.
[3] Truman Committee, *Hearings*, pt. 23, pp. 10011-13.

same time two other Liberty ships cracked. Then on March 29, 1943, the tanker, *Esso Manhattan,* split in two as she left the entrance to New York Harbor, the weather being clear with a mere force 2 northeast wind and a slight ground swell.[4] See Plate XX. These and other less sensational fractures during the opening months of 1943 raised doubts whether the 24 million tons of merchant shipping were being soundly built.

Publicity focused on the cracking of the *Schenectady.* The record-breaking speed of the Kaiser yards had been good copy; so was news of the fracture at Kaiser's Swan Island Yard. Having been exuberantly praised, Mr. Kaiser now found himself loudly mocked. Actually, the serious crackings occurred in ships built in many yards, old-liners as well as newcomers. The *Esso Manhattan* had been built at one of the most experienced of all tanker yards, that of the Sun Shipbuilding and Dry Dock Co. But its fracture and some others were kept from the public for " security " reasons. The *Esso Manhattan* had not yet cleared the area of minefields, and since she split with a loud bang, the crew assumed she had hit a mine and abandoned ship. The two parts continued to float (ultimately they were towed to a repair yard and rejoined). Later examination showed that the damage was different from that which would have been caused by an explosion.[5] This and other casualties at sea were not fully reported in the newspapers, but the cracking of the *Schenectady* occurred at the shipyard. It was headlined in the Portland newspapers the next morning and broadcast over the country.[6]

The break of the *Schenectady* was certainly extraordinarily complete and without apparent provocation. The vessel broke across the deck

> . . . just abaft the after end of the bridge about amidships, the fracture extending down both sides to the bottom shell plating which remained intact. All deck, side and bottom longitudinal frames fractured as did also the plating of the corrugated longitudinal bulkheads and the centerline deck and bottom girders, thus constituting a complete structural failure except for the flat

---

[4] Tyler, " Welding," pp. 60-61.
[5] Details of the failure are in the ship casualty files of the Division of Security and Communications, WSA and in MC gf 606-65, Apr. 20, 1943. For the resumption of operations by the *Esso Manhattan,* see New York *Herald-Tribune,* June 23, 1944, cited in Maritime News Digest, Daily Summary, June 23, 1944.
[6] E. g., *New York Times,* Jan. 18, 1943.

portion of the bottom shell plating. In no case did the fractures occur in the transverse welds.[7]

Some of the comment which this extraordinary event caused in shipping and government circles is reflected in an article appearing in the New York *Journal of Commerce* of February 1, 1943. Remarking that few cases of serious cracking had yet occurred, compared to the vast size of the shipbuilding program, this conservative journal suggested that they should serve as a warning to the Commission.

For the last year the Maritime Commission has used the construction records of the Kaiser yards as a sort of whip with which to goad other of the nation's yards into speedier construction. No one will deny that speed is needed in the construction and delivery of ships. However, no matter how speedily a ship is delivered its worth is practically nil if its plates crack, or for any reason the vessel must spend thirty to sixty days in a repair yard after one or two trips.[8]

Admiral Vickery had said that only 80 per cent of the capacity of the shipyards would be needed to produce the 1943 goal. The *Journal* suggested that the whole capacity be used with a schedule which would ensure sturdy as well as rapidly constructed vessels.

Fully realizing the seriousness of the *Schenectady* incident, Admiral Vickery immediately went to the scene himself, flying out in a snowstorm.[9] At the same time the Commission requested the American Bureau of Shipping to appoint a special subcommittee to investigate the causes of the *Schenectady's* failure.[10] The reports of this committee of experts, dated March 11, 1943, gave a diagnosis based on theories of locked-in stresses very different from the

[7] Findings of the American Bureau of Shipping subcommittee, in *Marine Engineering and Shipping Review*, Apr. 1943, p. 207.

[8] *Journal of Commerce*, Feb. 1, 1943, p. 18.

[9] Information from interview of D. B. Tyler with James W. Wilson, Aug., 1946 and Miss Irene Long, Apr. 1947. Admiral Vickery was away on this mission for 16 days.

[10] Findings of the American Bureau of Shipping subcommittee in *Marine Engineering and Shipping Review*, Apr., 1943, p. 207. This committee was made up of A. G. Bissell, senior welding engineer, Bureau of Ships, Navy Dept.; E. D. Debes, welding engineer, Bethlehem Steel Co.; La Motte Grover, structural welding engineer, Air Reduction Sales Co.; A. A. Norton, hull superintendent, Sun Shipbuilding & Dry Dock Co.; H. W. Pierce, welding engineer, New York Shipbuilding Corp.; Prof. S. H. Graf, consulting materials engineer and metallurgist, Oregon State College; and, as ex-officio members: J. L. Luckenbach, president of American Bureau of Shipping, D. Arnott, vice president, and chief surveyor, and J. L. Wilson, assistant chief surveyor.

theories the experts were to hold a few years later. The report said that the principal cause of the failure of the *Schenectady* was:

an accumulation of an abnormal amount of internal stress locked into the structure by the processes used in construction together with an acute concentration of stress caused by defective welding at the starboard gunwale in way of the abrupt ending of the bridge fashion plate, augmented by the hogging stress due to the ballasted condition; this accumulation and concentration of stresses caused a tensile failure at the starboard sheer strake which was formed of steel of sub-standard quality, all of which was aggravated to some degree by the drop in atmospheric temperature.

Among the contributing causes mentioned were: departure from fundamentals of good welding in an effort to speed up construction; lack of trained, experience welders, shipfitters, and supervisors; failure to adhere rigidly to established welding procedures and sequences; excessive forcing of poorly fitted subassemblies with consequent excessive welding; and serious accumulation of shrinkage stresses due to automatic machine welding.[11] The investigation disclosed that the temperature had dropped 15 degrees to 23° F. during a period of $8\frac{1}{2}$ hours preceding the break, while the water remained at about 40°. It was felt that this contributed to the brittleness of the steel. But the committee found no basis for believing that the alleged shock resulting from the breaking of an anchor cable during the trial trip, the alleged silting of the river under the mid-section of the vessel producing a straining effect, or an alleged shock caused by an earth tremor could have had any bearing in this case.

The *Schenectady* "jack-knifed" after the break so that the two ends settled into the silt while the mid-section rose as the deck parted with a 10 foot gap. She was refloated, repaired, and given severe sagging and hogging tests before the American Bureau of Shipping committee made its report. This enabled the committee to say that the defective welding had been taken out and that the vessel had passed successfully the severe tests imposed after repairs were completed. The committee felt that the break had relieved the *Schenectady* of its locked-in stresses and was confident that "closer control of welding procedure in which the builders have already been instructed will prevent a recurrence of such major failures. It is also the opinion of the committe that

[11] Findings of the committee, in *Marine Engineering and Shipping Review*, Apr. 1943, pp. 207-08.

the probability of fractures resulting from residual stresses in welded construction will decrease with the length of service." [12]

Admirals Land and Vickery agreed with the committee of experts of the American Bureau of Shipping in believing that stresses which were locked in during the process of welding formed one of the main causes of the disturbing fractures,[13] and the Commission took corrective measures based on this assumption. The magnitude of locked-in stresses could be minimized by improving welding practices in shipyards, but effecting this improvement required a time-consuming campaign of instruction and inspection. Meanwhile a certain number of specific corrective measures could be taken. One of the first things Admiral Vickery did was to prohibit the use on the ways of automatic welding machines on main strength members at key points, but they were still used in subassemblies. It was felt that the use of these machines either made a weaker weld or involved an application of heat which increased the locked-up stresses.[14] To control the effect of stresses, large girders were added as reinforcement beneath the decks of T2 tankers.[15]

Many small but important changes were made in design. On both tankers and Liberty ships the bilge keels were to be scalloped or serrated. Another change in the Liberty ships provided for separating the bulwarks from the top of the sheer strake and bridging the gap with riveted stiffeners. This separation would prevent a crack that started in the rail or bulwark from continuing down the side. Also on the Liberty ship, the cut in the top of the sheer strake for an accommodation ladder was eliminated since it was found that there was an intense concentration of stress at this break in the plate line. These important changes in Liberty

[12] *Ibid.*, p. 208.

[13] For Vickery's opinion given to the Iron and Steel Institute Advisory Committee Apr. 18, 1943, see *Congressional Record*, Vol. 89, pt. 3. p. 2780; and for repetition of the same emphasis in Jan. 1944, see Washington *Evening Star*, Jan. 20, 1944, quoted in Maritime News Digest, Jan. 21. For Admiral Land's judgment see his letter to Congressman Bland Feb. 24, 1943 in Land file, Historian's Collection and the *Journal of Commerce* (New York), Jan. 5, 1944 (Maritime News Digest, Jan. 5).

[14] House, *Independent Offices Appropriation Bill for 1945, Hearings* before the subcommittee of the Committee on Appropriations, 78th Cong. 2d Sess., Dec. 9, 1943, p. 797; and statements to me by E. E. Martinsky, Feb. 7, 1949.

[15] "Reinforcing the Liberty Ship" in *Marine Engineering and Shipping Review*, Sept. 1944, pp. 159-63, 216, an article based on a report entitled, "The Structural Reinforcement of Liberty Ships" in Proceedings of the Merchant Council, USCG, June 1944.

CRACK ARRESTORS

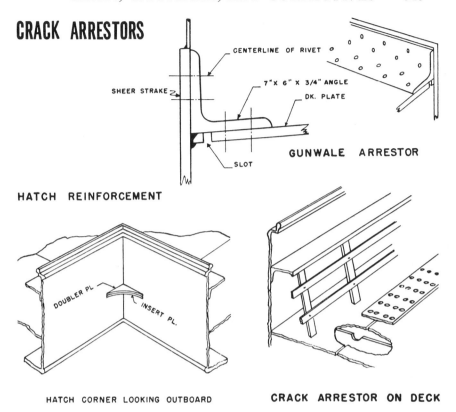

FIGURE 36.

ship design were issued to the yards in February 1943. Changes in design of tankers were also made to eliminate "notches," such as that formed by the abrupt ending of the bridge fashion plate.

Somewhat later, measures were taken to permit better distribution of the concentrated stress at square hatch corners. In June the Navy informed the Maritime Commission that it was reinforcing these corners on the Liberty ships it was operating, because the Navy's extensive experience showed the danger of cracks originating in sharp-cornered deck openings.[16] In July the Maritime Commission authorized the fitting of rounded plates to reinforce the hatch corners at No. 2, No. 3, and No. 4 hatches,

---

[16] Admiral E. L. Cochrane, Chief, Bureau of Ships, to Land, June 7, 1943, in MC gf 506-22-1.

upper deck, of Liberty ships in operation. On August 12, 1943, instructions and plans for these changes were sent to the Regional Offices for application to new construction.[17] In the light of later expert analysis it seems that these design changes were among the most important corrective measures.

While the Maritime Commission was going ahead during the first half of 1943 with needed changes in shipyard practice and ship design, it decided against approaching the problem through insisting that the steelmakers deliver better steel. That the quality of the steel was part of the problem was recognized in the report of the American Bureau of Shipping and was considered the main cause in a report to the Bureau by Professor Graf of Oregon State College, who had analyzed a sample taken from the defective plate on the *Schenectady*.[18]

At the time when the Truman Committee made Professor Graf's opinion public, other hearings of this committee headlined the fact that about 5 per cent of the tensile tests of samples made at the Irvin Works of the Carnegie-Illinois Steel Corp. had been falsely recorded (because of the methods used for cooling plate after this plant was changed over in 1942 from making automobile strip-sheet steel, for which it was designed, to making plate).[19] When the president of the Carnegie-Illinois Steel Corp. was called before the Truman committee in March 1943, he deplored the condition which had been brought to light, but pointed out that the shipyards had found the steel satisfactory and that the plate used in the *Schenectady* had come from the Homestead, not the Irvin plant.[20]

Actually, not only did steel plates produced by various mills have different properties, but various runs at the same mill differed. How much these differences were of a kind to affect the crackings was not known.

[17] J. L. Bates to Division of Public Relations, Att. S. W. Richards, Feb. 10, 1944, in MC, Historian's Collection, Research file 204.1.3a, Tyler, " Welding," pp. 38-39. Statements by Martinsky in Jan. 1949, and Esmond in Mar. 1949. Full details on these reinforcements are in *The Structural Alterations on Liberty Ships,* CG 140, published by U. S. Coast Guard, Treasury Department.

[18] Truman Committee, *Hearings,* pt. 18, Mar. 23, 1943, pp. 7179-80. (See also *ibid.,* p. 7194.)

[19] Truman Committee, *Additional Report,* No. 10, pt. 7, 78th Cong., 1st sess., Apr. 16, 1943, pp. 1-6; Truman Committee, *Hearings,* pt. 18, Mar. 23, 1943, pp. 7148, 7159, and Mar. 22, 1943, pp. 7221-78.

[20] Truman Committee, *Hearings,* pt. 18, Mar. 23, 1943, pp. 7168-69.

CRISES, CRITICISMS, AND CORRECTIVES 551

So long as there was uncertainty as to the main cause of crackings —defective steel, defective welding, or something else—both steel-makers and shipbuilders, particularly the Kaiser interests, were on the defensive. John Green, president of the Industrial Union of Marine and Shipbuilding Workers of America (CIO) testified before the Truman Committee that Kaiser's " admitted achievements are blown up into miracles and his failures are played down or suppressed." He asked: " Has the Maritime Commission revealed all of the instances of Kaiser-built ships cracking up? " [21] In reply Mr. Kaiser told newspaper reporters in New York that Mr. Green knew that the question of cracking ships involved many yards. " I am informed," he said, " that the shipbuilding companies owned by the steel companies themselves have had major cracks in ships, and we likewise have had some others, which have been minor ones." [22]

The total result of the controversy among steel-makers and shipbuilders was a general feeling of insecurity and doubt. Within the Maritime Commission, this feeling and a desire for more vigorous action was expressed by Commissioner Carmody, who had once been an inspector in a steel plant. He was particularly concerned about the question of the quality of steel plate. He had visited the Swan Island shipyard shortly before the *Schenectady* failure and had noticed some laminated and badly blistered plate.[23] In a memorandum to the Commission dated March 31, 1943, he recommended: (1) that a list be made of the serious fractures, such as " the *Schenectady*, the two ore vessels that shipping men are talking about and similar jobs that are now a matter of gossip in the trade "; (2) that the General Counsel be requested to advise whether the Commission had the basis for a claim against the Carnegie-Illinois or any other steel company; (3) that, in view of John Green's and Henry Kaiser's testimony before the Truman Committee, they be asked regarding their knowledge of steel failures in other shipyards, and (4) that data be collected regarding material inspectors with a view to seeing what the Commission could do to assure better steel. Said Commissioner Carmody, " In my judgment the Maritime Commission has an

---

[21] Truman Committee, *Hearings*, pt. 18, Mar. 25, 1943, p. 7350.
[22] MC, Maritime News Digest, Daily Summary, Mar. 26, 1943, vol. 7, No. 73, citing Washington *Post*, Mar. 26, 1943.
[23] Carmody to Vickery, in Carmody Reading File, Jan. 19, 1943.

obligation not only to recover appropriate damages but to participate in cleaning up testing and inspection practices on all Maritime steel." [24]

Against Mr. Carmody's inclination to have the Commission send its own inspectors into the steel mills,[25] was the reliance that Admirals Land and Vickery placed on the American Bureau of Shipping, the time-honored custodian of standards of the "shipbuilding fraternity." Mr. Carmody's memorandum was "laid over" for several weeks, until on April 20 it was merely "noted," along with a memorandum from the General Counsel advising that there was little basis for any suit against a steel company.[26] The Commission did not send its own inspectors into the steel mills; instead it continued to rely on those of the American Bureau of Shipping.

Far from "cracking down" on the steel mills, Admiral Vickery showed concern lest worry over high specifications slow down the production of steel plate and so delay ship construction. While Mr. Carmody's memorandum was being laid over by the Commission, Admiral Vickery spoke at the meeting of the Iron and Steel Industry Advisory Committee of the War Production Board when it considered the report on the *Schenectady* on April 8. He said the Maritime Commission would have accepted the steel from the Irvin Works even if the deviation from specifications had been reported.[27] The main problem was how far quality should be sacrificed to speed. Dr. Rufus E. Zimmerman of the U. S. Steel Corp. exhibited a letter sent by the American Bureau of Shipping in 1942 authorizing their inspectors to waive tests of specimens when in their opinion " delivery can be expedited and the serviceability of the ships for the present emergency will not be adversely affected." The latter said, " It must be recognized . . . that under present circumstances early completion of serviceable ships is of greater national importance than the high measure of perfection required for durability in peacetime." [28] The same attitude was expressed by the Chairman of the War Production Board on April 15, 1943 in a long telegram to producers of steel plate.

[24] MC Minutes, p. 24922.
[25] Statements by Mr. Carmody in interview.
[26] MC Minutes, p. 24759 (Apr. 2, 1943), and pp. 24918-26 (Apr. 20, 1943).
[27] *Congressional Record*, vol. 89, p. 3780 (Apr. 29, 1943).
[28] *Ibid.*, p. 3780-82. Dr. Zimmerman did not mention the American Bureau of Shipping by name, but it is obvious that this was the " Bureau " involved.

Fearful that the Congressional investigation might have the effect of slowing down production, he pleaded for inspections and tests which would meet two essential objectives—plates kept up to specifications and volume at the highest possible level. He called for " honest good judgment in testing and inspection " and at the same time, for avoidance of delay in the program due to a " blind, unthinking insistence on unattainable perfection." [29]

By July 1943 agitation over the testing of steel was dying down, but fresh news and rumors of crackings reawakened public suspicion of war-built ships. Although the most sensational earlier failures had been in tankers, attention now focused on Liberty ships, and again ships built by Kaiser were especially suspect. Doubts about Liberty ships were stirred by an announcement on crackings issued by Congressman Henry M. Jackson, Chairman of the subcommittee of the House's Committee on Merchant Marine and Fisheries which Judge Bland had appointed immediately after the *Schenectady* failure. Congressman Jackson emphasized that only two out of over a thousand Liberty ships had been lost through defects in structure, but he gave their names and verified the suspicion that there had been actual losses of that kind. Neither of the ships in question, the *Thomas Hooker* and the *J. L. M. Curry*, had been built by Kaiser. But through a misreading of explanation of asterisks in a Maritime Commission report, a member of the Merchant Marine and Fisheries Committee wrongly stated that four Kaiser-built Liberty ships had cracked up at sea. Upon Mr. Kaiser's demand, a public retraction was made when the mistake was discovered.[30]

When the winter months of 1943-1944 came around, cracking of ships became serious again. Winter, especially in the North Pacific, supplied the cold and storms in which fractures were most likely. Kaiser-built ships were those most affected that winter. A fact not always remembered in this connection is that his yards were near the North Pacific and a relatively large proportion of his output was assigned to those cold rough waters. Admiral Vickery came to the support of Kaiser and of the Liberty ship, but the casualties were numerous enough to cause worry in the Commission and shrill cries of alarm in newspapers and in Congress.[31]

[29] *Ibid.*, p. 3782 (Apr. 29, 1943).

[30] Tyler, " Welding," pp. 19-21. The *Curry* was constructed at Alabama Ship, the *Hooker* by South Portland Shipbuilding Corp.

[31] Tyler, " Welding," pp. 19-37.

The public outcry had the intensity and inaccuracy which through the centuries have always accompanied reports of losses in wartime. Each "cracking" was easily assumed to be a Class I fracture, that is, a complete break. One skipper was quoted as saying: "You can hear them crack like gunshots. And the cracks, once started, run like a woman's stocking. . . . These ships stand on a crest of a wave, with both ends shaking like jelly." [32] Congressman Magnuson of Washington named as many as fourteen ships that had cracked. With censorship on military information, the public could wonder whether many cases were being hushed up. The Truman Committee gave general endorsement to the Liberty ship as "the truck horse of the fleet," but by advising against its use as a transport left room for doubt.[33] All the excitement focused on welded ships; although cracks had occurred also in riveted ships, of which a few had broken in two, the object of suspicion was the welded ship.[34] Sober statements were offset by headlines which blared: "Another Ship Falls Apart." [35]

Even if the crackings were not as bad as the headlines might lead one to believe, they were in fact serious enough to intensify a search in January and February 1944 for more corrective measures. On the theory that several vessels had buckled at the bottom before they cracked, the West coast office of the Maritime Commission advocated reinforcement of the bottom plating. After considerable discussion, it was decided that the bottom reinforcement would not reduce stress on the strength deck to any extent and that there was no assurance whether the buckling was cause or effect. Yet in Libertys fitted as troop ships, additional girders were fitted to the inner bottoms.[36]

The question of ballast was also involved, for ships in the North Pacific always went out heavily laden and returned very light with water ballast fore and aft tending to cause a hogging stress. Since operators were very reluctant to give up cubic space to

[32] Washington *Evening Star*, Jan. 4, 1944, quoted in MC, Maritime News Digest.
[33] Truman Committee, *Hearings*, pt. 23, p. 9940; *Additional Report*, Report No. 10, pt. 18, 78th Cong., 2d Sess., June 23, 1944, p. 27.
[34] Admiral Land compiled a list of structural failures in riveted ships, including two instances in 1920 in which 59 lives were lost. See Martinsky's files.
[35] Chicago *Tribune*, Feb. 7, 1944.
[36] M. H. Fidalgo to Director of Technical Division, Feb. 24, 1944 in Wanless file, EC2-S-C1, and Martinsky Report, "Liberty Type Vessels, Structural Failures," Fall 1946-1947, p. 12, in Martinsky's files.

permanent ballast, the Technical Division of the Maritime Commission made an effort to develop portable tanks for water ballast amidships, but the war ended before experiments along this line were completed.[37]

The main corrective measures carried through in 1944 were the installation of various riveted "crack stoppers." At the top of the sheer strake this was provided either by a riveted gunwale angle bar, which was ordered for all new construction, or, on existing vessels, by cutting a slot in the sheer strake and covering it with a riveted strap. The orders to install these were issued January 21, 1944 and February 15, 1944, respectively. At the later date, orders were also issued to install on existing vessels intended for wintry waters slots cut in the upper deck for some 300 feet in length and fitted with a covering strap. But this crack arrestor on deck (Figure 36) was provided on very few ships; the War Shipping Administration, with its slogan of " Keep 'em Moving," was under very heavy pressure to keep ships in use and postpone repairs. Priorities in making alterations were given first to the reinforcements for hatch corners, second to the gunwale bar or sheer strake slot, and finally to the deck slot and strap.[38]

By 1944 the making of alterations on ships afloat was the main problem regarding Libertys, for many of the yards were being shifted over to the new type, the Victory, which incorporated in its original design features to prevent dangerous cracking.[39]

II. Efforts to Improve Welding

The *Schenectady* failure inaugurated in January 1943 a period of doubt when it was feared from day to day that some new serious failure would be reported. After the first corrective measures came a summer lull, followed by the series of crackings among Liberty ships in the winter of 1943-1944. During all this time public concern over ships that cracked made more urgent the need to improve welding practices in the shipyards.

A riveted angle bar or some other crack stopper could be added

[37] Tyler, " Welding," pp. 43-44, 106, and " Construction and Operation of Liberty Ships."

[38] Memorandum of Bates to Richards, Feb. 10, 1944 above cited, and notes by Martinsky on a draft of this chapter, filed in MC, Historian's Collection, Research file, 204-1-1.

[39] James L. Bates, " The Victory Ship Design," in *Marine Engineering and Shipping Review*, Apr., 1944, p. 158.

to new construction by a change in specifications, but better welding could not be suddenly obtained by decree. Research and training were both needed. It required scientific research, for the effects of different methods of welding were still relatively unknown; and above all it required steady effort by everyone concerned—the Commission, the American Bureau of Shipping, the yard managements, and the workers—to improve skills and welding procedure.

At the beginning of the emergency program, the Commission had left to the yard managements many decisions about how much use should be made of welding instead of riveting, how much to use automatic welding machines, how to organize the work, and how to train the raw recruits who were being so rapidly turned into "first class mechanics." The Commission's staff was not well prepared to advise or direct in detail on welding procedures. That was one of those things for which management was being paid. Only gradually and under the impulse of the alarm created by the sensational fractures of early 1943 did the Commission take more and more responsibility.[1]

The first general report made for the Commission on welding was by James W. Wilson, formerly a boiler inspector in the Bureau of Marine Inspection and Navigation, who was employed by the Commission to make a survey of the yards. In the process of preparing the report that he delivered to Admiral Vickery in June 1942, Mr. Wilson visited 35 shipyards and conferred with welding superintendents and field representatives of the Commission. He found that in general welders were certified for production work, under supervision, after 100 to 150 hours of training. In his opinion there were sufficient training facilities in the yards or in their immediate vicinity, but he recommended that training methods and welding procedure be standardized and that at least one welding inspector be provided for each shipyard.[2]

The inspection problem was complicated by the fact that it was carried on by four different agencies. The Commission's inspectors were in the shipyards, as representatives of the owner,

---

[1] Tyler, "Welding," pp. 152-54.

[2] Technical Division Files, Miscellaneous A-9, pt. 1, Wilson to Vickery, June 26, 1942. Judging by marginal notes penciled on this report, Admiral Vickery's advisers felt that, in the matter of training, the Commission should supplement rather than parallel the existing set-up under the U. S. Office of Education and War Manpower Commission.

to see that plans and specifications were carried out and that the work was complete. The American Bureau of Shipping inspectors were there to verify the seaworthiness of each vessel. The Bureau of Marine Inspection, which was under the Coast Guard after March 1, 1942, was concerned with the safety of passengers and crew. The shipyard had inspectors also, of course, to see that the builder's product satisfied the other inspectors and would be accepted by the owner.[3] Welding decisions were made in some yards by one official, in other yards by another.[4]

The most difficult problem was that of obtaining good supervisors and good inspectors in the shipyards. When, in April 1943, a series of meetings was held with shipyard personnel for the purpose of working out training procedures, W. H. Mackusick, welding engineer at California Shipbuilding Corp., asked that inspectors be included. He said:

> An Inspector for one of the Shipyard Inspection Bureaus who does not have the proper training background can cause more lost man hours by promiscuous use of his marking chalk, than any other single individual involved in the Shipbuilding Program. It is also evident that many Maritime, A.B.S., and Coast Guard Inspectors are being added to the Inspection Staffs under emergency conditions, and frequently a man with inadequate experience or training is put in a position of authority whereby a considerable amount of unnecessary repair work is done, while the same type of Inspector often passes over items that need attention.[5]

On the other hand, the inefficiency of the supervisors was emphasized by Mr. Wilson's report on the kind of conditions he found in some of the new yards in the spring of 1943. In the yard at Brunswick, Georgia, he came across a woman welder attempting to weld in the overhead position in the deep tank hatch coaming. She was using a rod of a type and size not intended for such work. After three-quarters of an hour's search, he found the woman's leaderman, whose job it was to check the rods as they were issued to workers under his supervision. The leaderman professed ignorance of the fact that the rods in question were not suited to the work. Mr. Wilson gave the following description of the yard as a whole:

[3] In the opinion of the Commission the work of these various inspectors was so different that their duties could not be combined. Land to WPB, Dec. 14, 1942 (dictated by Schmeltzer and countersigned by Vickery), MC gf 506-1.
[4] J. F. Lincoln to Vickery, Aug. 23, 1943 in gf 130-868.
[5] W. H. Mackusick, Welding engineer, Calship Corp., to Jack Wolff, May 8, 1943, Tech. Files, Miscellaneous, S29, Part 2.

As I walked through the yard, I found group after group of men whose insignia on their hats indicated that they were keymen, such as quartermen, leadermen and supervisors standing together discussing the war, ball games and other matters not pertinent to production. Loafing appears to be chronic among these men, and it is obvious they cannot perform their duties away from their official stations. This practice has the effect not only of reducing production, but also demoralizes the morale of their subordinates. This matter was discussed with the management and I found that our inspectors and the American Bureau of Shipping representatives were fully cognizant of the fact and had complained about it. I understand the management is about to insist that this group be made to give practical evidence that they are at least competent welders by having to prepare test specimens of their work to be tested by the American Bureau of Shipping, and those who fail to pass their tests will be terminated. This will probably have a wholesome effect on those who remain and also on the practical welders.[6]

Since bonus wages were paid for fast work, intentionally defective welding was done in a few instances. Mr. Wilson was called to Baltimore in April 1943 as an expert witness for such a case being tried in civil court. It was the trial of one of nine workers from the Bethlehem-Fairfield yard who were being held on charges of placing unfused electrodes and slugs of iron and steel in the plate grooves and then covering them with a weld. This was known in welding circles as "slugging." One defendant in this case was convicted of "making war material in a defective manner with the intent that his act would hinder, obstruct and interfere with the United States Government in preparing for and carrying on the War." Being a minor, he was sentenced to eighteen months in a reformatory. This and another similar case were publicized in the hope of deterring others from attempting the same means of making a lot of money fast.[7]

Poor work involving this kind of dishonesty was relatively infrequent; much more common was the poor work resulting from ignorance and lack of supervision. One immediate reaction to the splitting of the *Schenectady* was the completion in February 1943 of plans whereby instructors, superintendents, and foreman were to be sent from the various yards to the welding school conducted by the Lincoln Electric Co. at Cleveland, Ohio. They were to take a one-week course there at the expense of the Commission.[8] In March 1943 the Commission's staff began work on a

[6] Wilson to Vickery, June 15, 1943, in Tech. Division, Vickery file, Wilson Welding Reports.
[7] Tyler, "Welding," p. 72.
[8] Vickery to Regional Directors, Feb. 11, 1943 in gf 606-65.

welding manual which would combine in one authoritative document the best procedures. Such a "practical, basic, training manual" was suggested by Daniel S. Ring, Director of the Division of the Shipyard Labor Relations. He proposed that it be a collaborative effort by J. W. Wilson, other welding experts, and Jack Wolff, supervisor of shipyard training. Mr. Wilson approved of the plan but was occupied in making a survey of the shipyards for Admiral Vickery, and the technical work of preparing and editing the manual was done by E. E. Martinsky. An effort was made to secure and reconcile the opinions of men who were recognized as authorities in the matter of shipyard welding. The first edition of the manual appeared in August 1943 under the title "Welding Instructions for use by Welding Supervisors, Leadermen, etc., of All Crafts Concerned with Shipyard Welding," and it was re-edited in 1944. Its content was approved by both the American Welding Society and the American Bureau of Shipping. Widely distributed, the manual served its purpose by helping to standardize good welding practice.[9]

Since the Commission had started with the policy of giving the shipyards their head and relying on them to see that their workers were well trained and supervised, introducing new procedures was not easy now that the program was underway. This was particularly true of the sequences used in welding and in erection.

These sequences were believed to be intimately connected with the locked-in stresses and so a cause of fractures. The intense heat of welding produced expansion in the neighboring steel followed by contraction as the steel cooled. Consequently it was recognized that the shell, deck, and other plates should be put together starting from midship and working towards the two ends of the vessel. This permitted expansion and limited buckling. It was also important that erection should proceed simultaneously on opposite sides of the structure, so that expansion on both sides would correspond and a twisted hull be avoided. In the beginning of the emergency program Gibbs & Cox prepared an erection sequence plan, but not everyone thought it a good sequence, and practice varied from yard to yard.

The sequence of welding individual plates and sections was a more difficult matter to plan and to control, since often the

[9] Tyler, "Welding," pp. 67-68.

individual welder made the decisions. Not only did welders frequently have no knowledge of the best sequence, even experts disagreed. In this matter also Gibbs & Cox had prepared a drawing as a guide which the shipyards were told to follow, but shipyards were permitted to submit their own plans to the American Bureau of Shipping for approval. As a result a great variety of procedures were followed.[10] The testimony of John A. McCone, president of the California Shipbuilding Co., given before the Truman Committee explained:

> From that guide we developed our own welding procedure, which was reviewed with representatives from both the Maritime Commission and the American Bureau of Shipping. We improved that from time to time. We necessarily had to change it, as our methods of subassembly changed, because that was an evolutionary process, as you know, and tracing our file will show that there are innumerable prints and directives issued from time to time in connection with welding sequences.[11]

David Arnott of the American Bureau of Shipping stated that very few yards kept to the original subassemblies or welding procedure.

> Different methods of erection have been developed owing to differences in available crane capacity, shop equipment, electric power and skilled labor. The methods and details of welding also differ especially as regards the use and extent of automatic welding. Rigid adherence to so-called 'Codes' at this stage of progress of ship welding would handicap sound development and be quite undesirable and in this connection it should be said that the American Bureau has not attempted to dictate welding procedures or to reject proposed methods submitted for approval except where our experience has shown that they are not likely to achieve satisfactory results.[12]

In view of the fact that both the Maritime Commission and the American Bureau of Shipping had been encouraging experiment and diversity in the yards, intense difficulty was encountered when in 1943 and 1944 attempts were made to impose on them certain common standards. One good example arose on the West coast after the use of "unionmelt" automatic welding was restricted. The yards using these machines had found they speeded production and saved 30 per cent to 40 per cent on electrodes,[13]

[10] Tyler, "Welding," pp. 55-57; Sanford to All Principal Inspectors, Sept. 18, 1941, in gf 507-3-1.
[11] Truman Committee, *Hearings*, pt. 23, p. 10219.
[12] D. Arnott, "Some Observations on Ship Welding," p. 330.
[13] G. F. McInnis to Bates, Oct. 28, 1943, in gf 606-65.

at one time an important consideration because of shortage of supply. The machine made a deep or full penetration weld, a weld which the American Bureau of Shipping approved. After the ban on automatic machine welding of deck to shell on EC2's, most of the yards returned to the original manual double fillets. In December 1943 the American Bureau of Shipping inspector at Richmond No. 1 telegraphed the New York office saying that the Commission's West Coast Regional Office would not permit a return to double fillet weld on the grounds of workmanship. The somewhat agitated inspector informed Mr. Arnott that the Regional Office "insists they will fight approval of this change even by Washington or Bureau so suggest you follow this up direct to Washington." Mr. Arnott proceeded to do this, writing Admiral Vickery as follows:

> Our experience has not shown that either one of the two types of joints is to be preferred to the other to the extent that we would be justified in prohibiting one type in favor of the other and as a result this Bureau is still willing to accept either arrangement. As a matter of fact, if any conclusions can be drawn from our experience, these would rather lean in favor of the double fillet weld since the two types of vessels with which most of the serious difficulties have occurred, i. e., the EC2 and the T2 vessels, are those in which the full penetration welds have been used. However, in view of the stand taken by your West Coast Regional Office, we are at a loss as to just how to deal with the details as submitted to us by the different builders on the West Coast.[14]

This caused Admiral Vickery to write to the West Coast Regional Director, ordering a personal investigation to see "that instructions sent out from this office are carried out."[15] The attitude of the Regional Director was that he had told his inspectors it was not their responsibility to pass on how the weld was made, but it was their duty "to make sure it was a good weld"; and if those at Richmond felt that double fillet did not give a sound weld he would back them up, until the American Bureau of Shipping gave definite preference to one type over the other.[16]

The difficulty of standardizing welding techniques was also one of the obstacles encountered by the Lincoln Company when it was given a contract to act for the Commission in instructing

[14] Arnott to Vickery, Dec. 29, 1943 in gf 606-65.
[15] Vickery to Flesher, Jan. 1, 1944 in gf 606-65.
[16] Record of telephone conversation, Flesher with B. A. MacLean, in Flesher's files, Conferences, 1944, in MC, Historian's Collection.

shipbuilders how to improve their welding. Hiring a firm of welding engineers to do this kind of work was tried out in May 1943 in the Richmond yards, using the G. S. May Co., and at Pusey and Jones, where the Lincoln Company did a "pilot job" under contract to the Commission. Its work was considered so successful that in July 1943 the Lincoln Company was given a three-months contract: "1. To provide welding engineers to assist the shipyards to do faster, more economical, and more reliable welding.... 2. To furnish training facilities at Cleveland for supervisors and instructors. 3. To send monthly progress reports to the Commission." Mr. Wolff, the Commission's supervisor of training, was enthusiastically for it, and Admiral Vickery gave the company a list of the yards in which they should start work. Immediately James P. Lincoln reported that the differences in established procedures were an obstacle. Admiral Vickery wrote that the procedures advocated by the Lincoln Company agreed with those in the Commission's "Welding Instruction"; but the American Bureau of Shipping felt otherwise. In a letter dated August 26, 1943 the Bureau questioned some of the data in the Lincoln manual and also criticized it for over-emphasizing speed. The principal innovation advocated by the Lincoln Company was the use of their "Fleet Fillet" process which meant the use of electrodes of large diameter permitting greater current and producing more molten metal, and, consequently, faster welding. It was applicable only to downhand welding. Slight variations in the technique used—angle of welding rod, speed of travel, distance of rod from weld, etc.—could result in unsound results.

Since the main contribution of the Lincoln Company lay in its ability to speed up construction, the conflict between it and the American Bureau of Shipping was hard to reconcile. When crackings occurred in February 1944, the Bureau felt it necessary to warn its inspectors that the Lincoln manual overemphasized speed and to watch carefully to see that requirements regarding beveling of edges and back chipping of deep fillet welds were fulfilled.

The Commission itself disagreed with Mr. Lincoln as to the causes of the crackings. Mr. Lincoln was positive that locked-in stresses had not caused the splitting of the *Schenectady*. To his way of thinking, "These stresses obviously decrease continuously as soon as the weld is cold." He thought they would have to cause a failure immediately or not at all. The Commission felt that the American Bureau of Shipping report, which included locked-in stresses among the causes, was substantially correct.

These differences of opinion formed one reason, but not the only reason, why the Lincoln engineers were looked upon with suspicion in some of the shipyards. James W. Wilson was asked by many of his welding inspector friends how the Lincoln Company " came into the picture " and why other companies had not been consulted. They suspected the Lincoln people of " feathering their nest " by promoting their own products. Mr. Wilson explained that the Lincoln Company had offered to train welders and to make available its staff of engineers at a very reasonable rate, and that it was recommending the use of competitors' rods as well as its own.

As directed by the Commission, the Lincoln engineers tackled first New England, Walsh-Kaiser, Sun, Alabama, J. A. Jones, Todd-Houston, Richmond No. 3, Swan Island, Leatham D. Smith, and Barnes-Duluth. They came at the expense of the Commission and only after the yards had been informed of their purposes. Even so, they were frequently looked upon as meddlers and salesmen for Lincoln products. In one or two yards they made no headway. In others, Lincoln procedures were accepted with little difficulty, although it was not easy to enforce them permanently. Mr. Lincoln reported that Richmond No. 3 and New England had adopted all of his recommendations and that Houston, by adopting them, had increased welding speeds 100 per cent. His respresentatives' reports from Walsh-Kaiser, at Providence, were so uniformly favorable that he professed to be a little suspicious and asked the Commission to check on them. Mr. Wilson did so and found production greatly increased. In September and again in November 1943 he reported to Admiral Vickery that, in his opinion, the Lincoln engineers had also been very useful at Sun, where their efforts had been confined to the No. 4 Yard, that in which some friction had developed between the Negro workers and their white supervisors. Inspectors and superintendents at Portland, Maine, and at Bethlehem-Fairfield held varied opinions as to the usefulness of the Lincoln engineers. The more experienced superintendents were inclined to think that they did an excellent job in the new yards but were not needed in the old established ones.

In the Regional Offices there was skepticism regarding the value of these services. Although Mr. McInnis, director of the East Coast Construction Office, was one of the first to recommend the " Fleet Fillet " process, he also wanted to have it understood that welding had improved before the Lincoln Company started operations and that credit for improvement was, to a considerable extent, due to the Commission's own staff. Mr. Flesher, West Coast Director, reminded Admiral Vickery that his problem was " one of better loft work, more accurate layout and burning in the plate shop, more accurate slab work and subassembly work and finally better shipfitting on the ways." The best welders could not overcome the poor work of shipfitters.

The Lincoln contract was renewed for the last time in May 1944, and expired on August 28, 1944. There was less use of the " Fleet Fillet " process during these last months. In one of his last reports Mr. Lincoln estimated the extent to which his company's welding procedure was followed in thirty different yards. The " Fleet Fillet " technique had been dropped almost entirely " perhaps properly, in many cases," because of cracking difficulties. " It has been a very valuable producer," he said, " and could be made valuable still if properly applied with low carbon, low sulphur metal. It may be that the difficulty in getting proper parent metal is greater than the saving represented by the Fleet Fillet technique." [17]

The activity of the Lincoln Company was only one part of the attempt to improve welding. At the same time, Jack Wolff, the supervisor of training in the Commission's Division of Shipyard Labor Relations, was urging that the way to prevent bad welding was to require that welders be more adequately trained before they were graded as " journeymen " and paid the $1.20 per hour of that grade. As explained in chapter 13, the Commission's auditors checked over payrolls before reimbursing the companies, but the unions had arranged for rapid upgrading. The Commission hesitated to upset relations between the companies and the unions by insisting that promotion in grade come only after more thorough training. The result, as Jack Wolff saw it, was that thousands were made " journeymen " without

---

[17] Tyler, " Welding," p. 98. For the whole story of the Lincoln contract, see *ibid.*, pp. 79-100. On the personality of James P. Lincoln, see *Saturday Evening Post*, July 24, 1943, pp. 16-17, 51-52.

proper training. "The criterion is no longer the ability that a qualified welder should have—it is merely, ' I can weld as good as Joe Magee does and he gets a dollar twenty. So you have to pay me a dollar twenty.' Furthermore, that welder is no longer interested in extra efforts to increase his own welding skill because he is already receiving the highest rate of pay, and furthermore, the taking of additional training would be an admission on his part that he is not qualified to earn the rate he is already receiving." [18] Mr. Wolff made many recommendations for correcting the situation, but the Commission never applied drastically its power to refuse reimbursement in a way that would force the shipbuilding companies to improve training.

Something was accomplished by exhortations, to be sure, and the Commission's welding manual was a substantial help. Efforts to raise the standards in qualifying tests, in training methods, and in the acceptability of finished welds continued. A committee having fully the authority that comes from expert knowledge began visiting the yards in August 1944 to make concrete suggestions. This Welding Advisory Committee contained representatives of the Coast Guard, the Navy, the American Bureau of Shipping, and the Maritime Commission, and reported to a joint Board of Investigation whose most important function was research.[19] Need for more knowledge was one of the two basic difficulties in the way of the efforts to improve welding practices. The other difficulty was the lack among welders and welding supervisors of thorough training even in what was known.

III. Research and Retrospect

Scientific research required time, more time than was granted to the committee of experts which reported to the American Bureau of Shipping on the splitting of the *Schenectady*. That Committee had a preliminary report ready in about a month and even its formal report was made within two months. Consequently its report was only an analysis of the *Schenectady* failure by the application of existing knowledge, not as a result of new research.

The need for more scientific knowledge was driven home by the cracking of the *Esso Manhattan*. Why should a crack, once

[18] J. Wolff to Director, Div. of Shipyard Labor Relations, July 3, 1943 in MC Div. of Shipyard Labor Relations file, Training in Shipyards.
[19] Tyler, "Welding," pp. 111-12.

started in a defective weld, split the ship entirely in two? In addition to that pressing question, there were many others which scientists could not answer. Which of the qualities that differentiated the steel of various mills, or various heats, made the steel more or less liable to fracture? What changes took place in steel when subjected to the intense heat of an electric arc, or to a sudden drop in temperature, or to the stresses of a ship at sea? Search for this necessary information was undertaken by a committee entitled "Board of Investigation To Inquire Into the Design and Methods of Construction of Welded Steel Merchant Vessels, convened by order of the Secretary of the Navy." It represented the Coast Guard, the Navy, and the American Bureau of Shipping as well as the Maritime Commission. Since the Bureau of Marine Inspection and Navigation had been placed under the Coast Guard in March 1942, Admiral R. R. Waesche, Commandant USCG, felt that the Coast Guard had a direct and primary responsibility in the matter of cracking. On April 10, 1943, he requested his superior, the Secretary of the Navy, to set up an investigating board.[1] After consulting with Admirals Land and Vickery and having received their approval, Secretary Knox established such a board headed by Rear Admiral Harvey F. M. Johnson, USCG, Engineer-in-Chief, and having as its other members Rear Admiral E. L. Cochrane, USN, Chief of the Bureau of Ships, Rear Admiral H. L. Vickery, and David Arnott, Vice President and Chief Surveyor of the American Bureau of Shipping.[2] At its first meeting, April 28, 1943, the Board of Investigation agreed that "the failure of vessels and the rate of production should be considered as one problem and analyzed on the basis of calculated risk."

On the nineteenth of May a sub-board or "Working Committee" was set up to carry out the actual work of collecting information and making tests and investigations. The members of the board itself had no time for this kind of activity. The sub-board members were, for the most part, men who were already working on the problem. For example, Commander (then Lt. Comdr.) Philip A. Ovenden, USCG, had investigated and reported on the cracking of the *Esso Manhattan*. The Maritime Commis-

---

[1] Waesche to Secretary of the Navy, Apr. 10, 1943, gf 606-65.

[2] *Final Report* of this Board of Investigation (issued by the Navy Department, March 1947), p. 1; Tyler, "Welding," pp. 65-66.

sion members were J. L. Bates, J. W. Wilson, John Vasta and E. E. Martinsky. Commander R. B. Lank, Jr., USCG, became secretary for both the board and the subcommittee, while Captain C. D. Wheelock, USN, became chairman of the latter.[3]

By the middle of June the board and its subcommittee had worked out a plan for coordinating study by member agencies from six angles, as follows: (1) collection of data on casualties, to be done by Coast Guard, which had already begun this work; (2) review of design features and recommendation for improvements, primarily, the work of the Maritime Commission and American Bureau of Shipping with the assistance of studies to be undertaken at the Navy's David Taylor Model Basin; (3) review of steel specifications and inspection and test procedures, left to the Navy, Bureau of Standards, and steel manufacturers; (4) recommendations regarding erection procedures, to be made by a joint Maritime-Navy Committee; (5) consideration of means to improve the quality of workmanship, supervision, and inspection, Maritime Commission and American Bureau; (6) recommendations regarding the loading and ballasting of vessels, Navy and Coast Guard.[4]

As basic as any of these six procedures was a seventh, a series of controlled experiments designed to discover what occurred in steel during welding under various methods. The Board had no funds of its own for such research but the subject was being investigated by the War Metallurgy Committee of the National Academy of Sciences for the Office of Production Research and Development.[5] G. S. Mikhalapov, supervisor of welding research for the War Metallurgy Committee, became also chairman of the Research Advisory Committee to the Board of Investigation. His research organization gave coordination to a kind of research which had previously been going on somewhat haphazardly. The Welding Research Committee, sponsored by the American Welding Society and the American Institute of Electrical Engineers, had been active for some time, and in April 1942 the Commission had contributed to its work. From this Committee developed a

[3] Tyler, "Welding," pp. 101-02; *Final Report*, pp. 12-13.
[4] Tyler, "Welding," pp. 102-03; Harvey F. Johnson, USCG to Vickery, June 11, 1943, in MC gf 606-65.
[5] Senate, *The Government's Wartime Research and Development, 1940-44, Report* from the subcommittee on war mobilization of the Committee on Military Affairs, 79th Cong., 1st sess., subcommittee report No. 5, pt. 1, Jan. 23, 1945, pp. 227-30.

Welding Research Council on which Mr. Bates or Mr. Vasta was to represent the Maritime Commission. The Council conducted experiments on box girders and made other investigations. Research on welding was also conducted by some shipyards and by manufacturers of welding equipment and of steel. Gradually, through research conducted by many organizations but especially by Mr. Mikhalapov for the War Metallurgy Committee, various false notions were disproved and a firmer basis laid for the practice of welding. The corrective measures taken by the Commission in the summer of 1943 and the beginning of 1944 followed the lines laid down by the interim reports of these experts, and as soon as practicable their findings were also incorporated in the instructions regarding welding practices sent to the yards.[6]

Another series of tests, the most important, were devoted to stresses. The problem of residual or locked-in stresses involves the ability of steel to adjust itself to stresses by plastic flow. What is called biaxial or triaxial stress prevents plastic flow at points of high stress. Although this was known in a general way, the physical changes in steel under stress were not known with any precision. In an attempt to supply this need, G. S. Mikhalapov planned a series of tests on ships under construction. In approving of this project the Advisory Committee of the Board of Investigation asked that it be extended to include the history of stresses not merely during construction but also until the completion of the vessel's first voyage. The Maritime Commission approved this addition to the cost of construction of two tankers and two cargo vessels and agreed to assist in the taking of stress measurements.

In January 1944, the Advisory Committee recommended that the project be broadened to include the taking of strain measurements on many more vessels than originally contemplated. The Commission's assistance was again required since this plan involved the attachment of gauges for a period of about five or six hours just prior to delivery and again, on the same vessels, after they had been in service for some time.[7]

After these controlled experiments on welded ships, after many studies of steels and of welding, and after full statistical analysis

[6] Tyler, "Welding," pp. 102-23.

[7] Mikhalapov to Bates, July 3, 1943 and Vickery to National Academy of Sciences, Attention Mr. Mikhalapov, July 23, 1943; Mikhalapov to Bates Oct. 12, 1943 and Vickery to Flesher, Nov. 6, 1943; Vasta to Bates, Jan. 17, 1944; H. F. Johnson to Land, June 8, 1944 and Land to Johnson, Aug. 1, 1944, all in MC gf 606-65.

of all recorded fractures, the Board of Investigation released a report in March 1947. Its conclusions then were in marked contrast to the statements formulated in March 1943 by the committee of experts of the American Bureau of Shipping. The Board's *Final Report* makes the flat statement, " Locked in stresses do not contribute materially to the failure of welded ships." [8] The word " materially " covered a compromise. No one denied that locked-in stresses were part of the picture; the reason for saying that they did not contribute materially was that locked-in stresses were present in all ships, in those which suffered no casualties as well as in those that split.

The irrelevance of erection sequences was also stated baldly by the committee. " The magnitude of these stresses it was found, was generally unaffected by variation in the welding procedure or assembly sequence." [9] There were general reservations in the minds of at least some of the experts, as is indicated by a footnote in an appendix, which reads: " It should be noted that it is impossible with present strain gage techniques to determine locked-in stresses at structural discontinuities, e. g., hatch corners." [10] But instead of framing the problem so as to focus attention on residual stresses, the report focused on structural discontinuities and on a quality in steel which is called " notch sensitivity."

If there is a notch in the edge of a strip of steel and the steel is subject to strain, there is a concentration of stress at the notch which may produce cracking at that point. Lower temperature increases the notch sensitivity. In many of the cracked ships a poor weld or some feature of the design—notably the square hatch corners on the Liberty ships and the cut in the sheer strake—or a combination of one of these mistakes in design with a poor job of welding, had formed a notch at which stress had concentrated beyond the strength of the steel to resist.[11] A crack once started had the effect of a deeper notch. Looked at in this way, the problem had two main solutions: eliminating from the design the structural discontinuities, that is, features which in effect formed notches; and securing steel which was not notch sensitive within the temperature range of ship operation.

[8] *Final Report*, p. 10.
[9] *Ibid.*, p. 7.
[10] *Ibid.*, p. 138, note c.
[11] *Final Report*, p. 4. Defective welds were the starting point of the cracks on the *Esso Manhattan* and the *Schenectady. Ibid.*, p. 57.

In full, the conclusions of the Board read:

(a) The fractures in welded ships were caused by notches and by steel which was notch sensitive at operating temperatures. When an adverse combination of these occurs the ship may be unable to resist the bending moments of normal service.

(b) The serious epidemic of fractures in the steel structure of welded merchant vessels has been curbed through the combined effect of the corrective measures taken on the structure of the ships during construction and after completion, improvements in new design, and improved construction practices in the shipyards.

(c) Locked-in stresses do not contribute materially to the failure of welded ships.

(d) Existing specifications are not sufficiently selective to exclude steel which is notch sensitive at ship operating temperatures.

(e) A tendency for certain ships to incur repeated casualties can be measured but the trend is not great and the effect is not significant.

(f) The basic analytical method used in calculating nominal stresses in the main hull girder under a known bending moment is valid.

(g) The overall strength of the Maritime Commission ships is satisfactory.[12]

This report carries the signatures of Admirals H. F. Johnson, USCG, and E. L. Cochrane, USN, Captain T. L. Schumacher, USN, Technical Assistant to the Chairman, USMC, and D. Arnott, Vice President and Chief Surveyor, American Bureau of Shipping, and it was approved by James Forrestal, Secretary of the Navy. The actual editing was done by Commander R. D. Schmidtman and Lt. Commander E. M. MacCutcheon of the Coast Guard. For the most part it was a cooperative effort of the sub-board and represents, therefore, the composite of majority opinion of the experts who worked on welding problems.[13] It did not analyze the Navy's experiences with welding nor instances in which naval vessels broke in two (the reason for secrecy concerning warships is obvious) ; but a Board composed so largely of naval officers must, in its thinking, have taken account of welding practices in yards working for the Navy and of the performance of welded naval vessels.

[12] *Ibid.*, p. 10. A reprinting of this report, "Final Report of a Board of Investigation Convened by Order of The Secretary of the Navy to Inquire into The Design and Methods of Construction of Welded Steel Merchant Vessels, 15 July 1946," is in *The Journal of the American Welding Society*, vol. 26, no. 7 (July 1947), pp. 569-619.

[13] Interviews D. B. Tyler with Messrs. E. E. Martinsky and J. Vasta, Apr. 1947.

Surveying the record of the welded steel merchant vessels built by the Maritime Commission and considered in its report, the Board found that of the total of 4,694 only 25 sustained a complete fracture of the strength deck or bottom. Of those 25, 8 were lost at sea and 4 additional vessels, in which are included of course the *Schenectady* and the *Esso Manhattan*, broke in two but were not lost. The total lives lost due to structural failures of welded ships was 26.[14] There were no serious fractures of Victory ships nor of Liberty ships on which the corrective alterations had been made to provide crack stoppers. Compared to what one would expect from reading headlines about crackings during the winter of 1943-1944, the loss through structural failure of only 8 vessels out of 4,694 is a surprisingly good record.

The Board's conclusion after a full statistical study of the record of ships built in different yards was in direct contrast with the tendency in 1943-1944 to point the finger of scorn at particular shipyards. The Board found no marked correlation between fractures and the construction practices of the shipyards in which they were built. A higher than average incidence of fractures was shown by certain yards on the Gulf and West coast, but in view of the different conditions of service of different ships, conclusions from these averages would be questionable. Although it attributed little causal importance to the general system used from one yard to another, the final report asserted emphatically the importance of good welding. A defective weld at a point of stress due to structural discontinuity was found in case after case to be the starting point of the fracture.

Since much research remained to be done on welding as well as on the notch sensitivity of steel, the Board recommended continuation of the research it had been conducting. It recognized fully that its " final report " was not the final word on the subject of cracks in welded ships; its research functions were taken over by a new Ship Structure Committee.[15]

Foolhardy as it is for anyone not an expert in these technical questions to attempt a summary, the economic historian accus-

[14] *Final Report*, pp. 3 and 21.
[15] Tyler, " Welding," pp. 124-29. Study of welding stress was among the research projects to which the Maritime Commission was contributing in 1946-47, but it spent very little on research of any kind. See U. S. President's Scientific Research Board, John R. Steelman, Chairman, *The Federal Research Program* (Washington, 1947), I, 295-97.

tomed to hear our times described as an age of scientific technology can hardly conclude without remarking on the relation in this story between action and knowledge. When the emergency building began in 1941, relatively few all-welded ships had seen service.[16] An all-welded coaster 150 feet long was built in Britain in 1920, and welding was used increasingly in the twenties on larger ships, notably to save weight on the cruisers limited by treaty to 10,000 tons; but the tanker *J. W. Van Dyke* launched in 1937 is generally counted the first large welded merchant ship. She was 521 feet long and was riveted at the extreme ends.[17] In 1940-1941 the welded ships built for the Commission's long-range program were just beginning to see service. In spite of such slight experience with the performance of welded ships, that method was adopted for most of the 50-million-ton program. It became an essential in the methods of multiple production which turned out ships with phenomenal speed.

All this was done with almost no scientific knowledge of what happened to steel under the intense heat of the electric arc or what strains arose in a ship which by welding was made all one piece of steel. To explain the sensational crackings of the tankers *Schenectady* and *Esso Manhattan*, the experts talked about "locked-in" stresses which were "relieved" by the crackings—figures of speech used to describe the unknown, just as psychiatrists describe the mysteries of human personality by talking about the need of relieving inhibitions. Small wonder if at that time those responsible for the program had had nightmares about the sailings of schizophrenic ships likely to burst at any moment to relieve their split personalities! More soberly, it may be said that so long as the cause of crackings remained unknown there were grounds for worry over what would happen next.

But action could not wait for knowledge. Production reached its peak before there was any general agreement on the reasons why, in a few startling instances, two or three ships broke completely in two in calm weather. During this period of feverish construction, programs for improving welding practices could go only a certain distance; and then they bogged down in contro-

---

[16] In this connection, see the testimony of S. K. Smith, surveyor for the American Bureau of Shipping in Truman Committee, *Hearings*, pt. 23, pp. 10111-14.

[17] Tyler, "Welding before World War II," a typescript in Historian's Collection, Research file, 204.3.

versies arising from ignorance about what welding did to steel. Nevertheless, in spite of the lack of scientific explanations, practical measures were taken. The girders which reinforced the decks of tankers, the removing of the cut-out in the sheer strake of Liberty ships, and the rounding reinforcements of hatch corners, the angle bars, slots, and riveted straps which acted as crack stoppers provided practical remedies. These " crack stoppers " represented a slight return, but only a very slight return, towards the use of riveting. The final report of the Board of Investigation admitted " the mechanism of fracture is still not clearly understood," but looking to the future, it announced its conviction that crack arrestors could be dispensed with and that it would be possible, given proper design in all details, high quality of workmanship, and steel of low notch sensitivity, to build without them a satisfactory all-welded ship. Looking to the past, it said that " if welded construction in the building of both merchant and naval vessels had not been adopted at the outset of the program, the extraordinary results in speed and volume of construction would have been impossible of accomplishment." [18]

[18] *Final Report*, p. 10.

CHAPTER 17

THE VICTORY SHIP

I. THE DESIRE FOR FASTER CARGO CARRIERS

IN 1943, while the yards were setting their all-time record in the production of Liberty ships, the schedulers and designers in the Maritime Commission were making preparations for the production of various new types, of which the most important was the Victory ship, a faster type of cargo carrier. It was to be constructed in the emergency yards. Plans for the new type were well advanced even before crackings placed the Liberty ship under a cloud and they were not affected substantially by the rumors which maligned the Libertys. The plans were not put into effect immediately because they were opposed by the War Production Board. Its opposition led to a violent controversy between the Maritime Commission and the WPB, a controversy finally settled by the Joint Chiefs of Staff. At the end of 1943 the Commission's program had become more diversified and more closely meshed with specific military needs.

The extent to which standardization, the dominant thought at the inception of the emergency program, had been compromised even before the end of 1942, is shown by the Commission's program No. 11 formulated on December 17, 1942. It scheduled for delivery in 1943: 230 standard-type of dry-cargo carriers (C1, C2, C3), 209 tankers, 16 ore carriers, 69 frigates, 50 escort aircraft carriers, 54 landing ships, a number of types of Army and Navy transports, and 1,262 Liberty ships, in addition to minor and very special types such as the concrete and wooden ships. The effect of the original emphasis on standardization showed in the predominance of Liberty ships.[1] Deliveries of the new Victory type would not come until 1944 or possibly late 1943. By that time the Liberty ship would no longer bulk large in the Com-

---

[1] Program No. 11, in MC, Historian's Collection, Research file 117, from Production Division, Statistical Analysis Branch.

mission's program and standardization would to that extent be sacrificed.

One reason for this change was the steel shortage. The fastest yards were going to be forced to slow down anyhow in order not to run out of steel. Consequently, the Commission planned to put the steel allotted it to more efficient use by building faster ships.[2] When lack of steel limited the program for 1943 to 16 million tons, the Commission's program called for building as much as possible of that tonnage during the first half of the year, for the carrying capacity was needed as soon as possible.[3] During the later months of the year, when production would have to slow down anyhow in order not to use up more steel than had been promised for the year and in order not to exceed the Presidential directive, yards could afford the time to change over to a new type. The planning for this change began as early as August 1942. At that time standardization had already fulfilled its main purpose by enabling facilities and labor force to expand to such an extent that shipyards could use all the steel available and more.[4]

Another reason for the change, a reason that might have caused diversification in the program even if there had been no steel shortage, was the cumulative effect of demands for faster ships. Many demands for specific faster types were made by the Navy, not only for definitely military types such as the escort aircraft carriers, but also for C2 or C3 cargo ships to be converted into tenders or some other kind of auxiliaries.[5] For use as cargo carriers also, faster ships were being demanded, both for wartime use and for postwar competition. The 11-knot Liberty ships never would have been built at all if propulsion machinery could have been secured for faster vessels. As Admiral Vickery said, " It was a case of those ships or none at all. We are hopeful that expansion of turbine and gear production facilities may relieve the situation somewhat and we will be able to build more of the better ships." [6]

Newspaper opinion reflected and added to the criticism of

[2] Land (written by Searls) to Nelson, Aug. 25, 1942 in MC Historian's Collection, Douglas file, " Construction," and above pp. 332-33.
[3] Statements by G. J. Fischer. July program minus military and minor types.
[4] See above, chap. 10, sect. II, and Admiral Vickery's retrospective statements in House, *Independent Offices Appropriation Bill for 1945, Hearings*, pp. 705-06 (Dec. 9, 1943).
[5] The Land file in MC, Historian's Collection contains many such requests. See chap. 18.
[6] MC Press Release No. 991, Aug. 14, 1941.

the slowness of the Liberty ship. It became somewhat strident as submarine and bombing attacks became more menacing. An article in the Washington *Star* by Richard L. Stokes was headlined, " Held Ideal Target for Stukas; Maritime Commission, Having Acted in Haste, Now Has Leisure to Wonder." [7] Shipping interests were also unhappy about the construction of slow vessels in large numbers. J. Lewis Luckenbach, President of the American Bureau of Shipping, voiced this feeling in his semi-annual report given on July 29, 1941 in the presence of Admirals Land and Vickery. He contended that the Liberty ship was not the most suitable type for the emergency. He supported this opinion by quoting from editorials in the British publications *Motorship* and *Marine Engineer* which asserted that England would lose the peace if she overproduced slow and uneconomical vessels. *Marine Engineer,* asserting that 16-17 knot vessels would do twice the work of slower ones, quoted Lord Rotherwick's plea, " Now that we have got into our war stride the advantage of the fast cargo ship as against one of the ' box ' type cannot be gainsaid." [8] Mr. Luckenbach emphasized that he was quoting " from the press of a country more involved in this war than we are and where it is agitated that ships of a faster type be constructed. Apparently others are thinking ahead and I strongly recommend that we do the same as soon as facilities under construction for the production of turbine machinery will permit." [9]

When toward the end of August 1942, Sir Amos Ayre, Director of Merchant Shipbuilding in the United Kingdom, came to America and made a tour of the shipyards,[10] Admirals Land and Vickery learned about the fast standard cargo vessels then under construction in England.[11] The British decision to build faster ships was preceded by a careful consideration of such factors as deadweight, labor and material requirements, available berths, routes, number of voyages, and so forth.[12] They found that vessels of from 12 to 15 knots were frequently put into 10-knot convoys,

[7] Washington *Star*, July 2, 1941.

[8] *The Journal of Commerce*, July 30, 1941.

[9] *New York Times*, July 30, 1941.

[10] Vickery to Clay Bedford, Aug. 28, 1942, MC, Historian's Collection, Weber reading file.

[11] Land to Lane, May 11, 1949.

[12] Sir Amos L. Ayre, " Merchant Shipbuilding During the War," in Institution of Naval Architects, *Transactions*, vol. 87 (1945), p. 3.

which meant that their extra speed was wasted. Therefore, the British concentrated on a 15-knot vessel of about 10,000-12,000 tons deadweight with single screw and Diesel engine. Later on, two such engines with twin screw were used on larger vessels, giving a speed of 17 knots. The production of geared-turbine sets was increased through simplification and standardization.[13]

Shortly after Sir Amos Ayre's visit, Admiral Vickery had the Technical Division start work on the design of a new vessel of 15 knots. The approach to this new design was from the Liberty ship rather than from the C-type ship because the simplicity and standardization of the former type would facilitate production in quantity. From the start, it was decided to have the new vessel roughly equal the 10,000 ton deadweight of the Liberty ship.[14]

At this time Mr. Bates received a letter from Mr. Luckenbach, President of the American Bureau of Shipping, enclosing plans of a ship which he proposed as a substitute for the Liberty (EC2). The lines were those of the Commission's C1-B vessel with 12′ added in the middle. If powered with "Uniflow" engines, which he considered the best then available, Mr. Luckenbach thought the vessel would do 13 knots with capacity for an extended overload in case of emergency and would be more economical than the ordinary reciprocating engine at reduced speed. In closing, he said, " I hate to see the country loaded up with too many EC2's—all duplicates." [15] When informed that the Commission was at work on plans for a 15-knot vessel Mr. Luckenbach replied that he had not contemplated one of that speed being " afraid we could not get away with it from the machinery angle but the bigger and faster they are the better and more economical they would be for the work intended." [16]

II. The Hull Design of the Victory Ship

While the main problem, propulsion machinery, was still unsolved, the Preliminary Design Section of the Technical Division worked out a basic design for the new vessel, allowing for slight modifications in case reciprocating engines, Diesels, or turbines

---

[13] *Ibid.*, pp. 13-14.
[14] Bates to J. Lewis Luckenbach, Sept. 18, 1942, in MC, Technical Division files, A.B.S., pt. 3.
[15] Luckenbach to Bates, Sept. 8, 1942, same file.
[16] Luckenbach to Bates, Sept. 25, 1942, same file.

were to be used. From the start it was clear that no such box-like design as that of the Liberty would serve for a ship designed to make 15 knots. Assuming a reciprocating engine of 5600 horsepower and about 100 revolutions per minute, the basic dimensions of a design called the AP1 were worked out in September 1942, as follows:

LWL	445'
Beam	63'
Depth	38'
Designed draft	28'
Designed speed (80 percent of designed power)	15 knots
Designed power	5800 ihp.
Number of holds	5
Type of accommodation	Same as EC2-S-C1

After the plans had been worked on for a month Mr. Bates, Director of the Technical Division, showed them to John E. Burkhardt and James B. Hunter at the Quincy shipyard of Bethlehem Steel. The Bethlehem people agreed to assist in the preparation of the new design and, if their work-load permitted, to prepare working plans. They also agreed to make a plating model while the Commission prepared the lines, secured Model Basin results, and made calculations of weight, stability, and so forth. Mr. Bates commented after this meeting, "We must have a simplification party as soon as may be. I can see the design 'improving' right along already."[1]

At a conference with his staff on November 21, Mr. Bates reviewed the characteristics agreed upon and repeated the guiding principles laid down by Admiral Vickery. Deadweight must equal that of the Liberty ship and the speed must be 15 knots. As regards improvements, "they are not only not desired, but are not to be accepted, unless it can be shown that they are essential to acceptable performance, and at the same time that they can be incorporated without prejudice to prompt procurement." The design should "be generally such as to utilize to the greatest possible extent the principles found effective in the production of the present Liberty ship." One important change was made in

[1] Bates report of trip to Boston, Nov. 4, and New York, Nov. 5, 1942, in MC, Technical Division, Preliminary Design Section Correspondence, file VC2-S-AP1, 1942-43. Also David B. Tyler, "The Victory Ship Design," pp. 18-19, in MC, Historian's Collection, Research file 111. Tyler's research is the main basis for this chapter.

the vessel's dimensions. Its beam was changed from 63′ to 62′.[2] This difference of one foot may appear to have been inconsequential, but the change affected the other dimensions and also involved the question of stability. It was determined by the discovery that several of the yards, particularly Bethlehem-Fairfield, had ways that were too narrow for construction of a ship with a 63′ beam.[3]

It was not easy to accept the ban on improvements. The War Shipping Administration asked that the new design provide for greater loading on deck than was permissible on the Liberty ships. This required the use of steel shapes other than those which the steel companies were accustomed to supply. The objections of the steel companies were finally overcome by Admiral Vickery, however, after discussions with the officials of WPB. Provision was made thereby for larger permissible deckloads.[4]

As discussion of details progressed, the need for improvements became more insistent. It was a natural desire of the designers and planners to have the new vessel move away from the Liberty toward the C2; they thought it desirable to add such items as searchlights and gyrocompasses, for example, if they were avail-

[2] Notes on conference on new EC2 design, Nov. 21, 1942, in MC, Technical Division, Preliminary Design Section Correspondence file, VC2-S-AP1, 1942-43.

[3] Wanless to Vickery, Sept. 23, 1942, in Preliminary Design Section Correspondence file, EC2-S-C1.

[4] Ivan J. Wanless to F. C. Lane, Oct. 29, 1950 and Nov. 16, 1950. After consultation with Arthur C. Rohn, Mr. Wanless wrote, concerning the request for greater permissible loading on the decks: "This request forced the Preliminary Design Section to an angle larger than accepted steel mill practice or to split I beams to T's. This would entail considerable manpower expenditure for splitting and straightening, either at the mill or in the building yard. The new shape in question was a 9″ x 4″ angle. The steel companies protested that the cost of the rolls would be excessive, and that for a few vessels it would be far cheaper for the Government to order I sections and have them split and straightened. When the final arguments of steel conservation, manpower utilization, and probable number of ships that could be built were consolidated, the case was presented to the steel companies, with the request that the new 9″ x 4″ angle be provided. This request was again refused.

"Discussions between Sandmeir and Wanless developed that the steel companies would need facilities and manpower for straightening the economical I sections required to conserve steel. The case was then presented to Admiral Vickery for presentation to WPB. After discussion with the WPB on these two points, the steel companies agreed to provide the new rolls, to meet production requirements, and to absorb the cost. They were assured by the Maritime Commission that the orders for the Victory ship program would be of sufficient volume to insure no loss to them, and that this new shape would be of commercial value for both shipbuilding and other structural projects in the future."

able. In January 1943 Admiral Vickery approved the general policy of improving on the EC2 design whenever conditions of manufacture and design were such that no delay in procurement would result. Longer booms and improved winches were decided upon on this basis. Other features were dictated by military considerations—for example, lifeboat davits capable of handling landing craft.

While the form of the hull was being worked out, the characteristics of the propulsion machinery were still in doubt, as will be explained in the next section. The hull lines had been planned on the assumption that the ship would have an engine of 6000 shp. and a speed of 15 knots. It became apparent, however, that turbines of 8500 shp. might be available for some ships of the new type. Consequently, Admiral Vickery asked that basic plans for the machinery space provide that the bolting arrangements be the same no matter which engine was used, and that the hull lines be such that the ship could also operate economically at the higher speeds appropriate to the more powerful propulsion machinery. Designing the lines of the hull so as to provide for this range in the power of the engine required some sacrifice in optimum performance. As finally developed, the Victory Ship used slightly more power at 15-knot speed than did the C2; but, in spite of its greater deadweight tonnage, it used less power than the C2 at speeds of 16 knots or more.[5]

For working plans for the new type, the Commission turned first to the Shipbuilding Division of the Bethlehem Steel Co. Late in December 1942 the Commission's staff was told it was expected that 1,600 of the new type would be built with plan approval and procurement in Washington and with working plans being prepared by Bethlehem. In January it was announced that Bethlehem was too busy with its work on the troopships being built at Alameda, and that George G. Sharp's office would prepare the working plans for the new type. At a meeting in that office on January 20, 1943, with representatives of several yards, Mr. Bates explained the reasons for the new design and expressed the desire of the Commission to secure basic agreement on working plans to be developed by Sharp, it being understood that the individual yards could modify the plans to suit their particular methods of construction.

[5] Wanless to Lane, Oct. 29, 1950, and Nov. 16, 1950 in MC, Historian's Collection.

As might have been expected, since this was just four days after the spectacular cracking of the *Schenectady*, there was a great deal of discussion on the subject of welding. Mr. Aldrich of Bethlehem said his company still preferred to rivet shell seams while the others preferred to weld them as well as the frames. It was decided to shorten the bilge keels or eliminate them altogether if model tests indicated that could be done. Mr. Hunter of Bethlehem suggested a riveted stringer angle but Mr. Arnott of the American Bureau of Shipping thought it unnecessary, since the plates were not excessively thick, and since it would indicate a defeatist attitude towards welding. Immediate plans were made to eliminate the cut in the sheerstrake at the accommodation ladder. It was decided to rivet bulwark plates to sheerstrake and to exercise great care in the welding sequence used on stringer plate connections.

It was agreed that the design agent should prepare drawings for sections to accommodate the smallest crane capacities since the better equipped yards could put together and handle multiples of these small sections.

The yards' representatives, in the interest of uniformity, expressed a general desire to have all changes, after construction had begun, come direct from Washington. They also suggested that an effort be made to correlate the requirements of the Bureau of Marine Inspection and Navigation and the Navy.

By the end of March 1943 the characteristics and specifications for the new emergency vessel were settled.[6] Since the new vessel grew out of the Liberty ship, the basic idea being a vessel of the same deadweight but with 15-knot speed, it is interesting to compare it and the Liberty (Table 2). They were both full scantling ships with the machinery space between holds No. 3 and No. 4, and a total of five holds. Cargo handling gear was also similar, consisting, on the new ships, of fourteen 5-ton booms, one 30-ton, and one 50-ton boom. Unlike the Liberty ship, it had an extra deck in holds No. 1, No. 2, and No. 3 for the stowage of package goods, and used electricity rather than steam for anchor windlass, for steering, and for some pumps.

In appearance the ships were similar, with the single large house amidships. However, on the new ship, in order to simplify construction, a straight line sheer was adopted forward and aft

[6] Tyler, "Victory Ship Design," pp. 19-25.

and a forecastle added to prevent the sheer from being excessive forward and to make the vessel more seaworthy. Another change, in the interest of decreasing construction manhours, was the spacing of frames at 36" instead of 30". While this reduced the number of frames it required the use of heavier ones and resulted in a slight increase in steel weight.

The deep tank in No. 1 hold on the Liberty ship had never worked out very well for ballast or other purposes and was eliminated on the new ship. Its volume was added to the two large tanks at the forward end of No. 4 hold. These tanks could be used for fuel oil, dry cargo, or salt water ballast. Wing tanks at No. 5 hold were retained for use with fuel oil or salt water ballast only.

The crew was fixed at 51 as against about 44 on the Liberty and in both types of ships they were berthed in the midship house. Twenty-nine gunners were berthed in the poop deck house which also contained the hospital. On the Liberty they were all amidships. Profiting by experience in constructing the EC2, the houses were arranged in four complete sections so as to assist prefabrication.

Armament consisted of a 5" gun aft and a 3" gun forward. The latter was on an emplacement capable of supporting another 5" gun if the guns should be available. There were also eight 20 mm. guns—four amidships, two on the forecastle, and two on the poop deck.

In this new emergency ship, one-compartment flooding and damage stability were achieved without the use of fixed ballast. With a view to its use after the war, an effort was made to achieve a balance such that the vessel would not be too stiff after the removal of defense features. In this connection it will be remembered that the Liberty ship was designed before it was decided to add defense features.

The increased dimensions and speed of the new vessel called for different lines. The bow was semi-V-shaped, the stern was the "cruiser" type, and there was a 15 per cent parallel middlebody. As in the case of the Liberty, construction was facilitated by taking special care to avoid furnaced plates.[7]

---

[7] Tyler, "Victory Ship Design," pp. 25-27.

III. The Search for Engines

The propulsion machinery most satisfactory for a hull of this type would be turbines, provided the supply were adequate. The foresight which the Commission had shown in June 1941 in building new plants for the manufacture of turbines was showing results in the second half of 1942. As early as May 22, 1942, General Electric had begun delivery of C3 turbines from a site which had been only a barren swamp a year before. But most of the turbines from the new plants were earmarked for standard-type freighters or tankers. Yards which had been first planned for Liberty ships were changed over to faster types as rapidly as turbines were available.[1] To meet military and shipping needs, Alabama Ship, Marinship, and Swan Island became tanker yards.[2] Then, when more turbines and gears from the new plants became available, North Carolina Ship was changed over from Libertys to C2's in accordance with the original plan that turbines from the new factories would be used to build more of the standard types.[3] Some turbines of the C3 type might be available for use in the new type but not nearly enough to equip the 500 to 800 ships a year which were being planned.

Accordingly, there was no prospect of having enough turbines for all the ships of the new design; and when on September 9, 1942, Mr. Schmeltzer told the Engineering Section that they must find a way of raising the 2500 hp. of the Liberty ship to 5600 hp. for the new standard vessel, he said it would be necessary to stick to the reciprocating engine. Drawings must be prepared and a pilot engine built and tested; they would then be ready to go into quantity production. The matter was most urgent.[4]

Obtaining a satisfactory reciprocating engine of 5600 hp. was not going to be easy. The Skinner Uniflow, a 5600 (or 5400) hp. single expansion reciprocating engine which had been invented about the beginning of the century and had reached its greatest

---

[1] Vickery's memorandum on the Victory Ship cited below on p. 593, note 18; Tyler, "Victory Ship Design," pp. 6, 9-10.

[2] *Ibid.*, and Reynolds' Way Charts; for Swan Island, MC Minutes, pp. 22180-82 (June 11, 1942).

[3] MC Minutes, p. 23789 (Dec. 17, 1942).

[4] The following account of the development of the Lentz engine is taken from Tyler, "Victory Ship Design," which is based in part on interviews with Alan Osbourne.

popularity about 1910, was considered. The objection was made that it had lubricating troubles when used in salt water. This difficulty was eventually overcome, but the Skinner Uniflow being used for the twin-screw " baby flat tops " had 160 revolutions per minute, a speed much too high for use on a large single-screw cargo ship. The Sun Doxford Diesel used in the 10,000 ton C2-SU vessels was also considered. In the end, it was decided that the German Lentz engine was best suited for quantity production. It was no easy solution because no Lentz engine of the size required had ever been built. An important advantage was that, once production was under way, the Lentz engine would take fewer manhours to build than other engines of equal power.

This engine had been developed in Germany as an answer to the turbine. It is a "double compound" engine, consisting of two halves, each forming a complete and self-contained two-cylinder two-crank compound engine, the halves being connected only by the crankshaft. The two high-pressure cylinders are in the center and the two low-pressure ones forward and aft. Its economy and efficiency were proved. Although little used in this country, it had been used in a number of vessels built in Europe, and one of these had been purchased abroad by the Standard-Vacuum Co. The American Ship Building Co. of Cleveland was the only American company licensed to build these engines. Some were being put in the Great Lakes ore carriers being built for the Commission. Previously the American Ship Building Co. had done no more than "Lentzify," that is modify American vessels by installing Lentz cylinders and valves.

Because of the engine's German origin and very limited use in this country, the Commission could not avoid delays and difficulties in production. This was true even in the initial stage of making drawings. Since the American Ship Building Co. did not have enough men to spare for the job, it was decided to borrow draftsmen from other companies, especially those companies which would undertake to build the engine later on. The preparation of drawings was put under the direction of two engineers from the American Ship Building Co., one of whom, Paul Meidlich, owned some of the rights to manufacture. A new subsidiary company, the Lakeside Engineering Co., was organized for this particular work.

The legal question proved to be so involved that the contract with the Lakeside Company was not finally approved by the Commission until November 27. It permitted the company to retain ownership of the plans, subject to their use by the Commission. The license of the American Ship Building Co. extended only to the Great Lakes area and the patents involved were vested in the Alien Property Custodian. These technicalities caused delay as did the difficulties involved in setting up a new organization composed of men from various companies. Furthermore, the first engine had to be manufactured at still another establishment, the General Machinery Co.

The engine specifications as determined at conferences of members of the Technical Division held early in November provided that the engine should be of 5500 hp., 85 rpm., and have a 59″ stroke. This permitted the crankshaft to be lined up in other than the largest lathes available and kept the weight and dimensions of the bedplates from being too great. After the conference Mr. Bates made the following comment:

> It was accepted by this conference that such an engine would be a reasonable manufacturing possibility. It was also accepted that it would give reasonable expectation of securing the 15 knots sustained sea speed with an acceptable margin, although this margin would not be equal to that which the Commission has been securing in its recently designed ships.[5]

In December arrangements were made with the Navy to carry out tests of the first engine. Mr. Grant hoped that an exhaust turbine would increase the power to 7000 ihp. thus raising the speed to 16 knots.[6] It was expected that the tests would be completed in March 1943. Early in that month production studies were begun at the Bartlett-Hayward Division of the Koppers Co., at Joshua Hendy Iron Works, the General Machinery Co., Willamette Iron and Steel Corp., and the Filer & Stowell Co.[7]

While planning for the use of Lentz engines in most ships of the new type (the EC2-S-AP1), the designers also completed plans for the use of other engines. The designation EC2-S-AP2 was given to a design providing for a 6000 shaft hp. C2 turbine; the EC2-S-AP3 was the same vessel when provided with a 8500

---
[5] Bates, Memorandum for Files, Nov. 11, 1942, in Preliminary Design Section Files, Correspondence, VC2-S-AP1, 1942-43. These conferences were held on Nov. 2 and 3 and were attended by Bates, Schmeltzer, Grant, Osbourne, and Rohn.
[6] Wanless to Bates, Jan. 16, 1943 in file just cited. Tyler, "Victory Ship Design," p. 22.
[7] Ibid., p. 18.

shaft hp. C3 turbine; the EC2-S-AP4 a design using Diesel propulsion. The final separation of the new ship from the Liberty ship in the minds of its creators was signalized by changing its design designation from EC2 to VC2 on April 28, 1943, and by general adoption of the name " Victory Ship." [8]

By that time contracts had already been awarded for construction of the new type: on April 20, 1943 to Oregon Ship and Calship for deliveries of AP3's, with the C3 turbines, to begin in April 1944; [9] and on April 22-May 6, to Bethlehem-Fairfield, Richmond No. 1 and No. 2, Delta, Houston, and Southeastern, for AP1's, with the Lentz engines.[10]

Before the Commission's plans had reached this point, they collided with the plans for standardizing and scheduling which were being formulated in WPB. From January to May 1943 the preparation of working plans for the Victory was going forward in the office of George Sharp, the firm of naval architects which had been most closely connected with the Commission's designs of standard types—but not with the Liberty ship, for which, it will be recalled, the working plans were provided by Gibbs & Cox. Propulsion machinery was to be provided by turbines and Diesels so far as they were available, but since it was believed they were not available, chief reliance was placed on the Lentz engine, a type of reciprocating engine which was thought by some to have possibilities, if once developed, of competing with the turbine, but was as yet untested. This was the situation when William Francis Gibbs and the WPB began to interfere.

One result of that intervention was abandonment, as will be explained shortly, of plans to use the Lentz engine. Therefore its story is best completed here. The decision not to use it was made long before one had been completed for testing. Therefore its testing was delayed. The pilot or test engine reached the Naval Boiler and Turbine Laboratory at Philadelphia in September 1943 and by the time the tests were nearing completion the first Victory ship was delivered. After completing 85 hours of a test with only 4 more to go, the Lentz engine suffered a " local " break.[11] The foundrymen at the General Machinery Co. were not

[8] *Ibid.*, pp. 27-28.
[9] Minutes, pp. 24909-11.
[10] Minutes, pp. 24968-69, 25003, 25084.
[11] Grant to Lakeside Engineering Co., September 28, in MC, Technical Division

PLATE XXI. Above, Victory ships crowd the outfitting docks at Calship while being completed as attack transports. Center, the Victory cargo ship (VC2-S-AP2). Below, a Victory transport (VC2-S-AP5) under way.

PLATE XXII.

Among the 125 ships built by the Kaiser-Vancouver yard (left), across the Columbia River from Portland, Oregon, were 30 LST's (design S3-M2-K2), tank-carrying landing ships like that shown above just after launching, and 50 escort aircraft carriers (design S4-S2-BB3) like that shown below on its trial run.

experienced in the kind of castings required. After the break it was found that the cast iron had the normal grey iron flaky distribution of free carbon instead of the required modular distribution as in high strength cast irons, and that this was the cause of the break.[12] In regard to the potentialities of large-sized Lentz engines, it is well to note that they were built by the Germans during World War II for 9000-ton ships.[13] How satisfactory Victory ships powered with properly built Lentz engines might have been was not put to the test, for the controversy over the Lentz engine between WPB and the Maritime Commission created a situation which put the turbine companies under pressure to find a way of producing turbines for the Victorys.[14]

IV. COLLISION WITH THE CONTROLLER OF SHIPBUILDING

The chief problem of WPB during 1942 had been materials and their allocation; in 1943 it was production scheduling. A dominant role in WPB was being taken by the Production Vice Chairman, C. E. Wilson. His prominence in WPB reflected the emphasis on scheduling.[1] As part of his program for increasing production through standardizing components and smoothing the flow by careful scheduling, he brought into WPB to take charge of turbines and ship production the celebrated naval architect, William Francis Gibbs, who on December 19, 1942 became Controller of Shipbuilding. While Mr. Gibbs was the protagonist in the ensuing contest with the Maritime Commission, Mr. Wilson was well informed on the industrial matters involved since he was President of the General Electric Co., one of the leading manufacturers of turbines.[2]

files, Miscellaneous, S41; also, statements by J. E. P. Grant to D. B. Tyler, see Tyler, "Victory Ship Design," p. 55.

[12] Memorandum of A. Osbourne to F. C. Lane, May, 1949, inserted in Tyler's "Victory Ship Design," p. 55.

[13] Tyler, "Victory Ship Design," p. 88; *Shipbuilding and Marine Engineering in Germany during the period 1939-1945*, British Intelligence Objectives Sub-Committee Overall Report No. 2 (London: H. M. Stationery Office, 1948), p. 3.

[14] See below, sect. IV. In a letter of May 16, 1949, Alan Osbourne writes me: "Undoubtedly, the American turbine builders were not so much afraid of the Wolf cycle Lentz engine itself as that the shipyard people might get used to modern and moderate-powered types of reciprocating steam engines, particularly as it was known that this development could lead to the production of a reciprocating engine, both cheaper and of higher efficiency than steam turbines of comparable power."

[1] *Industrial Mobilization for War*, I, 263-64, 546, 581-82.

[2] *Ibid.*, I, 260, 682-84; Chaikin and Coleman, *Shipbuilding Policies*, p. 90.

Before the Gibbs appointment, both Admiral Land and Admiral Vickery voiced their opposition on the grounds that the firm of Gibbs & Cox, with which Mr. Gibbs was not severing his connection, was too deeply involved in Navy work for him to act impartially.[3] After arriving in Washington at his new post, Mr. Gibbs negotiated with Admiral Vickery in regard to the services of his firm in preparing plans for converting some EC2's to tankers. Mr. Gibbs considered the fee proposed insufficient and said he preferred to have the work done elsewhere if possible. He asked Admiral Vickery if he found it embarrassing in view of his, Gibbs', position in WPB to have the work done by his firm. Admiral Vickery replied that he did not, since Gibbs & Cox was already doing considerable work for the Commission under other contracts.[4]

The indefiniteness of Mr. Gibbs' authority was disturbing. His responsibilities and powers were established by an exchange of letters between WPB and the Navy Department,[5] and within WPB he assumed authority over its Shipbuilding Division, of which the functions were to " coordinate and facilitate the efforts of the Army, Navy, Maritime Commission, Coast Guard, Lend-Lease Administration, and other war agencies in the construction of ships on schedule and in the production of materials and parts. . . ."[6] The work of this division in increasing the supply of critical components such as turbines and coordinating their allocation has already been described in general terms (chapters 11 and 12). It had been under Admiral Land when part of the National Defense Advisory Council, and then came under Capt. T. O. Gawne who remained its Director for the entire war period but was subordinate to the Controller of Shipbuilding while that office lasted.[7]

The main problem tackled by Mr. Gibbs and the main reason for his appointment was the gap between the lagging construction of escort vessels and the Navy's intense need for these ships in its battle against submarines. That was primarily a problem of the Navy, but would affect the Maritime Commission if the remedial

---

[3] Land to Files, Jan. 26, 1943 in MC Historian's Collection, Land file.
[4] Vickery to Gibbs, Jan. 15, 1943 in MC gf 900-01.
[5] Chaikin and Coleman, *Shipbuilding Policies*, p. 90, note 169.
[6] *Ibid.*, p. 29.
[7] *Ibid.*, pp. 29-30, 90-92.

measures proposed were of such kind as to interfere with merchant shipbuilding. Before Mr. Gibbs had been Controller of Shipbuilding for many weeks, Admiral Land began to hear of conferences at which the Maritime Commission was not represented. He saw evidence that, as he wrote to Harry Hopkins, " there may be some ganging up on the Maritime Commission and involved therein, either directly or indirectly, is the fine Italian hand of our friend Gibbs." Since it was Admiral Land's understanding that Mr. Gibbs was to act as coordinator with reference to priorities and allocations, he wrote him on January 21 a long letter arguing against the creation of a special overriding priority for corvettes, the type of escort vessels, later called frigates (PF), which was especially needed. What was worrying Admiral Land most at this time was the fact that others were using the " back door to the White House " so effectively he was afraid a Presidential decision might be quoted against him before he had been adequately consulted.[8]

The danger was if anything even more sweeping and direct. Mr. Gibbs proposed to carry out his ideas by the methods of production scheduling being generally applied by WPB in 1943, and he had ideas not only about military types but also about what types of merchant ships should be built. In short, he was moving to take control of the sphere of decision which would make him a " Shipbuilding Czar." When the appointment was discussed at a luncheon given by Admiral Leahy, Admiral Land said that if that was to be Mr. Gibbs' position, he would fight it to the limit.[9] In this spirit Admiral Vickery wrote Mr. Wilson on January 6, 1943: " I assume, therefore, that Mr. Gibbs' duties will not embrace the merchant shipbuilding program." [10] But Mr. Wilson replied refusing to recognize any such limitation on the role of the Controller of Shipbuilding, claiming Admiral Land had said, " Mr. Gibbs' impartiality would be unquestioned and that the Maritime Commission did not therefore desire to be considered in opposition to such an appointment." Mr. Wilson said Mr. Gibbs would exercise without limitation the authority

---

[8] Land to Hopkins, Jan. 21, 1943; Land to Gibbs, Jan. 21, 1943; and Land's memorandum for Files, Jan. 26, 1943; all in Land reading file. Cf. Chaikin and Coleman, *Shipbuilding Policies*, pp. 86-87 on the priorities of escort vessels.

[9] Land's memorandum for Files, Jan. 26, 1943.

[10] Vickery (by Scoll) to Wilson, Jan. 6, 1943, copy in MC, Historian's Collection, Land file.

of WPB relating to shipbuilding.[11] In the ensuing controversy one of the issues was indeed the extent of WPB's authority compared to that of the Maritime Commission.

Aside from the conflict for power, the main issue was WPB's opposition to the Maritime Commission's plans to build Victory ships in the yards that had been building Libertys. Mr. Gibbs opposed the Victory on two grounds: that the Lentz engine was untried and therefore unsuitable for mass production, and that any discontinuance of the construction of Libertys violated the principle of standardization and would decrease production. WPB asked for more standardization, not less. It proposed that a single design of fast ship (the C2 was favored in January 1943) be the only design of fast ship built. That would mean that yards building C1's would build Libertys instead, and yards building C3's or C4's would build C2's. The WPB suggested also discontinuing the construction of large troop transports, called P2's. These changes, claimed the WPB, would increase the deadweight tonnage produced from Maritime Commission yards by 14 per cent. At the same time it would facilitate the standardization of turbine production in a way that would make possible the construction of more escort vessels for the Navy.[12]

When the Controller of Shipbuilding set forth these sweeping ideas at a conference on January 27, 1943, opposing arguments were presented by Admiral Land, who attended the conference without Admiral Vickery since the latter was at this date in Oregon inspecting the *Schenectady*. In regard to the escort vessels and to the C1 designs, the Commission was open minded, he said, but if an attempt was made to abandon either the C2 or the C3 there would be a fight.[13] His general attitude, maintained by the Commission throughout the controversy, was vigorously stated in a memorandum he wrote for the files on the eve of this conference.

> ... Neither Vickery nor I yield to Mr. Gibbs on the question of merchant ship design as it is a well-recognized fact that we are more familiar and have had greater experience with this subject than has Mr. Gibbs. If anyone

---

[11] Wilson to Vickery, Jan. 20, 1943, with routing slip showing that this carbon went to Commissioners and others, in Land file.

[12] Nelson to Leahy, Jan. 31, 1943 in WPB file 324.1041, in National Archives.

[13] Notes on Conference, Jan. 27, 1943; Tyler, "Victory Ship Design," p. 43; Land to Leahy, Feb. 17, 1943 in reading file. A copy of the reply was sent to Harry Hopkins,

has any doubt about the preceding statement, it is suggested that the records be looked into and that the matter be taken up with the National Council of American Shipbuilders, the American Bureau of Shipping, or any other agency familiar with merchant marine design and construction throughout the United States.

One of the first reports that came back to me through "the back-door route" was to the effect that as it takes longer to build the Maritime Commission's standard designs (C-types) than it does to build Liberty ships, it would be advisable to drop these standard designs and build Liberty ships.[14] The originator or the orginators of such an idea need have no delusion that he or they is or are Christopher Columbus for the idea was thoroughly well-known to Vickery and me, thoroughly explored and discarded for the most vital and important reasons, one of which is the question of speed and another is that we have plenty of shipbuilding yards concerned with building Liberty ships and plenty of additional capacity for additional Liberty Ships if the necessary material, particularly steel, is allocated to the Maritime Commission.

* * * * * *

Following the foregoing, the question was raised in various places "that the Maritime Commission has too many designs of its C-type ships," with the inference that if we concentrated on fewer designs that increased production would obtain. Anyone who has complete knowledge and a reasonable understanding of the situation should realize that while the idea of reducing numbers of designs has some theoretical merit, any careful analysis of our building program will readily demonstrate that as a practical matter the theory cannot be transformed into practice without serious maladjustment, delays in production and an over-all interference with the scheduled programs and no gains in any way commensurate with the losses involved.

* * * * * *

The Maritime Commission maintains that the number of designs have been kept to a minimum; that the distribution of contracts to given yards has permitted uniformity, standardization and repetitive work with the idea of obtaining maximum of production. It would be a serious mistake to shift over these yards from one type (design) to another as it would involve serious delays, especially when one considers that they are tooled up, jigged up, died, mold-lofted and experienced in the design and type of construction they now have under way. Full and by, such proposals just don't make sense when analyzed properly.

* * * * * *

As Director of Shipbuilding for the National Defense Council, I admit to having had some experience with the problems facing Mr. Gibbs. I

---

[14] See in this connection J. J. Friedman to R. R. Nathan, May 29, 1942 in WPB file 324.10413, which begins "By prohibiting the new construction of standard cargo vessels after July 1, 1942 and building Liberty vessels instead, we could complete 500,000 tons of merchant ships more this year than are now scheduled, and deliver escort vessels for their protection at least 6 months sooner than now planned,"

sympathize with him but I don't admit for a minute that, as a major drafting-room contractor for the Navy Department and a minor drafting-room contractor for the Maritime Commission, he has either the experience or the ability to be the "Czar of Shipbuilding" for these United States, and while he may be eminently fitted through years of experience with the United States Navy to direct and control some of their design and construction work, I find nothing in his record to justify any claims that he is equally fitted to do the same for the Maritime Commission.

Mr. Gibbs is one of my very best friends. Admiral Robinson and I are responsible for the first contracts he ever had with the Navy Department. It may be stated as a fact that I am more responsible than anybody else in the United States for his entrance into Navy contractual work. Our personal relations have always been most friendly and pleasant. Our professional relations have been likewise although there have been many differences of opinion, contractually and otherwise, though the final solution has been, to the best of my belief, generally satisfactory to both sides.

Finally, I think it is just plain stupid to precipitate a series of rows between important Departments of the Government which are concentrating on winning at the earliest possible date an all-out war. These rows can be avoided and proper procedure followed if proper methods of consultation, cooperation and coordination are carried out by the personnel concerned. They can not be avoided by high-handed methods, by playing Hamlet without the ghost, or by following circuitous routes to accomplish ends that may appear desirable to individuals concerned but seriously conflicting with proper efforts being made by equally patriotic and hard-working personnel whose knowledge may be greater and whose picture may be broader than those individuals originating or presenting the ideas.

The Maritime Commission is perfectly willing to "play ball" with all the cards on the table face up. We are entitled to be consulted whenever the shipbuilding program of the Maritime Commission is concerned. With an already overworked staff, we do not relish being put in the position by a "prosecuting attorney" of furnishing a lot of extraneous evidence for the purpose of establishing points already considered and discarded.[15]

The channels through which Mr. Gibbs formally pushed his ideas were the various "Combined" boards on which were both Americans and Britishers. A Combined Shipbuilding Committee (Standardization of Design) was created in March by the Combined Chiefs of Staff on the suggestion of the Combined Production and Resources Board. Admiral Leahy informed Admiral Land on March 6 that the Chairman would be a representative of the U. S. Navy. When the Committee was organized on March 21, 1943 Mr. Gibbs was made Chairman at the suggestion of Donald Nelson. Admiral Land named himself as the Maritime Commission's representative with Admiral Vickery as his alternate.

[15] Memorandum for Files, January 26, 1943, Land reading file.

This Committee undertook to say what type of fast ship should be built.[16]

In the Combined Shipbuilding Committee, the dispute centered on two questions: (1) Should Victory ships be built instead of Libertys? (2) What engines should be built for the Victorys and the C2's? Plans for building the Victory ship were far advanced by this time, and on March 25, 1943, just four days after the organization of the Combined Shipbuilding Committee, Admiral Vickery made a speech to professional societies of the shipbuilding and engineering fraternities announcing these plans and calling the vessel by its new name, the "Victory Ship." He said the Liberty ship had been redesigned for the greatly increased speed which the new Lentz engine would make possible.[17] In fact he was thinking at that time of constructing in 1944 about 524 Victory ships of which 347 would have the Lentz engines and 177 would have turbines designed for the C3. Production of the latter type of turbine was coming along faster than they were needed for the C3 hulls, so Admiral Vickery planned to use the surplus in Victory ships. At the same time he planned to place orders calling for C2 turbines in the future and use the turbines in Victory ships. With the C3 turbine the Victory was expected to do 17 knots, with the C2 turbine 16 knots, and with the Lentz engine $15\frac{1}{2}$ knots.[18] The Commission was on the point of awarding contracts to carry out this plan, but Mr. Gibbs took the position that it required the approval of the Combined Shipbuilding Committee.

Mr. Gibbs asserted his claim to authority by having Mr. Wilson direct the officials of WPB to refuse authorization for facilities or materials in connection with the Victory ship until the Combined Shipbuilding Committee had approved it.[19] When the Maritime Commission extended priority ratings to components for

[16] Tyler, "Victory Ship Design," pp. 45-47.
[17] Ibid., pp. 47-48. The address is in the Marine Engineering and Shipping Review, Apr. 1943, pp. 183-84.
[18] In long hand on the back of letter paper bearing his home address, Admiral Vickery wrote out a summary of his planning for the Victory ship and how it had been interfered with. This document which is without title is in the Historian's Collection, Vickery file, "Victory Ship Program." It is cited hereafter as Vickery's memo on Victory Ship.
[19] C. E. Wilson to Ralph Cordiner and J. A. Krug, Mar. 24, 1943, in WPB file 045.12 and MC gf 506-22-1. Letters to manufacturers dated Apr. 21, 1943 revoking preference ratings are in WPB file 324.10412.

Victory ships, the WPB acted to block delivery. Admiral Vickery declared that the WPB did not have the right to do this. He stated in the April 21st meeting of the Production Executive Committee of WPB, " The Maritime Commission is of the opinion that it has the authority to determine which types of merchant ships should be built. Moreover, he was under the impression that WPB has little or no jurisdiction over the ship construction program, except for the allocation of materials. . . ." Admiral Vickery was unmoved by Mr. Gibbs' arguments against the Victory ship, insisting that " since the Maritime Commission is using its best judgment, no other agency should interfere with its decisions." [20]

While the deadlock continued, Mr. Gibbs held conferences with the manufacturers of turbines to see if they could not increase output by standardizing production. Obviously, if the output of turbines could be sufficiently increased there would be less argument for using the untried Lentz engine. Representatives of the turbine manufacturing industry met together to work out details for quantity production of a standard turbine and reported that the prospects were good. This was the basis for an agreement on one of the points at issue. At a conference of Messrs. Wilson and Gibbs, and Admirals Land and Vickery on April 30, 1943, the Commissioners consented to drop the Lentz engines if enough turbines could be produced, and WPB agreed to drop its opposition to the Lentz engine if they could not.[21] Thereafter the Lentz engine was no longer an important issue.

With turbines the issue of the moment, Mr. Gibbs tried unsuccessfully to persuade the Commission to accept either those developed for escort vessels or the C1 turbine stepped up from 4000 to 6000 hp.[22] The path towards a solution was found rather in the plans for standardization and the various alternatives of increased production worked out by a group of consultants drawn from the companies manufacturing turbines. They developed plans for standardization in building a simplified type of 6000 hp. turbine (6600 shaft hp. max.) and presented them at a meeting

---

[20] Chaikin and Coleman, *Shipbuilding Policies*, pp. 171-72; Tyler, " Victory Ship Design," pp. 49-50.
[21] Memorandum of Conference held in the office of Charles E. Wilson, Apr. 30, 1943, in Land file; Vickery's memo on Victory Ship, cited above.
[22] *Ibid.*

on May 8 and 9, presided over by Mr. Gibbs. The general characteristics of the "Victory" turbine were agreed to at this meeting, and starting with these characteristics each manufacturer was to proceed with his own design. The gears of Westinghouse and Falk companies were to be of the nested type, and those of General Electric, Allis Chalmers, DeLaval, Farrel Birmingham, and Joshua Hendy were to be of the articulated type. All turbines should fit the same ship's foundation without structural change and all steam, drain, and lubricating connections should be identical. During the development of these plans, which amounted to a single composite plan, the industry was to arrange for close and current interchange of design information, in order to expedite matters as far as possible. The Commission agreed to give firm approvals at the outset, that is, before approval of detailed plans, so that materials could be ordered without delay. Enough progress had been made to indicate that a standard turbine could be produced in quantity and reasonably soon.[23]

A decision which made it easier to supply the necessary gears was the stepping up of the normal speed of the propeller to 100 revolutions per minute. On the C2's, comparable in size and in their turbines, the normal speed of the propeller had been set at 92 rpm. largely because of fear that a higher speed would cause hull vibration. Experience had indicated little or no danger of hull vibration, and although the higher speed would mean some loss of efficiency, the loss would be slight. The advantage of higher speed in the propeller was that it changed the design of the reduction gear so that gear-cutting machines not otherwise available could be used and gear cutting done more rapidly.[24]

On the basis of the plans of the turbine manufacturers the Maritime Commission agreed definitely on May 12, 1943 to drop the Lentz engine, and the WPB on its side authorized approval of facilities and components for Victory ships. Just how many yards were to be changed over to Victorys or at just what rate continued in dispute, however.[25] For Oregon Ship, WPB's approval of necessary facilities was held up until July 1943.[26]

[23] Tyler, "Victory Ship Design," pp. 49-54. A long record of this meeting is in the Vickery file in the MC, Historian's Collection.

[24] Statements by A. Osbourne, June 22, 1949 and James L. Bates, June 21, 1949, and Ivan J. Wanless, Oct. 24, 1950.

[25] Chaikin and Coleman, *Shipbuilding Policies*, p. 172; C. E. Wilson to Ralph Cordiner, May 13, 1943; J. A. Krug to Staff, May 14, 1943 in WPB file 324.10412.

[26] Vickery's memo on Victory ship; Tyler, "Victory ship Design," p. 74. Wilson

The new schedule for ship production which the Maritime Commission prepared May 15, 1943 on the basis of using turbines instead of the Lentz engine provided only 260 Victory ships to be delivered in 1944, instead of the 524 in the April program. Partly because the conflict with WPB had caused delay in equipping yards, and partly because the turbines would not be ready until later dates than those at which it had been planned to have the Lentz engine available, the May program scheduled fewer Victorys than had been planned in April and made up for it by scheduling 611 more Liberty ships. To that extent the Controller of Shipbuilding was succeeding in his campaign to continue production of Liberty ships, but the Commission kept trying to change over as much and as rapidly as possible from Libertys to Victorys and accordingly kept pressure on the turbine manufacturers. Its June program estimated that during 1944 there would be 177 C3 geared turbine units available and 207 Victory units, a total of 384. In conferences with the turbine manufacturers it expressed hope of getting back to its April goal of 524 Victory ships in 1944 and of getting as much as 90 units per month in 1945.[27] Its July program called for delivery of 340 Victory ships in 1944 (Figure 37).

After the propulsion unit for the Victory ship seemed to be settled, controversy over turbines continued because Mr. Gibbs was still pressing for more standardization. The turbines being built for C2 cargo vessels became for a time the central issue in the contest between the Controller and the Commission. Mr. Gibbs wanted to stop production of that type of turbine and to use in the C2 hulls the recently agreed-to Victory turbine. " I cannot justify," he said at a meeting June 10-12, " constructing two 6600 hp. plants side by side, one much cheaper than the other, much simpler than the other, with a slight diminution in economy, in a wartime development of this kind, . . ." [28]

Mr. Gibbs said he understood that in view of the saving in

---

to Land, July 3, 1943 (above cited) defends this refusal to approve facilities for changing yards from Liberty to Victory ships. Transcripts of telephone conversations between Vickery and Wilson, July 27, and Vickery and Murphy, July 28 in MC, Historian's Collection, Vickery file.

[27] Fischer, " Programming," p. 29 and Table IV; Tyler, " Victory Ship Design," pp. 54-57.

[28] " Memorandum of meeting held June 10-11-12, 1943 . . .," p. 40, in MC, Historian's Collection, Vickery file.

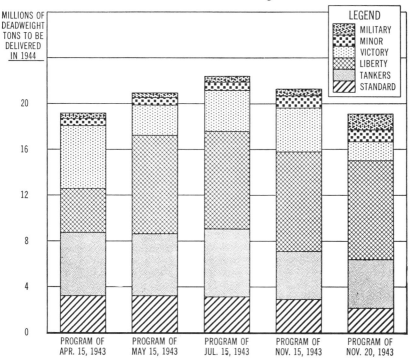

FIGURE 37. Source: Fischer, *Statistical Summary*, Table A-4.

money and critical materials, the Maritime Commission would consider the use of Victory equipment in C2 ships. According to the minutes of the meeting, Admiral Vickery replied with emphasis that such was not the case. The Commission had reached its own decision in the matter which it did not intend to alter. Vickery did not agree that the question of the type of turbine to be used in C2 ships was a matter for consideration by the Committee, and therefore he declined to make any explanations.[29]

At this point WPB undertook to assert its authority. Mr. Gibbs, with Mr. Wilson's approval on June 23, 1943, directed the Shipbuilding Division, WPB, to advise manufacturers to stop making C2 turbines and units beyond a minimum of 153 to be delivered

[29] Chaikin and Coleman, *Shipbuilding Policies*, p. 173 note 22.

from June 1943 to October 1944.[30] At the same time he wrote Admiral Land that in view of these considerations WPB was directing the General Electric Co. and the DeLaval Steam Turbine Co. to " cease and desist from the construction of further C2 turbines and gears beyond the numbers determined as the minimum to avoid a drop in production." [31]

The Maritime Commission refused to yield. At the regular meeting of the Commission on June 24 Admiral Land submitted Mr. Gibbs' letter and questioned its accuracy, stressing that it had previously been agreed, by the Navy, WPB, and the Commission, that the C2 and C3 turbines were not to be affected, and that the Commission had given up the Lentz engine and agreed to eliminate the C1 design for most of the 1944 program. The Commission reaffirmed its exclusive jurisdiction over the construction of the C2 turbines arising out of its contractual agreements with the manufacturers. In the opinion of the General Counsel the action taken by WPB had little authority in law. Part 3208 of its Order No. M-293 of February 26, 1943, applied to critical common components for defense, for private account, and for export. This did not cover the case in question and, moreover, was not supported by any statute or executive order. The authority of WPB, as established by arrangements with the service forces, was definitely limited, he argued.[32] By unanimous vote, the Commissioners (Vickery absent) authorized the transmittal of the following message to the General Electric and DeLaval companies:

> As Legal Contractual Agency the Maritime Commission requests you disregard any instructions relative to manufacturing C-2 turbines and gears for C-2 vessels unless such instructions are approved and confirmed by the Maritime Commission.[33]

Being thus defied by the Maritime Commission, the WPB took the offensive on a broader front, and resumed its fight to maintain

---

[30] *Ibid.*, p. 173, note 23.

[31] Gibbs to Land, June 23, 1943, in MC, Historian's Collection, Richards file. (That is, beyond a total of 145 for General Electric and of 8 for DeLaval.) Also Westinghouse, Falk, Allis Chalmers, Joshua Hendy, and Farrel Birmingham companies were being advised not to build C2 plants.

[32] General Counsel to Chairman, June 25, 1943, in MC gf 107-55.

[33] MC Minutes, June 24, 1943, p. 25551-52. A copy of the WPB telegram was telegraphed to T. H. Shepard, Jr. of the General Electric Co., by E. K. Henley on June 24, and is in MC, Historian's Collection, Richards file. The Commission's instructions were transmitted by telephone.

construction of Liberty ships. In a letter to Admiral Land on July 3, 1943, Mr. Wilson proposed that the construction of Liberty ships be continued in yards building that type, and that just one type of fast ship be built, namely, Victory ships, powered by Victory turbines. He proposed that during 1944 yards building C-types should change over to building Victorys.[34]

A complete deadlock was thus reached. The Maritime Commission and WPB were basically at odds over what ships to build and over who had power to decide. Before considering further the contest for power between the agencies, let us look at the substance of the issue.

The WPB argued in favor of the Liberty ship that it was the easiest to produce in quantity. The Maritime Commission's rebuttal was that the speed of the Victory would enable it to make more round trips per year and that the true basis of comparison for the different types was annual cargo-carrying capacity, that is, the amount of cargo which could be delivered within a year across a given route such as that to Europe. The Commission argued that the WPB program would require 12 per cent more steel, 7 per cent more manhours in ship construction, and 9 per cent more manhours in the operation of ships per unit of annual cargo-carrying capacity than the Commission's program.[35] The WPB plan would use more steel than the Commission's program, or else would slow down the rate of production in Liberty ship yards to 20 per cent under capacity. The Commission on the other hand proposed to schedule work for the yards up to about their capacity, but build more complicated, better ships. Although savings in manpower was debated, it was assumed that steel set the limit on production. Alternate programs were presented using the same amount of steel. The Commission proposed to use steel more efficiently by turning it into more efficient ships.

No doubt the Commission was right in theory in maintaining that cost per ton-mile of carrying capacity was the best basis of

[34] Wilson to Land, July 3, 1943, copy in WPB file 325.633 " Steam turbines and gears—Standarization and Simplification." The original, in MC, Historian's Collection, Research file 111, is covered, paragraph by paragraph with penciled marginalia in which Admiral Land commented alternately, " false," " any jackass knew all this over 2 years ago," " who says so," and concluded, " Taken over by Judge Byrnes so I think will stand pat."

[35] MacLean to Land, July 5, 1943 and especially G. J. Fischer to Vickery, July 28, in MC, Historian's Collection, Vickery file, and Research file, 111, Victory Ship.

comparison. The annual carrying capacity of each type was not known, however, with any certainty. In addition to other statistical difficulties, there were involved such variable military questions as whether 15-knot vessels could sail outside of convoys (or, on what routes they could sail without convoy) and what could be the speed of convoys. The Maritime Commission emphasized that fast ships required less escort protection; the WPB maintained that they would have to sail in convoys and would be slowed down to the speed of the slowest ship of the convoy. In the immediate future, at least, there would certainly not be enough Victorys to organize them in separate convoys.

The WPB arguments assumed that the largest possible delivery of new tonnage before June 1944 was the primary strategic consideration. The Maritime Commission, granting of course that the transition from Libertys to Victorys would reduce output temporarily and that the benefits of building the faster ship would not be fully felt until 1945, urged that there was no assurance that 1945 would be less important than 1944 in military operations.

While the relative usefulness during the war of the two types was the main question argued in the memoranda of June and July 1943, the Commission also sought support for its program by reference to postwar conditions. Fifteen-knot ships in the American merchant marine would be able to compete with those the British were building, 10-11 knot ships would not. The WPB argued that this concern with competitive position after the war was unsound; the United States would have more fast ships than it could use anyhow.

The argument for the Victory turbines or the C2 turbine followed somewhat the same general pattern. The WPB maintained that the same type of turbine should be used in both the Victory and the C2 on the grounds that this would permit the turbine manufacturers to increase their total annual output of turbines. Mr. Gibbs argued that the Victory turbine used less material and was more rugged. When the Commission objected that putting Victory turbines in C2 hulls made a ship in which there was lack of balance between the propulsion unit and other design features, WPB proposed that Victory hulls be built for the Victory turbines—ignoring the decline in productivity which would accompany such a change in the yards building the C-types.

Against this attack on the C2, the type which had been the

pride of the Commission from its start, Admiral Land advanced a whole battery of arguments. Yards were already equipped for the production of particular types; changing established plans would slow down production in the shipyards even if it would help the turbine manufacturers, which Admiral Land seemed to doubt. The C2 was too good a ship, too well proved by experience, to be powered by the untried Victory turbine or replaced in the program by Victory ships. The C2 turbine was not only the more efficient, it was also, Admiral Land stoutly maintained, the more rugged. The C2's were desirable for the postwar period because of their efficiency and were more desirable for the war in order to meet the repeated requests from the Army and Navy for ships suitable for conversion into auxiliaries.[36]

In short, the Victory was a better ship than the Liberty, and the C2 with its own turbine was a better ship than the Victory. On both issues the WPB was arguing for quantity at the expense of quality. The Maritime Commission was arguing that, both for postwar use and for use in war, it was better in each case to build the better ship.

V. Resolving the Conflict with WPB

The deadlock between the Maritime Commission and the WPB was expressed by their contradictory instructions to the turbine manufacturers and by the earlier conflict, not yet entirely resolved, between the Commission's contracts with shipbuilders for Victory ships and the WPB's refusal to grant the approvals and priorities so that the contracts could be carried out. A decision by some higher authority was necessary to break this deadlock.

At an earlier stage in the conflict, Admiral Land had appealed to President Roosevelt. In April when the use of the Lentz engine was still an issue, a Truman Committee report described the controversy between the WPB and the Maritime Commission in terms favorable on the whole to the Maritime Commission, and declared the dispute " most detrimental to the war program." [1]

---

[36] This exposition pro and con is based on the memoranda which are cited above elsewhere in this chapter and which are summarized with close attention to chronology in Tyler's "Victory Ship Design," pp. 64-78 and Chaikin and Coleman, *Shipbuilding Policies*, pp. 170-177. Interesting for comparison is Great Britain, House of Commons, *Sessional Papers*, 1940-1941, " Twentieth Report from the Select Committee on National Expenditures," ordered printed Aug. 6, 1941, pp. 4-5.

[1] Truman Committee, *Report* No. 10, 79th Cong., 1st sess., pt. 8, p. 9, Apr. 22, 1943.

Admiral Land sent the President a memorandum quoting this report and saying: "the 'big dispute' appears to have been 'planted' with the Truman Committee. It is earnestly recommended that you sign the enclosed memorandum to the War Production Board in order to settle the matter without a lot of useless controversy, which is nothing more or less than the subject you discussed with Vickery when you sent for him a short time ago."[2] The memorandum referred to was signed by the President and read:

> The Maritime Commission's shipbuilding program for 1943-44 was approved by the Budget, approved and appropriated for by the Congress and approved by me.
> Contracts have been awarded for a large majority of ships in this program by the unanimous action of the U. S. Maritime Commission.
> It is directed that no interferences be placed in the way of carrying out the construction of this program and that no changes be made therein unless they are submitted for my approval.
> The directives given by me to the Maritime Commission for this vast tonnage of merchant ships will not permit any delays in carrying out the program so vital to the war effort.[3]

When the whole issue was reopened in July, Admiral Land did not wish to bother the President again.[4] Who else would decide the conflict was not crystal clear, however. It might be the Combined Chiefs of Staff containing both British and American members, or the Joint Chiefs of Staff (American only), or the Office of War Mobilization.

Appeal by WPB to the Combined Chiefs of Staff was the natural first step, for Mr. Gibbs had been voicing his program as Chairman of its advisory committee, the Combined Shipbuilding Committee (Standardization of Design). In his letter of July 3 to Admiral Land, Mr. Wilson concluded: "When the views of the Combined Chiefs of Staff are known, it will enable you and me to consider this matter further." The Commission, on the other hand, while denying that the Combined Shipbuilding Committee had any jurisdiction in the matter of geared turbines for the C2's, had no wish to recognize the authority of the Com-

[2] Land to the President, Apr. 22, 1943, in Land reading file.
[3] MC,Program Collection, pt. I., Apr. 30, 1943.
[4] Draft July 5 of memorandum of Land to Leahy (rewritten and sent July 6, 1943) in Historian's Collection, Richards File, " Gibb Action," [sic] with attachment intended for Harry Hopkins saying, "I want to settle the matter, if possible, without involving the President."

bined Chiefs either in such a matter. Admiral Land wrote Admiral Leahy stating the Maritime Commission's objections to Mr. Gibbs' proposals, and saying he would like to have an opportunity to appear before any committee under the Combined Chiefs which might have the matter assigned to them; but he also expressed the opinion that the Combined Chiefs of Staff were not concerned over controversies having to do with installations on merchant ships.[5] In effect the Combined Chiefs did decline to act. Admiral Leahy, who was chairman of both the Combined Chiefs and the Joint Chiefs, replied " For the Joint Chiefs." He told Mr. Wilson that the questions involved were considered purely technical and not within their province, and since it had to do with a difference of opinion of two agencies of the United States, it was considered unwise to " bring the British Chiefs of Staff into the picture." [6] When Mr. Wilson appealed again asking for a statement on strategic requirements, Admiral Leahy wrote him on July 16, 1943 that the Joint Chiefs had already approved the shipbuilding program for 1943 and would make a study of the program for 1944. He reiterated that it was better for the American agencies to agree before submitting the question to the Combined Chiefs.[7] The secretary of the Combined Chiefs forwarded copies of these letters to Admiral Land and informed him that the matter had been referred to the Joint Military Transportation Committee for study and recommendations.[8]

At the same time that it was appealing to the Combined Chiefs, the WPB was also putting its case before James F. Brynes, who, as Director of the Office of War Mobilization, had been delegated by the President to unify the activities of Federal agencies concerned with production.[9] Mr. Nelson wrote Mr. Brynes that the essence of the disagreement was as follows:

> The War Production Board has found it possible, by technical alterations in the existing program for future construction of large cargo ships, to conserve materials, manpower and money, while meeting the requirements of the Combined Chiefs of Staff. The Maritime Commission is strongly opposed to the alterations and has indicated their unwillingness to comply.

---

[5] Land to Leahy, July 6, 1943, copy in Land reading file.
[6] Leahy to Wilson, July 13, 1943 in MC Historian's Collection, Richards file.
[7] Wilson to Leahy, July 15, 1943; Leahy to Wilson, July 16, 1943, both in the Richards file.
[8] Deane to Land, July 17, 1943 and see also Deane to Land, July 14, 1943, both in the Richards file.
[9] *Industrial Mobilization for War*, I, 554.

He contended that the Maritime Commission had failed to respond to " our repeated requests for a factual justification of their attitude " and therefore he had no alternative but to enforce the WPB decision; but concluded that he was most anxious " to resolve this matter with the understanding and agreement of the Maritime Commission, for whose work to date I have a high respect. The only requirement to this end, so far as I can see, is that they consent to discuss the real issue involved." [10]

When Mr. Byrnes sent a copy of this letter to Admiral Land for his comment, the Admiral replied on July 30 by setting forth the Commission's arguments at some length. At the same time he made sure that the Commission's arguments were known to Bernard Baruch, who was an influential adviser to both Mr. Byrnes and Mr. Roosevelt regarding industrial mobilization. But Admiral Land's letter to Mr. Byrnes was at pains to point out limits to the authority of the WPB in the matter:

> Whereas Mr. Nelson states that the sole question at issue is simply whether or not the War Production Board's " decision to make a change in the ship construction program is sound in terms of its effect on the prosecution of the war and the interests of the nation," it would appear that these decisions are policy considerations which are the responsibility of the Chiefs of Staff and the Maritime Commission respectively, and therefore outside the province of the War Production Board; . . .[11]

Thus, the independent jurisdiction of the Maritime Commission in regard to the long range interest of the nation in shipping was reiterated, and the Commission recognized the authority in the controversy of one other agency only, the Joint Chiefs of Staff.

In fact the Joint Chiefs decided the issue. On July 29, the day before he sent his reply to Mr. Byrnes, Admiral Land attended with Admiral Vickery a meeting of a subcommittee of the Joint Military Transportation Committee of the Joint Chiefs of Staff. The chairman of this subcommittee was Brig. Gen. John M. Franklin, president before and after the war of the United States Lines. He had been appointed by Admiral Leahy a member on the Combined Shipbuilding Committee, and to him also the

---

[10] Nelson to Byrnes, July 6, 1943, in MC, Historian's Collection, Richards file; Chaikin and Coleman, *Shipbuilding Policies*, pp. 175-76.

[11] Land (by Vickery) to Byrnes, July 30, 1943 in MC, Historian's Collection, Vickery file. See also Land to Baruch, July 24, 1943, in Land reading file, pt. 10. Admiral Land sent copies of his memorandum to Admiral Leahy of July 5 to Mr. Byrnes and Mr. Baruch, as well as to Mr. Hopkins.

Commission fully presented its arguments, especially those against the slow Liberty ship.[12]

On August 9, Admiral Leahy sent the decision of the Joint Chiefs to Mr. Wilson. It was to the effect that shipping would not in the near future be a bottleneck and that the production of other war materials would be more pressing. Therefore the building of great numbers of Liberty ships became "less compelling." The decision read:

> Since the beginning of the shipbuilding program, the fast C Ships, particularly the C-2 and C-3, have best met the strategic needs and have been seized to meet Army and Navy needs, often as combat ships. They are even now being rapidly converted as combat loaders and as combination passenger and cargo to fill vital military needs. The C-4 is building in direct response to Army requirements. The increased speed of these faster ships reduces the danger of loss of troops and cargo from submarine attack, shortens the time of turn around, decreases the requirements for escorts and saves crew manpower; advantages which it is held have justified their construction in the past and now with greater force will do so in the future. This experience leads to the conviction that our strategic needs in 1944 will best be met by the maximum number of fast ships.
>
> It is therefore the considered opinion of the Joint Chiefs of Staff, that the Maritime Commission No. 20 program [that of July 15, 1943], with its greater number of fast ships, better meets the strategic requirements of the war effort in 1944 than does any of the possible alternatives outlined by the War Production Board. Contingent to the understanding that slight changes may be requested, particularly in the field of smaller ships, the Joint Chiefs of Staff endorse the No. 20 Maritime Commission Program.[13]

Even before Mr. Gibbs' proposals had received this *coup de grâce*—but after Admiral Leahy had made clear that the decision was going to be by the Joint Chiefs, not the Combined Chiefs—the WPB had begun to give ground. At a conference on July 22 which included Nelson, Wilson, Land and Vickery, the WPB agreed to reinstate 39 C2 turbines it had attempted to cancel by its order of June 23,[14] and also to approve facilities for Oregon Ship which had not yet been cleared by the Facilities Clearance Board. There was some delay on the part of WPB in carrying

---

[12] Land to Franklin, no date, in Fischer reading file. Statement by Gerald J. Fischer.

[13] In MC, Historian's Collection, Vickery file, "Victory Ship Program" and in Land file.

[14] Chaikin and Coleman, *Shipbuilding Policies*, pp. 176-77; Land, Memorandum of Conference with WPB, July 22, 1943, in MC, Historian's Collection, Vickery files, "Victory Ship Program."

out this part of the agreement; the Chairman of the WPB Industrial Facilities Committee and Mr. Wilson misunderstood each other in the press of business, and Admiral Vickery had to go after them on the telephone before anything was done.[15] On its side the Maritime Commission yielded to the extent of agreeing to stop construction of C2 vessels as soon as existing contracts were completed, or not later than December 1944. These terms of settlement were confirmed by Mr. Gibbs and Admiral Land on August 12 on the basis of the decision of the Joint Chiefs. The WPB agreed to implement the Maritime Commission July program, and the Commission agreed to stop building C1 steam-propelled, C4 and P2 ships beyond those on order.[16] By these terms the Commission made no concessions of any practical importance. Long before the end of 1944 the Commission's program would be reviewed many times by the Joint Chiefs, and the Chiefs would decide then, as the occasion arose and on the basis of quite specific military needs, whether the Commission's program should include these special military types.[17] But the decision of the Joint Chiefs in August 1943 settled the heated controversy over building a faster type of merchant vessel.

How much the intensity of this conflict was due to the nature of the issues and how much to personalities is one of those questions which documents cannot answer conclusively. William Francis Gibbs retained throughout his service as Controller of Shipbuilding his connection with the firm of Gibbs & Cox, which had its own organization of naval architects and marine engineers. The Technical Division of the Maritime Commission, in contrast, took pride in the program developed by the long, full-time effort of Admiral Vickery and the staff he had developed and organized. They were intent on building the best ships they could. Mr. Gibbs' efforts to change the program seemed to Admiral Vickery and his staff unwarranted and unnecessary interference. On

---

[15] Vickery's memorandum on the Victory ship; and transcripts of telephone conversations between Vickery and Wilson, July 27, and Vickery and Murphy, July 28, in same Vickery file.

[16] Nelson to Land, Aug. 12, 1943, the original is in the MC, Historian's Collection, Land file. Chaikin and Coleman, *Shipbuilding Policies*, pp. 176-77, and Tyler, "Victory Ship Design," pp. 47-48 where this whole controversy is described in more detail.

[17] See below chaps. 18 and 19.

September 11, 1943, Mr. Gibbs resigned as Controller of Shipbuilding.[18]

One result of the controversy was that fewer Victory ships were built than had been programmed at the beginning of the year. Plans to build Victorys in the emergency yards in the South were abandoned. The Liberty yards of the West coast began their launching of Victory Ships, not in the closing months of 1943, as originally planned, but during the first half of 1944. Turbines of the C3 type were delivered before the Victory yards were ready to receive them and in November 1943 the Commission had to arrange for storage at Toele, Utah, of turbines on their way from New England to the West coast.[19] The first Victory ship to be built was delivered by Oregon Ship on the last day of February, 1944; and only 15 were delivered before May, 1944: 11 from Oregon Ship and 4 from Calship. Later, Victory ships were also built at Bethlehem-Fairfield, Richmond No. 1 and No. 2, and Vancouver. All the Victorys built at the latter yard were the modified form of the Victory design which began in May 1944 to fill the shipways of the West coast Victory yards: namely, the VC2-S-AP5, the attack transports (Plate XXI). Altogether the Maritime Commission built 414 Victory cargo ships and 117 Victory transports.[20] As matters worked out, the best justification for introducing the Victory design was its adaptability to the need for transports. The story of the carrying through of the Victory program ties into the story of building military types.

---

[18] *New York Times*, Sept. 12, 1943; Chaikin and Coleman, *Shipbuilding Policies*, p. 91.

[19] MC Minutes, pp. 26749-50, Nov. 9, 1943. Quarterly report of the Survey and Schedule Branch of the Production Division, Oct. 26, 1943.

[20] *Official Construction Record*, and Fischer, *Statistical Summary*, Table B-3.

CHAPTER 18

MILITARY AND MINOR TYPES

I. LANDING SHIPS AND ESCORT VESSELS

MILITARY TYPES occupied 103 of the 251 building berths in the Maritime Commission's Class I yards in August 1944 and formed 24 per cent of the total production of that year measured in displacement tons.[1]

Why were ships for military use being built by the Commission instead of by the Navy? A partial answer has been given in describing the expansion of the program after Pearl Harbor.[2] When the possibilities of multiple production became manifest in the summer of 1942, it was decided that the Maritime Commission yards could find room to build landing craft and escort vessels and also produce enough merchant ships to meet the goal of 24 million tons by the end of 1943. Fifty aircraft carrier escorts and 60 LST's were built in yards which in February had been planned for Liberty ships. In many instances military types were built in yards originally planned for merchant ships, either because vessels begun as C-types in the long-range yards were finished as military types, or because yards which finished their contracts for merchant ships received new contracts for military types to be delivered in 1944. In short, the main reason why military types were built, not in yards under the Navy, but in yards under the Maritime Commission was that there was more capacity available in the Maritime yards.

Once a yard was committed to building military types, arguments could be advanced in favor of having the Navy take over the contract for these types and supervise their construction. But this would have disrupted working arrangements. A complete reorganization of the inspecting staff would have been necessary, and the shipyard would have had to learn to meet a new set of standards. Moreover, some of the military types were really half military,

[1] Fischer, *Statistical Summary*, Tables E-8 and B-5.
[2] Chap. 6, sect. I.

half non-military—either in the sense that they were attempts to build ships which were in the main like merchant vessels in their construction, although purely for military use, or in the sense that they were designed, as were some of the transports, with some consideration being given to their postwar utility as merchant vessels.

The construction of military types under the supervision of the Commission was beset by special difficulties. Last-minute changes in design to take advantage of military experience, as well as other delays in the preparation of plans, made it hard for schedulers and shipyard managers to plan ahead. Consequently, the worst cases of "hoarding" labor and accumulating large inventories occurred in connection with building military types, as did some of the worst cases of bad management.

So many different military types were built under Maritime Commission contracts that it does not seem practical to try to discuss them all. A few details about the landing ships, the "baby flattops," the frigates, and various types of transports will show the difficulties involved in this aspect of the Maritime Commission's shipbuilding and will emphasize its importance to the Armed Forces.

A large landing-craft program was decided on in April 1942 as part of the plan for taking the offensive in Europe in the summer of 1943.[3] One of the largest vessels in the landing craft program was the tank carrying type, the LST, which was 327 feet, 9 inches long overall. With very shallow draft, $7\frac{1}{2}$ feet, powered with Diesel engines, it was designed to make 10 knots with twin screws. Although its deadweight tonnage, 2,286 tons, was about one-fifth of the 10,600 deadweight tons of the Liberty ship, its construction was more complicated and required more manhours per ship than a Liberty.[4] The design of the LST had been discussed by representatives of the British Admiralty with the Navy as early as November 1941, and it had been decided in December that the Maritime Commission should work out the plans.[5] But

[3] Dwight D. Eisenhower, *Crusade in Europe*, pp. 38-39; George E. Mowry, *Landing Craft and the War Production Board, April 1942 to May 1944*, [Historical Reports of War Administration, War Production Board Special Study no. 11] (Washington: Civilian Production Administration, 1944, reissued 1946) pp. 6 ff.

[4] Fischer, *Statistical Summary*, Tables A-1 and H-1.

[5] Letters in MC Technical Division files, S1, Nov. 21, 1941 and Dec. 11, 1941, cited in Helen E. Knuth, "The Building of LST's by the Maritime Commission." Miss Knuth's research is the basis of most of this account of building the LST's.

after the Maritime Commission had developed the plans and entered into negotiation to make contracts with shipbuilders, strategic planners set production schedules calling for the utmost speed. Admirals Land and Vickery decided they could not assume responsibility for meeting these production schedules unless they had control of the procurement of the machinery, namely, Diesel engines. Allocation of these engines was controlled by the Navy which needed them for other craft also and was unwilling to give up control. Consequently the Navy Bureau of Ships was placed in charge, and the drawings and specifications for the LST developed in the Preliminary Design Section of the Maritime Commission were turned over to the Navy.[6] By May 1942 the Navy had assigned contracts for the construction of 300 LST's to be completed before June 1943.[7]

The Navy awarded contracts to the yards with which the Maritime Commission had been in negotiation for the building of LST's; and, in order to reach the required total of 300 by June 1943, the Navy also called on the Maritime Commission to build 90 LST's in yards already under contract to it for other types. Admirals Land and Vickery were reluctant to use for the construction of LST's any yards that were building Liberty ships, for it would badly upset the production schedules. They agreed to do so, however, because the only alternative, in view of the decision at the highest level of command, seemed to be that the Navy would give contracts to these yards. In that case the Maritime Commission might never get the yards back for the Liberty program.[8]

Bethlehem-Fairfield was told officially on May 28 that it was being given a contract for 45 vessels [9] (reduced in October at the Navy's request to 30) [10] and consequently disrupted its work on Liberty ships in the fashion which has already been described.[11] In the outfitting of these LST's use was made also of two repair yards in Baltimore harbor, the Maryland Drydock Co. and the Bethlehem Key Highway yard. Delivery of the 30 vessels was completed in February 1943 nineteen days behind the contract

[6] Ivan J. Wanless to F. C. Lane, Nov. 16, 1950
[7] Knuth, "The Building of LST's," pp. 2-3, 7.
[8] Statement by William A. Weber, recorded May 7, 1946.
[9] Land to Bethlehem-Fairfield in MC gf 506-15-3.
[10] N. L. Rawlings to Land, Sept. 23, 1942, in MC gf 506-15-3; MC Minutes, p. 23092 (Oct. 1, 1942).
[11] Chap. 14.

schedule but ahead of the completion of the Maritime Commission's other contracts for LST's.[12]

Construction of the other 45 LST's was assigned on May 28, 1942 to the Kaiser Co., Inc. to be built in new yards from which no ships had yet been delivered.[13] The prospect of having to interrupt the multiple production of Liberty ships had brought spirited protests from Henry and Edgar Kaiser.[14] Their protests had some success, for the LST's were kept out of the existing yards at Richmond. Instead a new yard was built. Because it was attached to Richmond No. 3 and used some of the same facilities it was at first called Richmond No. 3A, later Richmond No. 4. It built 15 LST's; the other 30 built by Kaiser were constructed at the Vancouver yard near Portland, which broke off its work on Liberty ships after launching only two and clearing away two which were only half built. Although the original understanding with the Navy was that all the LST's would be delivered by February 1, 1943, Vancouver did not complete deliveries until March 20, 1943 and Richmond No. 4 until June 16 or 24, 1943.[15]

The chief reason for delay given by both Bethlehem Fairfield and the Kaiser yards was late arrival of the Navy-furnished materials, and they sent frequent protests to both the Maritime Commission and the Navy.[16] In October the Navy authorized the yards "to manufacture any items on LST's which will not be supplied by vendors in time to meet ship delivery schedules if items can be manufactured locally in the yards." At the same time Maritime Commission inspectors were given the authority to approve substitutions of materials and alterations in design which were in accordance with good marine engineering and would fulfill the purpose of the original parts.[17] The LST's were part of "one of the most outstanding high-speed efforts of the whole war effort," in which the Navy encountered very severe problems in trying to build on exceedingly short notice. It disrupted much of their building program, especially work on destroyer escorts.[18]

[12] Knuth, "The Building of LST's," pp. 17-19.
[13] Land to Kaiser Inc., May 28, 1942, in MC gf 506-15-2.
[14] Edgar Kaiser to Vickery and Henry Kaiser to Vickery, both May 16, 1942 in MC gf 506-15-2.
[15] Knuth, "The Building of LST's"; Reynolds' Way Charts; and MC Program of July 25, 1942 in Program Collection, Part II.
[16] See letters from the yards in MC gf 506-15-2 and 506-15-3.
[17] C. A. Jones, Bureau of Ships, to Vickery, Oct. 10, 1942 in MC gf 506-15-2.
[18] Mowry, *Landing Craft*, pp. 5-6, 23-24.

Its effect on the Maritime Commission program was estimated by Admiral Land to have meant the loss of 75 Liberty ships.[19]

While planning and preparing in 1942 to take the offensive on the other side of the Atlantic, the Anglo-American alliance was suffering very heavy shipping losses in American waters, even within sight of the coast, and in the Pacific was barely able to stop the Japanese in the Coral Sea and at Midway. These problems intensified the need for escort vessels of various kinds, and again the Navy turned to the yards of the Maritime Commission. Aircraft-carrying escort vessels constructed by converting merchant types had been under consideration by the Navy for some time. A Moore-McCormack C3 was converted to an escort aircraft carrier, at Newport News in March-June 1941. Other conversions followed, and in May 1942 work on this type concentrated at the Seattle-Tacoma yard, which had been working on C3's under Maritime Commission contracts but which took Navy contracts and came accordingly under the supervision of the Bureau of Ships thereafter. And a total of 8 of the C3-S-A2 escort aircraft carriers, intended for delivery to the British, were completed in various yards under the Maritime Commission in 1942-1943.

The main escort aircraft carrier program of the Commission consisted of the 50 built by Kaiser. Contract plans and specifications for this design (S4-S2-BB3) were developed by the Maritime Commission's Technical Division using its P1 design as a basis for the hull form and were finished with the cooperation of the Navy, which aided especially in designing the flight deck. George G. Sharp, acting as design agent for the Kaiser company, developed the working drawings, and the hull lines were drawn under the supervision of James L. Bates. Construction was in accordance with accepted standard marine commercial practice for both hull and machinery, with the Navy specifications for C3 conversions applied where appropriate. Because the output of turbines, gears, and Diesels was all pre-empted for other designs already in production, the design provided for reciprocating steam engines of the five-cylinder Skinner uniflow type capable of developing 5,400 horsepower on each of two shafts. They would give a speed, it was hoped, of 20 knots. The length overall was 490 feet, the light displacement 6,890 tons. The flight deck was longer than on the converted C3's, and the fact that the S4-S2-BB3

---

[19] Land to W. F. Gibbs, WPB, Jan. 21, 1943 in MC gf 107-55.

LANDING CRAFT AND ESCORT VESSELS   613

was designed from the beginning for its specific function gave it a number of other advantages, such as its two screws and separate machinery spaces. On the other hand it lacked certain features which would have been incorporated as a result of war experience but for the fact that its design was "frozen" in order to speed deliveries by the methods of multiple production for which Henry Kaiser was famous [20] (Plate XXII).

The decision to have these escort aircraft carriers built by Kaiser was stated by President Roosevelt to the Navy and the Maritime Commission at a conference at the White House on June 8, 1942. Kaiser's plan was to build them in the 12-way Vancouver yard as soon as that yard had finished its 30 LST's. Why then did not the Navy let the contract and take over the supervision of this yard as it had taken over the supervision of the C3 aircraft carrier escorts being built at Seattle-Tacoma? The Navy wanted the ships but did not want Kaiser. Accordingly, Admiral Land and Colonel McIntyre presented the case at the White House; Henry Kaiser also saw the President; and the construction of the 50 escort aircraft carriers of new design in the Kaiser Vancouver yard was agreed to on the basis that the construction of these, as of all other Kaiser-built ships, would be supervised by the Maritime Commission's staff, and the Maritime Commission should develop the design.[21]

When the contract was awarded, Henry Kaiser underestimated the time required in preparation before multiple production could begin for a ship of this type. Because of Kaiser's optimism,[22] the Commission's schedule of July 25, 1942 envisaged the delivery of 4 in February 1943 and completion of 50 by the end of the year. The first delivery was July 8, 1943. After the first round was off the ways, construction picked up speed and in the fourth round 6 were delivered in one month, June 1944, and the contract completed July 8, 1944.[23] In view of the size and the amount of complex equipment involved in a vessel of this type, it was a notable achievement in multiple production. The facilities of the

[20] Reports in MC, Historian's Collection, Research file 111; confirmed by interview with James L. Bates, June 21, 1949.
[21] *Ibid.*, and statements of Admiral Land, Mar. 22, 1949; MC Minutes, p. 27251 (Jan. 4, 1944), getting this contract is part of the Kaiser legend, see e. g. *Fortune*, Oct. 1943, p. 255.
[22] Statement by J. L. Bates, June 21, 1949.
[23] Program Collection, pt. II, and Reynolds' Way Charts.

Vancouver yard were increased substantially for the purpose so that it became one of the best of the emergency yards in equipment as it was in its lay-out.

"Baby flattops" were part of the answer to the submarine danger; escort vessels of the type which the British called corvettes were another. It was proposed in June 1942 at the peak of the controversy over steel that some of these be built by the Maritime Commission, but considering the facilities and steel available the Maritime Commission took no action then.[24] On December 8, 1942, however, acting on a verbal directive from the President,[25] the Commission awarded contracts for the construction of 69 corvettes. One purpose of these awards seems to have been to make use of the yards on the Great Lakes which would soon finish the coasters (and other special types) which had been awarded them in the summer of 1941 and for which many new yards had been built.[26] The five Great Lakes yards in the corvette program were: Walter Butler, 12 ships; Globe, 8 ships; Leatham Smith, 8 ships; American, 7 ships; Froemming, 4 ships. But the largest contracts went to the West coast: 18 ships to Consolidated at Wilmington, California, and 12 to Richmond No. 4. Moreover, Kaiser Cargo, Inc., was given a contract to develop the specifications, working plans, and requisitions and to place the orders for material under the supervision of the West Coast Regional Office.[27] Plans for handling in this way the design and procurement for the corvettes had been worked out at a conference at the West coast office on November 7, not without protest from the shipbuilders of the Great Lakes who were called on to furnish draftsmen and naval architects to go to the West coast and work with the Kaiser organization.[28] About fifty men were furnished by the Great Lakes Office and shipbuilders of that area.[29] Kaiser's control of procurement was protested by businessmen from Duluth before the Small Business Committee,[30] and the erection sequences developed

---

[24] W. L. Batt to Land, June 25, 1942, in MC gf 506-1-1-21.

[25] Land to Secretary of Navy, Dec. 24, 1942, in Program Collection.

[26] Statements to me by W. E. Spofford, Apr. 11, 1949 stressed the pressure from the Great Lakes to have work placed in that region. There was labor available there.

[27] MC Minutes, pp. 25696-700.

[28] Stenographic notes of "Conference held in West Coast Regional Construction Office, November 7, 1942—Subject: 'Corvettes'"; in Conference Reports, West Coast Office, in MC, Historian's Collection, Cabinet 7. It includes data on simplifying the design.

[29] Statement by W. E. Spofford, Apr. 11, 1949.

[30] See above, p. 407.

on the West coast, where they were accustomed to cranes able to lift forty or fifty tons, had to be done over by the Great Lakes Regional Office to fit its yards, few of which had cranes able to handle more than ten tons.[31] These were a few of the many headaches involved in building the escort vessels.

When the program started they were called corvettes but were later known as frigates. The temporary use of the name corvette was due to the fact that the Canadian design for vessels of that name was taken as the starting point by Kaiser and the West coast office in developing specifications.[32] They were light vessels of shallow draft with twin screws powered by reciprocating engines. Changes such as providing bunks instead of hammocks were necessary to fit them for American crews, and new specifications for electrical work and steel gauges had to be made to fit American industrial equipment.[33] Moreover, the Canadian design was modified by adding a foot to the beam to increase stability, and the length was increased by adding in the middle of the ship to relieve cramped conditions in the machinery space.[34] Therefore the frigates (S2-S2-AQ1) may be considered a new design (Plate XXIII).

A special problem encountered in building frigates on the Great Lakes was that of getting them to the sea. Those built in the three yards on Lake Superior could not pass Sault Ste. Marie between November 15 and April 15. All the frigates, whether built on Lake Superior, Lake Michigan, or Lake Erie, had to go down the Mississippi. They were too long to pass through the locks of the Cardinal and LaChine ship canals and down the St. Lawrence, as had the coasters and V4 tugs built in the Lakes. Therefore they had to go through the Chicago Drainage Canal down the Illinois River to the Mississippi. For the ships to pass under the bridges over the canal, their masts had to be taken down; and the outfitting was completed in Gulf yards.

Another difficulty was that the frigates had a draft of 13 feet and the channel down the Mississippi was normally only 9 feet and might be less in case of low water. Consequently the Great Lakes office worked out arrangements to attach four pontoons to

---

[31] Statement by W. E. Spofford, Apr. 11, 1949.
[32] Statement of J. L. Bates, Apr. 12, 1949.
[33] Statements by W. E. Spofford, Apr. 18, 1949.
[34] I. J. Wanless to F. C. Lane, Nov. 16, 1950,

each frigate so as to reduce its draft to only 8 feet $1\frac{1}{2}$ inches. One frigate was sent down the river without pontoons at a time of high water, but the rest required this equipment, and all had to be fitted with channel irons over the stern to guard against damage from contact with other vessels in the tow. The first five frigates were manned by hastily collected " scrub " crews which did so much damage that when the Navy took over the vessels they had a lot of outfitting and overhauling to do. Thereafter the Navy and Coast Guard supplied crews.[35]

As a whole the construction of frigates lagged several months behind schedule. When the contracts were awarded in December 1942, 69 frigates were scheduled for delivery in 1943, and deliveries did not total 69 until June 1944. By that time the Joint Chiefs had approved additions; in October 1944 the Commission's yards completed delivery of 96 frigates altogether.[36]

Among the causes for the delay was difficulty with lining up the engines. Doing it cold, as was originally recommended by the Commission or the West coast office, led to trouble with the bearings. It was found more advisable to line up the engines hot, as was done at Richmond No. 4 and Consolidated's Wilmington yard, and then elsewhere.[37] The fastest building of frigates was done by the Consolidated Steel Corp.

This delay in completing frigates led to an especially bad situation in the yard at Providence, Rhode Island, which had been begun by Rheem and was taken over by Walsh-Kaiser. The new management ran into difficulties which were attributed to late delivery of material and inexperienced management.[38] But in part the poor record of this yard even in 1944 is to be explained by the uncertainties and discontinuities connected with building for the military. In June 1943, a few months after the new management had taken over and begun work on the frigates, Walsh-Kaiser was awarded a contract for 32 cargo attack ships for the Navy (AKA) known as S4-SE2-BE1's. Later, because it appeared that the Navy would not have its designs for these BE1's ready in time,

---

[35] Statements by W. E. Spofford, Apr. 11, 1949. See also MC Minutes, pp. 28188 (Apr. 18, 1944); p. 29429 (Aug. 15, 1944).

[36] Comparison of *Official Construction Record* of deliveries and the schedules in MC Program Collection.

[37] House, *Walsh-Kaiser Co., Inc. Hearings*, pp. 251, 262, 272; and statements by W. E. Spofford.

[38] Land to A. J. Horne, June 23, 1944, in Land reading file.

Walsh-Kaiser was given a contract for 6 Liberty ships.[39] Consequently, when trouble developed with the frigates and their outfitting dragged on into the midsummer of 1944, work was going on in the yard on the frigates, the Liberty ships, and the attack cargo ships all at the same time. This was the situation in April of 1944 when the yard was investigated by the House Committee and charged with mismanagement and labor hoarding.[40]

II. TRANSPORTS

Whereas the LST's, the escort aircraft carriers, and the frigates were built with no non-military uses in mind, most transports were potential passenger ships, or combination passenger and cargo ships. In some cases that affected the design, and it may have been one reason why so many different kinds of transports were built. Also the Army and Navy had various and changing ideas about what kinds of transports they wanted. The consequent difficulties are fully illustrated in the history of the C4's.

The C4 design originated in the plans developed by the naval architect, George Sharp, in 1941 for a cargo ship for the American Hawaiian Lines and in the plans for its conversion into a troopship prepared at the same time so that it would be possible to proceed with the building of either type without delay. In his design, changes for conversion into a troopship were kept at a minimum because of the thought that these vessels would be turned over to the American Hawaiian Steamship Co. for use as cargo ships after the emergency.[1] In the latter part of October 1941 the Maritime Commission took over Sharp's design [2] and subsequently developed from it a number of designs both for cargo carriers and troop ships.

Two shipyards were constructed to build C4's. The first to be authorized was the five-basin Richmond No. 3 yard of Kaiser

---

[39] Later reduced to five. MC Minutes, p. 29199 (July 27, 1944).

[40] House, *Walsh-Kaiser Co. Inc., Hearings*, pp. 251-52, 262, 272 and *passim*. On the award of the frigates (corvettes) see MC Minutes, pp. 24461-62, 24547. On the work in the yards see the Reynolds' Way Charts.

[1] Memorandum from Ivan Wanless, Chief of Preliminary Design Section to Chief of the Final Design Section, Oct. 28, 1941, in MC gf 506-14-1. Cited in Helen Knuth, "The C-4 Shipbuilding Program," a typescript prepared under my direction. Miss Knuth's research is the basis of this section.

[2] Memorandum from Wanless to Chief, Hull Final Design Section, Oct. 28, 1941, in MC gf 506-14-1, and other memos in the same file.

Co., Inc. In building the yard the contractor encountered two unexpected developments which slowed up the work and raised the cost enormously. The first of these difficulties was the earth slippage which occurred at the east margin of the yard as the result of the instability of underlying strata. To correct this it was necessary to flatten out the slope of the embankment under the outfitting dock, to increase the width of the dock, and to build the dock out of concrete instead of lumber as had originally been planned. It was also necessary to rearrange certain buildings, railway tracks, and yard utilities in order to insure them against destruction from earth slippage. The cost of excavating the basin site and partially removing the hill behind the yard was increased by the unexpected encounter of a hard rock core.[3]

The second yard destined to build C4's was an extension of the Sun Shipbuilding and Dry Dock Co. at Chester, Pennsylvania, which was called Sun No. 4. When the construction of this yard was first planned, it was expected to build, not C4's, but seatrains —namely the 50 seatrains agreed to at the conference in President Roosevelt's bedroom on February 19, as described in chapter 5. The seatrain, as its name suggests, was a ship designed to serve as freighter for loaded freight cars. This specialized type of ship was developed for ocean service between Havana, Gulf ports, and New York. In 1941 both the Navy and Army had requested that several of these be secured from the operator, and in February 1942 the Army put in a request that 50 be built. Sunship submitted estimates and, after the White House conference, were given a contract (awarded February 24). But by March 13, 1942 the Navy became opposed to these vessels because of their vulnerability, emphasizing that they "could not possibly survive one torpedo hit," and that even a minor caliber projectile could jeopardize both ship and cargo if it hit along the water line. On April 27, 1942 the Army recorded its agreement that the seatrain was not the best ship for its purposes and asked for the construction of C4's instead at the Sun yard No. 4. Accordingly a new contract was made with Sun, awarded August 18, 1942, for 50 C4's.[4]

Plans for the construction of Sun No. 4 had been going forward

---

[3] Carl W. Flesher, Director, West Coast Regional Office, to MC, May 2, 1942, in MC gf 503-80; Flesher to Commission, July 10, 1944 (MC Minutes Aug. 1, 1944) in Historian's Collection, Weber file, "C-4's."

[4] Knuth, "The C4 Shipbuilding Program," pp. 2-8.

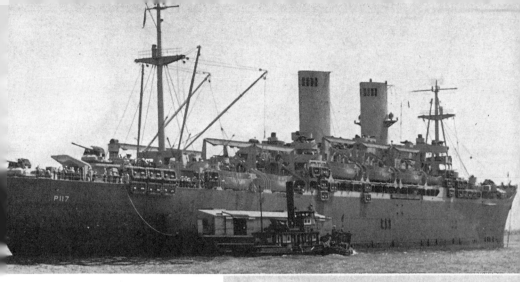

Plate XXIII.

Above, one of the army transports of the type built at Federal and Bethlehem-Alameda (a P2), the biggest ship built for the Commission during the war. Right, an army transport of the type built at Richmond No. 3 and Sun Ship (a C4). Below, the frigate, an anti-submarine escort vessel resembling the Canadian corvette (design S2-S2-AQ1). Many of these were built for the Maritime Commission in yards on the Great Lakes.

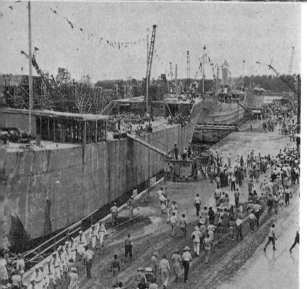

PLATE XXIV.

Above, the coastal cargo ship (C1-M-AV1) of which 218 were delivered in 1944-1945 for the needs of the Navy in the Southwest Pacific. Left, concrete ships being launched from basins at Tampa, Florida. Below, the powerful seagoing tug (V4-M-A1) of which 54 were constructed for the Maritime Commission. They formed part of the tug armada that towed concrete caissons across the English channel to form the artificial harbors off the Normandy beaches.

meanwhile. It was to be an 8-way yard constructed on land already owned by Sun lying directly north of the North Yard. It was estimated to cost $10.4 million and finally cost $12.6 million.[5] Delay occurred in its construction also, not because of difficulties of the site, but because shipway cranes were slow in arriving. It was May 1943 before keels had been laid on all eight ways.[6]

After the delays in constructing and equipping the two yards—Richmond No. 3 and Sun No. 4—came delays because of changes in the designs and specifications of the ships.

The contracts awarded Sun in August 1942 called for the C4-S-B1 design, a cargo ship which was primarily a tank carrier. Much time was spent developing a design to make possible the self-loading and discharge of tanks through interior and exterior ramps, but only one ship was finished on this design.[7] In September 1943, the Joint Chiefs decided that the 49 other C4's at Sun should be converted to Army troopships in order to fill the requirements for transportation of troops overseas in accordance with the decisions of the Quebec Conference.[8] Even 7 unfinished first-round ships were to be changed to the transport design (C4-S-B2).[9] Actually, 6 of Sun's C4's were turned over incomplete to the Navy to be converted into hospital ships and 8 were finished as troop ships.[10] In 1944 or 1945 there was another change, and the last 5 C4's built by Sun were again cargo ships (design C4-S-B5).[11]

At Richmond there was never any doubt that the C4's would be transports. But within the same basic transport design a great many modifications were possible. Both at Richmond and at Sun, construction was impeded by a larger number of modifications of this kind. Some arose from problems of armament and

[5] *Ibid.*, pp. 13-14; MC Minutes, p. 21271 (Mar. 3, 1942) and C. H. Lundegaard to MC, June 16, 1945 in Weber file on C4's.
[6] Reynolds' Way Charts; John G. Pew to Vickery, Sept. 1, 1942 in gf 120-124-2.
[7] Knuth, "The C4 Shipbuilding Program," pp. 20-21.
[8] Leahy to Land, Sept. 17, 1943, in MC gf 506-14-3.
[9] Reynolds' Way Charts; attachment to Robert Haig to Vickery June 17, 1944 and McInnis to Haig, Sept. 3, 1943, in gf 506-14-3.
[10] Minutes, p. 28745 (June 13, 1944) for completion of 8 as troop ships; Minutes, pp. 28943-44 (July 4, 1944) for delivery of 6 incomplete to the Navy. In Fischer's *Statistical Summary*, Table C, the C4-S-B2's turned over to the Navy for completion are listed as cargo ships because they were not delivered to the Commission as military types.
[11] Reynolds' Way Charts.

stability, desire to increase the troop-carrying capacity, and efforts to increase resistance to battle damage.[12] Only one aspect can be considered here in any detail—that connected with the uncertainty over what branch of the government should operate the ships.

There was a struggle going on among the Navy, and Army, and the Maritime Commission's Siamese twin, the War Shipping Administration, for control of shipping. This contest over who would operate the ships affected various features of their design and especially the provision for accommodations for crews, since the Navy's requirements in this respect were quite different from the Army's or the civilian standards followed by the War Shipping Administration.

When plans for the C4 transports were first drawn, the Maritime Commission proceeded on the assumption that they would be operated for the Army by the WSA or by the Army with civilian crews and that therefore the Navy need not be consulted about their design.[13] The Navy, on the other hand, kept in mind the possibility that the C4's might some day be turned over to them to be manned by naval personnel and for that reason strove to bring the ships as close to Navy standards as possible. Other questions besides that of accommodation for the crew were involved in the suggestions made by the Navy in the summer of 1942, but Admiral Vickery's instructions to the Commission's Technical staff were that since " the arrangements lay entirely between the Maritime Commission and the Army, the desires of the Navy Department should be ignored and the matter allowed to simmer." [14]

The " simmering " threatened several times to come to a boil, and a new arrangement was made in February 1943 which provided that crew spaces on C4 vessels " should be altered to their maximum capacity " so that if the Navy should be called on to man the vessels at the last moment " the vessels could be blown up to meet naval personnel requirements." [15] In April the matter

[12] Knuth, " C-4 Shipbuilding Program," pp. 19-23, based on the many letters on the subject in MC gf 506-14-2, 506-14-1, 506-14-3. Richmond built C4-S-A1's.
[13] Report of Conference held with Bates, McPhedran, Wanless, Haynes, Smith, Vickery, Flesher and Shulters, Nov. 19, 1941, in MC gf 506-14-1; House, *Production in Shipbuilding Plants, Executive Hearings*, pt. 3, June 23, 1943, p. 899.
[14] Bates, Memo. for Files, June 27, 1942, in MC gf 506-7-1.
[15] Bates, Memo for Files, Aug. 4, 1942, in MC gf 506-14-1 and Huntington Morse to Land and Vickery, Feb. 27, 1943 in MC gf 507-14-2.

really did boil over. On the sixteenth, John Pew, President of Sunship, wrote to Vickery that the preceding night 50 officers and 400 Coast Guardsmen had descended upon the town of Chester and stated that they were to work at the shipyard. Mr. Pew, who was obviously more than a little confused by the turn of events, reported that the Coast Guardsmen seemed to think that Sun was building ships for the Navy or the Coast Guard, although he had assured them that he was only building for the Maritime Commission and for private owners.[16]

But Mr. Pew was building for the Navy even though he did not know it, nor, apparently, did the Maritime Commission at that time. Late in the previous month, in March 1943, the Army had agreed reluctantly to have Army transports manned by naval personnel. Official notice came to Admiral Land in April in a letter setting forth the recent decision to assign the C4 ships to the Navy under bareboat charter, and stating that the Bureau of Ships would review the plans of the ships to see what changes would be necessary.[17]

Throughout the remainder of the spring and summer the Navy suggested modifications and changes.[18] In the latter part of October, Carl Flesher, Director of the West Coast Regional Office, telegraphed Admiral Vickery: "If you can find some way to stop the Navy Department from requesting so many changes, it will assist us greatly in new construction and conversion work in San Francisco area."[19]

Unfortunately this conversion to naval transports was not the last change these vessels were to undergo. The following summer the Commission was informed by the Naval Transport Service that the Navy was not in a position to man the C4 transports and could not therefore accept delivery of the vessels.[20] This left the Commission with the problem of converting the vessels for

[16] Pew to Vickery, Apr. 16, 1943, in MC gf 506-14-3.
[17] Knuth, "C4 Shipbuilding Program," pp. 25-26; Horne to Land, Apr. 30, 1943, in MC gf 506-14-2.
[18] Wanless Memo for Files, May 7, 1943, in MC gf 506-14-1; Report of Conference in office of George Sharp, May 26, 1943, in MC gf 506-14-1; Land, Memo for Admiral King, CNO, Sept. 15, 1943, in MC gf 506-14-2; Admiral James Pine to Land, Oct. 26, 1943, in MC gf 506-14-1.
[19] Flesher to Vickery, Oct. 28, 1943, in MC gf 506-14-1. For award of contract for conversion to Navy design see Minutes, p. 28636 (June 1, 1944).
[20] Bates to Vickery, July 6, 1944, in MC gf 506-14-1.

merchant crews to meet Coast Guard standards for merchant ships.[21]

There were other difficulties encountered by the Commission in the design of the C4's. For example, a great deal of time was spent in redesigning the first Army troopship because of a misunderstanding with the Army concerning the requirement for hospital space. The Maritime Commission apparently thought that the design had been checked and cleared by all cognizant sections of the Army and had commenced work on the ships when the Surgeon General's Office complained that they had not approved and could not approve of the plans.[22] Another difficulty was the matter of boatage. It was not possible to reconcile naval requirements for life boats or landing boats with Coast Guard standards for merchant ships. The correspondence on this subject alone was enough to fill two or three volumes in the files.[23]

Changes in the design of the vessels did not account for all the difficulties encountered in the construction of the C4's. Not only were the contractors delayed in completing the facilities, as has been explained, but they were slowed down by deficiencies in their labor force and supervisory personnel. In the Richmond yard, work was hampered by a shortage of skilled coppersmiths, electricians, sheet metal workers, and pipefitters; and it was in this yard, it will be recalled, that Kaiser had outstanding difficulties in fitting his " sand hogs " and the old-line shipbuilders like E. W. Hannay, Sr., together into one smoothly operating organization.[24] Another hindrance was the construction of the adjoining and closely connected yard known as Richmond No. 3A or 4, in which the LST's were being rushed to completion. The amount of progress made on the C4's by April 1943 seemed to Admiral Vickery so unsatisfactory that he wrote Henry Kaiser comparing his yard unfavorably with the Sun yard, which seemed to be doing better, he said, in spite of its being " completely manned by colored labor which had no experience in shipbuilding." Of

---

[21] *Ibid.*, also Green, Asst. Supt. Vancouver Yard to C. H. Johnson, Tech. Asst. to Vickery, July 11, 1944; Vickery to Kaiser Co., Richmond, Aug. 12, 1944, in MC gf 506-14-2.

[22] Surgeon General Magee to MC, Mar. 22, 1943, in MC gf 506-14-1; Colonel Hicks to Rohn, Sept. 18, 1943, *ibid.*; Bates to Director ECRO, Sept. 27, 1943, *ibid.*; Vickery to Somervell, Oct. 12, 1943, *ibid.*

[23] The series having the most material on this are: MC gf 506-14-3 and 506-7-1.

[24] See above, chap. 14, sect. 1.

Richmond No. 3, he said, "There seems to be something radically wrong, and I think before this matter becomes the subject of adverse publicity for you, you should take steps to correct it."[25] When the unfavorable publicity did come and led to the Congressional hearings in June of 1943, Admiral Vickery presented Kaiser's achievements in a favorable light and stressed the difficulties over design, especially those connected with armament.[26] In general, it was Admiral Vickery's practice to back up his builders in public while needling them in private.

One year later it was Sun's construction record about which Vickery was complaining. By April 17, 1944, 26 vessels should have been completed, but only two deliveries had actually been made. This was the result, Vickery concluded, of "a dilution of management, necessary dependence upon inexperienced personnel, and a lack of sufficient personnel." Since he believed that not more than 20 vessels could be completed by the end of 1945, and the facilities at Sun were needed to build tankers, he suggested that the balance of the ships be constructed elsewhere,[27] namely, 20 at the Kaiser Vancouver yard and 10 at Richmond No. 3. Contracts were made accordingly; but since labor shortage and the pressure of other construction (AP5's) was by the fall of 1944 slowing up work in the Richmond area, only 5 out of 10 were built at Richmond No. 3. The total output of C4's from that yard was 35. The other 5 were assigned to Vancouver and then cancelled, so that the final record on construction of C4's was as follows: [28]

Troop transports built at Richmond No. 3	35
Troop transports built at Vancouver	10
Troop transports built at Sun (including incomplete hospital ships)	14
Tank carrier built at Sun	1
Cargo ships built at Sun	5
Total	65

No other transports (perhaps no other ships) had as confused

---

[25] Vickery to Kaiser, Apr. 16, 1943, in MC gf 506-14-3, and Kaiser to Vickery, Apr. 18, 1943, in MC gf 503-80.

[26] House, *Production in Shipbuilding Plants, Executive Hearings*, pt. 3, pp. 890-918.

[27] Minutes, pp. 28194-95 (Apr. 18, 1944) and pp. 28328-30 (May 2, 1944).

[28] Two of the transports included in the 35 built at Richmond No. 3 completed their outfitting at Vancouver. Knuth, "C4 Shipbuilding Program," pp. 15-17, and Reynolds' Way Charts.

a record of changes in design and contracts as did the C4's. The record of the construction of P2's, at Federal and Bethlehem-Alameda, is contrastingly simple, although their construction was long drawn out. They were indeed big ships, the biggest built by the Commission during the war. With a length overall of 610 feet they were 10 feet shorter than the ore carriers, but had nearly twice the lightship displacement, being about 10,000 tons. Their design was made in the Maritime Commission late in 1941 under instruction from Admiral Vickery that the probable need for transports be given first consideration, but with some thought that the ships to be built at Federal would be operated by the Grace Line or Moore-McCormack to South America; and those to be built at Alameda would be operated by the American President Line, to the Orient. The former had 25 foot draught, the latter 29 foot, as was appropriate for the two routes. Each had two turbines and twin screws, with divided machinery space, which was necessary for safety in a transport but less economical in commercial operation. At Federal, standard C3 geared turbines were used, which gave 18,000 shaft horsepower; the ships built at Alameda had turbo-electric drive and 19,000 shaft horsepower. Both had a design speed of 19 knots.[29] See Plate XXIII.

Contracts were awarded on January 16, and February 3, 1942: 10 ships to Federal (design P2-S2-R2), where they filled ways which would otherwise have been used for C2's, and 10 ships to Bethlehem Steel Co. (design P2-SE2-R1), which then began building a new yard at Alameda, California, devoted entirely to these transports.[30] The delivery of the first P2 by Federal in July 1943 represented, said Admiral Vickery, " the shortest time for the development and construction ever achieved for a vessel of this type and size." [31]

The possible use of these vessels as passenger ships came under consideration late in 1944 when the end of the war appeared in sight. At the close of the year Federal had delivered 9 of its 10 and nearly finished the other one. Bethlehem-Alameda had

[29] Statements by James L. Bates, Apr. 11 and June 21, 1949, and A. Osbourne, June 22, 1949. Fischer, *Statistical Summary*, Table A-1.

[30] MC Minutes, p. 20637 (Jan. 16, 1942) —Federal; p. 20957 (Feb. 3, 1942) —the facilities contract with Bethlehem Steel Co. for this purpose.

[31] H. L. Vickery, " Commission Ships for Postwar Service," *Marine Engineering and Shipping Review*, May 1945, p. 146.

delivered 3, had launched 3 others, and had 4 on the ways.[32] Plans were made to have Federal build 3 more P2's on a slightly modified design (P2-S2-R4) for the United States Lines as passenger ships adapted to the needs of Atlantic travel, and to have Bethlehem-Alameda finish the last two of its vessels (P2-SE2-R3) equipped as passenger ships for the American President Lines. Only in the Alameda yard were these postwar plans carried through;[33] the ships at Federal were cancelled when it appeared that their success in the merchant service was problematical.[34]

Whereas the P2's were potential passenger ships in spite of their handicaps, the 32 Navy transports built by the Commission with the design S4-SE2-BD1 were a purely military type. They were much smaller, with shallow draft, provided also with divided machinery compartments, and were arranged to carry large amounts of combat equipment so that they were rated by the Navy as APA's, attack transports. They were built by the Wilmington yard of Consolidated Steel Corp., which finished its deliveries in April 1945 three months ahead of contract schedule. This achievement of Consolidated stood out the more by contrast with the difficulties of Walsh-Kaiser, which was building for the Navy attack cargo ships of the same dimensions (S4-SE2-BE1). The first keels for these two variations of the S4 type were laid at about the same time, but Walsh-Kaiser, hampered by having frigates and Libertys in the yard, did not finish its BE1's until August 1945.[35]

More numerous than any other type of transport were those built on the Victory ship design. Working plans, specifications, and procurement for the Victorys were pushed rapidly as soon as the controversy with WPB was settled in August 1943.[36] In October

[32] Reynolds' Way Charts. Federal delivered another P2-S2-R2, June 29, 1945, making 11 its total production of this type.

[33] H. L. Vickery, " Commission Ships for Postwar Service," *Marine Engineering and Shipping Review*, May 1945, pp. 146-49, also in *Nautical Gazette*, vol. 135, May 1945, pp. 60-61.

[34] Statements by James L. Bates, June 21, 1949; MC Minutes, pp. 32899-32901 (Aug. 17, 1945), just after V-J day. The cancelled ships were from 12 to 50 per cent complete.

[35] MC, " Permanent Report of Completed Ship Construction Contracts "; House; *Walsh-Kaiser Co., Inc., Hearings*, p. 262; Reynolds' Way Charts.

[36] C. H. Johnson to Technical and Procurement Divisions, Sept. 3, 1943 and Vickery to various suppliers, Sept. 3, 1943 in MC gf 506-22-1, and Tyler, " Victory Ship Design," p. 81.

of 1943 Admiral Vickery, while on a visit to London, stirred up controversy by some remarks about the commercial possibilities of the Victory in postwar competition;[37] but their suitability for wartime needs came to the forefront of attention again in November 1943 when the Joint Chiefs informed the Maritime Commission that 130 attack transports, as well as a number of other military types, should be added to its program. Obviously this required a reduction in the program for merchant shipbuilding so that the yards could be used for these military types, and the Commission's program of November 20, 1943 revised accordingly the schedule for 1944 and 1945.[38] Some of the new military needs were met by converting C-types, but the demand for 130 attack transports was met mainly by a modified form of Victory ship, VC2-S-AP5 design.

The Victory transport was enough like the Victory ships already planned so that there was no violent disruption of the continuity of production and procurement when yards which were constructing Victory freighters, or preparing to, built Victory transports instead. Richmond No. 2 changed over directly from Liberty ships to Victory transports in May and June of 1944. Calship and Oregon which had laid keels for Victory freighters in November 1943, launched two to four rounds of this design and then changed over to the Victory transports. Another West coast yard organized for multiple production, Kaiser-Vancouver, was assigned Victory transports in place of the tankers it had expected to build when it completed the aircraft carrier escorts, and it too was laying keels of AP5's in April and May of 1944. Altogether the Commission built 117 transports of this one design, VC2-S-AP5.[39] In constructing these transports the Commission was able to use fully the methods of multiple production, although in procurement and in labor force it encountered difficulties which will be examined in the next chapter.

Many other military types could be mentioned—Navy tenders,

[37] Tyler, "Victory Ship Design," pp. 82-87.
[38] Joint Chiefs of Staff to Land, Nov. 9, 1943, and schedule of Nov. 20, 1943, in MC Program Collection; and below, chap. 19.
[39] Kaiser-Vancouver began work on its C4's only after finishing 31 AP5's—Reynolds' Way Charts. On the contracts for these Victory ships see below, p. 679 and n. Vancouver, between the completion of its own AP5's and its building of C4's, outfitted 5 of the AP5's built at Calship, and Moore Dry Dock outfitted 4 of the AP5's built at Calship in order to finish them within the time limit set by the Joint Chiefs.—MC Minutes, pp. 31793-95 (Apr. 26, 1945).

five designs of Navy oilers, gasoline coastal tankers, other cargo attack ships, hospital transports, other troop transports, and attack transports.[40] Frequently the military types were conversions of ships begun as C-types. In instances in which change of type took place after work was far advanced, or the specifications were changed substantially while work was in progress, the tearing out of what had just been built seriously damaged the morale of workers. Congressman Bradley told his committee investigating shipyards of one instance in which a C3 had, after launching, been converted into a refrigerator ship, and when 85 per cent complete was to be reconverted, it was reported, into a combined cargo and refrigerator ship, which meant tearing out four refrigerator holds and reconstructing them. If that was done, the workers threatened en masse to stop buying war bonds. Workers suspected all this converting and reconverting was arranged by managements to make more money. Admiral Vickery, on the other hand, said that management was always against changes,[41] and I do not see how management could have derived any benefits, although they certainly must have had many headaches, from the innumerable changes made in military types during 1943 and 1944. But it was impossible to explain to the workmen the military reasons back of the changes. Even in retrospect only a technical military historian could pass any judgment on them. They upset both scheduling and morale and were one reason why building for the military increased the administrative and industrial problems out of all proportion to the total tonnage produced.

III. Concrete Barges and Other Minor Types

Very much less important than the military types but involving also notable engineering and administrative difficulties were the minor types built by the Commission. The construction of coastal cargo ships, tugs, and barges became part of the Commission's program during the third wave of expansion, as has been described in chapter 2. Additions were made from time to time, primarily to meet specific requests from the British and from the American Armed Forces. Out of a total production of minor types costing (before renegotiation) about $990 million, $620

[40] Fischer, *Statistical Summary*, Table A-1 of Form 106.
[41] House, *Production in Shipbuilding Plants, Executive Hearings*, pt. 3, pp. 894-95, June 28, 1943.

million was for coastal cargo vessels of steel construction. Most numerous in this category were the N3's built for the British on the Great Lakes, and the C1-M-AV1's built mainly in 1944-1945 at the direct request of the Joint Chiefs. The category next in importance in terms of total cost were the concrete ships and barges into which were poured about $151 million plus about $16 million for facilities.[1] Although that is not a large item compared to the total spent by the Maritime on ships ($13,311 million), it is a conspicuous example of an expenditure which produced relatively little. Moreover, construction of concrete ships gave rise to special problems which deserve attention, whereas the problems of the construction of steel coastal vessels were part of those discussed elsewhere in general terms. Altogether the Commission contracted for 142 concrete vessels of all types, of which 38 were cancelled, 104 built.[2]

The experiment of building concrete ships had been tried during World War I with results which moved Admiral Land to declare to a House committee in January 1941: "Personally, you are not going to get any from me as long as I am Chairman of the Maritime Commission. If you gentlemen start building concrete ships, as far as I am concerned, I am through."[3] He abandoned this attitude under the pressure of war. While doubtful of the wisdom of building concrete vessels he considered it a minor matter, certainly not a sufficient reason for resigning his post in time of war.[4] He and the Commission yielded to the various pressures which gradually edged them into it.

The opening wedge was inserted during the first scare over the shortage of steel plate, in June 1941. Something had to be done to increase the movement of petroleum products from Texas to the Northeast. The proposal to build a pipeline was not favored by the shipbuilding fraternity; the barge program provided an argument against building the pipeline.[5] In view of the shortage of steel, the Technical Division of the Maritime Commis-

---

[1] Fischer, *Statistical Summary*, Table H-5; House, *Independent Offices Appropriation Bill for 1946, Hearings*, p. 581.

[2] *Ibid.*, p. 580; Fischer, *Statistical Summary*, Table B-3.

[3] House, *Emergency Cargo Ship Construction, Hearings*, p. 15, punctuation changed to make sense.

[4] Statements to me by Admiral Land.

[5] Land to R. K. Davies, Deputy Pertoleum Coordinator, Sept. 5, 1941 in Land reading file. On the situation at that time, see above pp. 63-65, 315-18.

sion recommended that tow barges be built of reinforced concrete. It envisaged the construction of 50 or 60 during the first six months of 1942. It recognized that concrete ships would " require at least half the steel of an all steel hull," but the steel would be in the form of bars of low carbon content, and would thus avoid the wartime shortage of plate or strip.[6] The Director of the Technical Division, James L. Bates, supported the building of concrete ships primarily because he hoped to avoid another horrifying attempt, like that of World War I, to build those " beautiful wooden ships " for which President Roosevelt and other yachtsmen were openly enthusiastic.[7]

On these considerations, negotiations were opened with interested contractors, and conferences were held with the engineers who had worked on the many technical problems involved and were enthusiastic about the possibilities of their speciality.[8] Mr. Bates reported that the Technical Division was receiving " hearty cooperation from a number of strong companies," and also " a somewhat embarrassing interest from regional and political sources." [9] In November, contracts were awarded to three companies to build five barges each.[10]

When the first plans were made it was thought that no quarters for crews need be provided, but the Bureau of Marine Inspection and Navigation decided the barges must have crews to prevent them from becoming a menace to navigation. Provision for crews made it necessary to add not only a deck house but other required appurtenances including life-saving appliances and wartime defense features.[11] Then it was decided that the vessels should be classified by the American Bureau of Shipping, which therefore applied its rules concerning the strength of the structure and consequently informed the builders in January and February 1942 that their

[6] Report in MC gf 506-1 Exhibit 3. Cf. A. D. Kahn, " History of the Concrete Ship and Barge Program, 1941-1944 " (prepared in MC Technical Division, June 1, 1944, copy in Historian's Collection, Research file 111), p. 52.

[7] Statement by James L. Bates, June 21, 1949.

[8] J. Vasta, " History of Structural Design of Concrete Ships " (prepared in MC Technical Division, Feb., 1945, copy in Historian's Collection, Research file 111), pt. I, pp. 1-5.

[9] MC Minutes, p. 19326 (Oct. 15, 1941). See also Minutes, pp. 18491-93, memo from Bates of Aug. 13, 1941.

[10] Minutes, pp. 19658-68 (Nov. 8, 1941).

[11] Vasta, " History " above cited, pt. 1, p. 4; Minutes, p. 19326 (Oct. 15, 1941).

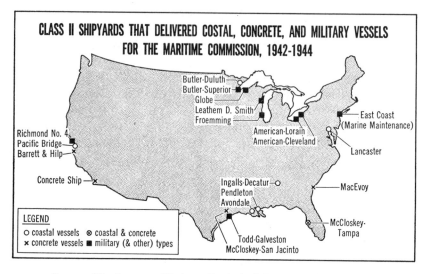

FIGURE 38. Source: Fischer, *Statistical Summary*, Table C-1.

designs must provide for more steel to reinforce the concrete.[12] Meeting the requirements of the Bureau of Marine Inspection and of the American Bureau of Shipping increased the overall utility of the vessels, but also increased the cost, the construction time, and the amount of steel consumed.

In spite of these complications, the Commission sharply increased the program in May and June of 1942. The three initial contractors—one working at Savannah, Georgia, one at Houston, Texas, and one at National City, California—were awarded new contracts. Also Barrett & Hilp, South San Francisco, received a contract to build 26 dry cargo barges to carry bauxite; and McCloskey & Co., with a site near Tampa, Florida, was given a contract to build 24 self-propelled concrete ships to transport sugar.[13] These awards were made at the time when submarine sinkings and steel shortage were both most acute. Reviewing results in April 1943, the Truman Committee reported that the steel used in concrete ships could have been put to better use

---

[12] Vasta, "History" above cited. Section on design B7-A1, pp. 5-6, 13; section on design B7-A2, pp. 5, 6, 11. See also Truman Committee, *Report No. 10*, 78th Cong., 1st sess., pt. 8, pp. 36-37, 48.

[13] Minutes, pp. 22338-43 (June 30, 1942).

in constructing steel producing facilities.[14] At the time he voted for the awards, Admiral Land gave as justification the need of grasping at straws. He wrote for the record: " While I am opposed to this construction under anything except great emergency conditions, in view of the seriousness of the submarine menace, the prospects for satisfactorily solving this problem, and the overall war picture, I vote ' aye ' on the basis of insurance to the U. S. A." [15]

Of the five contractors who undertook to build concrete vessels, the first to deliver—in February 1943—was Concrete Ship Constructors of National City, California, and they turned out many more than any of the other yards. They built in basins which were flooded to " launch " the finished hulls, and in which sections could be lowered into place without having to hoist them high over the sides of the hull.[16] Their initial design (B7-A2) provided one centerline longitudinal bulkhead and featured a typical transverse framing system, with frames closely spaced.[17] Later they built also 265-foot concrete lighters which were towed without crews and were used by the Army as floating warehouses. These lighters were considered by some experts " the outstanding contribution of the present program to concrete shipbuilding," because of two main points of difference between their design and that of all others previously used for concrete ship construction: " (1) the elimination of frames and beams and the substitution of a flat slab design suited to quick and economical building; and (2) the accommodation of the design and arrangement of reinforcing to pre-casting of bulkheads." [18] Records in rapid construction were made with this design.[19]

The other two firms that contracted in 1941 to build oil barges undertook to build them at Savannah and Houston according to a design (B7-A1) which was being developed by Col. P. M. Anderson, an enthusiast with much experience in concrete ship-

---

[14] Truman Committee, *Report* No. 10, 78th Cong., 1st sess., pt. 8, p. 48. By that time it was clear that there was shortage in the basic supply of steel. Cf. chap. 10.

[15] Memorandum for the Secretary and General Files, June 30, 1942, in Land reading file.

[16] Kahn, " History " above cited, p. 44.

[17] Vasta, " History " above cited, pt. I, p. 5.

[18] Kahn, " History " above cited, p. 19. For the history of this design see also Vasta's " History " above cited, section on design B5-BJ1.

[19] Delivery 42 days after keel laying, Vasta, " History," in section above cited, p. 10.

building in World War I. He died, however, in January 1942.[20] His design provided for a hull subdivided by two longitudinal bulkheads, and featured by a typical longitudinal framing system. From a structural point of view it was favored by the Technical Division over the B7-A2 design, but it proved " the most difficult design from the point of view of construction, as it was accompanied by difficult form work and objectionable but unavoidable steel congestion." [21] Technical difficulties in completing the design and working out construction methods were complicated by managerial changes, quarrels, and inefficiency in the two yards using it, MacEvoy and San Jacinto.

MacEvoy Shipbuilding Corp. was organized expressly for building concrete ships at Savannah by a contractor who was also active in building housing projects in New Jersey. His activity in both fields was investigated by the Truman Committee, which reported that " both cases show rapacity, greed, fraud, and negligence." [22] Clifford MacEvoy, the president of the corporation, and others connected with it were indicted for fraud.[23] The company finally delivered 7 barges between January and June 1944, and the contract for the others was cancelled.[24]

San Jacinto Shipbuilders, Inc. had a yard at Houston, Texas, which became the scene of so many personality clashes that it was referred to locally as the Second Battle of San Jacinto. There was " funny business " here also in the award of subcontracts and in the real estate arrangements. A change for the better was hoped for when the company was sold to McCloskey & Co. in September 1942, and new, more favorable contracts were awarded, but improvement was not rapid.[25] Four barges were delivered in November and December 1943, and the contracts for the others cancelled.[26]

---

[20] Truman Committee, *Report* No. 10, 78th Cong., 1st sess., pt. 8, p. 36.
[21] Vasta, " History " above cited, p. 5.
[22] Truman Committee, *Report* No. 10, 78th Cong., 1st sess., pt. 8, Apr. 1943, p. 17.
[23] *Ibid.*, p. 17, and for detail its Appendix III.
[24] Kahn, " History " above cited, p. 4. Vickery reported to the Commission on June 3, 1943 that notice of cancellation of contract had been given MacEvoy (MC Minutes, p. 25345) but complicated negotiations followed. See Minutes, pp. 25362, 25882-88, 25956, 26032.
[25] Truman Committee *Report No. 10*, 78th Cong., 1st sess., pt. 8, Appendix IV; MC Minutes, pp. 22935-37 (Aug. 27, 1942).
[26] Kahn " History " above cited, p. 4; Commission action dated Aug. 3, 1943 and Aug. 6, 1943.

The main activity of McCloskey & Co. in the shipbuilding program was at Tampa, Florida, where they developed a yard containing thirteen building berths in four basins. Concrete vessels were "launched" three at a time by flooding the basins. The cost of facilities was first estimated as $2.7 million and proved to be over $7.5 million, including housing.[27] Twenty-four ships, powered with steam reciprocating engines intended to drive the vessel at 10 knots, were delivered November 1943-December 1944. They cost $48 million which was $20 million over the estimated or "contract" price.[28] See Plate XXIV.

The delays, difficulties, and frauds encountered in some of the yards building concrete vessels give emotional backing to the feeling that the whole program was a mistake. The yards and the ships were built under contracts that guaranteed reimbursement of costs and payment of minimum fees, even when, as happened under McCloskey's contract, the cost proved much more than double the contract price.[29] In such a case clearly the government paid an outrageous amount compared to what it received. The West coast builders, Concrete Ship Constructors and Barrett & Hilp, produced somewhat comparable ships for about half the cost.[30] But even their costs were high compared to the utility of the vessel produced. The engineer with a scientific interest in solving problems can look back on the building of concrete ships with some satisfaction.[31] On the other side must be put the practical considerations summarized in Table 11.

The comparison is not meant to imply that the concrete barge was anywhere near comparable to the Liberty ship in usefulness. Quite the contrary. And none of the concrete barges, except the lighters, was found useful for the purpose originally intended. The gasoline shortage of the Northeast was relieved when a pipeline was finally built, and the 33 concrete barges constructed by the Maritime Commission for the transport of oil were taken by

[27] Kahn, "History" above cited, p. 44; Vasta, "History" above cited, pt. I, p. 8; Land to Sanford, May 29, 1942 in Land reading file. Finance Division record of vouchers paid under contract MCc7551. Exclusive of housing, the cost to July 11, 1944 was $5.6 million. Minutes, p. 29022.

[28] Finance Division record of vouchers paid under contract MCc7552.

[29] Senate, Truman Committee *Report No. 10*, 78th Cong., 1st sess., pt. 8, p. 39; Fischer, *Statistical Summary*, Table H-5.

[30] Fischer, *Statistical Summary*, Table H-5.

[31] This attitude naturally pervades the technical histories above cited by A. D. Kahn and J. Vasta.

the Navy which operated them in the southwest Pacific as floating storage tanks at advanced bases. The 20 dry cargo barges originally intended for bauxite were taken by the Army and 17 were used in the southwest Pacific for storehouses. Of the 24 steam cargo concrete vessels, 17 were converted by the Army into floating storehouses. Of the others, which were operated for a short time by private lines and then withdrawn, 5 were used by the Army as training ships and 2 found an honorable end when sunk to form

TABLE 11

CONCRETE BARGE COMPARED TO LIBERTY SHIP IN COST AND CAPACITY

	Concrete Barge B7-A2	Liberty (average of all yards)
Cost (thousands)*.....................	$1,326	$1,822
Steel used (short tons)		
For structural reinforcement†.........	1,600	.....
Hull steel§........................	.....	3,132
Carrying capacity in deadweight tons‡....	6,635	10,793
Speed under its own power (knots).......	none	11

   * Source: Fischer, *Statistical Summary*, Table H-5.
   † Source: Vasta, "History" above cited, section on design B7-A2, p. 15.
   § Source: Fischer, "Statistical Analysis, Steel," p. 1.
   ‡ Source: Fischer, *Statistical Summary*, Table A-1.

part of the breakwater protecting the American landing in Normandy at Omaha beach.[32]

Wood construction also deserves brief special mention. Altogether 1,164 wooden vessels were contracted for by the Maritime Commission and 1,131 were built. But of these only 41 were large barges and 20 were sea-going tugs; the rest were small tugs and barges which were ordered by and delivered to the British under lend-lease, most of them being small barges intended for use on the Tigris River in facilitating deliveries to Russia via Bagdad. That part of the barge program served the purpose for which it was intended.

Notably unsuccessful, however, was the program for building large wooden barges. They were planned as part of the effort to relieve the shortage of coal and other commodities in the Eastern seaboard cities. Although the main task of building barges for

---

   [32] A. D. Kahn, "Concrete Ship and Barge Program, 1941-1944," and House, *Independent Offices Appropriation Bill for 1946*, pp. 579-80.

use on inland waterways was shifted to the Office of Defense Transportation, which built both steel and wooden barges, the Maritime Commission was instructed to let contracts for wooden barges—some 180 feet long, others of 274 feet—which were represented as suitable for the Mississippi waterway, the inside passage along the Atlantic coast, and Gulf transportation.[33] Congressional pressure, especially from Maine and Washington, favored the barge program,[34] and President Roosevelt was full of sentimental enthusiasm for the wooden shipbuilders of Maine.[35]

Awards were made in April, June, and August of 1942, but progress in construction was very far from satisfactory. The crisis in the fuel supplies of the Northeast was solved in other ways, and in August 1943 the Office of Defense Transportation advised that the large barges would not be needed.[36] In September the Commission advised all the builders that it " would not accept delivery of any barge subsequent to December 31, 1943, and then cancelled construction of all wooden barges not completed to that date." [37] But the Commission could not so easily cancel the administrative headaches involved in winding up the contracts. None of the four Maine firms which had undertaken to build wooden ships were financially able to complete the work under the original terms, which were lump-sum contracts. An attempt was made to help them out, first by changing to price-minus contracts at higher price, then lifting the price again, and finally by waiving the claims to which they were subject because of defects. Although there was no doubt the defects existed, the Commission voted not to hold the builders financially responsible because they were all losing money on the work. The defects were said not to be the fault of the contractors but to be due to factors outside their

---

[33] " United States Maritime Commission History of the Wood Ship and Barge program, 1941-1944," by A. D. Kahn, Acting Chief Concrete Control Subsection, Technical Division, MC, prepared Dec. 15, 1944, pp. 7-11. Copy in Historian's Collection, Research file 111. Land to Senator McKellar, May 1, 1942 in *Congressional Record*, May 6, 1942, vol. 88, p. 2995; ODT, *Civilian War Transport, 1941-1946*, pp. 182-184.

[34] Truman Committee, *Hearings*, pt. 5, pp. 1283-84; *Congressional Record*, May 6, 1942, vol. 88, p. 2995; Land to Nelson, March 24, 1942, Land to Senator Bailey, July 17, 1942, Land to President, July 20, 1942, in Land reading file.

[35] Statements by Admiral Land in interview, Sept. 30, 1946.

[36] Kahn, " Wood Ship and Barge Program," above cited, pp. 9-10, 26-28.

[37] MC Minutes, p. 28512 (May 25, 1944).

control such as deficiencies in the lumber assigned the yards by the joint board which allocated timber supplies.[38]

Linked to the construction of barges was that of tugs, some of which were also built of wood as part of the effort to save steel. Especially notable were the ocean-going steel tugs, the 49 V4's, powered with Diesel engines. Many were built on the Great Lakes, and nearly all of them were finished in 1943. Although originally ordered to tow concrete barges, they proved very useful in other connections, especially in towing the sections of the artificial breakwater constructed off Omaha beach.

Commenting on the number of different kinds of tugs and barges contracted for by the Army, Navy, and Maritime Commission, the Truman Committe suggested it would have been preferable if this work had been channeled through one agency where experienced personnel was concentrated and which could effect a maximum in simplifying and standardizing design.[39]

In this discussion of minor types, attention has been given mainly to the ventures into concrete and wood construction because mention is made elsewhere of most of the other minor types, among which were: the 21 ore carriers (so large that their classification as a " minor type " is really an anomaly), 75 harbor tugs, 17 refrigerator ships, 5 Dutch motor coasters, 314 dry cargo coasters of steel construction, and 64 coastal and inland tankers. Although the total tonnage involved was small, the number of ships was large—and large correspondingly were the administrative problems. The Director of the East Coast Regional Construction Office wrote to Admiral Vickery: " The tugs and barges have given us more perplexing problems and a far greater number of aggravations than the equivalent number of large ships regardless of the tonnage differential." [40]

---

[38] Minutes, pp. 28512-15 (May 25, 1944). On the change to price-minus, see Anderson to Commission Jan. 28, 1943 in Anderson reading file. On the pressure from Maine, see notes on telephone conversation of Jan. 25, 1943 in East Coast Regional Office, Director's files, East Coast Investigation, in MC, Historian's Collection.
[39] Truman Committee, *Report No. 10*, 78th Cong., 1st sess., pt. 8, p. 16.
[40] McInnis to Vickery, Jan. 9, 1943 in MC gf 201-3-6.

Chapter 19

THE CONTRAST BETWEEN 1943 AND 1944

I. Programs and Production in 1943 and 1944

IN STANDARDIZED multiple production, 1943 was the year that broke all records; 1944 was to be of a different character. Whether the achievement in production was greater in 1944 than in 1943 is debatable, but it was certainly different in 1944 and it was beset with distinctive difficulties. The schedules set at the beginning of the year were not met so successfully as in 1943 for four reasons: (1) there was some let down after the record-breaking efforts of the last months of 1943; (2) the ships built in 1944 were of more complex construction and the change-overs from one type to another decreased productivity; (3) delivery of components lagged behind shipyard schedules; and (4) shortage of manpower became a more serious problem. The difficulties with labor supply deserve most attention, for they brought into question some of the basic policies of the Commission. They will be described in the next chapter. This chapter examines the other difficulties of 1944, and first of all the contrast between the years 1943 and 1944 in program and output.

The later months of 1943 witnessed an intense drive to make the highest possible record of ships delivered. In September 1943 Admiral Vickery sent out telegrams to the regional directors giving them high quotas of ships to be delivered before the end of the year in order to meet the President's directive.[1] By special effort yards delivered far more ships in December than they had in any previous month, although at the cost of delivering fewer than usual in January 1944 (Figure 40). Delta for example delivered 9 in December 1943 and only 3 in January 1944. Richmond delivered 27 in December 1943 and then dropped to 20.[2] In the yards as a whole there was an unusual amount of overtime

---

[1] Vickery (by Fischer) to each of the regional directors, Sept. 10, 1943, in Fischer file, Vickery, Sept. 10, 1943, Quotas.
[2] *Official Construction Record.*

in November 1943.[3] By these concentrations of effort, deliveries in December 1943 were raised to the record breaking total of 219 vessels.[4] In January 1944 the Maritime Commission proudly announced that the President's directive for 1942 and 1943 of 24 million tons, which seemed " impossible " when announced, had been exceeded by more than 3,325,000 tons.[5]

The year 1944 started with a month of relatively low production, while the yards were recovering their breath. In spite of warnings against complacency, there was a let-down. Moreover, the goal set for 1944 did not have the urgency and the definiteness of the goals for 1942 and 1943. The race between new construction and shipping losses had been won. Naval auxiliaries were now being inserted in the program in a way which prevented setting a fixed figure as a target. In testifying on December 9, 1943 concerning the Commission's program for 1944, Admiral Land emphasized the uncertainty and stressed how much depended on the way the war went. " We are merely the servants of the military on this program," he said.[6]

The President told the Commission to plan the budget as if they were going to produce 22 million deadweight tons, but named 16 million tons as the probable total production of merchant shipping.[7] Actually only 14 million tons of merchant shipping were produced in 1944.

Military types gradually filled more and more of the shipways. As a result of the military requests described in the last chapter, nearly all the West coast yards and many elsewhere were working on military types by the middle of 1944. Victory transports were being built in Calship, Oregon, Richmond No. 2, and Vancouver. To provide the total of 130 attack transports, four C3's were being finished in that form at Western Pipe and Steel, and seven at Ingalls. Moore Dry Dock at Oakland was building six attack

---

[3] As is shown by the difference between straight time and gross hourly earnings. The difference was .26 in July and Aug., .29 in Sept., .26 in Oct., .31 in Nov., .28 in Dec., and .23 in Jan. 1944. Fischer, *Statistical Summary*, Tables G-8, G-8.
[4] *Ibid.*, Table B-10.
[5] Background statement for press conference, Jan. 3, 1944, in MC gf 105-2. The final, revised figure for deliveries in 1942-43 was 27,254,518 deadweight tons. Fischer, *Statistical Summary*, Table A-3.
[6] House, *Independent Offices Appropriation Bill for 1945, Hearings*, p. 696.
[7] The President to Land, Nov. 12, 1943, in MC Program Collection.

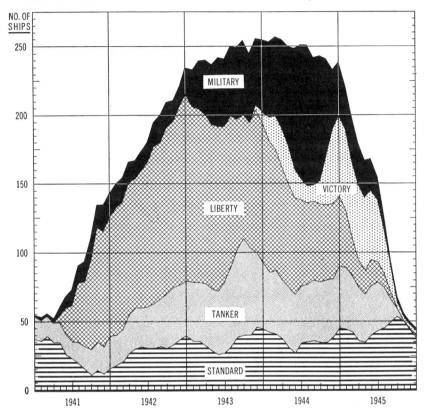

FIGURE 39. Source: Fischer, *Statistical Summary*, Table E-8.

cargo vessels.[8] On the East coast 24 more of these were being built by North Carolina, and Walsh-Kaiser was working on a special Navy design of attack cargo ships. The yards on the Great Lakes were finishing the frigates. On both coasts the slow building of the P2's (at Federal and Bethlehem Alameda) and of the C4's (at Sun and Richmond No. 3) was still continuing. In addition there were many military conversions of C-types and tankers. And many ships which were not specifically military types had been scheduled at the direct request of the Navy and Army and

[8] Reynolds' Way Charts; Land to the Secretary of the Navy, Dec. 6, 1943, and attachments to Bard to Land, Sept. 20, 1944 in MC Program Collection.

were desired by them in connection with specific operations in prospect in the Pacific. At their request the Commission had raised its program for dry cargo vessels of 5,000 tons (C1-M-AV1) from 100 to 200 [9] and placed orders for these not only with the Consolidated Steel Corp., which was already working on these types, but also in many yards in the Great Lakes, at Richmond No. 4, and later at Jones-Brunswick and Southeastern. Tankers too were urgently in demand, not only the special military types being built at East Coast Shipyards and Marinship but the ordinary T2's being built at Sun, Alabama, and Swan Island. Only in the Gulf yards did Libertys occupy more than a third of the ways, and even the Liberty yards of the Gulf turned out many "Liberty tankers" or other variations designed to respond to requests from the Armed Forces. On the West coast no Libertys were being built after August 1944. In the total production of the year, military types formed about 29 per cent, Libertys formed about 32 per cent.[10]

Since military types took so much longer to build, their importance is emphasized by an analysis of the use to which the Commission was putting its facilities. In August 1944 they occupied 103 out of 251 of the building berths in the Commission's Class I yards, whereas a year before they had occupied only half that many, 51 (Figure 39).

This change from one year to the next in the character of the ships being built makes difficult a fair comparison of the output of the two years. Output is frequently measured in the two kinds of tons: deadweight tons, which indicate the carrying capacity of the ship; and displacement tons, light, which indicate the weight of the ship. In terms of deadweight tons delivered, the output in 1943 exceeded that in 1944, as indicated in Figure 1. The better measure of industrial output is displacement tonnage, however, although the program was planned and publicized in deadweight tons to express its relation to the intense need for carrying capacity.

To describe production in displacement tons, two figures have been used: that for deliveries in a given period and that for

[9] MC Minutes, pp. 27580-81.
[10] Fischer, *Statistical Summary*, Table B-7, which gives production in terms of the standard manhours represented in the shipsworth produced. On deliveries by types see Figure 1 above.

PROGRAMS AND PRODUCTION 641

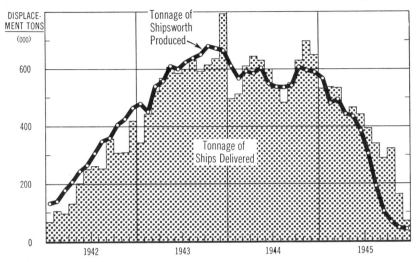

FIGURE 40. Source: Fischer, *Statistical Summary*, Tables B-2 and B-6.

shipsworth produced in a given period. The two figures are not the same because deliveries in one year are partly the result of production during the previous year. Some ships took longer to build than others. A measure of progress in construction month by month can be obtained by taking into account the number of months each vessel was on the ways. This measure is here called shipsworth produced per month. The comparison of shipsworth produced per month and deliveries per month is given on Figure 40. Yearly totals show that deliveries in 1944 exceeded deliveries in 1943 in the displacement tonnage (Figure 1), but shipsworth produced in 1943 exceeded the shipsworth produced in 1944. Many of the deliveries in 1944, as for example of C4's and P2's, were the result of construction during 1943 which did not result in deliveries until 1944.

A measure of output better even than the displacement tonnage is desirable however because some types, especially the military types, were more difficult to build than other ships of the same displacement. To allow for these differences in difficulty of construction, the Technical Division of the Maritime Commission prepared estimates of the number of manhours required to build each type. These estimates indicated that the standard T2 tanker should be built with 158 manhours per ton (light displacement), the Liberty ship with 184.5 manhours per ton, and the C-type freighters with 190.4 manhours per ton. The requirements estimated for the military types were from 219 manhours per ton for the Victory transport to 564 manhours per ton for the light but complicated frigates. Multiplying by these estimates the total displacement tonnage of the shipsworth produced of each type in 1943 and 1944 gives totals for the two years which make allowance for the fact that the ships produced in 1944 were more complicated, more difficult to build, than those produced in 1943. Even so, production in 1943 was larger than in 1944 (Figure 41).

One other method has been used for comparing the production of the two peak years. Figures prepared in the WPB and its successor, the Civilian Production Administration, give in standard dollars the " value in place " in ship construction month by month. They differ from tonnage figures of shipsworth produced chiefly in giving weight to work performed outside the shipyards in the production of components and materials (generally comprising 50 to 60 per cent of the value of the ship). The figures include this material as part of the " value in place " as soon as it is at the shipyard. In effect they count the manufacture of turbines, lifeboats, electric motors, steel plates, and so on as part of shipbuilding. They show the value in place for 1943 as $4,485 million and for 1944 as $4,569 million.[11]

All these figures show that the two years 1943 and 1944 stand out above the others as peak years. For our present purposes the most important figures are those that give the most accurate measure of production in the shipyards, namely, (1) the displacement tonnage of the shipsworth produced, and (2) the standard manhours represented by the shipsworth produced. According to both these measures, output in 1944 was less than in 1943.

The decline in output was accompanied by a decline in the labor force, a decline which was about 10 per cent. But it might be expected that a gain in the productivity of labor would occur to counterbalance the decline in the size of the labor force; for in 1944 the yards were fully equipped, and both management and labor had the benefit of a year or two of experience. In 1943,

---

[11] Fischer, *Statistical Summary*, Table B-9.

INDEXES OF PRODUCTION, MANHOURS, AND PRODUCTIVITY
For Shipbuilding Under The Maritime Commission (1942 = 100)

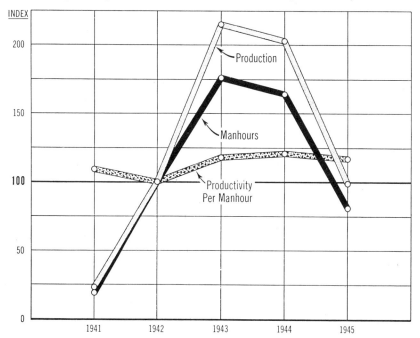

FIGURE 41. Yearly averages. Index of production weighted by standard manhour requirements. Source: Fischer, "Productivity," chart 4.

many of the yards were only just being finished and the labor force was just being brought up to its maximum. Nevertheless, for 1944 as a whole the productivity of labor was very little above what it was in 1943, 5.12 displacement tons per thousand manhours compared to 4.95. From the low level of approximately 4.2 tons in the second, third, and fourth quarters of 1942, productivity had increased to a high point of 5.4 tons in the final quarter of 1943. A slight decline followed and was not halted until the third quarter of 1944. Then it began to rise again and reached a new peak of 6.0 tons per thousand manhours in the second quarter of 1945. But the yearly averages were only 5.12 for 1944 and 5.35 for 1945. See Table 23.

Failure to increase labor productivity much in 1944 was due

LABOR PRODUCTIVITY IN MARITIME COMMISSION SHIPYARDS
Output In Displacement Tons Per Million Manhours, Quarterly, By Regions

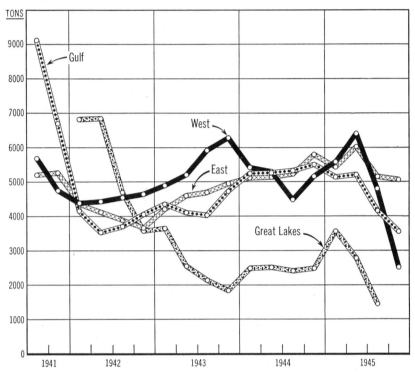

FIGURE 42. Quarterly averages from Fischer, *Statistical Summary*, Table H-4.

to the building of more military types, the diversification of the program, and the consequent interruptions in the continuity of production. This is shown by comparison of the trend of productivity in the four regions (Figure 42). On the West coast, where there were many change-overs and much work on military types, productivity fell from an average for 1943 of 5.66 tons to an average of 5.10 tons for the year 1944 (Table 23). In the Gulf on the other hand, most of the yards continued their established type of production (chiefly Libertys or tankers), and productivity rose from 4.29 for 1943 to 5.32 for 1944. A chart of productivity in individual yards in various locations shows also the importance of continuity in production of the same type (Figure 43). When

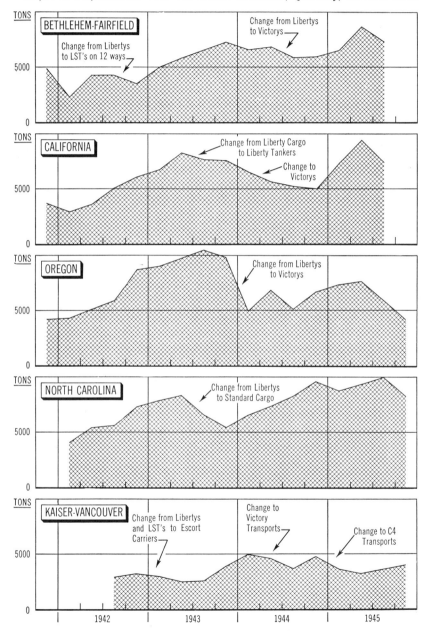

FIGURE 43. Source: Reynolds' Way Charts and Fischer, *Statistical Summary*, Table H-4.

a yard changed over to a new type it usually took a year or more to regain its former level of productivity. Three widely scattered yards made striking gains in productivity in 1944 by continuing production of the types which they had been producing in 1943— Alabama Ship, North Carolina Ship, and Swan Island.

Another way to locate the parts of the program that were being held back in 1944 by low productivity is to compare for each type the actual manhours on ships delivered in 1944 with the standard number of manhours estimated by the Commission's Technical Division to allow for the greater complexity of various types, especially military types. The comparison shows more failures to bring productivity up to the standard in the work on military types than in work on other types. Even this measure, which is weighted to allow for the greater complexity of the military types, indicates that productivity of labor was less in this kind of work.[12]

This comparison of 1943 and 1944 emphasizes how much output was adversely affected by interrupting a yard's continuity in production of a single type. The Commission and its staff knew, of course, that changes in the types of ships built would have this adverse effect on production, but decided to make the changes for the reasons which have been explained in the previous chapters. The schedules which they prepared for 1944 attempted to make allowance for the effects of change-overs and to estimate production accordingly. Nevertheless it soon became apparent that the schedules prepared at the beginning of the year were not going to be met. In April 1944 the Joint Chiefs noted that not as many ships as they had requested would be delivered in 1944.[13] The new schedule prepared by the Commission on April 15, 1944 showed that some of them had slipped over into the program for 1945.[14] In August, the Joint Chiefs again expressed concern at the slippage in the program;[15] the total scheduled by the Commission in September was down to 7 million displacement tons,

---

[12] Fischer, " Productivity," pp. 7, 9, and Table 2. For the sources of the figures of " standard manhours," see Fischer, *Statistical Summary*, pp. 35-36.

[13] G. C. Marshall to Land, Apr. 8, 1944, in Vickery file, Naval Auxiliary Requirements, in MC, Historian's Collection. Explanations of the drop are in Land to Leahy, May 15, 1944, in Land reading file.

[14] Fischer, " Programming," p. 42.

[15] William D. Leahy for the Joint Chiefs to Paul V. McNutt, Aug. 25, 1944, in Vickery's file above cited, n. 11.

whereas 8 million had been scheduled at the beginning of the year.[16]

Perhaps one reason for this slippage in the program was the letdown after record-making 1943 and the general effect of war weariness, merging, after the successful landing in France in June 1944, with the feeling that the war was almost won and the emergency shipbuilding program was nearing its end. More tangible difficulties which had not been allowed for in making the program were the shortages in critical components and in manpower. When they were overcome during the second half of 1944, there was no further slippage; schedules were met.

II. Supplying Components in 1944

In regard to components, the main problem was the fact that the schedules of the shipyards were ahead of the schedules of the manufacturers of components. After it was decided by the Joint Chiefs that the ships should be built, working plans had to be made and approved and orders placed with manufacturers of electrical equipment and other components. The interval between the dates when the Joint Chiefs asked for the ships and the dates when the Joint Chiefs wanted them delivered was so brief that when orders and specifications for components reached the manufacturers, they put these products into their schedules with delivery dates which were later than the dates at which the components were needed at the shipyards.[1] Consequently, the builders of C1-M-AV1's, for example, were " in a very bad position, over all, for lack of materials " in the spring of 1944, the most critical items being pumps, deck machinery, electrical equipment for these units, Diesel generator sets, and valves.[2] For the same reasons there were difficulties throughout the spring and summer of 1944 in supplying components for Victory ships of all types and especially for the attack transports and attack cargo vessels.

The increased complexity of the new types of ships added to the problem. All naval auxiliaries required much electrical

---

[16] Fischer, *Statistical Summary*, Table A-5; Land to Leahy, Aug. 9, 1944, in Land reading file, gives as the reasons "manpower . . . material . . . psychology."

[1] Quarterly Reports of the Branches of the Production Engineering Section of the Production Division, in Production Division Administrative file A9-3.

[2] *Ibid.*, report of Engineering and Scheduling Branch, Mar. 31, 1944. The same branch in its next report makes no special mention of the C1-M-AV1's.

equipment, and final approval of the plans was usually late. Even on the Victory cargo ships the scheduling of components was more complicated than it had been for the Liberty ship. For example, on the pumps and fans, the controls and motors, as well as the pumps and fans themselves, were purchased directly by the Commission. Then the motors and controls had to be allocated by the Commission to the pump and fan manufacturers. The motors and controls had to be matched, and in order to obtain the best use of spare parts, the equipment had to be allocated to ships in such a way that the motors on one ship would all be of the same make.[3] Even if the manufacture of components was all scheduled correctly and early enough, difficulties came from the failure of various manufacturers to meet their schedules.[4] Spreading more of the work among small business concerns as urged by the Small Business Act added to the problem of the Commission's Production Engineering Section, for while some of these did very well in converting to new products, others did not.[5] So intense was the demand in shipyards for components that many were shipped express by rail and or air. In July 1944 the Engineering and the Scheduling and Planning Branches reported there had been 94 shipments by air in the previous quarter, some weighing as much as 2,200 pounds, and that they expected to have to continue air shipments of electrical equipment and pumps. Since Bethlehem-Fairfield had slowed down its construction of Victorys, the difficulties in supplying components for this type were all in the yards on the Pacific coast. Their change-over first to the Victory cargo ships and then to attack transports caused large quantities of smaller items to be moved about among these Western yards.[6]

To the men in the Production Division of the Maritime Commission whose task it was to get the material to the shipyards when needed, the main cause of their difficulty seemed to be late ordering and the delays in approving plans. Insofar as the

[3] Reports cited in note 1, especially Director's report, for Jan.-Mar. 1944.

[4] The original reports of the Branches contain many specific references to the vendors who were and were not meeting their schedules. In the final, edited form, in which these reports are in gf 212-8, these references to companies are often omitted.

[5] Report of J. T. Gallagher, Chief of Production Engineering Section, in Production Division Administrative file A9-3; and statements by Mr. Gallagher, June 4, 1948.

[6] *Ibid.*, in report of Engineering Scheduling and Planning Branch, July 31, 1944.

lateness of plans was a necessary consequence of the lateness of the military authorities in making their decisions—to meet a rapidly changing military situation—there was nothing the Production Division could do about it. But, as a logical extension of the Division's scheduling activity, it began in the middle of 1944 to set up schedules of the dates when the requisitions should be written by the naval architects, the dates when orders should be placed by the Procurement Division, and the dates when they were placed. Thus they hoped to speed these preliminaries. The Director of the Division reported at the end of the year: " By expediting these functions and by calling attention to the design agent and the Procurement Division and the vendors that their functions are behind schedule, we have been able to cut down delay in starting production of components." [7] Scheduling the work of other divisions and of naval architects as rigidly as that of manufacturers looked like the long-run solution.

The immediate solution, which would affect deliveries in the middle of 1944, had to be sought through expediting and through priorities or special directives from the WPB. Since the Commission's expediters were finding it increasingly difficult during the spring of 1944 to keep the materials flowing on time, in view of the very tight schedules, the shipyards insisted that their own expediters be given a chance to try what they could do. Previously the expediters employed by the shipyards had been forbidden by the Commission, at least in theory, from doing more than calling their suppliers by telephone to learn of delivery dates. On their insistence they were allowed during the second quarter of 1944 to visit the vendors, that is, the manufacturers. The Maritime Commission itself assigned work to about 25 expediters from West coast shipyards. Whether they did any good was a question on which Mr. MacLean, Director of the Production Division, differed from the head of his Expediting Branch, except that they agreed that a group from Permanente Metals Corp. had done a good job of getting materials to the Richmond yards.[8]

---

[7] Report dated Dec. 16, 1944, for the quarter ending Sept. 30, in gf 212-8. This work was done by the Conservation Section for reasons described in their reports.

[8] Production Division report for the quarter ending Sept. 30, 1944, in MC gf 212-8, and the earlier draft of the Expediting Branch, Oct. 16, 1944, which says: " As a matter of history, the record shows that nothing was accomplished in this respect with the exception of Permanente. . . ." See also the Expediting Branch report for June 30, 1944, both in Production Division Administrative file A9-3.

In the summer and fall, more effective aid was obtained by higher ratings and from the Navy's large corps of expediters.

None of the Commission's program had top rating at WPB at the beginning of 1944. To be sure, the priority ratings of the kind which had figured prominently in the competition for steel and turbines in 1941 and 1942 were no longer very important. Nearly the whole program was AA-1. Priority AAA was really reserved for emergencies. But during 1943 new systems had been developed distinguishing the relative urgency of the many programs rated AA-1. First came President Roosevelt's specification of certain "must" programs for 1943. In the process of working out a way of implementing these "must" programs and reducing them to practical form, the WPB put the emphasis on production scheduling; and to guide in the scheduling it classified AA-1 products into seven groups according to the urgency with which they were needed. The "urgency rating," which was practically a priority rating under a new name, was made by the Production Executive Committee of WPB, a committee containing representatives of the procurement agencies. The Maritime Commission was represented by Admiral Vickery. This Production Executive Committee received from the President or the Joint Chiefs indication of military needs, since production plans were more and more closely connected with operational plans and experience as American forces became more fully engaged.[9]

Until the landing in Normandy in June, only one shipbuilding program was given an urgency rating in group I: landing craft, all of which then were being built by the Navy. To aid that program, as to aid other "must" items such as the atomic bomb (Manhattan) project and certain kinds of aircraft, the WPB issued directives to manufacturers changing their schedules of production or assigning to builders of landing craft scarce items, such as electric motors, valves, and pumps, which might have been originally scheduled for other shipbuilding.[10] Fear that this would delay the building of the ships scheduled by the Maritime Commission led Admiral Land to protest and cite the letter of

[9] *Industrial Mobilization for War*, I, 507, 516-17, 609, 755. This classification was done by a Production Urgency List which, in November 1943, indicated seven groups of relative urgency within the AA-1 category of priority.—Chaikin and Coleman, *Shipbuilding Policies*, pp. 97-98.

[10] *Industrial Mobilization for War*, I, 755; Chaikin and Coleman, *Shipbuilding Policies*, pp. 70, 135-36, 170.

SUPPLYING COMPONENTS 651

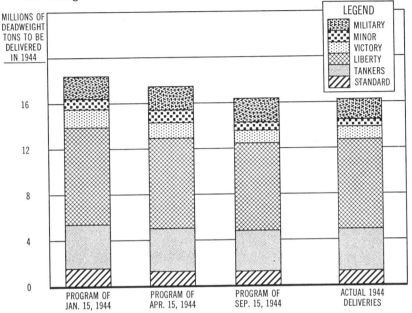

FIGURE 44. Source: Fischer, *Statistical Summary*, Table A-5.

President Roosevelt of November 12, 1943.[11] Consequently a proviso was made that when special directives were issued by the Special Rating Branch they would be limited in scope to a particular bottleneck and have the consent of the " key claimants," among which was the Maritime Commission.[12] Actually, the Maritime Commission's building also received some aid at the beginning of 1944 from the issue of directives from the Special Rating Branch, as well as from AAA priorities, to supply a few items especially badly needed;[13] and although the reports of the Production Division mention interference by the landing craft program, they put more stress on the late ordering and late plan approval.

[11] *Ibid.*, pp. 135-36.
[12] *Ibid.*, p. 137, n. 124.
[13] Reports of the Priorities Section, in Production Division Administrative file A9-3.

After the success of the landing craft program was assured, the attack cargo ships, attack transports, and the tankers being built in Maritime yards moved up to group I as a " must " program, and after D-day they rated above all other shipbuilding.[14] This made it possible to speed up the lagging production of components. In July and August an increased number of AAA priorities and directives from the Special Rating Committee speeded the supply of motors, electrical controls, and so on.[15] The problem of priorities, or urgency ratings, was then entirely different from what it had been at the beginning of the emergency program. In 1941 the Maritime Commission had been competing with the Navy for scarce items which the Navy wanted for combat ships and the Maritime Commission for cargo carriers. Now the Commission was building for the Navy and the Navy was urging that the very highest ratings be given ships being built in the Commission's yards.

The Navy not only helped raise the urgency ratings but also supplied expediters to follow through on their operation. In September 1944 the Chief of Naval Operations promulgated a letter to all Naval Districts and Navy Yards and the Bureaus of Ships, Ordnance, and Personnel stating that the completion of 114 auxiliaries being built in Maritime yards or converted in Navy conversion yards was of the utmost importance and should have precedence over all other building.[16] Maritime's Production Division created a Navy Liaison Group which prepared data on shortages and turned the information over to the Navy Department, which had a corps of inspectors in the plants of the manu-

[14] The General Production Urgency List approved by the Production Executive Committee on Nov. 3, 1943 placed combat loaders in Group III for labor referrals, and the Program Determination No. 500, Nov. 16, 1943, also from the Production Executive Committee giving the rating structure within AA-1, also placed them in Group III. In December the Navy recommended they be moved up but the Production Executive Committee refused.—Chaikin and Coleman, *Shipbuilding Policies*, pp. 178-79. Finally, in April the Production Executive Committee put combat loaders and tankers into Group I, the equivalent of " must " programs, and kept them there in June. Three weeks after D-day landing craft were dropped from the " must " list. *Industrial Mobilization for War*, I, 755-76.

[15] By the end of the year 731 requests for special directives had been received, and approval granted for 607. Chaikin and Coleman, *Shipbuilding Policies*, pp. 180-82.

[16] Ralph A. Bard, Acting Secretary of the Navy to Land, Sept. 20, 1944, with attachments, in MC Program Collection; Connery, *The Navy and the Industrial Mobilization*, p. 298.

SUPPLYING COMPONENTS

facturers to speed up shipments. " Through this system there has been placed at our disposal the large field organization of the Navy to insure material delivery," reported Mr. MacLean. " As a result, the schedules have been met, and the necessary components and spares have been aboard ship when delivered." [17]

Another help in expediting components for vessels to be turned over to the Navy was the Maritime Commission's readiness to pay for the overtime and other expenses incurred by manufacturers to accelerate deliveries. In October 1944 the Director of the Procurement Division was authorized to adjust contract prices upward to meet these extra expenses.[18]

During the last quarter of 1944 there was no further slippage in the Maritime Commission's program. Figure 44. Everyone from President Roosevelt down was putting on pressure to finish the combat cargo and attack transport vessels.[19] Higher urgency rating were being made effective to relieve the shortage of components. At the same time, similar higher urgency ratings were helping to solve the problem which loomed largest in 1944—the manpower shortage which will be discussed in the next chapter.

In a letter of commendation to the Maritime Commission, James Forrestal, Secretary of the Navy, reviewed the achievements:

> In 1944 the Maritime Commission was called upon to carry out for the Navy an extremely ambitious program of building high priority APA and AKA vessels critically needed in planned fleet operations. That this challenge was met and the assault vessel program completed in the time specified is a matter in which the Maritime Commission may take great pride. Not only did the Commission and its shipbuilding contractors meet the exacting delivery dates, but they cooperated with the Navy to effect introduction of recently developed and highly desirable military improvements in the later vessels of the program and to equip the vessels with adequate spare parts to permit them to go directly into service in the active war theatres. It is a pleasure to express the Navy's commendation and appreciation for excellent performance.

Land graciously replied: " As a matter of fact our accomplishments were only made possible by the excellent support given us here in Washington and in the field by Navy representatives in all the activities pertaining to the delivery of these vessels." [20]

---

[17] Report of Production Division, July–Sept., 1944, in MC gf 212-8.
[18] MC Minutes, pp. 30009-10 (Oct. 24, 1944).
[19] Land to the President, Oct. 13, 1944, and Land to the Joint Chiefs, Nov. 2, 1944, in Land reading file.
[20] Copies of both letters in Land reading file under Feb. 14, 1945.

CHAPTER 20

THE MANPOWER AND MANAGERIAL CRISIS

I. THE NEW MANPOWER PROBLEM OF THE SHIPYARDS

AMONG the difficulties encountered in meeting the schedule for 1944, the "manpower shortage" is of special interest because it called into the question the basic policy pursued by the Commission in regard to one of its major responsibilities, the managing of management. Emergency shipbuilding had begun in a time of unemployment, and its managers had developed a labor force of men and women who had had no previous shipbuilding experience. Thus they had broken the bottleneck that had been feared as a result of the shortage of skilled labor. In 1943, however, unemployment disappeared. In 1944 attention focused on manpower as the bottleneck restricting production, not only in shipbuilding but in the nation generally. Two lines of pressure immediately developed: one a pressure by the Maritime Commission's shipyards to maintain their labor force; the other a pressure on the Maritime Commission and the shipyard managements to get along with less labor so that more would be available for other needs of the nation.

Until July 1943 the number of workers employed in shipyards increased constantly in spite of the high rate of turnover. A separation rate of between 7 and 11 per cent was more than balanced by an accession rate of between 13 and 31 per cent. But in July 1943 the yards just broke even—they lost 12 per cent and hired 12 per cent. After this there was scarcely a month that did not show more workers leaving than taking jobs in merchant shipyards (Figure 45). The total labor force fell from 650,000 in July 1943 to 616,000 in January 1944. By July 1944 it had fallen to 560,000 (Figure 22).

The decrease in shipyard employment did not prevent achievement of the production schedule set for 1943, but early in 1944 it became a cause of serious alarm. By April 1944 it was clear that yards were falling behind the schedules which had been laid

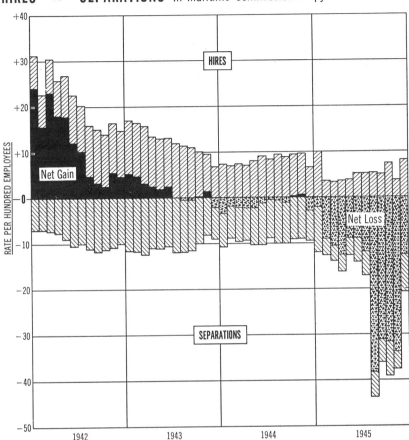

FIGURE 45. Source: Fischer, *Statistical Summary*, Table G-15.

down in January 1944, and a yard-by-yard analysis of this slippage showed on the whole a close correlation between failure to meet schedule and decline in employment. The Production Division reported to Admiral Vickery that the schedules for 1944 had been made on the assumption that employment would be maintained at the level of 625,000, as it was at the first of the year, and some way must be found to rebuild the labor force to that size if the slippage was to be overcome.[1] But both the decline in

[1] MacLean to Vickery, Apr. 22, 1944, in Fischer reading file.

employment and the slippage continued during the first half of 1944 (Figures 22 and 44).

Military service was a basic factor in this decline of the labor force, although after June 1943 the number of workers giving military service as the reason for quitting never rose above 1 per cent per month of the number employed in the shipyards. At its peak, September 1942 through March 1943, the number was between 2 and 2.7 per cent.[2] But many men may have left the shipyards in anticipation of being called, and the movement into the armed forces was a drain on the total male working population from which the shipyards were seeking replacements. Moreover it was a one-way movement out of the total labor force, whereas many of the other separations represented movement from one yard to another. The loss of skilled men was particularly serious and was protested by spokesmen of management.[3] The Maritime Commission's West Coast Director wrote concerning the loss of trained men: " It has been estimated we could effect a 20 per cent reduction in manpower and still maintain the same overall production rate if we could eliminate this enormous turnover . . . to the armed services." [4] In April 1943 Robert R. Nathan of the WPB had estimated that about half a million workers would be drafted from the aircraft and shipbuilding industry and would have to be replaced by women or men exempt from the draft.[5]

More numerous, however, than calls by Selective Service were lay-offs and discharges.[6] At particular times in particular yards large numbers of men were being discharged late in 1943 and 1944, although the shipbuilding companies as a whole were then complaining about " labor shortages." Assuming that the period had passed when speedy construction was the most important consideration, the big Bethlehem-Fairfield yard in Baltimore on

[2] Fischer, *Statistical Summary*, Tables A-5, G-2, and G-15.

[3] House, *Full Utilization of Manpower, Hearings*, before the Committee on Military Affairs, 78th Cong., 1st sess., Apr. 9, 1943, pp. 430-36, 441.

[4] Carl W. Flesher, " Draft Is Retarding Ship Production," *The Log*, vol. 38, Mar. 1943, p. 40.

[5] [Robert R.] Nathan to Ralph J. Cordiner, Apr. 2, 1943, in WPB files, document 232, Planning Committee.

[6] U. S. Department of Labor, Bureau of Labor Statistics, Employment and Occupational Outlook Branch, " Monthly Labor Turnover in Commercial Shipyards with Maritime Commission Contracts," in Production Division files; Fischer, *Statistical Summary*, Table G-15.

THE NEW MANPOWER PROBLEM 657

changing to Victory ships in 1944 cut its labor force down to what it considered an efficient size.[7] This sort of action led to protests from the labor union affected [8] and to criticism before the Truman Committee.[9] Early in 1944 West coast yards took advantage of the change to new types to weed out the least desirable of the unskilled, but at the same time clamored for more skilled workers such as electricians, sheet metal workers, and pipefitters. When the yards could not hire enough in those skilled crafts they were forced to lay off other workers so that a balance among the trades could be maintained.[10] Talk about lay-offs created uncertainty among the workers and many mistook necessary readjustments for a sign that the building program was about over. Although the situation varied from yard to yard, in some shipyards the "manpower shortage" of 1944 was not so much a general shortage as a shortage of the particular kinds of workers the shipyards needed at that time.

The largest element in the high turnover were "quits," which were 8.1 to 8.3 per cent per month at their peak in July-September 1943. After going down to 5.5 in November 1943 they were about 7 per cent throughout 1944.[11] Conditions in shipyards, especially in shipyard communities (See chapter 13), make quite understandable the relatively large number of quits each month throughout the war.

In addition there were some particular reasons why the rate remained high in 1944, when it might have been expected to decline in view of the efforts being put into improving conditions in shipyard communities. An intangible factor was the tendency to feel that the war was almost over. This not only caused quits but made recruiting new workers more difficult.[12] Eyes were focused on the expected landing in France in the spring of 1944,

---

[7] Chaikin and Coleman, *Shipbuilding Policies*, p. 162; Tracy to V. Lewis Bassie, July 22, 1944, in MC gf 508-1.

[8] Land to J. M. Willis and McInnis, Jan. 24, 1944 in Land reading file.

[9] Truman Committee, *Hearings*, pt. 26, pp. 12075-76, Dec. 8, 1944.

[10] Clipping from San Francisco *Examiner* attached to Henry W. von Morpurgo, to Felix Kahn, Feb. 5, 1944, in gf 508-1-12; MC, Manpower Survey Board [Reports] of Marinship, Jan. and May, 1944; Richmond, Jan. 1944; Swan Island, Jan. 1944; Vancouver, Jan. 1944; Oregon, Jan. and May 1944; California, Feb. 1944, in Historian's Collection.

[11] Fischer, *Statistical Summary*, Table G-15.

[12] Chaikin and Coleman, *Shipbuilding Policies*, p. 191; Morpurgo to Kahn, Nov. 22, 1944, in gf 508-1-12; Weyforth, *Manpower Problems*, p. 23.

and after the move across the Channel had been made successfully there came a feeling that the war in Europe would soon be over. Not only was that possibility seriously considered by the Joint Chiefs and the Maritime Commission in August 1944, it was commonly believed by the general public. Workers and managements became concerned about postwar jobs or postwar markets. In the WPB, problems of reconversion were being discussed as imminent.[13] There was a lot of talk about workers leaving shipyards, and Admiral Land felt it necessary to appeal to them to " stay in there and continue to pitch as you have done so handsomely in the past." [14]

Another factor of special importance in 1944 was the competing pull of repair yards and other plants in which the weekly " take-home " wage could be increased by overtime. Repair yards were extremely busy in 1944, especially on the Pacific coast; and they paid double time for all over eight hours a day.[15] Some other yards working for the Navy also offered opportunities for overtime. The drain of labor away from Maritime yards to those working for the Navy became so heavy in the Gulf area in the spring of 1944 that some Maritime yards went on nine- or ten-hour shifts.[16] Since the work on naval vessels and in repair yards required more skill on the whole than work in yards engaged in multiple production of merchant vessels, this competition in overtime pay was particularly important in that it attracted the more skilled and vigorous workers.[17]

A step taken by the Maritime Commission at the beginning of 1944 was alleged to have contributed to the decline of employ-

[13] Leahy to Land, Aug. 25, 1944, in MC Program Collection; *Industrial Mobilization for War*, I, 789 ff.

[14] Press Release, June 17, 1944, in gf 508-1.

[15] Pacific Coast Zone Standards, Apr. 23, 1941; Atlantic Coast Zone Standards, June 4, 1941; Gulf Shipbuilding and Repair Zone Standards Agreement, June 18, 1941; Great Lakes Shipbuilding and Repair Zone Standards, July 11, 1941; Amendments to the Pacific Coast, Atlantic Coast, Great Lakes and Gulf Zone Standards of the Shipbuilding Industry, effective July 19, 1942 (mimeographed) in MC, Historian's Collection, Research file 205.3.1; U. S. Department of Labor, Bureau of Labor Statistics, " Wartime Employment, Production, and Conditions of Work in Shipyards," *Bulletin No. 824* [1945], pp. 21-22.

[16] U. S. Department of Labor, Bureau of Labor Statistics, Employment and Occupational Outlook Branch, Labor Statistics for Private U. S. Shipyards, July 1944; Sanford to Vickery, Mar. 15, 1944, in MC gf 508-1.

[17] House, *Production in Shipbuilding Plants, Executive Hearings*, pt. 3, pp. 567-68, 595, 607, 614-15, 619, pt. 4, pp. 1096-97.

ment by its effect on both morale and overtime. In December 1943 Admiral Vickery ordered Maritime Commission yards to close on Sundays, unless they had permission from the regional directors to stay open for urgent reasons.[18] By reducing overtime, this measure decreased the ability of Maritime yards to attract and hold workers. Although workers were no longer paid double for working on Sunday *per se*, the questions of operating the yard on Sunday and paying overtime were closely connected. In part this was because the auditors had practical difficulties in checking the claims for overtime if the yard operated seven days a week.[19] Mainly it was because few yards except those on the West coast had so organized their labor force as to be able to work full tilt on Sundays without calling on some of their men to work a seventh day. Many old-line yards planned to use Sunday normally as a day for clean up and repair. If these yards did not work full shifts on Sunday, it meant paying double time; that was done to a certain extent in 1943 in the rush to meet the goals set for that year.

In explaining his order against Sunday work to the staff of the Gulf Regional Construction Office, Admiral Vickery stated that the Navy had been agitating for it since the previous October and he had refused to go on the six-day week until the 1943 goals were met. Now he was ready for a stiffer attitude towards labor. There had been many compromises because, said Admiral Vickery, " we could not afford to have the work stop." He told the Gulf staff in December 1943 that he did not propose to be pushed around and was then willing to risk a stoppage rather than compromise on issues which he thought should not be compromised.[20]

Admiral Vickery's ban on Sunday work was severely criticized by the WPB and the War Manpower Commission on the grounds that it slowed down production and contributed to the feeling that the crisis was over. On the other hand, the Maritime Commission maintained it was acting to conserve manpower, to increase efficiency, and to save money, and that eliminating work on Sunday would not slow down production. It pointed out that 60

[18] Vickery to Flesher and other Regional Directors, Dec. 10, 1943, in MC Division of Shipyard Labor Relations file, Zone Standards, General.
[19] Statements by L. G. Kell, who was Regional Construction Auditor, East Coast.
[20] Minutes of Staff Meetings, Gulf Coast Regional Office, Dec. 17, 1943, in Mr. Sanford's files. The Bureau of Ships memorandum of Dec. 9, 1943 relative to Sunday work is referred to in Land to Flesher, Dec. 15, 1943, in Land reading file.

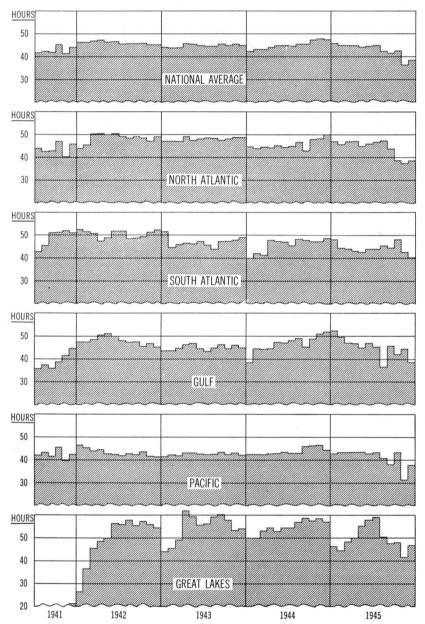

FIGURE 46. Source: Fischer, *Statistical Summary*. Table G-6.

per cent of its yards had never been on a seven-day week. Some of the yards, mainly those on the Great Lakes, had rarely employed anyone on Sunday. In others, the old-line yards and some of the emergency yards in the South, small groups were employed on that day to clean up and repair, or special crafts would be called upon from time to time to break bottlenecks. However, the large emergency yards had been on a 21-shift week, and in order to give individuals a day off they had been forced to hire more workers. The Commission wanted to save this one-seventh extra manpower. In yards where work done on Sunday differed from that done during the week it is likely that those who came in Sunday worked a seventh shift and thus earned double time. This had not been true in yards that were on continuous production. However, as manpower became scarce, these yards would have been forced to schedule seven working days for more and more individuals. The effect, as Admirals Land and Vickery saw it, would be either an increase in absenteeism or an increase in the cost of ships through payment of double time.[21]

The order had the immediate effect of cutting out Sunday work entirely in most of the yards that had been on continuous production, but by the middle of 1944 these yards had adopted a schedule like that of the old-line yards, working a full force six days and using Sunday to "clear decks."[22] The only area where the average number of hours worked per week per employee was appreciably lower in 1944 as a whole was on the Atlantic coast (Figure 46). In Pacific coast yards the average rose slightly. Nevertheless the Labor Division of WPB criticized the Maritime Commission for its action on the grounds that workers quit the yards because their pay was reduced and because they thought ships were no longer in urgent demand.[23] Coming at a time

---

[21] Chaikin and Coleman, *Shipbuilding Policies*, pp. 163-64; testimony of Ring in House, *Production in Shipbuilding Plants, Executive Hearings*, pt. 4, Jan. 27, 1944, pp. 1203-05; Washington *Star*, Dec. 12, 1943; Land to Flesher, Dec. 15, 1943; Land to Hon. Lindley Beckworth, Jan. 22, 1944; Land to C. E. Wilson, July 26, 1943 in Land reading file. Conclusions as to the number of yards working on Sundays and the percentage of the labor force employed were made after a study of BLS Production Schedules in MC, Historian's Collection. See also Memorandum, Horace B. Drury, Shipbuilding Stabilization Branch, to Paul R. Porter, "Effect on Shipbuilding Wage Costs of 1942 Amendments to Zone Standards Agreements, April 1, 1943" in MC Historian's Collection, Research file 205.6.1.

[22] See BLS Production Schedules.

[23] Joseph D. Keenan to C. E. Wilson, July 15, 1944, summarized in Chaikin and

when there was much discussion in the press of general cutbacks in war production, and being frequently cited as an example, the elimination of Sunday work in shipyards probably did have a psychological effect that was difficult to counteract.[24]

II. Controls of the War Manpower Commission

The attitude of the Maritime Commission on Sunday work showed that it was setting its face against one possible method of trying to maintain the labor force in its yards—namely, increased wages. The Commission's attitude was in accord with the general anti-inflationary policy of the government in 1944. Since the peacetime system of attracting workers—namely, offering higher wages—was not applicable, attention turned to the use of thorough government control and allocation of labor.

From the beginning of the war the question had been discussed whether workers should be drafted and made to stay in the shipyard or other plant to which they were assigned. The question came up repeatedly before Congressional committees, and in January 1944 President Roosevelt recommended such a system of national labor service after receiving a letter signed by Admirals Land and Vickery, as well as Secretaries Stimson, Patterson, Knox and Forrestal, explaining that the existing "manpower crisis" made it necessary.[1] Although in the early part of the war he opposed a labor draft, Admiral Land swung over in April 1943. He told a Congressional committee, " I have been a dally, linger, and wait bird on this thing for some time, but I am no longer." He said he was speaking for himself only, however, not for the Commission,[2] and neither then nor when testifying the following February in favor of the President's proposal did he say that the only way the Maritime Commission could meet its schedules was to keep workers in the shipyards by enforcing a labor draft.[3] He

---

Coleman, *Shipbuilding Policies*, p. 161. For Land's reply see Land to Wilson, July 24, 1944 in Land reading file.

[24] Examples of newspaper articles on cutbacks and reconversion are the following: New York *Herald-Tribune*, Dec. 22, 1943, in MC News Digest, Daily Summary, Dec. 23, 1943; Baltimore *Sun*, Dec. 12, 1943, New York *Journal of Commerce*, Dec. 3 and 6, 1943, in MC News Digest, Weekly Summary, Dec. 17, 1943.

[1] *The United States at War*, p. 450; Stimson, Knox, Forrestal, Land and Vickery to the President, Dec. 28, 1943, in MC gf 508-1.

[2] House, *Full Utilization of Manpower, Hearings*, Apr. 14, 1943, pp. 503, 600.

[3] *Ibid.*, Senate, *Manpower (National War Service Bill)*, *Hearings*, pt. 22, Apr. 15,

emphasized instead the general principle and the psychological effect: " The people of the United States accept and respect the will of Congress. If Congress will pass a national service act, there will be little need to enforce it by law." [4] Not finding the need sufficiently pressing, Congressional committees did not report out bills until 1945, and then the end of the war in Europe caused the matter to be dropped.[5]

Instead of drafting labor, the United States went through World War II under the methods of allocating labor which were developed by the War Manpower Commission in cooperation with the War Production Board and the procurement agencies. Some of the same procedures were used in allocating labor as in allocating materials: the procurement agencies argued the importance of their programs before the WPB, and the WPB issued " urgency " ratings for the guidance of the War Manpower Commission. However, allocations and urgency ratings were not so authoritatively applied to labor as to materials; no statutory penalties were attached to non-compliance.[6]

Starting with voluntary regional agreements, the War Manpower Commission worked out by 1943 a system under which new workers would be hired only through or with the approval of the United States Employment Service. This service, an arm of the Manpower Commission, would give jobs only to those who had a " certificate of availability " which showed they had been released by their previous employer. These " certificates of availability " were very easy to get. If an employer refused to release a worker he could appeal to a Review Unit of the U. S. Employment Service on any one of the following grounds: that his highest skill was not being utilized, that he was not being worked full time, that he was living an unreasonable distance from his present place of employment, or that he had " compelling personal

---

1943, pp. 934-35; Senate, *National War Service Bill, Hearings* before the Committee on Military Affairs on S. 666, pt. 2, 78th Cong., 2d sess., Feb. 10, 1944, pp. 167-76.

[4] Senate, *National War Service Bill, Hearings*, 1944, p. 168.

[5] *The United States at War*, pp. 451-54.

[6] " The Activities of the War Manpower Commission," n. d., prepared at the request of the Senate War Investigating Committee, p. 3, in MC, Historian's Collection Research file, 205.2.3; *Industrial Mobilization for War*, I, 710; " A Short History of the War Manpower Commission," Preliminary draft, mimeographed, by Technical Service Division, United States Employment Service, U. S. Department of Labor, June, 1948, pp. 11-12, 38-39, 113, 140-41.

reasons for wishing a change." [7] If all these appeals failed, a worker could usually manage to get himself fired and then be hired somewhere else. For example, an increase in crap games in the Calship yard in the spring of 1944 and the eagerness with which some workers pleaded guilty led the local Maritime officials to wonder whether " gambling is not being used as a means . . . to obtain an availability certificate." [8] Former farmers or housewives who wanted to return to their homes did not need any certificate. Consequently, the controls set up by the War Manpower Commission did not effectively bind men to their jobs.[9]

The looseness of these controls is shown in the rate of labor separations, but the separations did not worry shipyard managements as long as they could easily hire workers to take the place of those that left. In 1943 and 1944, however, the War Manpower Commission put into effect restrictions on the right to hire, and these restrictions vitally affected the efforts of shipyards to keep up the level of employment.

Under the system put into effect by the War Manpower Commission between September 1943 and July 1944, a ceiling, or maximum size of labor force, was established for each war plant through a committee under the chairmanship of the representative of the War Manpower Commission. Each Area Manpower Priorities Committee contained representatives of all the procurement agencies.[10] The Maritime Commission appointed as its representative in the areas of the Pacific coast executives of the shipbuilding companies.[11] In other regions the Commission was represented by its regional directors of construction. Only in

---

[7] " Short History of the War Manpower Commission," pp. 101-05; *The United States at War*, pp. 435-41.

[8] MC, Manpower Survey Board [Report] California Shipbuilding Corporation, May 1944, in MC, Historian's Collection.

[9] " Short History of the War Manpower Commission," pp. 105-07.

[10] *Ibid.*, pp. 105-13.

[11] They were Carlos Tavares in San Diego, Alden Roach in Los Angeles, Felix Kahn in San Francisco, Edgar Kaiser in Portland, and Henry D. Isaacson in Seattle.— Minutes, pp. 31474-75. Mr. Kahn's position caused Commissioner Carmody to question whether he should retire, but the regional director urged he be kept because of his standing in the community and because he had been " invaluable to this office on numerous occasions in assisting me personally in dealing with management in the shipyards in which he happens to be Director."—Flesher to Schell, Oct. 21, 1944, Flesher reading file, 1944, Confidential. Mr. Kahn's very active alternate, Henry von Morpurgo was paid until 1945 by Permanente Metals Corporation.—MC Minutes, pp. 31252-53.

THE WAR MANPOWER COMMISSION   665

July 1944 were they authorized to hire delegates, 5 on the East coast and 5 on the Great Lakes, to represent them in various cities in which there were committees.[12] In Washington the Maritime Commission was represented directly on the War Manpower Commission by Admiral Vickery, who made Mr. Ring his alternate.[13]

In general, the Commission used its representation in these manpower agencies to support the requests of yard managements. They calculated their needs for labor by taking the number of ships they were under contract to deliver and multiplying by the number of manhours which were estimated as necessary per ship. The company's figures were then checked over by the regional offices of the Maritime Commission and at Washington in the Statistical Analysis Branch and by Admiral Vickery personally.[14] Representatives of the War Manpower Commission contended that these ceilings were often too high. In many cases they were well above the actual levels of employment. In shipbuilding at least, ceilings were set so high that shipyards could not find enough workers to bring employment up to the ceilings.[15]

Of more practical importance than the ceilings therefore in restricting hiring by shipyards were the urgency ratings. The urgency ratings of the plants were established in each area by an Area Production Urgency Committee which contained the same membership as the Area Manpower Priorities Committee but was under the chairmanship of a representative of the WPB. In all these committees also the Maritime Commission was represented as one of the major procurement agencies, and upheld the claims of its yards on the nation's labor supply.[16]

[12] MC Minutes, pp. 29150-53 (July 25, 1944); pp. 29206-08 (July 27, 1944).
[13] Land to the President, Dec. 8, 1943; the President to Land, Jan. 3, 1944, enclosing Executive Order No. 9409 signed Dec. 23, 1943; and Land to McNutt, Jan. 12, 1944, all in MC gf 107-78.
[14] Testimony of Clay Bedford in House, *Production in Shipbuilding Plants, Hearings*, pt. 3, June 18, 1943, p. 607; testimony of Vickery in House, *Walsh-Kaiser Co., Inc. Hearings*, Apr. 5, 1944, p. 269.
[15] House, *Walsh-Kaiser Co., Inc., Hearings*, p. 20; *Industrial Mobilization for War*, I, 849. For a comparison of ceilings with employment, see Progress Reports of Northern California Production Urgency and Manpower Priorities Representatives signed by H. W. von Morpurgo, especially those of Dec. 22, 1943, Apr. 8, 1944, Sept. 30, 1944; and Nov. 6, 1944, in gf 508-1-12; and Fischer, *Statistical Summary*, Table G-1.
[16] Chaikin and Coleman, *Shipbuilding Policies*, pp. 160-64; *The United States at War*, pp. 439-42; Progress Reports signed by Morpurgo cited above; telegram—Vickery to Edgar Kaiser, Dec. 7, (or 17) 1943, in MC gf 508-1-12.

Ceilings and urgency ratings were put into operation first on the Pacific coast in the fall of 1943.[17] That was the region in which the shipyards were shouting loudest about labor shortage, where turnover was highest, and where there was most construction of the military types for which the Navy was clamoring. But labor was needed in that region for other industries as well as for shipbuilding. In 1943 there was more question of using manpower controls to prevent the shipyards from hiring too many workers than there was of using these controls to help shipyards maintain their level of employment. Consequently the shipyards were not given the top urgency rating; the effect of the controls was to make it slightly more difficult for them to hire locally.[18]

The effectiveness of the urgency ratings depended on their application by the U. S. Employment Service, which was charged with enrolling workers and referring them to employers. While it thus helped recruitment by shipyards and other industries, it did not relieve shipyard managements of the primary responsibility for finding the labor they needed. No shipyard or other employer could hire without the approval of the U. S. Employment Service, but the U. S. Employment Service delegated the right to approve on its behalf to unions and employment agencies. Most hiring by the shipyards was done through an employment-union office, which was responsible for seeing that workers hired had "availability certificates," that ceilings were not exceeded, and that the ratings and local rules of the War Manpower Commission were observed. When a worker came to one of these employment offices, or that of the U. S. Employment Service, he was supposed to be told of the jobs open at the plants with highest ratings and urged to go there. But he was free to reject these suggestions and take a shipyard job so long as the shipyard was under its ceilings. Most recruits from the local labor market were hired through a plant's union-employment office, having been told to apply by a friend or relative in the yard.[19]

[17] "Short History of the War Manpower Commission," pp. 119-23, 109-10.
[18] Progress Reports signed by Morpurgo dated Dec. 22, 1943, Feb. 5, 1944; MC, Manpower Survey Board [Reports] of Richmond Shipbuilding Corporation, Jan. 1944; Marinship, Jan. 1944; and Oregon, March (?) 1944; *Industrial Mobilization for War*, I, 755-56.
[19] The U.S. Employment Service could delegate authority to unions or private employment agencies.—"Short History of the War Manpower Commission," pp. 110-14. A sheet dated Mar. 25, 1944 in MC, Production Division folder, Walsh-Kaiser, tells of the USES representative at the yard's employment office. The West Coast

When the yard managements tried to recruit from other areas, as they did increasingly in view of the difficulties of obtaining workers locally, they were at the same time more firmly controlled and more helped by the U. S. Employment Service. In-migration was the "life-blood" of critical areas like the West coast, and the shipyards and government both did their best to encourage it.[20] Whereas the War Manpower Commission would not allow shipyards to advertise for workers locally because other plants had higher urgency ratings, outside of the shipbuilding centers there were regions where there was no labor shortage and in these regions the shipbuilding companies could advertise. With the help of the U. S. Employment Service, the Kaiser organization had recruited 20,000 workers from other regions in the space of four months during the spring of 1943,[21] and it finally obtained the approval of the Commission for paying the expenses of workers brought from New York. New recruits came from every state in the union.[22] In seeking areas which had labor available, the shipyards were guided by the U. S. Employment Service, which canvassed the whole nation and divided it into regions according to the needs in each for labor in war industries.[23]

Until the spring of 1944 the controls of the War Manpower Commission were dominated by the pressures working to divert labor from shipyards to other industries in which it was believed to be more urgently needed. Many people in the War Manpower Commission and the WPB believed the high wages of the shipyards had enabled them to suck up far more than their proper share from the nation's reservoir of labor. Those who looked at the problem from this point of view felt dissatisfied with the controls

Manpower Survey Board Reports cited below leave no doubt as to the importance of their own recruitment efforts over those of the USES and of the position of the unions in the hiring process. MC, Manpower Survey Board [Reports], Richmond Shipyards, Jan., May 1944; Kaiser Company-Vancouver, Jan. 1944; Kaiser Company-Portland, Jan. 1944; Oregon Shipbuilding Corp., Jan. 1944; California Shipbuilding Corp., Jan. 1944, in MC Production Division files or Historian's Collection.

[20] Morpurgo to Kahn, Aug. 5, 1944 in gf 508-1-12; House, *Production in Shipbuilding Plants, Hearings, passim.*
[21] "Short History of the War Manpower Commission," p. 80.
[22] Edgar Kaiser to MC, atten. Vickery, June 14, 1943; Ring to MC, June 14, 1943; Anderson to MC, June 28, 1943; all in gf 508-1; testimony of Henry Kaiser in Truman Committee, *Hearings*, pt. 16, p. 7014, and of Edgar Kaiser, in House, *Production in Shipbuilding Plants, Hearings*, pt. 4, pp. 1098-1100.
[23] "Short History of the War Manpower Commission," p. 116.

because they did not make the shipyards disgorge "hoarded" labor.

After June 1944 the controls of the War Manpower Commission became of positive help to many shipyards. Tankers and combat loaders were on the "must" list, whereas after the successful Normandy landings, landing craft were not. Moreover the Production Executive Committee of WPB on June 28, 1944 placed 42 Maritime Commission shipyards on its Production Urgency List and thereby instructed the War Manpower Commission to give them top labor priority. The War Manpower Commission consequently approved interregional recruitment by these yards.[24] But the War Manpower Commission followed a supplementary rating system of its own in deciding where to send workers. Within the "must" category it gave plants three ratings: "AA," "A," and "B." Shipyards working on military and other "must" types received the "B" rating, and the U. S. Employment Service began sending workers to these yards rather than to employers who had only "C" rating. When, early in July, Edgar Kaiser asked for a higher rating, Admiral Vickery told him that in all the country there were only 20,928 job openings with higher rating, and of 49,000 openings with "B" rating, Oregon and Richmond had been allotted 20,000.[25] But later in the month Admiral Land was protesting to WPB that the "B" rating was not high enough. "The Production Urgency list of the Production Executive Committee and the National Manpower Priorities system of the War Manpower Commission do not jibe," he wrote. On the "must" list, programs were not rated one above another, but the War Manpower Commission divided them into "AA," "A," and "B,"and gave the Manhattan project and Boeing higher ratings. "To get on the A or AA requires administrative pressure under no stated rules of the game," objected Admiral Land.[26]

Even the B rating which resulted from being on the "must" program helped, however, both in securing labor locally and by enabling some shipyards to recruit labor from distant areas hitherto closed to them by the regulations of U. S. Employment

---

[24] Chaikin and Coleman, *Shipbuilding Policies*, pp. 161-64.

[25] Vickery to Edgar Kaiser, July 8, 1944 in MC gf 508-1-1; Ring and Van Riper to Vickery, July 7, 1944 in MC gf 508-1-12.

[26] Land to General Jaffe, Ex. Sec. Prod. Ex. Com., WPB, Aug. 10, 1944 in Land reading file.

Service.[27] Oregon Ship increased its labor force substantially, from 27,800 in June 1944 to 34,600 in November 1944, and the Vancouver yard, Calship, and Consolidated also increased employment, but at the Richmond yards, Swan Island, and Moore employment continued to fall off. Of course all the yards were at the same time losing men—separations on the West coast were between 10.1 and 12.9 per cent through 1944—and the main effect of the urgency ratings was to enable them to obtain replacements.

Whether employment was increasing or decreasing in the shipyards, they were continually being accused by the War Manpower Commission in 1943-1944 of hoarding labor—of not making good use of the labor they already had. To prevent this the War Manpower Commission proposed to send its representatives into the yards to make "manpower utilization surveys" with the idea of then cutting down the ceilings given the yards and having them disgorge the hoarded labor. It was empowered by Economic Administrator Byrnes in September 1943 to make such surveys, but the ability of its staff to do so with any understanding of the problems involved in shipbuilding was doubted by officials of the Maritime Commission; and in December 1943 Admiral Vickery instructed all regional directors to set up Manpower Survey Boards composed of members of their own staffs.[28] Commission officials then resisted efforts of the War Manpower Commission to make surveys independently both of it and of management, or criticized as incompetent those made.[29] In the cases where an immediate excess of labor did exist in a yard, it was defended on the ground that it was not hoarding but necessary provision for work scheduled for the near future. The War Manpower Commission on the other hand claimed that its efforts had the approval of managements,[30] and criticized the surveys made in some areas by the

[27] Approval for inter-area recruiting with an "A" rating was given to Calship and with "B" rating to Marinship, Moore, Ingalls, and Sun. Manpower Bulletin No. 1, issued by the Division of Shipyard Labor Relations, Sept. 23, 1944 in MC gf 508-1-12.

[28] *The United States at War*, p. 440; Ring to Vickery, Feb. 10, 1943; and Vickery to all Regional Directors of Construction, Dec. 4, 1943, in gf 508-1.

[29] Telegram, Flesher to Land, Dec. 19, 1944, in gr 508-1-12; Vickery (by D. R. Dorn) to Bewkes, Apr. 17, 1944, in gf 107-78; Dorn to Weber, Apr. 20, 1944, in gf 508-1; testimony of Vickery in House, *Walsh-Kaiser Co., Inc., Hearings*, pp. 257-58.

[30] Bewkes to Watt, Dec. 22, 1944; Bewkes to Watt, Dec. 30, 1944; Bewkes to Hunter, Jan. 1, 1945, and Bewkes to Watt, Jan. 1, 1945, all in gf 508-1-12.

Commission's staff. In a final statement of the Maritime Commission's official attitude Admiral Land wrote to all Regional Directors of Construction on January 2, 1945: "The Maritime Commission is not opposed to such surveys [by the WMC] but considers that proper utilization of labor is primarily a responsibility of management. Accordingly whenever management has agreed with the War Manpower Commission that a labor utilization survey should be undertaken, the Maritime Commission will cooperate fully with the War Manpower Commission and Management in making such a survey." The final say in accepting or rejecting the recommendations was to rest with the management.[31]

In at least one case the Manpower Commission's consultants made a survey which they considered thorough and impressive in its results—in the big Sun yard at Chester, Pennsylvania, where they reported it was possible to cut the existing work force about 10 per cent and still exceed the production schedules.[32]

The relations of the Maritime Commission with the War Manpower Commission may be summarized by saying that the Maritime Commission sought and obtained sufficient representation on the manpower committees so that its shipyards were allowed adequate ceilings; but employment was not always at the desired level, either because of a shortage of applicants for jobs, or because the WPB fixed ratings so that the U. S. Employment Service made few referrals to the shipyards. At the same time the War Manpower Commission and the WPB were constantly criticizing the Maritime Commission for not making shipyards use their labor more effectively.

III. Fumbling with Functions of Management

Efficient use of labor is at the very core of the managerial function. Wartime controls had limited management's freedom of action in many spheres where it had once had authority—procurement, labor supply and wages, even executives' salaries. From the beginning of the emergency program the Commission took a vigorous part in making many decisions ordinarily left to private managements—decisions about the selection of shipyard sites, the

---

[31] Land to All Regional Directors of Construction, Jan. 2, 1945, in gf 508-1.
[32] "Short History of the War Manpower Commission," p. 58.

layout of the yards, the construction procedures, and the amount to be invested in equipment. Now shortage of manpower put pressure on the Commission to take yet another step and assume some of the managerial function of seeing that labor was utilized efficiently.

Although the reduction in the labor force was one factor in the failure of yards which did not meet their schedules in 1944, it was not the only factor. A yard-by-yard comparison shows some correlation between the extent to which they fell behind and the extent to which employment declined during the first half of 1944, but there were exceptions. Some yards, notably Walsh-Kaiser, fell behind schedule in spite of the fact that employment was maintained and reached the ceiling in July 1944. Other yards, notably Kaiser-Vancouver and North Carolina, practically made their schedules in spite of severe drops in employment, 9 per cent in the one case and 22 per cent in the other.[1] Those two yards were building military types and so had the advantage of high urgency ratings to help in keeping the flow of components on schedule; but the contrast between their performance and that of some other yards also building military types indicates that decrease in the labor force could be overcome by more effective use of the labor available.

Therefore, although public discussions of the difficulties being encountered in 1944 spoke loudly of " the manpower problem," it might equally well have been called " the management problem." To be sure the change which created a new situation was the shrinkage in the supply of labor; but since in a nation at war this shrinkage was unavoidable, the question was how to train and use more effectively the smaller supply. There were two possible ways for the Commission to act on that problem. It could try to act through yard managements, or it could try to take over from the managements functions which the private managers were not performing satisfactorily. As already explained, this was the central issue in the whole problem of managing managements that had been taken on by the Commission when launching a program much larger than the capacity of the existing industry.

Admiral Land and Admiral Vickery did not want to take over the problems of management any more than necessary; their

[1] Compilations by Gerald J. Fischer and Blanche Coll in Historian's Collection, Research file 205.2.6.

policy was to drive the managers into finding solutions. Admiral Vickery was still going from yard to yard on inspection tours, finding what was wrong and what could be done on the spot to correct it, and leaving behind doses of his own special brand of "Victory vitamins." Loving hard work and feeling that the program depended on him, he had been driving himself at a terrific pace all through 1941, 1942, and 1943, and in 1944 was still flying from coast to coast keeping his "needle" in the builders. A week at the yacht races at Marblehead in July 1944 was practically his first vacation since the program began. On September 25, 1944 he suffered a heart attack and was forced to stop working.[2] Technical assistants could do some of his tasks, but until Admiral Vickery returned in the middle of February 1945 (and immediately left for a long inspection trip), Admiral Land added to his duties as Chairman of the Commission and Administrator of WSA the handling of many matters which had been in Admiral Vickery's field.[3] There was no abandonment during 1944 of the effort to manage management by personal relations, but it was making a heavy drain on the men who carried the burden.

Although in general the Commission followed the policy of putting responsibility on the management, on some points it had compromised this principle. It sought to find solutions through experts outside the shipbuilding companies and then to tell the managements what to do. When faced in 1944 with the need for more efficient utilization of labor, the Commission moved in three ways towards telling management how to solve the problem: through manpower surveys, through use of consulting engineers, and through the training program.

Manpower surveys were an opportunity to find out how well management was doing its job and to tell it how it could do better. Some of the regional directors of the Maritime Commission took Admiral Vickery's instruction of December 8, 1943 in this sense. In the Gulf region, Mr. Sanford at once appointed the chiefs of his Production and Construction Sections to a Regional Manpower Survey Board under the chairmanship of his Industrial Relations Advisor and assigned alternates who were to spend full time in field work. Before the month was out they were surveying in detail the yard nearest the regional office. Their report listed

---

[2] Record of his leave and of his travel kept by Miss Irene Long.
[3] Land's Memorandum for Executive Director, Sept. 28, 1944, in Land reading file.

the number of men in each department, how efficiently they were used, how their number compared with the manhours actually expended on each ship and with the standards of low manhours per ship set by the more efficient builders. It suggested places where labor could be saved by more machinery, or more efficient arrangement, or different procedures, and reviewed the whole supervisory organization as well as the programs for training and the efforts to reduce turnover and absenteeism. While recognizing that the overstaffing came from producing beyond the yard's normal capacity (as was done in December 1943), the report recommended a stabilization of the labor force of 3,500 below the number actually employed. A resurvey nine months later stated that many of the suggestions had been followed and the force had been reduced nearly 3,000, yet there were still idle workers. This board surveyed similarly other yards in the Gulf region and made specific suggestions, but in most yards it focused its recommendations, not on decreasing the number employed, but on increasing the production by better training and utilization.[4]

In the Great Lakes region also a Regional Manpower Survey Board was at once created by the Regional Director, Mr. Spofford, and some of the staff was assigned to work full time on its functions. Although it analyzed the number of workers in different categories and the number of manhours per ship, it focused more on the labor situation than did the survey made in the Gulf, and did not go into analysis of managerial organization or purely engineering questions. At the Superior Yard of Walter Butler Shipbuilders, Inc. a labor utilization survey had already been undertaken during the month of November 1943, at the request of the company's president, by the office of the Regional Industrial Advisor together with technicians of the Division of Manpower Utilization of the War Manpower Commission. This was a result of allegations by the War Manpower Commission that the yard was overmanned, and yet, by its continued recruiting, was creating a critical labor shortage in the Duluth area. When the Regional Manpower Survey Board of the Maritime Commission, also headed by the Regional Industrial Advisor, took over, it utilized the field work of the Manpower Utilization Staff. The report con-

[4] Gulf Coast Regional Construction Office, Regional Manpower Survey Board, Confidential Survey Report[s] on Manpower Utilization, in Historian's Collection, Research file.

tained definite recommendations concerning the organization of the personnel department, safety, the cafeteria, and the proportion of men to be employed on day shift and night shift. If the latter could be more evenly averaged, it would not be necessary to increase the number of workers in the yard, the report concluded.[5] Shortly thereafter, the Great Lakes region was combined with the Gulf and Mr. Sanford made Regional Director of both. One reason for this was the plan for finishing in the Gulf the frigates begun on the Great Lakes, but Mr. Spofford attributed his removal to Mr. Butler's resentment at having had his manpower ceiling cut, and the Commission action removing Mr. Spofford said he " no longer has the confidence or cooperation of the shipyards." [6]

On the West coast the response to Admiral Vickery's directive was very different. A questionnaire was made out and answers to it were written either by the industrial relations representative of the Maritime Commission or by company officials. In neither case did these reports enter into managerial or engineering problems or consider department by department the way men were utilized; they discussed the number of additional workers wanted in various crafts and the difficulties of obtaining them. The manpower surveys from southern California contain many criticisms of the way labor was being handled in the yards, but those from northern California and Oregon expound the company's need for more recruits and sometimes turn into pleas to be allowed to pay more for overtime. That from Marinship for January concluded: "... may we quote a remark made by Admiral Vickery at a luncheon at the Cafeteria at Marinship on January 27, 1944 as follows: ' Our men are not here in the yard to find what is wrong with things, but rather to find a way to get things done ' (or words to that effect)." [7] For the East coast, formal manpower surveys cannot be found and the Chief of the Inspection Section does not recall that there were any.[8]

Another way of telling managements how to save labor was

[5] Great Lakes Regional Construction Office, Regional Manpower Survey Board, Confidential Preliminary Report on Manpower Utilization for December 1943 at Walter Butler Shipbuilders, Inc., especially p. 31, in Historian's Collection, Research file.

[6] Statement by W. E. Spofford. MC Minutes, pp. 28331-32 (May 2, 1944).

[7] Manpower Survey Board, MC, [Reports], in Historian's Collection, Research file.

[8] Statement by W. H. Blakeman. In a memorandum to Commissioner Carmody, Jan. 2, 1945, Mr. Ring mentioned surveys received in his office from the three other regions but none from the East coast.

through the employment of firms of advisory management engineers. Commissioner Carmody wished to urge on the shipbuilding companies the employment of such firms, or even to have the Commission employ them, while Admiral Vickery felt the question of whether or not to hire them should be left to yard managements. At the initiative of the yards themselves, more use was made of such firms in the later years of the program when more and more emphasis was being put on cutting labor costs. After the reorganization of the South Portland yards at the beginning of 1943, the newly formed New England Shipbuilding Corp. engaged on May 24, 1943 the firm of Thompson & Lichtner which worked in that yard until early 1945. In that interval the total manhours per Liberty ship was reduced from 760,000 to 410,000. The reduction was larger than that in comparable rounds of the ways at most yards.[9] Much credit for the improvement must of course go to the new management as a whole and to the way the problems of trackage and storage were eased by the addition of the North area; but Thompson & Lichtner could point to many savings which were their suggestions: a way of planning the use of trucks and railroad cars which eliminated the need for hiring trucks, systems of controlling inventory and scheduling the movement of materials and components in the yard so that they were at the right place when needed, simplifications of storekeeping and timekeeping methods so that the clerical force could be reduced, and an incentive system for welders based on poundage.[10] This consulting firm was particularly proud of developing its system of measuring the achievement of welders; by measuring it in pounds instead of feet a sounder basis was provided for incentive payments.[11] In addition to the work at New England Ship, this and other firms were employed to survey efficiency in a number of yards on the East coast and the Gulf.[12]

Whether the fees paid such engineering firms should be reim-

[9] At the West yards, rounds 5 to 18 were in this period. Compare Figure 16.

[10] Sanford E. Thompson to F. C. Lane, June 22, 1948, and July 15, 1948; John M. Carmody to F. C. Lane, June 29, 1948, and "A Few Accomplishments in Shipbuilding," a brochure put out by Thompson & Lichtner, all in Historian's Collection, Research file 203.3.5; also reports of Thompson & Lichtner seen in office of Plant Engineer at South Portland yard in August 1946.

[11] J. Saliba, "Welding Production Controls for Shipyard Operations," *Marine Engineering and Shipping Review*, March 1947, pp. 74-78.

[12] Correspondence with Thompson & Lichtner and with the Emerson Engineers, in MC, Historian's Collection, Research file 203.3.5.

bursed by the Commission to the shipbuilding company was a matter of dispute. The services of Thompson & Lichtner under their first contract with New England Shipbuilding Corp. were paid for by the Commission, and the contract was concluded subject to the Commission's approval and agreement to reimburse. Its services under subsequent contracts were paid for by the shipbuilding company without reimbursement.[13] In general, reimbursement did not seem justified. Admiral Vickery testified concerning efficiency experts hired by Walsh-Kaiser: "We would not allow that in cost. They pay for it simply to get it. Our contention is that management gets a fee for managing the yard. If they need some professional help to show them how to manage that yard they should pay for it themselves, because they are already paid for doing a job." [14] If the Commission reimbursed the shipbuilding company for the services of the consulting engineers, it might be said to be paying twice for the same thing. On the other hand, legality aside, from the point of view of the effect on the national economy as a whole in 1944, the extra expenditure of money would have been relatively harmless, the saving of manpower might have been quite important. If the labor shortage had grown increasingly severe it might have been economical to pay twice over for the sake of saving manpower.

Another approach to the manpower problem, which not only proposed to tell management what to do but even to compel compliance, was the training program championed by Jack Wolff in the Division of Shipyard Labor Relations. The " Basic Principles to be Observed in Establishing Production Training Programs in Shipyards " which the Commission had approved in 1942 called for the systematic training of workers before they were upgraded to " first class mechanics," and stressed the need for training supervisors, especially through supervisors' conferences. The changeovers in many yards during 1944 to new and more complicated types of ships made it all the more desirable to have workers who could do more than " one small detail of a job." [15]

In many yards the Supervisor of Training felt he was successful in his preaching of this program and in his on-the-spot attempts to

[13] Thompson to Lane, July 15, 1948, above cited.
[14] House, *Walsh-Kaiser Co., Inc., Hearings*, p. 274.
[15] Analysis of Shipyard Training Report for Month of Jan. 1944, in Division of Shipyard Labor Relations file, Shipyard Training Reports. Compare above chap. 8, sect. IV.

help put it into effect. In February 1944 he reported that only two yards had no in-plant training programs, and that twenty-nine shipyards had shown marked improvement during the past year both in the quantity and quality of the training given.[16] But in many places the training was not as he thought it should be, and the Commission was not using the power that it had through checking pay rolls and reimbursements to prevent the upgrading of those without proper experience and training.[17] Commissioner Carmody, to be sure, strongly favored the program. He advised in February 1943, " The yards will take the matter seriously and do a job only if Admiral Vickery makes it clear to them that this is not some phoney boondogling project but a matter of production that is being lost and dollars and cents that are being unnecessarily wasted. . . ."[18] Mr. Wolff drafted a letter for Admiral Vickery's signature to go to all regional offices and shipyards but the letter was not sent.[19] Instead of this personal approach, rules to be enforced in reimbursement were made jointly with the Navy and Army through the Shipbuilding Stabilization Zone Standards Committee. It issued new regulations looking towards a tighter policy in allowing costs from training and upgrading, but each seemingly stringent provision contained a loophole and was followed by modification.[20] By failing to regulate upgrading from the start, as explained in chapter 13, the Maritime Commission lost control of it, and consequently they were unable to use it to force better training.

Although skilled labor was most urgently needed on the West

[16] Wolff to Director, Division of Shipyard Labor Relations, Feb. 2, 1944, in Division of Shipyard Labor Relations file, Training in Shipyards.

[17] Letters from Wolff in file above cited and in gf 508-1-1.

[18] Carmody to C. R. Dooley, Feb. 19, 1943, in Division of Shipyard Labor Relations file, Training in Shipyards.

[19] Wolff to Director, Division of Shipyard Labor Relations, May 31, 1943, in the same file.

[20] Administrative Instruction No. 10, Jan. 13, 1943; telegram, Ring to C. W. Eliason, West Coast Industrial Relations Adviser, Feb. 18, 1943, in gf 508-1-1. A set of the Administrative Instructions is in Historian's Collection, Research file 205.6.

For example its regulation of Oct. 23, 1943 forbade upgrading in less than 180 days after assignment to productive work. There were immediate protests including one from Edgar Kaiser that the Commission was thereby upsetting the arrangements with labor unions which it had approved. Edgar Kaiser to Vickery, Nov. 14, 1943; Wolff to Director, Division of Shipyard Labor Relations, Nov. 14, 1943 and Dec. 8, 1943, all in gf 508-1-1. Actually the rule did not prevent workers moving to another yard at a higher grade and it was modified in April.

coast, that is where the Commission's program for training had least effect. Around Portland, Oregon, training had been from the first turned over to public schools, and in 1943 the Kaiser companies were still paying trainees to go to public schools for courses in spite of the fact that in-plant training was recommended by the Commission's "Basic Principles." [21] The West coast regional director preferred the view of the shipyard managers to those of Mr. Wolff, and protested against the method used by the latter in calling conferences at the shipyards without informing the regional office.[22] The Industrial Relations Advisor on the West coast objected to having Mr. Wolff visit the yards there, and he did not after August 1943. When the regional assistant supervisor of training resigned, having been critical of yard practices, the regional office replaced him with a man who, so Commissioner Carmody said, had no qualifications for the job.[23] In Mr. Wolff's opinion the West coast yards were "running more or less wild in applying uneconomic and highly unproductive training procedures." [24] Whereas Mr. Wolff's eyes were focused on training, the regional directors were of course responsible for results as a whole. Their attitude may have been influenced by the fact that the yards of which the training director was most critical were precisely those which had the outstanding production records. "The Kaiser Yards," Wolff commented, " can turn out lots of ships with a poor training set-up because they have room enough to employ one and one-half or two people for every one qualified person who might do the job . . . [and] their engineering and plant layout together with their national procurement has been so brilliant that it more than offsets personal inefficiency which would otherwise be fatal." [25] In September 1944 all the regional training directors were made subject to the regional directors functionally as well as administratively.[26]

The training program, the use of consulting engineers, and the manpower surveys are probably less important for what they

[21] Wolff to Ring, Jan. 29, 1943, in Division of Shipyard Labor Relations file, Training Survey in Maritime Shipyards; Theodore A. Hellenthal to W. H. Quarg, Apr. 24, 1943, in gf 508-1-1.
[22] Flesher to Vickery, May 15, 1943, in Flesher reading file, confidential, 1943.
[23] Wolff to Ring, Jan. 16, 1945, in gf 508-1-1; Carmody to Ring and Wolff, Jan. 15, 1945, in Carmody file.
[24] Wolff to Ring, Jan. 16, 1945, in gf 508-1-1.
[25] Wolff to W. J. Freeling, July 8, 1943, in gf 508-1-1.
[26] Vickery to McInnis and other regional directors, Sept. 16, 1944 in gf 508-1-1.

actually accomplished in 1944 than as indications of the direction in which the Commission might move under pressure for more efficient use of manpower. One cannot say how effective these measures could have been, for they were never vigorously pushed with backing from the top. If pushed, they would have splintered the responsibility for yard management. Since the shipbuilding contractors were receiving large fees as rewards for undertaking to manage the yards, the Commission clung to the principle that these contractors should have enough freedom of action so that they would feel their managerial responsibility.

IV. THE LUMP-SUM CONTRACTS IN 1944 AND 1945

If private managements were to be depended on to solve the manpower problem, a change to lump-sum contracts was the obvious way of putting pressure on them to use labor more efficiently. Contracts which assured reimbursement of costs did not, in spite of their variable fees, provide enough incentive to economy. As explained in chapter 14, the Commission had been attempting ever since June 1943 to induce builders to take lump-sum contracts, and had devised a special type of lump-sum contract, the selective-price contract, to enable the builders to minimize their risks. But only one yard (North Carolina) was persuaded in 1943 to change from a cost-plus-variable fee (or manhour) contract to selective-price, although a few old line yards—notably Federal and Ingalls—clung to fixed-price contracts all along. The construction of Victory ships in 1944 was begun under the cost-plus-variable-fee contracts awarded in April 1943 [1] and the change from Victory freighters to Victory transports was authorized simply by letters to the yards from Admiral Vickery.[2]

While the desirability of lump-sum contracts increased in 1944, so did the difficulties of imposing them, because many yards were then building military types. These were new types subject to late revision of plans, and estimating the cost in advance was difficult—in many cases impossible. Moreover the military types were urgently needed. It was not desirable to impose a contract which

[1] "Permanent Report of Completed Ship Construction Contracts," pp. 1B-21B. The award of Victory transports to Kaiser-Vancouver was made Dec. 14, 1943 but on the same terms as the awards of Apr. 20, 1943. See Minutes, pp. 27091-92 and for the Apr. 1943 contracts, pp. 24910-11, 24968-71.

[2] These letters are in Technical Division, reading file, Dec. 13, 1943.

might lead the builder to sacrifice speed for the sake of economy. When the time came to award new contracts to provide yards with a basis of operations in 1945, a new effort was made to return to lump-sum contracts. Awards for 174 Victory ships were made June 15, 1944 with the statement: "It will be desirable to have the contracts for the construction of these vessels on a fixed price basis. However, negotiations as yet have not been completed with the various shipbuilders with respect to prices and terms of the contract. In order that there will be no delay in the commencement of the program [of the Victory transports and freighters ordered by the Joint Chiefs for delivery before June 1945] pending the completion of such negotiations, awards of contracts should be made at this time to the various shipbuilders upon the understanding that the terms and conditions will be agreed upon at a later date."[3] Letters of intent were sent accordingly. On the basis of these letters of intent, and without any contract, the yards made preparations to construct more Victorys in 1945.

This period in 1944 when the manpower shortage was most acute was therefore a period in which contractual relations with the shipbuilders, especially in the emergency yards, were not of such a character as to put the builders under substantial pressure to economize. It was also the period in which Admiral Vickery was unable to maintain personally his pressure on the builders. His heart attack kept him in the hospital from September 25, 1944 to February 1945. Admiral Land was carrying both his duties as Administrator of WSA and the heaviest part of the responsibilities previously borne by Admiral Vickery. In December 1944 he undertook to place the Victory program for 1945 on some kind of a contractual basis. He proposed, and the Commission unanimously approved, on December 7, 1944, cost-plus-variable-fee (manhour) contracts for all the Victory yards.[4] Presumably the work, which had already been started had proceeded on the assumption by the builders that they would be reimbursed according to the terms of such contracts, and the Commission action of December 7, 1944 formally approved what had been an informal understanding under the letters of intent.[5]

[3] Minutes, pp. 28760-61. Similar letters of intent went to North Carolina (Minutes, p. 28763) and Alabama and Sparrows Point (Minutes, p. 28774).
[4] Minutes, pp. 30418-19 (Dec. 7, 1944), where it says the letters of intent had been sent in July or August.
[5] Confirmed by comment of Admiral Land.

All these contracts were concluded at a time when the Maritime Commission had no approved program for extending emergency ship construction much beyond the middle of 1945. The program which the Commission drew up September 15, 1944 (No. 34) envisaged the construction of only 10,254,000 deadweight tons in 1945. In approving this program in October 1944 the Joint Chiefs of Staff recommended that " no additional new construction for the war effort be programmed for 1945 at this time." It then seemed possible that Germany would be defeated at the end of 1944.[6] But in December it was clear that the war in Europe would continue into 1945, and consequently the Joint Chiefs asked for additional construction. Justice Byrnes, Director of the Office of War Mobilization and Reconversion authorized the Commission to " go ahead to avoid a lag in deliveries," and the Commission drew up schedules which would keep the Victory and tanker yards busy to the end of 1945 and produce 12,645,000 tons.[7]

The new awards necessary to fulfill this program were the occasion of a new successful attempt to make the yards take lump-sum contracts. Telegrams were sent out December 29, 1944 telling them that they would not have to close down as soon as had been expected. There would be more ships to build, but they must be built under fixed-price contracts.[8] To arrange these contracts the Commission's Committee on Awards, which had practically lapsed during Vickery's illness, was reconstituted on January 9, 1945 by adding young men who took a prominent part in later negotiations with the builders: W. S. Brown of the Legal Division; W. A. Weber, Admiral Vickery's aide; and R. P. Mills of the Technical Division. R. E. Anderson continued and acted as chairman in the absence of Admiral Vickery.[9] During January-March 1945 the reconstituted committee recommended a series of contracts which put all the yards on selective-price or fixed-price contracts.[10]

[6] Fischer, *Statistical Summary*, Table A-5; Joint Chiefs to Land, Oct. 16, 1944 in MC Program Collection.

[7] Land to Joint Chiefs, Dec. 22, 1944; Byrnes to Land, Dec. 23, 1944; Joint Chiefs to Land, Dec. 30, 1944—all in MC Program Collection. Also MC Press Release 2125, Jan. 3, 1945, and Fischer, *Statistical Summary*, Table A-5.

[8] House, *Investigation of Shipyard Profits, Hearings*, (1946), p. 643. Land (by Weber) to John A. McCone and others, Dec. 29, 1944 in Weber reading file.

[9] Minutes, p. 30709.

[10] Minutes, p. 30813 (Jan. 18, 1945); pp. 31184-86 (Feb. 27, 1945), Oregon Ship, and other contracts at later dates.

As part of these arrangements many cost-plus-variable-fee contracts for ships which had already been completed were converted to lump-sum contracts. These conversions were a way of settling conflicts over reimbursables and of avoiding the auditing difficulties which would arise if the same yard had both cost-plus and lump-sum contracts. To the unwary reader of the record of Maritime Commission contracts it might appear that many Victorys and tankers delivered before 1945 were built under lump-sum arrangements, since the contracts were dated back to cover the period of construction, but it is important to realize that the contracts were not actually agreed to until 1945—until after the ships were delivered. They did not in 1944 create incentives to economy.

Shipbuilding operations during 1945, however, were influenced by the agreements reached early in that year. Victory yards were converted to selective-price contracts, tanker yards were put on a lump-sum basis.[11] The negotiations with the tanker yards show how similar the system of pricing was under these two forms of contract. Estimates of the actual cost without profit of building T2's were: at Sun, $2.6 million; at Kaiser-Swan Island, $2.7 million; at Alabama, $3 million. Sun offered to build twenty tankers at the fixed price of $2.7 million. The Commission approved setting that figure as the " normal " price in selective price contracts in which the contractor could make 4 per cent profit. If he picked a higher price, his allowable profit would be reduced, down to $1\frac{1}{4}$ per cent, if a lower price, it would be raised, up to 7 per cent. But Sun preferred a fixed-price contract and secured it on terms which permitted the company to retain a maximum of $245,000.[12] Marinship said it too preferred a fixed-price contract and the Commission approved one naming $2.8 million as the price and permitting retention as profit of only $200,000. The same principle was applied in awarding four Navy oilers to Marinship. The Technical Division of the Maritime Commission estimated the cost of building should be $3.7 million. Marinship wanted a fixed price of $4.45 million. A contract was approved at $4.3 million, but retainable profit was fixed relatively low, at $150,000.

[11] According to the card file of contracts prepared for me by the Finance Division, all contracts awarded after August 1944 were selective-price or fixed-price, or were converted to fixed price, except a few price-minus contracts of Sunship.

[12] Minutes, pp. 30785-87 (Jan. 16, 1945). For earlier efforts to arrange fixed prices for tankers see Minutes, pp. 30128-29 (Nov. 2, 1944); p. 30386 (Dec. 5, 1944).

The Committee on Awards commented: "Thus, as in the case of the T2 tankers, the contractors' recapture ceiling has been materially reduced due to the shipbuilder's unwillingness to accept a price which the Commission estimates should be charged for these vessels." [13] As the instance shows, the same principles were applied in fixed-price contracts as in selective-price contracts. The difference was that selective-price contracts named a schedule according to which higher price meant less chance of profit, and it empowered

TABLE 12

WAY TIME AND NUMBER OF EMPLOYEES IN SELECTED YARDS, JANUARY-MAY 1945

	Jan.	Feb	March	April	May
Alabama					
Way Time (days)	65	63	61	67	71
Employees (hundreds)	221	176	146	130	120
Swan Island					
Way Time (days)	38	39	57	68	61
Employees (hundreds)	210	158	135	119	102
Beth-Fairfield					
Way Time (days)	54	61	55	49	49
Employees (hundreds)	270	266	257	245	232
Richmond No. 1 and No. 2					
Way Time (days)	57	52	47	44	39
Employees (hundreds)	410	383	341	293	243

the contractor to pick at any time prior to keel-laying the price he preferred, whereas in negotiating fixed-price contracts the same principle was applied in advance of signing the contract.

Although this return to lump-sum contracts came too late to affect labor utilization at the time when the pressure was greatest, it is interesting to inquire whether the productivity of labor did increase when yards changed from cost-plus-variable-fee to lump-sum contracts. Did the shipyards then disgorge "hoarded" labor? At the time the change was made so many yards were changing production from one type to another that only in very few is it at all possible to compare their record of labor productivity before and after the new types of contract went into effect. It has been attempted for four yards—two building tankers, Alabama and Swan Island, and two building Victory ships, Bethle-

[13] Minutes, pp. 31235-36, 31289-90 (Mar. 6, 1945), p. 31322 (Mar. 13, 1945). Commissioner Carmody's opposition to this contract will be discussed in chap. 23.

hem-Fairfield and Richmond No. 1 and No. 2 considered as one yard. In these yards the same type of ship was produced through the period of the change from one form of contract to the other. Table 12 shows some big drops in the number employed. The shipyards did disgorge manpower, but there was also some slowing down of production. At Bethlehem-Fairfield and Richmond the decline in employment had been going on all through 1944. Some of the Victory ships launched in January were the first round of the ways, some were the third, of that particular type, so that the increase in speed in these yards must be discounted to allow for the usual acceleration of successive rounds.[14]

In the yards as a whole labor productivity reached its peak in the second quarter of 1945 (Figure 42). Some of this increase in productivity in 1945 can be attributed to the new types of contract, but much was due to cutting down the third shift.[15] That step could of course have been taken by the Maritime Commission at any time without change in the form of contract, just as it had cut out Sunday work. Everyone knew that the output per worker was less on the "graveyard shift." But to have cut down the third shift would have delayed the delivery of vessels. Therefore that way of solving the manpower problem was not insisted on in mid-1944 when the Navy was clamoring for the auxiliaries being built by the Commission.

To be sure, the loss in speed seems to have been relatively slight compared to the saving in manpower. But how measure the value of time during a war?

An indication that costs were reduced under the lump-sum contracts is the size of the profits made under these contracts. By cutting costs on later deliveries, many yards earned the maximum retainable profit and in addition repaid the Commission millions in recaptured profits. These recaptures arose from the

[14] It is difficult to determine at exactly what date the new incentives in the new form of contract would show in the operations of the shipyard. Anticipation of the change, as well as the change itself, may have affected managerial policy. The date of the contract and the date on which it was executed were often quite different, and the date on which the parties orally agreed and made up their minds what was to be done is something else again. The price range of the selective-price contract for new building was affected by previous costs. Being unable to work out all these factors in detail for each yard, I present the record of the period when the policy decision, approved by the Commission action Jan. 18, 1945, was being put into effect.

[15] BLS data sheets show reductions in third shifts.

fact that the costs of production were below the lowest figure envisaged in the contract (Table 19).

But against these indications that the new form of contract was the cause of an increase in productivity must be set the evidence that continuity in production of a single type for many successive rounds of the ways was the real cause—that this continuity would have produced savings in manpower and money even if there had been no change in the type of contract. It is instructive to compare the reduction in manhours expended in producing successive rounds of Liberty ships, all of which were built under cost-plus-variable-fee contracts. Between the 3rd and 7th round of the ways, manhours dropped about 25 per cent; they dropped 10 per cent more between the 7th and 11th round, and another 10 per cent between the 11th and the 17th (Figure 16). There were similar gains in productivity in the continuous production of tankers (Figure 20).

To estimate the importance of continuity of production in the Victory yards it is necessary to describe in some detail the output of Calship and Oregon as examples. Between April and September 1944, Calship delivered 32 Victory cargo ships (AP3's), that is, an average of three rounds of the ways. Then they built 30 of the military type of Victorys, the attack transports (AP5's). In January 1945 they began delivering Victory cargo ships again (AP2's) and completed five rounds off each way (except one which built three rounds) before the yard closed down.

Activity at Oregon Ship followed much the same pattern. In February through July of 1944 it delivered 32 Victory cargo ships (AP3's), roughly three rounds of the ways. In August through December 1944 it delivered 30 Victory attack transports (AP5's). In January 1945 it again began delivering Victory cargo ships (AP3's) and made delivery that year of about eleven rounds of the ways, 67 ships.[16]

The construction of different types of Victorys makes it difficult to estimate exactly the benefits of continuity of production, but comparison with the reduction in manhours on successive rounds of Liberty ships would lead one to expect a reduction of at least 10 per cent in the cost on these later rounds of Victory ships, even if there had been no change in the form of contract.

In defense of the selective-price contracts, comparisons were

[16] Reynolds' Way Charts.

made between the prices which had been paid for Victory ships under the cost-plus-variable-fee contracts and the prices paid under the selective-price contract. For example, it was pointed out that $2,918,000 per ship had actually been paid for Victory cargo ships (AP3's) under the cost-plus-variable-fee contract, and that Oregon Ship would deliver them under its selective-price contract for $2,618,000.[17] The reduction is only 10 per cent, yet the price under the cost-plus-variable-fee contract was that for the first three rounds of AP3's, built in 1944, and the experience with Liberty ships would lead one to expect a reduction of more than 10 per cent in later rounds. Actually, Oregon Ship selected the price of $2,325,000 for the last 13 vessels.[18] Considering how many rounds of the ways had been produced by the time they made their last deliveries, the reduction is again no greater than occurred for Liberty ships under the price-plus-variable-fee contracts. This would seem to indicate that, in determining productivity, continuity was more important than the form of contract.

On the other hand, it is pertinent to emphasize that cost-plus-variable-fee contracts would probably have encouraged waste in 1945 far more than they did in 1943. In 1943 there was pressure to finish one contract quickly in order to begin on a new contract. In 1945 the situation was quite the opposite. In 1945 it was known that the shipyards would close down as soon as their current contracts were finished. Under these conditions, a cost-plus-variable-fee form of contract would have put a premium on dragging out the execution of the contract as long as possible.

The profits on the selective-price contracts will be considered in chapter 24 in connection with their exemption from renegotiation. Profits were not so large as they might have been, however, because some of the ships in these contracts were cancelled. As early as April 20, 1945, shortly before Germany surrendered, the Commission, acting on instruction from the Director of War Mobilization and Reconversion, ordered work on thirty tankers terminated.[19] Schedules for other types were based primarily on

[17] Unsigned memorandum dated Apr. 9, 1945 in Weber's files, "Selective price." Similar comparisons for Calship and a number of other yards, which show no greater reductions. Comparisons subject to the same considerations, although concerning the price of tankers, are given in House, *National War Agencies Appropriation Bill for 1946, Hearings* before the Subcommittee of the Committee on Appropriations, 79th Cong., 1st sess., Apr. 25, 1945, p. 402.
[18] House, *Investigation of Shipyard Profits, Hearings* (1946), p. 651.
[19] MC Minutes, p. 31737 (Apr. 20, 1945).

the needs of war in the Pacific and continued in force until the end of hostilities with Japan. Then, on August 17, 1945, the Commission cancelled the construction of 132 vessels, including 42 Victory freighters. By the end of 1945 monthly deliveries were declining rapidly and many yards were closing.[20]

Some features of the lump-sum contracts arranged in 1945 will be re-examined later. Pertinent to the problem in managing management which we have discussed are the following conclusions: (1) The return to lump-sum contracts was carried through only after the pressure for speed was relaxed; it was not put into effect, and in my judgment could not have been put into effect, until after the rate of production was allowed to taper off. (2) Even then, it was carried through on terms which permitted large increases in private profit. (3) It was followed by substantial savings in manpower. (4) But these savings are not by any means to be attributed entirely to the change in the form of contract. To a large extent they would have occurred under a carefully audited cost-plus contract as a result of (a) the savings attendant on continuity in production of the same type in successive rounds of the ways and (b) the elimination of the relatively unproductive night shift and the laying off of the least productive workers, steps which were accelerated no doubt by the new form of contract, but steps which could have been taken also under cost-plus contracts as soon as emphasis was put more on economy than on speed.

In industrial mobilization as a whole as well as in shipbuilding, problems came in a certain rhythm. First it was facilities, then materials, then smoothing production through scheduling and standardization, and finally manpower. The war ended before the pressure on manpower had become as intense in the United States as it was in countries more severely tested, such as Germany and Britain. The United States was therefore never pushed into making a real trial of methods for assuring efficient use of labor. Overall, the controls proposed by the War Manpower Commission were never given statutory force; there was no labor draft. Within its narrower sphere of action, the Maritime Commission was not pushed into making a thorough trial of either of the two methods by which it sought to secure more efficient use of labor in wartime shipbuilding. It neither forced the companies to operate in 1944

[20] Fischer, " Programming," pp. 47-50; Reynolds' Way Charts.

under lump-sum contracts, nor used the Commission's power over reimbursement to force on them its recommendations for efficiency. Like the nation as a whole, the Commission never had to take all-out measures in dealing with manpower.

The pressure was enough in 1944, however, to indicate what kind of steps would have been taken next had the pressure continued. One possibility was a labor draft by an act of Congress. That might have enlarged the total labor force and have reduced turnover, but it hardly seems likely that conscripted labor would have been used with more efficiency. A return to lump-sum contracts was an unattainable alternative; the heads of corporations would not sign contracts which might, because of circumstances which they felt to be beyond their control, prove unprofitable or even ruinous. If they had been forced to sign such contracts under duress, that would have produced more bitterness than industrial efficiency. The most likely alternative was an increase in bureaucratic controls: more use by the Commission's officials of their power to disapprove expenditures. This trend was indicated by the manpower surveys, but it was checked by the fact that the Commission was against splintering the functions of management.

The extension of bureaucratic controls might have had a deadening effect. Certainly it would have placed a heavier burden on the Commission's regional offices, perhaps more of a burden than they were capable of carrying.

CHAPTER 21

ADMINISTRATIVE PROBLEMS:
(A) THE REGIONAL OFFICES

I. THE FOUR REGIONS

ALTHOUGH IN ITS relations with the shipbuilding corporations the Commission depended heavily on direct personal contact between the managers of the corporations and the top officials of the Commission, the necessary checking to see that the ships were soundly built and that the money was being spent legitimately depended on the detailed work of inspectors, auditors, and other officials. This staff also made important contributions at its level in the drive for speed and efficiency. How far the Commission's staff should take a hand in functions which the corporations felt were their managerial responsibilities was a continual problem. If more reliance had been placed on the rank and file of the Commission's staff and more attention given to organizing and administering their efforts, this staff could, at least so it seems to me, have done more in guiding yard managements into efficient methods, especially in 1944. But as it was they had plenty to do.

The portion of the staff which was in closest touch with the problems of shipyard management consisted of men reporting to a Regional Construction Office. In planning, procurement, and the formation of policies, decisions in Washington were of prime importance, but execution depended on " the field." The men in the regions felt that they built the ships.

When the four regional offices were established in March and April 1942 at Oakland, New Orleans, Philadelphia, and Chicago, the primary purpose was to speed up the answering of the host of questions about plan approval, inspection, and auditing which were creating congestion in Washington. The decentralization created some new administrative problems, but it relieved the congestion and also made it possible to modify the application of the Commission's supervisory functions in accordance with the different needs of different regions.

The most important regional office was that in Oakland, California, as is indicated by Table 13. The extent of each region was accurately indicated by its name except that shipyards on the Atlantic coast of Florida were placed under the Gulf regional office, since Jacksonville was on the railroad line to Tampa and Gulf had fewer yards to supervise than did East Coast.[1]

When decentralization occurred and the key men for the regional offices were picked, it seemed possible that the Gulf office would

TABLE 13

DELIVERIES, BUILDING BERTHS, AND EMPLOYEES BY REGIONS

Region	Deliveries 1939-45 (millions of displacement tons)	Building Berths		MC Employees Functionally under Regional Director in 1944
		In Class I Yards	In All Yards	
West Coast....	10.2	101	157	1656
East Coast....	7.7	104	138	962
Gulf..........	3.9	62	104	832
Great Lakes...	.5	0	60	341

Sources: The figures on deliveries are derived from Fischer, *Statistical Summary*, Table B-1, but since Fischer's figures are for geographic areas, these for the administrative regions have been computed by adding or subtracting the deliveries from St. Johns River Shipbuilding Co. and Arlington Shipbuilding and Engineering Corp. of Jacksonville. Figures on building berths are similarly derived from Tables E-1 and E-3. The figures for MC personnel are from MC, "Justification and Schedules for Budget Estimates for Fiscal Year, 1945, being estimates of the actual number in Fiscal 1944," corrected in pencil. Compare totals in Table 16.

have the heaviest load, for the Higgins 44-way shipyard was then in the program. Even after the Higgins contract was cancelled, Gulf was relatively more important than statistics indicate. On the West coast the biggest yards were the new ones established by the Kaiser-Bechtel-McCone group, which had gotten off to a fast start and demonstrated that they could transfer their highly rated industrial " know-how " to the field of ship construction. On the East coast the biggest yards were managed directly or indirectly by strong experienced companies—Bethlehem, Newport News, Federal, and Sun. No need to teach them shipbuilding! In the Gulf, however, nearly all the yards were new and having difficulties; skilled labor and management were hard to find. More

[1] MC Minutes, p. 24128 (Jan. 21, 1943).

than in the other regions, the yards in the Gulf looked to the Maritime Commission's regional office for guidance.

In each region the staff was organized into three groups of sections, each group under an Assistant Director. Sections doing plan approval and cost analysis were under an Assistant Director who was in charge for the region of functions formerly performed by the Technical Division in Washington. The production engineers, the shipyard inspectors, and the field force generally were under a second Assistant Director who was in charge for the region of the functions previously performed by the Construction Division. Under him also was a regional Trial Board which inspected finished vessels when they made their final runs at the time of delivery. These two Assistant Directors and the staffs under them were functionally responsible to the Regional Director, that is, they received all their instructions from him. Many other Maritime Commission employees in the regional offices were functionally responsible to division directors in Washington—for example, the Regional Attorneys were functionally responsible to the Legal Division—but were administratively under the Regional Director. Their office space, vacations, and personnel problems were handled, as were similar problems for the whole regional staff, by a third, less important Assistant Director in charge of administration.[2]

This was the general plan, but there were variations from one region to another. In Philadelphia, for example, a fourth Assistant Director, and one of the most important, was appointed in February 1944. He was the Regional Attorney, William R. Burns. With the title of Assistant Director for Departmental Activities, he handled the relations between the Director and a number of officials who, like himself, were responsible functionally to Washington—the industrial adviser, the training superviser, the safety and health consultants, the construction and disbursement auditors, the public relations representatives, and the consultant for smaller war plants.[3] This list shows the wide variety of functions which the Commission had assumed in connection

[2] Office orders and charts, especially that accompanying Vickery to Regional Directors, June 13, 1942 in gf 201-3-6, 201-3-3, 201-3-4, 201-3-8; and in Historian's Collection, Research files 212.2, Admin. Order No. 37, Suppl. 56.

[3] Regional Admin. Order No. 1, Suppl. 1, Feb. 1, 1944, in Historian's Collection, East Coast file, Director, 1944.

with its shipbuilding and indicates how many different kinds of problems flowed through the regional offices.

Although there was much similarity in the formal organization of the four regional offices, there were notable differences in the way in which the organizations worked in practice, for both the personalities of the directors and the problems they faced differed. These problems were considered in choosing the directors. For example, Mr. Flesher's previous experience had been mainly with engines and with priorities, and these seemed likely to be the most pressing problems on the West coast. Mr. Sanford's long experience, which included both building new shipyards and inspecting construction, was qualification for the large organizational job to be done in the Gulf. The nature of the shipyard managements in the region and the distance from Washington were of prime importance in the administration of the regional offices.

II. Relations with Washington

Dominating all other factors in the relations between the regional offices and Washington was Admiral Vickery's direct relation to the four regional directors. He held them responsible for getting out the ships, and told them in no uncertain terms what was expected of them. Just as the yards were raced one against another, so were the regions; the regional directors were started off with the injunction, "get on your horses and ride." To encourage their rivalry, Admiral Vickery sent every month telegrams like the following: "Congratulations to West coast director who with 40 ships and rating 105 takes top honors for September [1942]. Congratulations also to East coast director with 37 ships and rating of 103, and to Gulf coast director with 15 ships and rating of 100. To Lakes director with one ship and rating of 34, . . . do you desire a set of spurs?" There followed the quota of ships expected from each region in October. From the Gulf, Mr. Sanford, who had scored 100 but not top honors, wired back: "Appreciate congratulations. We salute East and West coasts. Spurs useless without oats, a starving horse cannot be . . . ridden. Give us material and we will give you ships. Re October quota of 15 ships, 12 is all there is in sight. No material, no ships."[1]

[1] Gf 201-3-3, Oct. 6, 1942.

This was the general pattern which the telegrams followed month after month with first one region and then another rating tops. Admiral Vickery, or during his illness Admiral Land, called for quotas which the regions thought too high but tried to meet.[2] Occasionally the monthly telegram was all commendation. On January 1, 1944, Admiral Vickery wired: " The President's goal has been achieved. . . . The December rating for each of you is 100%. . . . On my team you are the four horsemen of an all-American backfield. My heartiest congratulations to you, to each member of your staffs and to the shipbuilders in your region, all of whom have turned to and done a bang-up job. . . . January quotas follow in a few days."[3]

The reliance which Admiral Vickery placed on his regional directors strengthened their hands in their continual tug of war with the division directors in Washington. Defining exactly the functions and authority of regional office on the one hand and Washington staff on the other was bound to be difficult. Each had important responsibilities and strove for the powers which it felt were necessary to meet those responsibilities.

One of the first and most pressing issues concerned the hiring of urgently needed personnel when the regional offices were being set up. In March 1942 on the eve of the decentralization, Mr. Flesher proposed decentralization of the authority to appoint personnel in the regions so that the regional directors could put inspectors to work immediately and secure approval of their positions and qualifications from Washington later. Mr. Schell, the Executive Director, who took a keen interest in questions of personnel, objected and persuaded the Commission to approve a procedure that he worked out which retained a centralized control over appointing personnel. By this procedure positions at previously established grades were created and approved first. Regional directors could then make appointments, after clearing the qualications of the appointees with the district offices of the Civil Service Commission, but final approval of the appointments was routed through the Personnel Division in Washington, to which the personnel officers in the regions were

---

[2] In addition to the copies in gf 201-3-3, the series is in Historian's Collection, Weber files, Miscellaneous Correspondence, Regional Directors. Attitudes towards these quotas are shown in Minutes of the Staff meetings, Gulf region (in Historian's Collection) , e. g. Aug. 8, 1942, Aug. 12, 1944, Nov. 11, 1944, Dec. 16, 1944.

[3] Gf 201-3-3, Vickery (by Weber) to all Regional Directors.

functionally responsible, and appointments were checked over by the Executive Director before coming to the Commission for final approval.[4] Once this system was in operation, regional offices could hire fast, but the first pay checks were slow in coming.[5]

Another issue which came up immediately was how far the regional offices could go in approving the changes which were made in working plans and specifications to fit them to each yard's own equipment, supplies, and methods of construction. The main reason for the decentralization was to speed up this kind of approval " on the spot." The regional offices were formed by moving out men who had been doing this kind of work in Washington. But the Technical Division left in Washington was still responsible for the " design " of the ship and had authority to review changes which would affect " design." [6] The difficulty was in drawing the line between changes which did or did not affect " design," or, as the regional directors usually put it, " basic design." Were ventilation systems, dining-room plans, or joiner work part of " design " or " basic design "? The regional offices thought the answer was " No." [7] They wished power to approve changes in detail here and there when necessary to get the ships out. On the other hand, the Technical Division in Washington argued: " Appreciable departure from design characteristics are seldom the result of one single variation from specified or assumed requirements. . . . More often . . . departures from design characteristics are the result of the cumulative effect of a number of small changes." " The experience of the Commission with ships already built shows that the greatest discrepancies between design and construction seldom, if ever, occur in the form characteristics, but rather in structural arrangements, scantlings, joiner work, heating and ventilation, etc., and the resulting discordance is mostly confined to lightship weight and position of center of

[4] Minutes, pp. 21524-26 (March 31, 1942) and Admin. Order No. 44, Suppl. 3, May 1, 1942 and Suppl. 4, July 16, 1942.

[5] Statement by E. G. Montgomery, March 14, 1948. The issue was elaborated in Admin. Order No. 44, Suppl. 8, March 23, 1943 and Suppl. 9, Feb. 25, 1944. The latter reaffirmed that field personnel offices were functionally responsible to the Division of Personnel in Washington.

[6] Admin. Order No. 37, Suppl. 73, June 26, 1942.

[7] R. M. Smith to Sanford, June 24, 1942, and Sanford to Vickery, July 4, 1942, in gf 201-3-3.

RELATIONS WITH WASHINGTON

gravity; the basic characteristics affected thereby are mostly those of deadweight and stability." [8]

The meaning of " basic changes in design " was restated at a meeting attended by the regional directors in Washington August 17, 18, and 19, but only in general terms.[9]

The stability of the Liberty ship was a matter of particular concern to the Technical Division since they had reason to believe that some late features of the design, notably the guns and armor above deck and the use of welding, would so raise the center of gravity as to make the ship liable to capsize if it received any battle damage. The Technical Division ordered several hundred tons of permanent ballast placed in new ships, but regional directors used their own judgment in carrying this direction into execution.[10]

Conflict of jurisdiction in regard to changes was avoided entirely for some of the new types of ships introduced later in the program, since working plans and specifications were prepared originally under supervision of the regional office most concerned with their construction. An example already mentioned is the preparations of the frigate design in the West coast office.

The directors and section chiefs of the regional offices were drawn partly from the Technical Division and partly from the short-lived Construction Division. Their relations with that part of the Construction Division which remained in Washington and became the Production Division were very close, and the problem of defining the functions and authority of the regional offices on the one hand and the Production Division on the other was as difficult in this case as in regard to the Technical Division. Theoretically the line was easier to draw. The Production Division was to supply materials and components, the regional offices were to supervise construction. Procurement was centralized, construction decentralized. In practice it was not so easy.

From the point of view of the Production Division, supplying materials depended on accurate scheduling, and scheduling required accurate information from the yards concerning both the

[8] Asst. Chief, Preliminary Design Section to Director, Technical Division, July 25, 1942, in gf 201-2-11.

[9] Minutes of the meeting are in Historian's Collection, West Coast files, Conference Report No. 2 in Conference Reports, Series of 1942.

[10] D. B. Tyler, " Construction and Operation of Liberty Ships from the Technical Point of View," Oct. 23, 1946, section V, in Historian's Collection, Research file, 111.1a.

materials on hand in each and the stage of completion of the ships. Therefore Admiral Vickery was persuaded to issue Office Order No. 2, May 28, 1942, creating in each yard a Material Coordinator who would report directly to the Production Division in Washington. The regional directors protested vigorously, and Mr. Sanford emphasized especially their objection to having the material coordinators report on progress in construction and estimated deliveries. He considered the estimating of future deliveries one of the main functions and one of the most difficult functions of the regional offices, and said, " independent information going to Washington in that connection would result only in chaos. . . ." [11] Coming at the same time as the controversies over authority to hire personnel and to approve changes in plans and specifications not affecting design, Office Order No. 2 seemed like an attempt to rob the regional offices of independence and make them report through the Production Division.[12] Fuel was added to the fire by the fact that the Production Division hired and sent out material coordinators without in all cases previously informing the regional offices.[13] The whole issue was threshed out at the regional directors' meeting in Washington with Vickery on August 17-19. Material coordinators were made responsible to the regional directors and told to correspond with the Production Division only through the regional offices.[14] The Production Division received by this route the reports on steel receipts and consumption, but the reports on percentage completion of hulls and expected dates of delivery were made by the regional directors themselves on the basis of reports from their hull or machinery inspectors. Although the Statistical Analysis Branch of the Production Division did some checking of these estimates, the most important review and the basis for the monthly quotas allotted the regions was Admiral Vickery's personal study of the records in his chart room and his frequent visits to the yards.

[11] Sanford to Rockwell, July 28, 1942, in gf 201-3-3.
[12] See above, chap. 5, sect. V, and Spofford to Vickery, June 24, 1942 in gf 201-3-31; Sanford to Vickery, July 24, 1942 in gf 201-3-3; and McInnis to Sanford, July 30, 1942 all in East Coast file, Director, 1942 in Historian's Collection. A copy of Office Order No. 2, and of " Instructions to Material Coordinators, July 16, 1942 " is in Historian's Collection, Research file 212, 1, 1.
[13] Rockwell to Regional Directors, Aug. 21, 1942 in gf 201-2-31.
[14] MacLean to Vickery, Aug. 19, 1942, in gf 201-2-31; Minutes of the meeting in West Coast Conference Reports, Series of 1942, in Historian's Collection, West Coast files.

This settlement gave the Production Division less complete control over schedules than they thought they needed, and only round-about information concerning inventories. At a later date, October 1943, material coordinators were permitted to correspond directly with the Production Division,[15] but it still felt inadequately informed and depended largely on the contractor's expediters to determine the needs of the yards.[16]

On the other hand, the regional offices received intially less control than they found they needed over allocation of materials and components. In theory all allocation was to be done from Washington, but in practice when a shipyard desperately needed something to finish a ship it made "emergency purchases" or "borrowed" from anywhere it could, which usually meant some other yard in the same region, or a shipment on the way to some yard, with a promise to pay back when its own allotment came in. For example, the material coordinator at Southeastern wrote in December 1942 that in the last two months two thousand items "ranging from rivets and globe valves to anchor chains" had been transferred. "The authorizations for these transfers were made on a very haphazard basis, embracing telephone calls, wires and letters, originating both from the Washington office and the Philadelphia office, and quite often handled by several persons." The Production Division could not keep up with these transfers and told the East coast regional director that in such cases he was responsible for protecting any yards which might suffer, a responsibility which the regional director readily accepted.[17] Gulf and West coast did even more switching around of material than did East coast.[18] Each regional director organized in the regional offices a section or sections to help the yards obtain the materials they needed. In the Gulf, the regional director created a Material Coordinator Section which he told the Production Division was for the purpose of supplying them with the reports they needed. He issued elaborate instructions to the coordinators stationed in the

---

[15] Admin. Order No. 37, Suppl. 92, Oct. 26, 1943.
[16] Statement by J. T. Gallagher, June 16, 1949. See above chap. 11, sect. V.
[17] William L. Jones, Material Coordinator at Southwestern to Kirkpatrick, Dec. 4, 1942; H. R. Shea to Blakeman and Kirkpatrick, Dec. 3, 1942; Rockwell to McInnis, Dec. 4, 1942; McInnis to Rockwell, Dec. 16, 1942, all in Historian's Collection, Kirkpatrick's files, Ship Construction. There is much more in this file on the need for the material coordinators and their coordination.
[18] Statement by J. T. Gallagher, June 16, 1949.

yards and in practice seems to have controlled their appointment.[19] The East coast regional office planned close supervision of the material coordinators in its organization of a new Production Section in March 27, 1943.[20]

On the West coast the issue came up in October 1943 when the Production and Procurement Divisions made new efforts to increase their authority in the regions. The administrative order re-establishing their direct contact with the material coordinators also restated the Productions Division's control of allocation and gave it authority to dispose of surplus accumulated at any yard.[21] On the other hand, the West Coast Director created at this time a Material Control Section in his regional office. Its chief described the pre-existing situation as one in which there was a great lack of coordination among the material coordinators. He worked up a table of organization and complete manual of instructions, but when he tried to put the table of organization into effect, he had trouble putting the necessary steps through the Personnel Division in Washington because the Executive Director said that it would require Commission action on new administrative orders " which would show the relation of the various Washington Sections with this proposed Regional Section." [22] Mr. Flesher urged the need of permitting the regional office to carry through this reorganization and tighten its control over material coordinators, because of what he called the " inefficient way in which material has been handled by the Production Division." He blamed that division for having created the surpluses.[23] On the other hand, from the point of view of the Production Division, the surpluses—namely, the excessive inventories at certain yards—had been created by the regional office in permitting the shipyards to make " emergency purchases " of material which had already been allocated to them from Washington, or by switching steel and components from one yard to another without informing the Division in

---

[19] Sanford to Rockwell, Sept. 24, 1942 and Dec. 24, 1942, Sanford to all Material Coordinating Section Personnel in Gulf Coast Region, Sept. 1, 1943; all in gf 201-3-3.

[20] W. H. Blakeman to J. F. McInnis with attachments in Kirkpatrick file, Ship Construction Organization Charts, in Historian's Collection. Later the Material Coordinators were in the Cost Review and Material Control Section. Organizational Manual, April 1944 in Historian's Collection, Files from Regions.

[21] Admin. Order No. 37, Suppl. 92, Oct. 26, 1943.

[22] L. T. McCarthy to C. W. Flesher, Nov. 24, 1943, in Flesher reading file, Confidential, 1943.

[23] Flesher to W. A. Weber, Nov 24, 1943 in Flesher reading file, Confidential.

Washington.[24] The regional director justified the emergency purchases by saying they were the reason the West coast had such a good record for turning out ships.[25]

In December 1943 the Commission laid down new rules governing "emergency purchases." To avoid duplicate purchases and the formation of surpluses, notification of each emergency purchase was to be furnished by the shipyard to the Procurement Division, which was enjoined to either approve or disapprove the purchase within forty-eight hours. Procurement was to give notification of approvals (a) to the contractor to go ahead and buy; (b) to the Production Division so that they could adjust allocation; and (c) to its own employees so that they could cancel any order which would be duplicated.[26]

Actually it seems that Washington and the regions each worked independently in helping the contractors get materials. The regional offices improved their supervision of the material coordinators, and evened up distribution among the yards of the region.[27] The Production Division in Washington secured from the shipyard expediters more detail of just which kinds of steel, or other material and components, they had on hand; and instead of allocating a complete bill of material for the ships, allocated only the kind of steel or other material needed.[28] The ships did get built, but it cannot be said that inventories were kept at a minimum, nor that the administration of materials was smoothly integrated.

Another point of conflict between Washington and the regional offices was the Purchase Controller. This official in each yard was

[24] Statement by J. T. Gallagher, June 16, 1949.
[25] Flesher to Vickery, Apr. 16, 1943, in Flesher reading file, Confidential, 1943. Notices of all these purchases had been sent to Washington, he said.
[26] Admin. Order No. 37, Suppl. 83C, Dec. 23, 1943.
[27] As above, notes 19 and 22. The organization of a Material Coordinator's Section was carried through in West Coast. West Coast Admin. Order No. 33, Suppl. 19, Oct. 12, 1943 and statements of E. G. Montgomery and C. L. Miller, June 17, 1949. For its duties in the disposal of surplus or excess materials see West Coast Regional Office, Office Order No. 13 series of 1943, May 13, 1943. When Admin. Order No. 33, Suppl. 19, Oct. 12, 1943 put it up to Regional Directors to see that reports on material at the yards were sent to Washington, the West Coast Director issued his Office Order 27 series of 1943 on Nov. 8, 1943, saying that "the jurisdiction and responsibility for determining that Material Received Reports for Vessel Construction Material *only* are prepared by the Shipyard" fell on the material coordinators. Up to that time on the West coast these duties were being carried out by auditors. Cf. chap. 22, sect. III.
[28] Statement by J. T. Gallagher, June 16, 1949.

THE REGIONAL OFFICES

in charge of approving the purchases made by the shipbuilding contractor (whereas the material coordinators were concerned originally only with Commission-furnished materials). Each shipyard bought regularly for stock such things as welding rod, paint, nuts and bolts, rivets and gaskets. In the decentralization, the authority to approve such purchases for reimbursement was given to the regional directors and exercised in the shipyards at first by inspectors and later by purchase controllers. The division of duties between inspectors, material coordinators, and purchase controllers evolved gradually and was not the same in all regions. In the Gulf in 1942 the purchase controller appears as a kind of clerical assistant of the Principal Inspector.[29] On the West coast, purchase controllers were charged in June 1942 not only with approving yard purchases but also with keeping track of Commission-purchased material.[30] To bring yard purchasing in line with that of the Procurement Division, it was proposed in April 1943 that the purchase controllers be placed under the Procurement Division, but this was not done.[31] They remained under the regional directors.

An aspect of supply over which the regional directors had very little control until 1944 concerned the allowance-list items. At the beginning, these were purchased and supplied to the yard by the Division of Purchase and Supply which was later absorbed into the new Procurement Division. It sent to the yards a Supply Officer who was not under the regional director. But the work of supplying new ships became larger and larger, and, in the opinion of one official in the Gulf office, got worse and worse until "in many cases the ship has to sail without even a wrench to tighten a bolt." In March 1944 the supply officers were abolished by an administrative order of the Commission, and their functions were taken over by the material coordinators or purchase controllers.[32]

In general the trend was towards increase in the authority

---

[29] Minutes of Gulf Staff Meetings, July 4, 1942 and Aug. 1, 1942.

[30] West Coast Regional Construction Office, Office Order No. 3, series of 1942.

[31] Flesher to Vickery, Apr. 16, 1943 in Flesher reading file, Confidential, 1943, with a proposed Admin. Order No. 11, Suppl. 2, attached, but the approved Admin. Order of that number is on another subject.

[32] Admin. Order No. 33, Suppl. 20. On the background, Warren Adams to Weber, Feb. 4, 1944 and Vickery to MC, Feb. 15, 1944, in gf 201-2-43. Minutes of Gulf Staff Meeting, Aug. 28, 1943.

of the regional directors. Their recommendations were sought and followed in regard to the host of community problems in which the Commission was asked to help—housing, busses or ferries or other forms of transportation, health and safety measures, and provisions for in-plant feeding. As one special problem after another arose, the Commission appointed an increasing number of officials in the regional offices who were partly supervised by the regional director and partly by a division in Washington: the industrial adviser for labor relations,[33] the assistant training supervisors,[34] safety and health consultants,[35] and public relations representatives.[36] The story of their activities has been told in earlier chapters, and they show both cooperation and conflicts between the divisions in Washington and the regional directors. If differences arose, the rulings of the regional directors usually prevailed. This triumph of the regional directors was formally expressed in the case of the training supervisors by issuance on September 1, 1944 of an order making them functionally as well as administratively subject to the regional offices.[37] The one outstanding exception from the tendency of the regional director to dominate all the Maritime Commission staff in the shipyards is the position of the auditors, which will be discussed shortly.

Mr. Sanford set forth the regional director's ideal by citing a letter sent April 25, 1942 by Admiral Vickery advising the shipyards that all correspondence hitherto addressed to the Commission for the attention of Commissioner Vickery should thereafter be addressed to the regional directors. "Pursuant to those instructions," declared the Gulf Regional Director, "this office endeavored to have them carried out literally but experienced considerable difficulty in accomplishing this objective due to

[33] Created by Admin. Order No. 37, Suppl. 60, Apr. 3, 1942, he was appointed by the Commission, but "directly responsible to the Regional Director" and communicating with the Commission only through him.

[34] See above, chap. 8, sect. IV.

[35] When created, they were responsible through the Chief Safety Consultant and Chief Health Consultant to the Director of Shipyard Labor Relations, but could only inspect and recommend. Enforcement was "reserved to the office of Regional Director." Vickery to Regional Directors, Apr. 23, 1943, in gf 503-1-6. On June 13, 1944, they were transferred to the regional offices, Ring to MC, June 13, 1944 in gf 503-1-6.

[36] Functionally responsible to the division in Washington. O'Dea to McInnis, Dec. 4, 1942 in gf 201-3-6.

[37] Vickery to McInnis and other regional directors, Sept. 16, 1944 in gf 508-1-1. Cf. above chap. 20, sect. III.

a tendency both on the part of the shipbuilders and on the part of certain individuals and certain divisions of the Maritime Commission in Washington to correspond directly either over the head of the Regional Director or around the Regional Director. It was common practice for many shipyards, and still is in some cases, to attempt to obtain favorable decisions from Washington after having obtained unfavorable decisions from the Regional Director or his office." This he considered embarrassing as it led to decisions which were not based on knowledge of all the facts. " Fortunately," he went on, " in most cases the relations of this office with the shipbuilder in this connection are most satisfactory but it was only by the usual slow process of education that this objective was achieved." [38]

Obviously the question of who could go over whose head to the higher-ups cannot be given a sweeping and definitive answer. A particularly difficult aspect of that question concerns the use of political influence. The written record is small, the quantity of rumor large. Appeals to the White House and the Congressional representatives have been mentioned in earlier chapters in various connections, some of them being of a kind quite in accordance with the American ideal of democracy.[39] There may have been appeals which affected the distribution of materials to this or that shipyard. In the selection of the headquarters for the regions, political pressure is said to have influenced the choice of Philadelphia over Baltimore on the East coast,[40] but the selection of Oakland instead of San Francisco on the West coast was made by Mr. Flesher in spite of strong protest from the Congressman representing the latter city.[41] The selection and promotion of inspectors was under civil service, and government jobs were so easy to get in this period that appeals to political influence,

---

[38] Minutes, p. 29032 (July 11, 1944) with reference to the activities of the Shipyard Advisory Committee (created by Admin. Order No. 78, May 2, 1944) asking that they be routed through regional offices.

[39] E. g., those from South Portland about the railroad. For other examples of appeals through Congressmen, Senators, and the White House, see chap. 5.

[40] Statement of S. Schell.

[41] And also from the District Manager for shipping operations who had his office in San Francisco. Mausshardt to Schell, Apr. 18, 1942 with enclosures; Schell to Flesher, June 15, 1942. The objections of Congressman Welch were the more important because he was a member of the House Committee on Merchant Marine and Fisheries. Flesher said he chose Oakland because it was nearer to most of the yards, Flesher to Vickery, Apr. 8, 1942. All in gf 201-3-4.

although not unusual, were relatively unimportant. In one case the Maritime Commission supervisor dealt so roughly with a Congressman to whom an inspector appealed in seeking promotion that the Congressman wrote Admiral Land, who referred the letter back to the regional office, and after receiving an explanation wrote to the supervisor personally: " I am not at all surprised at your indignation but I must invite your attention to the fact that appeals to Congressmen are rather common in this country, frequently cultivated by certain types, and are something that all of us have to live with whether we like them or not." He asked that they be received with courtesy and tact " whether the ultimate answer be ' yes ' or ' no '." [42] The staff had instructions not to be guided by political influence in their appointments or promotions.[43]

III. COORDINATION WITHIN THE REGIONS

Whether the thousands of Maritime Commission officials stationed in the field were functionally responsible to the regional office or merely attached to it " administratively," the coordination of their activities was in practice one of the chief tasks of the regional director. This coordination was needed at two levels, in the shipyard and in the regional office, and in the communication between the two levels.

In each shipyard there was a group of Maritime Commission employees headed by five or six principal officials each of whom reported to a separate supervisor. In the Bethlehem-Fairfield yard, for example, the Maritime Commission staff in 1944 consisted of 150 people distributed as follows: Resident Auditor and staff, 60; Principal Hull Inspector, and Principal Machinery Inspector, with 60 other inspectors and a few clerks, making the total inspection staff 70; Resident Plant Engineer and staff, 5; Material Coordinator and staff, 4; Purchase Controller and staff, 6; Supply Officer and staff,5.[1] Since there was no official among them who outranked all the others, an attempt to coordinate at this local level took in some regions the form of creating a Yard Staff Committee with rotating chairmanship. It contained the Resident Auditor, the

---

[42] Land to McInnis, Jan. 25, 1944 and Land to Kirkpatrick, Feb. 1, 1944, in Land reading file. Land ended his letter, " Just forget the whole thing."
[43] Statement of E. G. Montgomery, June 1949.
[1] Payroll for personal services, MC, East Coast Regional Construction Office, Feb. 16, 1944–Feb. 29, 1944 (standard form 1013 rev.) .

Principal Hull Inspector, the Principal Machinery Inspector, the Resident Plant Engineer, and the Material Coordinator or Purchase Controller or both.[2] These committees were formed on the East coast in the summer of 1943 [3] and in the Gulf that fall.[4] What they discussed and the uses made of them varied much from yard to yard. In many instances they discussed such matters as their own office space, parking privileges, and cafeteria service, but occasionally more weighty matters appear in their records. The head of the field force in the East coast found them useful in dealing with disciplinary problems—to reprimand and reform or to discharge an employee who drank too much. A general report on the management of the yard was asked from them by the regional office in at least one case, and occasionally the Yard Staff Committee was called to a special meeting to see what could be done about some difficulty which the management was not taking care of—such as the meeting the Yard Staff Committee at Bethlehem Fairfield called on April 13, 1944 because there were " eight steam turbines in the Weyerhauser field, without protection." On the other hand, the yard staff committees were on occasion told not to spend their time criticizing the yard managements.[5]

The main center of coordination was of course the regional office. A constant stream of inquiries, instructions, and reports went from the Maritime Commission's officials in the yards to the appropriate sections in the regional office, and the chief officials of the regional office made frequent trips to the yards to clear up difficulties on the spot. Instructions to inspectors were codified by the Gulf regional office into an inspector's manual, topically arranged and indexed. As it was in loose-leaf form, old instructions could be removed and replaced by new rulings. T. E. Shaw, the Assistant Director who was responsible for it, found that this codification saved much confusion and cut down markedly the number of inquiries going to the regional office.[6]

[2] Whether the Purchase Controller was the same person as the Material Coordinator, or subordinate to him, varied from time to time and region to region. The six named appear in the East Coast Yard Staff Committee cited below.
[3] Minutes of the committees in the East coast yards, among files of W. H. Blakeman, Assistant Director—Field.
[4] Minutes of Staff meetings, Gulf Regional Office, Sept. 11, 1943, Apr. 7, 1945.
[5] Minutes in Blakeman's files, above cited.
[6] Copy in Shaw's files. East Coast inspector's manual, from Blakeman's files, in

COORDINATION WITHIN THE REGIONS 705

Quite as important as the activity of the regional office in guiding the staff in the region were its direct relations with shipyard managements. A large part of the management of management, especially of the smaller firms and of those on the Gulf and West coast, was done in the regions and by the regional directors personally. Their recommendations to Admiral Vickery and to the Commission had influence on many financial and contractual matters which the regional directors examined on the spot.[7]

Engineering problems, however, were the main theme in the relations of shipyards to the regional office at the beginning of any new program. When yards were getting started or construction of a new type of ship was in prospect, a large conference of naval architects and engineers from the contracting companies and from the regional office assembled to work out the details— how the working plans were being made and routed, how this or that detail of construction could best be handled.[8] Labor difficulties and material shortages also were the theme at many conferences with particular companies.[9] A series of regular bi-monthly conferences with shipyard managements was scheduled by the East coast director in June 1943, and section chiefs were instructed to prepare to answer questions and bring up matters they wished discussed.[10] Conferences with management were more important than any other type of procedure in coordinating the work of the region, at least on the West coast and East coast.

To tie together the staff in the regional office itself and define their functions, there were various kinds of office orders or office memoranda, but no single system of such administrative orders. The directors were instructed to hold staff conferences. Only in the Gulf region, so far as I know, were they of major importance; on the East coast and West coast, conferences with management

Historian's Collection. Both were based on the instructions sent out by the Construction Division before decentralization.

[7] In addition to examples, given elsewhere in this history, note Anderson's request to Sanford for "vigorous personal cooperation" to settle claims of Louisiana Shipyards, Anderson reading file, Feb. 22, 1944.

[8] E. g., the conference on the corvettes, at West Coast regional office, Nov. 7, 1942 and that about conversions at Western Pipe and Steel, Nov. 27, 1942, in Historian's Collection, West Coast regional files, Conferences.

[9] For example the conferences with Marinship, Nov. 30, 1942 and June 5, 1944 in the above series.

[10] Mr. Kirkpatrick's file, "Bi-Monthly Conferences," now in Historian's Collection, Regional Offices.

called for specific purposes and attended by a large number of the regional office staff may have served many of the same purposes.[11] The staff conferences of the Gulf regional office were held every week or two except when Mr. Sanford was away from New Orleans. They were attended by fifteen to twenty persons, branch heads as well as section chiefs, and gave opportunities for each to report on his phase of activity and ask questions. Mr. Sanford reported the latest news from Washington, and discussion frequently centered on the relation between the quota of ships fixed by Admiral Vickery and the expectations of the men in closest touch with the yards.[12] In the Gulf also, quarterly reports were requested from section chiefs and an annual report, at least in 1944, was prepared for the region as a whole,[13] but in general the emphasis was not on any system of paper work, the main efforts of the regional offices being focused on meeting the problems of the shipbuilders as they arose.

The work of the auditors created the most difficult problem in coordinating the activities of the regional staff. The auditors were not functionally responsible to the regional director but functionally responsible to a Regional Auditor who reported through the General Auditor of Construction to the Director of the Finance Division in Washington. Although the auditors had no authority (outside the field of accounting) to make rules about what the contracting shipbuilder could or could not do, they had authority to enforce the rules. They did so by refusing to approve reimbursement of the expenditures of the shipbuilders who were operating under contracts providing for reimbursement of cost, as most contracts did. In practice the distinction between making rules and enforcing them was often blurred. Many specific expenditures by the contractor required the approval of the regional director or his engineering staff, but the auditor had to decide whether a parti-

[11] Among the minutes filed under "Conferences" in the West Coast director's files, that for Oct. 17, 1942 is of a staff meeting, but these minutes do not show any regular series of meetings. Instead they show special conferences of varying membership, seldom regional staff alone. West Coast staff whom I have asked do not recall staff meetings.

[12] Minutes of the conferences in Historian's Collection, Regional files. Mr. Shaw says these minutes were distributed to the staff. Similar staff conferences began to be held on the East coast when Mr. Sanford was placed in charge there also. Kirkpatrick's file, Bi-monthly conferences.

[13] "Report covering the Activities of the Gulf-Great Lakes region (New Orleans Area) July 1, 1943-June 30, 1944" in Historian's Collection, Regional files.

cular expenditure was of the kind requiring this approval, and whether an approval which had been given covered all for which the contractor was asking reimbursement. Conflicts developed because the auditors had a different point of view from that of the inspectors and regional directors. The auditors thought in terms of formal correctness and of saving money, the inspectors and regional director were mainly concerned with getting ships finished as soon as possible.

This conflict and other problems of coordination became an issue in the West coast region in 1943 as one of the results of Edgar Kaiser's plea, described in chapter 14, for more " freedom of management." The special committee which was sent out from Washington to look into Kaiser's complaints submitted a report which seems worth quoting at some length because it sets forth in specific terms the kind of difficulties which arose in coordinating the activities of the regional staff. It said:

> The major portion of such suspensions of reimbursement for expenditures by the shipbuilding companies appear to have been caused by the failure of the contractor to obtain required approvals from the Regional Office or the Commission in Washington. Although the facts considered showed in many instances undue delays in the regional office or in the offices of the Commission in Washington in granting approvals, considerable responsibility for the situation would appear to rest upon the contractor itself due to its failure to follow up its original request for approval. On the other hand, in certain instances, such as the approval of the shipyard paper which, under Commission procedure, requires the approval of the Regional Director, West Coast Region, the contractor has made a number of attempts to obtain action from the regional office, all such attempts having been futile.
> It was also pointed out by the shipyard that a great deal of confusion was caused due to the fact that various officers in the regional office purported to have authority to approve or disapprove matters on behalf of the Regional Director. Nowhere did the authority of such officers appear to be defined, and in certain instances conflicting instructions were issued from the regional office to the shipyard. It was also apparent from the discussions held that there was considerable doubt as to the exact extent of the authority of the Resident Plant Engineer.
> Although the disallowances and suspended items are of extremely minor character and could not be reasonably expected to delay shipbuilding operations, considerable annoyance appears to be occasioned in all the shipyards by the failure of the regional office or the Commission to clearly define the duties of the various officers in the shipyards and the regional office. This situation is most serious in connection with the scope of authority of the Regional Auditor and the shipyard auditors as opposed to that of the administrative officers of the regional office in the shipyard but also exists as

between various administrative officers. It is, therefore, recommended that definite instructions be issued defining the scope of the duties and authority of the Regional Director, the Regional Auditor, and the various section heads in the regional office, the Resident Plant Engineers and the shipyard auditors. If this is done, it is believed that considerable confusion may be avoided.

This committee was impressed by the lack of cooperation within the regional office and the fact that the Regional Director, the Regional Auditor and other responsible officers in such office conferred with each other infrequently. The situation might be alleviated if the Regional Director were to hold staff meetings at stated intervals and conduct Maritime Commission staff meetings at each shipyard that he visits.[14]

To comply with all these recommendations was beyond the powers of the regional director. Only the Commission could redefine the line between his authority and that of the auditors. Admiral Vickery sent the report to the West coast director without immediate comment except to ask for his suggestions on that point.[15] Before he replied, the auditing staff on the West coast was reorganized, although without any re-definition of its authority. Mr. Quarg, who had been regional auditor there since the beginning of the program, became an Assistant Director of Finance, and the resident auditors in the shipyards were raised in pay and rank, so that they received more than inspectors.[16] Mr. Flesher's reply, October 2, 1943 protested that " our principal Inspectors in our larger yards have just as much responsibility as the auditor," and should have as much pay. He criticized auditors for doing too much checking—for example, asking to see the driver's licenses of men paid as truck drivers—and in general for their questioning in matters outside their competence, such as whether the supervisory force in a yard was too large. Many auditors, he said, took an antagonistic attitude, and continued:

That was one of the first things I had to overcome with my inspection force when the Regional Office was formed. They were of the opinion that they were not doing their job unless they showed their authority. However,

---

[14] Henry E. Frick, William Slattery, and Walston Brown to the Chairman, Aug. 21, 1943 in Carmody file, Br-Bz and in Flesher's reading file, Confidential.

[15] Original in Flesher's reading file, Confidential, Sept. 9, 1943.

[16] MC Minutes, pp. 26151-54 (Sept. 2, 1943). The stated purpose of the reorganization was to enable shipbuilding companies to bring their accounts up to date. A Regional Construction Accountant, charged particularly with that task, and other intermediaries were created between Mr. Quarg and the resident auditors, who were raised to CAF-13. Principal inspectors were at grades corresponding to CAF-12. Flesher to Vickery, Oct. 2, 1943 in Flesher reading file, Confidential.

after numerous transfers, dismissals, and a continual educational program, that problem has been overcome, and without curbing, but rather increasing the efficiency of the inspection force. In other words, the Principle [sic] Auditor can be selected and educated to the point of cooperating 100% with the management without relaxing from the rules and regulations by which he is bound.

If the function of selecting and transferring auditors were delegated to the Regional Director as an administrative function, I believe this would be another factor in assisting management greatly with some of their present problems. In other words, I personally would attempt to select auditors for the yards from a personality standpoint, while the Regional Auditor would pass on the man's qualifications as an auditor.[17]

His proposal that the regional directors be empowered to select and transfer auditors was not carried through. In November 1943, Admiral Vickery wrote Mr. Flesher suggesting that he meet the criticism about lack of coordination in the regional office by holding regular staff meetings and specifically outlining functions and jurisdiction in office orders.[18] The regional director had indeed ordered regular staff conferences earlier, but actually the system of conferences with company representatives about specific problems prevailed.[19]

The general satisfaction of Admiral Vickery with his four regional directors is attested by the fact that none of them was ousted to be replaced by a new man. The first to go was Mr. Spofford, director of the Great Lakes office, for in May 1944 the Great Lake and Gulf regions were combined into one under Mr. Sanford. There were three reasons for making this change at that particular time. One was that Mr. Spofford was then ill with pneumonia.[20] Another was that, as recorded in the Commission's minutes, " the Great Lakes Office no longer has the confidence or cooperation of the shipyards under its jurisdiction, and has lost its effectiveness as evidenced by the lack of satisfactory production of such shipyards." [21] What lay behind this statement, according to Mr. Spofford, was the hostility of Walter Butler,

[17] Flesher to Vickery, Oct. 2, 1943 in Flesher reading file, Confidential.
[18] Vickery to Flesher, Nov. 26, 1943, in Weber reading file.
[19] West Coast Regional Office Order No. 8, Series of 1943 (Apr. 8, 1943), called for meetings at 3 p. m. the 15th and last of every month of 22 specified staff members, but statements to me on June 17, 1949 by individuals who would have attended— E. G. Montgomery, C. L. Miller, and B. F. Carter—indicate the order was not executed.
[20] Statement by W. E. Spofford, Apr. 11, 1949.
[21] MC Minutes, pp. 28331-32 (May 2, 1944).

a large contractor active in the Democratic Party in Minnesota, whose yards in Duluth had been given by Mr. Spofford a manpower ceiling lower than Mr. Butler wanted. In addition, Mr. Spofford felt that the relatively poor showing of the Great Lakes in meeting its quotas was due to lack of understanding in Washington of the special problems of that region, such as the necessity of closing in the ships before winter so that work could go on inside them during cold weather, and the need of having materials and components delivered accordingly.[22]

The third reason for combining the Great Lakes office with the Gulf office was that the main construction in that area was of frigates and coastal cargo vessels which had to be brought down the Mississippi and reassembled or re-outfitted in Gulf yards. Arrangements for delivering the frigates in this way had already been made (the frigates were delivered mainly in May and June of 1944).[23] Mr. Sanford made arrangements to have the reassemblage of the coastal cargo ships done in Gulf yards under lump-sum contracts,[24] and moved as many as he could out of the Lake Superior yards before navigation closed for the winter of 1944-1945 (December to April). He gave as a second reason for this move the fact that progress in the Butler yard had not been satisfactory and he wished to impress the fact on the management.[25] He also reorganized the Great Lakes office, in which he declared there was a lack of coordination and control, so that it fitted more nearly into the pattern of the Gulf office, but retained in authority there a number of the key men.[26]

For about a year there were the three regional offices—Gulf-Great Lakes, East Coast, and West Coast. Then, when Messrs. McInnis and Flesher resigned to enter private industry, in July 1945, Mr. Sanford was made the Director of Eastern Region including all except the West coast, and in August he was named the Director of a Division of Construction into which all the regional offices were absorbed.[27] Postwar reorganizations then abolished that division also and scattered most of its staff.

[22] Statements by W. E. Spofford, June 29, 1948, and Apr. 11, 1949.
[23] *Official Construction Record,* and above, chap. 18.
[24] MC Minutes, pp. 29993-95 (Oct. 17, 1944); also pp. 30073, 30131-32, 30332-24, 30407-10.
[25] MC Minutes, pp. 30295-300 (Nov. 21, 1944).
[26] Gulf Regional Office Staff Meetings, Minutes, May 20, 1944.
[27] Admin. Order No. 37, Suppl. 56C, 56E.

Whether the decentralization carried through in the spring of 1942 by the creation of the regional offices was a mistake is a question on which there is difference of opinion. I am inclined to think the decentralization was wise, in spite of the conflicts that arose between regional offices and the divisions in Washington. In the comparable part of the staff of the Navy Department there were sharp rivalries also, although it was very differently organized. The Inspectors of Naval Material, the Material Coordinating Agency, and the Bureaus fought over the responsibilities for expediting, scheduling, and distributing materials and components.[28] The conflicts that arose within the Maritime Commission, such as that over the material coordinators, should have been ironed out more quickly, but such problems were sure to arise under any administrative structure.

[28] Connery, *The Navy and the Industrial Mobilization*, pp. 319-339.

CHAPTER 22

ADMINISTRATIVE PROBLEMS:
(B) THE FLOW OF MONEY

I. Appropriations

IN ADMINISTERING the flow of money from the government to the shipbuilding corporations, their suppliers, and their workmen, the Maritime Commission aimed above all at making the flow ample and rapid so that there would be no delay in production. Next in importance was the auditing of payments at the time they were made in order to make sure they were authorized expenditure. Recording the outward flow in a system of accounts which would summarize the record of expenditures was an obligation of the Commission, but was treated as less urgent.

Congressional appropriations were of course the fountainhead. Most of the money was appropriated by Congress directly to the Commission but a part came in transfers from other departments and in allocation by the President of money which Congress placed at his disposal. Of the appropriations to the Commission, the bulk went into its "Construction Fund." This was a revolving fund created by the Merchant Marine Act of 1936 which provided that funds from various sources connected with subsidized merchant marine and the activities of the Shipping Board Bureau should be deposited in the U. S. Treasury and "there maintained as a revolving fund, herein designated as the construction fund, and ... be available for expenditure by the Commission in carrying out the provisions of the Act."[1] All receipts and disbursements under that act were paid into and out of this fund, administrative expenses as well as those for ship construction. The amount in the fund when the Commission started was sufficient so that no additional appropriation was needed until for fiscal 1940 the sum of $100 million was added.[2] The Commission presented its budget

---
[1] Sec. 206.
[2] 53 Stat. 524, Public Law 8, Mar. 16, 1939, 76th Cong., 1st sess.

to Congress and justified its intended expenditure every year, however, even for years in which the balance in the fund and the receipt from such items as the operation or sale of ships made it unnecessary to ask for more money.[3]

The emergency need for ships created by the war in Europe led to several appropriations for restricted, specified purposes and then, in August 1941, the floodgates were opened wide. The earlier, restricted appropriations seem small compared to those of the war years, but in 1940 and 1941 they seemed large indeed. First came additions to the construction fund to take care of the speeding up of the long-range program: $144,500,000 was provided for this purpose on April 18, 1940 [4] and $160,000,000 on April 5, 1941 in the regular appropriations bill.[5] The later bill was based on figures discussed with the Bureau of the Budget in the fall of 1940, since the preparation of the Commission's budget was normally initiated about ten to twelve months before the beginning of the fiscal year to which it applied.[6] Emergency measures moved very much faster, and consequently, even before the regular appropriation bill for 1942 had been passed, Congress made, on February 6, 1941, a special appropriation to start the Liberty ship program. In doing so it set up a new fund, the "Emergency Ship Construction Fund," and appropriated to it $314,000,000 for 1941.[7] This was the fiscal basis of the first wave of expansion in the Commission's emergency program. When the second wave was decided on in April 1941, it also was financed out of a special appropriation, the Defense Aid Supplemental Appropriation Act, 1941, commonly called the lend-lease appropriation, which had been approved March 27, 1941. From the sum placed at his disposal under the terms of that act, the President on April 14, 1941 allotted to the Commission $550,000,000.[8] Thus before the middle of 1941 the Commission had received large appropriations into three different funds for the construction of ships and ship-

---

[3] House, *Navy Department Appropriation Bill for 1947, Hearings*, p. 43.

[4] 54 Stat. 111, Public Law 459, Apr. 18, 1940, 76th Cong., 2d and 3d sess.

[5] 55 Stat. 92, Public Law 28, Apr. 5, 1941, 77th Cong., 1st sess.

[6] Hymen Ezra Cohen, "The Budgeting Process During World War II," a typescript in Historian's Collection, Research files, 106.25.

[7] 55 Stat. 5, Public Law 5, Feb. 6, 1941, 77th Cong., 1st sess. Out of the total given above, $500,000 had been allocated from the Emergency Fund for the President.

[8] 55 Stat. 53, Public Law 23, Mar. 27, 1941, 77th Cong., 1st sess.; the President to Land, Apr. 14, 1941, in Maritime Commission Program Collection.

building facilities—the Defense Aid Fund, the Emergency Ship Construction Fund, and the Construction Fund. The legal authorization of each fund was different, and each was, or might be at Presidential discretion, usable only for certain kinds of ship construction. The "construction fund" was still limited by the provisions of the Act of 1936, and during the first half of 1941 it was still an "iffy" question whether or not all the Liberty ships would be turned over to the British; so that each fund required a separate accounting.

In August 1941 both the terms of the appropriations and their size assumed wartime characteristics. In explaining the need for funds under the First Supplemental National Defense Appropriation Bill for 1942, Admiral Vickery stated the Commission's intention " to load the shipbuilding industry with all the ships it can absorb as fast as it can absorb them, as long as the war lasts." [9] That appropriation, approved August 25, 1941, provided the added money for the Commission's third wave of expansion, and for the continuation of building as much as possible through fiscal 1943. It made available $698,650,000 and in addition authorized contracts for $1,296,650,000.[10] After Pearl Harbor the principle of building as much as possible continued to be the basis of appropriations and they were on an expanded scale as follows:

On April 5, 1942	$1,502,000,000 [11]
June 27, 1942	980,080,000 [12]
March 18, 1943	4,000,000,000 [13]
June 26, 1943	1,289,780,000 [14]
June 27, 1944	6,766,000,000 [15]

All these appropriations, including that of August 25, 1941, were made to the Construction Fund. There were some additional

[9] House, *First Supplemental National Defense Appropriation Bill for 1942, Hearings*, pp. 409-10.

[10] 55 Stat. 669, 671, 681, Public Law 247, Aug. 25, 1941, 77th Cong., 1st sess.

[11] Fifth Suppl. Nat. Def. Appr. Act, 1942, 56 Stat. 130, Public Law 474, Mar. 5, 1942, 77th Cong., 2d sess.

[12] Ind. Off. Appr. Act, 1943, 56 Stat. 418, Public Law 630, June 27, 1942, 77th Cong., 2d sess.

[13] First Def. Appr. Act, 1943, 57 Stat. 25, Public Law 11, Mar. 18, 1943, 78th Cong., 1st sess.

[14] Ind. Off. Appr. Act, 1944, 57 Stat. 191, Public Law 90, June 26, 1943, 78th Cong., 1st sess.

[15] Ind. Off. Appr. Act, 1945, 58 Stat. 380, Public Law 358, June 27, 1944, 78th Cong., 2d sess.

allotments by the President to the Defense Aid Fund and a small addition to the Emergency Ship Construction Fund,[16] but appropriations to all other funds were dwarfed by those to the Construction Fund, for which the cumulative total to June 30, 1945 was $15.7 billion and for the fiscal years 1941-1945 alone was $15.5 billion.[17]

The uses of the fund changed markedly while it was being thus expanded. For one thing, the creation of the War Shipping Administration on February 7, 1942 led to the presentation by Admiral Land in June 1942 of a separate budget for that agency. Its expenses from its establishment until June 30, 1942 were taken care of by the transfer to the new agency of $145,090,000 out of the Maritime Commission's Construction Fund.[18] Then, from July 1, 1942 until July 1947 expenses directly connected with operation, those for personnel as well as those for requisitioning or buying ships, were paid out of separate appropriations to the War Shipping Administration.[19] To be sure, some expenses which might be called those of operation were still paid for from the appropriations to the Maritime Commission, for it paid the personnel of certain divisions which performed services for the War Shipping Administration as well as for the Maritime Commission. These were the Personnel, Procurement and Public

[16] USMC, *Report to Congress*, 1947, Appendix L. Because of the number of allotments and transfers some of these figures are questionable and it is difficult to summarize them clearly and accurately. For the Emergency Ship Construction Fund, Public Law 5 cited above appropriated $36 million for fiscal 1942, and $125 million was transferred from the Construction Fund by Commission actions on Sept. 26, 1941 and Apr. 2, 1942 (See MC Minutes, pp. 19084-85, 21546-47). But the *Report to Congress*, p. 57, gives $161 million under appropriations for 1942.

[17] *Ibid.*

[18] House, *First Supplemental National Defense Appropriation Bill for 1943, Hearings* before a subcommittee of the Committee on Appropriations, 77th Cong., 2d sess., pt. 1, pp. 266, 292; Land to Harold D. Smith, May 16, 1942 in MC gf 301-4.

[19] Public Law 678, 77th Cong., 2d sess., chap. 524, approved July 25, 1942, appropriated $1,100,000,000 to the WSA revolving fund, and other appropriations to it followed. Public Law 299, 80th Cong., 1st sess., chap. 414, approved July 31, 1947, transferred the balance of the WSA revolving fund into the Treasury, and appropriated to the Treasury a sum not to exceed $200 million to liquidate all obligations which were found by the GAO to have been incurred by the WSA prior to Jan. 1, 1947. Current income from the former WSA function of operating ships was made available to the Maritime Commission to pay for operating functions for a time. Then in June 1948 the appropriations made to the Maritime Commission included funds for operations and provided that receipts should go directly to the Treasury. Public Law 862, 80th Cong., 2d sess., chap. 775, approved June 30, 1948. Compare note 40 on p. 748.

Relations Divisions, the Office of the Secretary, and a large part of the Finance Division. They were paid out of the Maritime Commission's Construction Fund, but most of the expenses of War Shipping Administration were paid from its own appropriations.[20]

An even more important change in the use of the Construction Fund occurred when it was made available for all types of ships and thus freed from the clauses in the Merchant Marine Act of 1936 which restricted its use to the construction of ships wanted in connection with the long-range plan for improving the American merchant marine. The Appropriation Act approved in August 1941 said the money appropriated to the fund was for "the construction in the United States of merchant vessels of such type, size, and speed as the United States Maritime Commission . . . may determine to be useful for carrying on the commerce of the United States and suitable for conversion into naval or military auxiliaries; . . ." for their components and equipment; and for facilities for their construction; as well as for certain functions later taken over by the War Shipping Administration.[21] Later appropriation bills continued to authorize this broad use of the Construction Fund; [22] and when the Emergency Ship Construction Fund and the Defense Aid Funds were exhausted, all kinds of ship construction were charged to the one Construction Fund, and bookkeeping thereby simplified.[23]

As a revolving fund it was replenished from other sources as well as appropriations—from the sale of scrap by the shipyards, for example, and from the sale of ships. A cumulative tabulation of the income and outgo and the balances shows how adequately it provided financial foundation for the huge shipbuilding program (Table 14).

---

[20] House, *Independent Offices Appropriation Bill for 1944, Hearings*, p. 667; and budget estimates for 1944. On the status of Personnel Management and Procurement see next chapter.

[21] Public Law 247, Aug. 25, 1941.

[22] Cohen, "Budgeting Process," pp. 46-48.

[23] One incidental effect of this was the use of the phrase "long-range program" in a budgetary sense which was quite different from its original meaning and from the use in the Technical and Production Divisions. Budget statements used "long-range program" to refer to all ships being paid for from the Construction Fund and therefore included thereunder most of the Liberty ships, and some minor and military types, although they were not in any ordinary sense a part of the Commission's long-range program. For example, House, *Navy Department Appropriation Bill for 1947, Hearings*, pp. 282-83; MC *Report to Congress* 1941, p. 53.

APPROPRIATIONS 717

The existence of billion-dollar balances at the end of each fiscal year during the war shows a situation in which there was no danger that ship production would be held up by lack of funds although the budget estimates for all of those years were beset by uncertainties. The description in earlier chapters of the changeableness of the program shows how uncertain the budget estimates were bound to be. The procedure for making these estimates

TABLE 14

CONSTRUCTION FUND OF U. S. MARITIME COMMISSION

(in thousands)

	Appropriations, Collections, and Transfers	Disbursements	Rescissions	Cumulative Balance at End of Year
Oct. 1936-June 1938	$ 160,737	$ 48,362	....	$ 112,374
Fiscal 1939	25,425	65,681	....	72,118
Fiscal 1940	133,750	136,009	....	69,859
Fiscal 1941	301,711	197,410	....	174,160
Fiscal 1942	2,496,307	782,930	....	1,887,537
Fiscal 1943	5,375,393	3,332,655	....	3,930,275
Fiscal 1944	1,868,103	4,479,629	....	1,318,749
Fiscal 1945	7,706,333	3,911,225	$3,100,068	2,013,789
Fiscal 1946	477,050	941,447	796,568	752,824
Fiscal 1947	478,729	292,363	712,399	226,791
Total	$19,023,542	$14,187,761	$4,609,035	....

Source: Figures extracted from MC, *Report to Congress*, 1947, Appendix L. Figures rounded off so that columns may not add, and see chap. 22, sect. IV regarding the difficulties in using these figures.

was to have the Cost Section of the Technical Division figure the amount of money needed for the ships scheduled, and make a spread of the cost of the program over the estimated period of construction. When approved in the Technical Division and by Admiral Vickery this estimate went to the budget officer and was incorporated by him in the budget. Mr. Kirsch, who was in charge of preparing the budget, also received from each director of a division an estimate of the funds needed for personnel.[24] But these needs naturally varied with the tasks to be done, and the tasks changed as the program changed—not only because it grew larger but also because the building of a different type, such as Victory ships or concrete barges, increased the load of work.

[24] Cohen " Budgeting Process," especially pp. 6, 63, 153-56.

Under these conditions it was obviously impossible to forecast expenses accurately, and deficiency appropriations were necessary.

Equally difficult was forecasting the receipts which would be coming into the Construction Fund during the next fiscal year as a result of such activities of the Commission as the operation of housing units and of bus lines, as well as from the sale of scrap and various kinds of recaptures. In July 1944 the Budget Officer asked the Technical Division whether they could allow in their estimate of construction costs for such items as revenues from cafeterias, canteens, vending machines, and pay telephones. Actually, he had to prepare budgets with no nearly accurate knowledge about either past or future payments and receipts of this kind, which were going on under the direction of the regional offices.

In July 1944, just after D-day in Normandy, the Bureau of the Budget asked for four different estimates on four contrasting assumptions about the course of the war. The Maritime Commission asked to be allowed to delay making these estimates until after the meeting of the Joint Chiefs in October, but the Bureau of the Budget rejected the request.[25]

To appreciate fully the task of the Budget Officer it must be recognized also that he had to present to Congress estimates regarding many funds. Besides those used for ship construction which have been mentioned, there were many others set up by Congressional or Presidential action, such as various insurance funds, seamen and cadet officer training funds, state marine school funds, and so on. There were also transfer funds created by inter-agency payments. A statement on all these had to be gone over with the Bureau of the Budget and presented to the Appropriation subcommittees.[26]

The Executive Director, Mr. Schell, took a vigorous part in putting the budgets in final form. After all estimates and supporting documents were in the hands of the Budget Officer, they were tabulated and analyzed. Then the Executive Director conferred with the Budget Officer and the various directors of divisions and decided about many questionable items.[27]

---

[25] Cohen, " Budgeting Process," pp. 115-18.

[26] See the number of funds listed in MC *Report to Congress*, and the discussions in the various *Hearings* on appropriation bills.

[27] See comments written by Mr. Schell on the typescript of Cohen, " Budgeting Process."

When, after negotiations with the Bureau of the Budget, the Commission's needs came before Congressional subcommittees as part of the President's budget estimates, the lead in explaining them was almost always taken by the Chairman, Admiral Land, although the Executive Director, the Budget Officer, the General Counsel, the Legislative Counsel, and other members of the staff attended the hearings and answered questions of detail. The whole program and the general policies of the Commission were expounded at these hearings, and the liberal appropriations amounted to a general vote of confidence. The inevitable uncertainty of the estimates of future needs was recognized. When, for example, the House in January 1944 cut the amount for administrative expense from $40 million to $30 million, Congressman Woodrum, explaining that it had been done as an incentive to economy, admitted the possibility that the larger amount might prove to be needed. After Admiral Land explained the need to the Senatorial subcommittee, most of the cut was restored.[28] Moreover, under the pressure of war, Congress felt obliged to approve the deficiency appropriations.

Although Congress voted the money asked during the war, there were some signs of restiveness, symptoms of the attitude which was to dominate once hostilities were over. When the House was voting the deficiency appropriation for 1943, on February 25, 1943 Congressman Wigglesworth said: " The hearings reflect . . . a tendency to override the work of the regular subcommittees. . . . They reflect a tendency to come in before the Deficiency Subcommittee with a lot of personnel that was never authorized by the regular subcommittee, placing the Deficiency Subcommittee in the position of either approving the increase or compelling a lot of personnel to be severed from the rolls of the Agency." [29] He and other Congressmen, although they showed a dislike of having their hand forced, recognized the importance of giving the Commission adequate funds.

Earlier in February 1943 the whole record of the Commission and of the War Shipping Administration was debated when the regular appropriation bill was before the House. On the one hand, criticism was directed against the high prices paid for ships

---

[28] Cohen, " Budgeting Process," pp. 108-13; *Congressional Record*, vol. 90, pt. 1, p. 827, Jan. 27, 1944.
[29] *Congressional Record*, vol. 89, pt. 1, p. 1333, Feb. 25, 1943.

requisitioned by the War Shipping Administration, against the arrangements made regarding the shipyards in Tampa, Savannah, and South Portland, against the cancellation of the Higgins contract, and even against the building of those " ugly ducklings," the Liberty ships. On the other hand Admiral Land was highly praised for his ability and integrity.[30] Congressman Woodrum, Chairman of the subcommittee in charge of the appropriation said of him that " with all of the hot spots he had had to sit on in this great shipbuilding program " he had " performed a service in this war effort second to no man who wears a uniform." [31] The appropriation asked for was approved.

In connection with making appropriations, Congress also discussed how naval auxiliaries being built by the Maritime Commission for the Navy should be paid for, and the relative responsibility of the Navy and Maritime Commission in this kind of building. Congressman Carl Vinson proposed to make sure that all naval auxiliaries would be paid for by funds appropriated to the Navy Department, and an amendment to that effect was added by the House to the First Deficiency Appropriation Bill for 1943.[32] Much of the drive behind this rigid proposal seems to have been removed, however, by the passage of Public Law 76 which on June 17, 1943 affirmed the authority of the Navy over the design and construction of auxiliaries (see above chapter 18).[33] The opposite policy in regard to transfers of funds from the Navy was stipulated on December 17, 1943 by Public Law 204 which provided " That no sums expended by the Maritime Commission from funds appropriated to it for the construction of vessels which are transferred to the Navy shall be reimbursed from naval appropriations, except to the extent of agreements existing on the effective date of this Act." [34] The exception, however, was most important. The Maritime Commission received about $2 billion from the Navy for construction of 519 vessels, such as the escort carriers

[30] *Ibid.*, pp. 968-78, 1009-10, 1028-30, and 1033-34.
[31] *Ibid.*, p. 970.
[32] Cohen, " Budgeting Process," pp. 86-90, and *Congressional Record*, vol. 89, pt. 1, pp. 1360-63, Feb. 26, 1943. The previous plan of the Bureau of the Budget I do not understand.
[33] 57 Stat. 156, Public Law 76, June 17, 1943, 78th Cong., 1st sess.
[34] 57 Stat. 604, Public Law 204, Dec. 17, 1943, 78th Cong., 1st sess.

built at the Vancouver yard, and advances for conversion and delivery costs on 353 other ships.[35]

Including receipts from its operations, transfers such as those from the Navy, and appropriations to all its various funds, the amount made available to the Maritime Commission, 1936-1946, was about $23 billion,[36] of which some was transferred by it to other agencies. Not all of it was spent, to be sure; from the Construction Fund approximately $5 billion was transferred to the surplus funds of the U. S. Treasury or available as appropriation for fiscal 1947.[37] The total spent was about $18 billion.

II. Disbursements

In paying out the money appropriated the Commission adopted methods which assured speed and allowed certain problems of auditing and accounting to pile up trouble for it in the future. The central problem was the processing of vouchers, that is, the bills for services and materials submitted by shipbuilders and vendors. When certified by the authorized officers of the Maritime Commission, these vouchers were paid by disbursement officers of the U. S. Treasury, who, in turn, were supplied with the funds for the purpose by the requisitions which the Maritime Commission sent to the Treasury calling for the placing of appropriated funds in the hands of the Treasury disbursing officers to the account of the Commission. To explain the process by which the vouchers were certified and placed with the Treasury for payment requires some explanation of the organization of the Commission's Finance Division.

[35] Working funds were recorded on the Maritime Commission books from advances by the Navy Department to the Maritime Commission as follows:

Net advances, less rescissions and refunds, for the construction of 519 vessels, including cost of special facilities..........	$1,623,357,237
Net advances for the conversion and delivery costs on 353 vessels on a loan basis...............................	419,562,428
	$2,042,919,665

In addition to the above, the Navy advanced $13 million to the Maritime Commission for the cost of repairing 24 tankers. Figures furnished by J. F. Winter, Budget Division, MC, Aug. 9, 1948. All of this except $200 million was expended through the Construction Fund.—Statement of E. G. Montgomery, Apr. 25, 1950.

[36] MC, *Report to Congress*, 1946, Appendix A. But see note 16 above. Roughly, total receipts in all funds were $24.7 billion, transfers between funds $1.7 billion.

[37] See Table 14 above.

Before the war the organization of the Finance Division gave recognition rather simply to a basic threefold division in its functions. One section under J. A. Honsick was devoted to auditing ship construction. Another section under J. M. Quinn was devoted to auditing operations and the accounts of subsidized operating companies. A third section under W. U. Kirsch handled matters which concerned the Commission as a whole or concerned both construction and operation. This third section, originally called the Accounting Section, was responsible for maintaining the books of the Commission, as well as preparing budget estimates. Three sections, one for auditing construction, one for auditing operations, and one for the budget and general accounts made a simple logical picture.[1]

Expanding activities brought minor changes within these three sections in 1941, but the first major reorganization came in March 1942 at the time of general decentralization by the creation of the regional offices. Simultaneously the War Shipping Administration was beginning to take shape. To fit the new circumstances the Finance Division went through a reorganization which had three main features.[2]

(1) The Division was divided into two parts, each under an Assistant Director: Mr. Honsick was made Assistant Director of Finance (Construction); Mr. Quinn was made Assistant Director of Finance (Operations). Actually the part of the division under Mr. Quinn included more than the staff concerned solely with operations. It included the Budget and Accounts Section which continued to handle the bookkeeping for both operations and construction; in fact it included all that portion of the Finance Division which was in the Joint Service Organization performing functions both for the War Shipping Administration and the Maritime Commission.

(2) A new section, the Disbursement Section, was created, so

---

[1] Actually this neat trifold organization existed only from Oct. 6, 1938 to Jan. 12, 1939. Earlier there were four sections in the Finance Division because it included an Examining Section, which naturally disappeared when in 1938 an Examining Division was created to take over its function of reviewing all aspects of proposed subsidy agreements. Admin. Order No. 37; Admin. Order No. 47. After Jan. 12, 1939 there was a Collection and Securities Unit to which were transferred some of the functions of the Accounting Section. Admin. Order No. 37, Suppl. 2, Jan. 12, 1939.

[2] Admin. Order No. 37, Suppl. 55.

that vouchers could be processed more quickly. Although it was within the Joint Service Organization and Mr. Quinn's half of the Division, it was concerned in a very essential way with ship construction, for to keep the shipyards operating it was essential to speed the flow of vouchers into the United States Treasury and the flow of checks back to the shipbuilding companies and suppliers, Otherwise they could not have met their mounting pay rolls and bills for materials. Before the reorganization of March 1942, the audited vouchers had been sent by the resident auditors to a unit with the Construction Audit Section, which checked them and sent them to the Accounting Section for final approval and transmittal to the Treasury.[3] Prior to the fall of 1941 the main vouchers arising from ship construction were for advance progress payments under lump-sum contracts, a proportion of the price according to how far the work had progressed. But when the emergency program got under way, mail bags full of other kinds of vouchers began arriving in Washington—vouchers sent in by resident auditors for reimbursable costs under the manhour contracts and vouchers from steel companies or other vendors from whom the Commission was buying the materials or components with which they supplied the builders. A step to meet this situation was taken November 10, 1941, before the general reorganization. Authority to approve the vouchers for payment was given to each of the three sections, Construction Audit, Accounting, and Auditing and Financial Analysis (operations).[4] An episode at the time this arrangement was being put into effect gives some idea of the size of the problem. At the Finance Division staff meeting on December 26, 1941, Mr. Honsick said that his section was ready to handle construction vouchers in the future according to the administrative order of November, but had a few they would be sending along to the Accounting Section since they had been worked up for manifesting to that office. When 100 steel vouchers arrived next day, Mr. Kirsch thought these were the few referred to. But 700 more steel vouchers reached his section on the thirtieth and 400 more on the thirty-first, adding to the 1,500 steel vouchers already on hand.[5] By March the flood of vouchers had risen,

---

[3] Organization Charts dated Jan. 16, 1940 and Apr. 17, 1941.

[4] Admin. Order No. 33, Suppl. 9.

[5] Chief, Accounting Section, to Chief, Construction Audit Section, Jan. 1, 1942, in gf 201-2-5.

even if it was not yet at the crest of 88,000 a month attained later. The way of handling them proposed in the fall, by spreading the work among the three sections, was reversed, and the reorganization of March 19, 1942 concentrated the final approval for payment of all vouchers in the Disbursement Section.[6]

(3) Decentralization was applied to the Finance Division, for in March 1942 both its Construction Audit Section and the Disbursement Section were decentralized by the creation of four Regional Construction Auditors and four Regional Disbursement Auditors. According to the usual formula, they were said to be administratively under the regional directors and functionally under the appropriate section chiefs in the Finance Division in Washington.

After the reorganization ordered in March 1942 was completed, the flow of vouchers through auditing to final approval for payment was of two kinds. In the field, vouchers providing for the reimbursements of pay rolls and other costs in accordance with the contracts were approved by resident auditors at the shipyards and sent by them to the regional offices. When passed by the regional construction auditors and regional disbursement auditors they were presented for payment to nearby offices of the Treasury.[7] In Washington, vouchers for steel and other material or components purchased directly by the Commission flowed through for payment without ever passing through the Construction Audit Section. Like vouchers covering the administrative expenses of the Maritime Commission staff and the War Shipping Administration, they were placed by the Disbursement Section in line for payment by the Treasury.

For a while the vouchers providing payment for the purchases which had been made by the Procurement Section of the Production Division were supposed to be administratively certified by that Division,[8] which, according to an administrative order established a Voucher Certification Section.[9] But since the review

[6] Accordingly the Accounting Section stopped scheduling vouchers to the Treasury at the close of business Mar. 31, 1942. Chief, Accounting Section to General Auditor of Disbursements, Mar. 30, 1942, in MC gf 201-2-5.
[7] They were Authorized Certifying Officers to approve public vouchers for payment. Admin. Order No. 33, Suppl. 13, July 14, 1942.
[8] Director, Finance Division (written by A. C. Waller) to Director, Production Division, Apr. 20 and May 8, 1942, in gf 201-2-5.
[9] Admin. Order No. 33, Suppl. 16, Oct. 20, 1942.

DISBURSEMENTS

of vouchers by checking their contract authorization and matching them with purchase orders, bills of lading, and receipts was being done anyhow by the Disbursement Section, it was given full and sole responsibility in April 1943; and public vouchers covering materials received or services rendered under contracts entered into by the Procurement Section or Division, Washington, were routed directly to the Disbursement Section.[10]

As it settled to its work, the Disbursement Section grew in November 1943 to 340 employees in Washington organized into three units and 74 employees in the four regional offices.[11]

Although the purpose in creating a Disbursement Section was to speed the payment of vouchers, it was hard pressed to keep up with those presented to it. The number handled rose to 3,000 a day at the peak, and the average, 1943-1945, was about 1,400 a day.[12] The total number sent to the Treasury for the fiscal year ending June 30, 1943 was recorded as 342,669, of which approximately one eighth were for the War Shipping Administration.[13]

Vouchers which had been made out correctly and had the proper supporting documents were normally cleared through the Disbursement Section in five or six days, and the Treasury normally took about the same length of time to make the payment, although it would rush cases through in a day or two on special request. Vouchers requiring investigation might be very much delayed, however. In many cases they had been made out incorrectly by the vendor and had to be sent back to him, perhaps two or three times. Manufacturers who were unaccustomed to government orders had to be instructed how to make out the vouchers. After some of the staff of the section were sent on a trip to meet with manufacturers and explain the form to be filled out, there was a marked improvement.[14]

---

[10] Admin. Order No. 33, Suppl. 17; statements by A. C. Waller, July 29, 1948.

[11] On its organization: A. C. Waller to Kirsch, in MC gf 301-10, pt. I, and statements by A. C. Waller, July 29, 1948; E. L. Kohler to Lewis W. Douglas, Dec. 14, 1943, in House, *Accounting Practices of the War Shipping Administration and the United States Maritime Commission, Hearings pursuant to H. Res. 38* before the Committee on the Merchant Marine and Fisheries, 79th Cong., 2d sess., pt. 1, July 17, 22 and 24, 1946, p. 414. Cited hereafter as House, *Accounting Practices, Hearings*.

[12] Statements by Mr. Waller and C. W. Levis in July 1946, and E. L. Kohler to L. W. Douglas, Dec. 14, 1943, printed in House, *Accounting Practices, Hearings*, p. 415. Mr. Kohler reported that a count made during the week which ended Nov. 20, 1943 showed 6,847 vouchers sent to the Treasury of which one-fifth were for WSA.

[13] *Ibid.*, p. 415.

[14] Statements by A. C. Waller and C. W. Levis, July 29, 1948.

During the war, emphasis in the Disbursement Section was on moving the vouchers rapidly. After the war there was time for criticism of the whole system of handling vouchers. Some of the criticism concerned the vouchers relating to the work of the War Shipping Administration, and it must be remembered that they were a substantial part of the work being done by the Joint Service. From the point of view of ship construction, the center of difficulty was the processing of vouchers providing payment for Commission-furnished materials and components.

The system under which the government bought steel, engines, boilers, and so on directly from the manufacturers meant that the bills for these materials and components went to Washington and the materials themselves went to shipyards. When these bills, in the form of vouchers, arrived in Washington, how were they to be audited? The Commission had no inspectors regularly resident at all the manufacturing plants to inspect and accept delivery there of the product, as did many procurement agencies. To have left the vouchers unpaid until notice had been received in Washington that the goods had been received at the shipyards, or until the vouchers had been sent to the shipyards to be checked by the resident auditors against the shipbuilders' records of receipt, would have delayed payment so long that the manufacturers could not have kept up production. Consequently it was decided to pay the vouchers on the basis of the record of shipment, namely the bill of lading signed by the railroad. It was agreed that title passed to the government when the goods were shipped. Checking up on whether they arrived, whether they were defective, or whether, as in the case of steel, they might have been billed in excess of true weight was left to be done later. Then work piled up so fast that much of this checking never was done.

Plans for checking the receipt of material were not lacking. As early as March 1941 the Director of Finance, in proposing that payments for steel be made on the basis of the government bills of lading, had in mind that copies of these, together with a detailed schedule of the materials included in the shipment, be sent from the Finance Division in Washington to the resident auditors who would have them checked by the shipyard's receiving department and then report back any shortages or discrepancies. From Washington the Finance Division would then present its claims against the supplier.[15] The burden this would have placed on already overburdened resident auditors may have been the reason for its abandonment in favor of a different system. Joint Instructions issued by the General Auditor of Construction and the General Auditor of Disbursements, April 24, 1942, required the resident auditors to send a "receiving report" concerning any material which had been rejected by the shipbuilding company as damaged and defective. Of course the shipbuilder received a bill of lading with the goods. If the vendor was at fault he was to be notified; and "it is, of course, to be understood," so read the Joint Instruction, "that the vendor will not bill the Commission or the contractor for items to be shipped in replacement."[16] In October 1943, an administrative order provided: "A separate Material Receiving Report for each purchase order issued shall be furnished the Disbursement Section immediately upon receipt of such material by the shipbuilder." The shipyards were to prepare the receiving reports but the regional directors were to designate a member of the Commission staff at each yard, usually the material coordinator, to be responsible for determining that the reports were prepared.[17] Since "over, short, and damage reports" were naturally included, they contained the basis for innumerable claims of the Commission against vendors, but going through them and matching them against purchase orders, contracts, and vouchers was a time-consuming task.

In general, whatever could be put off had to be put off, it was felt, in order to get done those things which were necessary to keep production going.[18] Consequently, receiving reports were not checked during the war. In July 1946 there were "175 file cases—four drawer file cases—. . . filled with over, short, and damage reports" mixed up with correspondence, duplicate bills and other incidental papers which were being slowly worked through.[19] In May 1947, 30 per cent of the filing cases had been gone through, and from the over, short and damage reports alone $3,000,000 had been recovered for the government,[20] but in May

[15] Anderson to Vickery, Mar. 8, 1941, in Anderson reading file.
[16] Joint Instruction No. 2, in O. B. Vogel's collection of instructions.
[17] Admin. Order No. 33, Suppl. 19; Minutes, pp. 26502-03.
[18] Statements by A. C. Waller and C. W. Levis, July 29, 1948.
[19] Testimony of Mr. Slattery in House, *Accounting Practices, Hearings*, p. 175.
[20] Testimony of Mr. Quinn in House, *Independent Offices Appropriation Bill for 1948, Hearings*, pt. 2, p. 652, and explanation of C. W. Levis, Aug. 1948.

1948 it was estimated that there was a backlog of approximately 105,000 property removal notices, receiving and delivery tickets.[21] Completing the processing of these notices was closely connected with the accounting difficulties, of which much more will be said shortly.

III. Auditing Construction Costs

More complicated than the auditing of vouchers by the Disbursement Section in Washington, and coming closer to checking the real facts behind the papers was the auditing of the vouchers paid through the regional offices. This auditing was done in that part of the Finance Division which was under J. A. Honsick as Assistant Director (Construction) and contained four subdivisions: a small Administrative Unit, a small Special Studies Unit, a Construction Review and Analysis Section, which worked mainly on the final settlement of contracts, and the Construction Audit Section.[1] The latter contained 95 per cent of the personnel of the Finance Division outside of the Joint Service.[2] It audited the vouchers submitted by the contractors day by day as building progressed and accordingly most of its staff were in the field. At its head was the General Auditor of Construction,[3] under him were four Regional Construction Auditors, and under them in 60 to 70 shipyards were the Resident Auditors who were the foundation stones of the audit of construction.[4]

A resident auditor worked first of all from the prime records: pay rolls, subcontracts, and invoices and bills submitted by firms selling materials to the yard. His task was to see that all those

---

[21] Chairman Smith to Senator George D. Aiken, May 27, 1948, in MC gf 301-12.

[1] Admin. Order No. 37, Suppl. 55, Mar. 19, 1942; organization charts dated Mar. 19, 1942, and Dec. 23, 1942 (in MC, Historian's Collection, Research file 116 MC 23.10); and a booklet issued in elaboration of the administrative order, " Duties of the Assistant Director of Finance and Functions of Sections and Units Reporting To Him," obtained from Wesley Clark, all in MC, Historian's Collection, Research file 203.4.

[2] According to MS " Justification and Schedules for Budget Estimates, Fiscal Year 1944," pp. 11-14, in files of Budget Officer, MC.

[3] R. H. Mohler, the first General Auditor of Construction, resigned June 16, 1942, and F. L. Lynch became Acting General Auditor.—Instructions to Auditors, No. 43, in Construction Audit Section. On Nov. 2, 1942 W. L. Slattery was appointed General Auditor of Construction and served until Feb. 7, 1946. Mr. Lynch became Assistant General Auditor.

[4] Organization chart of Dec. 23, 1942, cited above, tabulates the number of positions. Slattery gave the total number under him at peak as 1,800. House, *Accounting Practices, Hearings*, p. 206.

AUDITING CONSTRUCTION COSTS 729

were properly certified by company officials and Maritime Commission inspectors. For example, after a purchase had been accomplished, the shipbuilder prepared a voucher certifying that he had received and paid for the material for use on work under the contract. The voucher then went to a resident plant engineer, or principal hull inspector, or purchase controller, one of whom, according to the nature of the purchase, certified that the material was needed for the contract. Then it went to the resident auditor who checked the voucher with the vendor's invoice and the company's record of receipt, checked on the way the purchase had been made, checked the approvals by inspectors, and checked the amount available for expenditure under the contract. The invoices were then put in form for transmittal to the regional office which made sure the papers were in order—signed in the right places by the right people, with no disparity in the figures—and sent the vouchers to the nearest Treasury disbursing officer for payment.[5]

In addition to checking from the prime records, the resident auditor and his staff, which might number 5, 20, or even 60 people according to the size of the yard and the nature of the contract,[6] went over the basic books of the contractors, at least in yards under cost-plus contracts. All contracts provided that he should have access to company books for needed information. Examination of the company accounts was a means of catching errors; the correctness of payroll figures could be checked there in the cash account. Moreover, access to the company's books was essential for determining how much overhead should be allocated to a contract and reimbursed.[7]

The difficulties encountered at East Coast Shipyards, Inc. will give an idea of what might be involved in the audit of the company's books. In justifying his need of having Senior Accountants, not Juniors, Mr. Tartter, who came to that yard on April 28, 1943, wrote: "I found that . . . the distribution such as it was of the current payroll was not in agreement with the payroll itself." After explaining how he had obtained the higher grade accountants needed to discover errors of this kind he described what

[5] This general picture is based on letters and testimony printed in the *Hearings* cited above, on GAO reports, and many letters I have read in Commission files, and has been checked by Messrs. Kell, Anderson, and others.
[6] House, *Accounting Practices, Hearings*, p. 158.
[7] House, *East Coast Shipyards, Inc., Hearings*, pp. 164-65.

happened when he assigned one of them to audit the "Operating (Overhead) Expense Ledger." "Almost immediately, we determined that the Operating Expense Ledger was incorrectly posted, incorrectly added both horizontally and vertically, the individual departments did not agree with the Summary and the grand total of the Summary did not agree with General Ledger Controls." Since a complaint to the company Comptroller produced no results, a member of the staff of the company's firm of Certified Public Accountants was called in. "At the end of two days' work, he threw up his hands and said that he could not fathom what the Shipyard's Accounting Department had done and would appreciate it if our staff would straighten out the Operating Expense Ledger." The result of a general straightening out, including checking of insurance payments, of over-charges by subcontractors, of improper payments from petty cash, and of responsibility for damaged material was a saving for the government of about $500,000 by disallowances. Mr. Tartter recognized that East Coast's lack of qualified accountants was connected with the very late date at which that company was organized. There is every reason to think that East Coast was an exceptional and extreme case, but just for that reason it illustrates vividly the kind of problem which came up in auditing the books of hastily organized shipbuilding companies.[8]

Although no general manual of instructions for the auditing of manhour and price-minus contracts was issued to the resident auditors, they had a general basis to work from in the *Regulations Prescribing Methods of Determining Profit* prepared before the war, in the "Uniform Classification of Accounts" issued near the beginning of the emergency program, and in the terms of the shipbuilders' contracts.[9] Specific day-by-day instructions on doubtful cases came to them from the regional offices, and more general instructions covering a whole category of cases were sent out from Washington.

Under the General Auditor of Construction, the Construction Audit Section in Washington was divided into units which made

---

[8] Edwin E. Tartter to Leroy G. Kell, May 7, 1945, in Resident and Regional Auditors' Replies to Surveys of Field Offices of the Construction Audit Section by Classification-Organization Section, Division of Personnel, in MC, Historian's Collection.

[9] See above in chap. 4, sect. V. Forms to be used in general financial statements by companies were prescribed in Instruction No. 9, suppl. 1, Nov. 17, 1942, and additional classifications of accounts were issued from time to time.

analyses of particular auditing problems. For example, the Labor Unit recorded the job classifications of the yards, the zone standards, and the methods of time keeping; the Materials Unit compared the prices paid for similar materials in different areas; the Overhead Unit compared the ratio of overhead costs in various yards; the Analysis, Cost Studies and Reports Unit made comparative studies of profits, and so on. Separate units were established later for insurance, travel, cafeterias, and housing.[10] Using the material worked up by more specialized units, the Audit Practice and Procedures Unit prepared the many " Instructions to Auditors " sent out by the General Auditor of Construction.[11] Towards the end of the war when the selective-price contracts were being introduced, a general manual was made for auditing contracts of that type and lump-sum contracts.[12]

One particular problem was how to notify the auditors in the field of the rulings by the Comptroller General pertinent to their duties. In answer to a request from a resident auditor for copies of the Comptroller's decisions, the Assistant General Auditor wrote him that there were twenty volumes of these decisions, but relatively few applied to shipyard work, and a digest of the more pertinent was being made for distribution. Of course new ones were coming out all the time.[13] One issued June 15, 1943 disallowed " the extra five cent charge which is usually made if a telephone call originates from a hotel." Sardonically calling it a " rather momentous decision," Mr. Honsick remarked that enforcement would require a corps of auditors who, if they found 200 cases, would recover a half of a day's pay.[14] But many other decisions of the General Accounting Office, like that permitting reimbursement for the publication cost of plant newspapers, were of immediate importance to the practical work of auditing.[15]

Another example of the kind of problem which had to be dealt with by instructions from the General Auditor for Construction

[10] The charts and booklet cited in note 1 above, and Job Classification sheets of various dates in files of Personnel Unit, Construction Audit Section.

[11] *Ibid.*, and statements by O. B. Vogel of that Unit.

[12] It was compiled by E. E. Hartline, who had been Regional Auditor at Chicago, after the closing of that office, working closely with the Audit Practice and Procedures Unit, and was edited by Mr. Honsick. See personnel file of J. A. Honsick where there is a copy. Another copy in MC, Historian's Collection, Research file 209.

[13] F. J. Lynch to C. J. Culleton, July 1 or 11, 1943, in MC gf 306-1.

[14] J. A. Honsick to W. L. Slattery, June 15, 1943, in gf 306-2.

[15] General Auditor of Construction to auditors, May 29, 1943, Instruction No. 78.

arose from the rental of equipment by the yards. The contracts originally provided that if the total rental paid equalled the value of the equipment, title to it should pass to the Commission. This worked a hardship particularly in the case of owners of trucks who depended on this equipment for their livelihood. Consequently the Commission authorized the regional directors to amend the contracts in such cases, subject to safeguards against abuse, and the General Auditor sent out an "Instruction to Auditors" explaining the change and the safeguards.[16]

While on the one hand the resident auditors had to keep up with the stream of decisions and instructions coming from Washington and from the regional offices, on the other they strove to keep track of what was actually going on in the plant. How far they went beyond merely matching up the papers which came over their desks, how far they inquired into what was back of the papers, varied from yard to yard. In some cases there was more need of it than in others. An auditor who not only formed a general opinion of how the yard was being managed but also talked about it publicly and critically was told it was none of his business.[17] But there was a borderline area in which his position was not clearly defined—namely, the control of inventories.

From the point of view of the auditing profession, control of inventory had been plunged in the center of attention by the hearings and writings on the subject which followed the failure of accountants to detect faked inventory records connected with the failure of the drug firm of McKesson and Robbins in 1938. After that episode and the statement of principle drawn up in consequence by the American Institute of Accountants, an auditor had reason to feel uneasy and frustrated if he could not make test checks to see whether the materials reported to him were actually physically present.[18] The resident auditors were not able to make many such test checks.[19] In some cases they did not have even adequate paper records against which to make a physical check.

[16] MC Minutes, Sept. 18, 1942, pp. 22981-82; Instruction No. 51, Oct. 10, 1942.
[17] House, *Production in Shipbuilding Plants, Executive Hearings*, pt. 3, pp. 907-08, Vickery's testimony, June 28, 1943.
[18] *New York Times*, Dec. 2 and 7, 1938; *The Journal of Accountancy*, vol. 67, pp. 16, 342 ff.; *Business Week*, May 13, 1939, p. 54.
[19] House, *Accounting Practices, Hearings*, pp. 66, 87, 107, 290, and especially 437. On *ibid.*, pp. 157, 188-190, Mr. Slattery defends the amount of inventory control by the auditors but his answers show how much they had to depend on the inspectors.

Two reasons for this inadequate control of inventory are apparent. One is simply that things were moving too fast. All the emphasis was on speed, on turning materials rapidly into ships. Slowing down merely in order to be sure to count everything before it went into the ships was not to be considered seriously in view of the large production being asked for, even from yards which were only just getting organized. Mr. Slattery, General Auditor of Construction, called the Commission's policy " construction of ships without regard to cost." [20]

The other reason was the lack of a common understanding as to who was responsible for control of inventory. As a matter of fact, most auditors could not inspect the material and equipment in the yard. They had to depend on the Commission's hull inspectors, resident plant engineers, machinery inspectors, or material coordinators. This staff of engineers was responsible for seeing that the material was up to specifications, and it was better placed to check the quantities actually received. Admiral Vickery said on one occasion that the inventory was not at all within the province of the resident auditor. To be sure, he was not speaking of Commission-furnished material; he was speaking of Richmond No. 3 at a time when Kaiser Co., Inc. was buying its own material for that yard, but being of course reimbursed for the cost.[21] Actually at that yard as at others, responsibility regarding inventory was divided between the resident auditor, who reported to the Finance Division, and the material coordinator, or purchase controller, who reported to the regional office and the Production Division, and through them to Admiral Vickery.

The auditor and the material coordinator came at the problem from different angles. The latter was providing reports which were the basis of allocations and of scheduling. His concern was the adequacy of the inventory for the production schedule—to check on whether the management was making provision to get what was needed to build ships, and secondarily to see that it was not getting more than its share. The material coordinator

[20] House, *Accounting Practices, Hearings*, p. 326.
[21] House, *Production in Shipbuilding Plants, Executive Hearings*, pt. 3, p. 897. Since in this testimony he was rebutting the testimony of a resident auditor who had in fact not only criticized the lack of inventory but also matters more clearly outside his province (*ibid.*, pp. 803-04), Admiral Vickery may have overstated how much he would restrict the auditor's role.

at Richmond No. 3 said that the card file of receipts into and removals from inventory gave a reasonably adequate record, although he admitted that during the night shift items were often taken by someone who forgot to make out a card.[22] The auditors, on the other hand, were concerned with accounting for the expenditure of government money, and from that point of view wished strict accountability to show what the yard received and on what contract it was used.

The difficulties over inventories would not have occurred, of course, if all shipbuilding companies had perfectly fulfilled their obligations to keep track of materials, either those which they bought or those which the Commission furnished and of which they were custodians under the manhour contracts. But the flow of materials was too large.[23] The main interest of the companies was in high production, not in economy, and the Commission's supplementary systems of checking were defective. From the point of view of allocations and scheduling more uniform methods of "keeping store" were much to be desired. From the point of view of procurement, the system of paying for Commission-furnished materials through Washington, although it facilitated the main goal of helping manufacturers keep up the supply, left loopholes because receipt and removal notices piled up unrecorded. In the yards the resident auditors lacked both the staff and authority to take stock of what was physically present. In addition, of course, the huge quantity and diversity of the materials and components going into 5,500 ships contributed to making control of inventory a weak point in the Commission's administration of its program.

In regard to payrolls, the resident auditors were more successful in probing the realities behind the figures. They could inspect the system of time clocks and the way timecards were punched and handled, and occasionally make test checks of the numbers of men at work or passing in or out.[24] To tell whether men were actually working at the kind of job for which they were being paid took further investigation. The most sensational case in which the Maritime Commission auditor uncovered frauds in the

---

[22] *Ibid.*, pp. 937-42; and above chap. 21.
[23] Charles R. Holton, "Procurement and Storekeeping," in Fassett, ed., *The Shipbuilding Business*, II, 136; MC Minutes, pp. 26151-54, Sept. 2, 1943.
[24] House, *Accounting Practices, Hearings*, pt. 1, p. 189.

AUDITING CONSTRUCTION COSTS

payrolls was at the Sparrows Point yard where counters certified inaccurate time sheets for the welders who, when they received the money, paid the counters a part of the overpayment.[25] Tipped off first by union officers and then by "outside sources," the Maritime Commission resident auditor in August 1942 conferred with the company on how to end the overcharges. The FBI was called, but the Department of Justice did not return any indictments against the welders and counters until April 3, 1944.[26] None was ever finally convicted.[27] But the Commission disallowed reimbursement of $470,947.83.[28] In many other cases in which not fraud but carelessness was charged, resident auditors disallowed small sums.

While the resident auditors were checking on the shipbuilding companies, they in turn were being checked up on by field investigators from the General Accounting Office. Judged by what the companies were saying about them, resident auditors were questioning every little thing, so much so as almost to hamstring the management. Judged, on the other hand, by what field investigators from the General Accounting Office were saying, Commission auditors were permitting a lot of unauthorized expenditure. The reports of these investigators were sent by the Comptroller General to the Chairman of the Commission and routed through the Finance Division back to the regional and resident auditors who prepared replies. The GAO investigators objected to some general policies of the Maritime Commission for which resident auditors had no responsibility, such as paying interest on loans by which a company secured a large part of its working capital. They picked up a few more cases of the same kind the resident auditors were customarily catching—improper assignment of overhead, payroll errors, and so on—and criticized the loose control of inventory.[29] They objected also to resident

---

[25] *Ibid.*, pp. 158-59 and 110.

[26] Van Riper and Barnes to MC, Minutes, pp. 33064-65 (Sept. 4, 1945); MC Minutes, p. 30035 (Oct. 26, 1944); Ralph C. Tucker, Resident Auditor, to Slattery, Aug. 9, 1944, in Carmody file, Bethlehem-Sparrows Point; *New York Times*, Apr. 4, 1944.

[27] Clippings from Baltimore *Sun* attached to Ward B. Freeman to Carmody, May 24, 1945 and Tucker to Freeman, Aug. 29, 1945, in Carmody file, Bethlehem-Sparrows Point.

[28] MC Minutes, pp. 33071-72 (Sept. 4, 1945).

[29] House, *Accounting Practices, Hearings*, pt. 1, pp. 107-11, the exhibit there referred to which is really 9F, pp. 402-08.

auditors or members of their families receiving launching gifts, dinners, use of automobiles, or other personal favors.[30]

According to the law, the resident auditors were required to make their records fully available to the auditors from GAO. Mr. Honsick instructed them to do so but not to reveal any technical data about the shipbuilding program.[31] There is evidence of resentment against GAO investigators. Even a resident auditor who was himself extremely critical of the company he was auditing did not feel like being especially cooperative with a man sent in from another agency to look for trouble.[32]

While the GAO was making one kind of a review of the work of the resident auditors, a review from the opposite point of view was provided by the contractors' appeal from the auditors' decisions. In addition to day-by-day referrals to a regional auditor and even to Washington for rulings on specific new questions as they arose, there was a system of review. This enabled contractors to state all their objections to a whole group of the disallowances made by the local auditors. These objections were considered first by a Regional Construction Cost Committee consisting of the Regional Director, the Regional Construction Auditor, and the Regional Attorney, or their alternates. Each committee designated a secretary to whom contractors could address their appeals. A contractor who felt aggrieved was granted a hearing. The minutes of the Regional Construction Cost Committee were sent by its secretary to a Washington Construction Cost Committee for yet another review.[33]

The Washington Construction Cost Committee, consisting of the Director of Finance, the General Counsel, and the Assistant Director of the Technical Division, could overrule even an unanimous decision of a regional committee if it was found out of line with general policy.[34] When a regional committee was not unanimous, decision was up to the Washington committee. The contractor could in either case request a hearing in Washington

---

[30] Land to Lindsay Warren, Feb. 28, 1944 in Anderson reading file.
[31] Honsick to Auditors, July 3, 1942, Instruction 46.
[32] Quarg to Honsick, Nov. 5, 1943—his answer to the GAO report on Marinship, in gf 503-86; also Minutes of Staff Meeting, Gulf Regional Office, Nov. 27, 1943, in Mr. Sanford's files.
[33] Admin. Order No. 37, Suppl. 89; MC Minutes, p. 25088.
[34] Not to be confused with the Review Committee (Construction Cost) which passed on profits of subcontractors. Minutes, p. 26275 (July 16, 1943).

by notifying the Director of Finance.³⁵ In the final audit of his contract he had yet further opportunities for appeal, first to the Commission which had to approve the action of the Washington Construction Cost Committee before it became final, and after that to arbitration as provided in the contracts, or to the Court of Claims. Consequently many of the biggest contracts remained unsettled even in 1948.³⁶

So varied were the expenditures which resident auditors disallowed as not reimbursable and for which reimbursement was in some cases provided for after appeal that it is impossible to do more than pick illustrations. The ordinary appeal concerned many items involving a few hundred or few thousand dollars each, although altogether they might total up to millions for a single contract. Disallowances for minor payroll mistakes were sometimes overruled on the ground that the company had used due diligence. Even the disallowance of the overpayments to welders at Sparrows Point, a result of fraud among the company's employees, was appealed on the ground that the company's freedom of action had been repeatedly interfered with by the labor union involved, the War Labor Board, and other government agencies. Allowing some weight to that plea, the Washington Committee cut the amount of disallowance to $300,000; but the Commission sent the matter back for further study.³⁷ Many appeals concerned overhead, bonuses or incentive payments, fees for professional services, taxes, and welfare, entertainment, or advertising expenses by the contractor, such as $2,509.27 for band rehearsals, and $12,938 for a training film.³⁸

In innumerable cases the original disallowance made by the resident auditor was upheld. But some cases were " compromised," to the disgust of the General Auditor of Construction. For example, the Institute on Human Science, or Human Engineering, Inc., made arrangements for its president to give in six shipyards a course of lectures which would have run up a bill of $290,000. Al-

---

³⁵ *Ibid.*, and many particular cases described by letters in Anderson reading file, for example, Anderson to Southeastern Shipbuilding Corp., July 14, 1943, and routing slip Anderson to Skinner and Mills, July 20, 1943. It was in this final review of contracts before settlement that the Construction Review and Analysis Section did most of its work.

³⁶ House, *Accounting Practices, Hearings*, pt. 1, pp. 193-94.

³⁷ Above note 26, and MC Minutes, pp. 30239-55.

³⁸ Anderson to Calship, May 2, 1944, in Anderson reading file and MC Minutes, p. 26272.

though questioned by the Finance Division, the expenditure had been approved by the Gulf Coast Regional Director, who said he had taken the course and thought it good. When the GAO objected, the amount already due was $50,000 and as a compromise $25,000 was allowed.[39] It is notable that the General Auditor of Construction, who by reason of his position was the Washington champion of the resident auditors, was not a member of the Washington Construction Cost Committee; the Finance Division was there represented by Mr. Anderson, and the other members represented the point of view of the engineers and the lawyers.

At a later date when Mr. Anderson had been ousted and Mr. Slattery was Director of Finance (Construction), the latter expounded publicly his feeling that the resident auditors had not been backed up by the Commission as much as they should have been.[40] He named fifteen who had resigned or been transferred under pressure from the contractors or because of their dissatisfaction with the contractors' methods.[41] Being backed up was not merely a question of being upheld in the Construction Cost Committee; more important probably were the day-by-day appeals by the companies to the regional auditors and especially to the regional directors as new issues came up for decision. The regional directors were all engineers, and they were all being hounded by Admiral Vickery to meet their quotas. Inevitably they had a different point of view from the conscientious professional auditor. Even so, the West Coast Regional Director protested in 1944 at the extent to which costs that had been disallowed by the regional office were later approved by the Washington Construction Cost Committee.[42]

Just before Mr. Slattery left the staff of the Commission he asked the regional auditors to report on the total amount which they had disallowed. They reported a total of $107,210,730.26, which included many items still in dispute because the contracts had not been finally settled. The West coast, in addition to reporting most of this figure, namely $78 millions of actual savings, reported savings of $59 million through changes they had made

[39] MC Minutes, pp. 31463-66 (Mar. 27, 1945); House, *Accounting Practices, Hearings*, pt. 1, pp. 161-63.
[40] House, *Accounting Practices, Hearings*, pp. 152, 160-63, 329.
[41] *Ibid.*, p. 455.
[42] Flesher to Vickery, May 12, 1944 in Flesher reading file, Confidential, 1944 in MC, Historian's Collection.

in the policies of managements, for example, through requiring a different system of time clocks. In sending the complete report to the Commission, Mr. Slattery emphasized also that " the moral effect of the Resident and Regional Auditors represent a further saving to the taxpayers which is beyond calculation." [43]

In the whole process—the resident auditor's checking of the companies' books and prime records, the checking of vouchers in the regional office, the investigations by the GAO, the examination of appeals by regional committees, the Washington committee, and the Commission—there was certainly a lot of checking up on figures, claims, and counterclaims. But back in the shipyards, where events moved more rapidly, there was not sufficient checking up on materials.

IV. THE ACCOUNTS

The foregoing sections have made clear that the billions of dollars spent by the Maritime Commission were accounted for in one sense: all payments were based on vouchers with supporting documents which stated who was getting the money and why. To a professional accountant, however, the words " properly to account for " have a different meaning: they denote that payments, receipts, and obligations have been recorded in a well-ordered system of ledger accounts. In this latter sense the accounting of the Maritime Commission was incomplete. Consequently the Comptroller General, using the words in their technical accounting sense, and Congressmen, using them with awareness of the political overtones, charged in 1946-1948 that the Commission had " not properly accounted for " its wartime expenditure.[1] The investigations which followed did not discover any fraud. They did demonstrate many mistakes and general confusion in the bookkeeping.[2] During the rush to get the ships built, accounting had been a secondary consideration.[3] It had been shoved aside and postponed to such

---

[43] MC Minutes, untyped, Oct. 31, 1946, with attachments which were probably scattered through general files once the minutes were typed.

[1] *Congressional Record*, vol. 92, pp. 278-92, 6115, and the Comptroller General's audit report in vol. 92, p. 7741. Also see below chap. 23.

[2] House, *Wartime Accounting Practices of the United States Maritime Commission and the War Shipping Administration, Report* No. 1, from the Committee on the Merchant Marine and Fisheries, 80th Cong., 1st sess., Jan. 3, 1947 (hereafter cited as *Wartime Accounting, Report*), and House, *Accounting Practices, Hearings*.

[3] Kirsch to S. Schell, Sept. 13, 1944 in House, *Accounting Practices, Hearings*,

an extent that it became impossible to catch up and give a complete and adequate accounting record.

The shoving aside was literal as well as figurative in that the Budget and Accounts Section and the Disbursement Section of the Finance Division were both crowded out of the Commerce Building to make room for the divisions more immediately concerned with getting ships built and for the divisions of the War Shipping Administration whose motto was " Keep 'em Moving." The accounting staff was shifted in April 1942 to a building at 1436 U Street. A year later this was declared by the Director of Finance to be congested and unsuitable, and when hot weather approached another move seemed imperative.[4] These sections were then moved to the Temporary Y Building, at the end of B Street, N.E., which provided more space but was difficult to get to, especially under the restrictions on the use of automobiles.[5]

The housing of these sections was one of several factors which prevented their hiring adequate staff. The location increased the rate of turnover in personnel; and clerks, both white and Negro, had to be trained. Much of the time of the experienced staff was wasted training newcomers who did not stay long.[6] Disbursement was given preference in assigning what staff was available, since payment of vouchers was essential to keep up production, and the accounting was relegated to last place.[7] Moreover, the rating which Civil Service would allow for accounting clerks was so low (CAF-3 or CAF-4) that persons really capable enough to do the operations that were the basic first steps in building up the books could obtain better jobs in the War or Navy Departments, or

p. 425 says: " Because of the inadequacy of numbers of employees, it became apparent to the Director of Finance over two years ago that the handling of vouchers for payment would have to be established in a section independent of the Accounting Section and accordingly the Disbursement Section was set up. Vouchers had to be paid more promptly—the accounting requirements were a secondary consideration."

[4] Anderson to the Secretary, May 4, 1943 and June 7, 1943, in Anderson reading file.

[5] House, *Navy Department Appropriation Bill for 1947, Hearings*, p. 149 and House, *Accounting Practices, Hearings*, pp. 424-25 and p. 324 where the location was called " so undesirable that it was difficult to obtain employees and of those obtained, clerical absenteeism as high as 33 percent was recorded. With all this there was a devotion to duty and long hours were spent on these jobs."

[6] Statements of A. C. Waller, July 29, 1948.

[7] Kirsch to Schell, Sept. 13, 1944, in House, *Accounting Practices, Hearings*, pp. 424-25.

elsewhere.⁸ Inability to expand the staff to an extent corresponding to the rate at which the task itself expanded was the chief reason given by the head of the Accounts and Budget Section for its shortcomings.⁹

The basic job for which Civil Service would not approve a rating high enough to attract the needed personnel was the coding of vouchers, that is, the indication for each voucher of the entries which should be made in the account books in order to record how payment of the vouchers affected the obligations of the Commission or the obligations of others to the Commission.

Coding of disbursement vouchers to the proper appropriation accounts became difficult as the number of funds set up by appropriation acts increased. For example, in 1941 a serious problem was the allocation of expenditure at the emergency yards between the Emergency Ship Construction Fund and the Defense Aid Appropriation. Facilities in some yards were to be paid for entirely from one fund or entirely from the other; but in a number of yards a certain percentage of the costs were allocated to each fund under an arrangement worked out after conferences with the Treasury, the Bureau of the Budget, and the General Accounting Office.¹⁰ Another example of difficulty in applying the terms of the appropriation acts was provided by the requests which Mr. Kirsch addressed to the General Counsel asking him to draw the line between contracts which could be charged to ship construction and those which would come under the limitation placed by the appropriation for contractual services. How about a contract for appraising properties taken in condemnation proceedings? Or for a survey of the electric power rates paid by shipyards? Or for working plans for wooden tugs? ¹¹ The Comptroller General's Audit Report for fiscal 1943 charged that approximately $30 million had been paid from the wrong appropriations.¹²

---

⁸ House, *Wartime Accounting Report*, pp. 62-63, report by Messrs. Wedeberg and Cissel. In recognizing the need for higher classification for coders and general ledger bookkeepers, they agree with statements by Messrs. Waller and Kirsch, cited above.

⁹ *Ibid.*, pp. 424-25, and statements to me by Messrs. Quinn, Waller and Kirsch; also the reason given in Admiral Smith to Senator Aiken, May 27, 1948, in MC gf 301-12.

¹⁰ Director of Finance to Director of the Bureau of the Budget, Apr. 10, 1941, in Anderson reading file.

¹¹ Budget Officer to General Counsel, Dec. 11, 1942, in gf 301-10.

¹² House, *Accounting Practices, Hearings*, p. 255.

Some of this may not have been due to mistakes at the clerical level but to decisions at higher levels on debatable matters of legal interpretation.

Much more, however, than merely showing the appropriation from which money was to be paid, was required in coding vouchers, in view of the considerable number of complicated bookkeeping operations in modern accounting for classifying and analyzing transactions.

Although some coding was done in advance of payment by the Disbursement Section, the main burden of coding and of recording vouchers fell on the Accounting Branch which until 1944 was in the Budget and Accounts Section.[13] In November 1943 its personnel totaled 156; and they were handling accounts of the War Shipping Administration as well as of the Maritime Commission. At one time as many as 88,000 vouchers came in during one month. A backlog piled up, and with the staff available and their methods of bookkeeping, it was not possible to keep the accounts abreast of operations.[14]

Of course, vouchers were not the only basic documents giving rise to bookkeeping entries. Among the others were the removal notices which recorded the shifting of property from one shipyard to another or to a warehouse. Such notices were made out by the resident auditors. In an instruction to them issued May 22, 1944, the General Auditor of Construction insisted that: " There are no exceptions to the necessity for priced Property Removal Notices when the contractor shifts his contractual responsibility as custodian." Record of the transfers, he said, would be made by the Accounts Section in Washington.[15] But this flood of papers was more than the staff of the Accounts Section could handle, and these notices accumulated unrecorded.[16] As explained above, even the resident auditors did not in all cases have accurate records concerning inventories, and in regard to many matters concerning

---

[13] After 1944 its position underwent many administrative reorganizations. See next chapter.

[14] Kohler to Douglas, Dec. 14, 1943, in House, *Accounting Practices, Hearings*, pp. 413-14; interview with Mr. Quinn, Aug. 12, 1947; Mr. Slattery's testimony in House, *Navy Department Appropriation Bill for 1947, Hearings*, p. 149, and in House, *Accounting Practices, Hearings*, p. 324.

[15] Instruction No. 96 (B20). On the Accounts Section, see below chap. 23, sect. III.

[16] *Supplement to Preliminary Report on Comptroller General's Audit Reports for Fiscal Years 1943 and 1944 and the Replies Thereto by the Maritime Commission and the War Shipping Administration*, Exhibit 10, pp. 35-36.

Commission property, the Accounts Section either was not informed of what was done under the regional offices or did not get it into the bookkeeping record. A glaring example of this was the fact that until 1944 the Accounts Section had no record of $2,663,186.22 of inventory on hand in a Commission-owned warehouse in Emeryville, California.[17]

Because so much of the record was incomplete and because the time or staff could not be found for a full accounting analysis of the vouchers, a very large number of payments were recorded by entries in "suspense accounts." These were catch-alls which it was hoped could be cleared up later by transfers to other accounts. But until they were cleared up they interfered with giving any clear, well-founded picture of the Commission's financial transactions as a whole. Efforts to produce general balance sheets from the incomplete and half-analyzed records required many adjustments which the General Accounting Office severely criticized.

In his reports on the Commission's accounts for 1943 and 1944 the Comptroller General said this audit disclosed that "the accounting methods and procedures employed were not of the kind or type to result in accurate recording of the financial transactions to the extent necessary to properly and completely disclose the results of its operations or financial condition . . ."; that at no time during those fiscal years "could the management of the Commission have been furnished current financial information from the accounting records"; and that "the financial interests of the government were not adequately protected. . . ." Many specific examples supported this attack.[18]

Contributing to the vigor of this criticism was the fact that the General Accounting Office had disagreed with the officials of the Maritime Commission ever since its inception as to what system of accounts it should use.[19] The question was partly a highly technical one involving one of the most ancient of accounting problems, voyage accounting,[20] and the fact that the Com-

---

[17] House, *Accounting Practices, Hearings*, p. 326.
[18] *Ibid.*, pp. 253 ff., and pp. 289 ff.
[19] *Ibid.*, pp. 80-81, 386-87.
[20] *Ibid.*, p. 215 and pp. 445-46; House, *Wartime Accounting Report*, pp. 22-23; F. C. Lane, *Andrea Barbarigo, Merchant of Venice* (Baltimore, 1944) pp. 164-69; and *idem.*, "Venture Accounting in Medieval Business Management," in *Bulletin of the Business Historical Society*, XIX (Nov. 1945), 5.

744   THE FLOW OF MONEY

mission's books were not set up fully on an accrual basis.[21] Mr. Reavis, who as investigator for the General Accounting Office specialized on the Maritime Commission, as he had earlier on the Shipping Board, testified that for the period before fiscal 1942 the system used, "since the accounts were not so numerous," while it "was not what I would want for an accounting system myself, it did work,"[22] but he felt that it should have contained, as it did not, a full system of budgetary control accounts. As long as the Commission had the one revolving fund, the Construction Fund, the lack of these controls did not bring serious trouble, and budgetary control could not be maintained in wartime, to be sure.[23] But clinging to a system of accounts other than that recommended by the General Accounting Office left the Maritime Commission in a weak position when its own system failed to cope with the situation, and in an even worse position after the war when, before the accounts had been disengaged from the wartime disorders, Congress did insist on budgetary controls.

The specialists in government accounting (that is, accounting for public funds) and the firms of Certified Public Accountants who audited commercial companies had widely different points of view. The former were entrenched in the General Accounting Office, but Admiral Land placed confidence in the latter group.[24] At the creation of the Commission, a special audit was made by Price, Waterhouse & Co.; and the Director of Finance from 1937-1939 was D. F. Houlihan, a member of Price, Waterhouse & Co.[25] Meanwhile the standing of that firm and the general reputation of Certified Public Accountants had been subjected to question as the result of failure to prevent the frauds in inventory which preceded the sensational failure of McKesson and Robbins.[26]

[21] House, *Accounting Practices, Hearings*, pp. 80-81, 424; House, *Independent Offices Appropriation Bill for 1948, Hearings*, pt. 2, p. 658.

[22] House, *Accounting Practices, Hearings*, pp. 80-81. See also the *Annual Report of the Comptroller General of the United States*, 1941, p. 66.

[23] Statements by T. H. Reavis, Aug. 6, 1948, and see above in sect. 1 of this chapter the many deficiency appropriations.

[24] House, *Hearings on H. R. 6790* before the Committee on Naval Affairs, Apr. 13, 1942, p. 2968.

[25] Blanche D. Coll, "1937 Audit of Maritime Commission Accounts," a typescript in Historian's Collection, Research file 106.23.

[26] On the scandal see *Time*, Dec. 19, 1938, pp. 48-50; Dec. 26, 1938, pp. 35-37; Jan. 2, 1939, p. 29; Jan. 9, 1939, p. 49; Jan. 16, 1939, p. 54. For discussions by professional accountants during 1939 of the attacks then being made on them and

THE ACCOUNTS 745

When the War Shipping Administration was created, Mr. Houlihan was brought back to be its Director of Fiscal Affairs. The Comptroller General wrote Admiral Land that Mr. Houlihan's position in the War Shipping Administration was " incompatible with the public interest," because of business done by Price, Waterhouse for War Shipping Administration, and although the Comptroller General later withdrew his objection, Mr. Tulloss, Chief Investigator of the General Accounting Office, expressed complete lack of confidence.[27] There was a feeling among government accountants that those working with commercials firms did not understand what was required in dealing with public funds, which had to be expended according to law.

Other government agencies fell behind in their accounting work during the war, as did even the General Accounting Office itself in its work of post-auditing the vouchers paid by the government.[28] The same kind of criticisms made of the Maritime Commission accounting were made of five other government agencies, particularly the Reconstruction Finance Corporation.[29]

Criticism of the Maritime Commission's accounts was particularly severe because it was part of a general attack by the Comptroller General on the Maritime Commission.[30] In addition to the transactions concerning the Tampa Shipbuilding Corp. which have already been described, many arrangements involving millions of dollars that were made by the Maritime Commission and War

the extent of their responsibilities for inventories, see *The Journal of Accountancy*, vol. 67, pp. 208-10, 375, and *passim*; vol. 68, pp. 16, 345.

[27] Comptroller General to Administrator, WSA, Oct. 26, 1942, and Nov. 20, 1942; memorandum of Conference held in Mr. Douglas' Office, Nov. 4, 1942, both in Douglas file, Fiscal Affairs.

[28] In fiscal 1941 it audited 20 million vouchers and had 4.5 million unaudited at the end of the year. In fiscal 1945 the GAO had a workload of about 81.5 million vouchers, and had 33.5 million on hand at the end of the year. In short its workload increased fourfold from 1941 to 1945 and the number of vouchers on hand at the year's end increased from 22.5 per cent of the year's receipts to 43.6 per cent of the year's receipts. Its staff meanwhile increased 250 per cent. *Annual Report of the Comptroller General*, 1941, pp. 50-51; *ibid.*, for 1945, pp. 20-21; House, *Independent Offices Appropriation Bill Hearings for 1946*, p. 1117.

[29] House, *Independent Offices Appropriation Bill for 1948, Hearings*, pp. 78-82; House, *Accounting Practices, Hearings*, p. 113.

[30] The " Annual Report of the Comptroller General of the United States for the Fiscal Year ended June 30, 1942," (typescript, in GAO library) is not critical of the Maritime Commission, although it mentions investigations in progress. The *Annual Reports* for 1943 (pp. 78-81) and 1944 (pp. 75-77) signalize the Maritime Commission for special criticism.

Shipping Administration in fixing charter rates and in requisitioning vessels were declared by the Comptroller General to be improper or even illegal. Since the operation of the merchant marine is not part of this history, there is no need to go into the complicated legal arguments made on both sides, but the general statements of the General Accounting Office about the whole accounting of the War Shipping Administration and Maritime Commission are better understood if it is remembered that the General Accounting Office claimed also that millions had wastefully and unnecessarily been paid to operating companies for their ships. Even when the Attorney General upheld the legality of the actions, the Comptroller General continued to believe them illegal.[31] As already explained in regard to the arrangements made to keep the Tampa shipyard going, the Commission adopted what looked like a practical solution to get production. The Comptroller approached all these problems from the point of view of rules and regulations. That was his duty. It is a commonplace among political scientists that the General Accounting Office is so set up by law that it feels no responsibility for what would be the practical results if all its rulings were followed.[32] Accounting was only one of many things which it felt duty bound to criticize in the activities of the agencies headed by Admiral Land, who was responsible for results.

In regard to the accounts of the Maritime Commission, many of the deficiencies pointed to by the General Accounting Office were admitted, although wartime conditions were cited to excuse them and their seriousness was disputed. It is difficult for anyone not an accountant to distinguish always between plain bungling and the accountants' quarrel over cash or accrual methods, but it does seem necessary to attempt some estimate of how serious were the practical consequences of the deficiencies.

The failure to keep the accounting abreast of the disbursements made it difficult to be sure that the same bill was not paid twice. The General Accounting Office found that overpayment was made by the Commission to such an extent that over $200,000 in erroneous payments were voluntarily repaid by contractors and vendors in fiscal 1943.[33] As a whole, the recording of government receiv-

---

[31] House, *Independent Offices Appropriations Bill for 1948, Hearings*, pt. 1, p. 64. On Tampa, above chap. 15.

[32] Harvey C. Mansfield, *The Comptroller General* (New Haven, 1939) pp. 3, 286.

[33] House, *Accounting Practices, Hearings*, p. 255.

PLATE XXV.

The Regional Directors, left to right: W. E. Spofford, Great Lakes; J. F. McInnis, East Coast; L. R. Sanford, Gulf; and C. W. Flesher, West Coast (at right). Below, the 14-way yard of the California Shipbuilding Corp. on Terminal Island. Planned as an 8-way yard at the beginning of the emergency program, it was enlarged horizontally along the Cerritos Channel that connects the Port of Los Angeles with the Port of Long Beach. It launched 306 Liberty freighters, 30 Liberty tankers, 101 Victory freighters, and 30 Victory transports.

PLATE XXVI. Upper left, bulkhead and forward section of tanker on ways. Upper right, lifting a forepeak section. Lower left, completing main deck of a Victory ship. Lower right, cutting scrap steel for use in welding schools.

ables was so delayed and confused that their collection was jeopardized. Lack of clarity and of completeness in the bookkeeping record placed the government in a weaker position in negotiating or pressing for settlement.[34] The inadequate controls of inventory were detrimental to the interest of the government in the same way. Failure to process the receiving reports meant that if shipments of steel, for example, were less than the amount billed, the shortages were not detected until years later, when indeed attempts were made to establish claims against producers and against railroads to recover overpayment of freight.[35] The amount of monetary loss which it was proved the government had suffered because of these various deficiencies, although it ran into millions, was relatively small compared to the billions expended; but the nature of the deficiencies and mistakes in the accounts make it impossible to say with assurance that larger sums may not have been lost.

A consequence of the failure to record currently the receiving reports and removal notices was that the Commission had no complete bookkeeping record of what material and components went into any particular ship. It could not therefore establish from the accounts the costs of individual ships or even of all those built under one contract; for although the amount paid the shipbuilder in the final settlement of the contract was clearly recorded, the cost of the Commission-furnished material and components was not. The engines, for example, were bought by placing with manufacturers orders which were called pool orders because when they were placed it was not known to which shipyards the engines would go. That might depend on last minute scheduling or spot allocation. The price of the engines varied somewhat from order to order. By going through all the receiving and removal notices, if they were as complete as they were supposed to be, it would have been possible theoretically to find out just which engine went in which ship and to allocate the cost accordingly.[36] But this was never done. In the judgment of both the Commis-

[34] House, *Wartime Accounting Report*, p. 22.
[35] *Supplement to Preliminary Report on Comptroller General's Audit Reports for Fiscal Years 1943 and 1944* . . . , Exhibit 10, p. 37; *Iron Age*, Sept. 4, 1947, pp. 111-112; House, *Wartime Accounting Report*, p. 20; House, *Accounting Practices, Hearings*, pp. 48-51; and later appropriation hearings.
[36] House, *Navy Department Appropriation Bill for 1947, Hearings*, p. 147, and Slattery to the Commission, May 24, 1946, approved May 31.

sion's accountants and those from the General Accounting Office, it was not worth doing.[37] For practical purposes a sufficiently good estimate of the cost of ships by yards and types could be made by statistical methods, using a standard bill of materials, and therefore this aspect of the loss of "accountability" did no serious harm.[38]

A consequence emphasized by some accountants was that the Maritime Commission accounts, not being current, could not serve as a tool of management; but these critics gave no specific illustrations of the kind of accounting data which could have been useful in making the kind of managerial decisions which confronted the Commission during the war.[39] The Finance Division of the Maritime Commission did develop ways of measuring the efficiency of the shipbuilding companies, as we have seen, in order to race them against each other. Useful figures of this kind were developed by applying statistical methods to the records of cost kept by the resident auditors.

Current accounts would, to be sure, have made much easier the budgeting of the future financial needs of the Commission. The deficiencies in the accounts had serious effects on the Commission's relations with Congress immediately after the war. Moved by the spectacle of past and present confusion in the financial statements of the Maritime Commission and War Shipping Administration, Congress, the first Republican Congress since the formation of the Maritime Commission, in 1947 not only cut the Commission's appropriations for fiscal 1948 but provided that the Construction Fund created in 1936 should cease to be a revolving fund. Any balance above the amount appropriated for specified purposes in fiscal 1948 was, as of June 30, 1947, to be transferred from the Commission to the Treasury.[40]

Among the many consequences of the delays and disorders in accounting, the most important seems to have been the possibility that fraud occurred and money was wasted in cases which cannot be itemized precisely because of the gaps or mistakes in the accounting record. None of the consequences was as serious,

---

[37] *Ibid.*, pp. 147, 609; statements of T. H. Reavis, Aug. 6, 1948.
[38] Statements by T. H. Reavis, Aug. 6, 1948, and see below in chapter 24 on costs.
[39] House, *Wartime Accounting Report*, p. 9.
[40] House, *Independent Offices Appropriation Bill for 1948, Hearings*, pt. 2, p. 666; 61 Stat. 603, Public Law 269, July 30, 1947, 80th Cong., 1st sess.

of course, as would have been a slowing down of ship production in order to improve the auditing or to let the accountants catch up. But the accounting could have been done better without any such interference with production if qualified personnel had been found. The two professional accountants engaged by Congress to pass judgment, Sivert M. Wedeberg and C. Wilbur Cissel, concluded: " It appears that if an intelligent effort had been made in the early days of the war to simplify accounting procedures and methods, what now appears to be ' accounting patchwork ' might have been avoided and many man-hours saved in its avoidance." [41] They criticized " ' blind ' acceptance of an industry's peacetime accounting methods for use during a war emergency " and said it indicated " a lack of foresight, planning, and comprehension of accounting under emergency wartime conditions," adding, " Fortunately for the Nation the actual construction of ships was not approached with this philosophy." [42]

This judgment was accepted by the Congressional investigating committee, and it also directed its criticism at the officials higher up. Noting that the " chaotic condition " of the accounting systems had been called to the attention of Admiral Land as early as 1943, and yet was not corrected, the Committee's report said: " This failure by the Commission and the War Shipping Administration to take adequate measures during the following years to rectify a known condition is not understandable." [43] Any effort to " understand " requires further study of general administrative problems of those two agencies and of their difficulties in finding qualified personnel.

[41] House, *Wartime Accounting Report*, p. 25.
[42] *Ibid.*, p. 23.
[43] *Ibid.*, p. 8.

Chapter 23

ADMINISTRATIVE PROBLEMS: (C) THE COMMISSION AND THE WAR SHIPPING ADMINISTRATION

I. Expansion of Personnel

PERMEATING ALL the administrative problems of the Maritime Commission was the difficulty of finding persons with the ability to do the many jobs which had to be done. The similar problem facing the shipbuilding contractors in finding managerial and skilled labor has already been discussed. The size of the problem within the government service itself is clearly shown by the number of employees: [1]

	June 30, 1940	Peak
Technical Division	351	691 in 1945
Regional Construction Offices	0	4,630 in 1943
Production Division	0	624 in 1944
Procurement Division	95	1,351 in 1944
Finance Division	197	3,103 in 1945
All Employees, MC and WSA	1,361	14,822 in 1945

The personnel was all brought under Civil Service classification except for a few contract employees in the higher brackets. Most of the staff added in 1941 came from Civil Service lists, but after Pearl Harbor these lists were quickly exhausted. War Service Regulations effective March 16, 1942 permitted government agencies to pass on the qualifications of those they hired and give them war service appointments. The Maritime Commission was also able to recruit some help from other civilian agencies. It was given priority in Group I, as were the Army and Navy, and consequently persons could transfer to it from civilian agencies in

---

[1] Figures derived from Table 16 in this chapter. Construction and Production were part of Technical in 1940. Procurement and Finance include persons concerned with both construction and operation of ships. The total includes many other divisions less important in ship construction.

750

Group II and retain right to re-employment after the war in the agency from which they transferred.[2]

Although in a favorable position in comparison with other civilian agencies, the Maritime Commission was at a disadvantage in competing with the Army and Navy, which could offer the attractions of uniform and military rank, and which benefited from Selective Service. Some key members of the Maritime Commission staff, for example, Lt. Comdr. William A. Weber, aide to Admiral Vickery, and Lt. Comdr. Walston Brown of the Legal Division, obtained Navy commissions and were detailed to the Maritime Commission. A few officers in service in the Navy Department were also detailed to help Admiral Land as Chairman of the Maritime Commission or as Administrator of the War Shipping Administration. But the Navy's own needs were so pressing that it would lend few of its officers.

In seeking and retaining its large corps of inspectors and engineers, the Commission met also the competition of the very shipbuilding companies it was supervising. They could offer salaries higher than those which the government paid in accordance with Civil Service ratings, and the companies could list these salaries among the costs for which they were reimbursed under the terms of their shipbuilding contracts.[3] After controls were instituted by the Manpower Commission, the Maritime Commission could use them for what they were worth to prevent its employees from leaving. In general, however, government service depends for attractiveness on its relative permanence. War-service appointees in a wartime agency could not be offered that kind of security. All things considered, the wonder is that the Commission retained as many good men as it did with private corporations paying higher salaries.

The personnel problems of the Maritime Commission were in many respects unique because it was not a new wartime creation like the WPB nor an agency maintained in peace but planned for war, like the Navy. It was like the Navy in having a nucleus of experienced professional career men around whom to build the wartime organization, but it had no reserve corps from which to draw reinforcements and no mandate in peacetime to lay the groundwork for wartime recruitment. There were other civilian

[2] Statements by C. L. Miller, June 22, 1949.
[3] See, for example, chap. 4, sect. V.

agencies of war administration which were linked to peacetime agencies, but none in which the peacetime agency itself had so large a part in industrial mobilization as the huge shipbuilding program gave the Maritime Commission.[4]

Because the Maritime Commission was an established agency, its expansion could be regarded as a problem in dilution. As Executive Director Schell saw it, the staff of the Commission had to go through the same process as did the staff of a new shipyard: a small number of experienced men had to be spread thin in key positions so that their "know-how" would season the mass of men with less or different experience.[5] But expansion could also be a process of invigoration by the injection of new blood. The effect of bringing in outsiders, whether it quickened or merely diluted, was most easily spotted at the top but was felt all down the line.

Most of the key men in the Commission's wartime organization —directors of divisions and regional offices, the section chiefs, and Admiral Vickery's personal assistants—were members of the staff which Admirals Land and Vickery had built up before the war. There were outstanding exceptions: namely, Charles E. Walsh, Director of the Procurement Division; Allen D. MacLean, Assistant Director and then Director of the Production Division; and Colonel W. F. Rockwell, the first Director of the Production Division. Many of the section chiefs and heads of branches under them were new employees.

Colonel Rockwell is the only instance of a Commission official who was what is commonly called a "dollar-a-year" man, meaning, not that he was paid at that rate, but that he continued to receive salary from the companies with which he was connected and to devote part of his time to their affairs.[6] Colonel Rockwell's

---

[4] The Office of Defense Transportation was linked to the Interstate Commerce Commission, and the War Foods Administration to the Department of Agriculture. They are therein comparable to the War Shipping Administration, which was linked to the Maritime Commission. The unique feature of the position of the Maritime Commission is that it not only had a large emergency organization attached to it but was itself a major agency for wartime industrial mobilization, ranking as a procurement agency second only to the Army and Navy. On the Navy's use of reservists in its inspection service, see Connery, *The Navy in the Industrial Mobilization*, p. 341.

[5] Statement by S. Schell.

[6] Henry E. Frick was similarly paid $25 per diem while retaining connection with his company, but did not have a definite administrative position. See Personnel files; also see above, chap. 14 and chap. 11.

companies did no business with the Commission,[7] but his double position was a source of some criticism because he was away so much; and it became a cause of embarrassment to the Commission when he testified publicly against the Renegotiation Act. He appeared before a Congressional committee in September 1943 to speak for the Timken-Detroit Axel Co., of which he was Chairman of the Board. He spoke for men who, as he said of himself, had a million dollars, wished to keep it, and felt that the Army's renegotiation threatened ruin. He denounced the whole system of renegotiation, part of which was being administered for shipbuilding contracts by members of the Maritime Commission.[8] In January 1944 Colonel Rockwell tendered his resignation, adding: "I should very willingly, for at least a time, continue to act in the same capacity as I have heretofore but only in an advisory manner due to the fact that my companies' needs will require that I devote almost my full time to their welfare." The Commission simply accepted his resignation.[9]

In spite of the difficulty the Commission had in recruiting help, there were charges that it was overstaffed in some divisions. The Production Division and its Controlled Materials unit especially were criticized by a Congressional committee's investigators both for overstaffing and for employing so many persons from Colonel Rockwell's companies.[10] The Commission's own Personnel Division was critical, as was their function, of the amount of upgrading in that division and others.[11] The list of those in the Maritime Commission and the War Shipping Administration who were raised two or more grades between 1940 and 1947 fills 36 pages of committee hearings. Quite a number with

---

[7] In letters at the time of his appointment he stated that he expected to continue his connections with his businesses, and that the Maritime Commission was the only governmental agency for which his companies were not delivering material. Rockwell file in Personnel Division.

[8] House, *Renegotiation of War Contracts, Hearings*, before the Committee on Ways and Means, 78th Cong., 1st sess. on H.R. 2324, 2698, 3015. On his company and ordnance contracts, see Rockwell to Land, Dec. 17, 1943, in MC Historian's Collection, Land file.

[9] MC Minutes, p. 27302 (Jan. 11, 1944).

[10] *Report on Abuses of Merit System and Misuse of Equipment and Funds, Maritime Commission* (Confidential Committee Print) [House Civil Service Committee] Staff report, Dec. 9, 1944, in MC Personnel Division files, Production Division.

[11] Personnel Division files. The Controlled Materials unit section numbered 65 in Dec. 1943; 17 in Dec. 1944. Montgomery to Schell, Dec. 14, 1944.

salaries of about $2,500 in 1940 received $5,000 by the end of the war.[12] As in other agencies the usual slow promotions of Civil Service were brushed aside in competing for personnel against the lures of private industries and other branches of the government.

II. WSA AND THE JOINT SERVICES

No analysis of how invigoration or dilution worked out can be pushed very far without considering at the same time the relation of the Maritime Commission (MC) and the War Shipping Administration (WSA).

In selecting Admiral Land to be Administrator of WSA while continuing to be Chairman of the Maritime Commission, President Roosevelt was reaffirming a decision taken in February 1941 when he made Admiral Land his adviser on ocean shipping to analyze needs and make plans for control of shipping operations.[1] To assist in carrying out that mandate, a Division of Emergency Shipping was formed within the Maritime Commission under H. H. Robson and other executives brought in from big shipping companies.[2] After WSA was created on February 7, 1942, Admiral Land wrote, ". . . it is my idea to disturb the Maritime Commission's work as a whole as little as possible, which, in reality, means setting up a War Shipping Administration unit inside the Maritime Commission circle."[3] At the upper levels this was carried out to the extent of appointing two Commissioners as Deputy Administrators: Captain Macauley became Deputy Administrator in charge of merchant marine personnel, and Admiral Vickery Deputy Administrator for new construction, a post which gave status but no additional staff beyond that which Admiral Vickery supervised within the Maritime Commission.[4] Also, the Executive Director of the Maritime Commission, Mr. Schell, was appointed Executive Officer of the WSA.[5] To that extent there was union at the top.

[12] House, *Independent Offices Appropriation Bill for 1948, Hearings*, pt. 2, pp. 856-91.
[1] MC Press Release 837, Feb, 12, 1941, in Land file. For the recommendations behind this move see W. A. Harriman to Stettinius, Dec. 20, 1940 in OPM file 324.1041.
[2] MC Minutes, p. 16378 (Feb. 25, 1941).
[3] Land to Mausshardt, Mar. 2, 1942, in Land file.
[4] WSA Admin. Order No. 2, Suppls. 3 and 4, June 9, and 17, 1942.
[5] WSA Admin. Order No. 4, Feb. 18, 1942.

But the most important Deputy Administrator of WSA, after Mr. Robson retired, was not a part of the Maritime Commission circle, he was not even a part of the shipbuilding and shipping fraternity. He was Lewis Douglas, President of the Mutual Life Insurance Company, a former Director of the Bureau of the Budget, and intimate of President Roosevelt. Suggested by Admiral Land to the President and given the President's personal

TABLE 15

TYPES OF ACTIVITY OF ADMINISTRATIVE PERSONNEL OF THE MARITIME COMMISSION AND WAR SHIPPING ADMINISTRATION FISCAL YEARS 1940 TO 1947

Type of Activity	Fiscal Year (Ending June 30)							
	1940	1941	1942	1943	1944	1945	1946	1947
Total, All Activities	1,361	1,926	5,444	11,616	14,318	14,822	9,488	4,874
Administrative and Miscellaneous....	268	317	812	891	1,042	1,025	736	466
Legal and Regulatory	117	101	155	172	264	285	264	211
Auditing, Accounting and Budget......	197	306	1,120	1,146	2,985	2,993	2,007	1,166
Supplies, Vessel Procurement and Disposal, and Renegotiation...........	95	204	475	679	1,493	1,289	2,268	570
Vessel and Facilities Construction.....	377	551	2,131	5,662	4,959	4,856	822	307
Vessel Operation...	150	280	466	2,063	2,425	3,071	2,612	1,972
Research, Statistics and Government Aids............	117	118	165	283	331	368	257	91
Training and Recruitment........	40	49	120	720	819	935	522	91

Source: Similar table prepared by MC Budget Division, Feb.-Apr. 1948.

promise to back him,[6] he became Deputy Administrator without closely limited functions under a plan of organization approved by the President May 16, 1942.[7] Using many members from the Commission's staff, many men brought in from the shipping industry, and many who had no previous connection with shipping, Mr. Douglas and Admiral Land shaped up the bulk of the WSA into an organization which was under Mr. Douglas as Deputy Administrator for Vessel Utilization, Planning, and Policies. At

[6] Statement by Admiral Land, June 17, 1949.
[7] Land to the President, May 16, 1942, with changes and O.K. in the President's hand, in Land file.

the beginning of 1943 he had four Assistant Deputy Administrators, one for Ship Control, one for Ship Operations, one for Tanker Operations, and one for Fiscal Affairs.[8] None of these held any position in the Maritime Commission simultaneously nor had they been connected with it before the war, except Mr. Houlihan, the Assistant Deputy Administrator for Fiscal Affairs, whose connection has been explained in the previous chapter, and who had been brought into the WSA by Admiral Land before Mr. Douglas took over.[9]

The separateness of WSA was accentuated by the appointment on November 12, 1942 of an Administrative Officer reporting to Mr. Douglas and performing for his part of WSA the same general functions which Mr. Schell performed as Executive Director for the Maritime Commission. The Administrative Officer for WSA, Amyas Ames, was made responsible for the preparation of administrative orders and their execution, for office space, for the review of budgetary plans and recommendations concerning personnel, and for coordinating the work of divisions to prevent duplication. He was instructed to work closely with Mr. Schell in the latter's capacity as Executive Officer of WSA, but only in regard to "matters requiring action by the Administrator" was consultation explicitly called for.[10] In the large field in which Mr. Douglas himself could act, he relied directly on Mr. Ames. The same trend was carried further by the creation on February 22, 1943 of a separate WSA Division of Personnel Management.[11] By fiscal 1944 this division had a staff larger than that of the Personnel Management Division in the Maritime Commission (Table 16).

In spite of the growing separateness of WSA, an overall view of the figures concerning the personnel of the two agencies shows how closely these Siamese twins were still connected. Out of a total of 14,318 employees in fiscal 1944, only 27 per cent were in the WSA, and about 73 per cent were in the Maritime Commission,

[8] They were, in that order: Franz Schneider, for Ship Control; John E. Cushing for Ship Operations, until Apr. 1943 when moved to west coast; D. F. Houlihan for Fiscal Affairs, from June 17, 1942 to Jan. 15, 1943 when he was succeeded by Richard W. Seabury (WSA Admin. Order No. 13, Suppl. 2) who was in turn succeeded in Oct. 23, 1943 by Percy Chubb (WSA Admin. Order No. 13, Suppl. 7) and B. Brewster Jennings for Tanker Operations (WSA Admin. Order No. 2, Suppl. 4).
[9] WSA Admin. Order No. 2, Feb. 17, 1942.
[10] WSA Admin. Order No. 2, Suppl. 9 and 10.
[11] WSA Admin. Order No. 40.

according to a breakdown for budget purposes (Table 16). Analysis in terms of functions shows that 22 per cent of the total of both agencies were directly connected with operations under Mr. Douglas or in the recruitment and training of seamen under Captain Macauley; 34 per cent were in divisions exclusively concerned with planning and supervising under Admiral Vickery the construction of ships and shipyards (Table 15). Personnel performing other functions (44 per cent of the total) were mainly on the rolls of the Maritime Commission. Compare Tables 15 and 16. Even in the Division of Ship Operations a few persons were on the Maritime Commission payroll. About a third of the total staff were in divisions or offices serving both construction and operations, so substantial was the connection between these Siamese twins.

A starting point for this intermingling of the two staffs was laid in the very first WSA Administrative Order. It authorized anyone on the Maritime Commission's rolls who in all or part of his time had been responsible for performing a function which had been transferred to the new agency to keep on performing it.[12] Certainly during the spring of 1942 there were many of the staff who neither knew nor cared which agency they were part of; they were too busy with tasks needing to be done immediately. When relations became more formalized, a part of the Maritime Commission staff was specifically recognized as acting for both agencies. Its size varied from time to time. It appears smallest in the estimates of the Budget Officer for fiscal 1944, where he designated as the Joint Service Organization the following: " the Office of the Secretary, the Division of Public Relations, and that portion of the Finance Division performing, for both the Administration and the Commission, auditing, accounting, disbursing and related functions other than construction auditing." He defined their position by saying that the Administrator had determined that " the interests of the Commission and the Administration will best

[12] " Effective immediately, and pending further orders, all employees and agents of the United States Maritime Commission are authorized to act in behalf of the Administrator, War Shipping Administration, in respect of functions, duties and powers conferred by law upon the United States Maritime Commission and transferred to the Administrator by the Executive Order Establishing a War Shipping Administration in the Executive Office of the President and Defining its Functions and Duties, dated February 7, 1942, to the same extent and with the same effect as they would have been authorized to act in behalf of the United States Maritime Commission if said Executive Order had not been issued."

TABLE 16

PERSONNEL IN ORGANIZATIONAL UNITS OF MARITIME COMMISSION AND WAR SHIPPING ADMINISTRATION AT END OF FISCAL YEARS 1940-1945

Organizational Unit	1940	1941	1942 MC	1942 WSA	1942 Total	1943 MC	1943 WSA	1943 Total	1944 MC	1944 WSA	1944 Total	1945 MC	1945 WSA	1945 Total
Total of Administrative and Operating Personnel	2,263	3,029	...	...	6,816	...	...	19,879	...	...	23,033	...	...	24,400
Administrative Personnel														
Total	1,361	1,926	5,206	238	*5,444	8,621	2,995	11,616	10,578	3,740	14,318	10,317	4,505	14,822
Office of Commissioners, MC	37	36	48	...	48	66	...	66	77	...	77	100	...	100
Office of Administrator, WSA				2	2	...	9	9	...	12	12	...	5	5
Legal Division	67	65	71	51	122	50	99	149	76	166	242	68	197	265
Regulations Division	50	36	33	...	33	23	...	23	22	...	22	20	...	20
Price Adjustment Board						22	4	26	73	42	115	79	57	136
Trials and Surveys Division	10	16	15	...	15	2	...	2	...	...	...	...	...	...
Office of the Secretary	173	202	554	...	554	408	...	408	421	...	421	416	...	416
Public Information Division	10	14	42	...	42	40	...	40	43	...	43	51	...	51
Personnel Management Division	48	65	136	30	166	211	157	368	227	262	489	321	132	453
Government Aids Division	11	11	11	...	11	8	...	8	5	...	5	5	...	5
Research Division	106	107	112	42	154	62	213	275	79	247	326	99	264	363
Finance Division	197	306	1,120	...	1,120	1,143	...	1,143	2,929	...	2,929	3,103	...	3,103
Office of Director of Fiscal Affairs							3	3	...	24	24	...	92	92
Budget Division									24	32	32	65	40	105
Procurement Division	95	204	475	...	475	647	...	647	1,351	...	1,351	1,000	...	1,000
Large Vessel Sales Division												7	...	7
Small Vessel Sales Division							6	6	...	27	27	...	52	52
Contract Settlement and Surplus Property Division												94	...	94
Ship Operations Division	82	213	231	55	286	15	267	282	16	726	742	14	1,183	1,197
Maintenance and Repair Division	51	50	117	...	117	15	672	687	3	695	698	1	1,050	1,051
Insurance Division	17	17	35	28	63	22	146	168	19	148	167	17	128	145
District Directors WSA							926	926	...	818	818	...	678	678

TABLE 16—Continued

PERSONNEL IN ORGANIZATIONAL UNITS OF MARITIME COMMISSION AND WAR SHIPPING ADMINISTRATION AT END OF FISCAL YEARS 1940-1945

Organizational Unit	1940	1941	1942 MC	WSA	Total	1943 MC	WSA	Total	1944 MC	WSA	Total	1945 MC	WSA	Total
Administrative Personnel (continued)														
Technical Division	351	517	441	...	441	462	...	462	613	...	613	691	...	691
Regional Construction Offices	...	...	1,127	...	1,127	4,630	...	4,630	3,692	...	3,692	3,321	...	3,321
Production Division	...	...	528	...	528	544	...	544	624	...	624	507	...	507
Shipyard Labor Relations Division	16	18	20	...	20	24	...	24	30	...	30	30	...	30
Recruitment and Manning Organization	...	...	...	30	30	...	493	493	...	541	541	...	627	627
Training Organization	40	49	90	...	90	227	...	227	278	...	278	308	...	308
Operating Personnel														
Total	902	1,103	...	1,372	1,372	...	8,263	8,263	...	8,715	8,715	...	9,578	9,578
Warehouses	12	20	...	32	32	...	86	86	...	287	287	...	757	757
Reserve Fleet	256	112	...	77	77	...	6	6	...	6	6	...	4	4
Terminals	165	154	...	96	96	...	91	91	...	105	105	...	111	111
Training Organization	469	817	...	1,167	1,167	...	8,080	8,080	...	8,317	8,317	...	8,706	8,706

Source: Similar table prepared by MC Budget Division, Feb.-Apr. 1948.
* War Shipping Administration created February 7, 1942.

be served by retaining those Divisions on the rolls of the Commission,—those Divisions, however, for the purposes of directive control, being primarily responsible to the Administrator." [13]

The Secretary of the Commission, W. C. Peet, Jr., was designated Secretary of the WSA at its creation,[14] and A. J. Williams succeeded him October 1, 1943.[15] Under them were about 400 persons performing the very essential but very little appreciated tasks of recording and circulating the minutes of all official actions, answering telephones, handling a staggering volume of mail and a system of general files which grew correspondingly, maintaining the library, providing stenographic, tabulating, and duplicating services for all divisions, maintaining records of travel expenses, and moving the staff from office to office. They worked as one unit for both Maritime Commission and WSA, and their tasks expanded with those of both agencies.[16]

The functions of the Division of Public Relations enabled it also to operate more easily as a joint service. Admiral Land and the other Commissioners for whom it furnished material for speeches were presenting to the public not only the achievements in building ships but especially the need for ships and the uses to which they were to be put. The same combination—the intense need and the efforts necessary to meet it—were emphasized in many other of the Division's activities: the feature articles placed in magazines, descriptive booklets, the Maritime News Digest circulated among the staff of Maritime Commission and WSA, radio and newsreel material, and the poster campaign which, has already been mentioned. From August 1941 to February 1944 the Director of this division was Mark O'Dea, president of a New York advertising agency, and in 1944-1946 Robert Horton. Its personnel grew from 10 in 1940 to 51 in 1945 and included representatives in Regional and District Offices.[17]

The other part of the Joint Service Organization, as narrowly defined, was the portion of the Finance Division under Mr. Quinn.

---

[13] MC, " Justification and Schedules for Budget Estimates, Fiscal Year 1944," pp. 9-10, in files of Budget Officer.
[14] WSA Admin. Order No. 4, Feb. 18, 1942.
[15] MC Admin. Order No. 37, Suppl. 90, Oct. 1, 1943, and Williams to Director of Personnel, Oct. 4, 1943, in WSA Personnel files.
[16] Organization charts in MC, Historian's Collection, Research file 106 MC 46.
[17] Blanche D. Coll, " Division of Information," a typescript in MC, Historian's Collection, Research file 106 MC 29.

Its functions are described in the previous chapter and its administrative reorganizations will be considered shortly. Also in fact acting jointly for both the Maritime Commission and WSA were a number of other divisions. Although the repair of ships came under WSA, the audit of ship repair contracts on the West coast was done under the Regional Auditor by construction auditors.[18] Closely connected with fiscal matters also was the Procurement Division, into which was absorbed in 1943 the earlier Division of Purchase and Supply.

The Procurement Division was the biggest joint service in the field. It maintained two sets of warehouses for the storage of material: one group for construction material and the other for material and equipment for operating, repairing and converting ships. In 1943 it had eleven warehouses altogether and handled stock valued at $15 million.[19] It did the buying for WSA under the general authority which the very first WSA Administrative Order gave to the staff of the Maritime Commission.[20] Although the WSA made many kinds of contracts for services, including contracts for repairs in which the contractor supplied the materials, it did no direct buying of materials except through the Procurement or Purchase and Supply Division of the Maritime Commission.[21]

Partly for this reason, the District Offices of the WSA were in a certain sense part of the joint services. Quite distinct from the Regional Construction Offices, the District Offices existed to handle the operation, maintenance, and repair of ships. The Commission's Division of Maintenance and Repair, like its Operations Division and Training Division, had been almost entirely transferred to the WSA, but even so the Maritime Commission personnel assigned to the District Offices in June 1944 were: 590 to the New York office, 108 to the Baltimore office, 119 to the Norfolk office, 195 to the New Orleans office, and 377 to the San Francisco office. The Maritime Commission personnel under the District

---

[18] Quarg to Slattery, Aug. 13, 1946, in Minutes Section, attachments Oct. 31, 1946.

[19] Director, Procurement Division to Secretary, Annual Report, Fiscal Year 1943, Nov. 5, 1943, in Historian's Collection, Cabinet 16.

[20] Buying by the Division of Purchase and Supply on behalf of WSA is assumed also in WSA Admin. Order No. 6, Feb. 25, 1942. See also MC Minutes, pp. 27162-64 (Dec. 28, 1943) and on transfer of WSA warehouse personnel to MC, see Minutes, p. 28313 (May 2, 1944).

[21] Statements by John G. Conkey and Harold E. Steffes, Aug. 13 and 17, 1948.

Director in New York came from the following divisions: 205 from the Procurement Division, 319 from the Division of Finance, 2 from the Division of Economics and Statistics, 54 from the Technical Division, and 10 from the Production Division.[22] As in the regional construction offices, this personnel was administratively under the District Director, and functionally responsible to the division directors in Washington.[23]

The double system of decentralization—Regional Offices for Construction and District Offices for WSA and some joint functions —led to situations which seem confusing. At San Francisco the procurement officials in that District Office bought all office supplies for the Regional Office located across the Bay at Oakland.[24] Sometimes the overlap contributed to the confusion in the accounts. Urgently needed repairs were made by drawing if necessary on the material in one of the construction yards;[25] and while it was covered by a removal notice, not all of these notices, as has been explained, got recorded in the accounts of the two agencies. That example indicates also the advantage of having the Procurement Division act as a joint service. In waging war, to secure rapidly the parts needed to keep ships operating was more important than accurate accounting.

Some other portions of the Maritime Commission were not so closely connected with the WSA's mission to keep ships moving, and yet were as much concerned with operations as with ship construction. Since the WSA was purely a wartime agency, the Maritime Commission continued to exercise the regulatory functions given it by the Merchant Marine Act of 1936. The Regulations Division continued to investigate and pass on rate structures; the Examining Division continued to study operating subsidies, although its activity was much reduced;[26] and the Director of Finance and the General Counsel still kept an eye on the Reserve Funds of the operating companies.[27]

[22] Both sets of figures are taken from the Annual Report for fiscal 1944, condensed from personnel chart by L. V. Holt, Sept. 29, 1944. See Historian's Collection Research file, 116.2 and Cabinet 16.
[23] Statement by H. E. Steffes, Aug. 17, 1948.
[24] *Ibid.*
[25] Statement by John G. Conkey, Aug. 13, 1948.
[26] MC, *Report to Congress*, for the period ending June 30, 1943, pp. 31-35, 43-51; MC, *Report to Congress* for the period ending June 30, 1944, pp. 15-25.
[27] Memoranda on these subjects appear in Anderson reading file, and after the middle of 1943 postwar problems fill about as much of the file as does war construction.

A larger unit was the Division of Economics and Statistics. This division was one of the most important in the Maritime Commission before the war—one of those seven which had more than 100 employees in the Washington office. Its main function was providing the facts on the basis of which the Commission determined what were essential trade routes and what level of subsidies was justified under the Merchant Marine Act of 1936, but in addition it made statistical reports as need arose on various other topics such as the number of shipways active or available and the prospective labor supply. During the war its research in regard to the bases for subsidies ceased to be of immediate importance, and for a while it expanded its work on special statistical reports. In no case did the division have direct access, however, to the sources of statistical data; it worked with material received from the Customs Bureau, or the American Bureau of Shipping, or the Bureau of Labor Statistics, or divisions within the Commission. With the growth of the divisions more closely connected with execution of policy, such as the Production Division (MC) or the portion of WSA controlling ship movements, these operating divisions developed their own statistical staffs, which were on the one hand closer to the sources of the statistical data and on the other hand more closely in touch with the purposes for which statistics were needed. In 1941 the Division of Economics and Statistics made reports on shipbuilding costs, ship production, labor requirements and wage rates; in 1943 regular reports on these costs were prepared in the Finance Division, on production by the Statistical Analysis Branch of the Production Division, and on wage rates both in the Division of Shipyard Labor Relations and the Finance Division. Similarly in regard to the size and movement of the merchant fleet, in 1943 a Division of Statistics and Research in WSA under the Assistant Deputy Administrator for Ship Control was compiling day by day more complete and up-to-the-moment reports than had ever been needed before. Although the Division of Economics and Statistics continued to make some special studies in 1943 on current problems, it turned more and more to analysis of postwar policy. Throughout, it kept card files recording so far as possible the basic facts about every ship in the world of 1,000 gross tons or over, and made overall surveys of the world fleet. Its varied activities before the war prepared it to be part of the joint services during some months

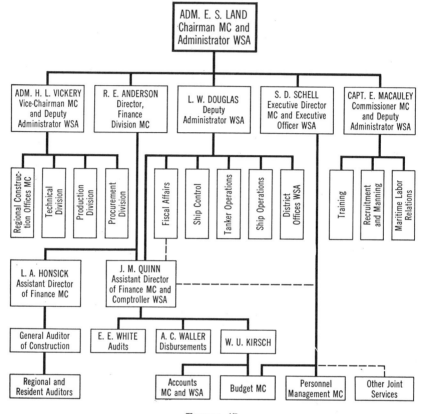

FIGURE 47.

in 1942; but it focused later on the long-range functions of the Maritime Commission.[28]

From this survey of their interconnections in 1943 it is clear that Admiral Land had neither wholly succeeded nor wholly failed to keep the WSA within the Maritime Commission circle.

[28] This analysis is based on the facts collected by Mrs. Marie D. Werner under the supervision of Hymen E. Cohen in her typescript, " An Administrative Study of the Division of Economics and Statistics of the U. S. Maritime Commission, 1941-1946," in Historian's Collection, Research file, 106 MC 17, and on an interview with the director of the division during most of the war, Henry L. Deimel.

Figure 47 is an attempt to show the realities in the most vital administrative relationships of the two agencies so far as realities of administration can be shown on a small chart. It fails to indicate the importance of the legal staffs, which advised directly the heads of both agencies. It diagrams clearly, however, the overlapping of supervisory authorities in regard to the units which caused most difficulty, those handling accounts.

III. REORGANIZATIONS IN THE FINANCE DIVISION

As the chart shows, there were two or three lines of authority over many individuals administering the finances. Mr. Quinn for example, acted in a dual capacity: as Assistant Director of Finance (Operations) he was an official of the Martime Commission; as Deputy Director of Fiscal Affairs and later as Comptroller he was an official of the WSA. He was also under the coordinating supervision exercised by Mr. Schell either as Executive Director of the Maritime Commission or as Executive Officer of WSA. Mr. Quinn's position had awkward possibilities. He declared himself ready to recommend the separation of the finances of the two agencies.[1] Attracted by offers from a shipping company, he resigned but later withdrew his resignation.[2] Throughout the war he retained his double title and capacity, an official in both the WSA and the Maritime Commission.

In WSA there was a certain amount of overlapping between Mr. Quinn and Percy Chubb, the Assistant Deputy Administrator (WSA) for Fiscal Affairs, although Mr. Chubb was mainly concerned with insurance, charters, requisitioning ships, and determining various fiscal policies.[3] When Mr. Quinn was Deputy Director of Fiscal Affairs, he was under the Assistant Deputy Administrator,[4] but after he was made Comptroller, he seems to have been responsible either to the Deputy Administrators or the Administrator.[5]

[1] Statements by Joseph M. Quinn, Aug. 12, 1947.
[2] On his resignation Quinn to Land, Mar. 3, 1943 and Land to Quinn, Mar. 23, 1943, in Personnel Division files and statements to me by R. Earle Anderson.
[3] Memorandum by Percy Chubb, June 14, 1944 with Chubb to Schell, June 15, 1944, in MC gf 201-2-5.
[4] WSA Admin. Order No. 13, Apr. 7, 1942.
[5] WSA Admin. Order No. 13, Suppl. 4, Feb. 18, 1943, creating him Comptroller did not say to whom he was to report. Suppl. 4-A to the same order, Feb. 5, 1944, says he is responsible " to the Administrator and Deputy Administrators " for his

His immediate superior in the Maritime Commission, Mr. Anderson, had no post in WSA, being a Maritime Commission official exclusively. Mr. Anderson's large role in the shipbuilding program as business adviser of the Commission did not prevent him from continuing to take an intense interest in ship-operating companies and the long-range development of the American merchant marine, but he gave his personal attention to the negotiation of contracts and the finances of the companies with which the Commission was doing business more than to auditing and accounting.[6] By early training Mr. Anderson was a naval architect. His Assistant Directors and Section Chiefs were men who had been dealing with audits and accounts all their lives, and they were all employees carried over from the Shipping Board Bureau to the Maritime Commission before Mr. Anderson became the Division Director.[7] To coordinate their activities, he did not depend on office memoranda but held staff meetings twice a week at which about sixteen persons attended.[8] He occasionally questioned the rulings regarding the auditing of construction contracts and was active on the Washington Construction Cost Committee,[9] but on the whole his role in the Commission was more like that of a company's treasurer than that of a comptroller.[10] His basic interests were much the same as those of Admiral Land with whom he had been friends since the years 1911-1913 when they both worked in the Bureau of Construction and Repair of the Navy.

Mr. Schell, on the other hand, was thoroughly familiar with all the details of auditing voyage accounts since he had worked at it years before, side by side with many of the men who were Mr. Anderson's Section Chiefs. Having been promoted through various

duties. The WSA organization chart for the administrative budget of 1944 shows him reporting to Mr. Douglas as Deputy Administrator and to Mr. Schell as Executive Officer.

[6] This general conclusion is based on study of his reading file and interviews with many members of his Division.

[7] Biographies furnished by Division of Information or Personnel Management Division.

[8] Anderson to Quinn, Honsick, Kirsch, and Haines, June 4, 1941, in Anderson reading file; and statement of Miss Sprott, Aug. 5, 1947.

[9] Anderson to Slattery, Mar. 29, 1943, Anderson to Honsick, May 18, 1943 and May 31, 1943, in Anderson reading file.

[10] This distinction, which he made in regard to his Regional Assistants (see chap. 14, sect. II, note 39, Anderson to Flesher, Aug. 11, 1942) seems to me to apply to Anderson himself.

positions involving ship operations, traffic, and accounting, he had become Chairman of the Board of Trustees and President of the U. S. Shipping Board—Merchant Fleet Corporation in the years before the Maritime Commission was created.[11] He therefore knew in detail better than did Mr. Anderson the work of those sections of the Finance Division which were under Mr. Quinn. As Executive Director, his duties regarding the budget required that he work directly with Mr. Kirsch, and in all financial sections his influence was greater than it was over the sections which reported to the Commission through Admiral Vickery.[12] Like Admiral Land and Mr. Quinn, he was both in the Maritime Commission and the WSA, and could act in a dual capacity in reorganizing the portion of the Joint Service Organization which was in the Finance Division.

A continual process of reorganization occurred in the first eight months of 1944. Three sections had been established under Mr. Quinn at the time of the decentralization in 1942—namely, Audit and Analysis for ship operations under E. E. White, Disbursements under A. C. Waller, and Budget and Accounts under W. U. Kirsch. On January 25, 1944 an administrative order called for the division of the Budget and Accounts Section. Mr. Kirsch was to remain in charge of budgeting only, and in that capacity was to report to Mr. Schell quite as much as to Mr. Anderson. The accounting branches of his old section were to be combined with the Disbursements to form a new Accounts Section headed by Mr. Waller. In short, bookkeeping and accounting functions were transferred from Mr. Kirsch to Mr. Waller, who was given the title of General Accountant. He reported as before to Mr. Quinn who signed as Comptroller.[13] The reason for this change was that the preparation of budgets was growing more difficult, and the Bureau of the Budget was complaining at the delays in their preparation.[14]

Creation of the new Accounts Section required separation of the personnel in the section which Mr. Kirsch had headed ever

---

[11] Statements by S. D. Schell, Aug. 13, 1948, and A. C. Waller, Aug. 3, 1948.

[12] This is my impression from talking to many of the persons concerned.

[13] MC Admin. Order No. 37, Suppls. 95 and 55A, and Executive Director and Director of Finance to MC, approved Jan. 25, 1944, MC Minutes.

[14] Statements by Mr. Quinn, Aug. 12, 1947. The 1945 budget was not ready to go to the Bureau of the Budget until Sept. 25 or 28, although asked for June 24, 1943. See Cohen, " Budgeting Process."

since the formation of the Commission. The separation was not completed before May at least, five months after the date of the administrative order.[15] In fact, Mr. Waller never did get his accounting staff fully organized;[16] other reorganizations were in the wind.

Mr. Douglas had been moving in the direction of separating more and more completely the staffs of the Maritime Commission and the WSA. The appointment of a separate Administrative Officer for WSA in November 1942 and the creation of a separate Personnel Management Division in February 1943 were followed in November 1943 by the appointment of a separate Budget Officer for Operations who reported to the Assistant Deputy Administrator for Fiscal Affairs, at that time Percy Chubb.[17] In November 1943 also, a survey of the whole financial administration of WSA was made for Mr. Douglas by Eric L. Kohler, an eminent certified public accountant who had won wide praise, including that of the General Accounting Office, by straightening out the accounting of the Tennessee Valley Authority.[18] The report which he submitted December 14, 1943 was highly critical, and was most severe on the Budget and Accounts Section. He recommended immediate separation of the fiscal and accounting staffs of the Maritime Commission and the WSA in order that responsibility be more clearly defined, and he urged that an outsider be brought in and given full authority as Comptroller for the WSA.[19]

On receiving Mr. Kohler's report and after conferring with Mr. Quinn, Mr. Douglas wrote to Admiral Land:

I myself feel that we are living on top of a volcano and that unless we proceed speedily to cure this situation there may be a pretty devastating eruption.

Mr. Kohler's recommendations are substantially endorsed. The first step,

---

[15] House, *Accounting Practices, Hearings*, p. 422; General Accountant to Price Adjustment Board, May 9, 1944, and Budget Officer to Director of Personnel Supervision and Management, concerning division of personnel, May 29, 1944, both in gf 201-2-5; interview with Mr. Quinn, Aug. 12, 1947.

[16] Statements by Mr. Quinn, Aug. 12 and 13, 1947, by A. C. Waller, Aug. 1948, and C. W. Levis, Aug. 18, 1948.

[17] Norman L. Johnson to Percy Chubb, Jan. 17, 1944, in Historian's Collection, Research file 106 WSA 82, from Richards file.

[18] House, *Independent Offices Appropriation Bill for 1948, Hearings*, pt. 1, p. 82.

[19] House, *Accounting Practices, Hearings*, pp. 410-15.

however, is to employ the ablest man we can find. I have two names that might fit this specification.

I am cautious about having the attached documents read because of the gossip they would give rise to and the feelings they would prematurely hurt, if too widely known. I have therefore limited their circulation to those who have zippers on their mouths.[20]

Mr. Kohler's report was not officially laid before the Maritime Commission, although its criticism, while stressing the part related to WSA, applied to the accounting being done for both agencies, and it is reasonable to assume that its content was known to many through the " grapevine." [21]

As a result of Mr. Kohler's report,[22] another accountant from the Tennessee Valley Authority, J. F. Stone, was brought in and " charged with the establishment and administration of the Division of Budget and Accounts of the War Shipping Administration." Pending the establishment of such Division," so read the WSA Administrative Order of February 5, 1944, " Mr. J. F. Stone will direct the activities of the Budget Office and Accounts Section, U. S. Maritime Commission, insofar as they relate to the activities of the War Shipping Administration." [23] Mr. Stone promptly made a report concerning the way he hoped to have the accounts organized,[24] but a half-year of uncertainty followed. The new division envisaged in February was not formally created until April 1, 1944,[25] and the assignment of personnel to it took more time.[26] Thus it was that while Mr. Waller was still trying to organize under the MC Administrative Order of January, Mr. Stone was trying to organize under the WSA Administrative Order of February, and both were finding it difficult to obtain from Mr. Kirsch's old section of Budget and Accounts the personnel they wanted, especially since there was constant complaint

---

[20] L. W. Douglas to Admiral Land, Jan. 17, 1944, in Douglas file, " Fiscal." Published in *Supplement to Preliminary Report on Comptroller General's Audit Reports for Fiscal Years 1943 and 1944 and the Replies Thereto by the Maritime Commission and the War Shipping Administration* (Prepared for the Merchant Marine and Fisheries Committee, House of Representatives).

[21] Statement by Admiral Land, June 1949.

[22] Statement by J. M. Quinn, Aug. 12, 1947.

[23] No. 13, Suppl. 4-A.

[24] J. F. Stone to Amyas Ames, Feb. 23, 1944, in House, *Accounting Practices, Hearings*, pp. 442-43.

[25] WSA Admin. Order No. 58, Suppl. 1, Apr. 1, 1944.

[26] Comptroller (written by Harford) to J. F. Stone, Special Assistant to Administrator, Apr. 1, 1944, in gf 201-2-5, and above, n. 15.

from all sides that qualified personnel for fiscal administration was inadequate.

Meanwhile the trend toward the separation of the WSA from the Maritime Commission was reversed. On March 3, 1944 Lewis Douglas resigned, effective April 1, as Deputy Administrator in WSA, giving as his reasons that the main task of organizing WSA had been achieved and that he could ignore no longer the doctor's orders that he take care of his sinus headaches.[27] He was succeeded as Deputy Administrator by Captain Granville Conway who had been Director, Atlantic Coast District, for the Maritime Commission and WSA before and during the war. Captain Conway had done outstanding work in directing shipping from New York and was regarded by Administrator Land as his " Number one adviser." [28]

After Mr. Douglas' resignation, his Administrative Officer, Amyas Ames, departed also; the WSA Personnel Management Division was joined to that of the Maritime Commission May 8, 1944;[29] and on July 6, 1944 Mr. Schell was designated Executive Deputy Administrator of WSA,[30] a change of title indicating that his position in WSA was thereafter to be more like what it had always been in the Maritime Commission.

In accordance with the new trend, the complete separation of the accounting of the two agencies was not carried out. A report which Mr. Stone submitted on May 3 calling for completion of the separation led to two new surveys of the situation—for Admiral Land by Mr. Anderson, and for Captain Conway, by a shipping man from outside the Commission, C. D. Gibbons.[31] Mr. Quinn was willing to recommend separation but he objected to parts of Stone's proposals which trespassed on his field, the auditing of voyage accounts.[32] Percy Chubb, Assistant Deputy Administrator for Fiscal Affairs, wrote a long memorandum of June 14, 1944 objecting to the existing confusion of responsibilities

---

[27] His letter of resignation, the original, dated Mar. 3, 1944, and copies of Admiral Land's acceptance, dated Mar. 24, and expressing warm appreciation for what Mr. Douglas had accomplished, are in the Land file in MC Historian's Collection.

[28] Land to F. C. Lane, June 8, 1949.

[29] MC Minutes, pp. 28398-99 (May 8, 1944). Mario P. Canaipi, Director of the WSA Division became Director of the MC Division.

[30] WSA Admin. Order No. 2, Suppl. 23.

[31] Statements by Mr. Quinn, Aug. 12 and 13, 1947, Mr. Waller, Aug. 1948, and Mr. Levis, Aug. 18, 1948.

[32] Interview with Mr. Quinn, Aug. 12, 1947.

between his organization and that under the Comptroller and asking that they be clearly defined.[33]

Mr. Gibbons, who had been with the Shipping Board Bureau and in the Finance Division of the Maritime Commission from 1922 to 1936,[34] summed up the argument against separation by saying that two separate divisions would require larger staff than the Joint Service, and added: " Many of the key auditing and accounting employees have had long service and they naturally feel that their futures with the USMC are more secure than with the WSA, a war emergency organization." If a separation was made, at a future reunion of WSA and Maritime Commission there would be severe personnel problems. Existing difficulties could more easily be cured by improving the organization of the Joint Service than by its separation, he believed.[35]

Action was taken August 4, 1944. Mr. Stone's position was made purely advisory and an Administrative Order drafted by Mr. Schell redefined the powers of Mr. Quinn as Comptroller.[36] Disbursements was now linked to Audits in a new section headed by Mr. Waller with the title of General Auditor. Accounts was to be a section by itself under Mr. White who had been previously the head of Auditing and Financial Analysis. By this reorganization of August 4, 1944, the Joint Service in regard to disbursing and accounting was firmly re-established and was not thereafter seriously threatened with division between Maritime Commission and WSA.

The reorganization shelved Mr. Stone, and he addressed to Admiral Land a parting blast full of " vitriolic " comment.[37] He not only criticized the financial statements but accused the persons charged with their preparation of having " no conception of their responsibilities." Of the accounting in general Mr. Stone said:

[33] Under June 14, 1944 in MC gf 201-2-5.
[34] Information from Personnel Division.
[35] I have read Mr. Gibbons' report in the copy given me by Mr. Quinn. I have not seen Mr. Anderson's, nor Mr. Stone's letter of May 3.
[36] Copies of WSA Admin. Order No. 64 dictated by Mr. Schell are in MC gf 201-2-5. I have not found this new organization of the Finance Division in the Admin. Orders of the Maritime Commission, but Suppl. 1 to WSA Admin. Order No. 64 named J. H. Yates Acting Assistant Accountant for the Maritime Commission, so it was probably thought of as a Joint Order. Mr. Stone was relieved of accounting responsibilities when WSA Admin. Order No. 2, Suppl. 24, made him a special assistant to the Administrator. See also Suppl. 25.
[37] House, *Accounting Practices, Hearings*, p. 424.

The facts are that known and ascertainable domestic transactions were not recorded, accounts receivable running into millions of dollars were not recorded or billed, accounts payable and commitment records were non-existent, the amount of simple administrative expenses had not been determined . . . No procedure exists for the systematic recording and billing of accounts receivable.[38]

Mr. Stone's letter was sent by Mr. Schell to Mr. Quinn who sent it back, saying he was not in that period responsible,[39] and then Mr. Schell sent it to Mr. Kirsch who while writing a reply himself sent it to the subordinates most immediately concerned and asked for detailed replies. The replies of Mr. Kirsch and of those under him, such as replied, stated that Mr. Stone never understood the system of accounts used in the Commission, that they were not misleading to those for whom they were prepared, who did understand them, and that Mr. Stone's own attempts to handle the accounting problems had been quite inferior. It was admitted that the recording of vouchers continually fell behind, but Mr. Kirsch vigorously maintained: "Everything that was humanly possible was done in the Accounting Section to bring the accounting to a more current basis, but the ever-increasing shipbuilding program, ever-increasing numbers of ships in operation, with no comparable increase in personnel, has made the job exceedingly difficult." [40]

Mr. Stone's criticisms did not lead to any further changes during 1944 in the organization of the Finance Division. But the August reorganization was not formally approved by the Maritime Commission until December 1945,[41] nor was it entirely carried out as Mr. Schell intended.[42] Mr. White considered the assignment of accounts to him as temporary,[43] but he organized the Maritime Commission accounting under Mr. Yates as General Accountant for Maritime Commission, and the WSA accounting under Mr. Hicks, who had been brought in by Mr. Stone and became General Accountant for the WSA.[44] They worked in

[38] *Ibid.*, pp. 416, 422.
[39] *Ibid.*, p. 422, and interview with Mr. Quinn, Aug. 12, 1947.
[40] Kirsch to Schell, Sept. 13, 1944, in House, *Accounting Practices, Hearings*, pp. 424-25.
[41] MC Minutes, pp. 34204-05 (Dec. 27, 1945).
[42] Statements by Mr. Schell, Aug. 13, 1948, and inference from the description of the organization given me by Mr. Levis.
[43] Statements by E. E. White, July 28, 1948.
[44] *Ibid.* Organization Chart of Mar. 19, 1945, and explanation of it by Mr. Levis, Aug. 16, 1948.

the Y building and Mr. White made trips over there in the shuttle bus from the Commerce Building.[45] So matters proceeded for a year and a half, until after the end of the war and a change, to be described shortly, in the membership of the Commission.

How this story of the reorganizations in the Finance Division makes more "understandable" what the Congressional investigating committee later called the "failure to take adequate measures . . . to rectify a known condition" may be left to the reader's judgment with the hope that the complexity of the narrative has not hidden general aspects of the situation. The men at the top of the Maritime Commission and War Shipping Administration all had much on their minds other than accounting. The pressure of having many weighty problems to solve immediately is impossible to measure, as is also the concern of this or that individual with his power or bureaucratic position. Everyone concerned offered solutions he was unable to have carried out, proposals which got lost. There were more of these than I have mentioned. Admiral Land made several serious efforts to obtain an expert accountant-paymaster from the Navy, but the Navy advised that none was available. No able outsider was secured and given a clear field. Admiral Land was inclined to discount the critics because he considered them candidates for the top job in a new set-up and because he distrusted the accounting of the TVA, the agency from which they came. After serious consideration and many consultations, Admiral Land determined "not to change horses in the middle of the stream." [46]

IV. COMMISSION ACTION

In the preceding history the words "Maritime Commission" have often been used to refer to the whole body of officials in that agency or any pertinent portion thereof. The stricter usage, which would apply the term only to the board of five men, each of whom had an equal vote according to the Merchant Marine Act of 1936, will be followed in this section, which will examine the relation of that board as a board to the activities of the staff and of the individual Commissioners.

---

[45] Statements by E. E. White, July 28, 1948.
[46] Land to F. C. Lane, June 8, 1949. Also, Anderson to Land, Aug. 14, 1944, and statements by Land and Schell.

Throughout the war the Commission met regularly on Tuesday, Thursday, and sometimes Friday morning, assembled frequently for special meetings, and took many actions "by notation," that is by circulating memoranda to the offices of the individual Commissioners in order to obtain their approval and then record their action in the official Minutes as action by notation at a special meeting. The regular meeting held twice a week usually lasted about two hours and approved a score or more of memoranda submitted by the staff. In the official minutes there are almost no instances of a majority of "nay"; when Commissioners objected to recommendations presented, the matter was generally "carried over" to a subsequent meeting, and if the objection was serious and sustained the recommendation was reshaped or withdrawn without ever coming to a vote. The agenda or "docket" for each meeting was placed in the hands of each Commissioner in advance, usually the day before. Most matters on the docket were routine—indeed the official minutes contain pages of details which it may have been useful to record there as a matter of formal record, but which could not have received personal attention from the Commissioners.[1] This does not mean that the Commission meeting as a whole had become a mere formality, however. Although the bulk of the "docket" was approved unanimously with little discussion, Commissioners who found in perusing the agenda proposals which they thought required examination often called in the appropriate members of the Commission's staff to clear up the points about which they were in doubt. If they wished, they could bring up the matter in the Commission meeting, and see that action was not taken without serious consideration.

The membership of the Commission remained unchanged from December 13, 1941, when Commissioner John M. Carmody took the oath of office,[2] until September 25, 1945, when Commissioner Thomas M. Woodward's second term expired.[3] The terms of Chairman Land, Commissioner Vickery, and Commissioner Macauley expired during the war but they were all reappointed for six-year terms.[4]

[1] This general picture of procedure is based on my reading of the Minutes for 1941-45, and confirmed by interview with A. J. Williams, the Secretary.

[2] He was appointed Nov. 19, 1941 to fill the unexpired term of John J. Dempsey who resigned effective July 7, 1941.

[3] He was reappointed after his first term of office expired Sept. 25, 1939 and, after Senate confirmation, took the oath of office for his second term on May 8, 1940.

[4] Admiral Land's first term expired Apr. 15, 1943, his nomination for a second

Although commissions generally act more slowly than a single administrator, I have found no evidence that the need of obtaining approval from this five-man board slowed down the action of the agency in carrying out the emergency shipbuilding program. Delays occurred of course, but they occurred in the administrative process in the divisions or regional offices, rather than in securing Commission approval. The delays in defining exactly the functions on the borderlines between the regional offices, the Technical Division, the Production Division, and the Finance Division were not the result of any need for bringing these matters before the Commission for settlement. If Admirals Land and Vickery could have arrived at solutions of these difficult problems, their decisions would have been recognized by the staff as authoritative.

The ability of the Commission to conclude its own business with speed was due in large part to the specialization of functions among the Commissioners and the general willingness of the Commissioners, prior to 1945, to follow on vital matters the advice of the Commissioner chiefly concerned. Nearly all memoranda on the Commission's docket had been approved previously by an appropriate division head, and many by a Commissioner, or a committee of division heads and Commissioners.

Actions vitally affecting the production of ships normally had the prior approval of Admiral Vickery who conferred regarding important questions with Admiral Land. They worked closely together and their recommendations were followed by the other Commissioners partly because of admiration for the task Admiral Vickery was carrying through and partly because it was known that he was acting under directions from the President, the Joint Chiefs of Staff, or the WPB, and that President Roosevelt wished the Commission to be governed in these matters by the judgment of Admirals Land and Vickery. Although the Merchant Marine Act of 1936 which created the Commission had given it powers of action that did not depend on Presidential approval, many of the powers exercised by the Commission during the war were dele-

term was confirmed by the Senate Mar. 3, 1943 and he took his new oath of office Apr. 15, 1943. Admiral Vickery, who had been appointed to fill the unexpired term of Edward C. Moran, Jr., and had taken his first oath of office Sept. 26, 1940, was reappointed Apr. 10, 1942 and took a new oath of office May 7, 1942. Captain Macauley, who was appointed to fill the unexpired term of Max O'Rell Truitt and entered on duty Apr. 3, 1941, was reappointed and confirmed by the Senate Nov. 21, 1944—Information supplied by MC Personnel Division.

gated to it by the President out of his war powers. He could at any time have transferred the supervision of the whole shipbuilding program from the Maritime Commission to the WSA—making use of the First War Powers Act under which the WSA was created.[5] The fact that Admiral Vickery was already Deputy Administrator for Ship Construction in WSA was a patent reminder of this possibility. If the Commissioners had blocked action which Admirals Land and Vickery thought essential, and they had appealed to President Roosevelt, he could have given Admirals Land and Vickery authority to run the shipbuilding program through the WSA without consulting the Commission. But the Commission never gave occasion for such action.

Admiral Vickery stated more than once his feeling that the President personally had made him responsible for the production of ships. The special responsibility of Admiral Vickery was recognized by the Commission in February 1942,[6] in electing him Vice Chairman (in place of Commissioner Woodward who had served longer on the Commission). When he was re-elected the following year, Admiral Land as Chairman sent the following memorandum to the secretary to be recorded in the minutes: "On March 19, 1943 I reported to the Commission that I had taken up with the President the question of continuing Rear Admiral Howard L. Vickery as Vice Chairman of the Commission. The President directed that Vickery be continued as Vice Chairman ' during good behavior.' "[7] I can find no authority for such Presidential action in the Merchant Marine Act for 1936, which provided only: "The President shall designate the member to act as chairman of the Commission, and the Commission may elect one of its members as vice chairman."[8] At the subsequent re-elections of Admiral Vickery as vice chairman in 1944 and 1945, also by unanimous vote, there is no record of Presidential wishes or "direction."[9] But so long as the Commission was subject to the President's war powers, the roles of individual Commissioners were vitally influenced, directly or indirectly, by the President's will.

In regard to many functions the individual Commissioners

[5] *The United States at War*, pp. 104-05.
[6] See above chap. 5, sect. IV.
[7] MC Minutes, p. 24626 (Mar. 19, 1943).
[8] Section 201 (a).
[9] MC Minutes, p. 27491 (Feb. 7, 1944); p. 31043 (Feb. 15, 1945).

became links in the chain of command leading down from the President. Approval by the Commission became only a formality in regard to such matters as the use of yards for the construction of military vessels. For example, after Kaiser-Vancouver had been awarded by the Commission a contract for 117 Liberty ships, work on that contract was stopped June 18, 1942 in order that the yard could build LST's and carrier escorts, although the Commission did not act to cancel the contract for Libertys until February 17, 1944.[10] Later, West coast yards were shifted from the construction of Victory cargo ships to the construction of Victory attack transports (vessels costing nearly twice as much) simply by letters from Admiral Vickery.[11] Many examples can be found in the events recorded in previous chapters, such as those concerning steel, in which the Commission's staff under Admirals Land and Vickery made decisions as part of the chain of command established for industrial mobilization. Decisions concerning steel quotas, abolition of Sunday work, or the conduct of manpower surveys were not voted by the Commission, although Admiral Vickery discussed them with the other Commissioners.

In the chain of command passing through the Commission and through individual Commissioners, Admiral Land was the topmost link—practically because of his close personal relations with President Roosevelt, and officially, because of his dual capacity as Chairman of the Commission and as Administrator of War Shipping. His actions as Administrator did not require even the formal approval of the Commission. Many decisions involving the WSA were brought by him before the Commission for approval nevertheless.[12] It was part of his general policy of tying the two organizations closely together. In spite of having expressed dislike for the commission-form of government, he used it.

A special and significant instance is the way contracts for Maritime Commission shipbuilding were renegotiated to eliminate excessive profits. The Renegotiation Act conferred the power of

---

[10] Fischer, *Statistical Summary*, p. 23.

[11] Vickery to California Shipbuilding Corp. and other companies, Dec. 13, 1943 in MC, Technical Division reading file. The change to AP5's was handled as if it were a design modification.

[12] Many administrative orders were issued jointly. The Administrator and the Commission acted jointly in the purchase of ships. Minutes, p. 22874 (Sept. 3, 1942). The decision on the Sword Line by the Administrator, WSA, was "Concurred in" by the Commission by vote. Minutes, pp. 31194-98 (Feb. 27, 1945).

renegotiating contracts, not on the Commission, but on the Chairman of the Maritime Commission.[13] Like the Undersecretaries of the War and Navy Departments, he set up a Price Adjustment Board to carry out the purposes of the act. The Price Adjustment Board of the Maritime Commission was created by action of the Commission, however, through an administrative order;[14] and Commissioner Woodward was named the Chairman of the Board. Originally there were two other members: one a representative of WPB, the other a representative of the Commission's Division of Finance, namely, Arthur G. Rydstrom. He was employed in May 1942 so that " he would be in a position to carry on most of the activities of that board and thus relieve the Chairman of much detail."[15] He resigned his position in March 1945.[16] Meanwhile, in 1944, John R. Paull had been added to the Price Adjustment Board and in September 1945, after Mr. Woodward's term as Commissioner expired, he became its Chairman.[17] In its investigations, the Board was authorized to call for help from the Commission's staff, especially from the Construction Review and Analysis Section of the Division of Finance, which the Commission designated to act as the Board's fact-finding agency.[18] The decisions of the Price Adjustment Board in renegotiation were submitted by Commissioner Woodward to the Maritime Commission for its formal approval.[19]

The way the Price Adjustment Board was organized illustrates not only Admiral Land's use of the Commission as a board of review but also his policy of fitting individual Commissioners into the chain of command by giving them specialized functions. The delegation might be made by the Commission, by Admiral Land formally or informally as its Chairman, or formally by Admiral Land as Administrator of War Shipping. Commissioner

---

[13] Section 403 of Title IV of the Sixth Supplemental National Defense Appropriation Act, 1942 (April 28, 1942). House, *Investigation of the Progress of the War Effort, Report No. 733* of the Committee on Naval Affairs, 78th Cong., 1st sess., pursuant to H. Res. 30, Oct. 7, 1943, pp. 2, 16, 20-21.

[14] MC Minutes, pp. 21828-29 (May 5, 1942); Adm. Order No. 64.

[15] MC Minutes p. 21831 (May 5, 1942); Anderson reading file, May 4, 1942. See also Adm. Order No. 64, Suppl. 1, Land to Nathan, May 19, 1942, and Anderson to Rydstrom, May 28, 1942 in Anderson reading file.

[16] MC Personnel Division files.

[17] Adm. Order No. 64, Suppl. 1-B, and information from Personnel Division.

[18] Adm. Order No. 64.

[19] MC Minutes, p. 24134 and *passim*.

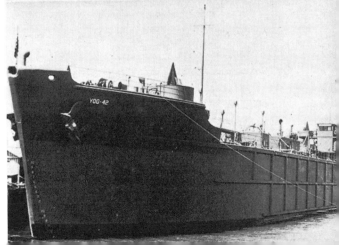

PLATE XXVII.

Above, a Navy attack cargo ship (S4-SE2-BE1) built by Walsh-Kaiser at Providence, R. I. Upper right, putting the finishing touches on the propeller for a tanker. Center, preassembly arriving by truck at the head of the ways at Marinship. Below, a concrete barge (B7-A2) used for oil and gas. Twenty-two of this type were built at National City, California, by Concrete Ship Constructors, the most successful builders of concrete barges.

PLATE XXVIII. Above, workers leaving Bethlehem Fairfield, Baltimore, largest of the emergency yards, with 16-ways. It built 30 tank-carrying landing ships, 94 Victorys, and 384 Libertys. Below, evening, a Liberty ship rides at anchor.

COMMISSION ACTION						779

Woodward was delegated special supervision not only over renegotiation but also over the screening operations of the Shipyard Site Planning Committee and over transportation problems.[20] Mr. Carmody, almost as soon as he joined the Commission, began countersigning recommendations for the acquisition of land by condemnation proceedings, and he supervised the work of the Solicitor, Mr. Page, after the Commission had reversed its earlier policy and was attempting to correct the disadvantageous position in which it had been placed by the leases connected with the early facilities contracts.[21] He was chairman of a number of committees including that which arranged for the selection of names for Liberty ships.[22] Captain Macauley spent most of his time on his duties as Deputy Administrator of WSA in charge of the recruitment, training, and labor relations of seamen and officers of the merchant marine. In his own sphere he made a most notable contribution, but the nature of Captain Macauley's special duties has caused him to receive far less mention in this history than he should receive in a general history of the Maritime Commission.[23] The staff commanded by Admiral Vickery was far larger than that directly under any other Commissioner, but each of the five had some special spheres of supervision, except Admiral Land, who headed the whole organization and spoke for it in its relations to other agencies, to Congress, and to the President.

In addition to their specialized functions, each Commissioner had the right and opportunity as a member of the Commission to exercise the function of reviewing the actions approved by other Commissioners and of registering dissent. The vast majority of Commission actions were unanimous or without recorded

[20] See above chaps. 5, 13, 15.

[21] The Commission minutes of the spring months, 1943 are full of recommendations concerning real estate countersigned by Carmody. On facilities contracts, see chap. 4.

[22] Cohen, "Commissioner John Michael Carmody."

[23] The Senate approved its committee's recommendation of Mar. 2, 1945 that Captain Macauley be raised to the rank of Rear Admiral, in spite of his being on the retired list, because of his extraordinary service in connection with training and manning. (Senate, *Report No. 80* on S. Res. 646 before the Committee on Military Affairs, 79th Cong. 1st sess.) When the resolution came before the House committee, Admiral Land testified, "So far as merit is concerned, I heartily endorse this," and the committee reported favorably, May 14, 1945.—House, *Hearings on S. Res. 646* before the Committee on Naval Affairs, 79th Cong., 1st sess., May 11, 1945; *Report No. 567*, May 14, 1945. But the House did not complete action.

dissent, but there were lively debates not recorded in the minutes.[24] Until 1945, "nay" votes were cast much more frequently by Commissioner Woodward than by any other member of the Commission, although other Commissioners also, even Admiral Vickery, occasionally voted "nay" on some particular measure approved by the majority. Commissioner Woodward's dissents concerned mainly financial arrangements with the shipbuilding companies, such as V-loans, advance payments of fees, and real estate leases—particular instances in which he felt the terms were too favorable to the companies. Commissioner Carmody registered his dissatisfaction sometimes by a "nay" vote but more frequently by sending memoranda to the Chairman or memoranda to the Commission, which were entered in its minutes merely as having been "noted." But prior to 1945 the minutes record no regular pattern of division between "yeas" and "nays."

Many of the activities of the two civilian Commissioners as well as those of the two Admirals have been described in various connections in the preceding chapters. To judge from the minutes, more time was spent in the meetings of the Commission in 1943 on such matters as housing, transportation, and in-plant feeding than on the award of contracts or the formulation of shipbuilding programs. This was a natural result of the fact that ship construction programs were being determined on technical grounds and in consultation with the Joint Chiefs of Staff, whereas in handling the problems of the shipyard communities the Commission had more freedom of action and the civilian Commissioners had more knowledge and more part in formulating policy.

Some other matters in which the civilian Commissioners were particularly active deserve attention. In the spring of 1942 they investigated the freight rates and other charges which the Commission was paying to the railroads for shipment of Commission-furnished material, primarily steel. They backed the work of Mr. Geoghegan, the Commission's transportation expert, in checking on railroad charges and in securing "land-grant" rates, which saved the Commission millions.[25] At the same time an investigation of how much the yards were paying for electric

[24] MC Minutes and interviews with the members and secretary of the Commission.
[25] Carmody to Land, Mar. 14, 1942 and Carmody to the General Counsel, June 25, 1942, both in Carmody reading file; MC Minutes, pp. 22345-47 (June 30, 1942); Cohen, "Commissioner John Michael Carmody," pp. 27-29.

power was initiated by Commissioner Carmody, who had been head of the Rural Electrification Administration in 1936-1939.[26] Admiral Land referred to Mr. Carmody in November 4, 1942 as the Power Procurement Officer of the Maritime Commission.[27]

In these and other activities Mr. Carmody revealed a spirit of discontent and distrust from which was to develop an important split in the Commission. In August 1943 Mr. Carmody wrote:

> When I came here some eighteen months ago I discovered no one was paying much attention to power contracts made by shipyards, in spite of the fact that all such costs either for installation or for energy are reimbursable. Contracts were made either by the contractors themselves or by our resident plant engineers, who knew little about this specialized phase of engineering. I persuaded the Maritime Commission to employ special engineering counsel to review these contracts and the rate schedules. Months have passed and the only tangible evidence I have seen of really squeezing the sugar out of these contracts is what has been done in the Walsh-Kaiser case in Providence R.I. . . . We can't expect much help from our contractors. Most of them haven't got a nickle invested in these yards. They are interested chiefly in their fees and in building themselves up in the public mind as great producers. We pay the bill.[28]

Commissioner Carmody's attitude, not only towards power rates but in general in regard to the shipbuilding costs, appears also in a telegram he sent the president of the New England Shipbuilding Corp.

> You don't pay these power costs. We don't pay these power costs. These abnormally high power costs find their way directly into the cost of ships. Every man and woman and child in Maine who pays income taxes or buys war bonds or stamps helps pay these costs. This includes shipyard workers in the very shipyards where these high cost are fastened on your operations. Frankly, I think power company officials are fully aware of need for readjustment but are stalling for time hoping we will tire out. We won't tire out. . . . [29]

Persistently, with Commissioner Carmody's assistant, Ward B. Freeman, keeping the matter in hand, power charges were reduced in one company after another. On October 19, 1945 Commis-

---

[26] Carmody to Chairman and other Commissioners, Apr. 6, 1942 in Carmody reading file.
[27] Cohen, " Commissioner John Michael Carmody," pp. 74 ff. Carmody's recommendations for surveying power rates were approved by the Commission, Sept. 22, 1942 and Nov. 24, 1942.
[28] Carmody to Basel Manley, Aug. 10, 1943 in Carmody reading file.
[29] Carmody to C. L. Churchill, Aug. 13, 1943 in Carmody reading file.

sioner Carmody reported total savings in electric power costs of $2,668,502.[30]

Commissioner Carmody's zeal for cutting down costs and hitting at what he felt to be war profiteering is shown also in his concern with the large amounts the shipyards were paying for insurance. In this matter too he acted in collaboration with Commissioner Woodward. Premiums paid for casualty and other kinds of insurance were of course reimbursed by the Commission, and the shipbuilding companies claimed reimbursement also for whatever they paid to "insurance advisers." Commissioners Woodward and Carmody were both dissatisfied with the provisions regarding these matters made by the Commission's Division of Insurance.[31] The particular arrangement which the director of this division recommended for the Higgins yards so aroused Mr. Carmody that he wrote as follows to the head of a private casualty insurance company chiefly involved:

> I have no idea what action the Commission may take, but on the basis of tentative cost figures. . . . I gag at the fees I understand will be paid to agents or brokers under the guise of being insurance "advisers" to Higgins. I understand Higgins is recommending these "advisers." . . . Specifically I understand fees of these "advisers" may run, under the proposed contract, to $50,000 or $60,000 per year. For what? We don't allow any shipyard to pay even its president or its top "know how" executive over $25,000. . . . We are asking shipyard workers to work overtime and take pay for it in war bonds. Kids are buying war stamps with pennies. I know you can't run a war without waste as we understand the word in peacetime economy. I do think however, the public officials have no right to hide behind war needs to encourage wasteful extravagance. Frankly, if I thought I were a party to paying insurance agents or brokers under any name $50,000 or $60,000 per year for *any service* they can render the Commission in connection with insurance coverage in the midst of a war crisis that compels our President to work twenty hours a day for approximately the same salary, I would never be able to walk through the Higgins shipyard and hold my head up or look an honest workman in the eye.
>
> I'm told agents have great political influence and that in some states

---

[30] MC Minutes, p. 33561 (Oct. 23, 1945). The saving was divided as follows: $1,808,571 from reduction in charges for electric energy; $759,931 from "recoveries of monies paid by contractors, in the early construction days, for power facilities that should have been installed at their own expense by power companies supplying the energy." An earlier report, Feb. 22, 1944, is in Minutes, p. 27624. For more detail see Cohen, "Commissioner John Michael Carmody."

[31] MC Minutes, pp. 28548-51 (May 30, 1944). Much more on the subject may be found in the minutes under "Insurance" in its index.

Commissions, set up ostensibly to protect the public, compel the U. S. Government to pay these fees or else. . . . Do you wonder that I gag? . . .

I have no idea how my colleagues feel about these fees; I can only express my own views.[32]

When approval of insurance contracts came before the Commission, the Division of Insurance was in fact overruled more than once. Mr. Carmody gave much of the credit to "the persistence of Commissioner Woodward, supported by Commissioner Macauley and myself."[33] Insurance is too complicated a subject to examine thoroughly here, however. Most significant for the study of the Commissioners is the way in which Mr. Carmody's letter reveals his lack of confidence that all the Commissioners would react as he did against suspected profiteering.

Although Mr. Carmody's "nay" votes were not actually as numerous as Mr. Woodward's prior to 1945, he seems to have been more definitely out of sympathy with the basic policy being followed by Admirals Land and Vickery towards the shipbuilding companies. As between the two horns of the dilemma which arose in the management of management—between on the one hand extending bureaucratic controls even at the risk of splintering the responsibilities of management in a way that might decrease production, and on the other hand emphasizing management's responsibility but being unable to force the managements to economize—Commissioner Carmody wished to move toward more control. He hoped to do so without diminishing the vigor of private management, but he felt that Admirals Land and Vickery shared too closely the same point of view as that of the shipbuilding contractors. He felt that they were permitting excessive costs— partly because of an interest "in building themselves up in the public mind as great producers," but partly also because they and their chief assistants were naval architects who although very competent in their specialty were relatively naive and unwary in their handling of many of the new problems being piled on them, such as electric power rates, insurance, real-estate values, housing, training, and labor relations. From Mr. Carmody's own experience with such matters, both in his earlier business career and more recently in the "New Deal" agencies of the Roosevelt

---

[32] Carmody to Stewart McDonald, Chairman, Maryland Casualty Co., June 29, 1942, in Carmody reading file.

[33] Carmody to Paul Scharrenberg, Aug. 28, 1946, in Carmody reading file.

administration, he had acquired a different approach from that of the Navy constructors.

In regard to the administration of the Commission's own staff also, Commissioner Carmody felt dissatisfied. He took a lively interest in appointments, promotions, dismissals, and the whole field of personnel management. He personally started many investigations into personnel actions, especially dismissals, seeking to correct injustices.[34] He felt it to be distinct improvement and a result of his efforts when in 1943 the Commission created a Division of Investigation, which provided an independent channel for looking thoroughly into charges of misconduct by members of the Commission's staff.[35] Under T. E. Stakem as Chief Investigator, a staff of lawyers checked up on the many complaints about bribery, discrimination, and kickbacks on purchases. Mr. Stakem worked closely with the Department of Justice on the one hand and with the Executive Director and regional directors of the Maritime Commission on the other. He reported to them cases that called for criminal or disciplinary action.[36] Commissioner Carmody felt that these reports were not always followed up in the right way, and persuaded the Commission in January 1945 to call for quarterly reports on the steps taken to correct conditions revealed by Mr. Stakem and his staff.[37]

Another expression of Commissioner Carmody's concern with personnel management was the Commission's Suggestion Committee, of which he was chairman. This was first created September 16, 1942 with small prizes privately donated, and was revived May 2, 1944 after larger cash awards for employee suggestions had been voted by Congress. In 1944 and 1945 a total of 1,377 suggestions were received and 100 cash awards were made, totaling $4,165 for suggestions which it was estimated saved $308,595. Mr. Carmody took a lively personal interest in the suggestions

---

[34] Cohen, "Commissioner John Michael Carmody," and notes of Hymen Cohen's interview with Carmody, Dec. 17, 1946, with the typescript in Historian's Collection, Research file 118.3.

[35] Statements by Carmody to Hymen Cohen, Dec. 17, 1946, notes in Historian's Collection.

[36] Statements by T. E. Stakem to F. C. Lane, June 16, 1949 and letter of June 19, 1946 in Historian's Collection, Research file 106.31. Mr. Stakem showed me the "Summary of all Division of Investigation Cases" in his office. An example of a case turned over to the War Frauds Division is in Flesher reading file, Confidential, 1944, Dec. 7, 1944.

[37] MC Minutes, p. 30675 (Jan. 4, 1945).

offered, commenting on them and writing encouraging letters to those who sent in suggestions.[38]

Although Commissioner Carmody could point to some successes in his efforts to reduce costs in shipyards and improve the handling of personnel in the Commission, his position did not enable him to take charge of the Commission's policy in any of these matters. In regard to the Commission's Personnel Division he wrote in February 1945, " I do know that we have done a bum job here. . . . They [its Directors] are not really directors; they are assistants to an old time executive director, who makes all important decisions for them and handles all personnel business with the Commission." [39] Mr. Carmody was occasionally able to make his point of view prevail, but in the main the Executive Director S. D. Schell managed personnel and Admiral Vickery managed the shipbuilding companies, each being backed of course by Admiral Land as Chairman.

Not wishing to restrict himself to a specific sphere and finding all the main spheres of the Commission's activities already being supervised by others, " Commissioner Carmody fulfilled what may be designated as a planning and review function. . . ." [40] He considered this " review function " to be the role a Commissioner ought to assume, saying: " I don't agree with my colleagues that the Commission business can be handled most advantageously by making Commissioners, as they come into office, division heads, as it were, only to wind up as specialists fighting for their special activities without regard to the balance that only men without vested interest in part of a program can achieve. . . ." [41]

In resisting the tendency to absorb the Commissioners into a chain of command by delegation of specialized functions, Commissioner Carmody was objecting also to the tendency to make some Commissioners superior to others by reason of their position as chairman or vice chairman. He did not consider himself a subordinate of the Chairman. When he proposed to attend a meeting of the WPB and was asked whether he had " cleared " with Admirals Land and Vickery, he replied, " Of course, as an independent member of this Commission I don't ' clear ' anything

---

[38] Cohen, " Commissioner John Michael Carmody," pp. 123-31.

[39] Carmody to A. J. Sarre, Feb. 17, 1945 in Carmody reading file.

[40] Quoted from Cohen, " Commissioner John Michael Carmody," pp. 18, 19 in which this review function is analyzed at more length.

[41] Carmody to Lydell Peck, June 19, 1943, in Carmody reading file.

with either of them," and explained, " I realize you and your colleagues in WPB are too busy to bother about the composition of the many Federal agencies you do business with but, after all, the Maritime Commission is a statutory body created by Act of Congress with five members, all appointed by the President and all subject to Senate confirmation. None of them has any authority over or responsibility for the vote or action or conduct of any other member with respect to Commission business." [42]

Unlike Admiral Land, Commissioner Carmody believed in the commission-form of government as did Commissioner Woodward.[43] Furthermore, Mr. Carmody believed in commissions in which the chairman was elected by the members from among their own body. He considered that procedure to be "the essence of democracy in public administration and management." Speaking of commissions in general, he said that when the chairman was appointed by the President, the result was that attention centered on " some one man thus thrown more sharply into the limelight than is good for his ego or for the fortunes of the enterprise." [44]

Although without experience in shipbuilding before becoming a member of the Commission, Mr. Carmody while a Commissioner spent more of his time visiting shipyards than did any other Commissioner except Admiral Vickery, and Commissioner Carmody always went first to the Commission's representatives in the yard before going to see the yard management.[45] The energy and idealism which he poured into his " reviewing function " led him to establish direct contact with members of the Commission's staff in many divisions in all parts of the country. For a while at least he had hopes of inducing the Admirals to see what he was driving at in his zeal for efficient administration. " When I have had a couple of more years with the Naval architects," he wrote in June 1943, " and the Naval architects have had a couple more years in the broader administrative field that they have been pitched into, each of us will understand the other's language better and get more done with less waste." [46]

Unfortunately there was no such increase in mutual under-

[42] Carmody to C. E. Wilson, July 10, 1943, in Carmody reading file.
[43] Note by T. M. Woodward on a draft of this chapter.
[44] Letter to the editor, Washington *Post*, Feb. 28, 1947.
[45] Interviews with various members of the Maritime Commission staff.
[46] Carmody to Lydell Peck, June 19, 1943 in Carmody reading file.

standing. On the contrary, clashes became more severe after victory was assured and sharpened after the war was over.

Considering both aspects of Commission action—on the one hand the decisions made as part of the chain of command in which Commission action was a mere formality, and on the other the decisions which were substantially affected by the need of receiving the approval of the Commission, or were reshaped as a result of the discussions in the Commission—the situation may be summarized thus: The Chairman and Vice Chairman were in command, backed by the Presidential war powers. They could, if they wished, decide many vital matters without fear of being overruled by the five-man Commission, but they brought important questions of policy before the Commission, and the Commission meetings forced them to listen to criticism from critics who were in a position to be well informed. The other Commissioners made their views effective on incidental matters, on details of ways and means rather than in formulation of basic policies.

V. The Disruption of the Wartime Commission

To recount fully and authoritatively the events which tore the Commission apart in the last year of the war and, after the resignation of Admiral Land in January 1946, led to the dominance in the Commission for a time of Mr. Carmody's point of view is impossible here. It would require more sure knowledge than I possess, and would require going into issues beyond the story of wartime shipbuilding. In August 1944, many contracts for ship construction were cancelled; in 1945 the emergency yards were closing down.[1] Planning for the postwar merchant marine and for demobilization became of increasing importance to all Commissioners. A brief summary of the denouement is in order, however, and in attempting it I will give particular attention to the issues arising out of the shipbuilding program itself.

In 1944, the strain of incessant labor and difficult decisions began to have effects. In late July 1944 and again in August and September there were some meetings of the Commission attended by as few as three or even only two of the Commissioners, the others being absent on account of illness.[2] Although thus reduced

[1] See above, pp. 686-87.
[2] MC Minutes, pp. 29177, 29184, 29456, 29502, 29496, 29788.

in numbers, the Commission was still able to function because of the proxies which other Commissioners gave to Chairman Land. In July 1944, Admiral Vickery was still combining inspection flights to shipyards with the innumerable other demands on his energy. The intensity with which he identified himself with the shipbuilding program appears in innumerable turns of phrase in the testimony he gave at various Congressional hearings. He took responsibility for all decisions, those which had proved wrong as well as those proved right, and many for which it would have been easy to shift blame.[3] But one result was to make the shipbuilding program appear his, rather than that of the Commission as a whole. At a hearing on East Coast Shipyards on July 12, 1944, when questioned about Commissioner Carmody's criticism of the management of the yard, Admiral Vickery blurted out: " I merely state this, that I don't think Mr. Carmody's judgment on shipbuilding is very good."[4] During this crowded period, when the 1944 program was slipping behind schedule and there was pressure from the Joint Chiefs to finish the military types on schedule, formal memoranda appear in the relations between Admiral Land and Admiral Vickery.[5] On September 25, 1944 Admiral Vickery suffered a heart attack which kept him in the hospital until the following February and from which he never entirely recovered, although in mid-February he was off again for a twenty-day trip inspecting shipyards.[6]

In the emergency created in September 1944 by Admiral Vickery's hospitalization, Admiral Land reorganized, without any formal Commission action, the lines of authority which Admiral Vickery had previously tied together. The Chairman sent to all concerned a memorandum instructing that no mail was to go out over Admiral Vickery's signature and that the following should report thereafter to the Chairman: (1) Mr. Bates, the Director of the Technical Division, (2) Mr. Ring, the Director of the Division of Shipyard Labor Relations, (3) J. S. Wilson and (4)

[3] House, *Investigation of Rheem Manufacturing Co., Hearings*, pp. 60-69, Oct. 7, 1943; House, *Walsh-Kaiser Co., Inc., Hearings*, pp. 267-78, Apr. 26, 1944; House, *East Coast Shipyards, Inc., Hearings*, pp. 375-424, July 12, 1944.

[4] *Ibid.*, p. 384.

[5] Land to Vickery, July 6, 1944, July 19, 1944, July 29, 1944, in Land reading file.

[6] Record of travel and leave maintained by Irene Long. On his efforts to remain on the Navy active list, see Land to Secretary of Navy, April 4, 1945, in Land reading file.

C. H. Johnson, technical assistants, and (5) Lt. Weber, Vickery's aide. Mr. Johnson had been engaged March 26, 1943 to replace Mr. Schmeltzer after the latter's death, and had been appointed "Technical and General Assistant to Admiral Vickery." He came from the Federal Shipbuilding and Dry Dock Co. of which he had been Assistant Chief Engineer.[7] Admiral Land now gave instruction that the directors of the Procurement and Production Divisions should report to Mr. Johnson, technical matters should be referred to Mr. Bates, other letters and memoranda addressed to Admiral Vickery including those from the regional directors should be sent to Lt. Weber, and mail previously signed by Admiral Vickery should be prepared for Admiral Land's signature.[8] The Chairman regarded the burden of supervision he thus assumed as temporary and sought assignment from the Navy of someone who could take over while Admiral Vickery was ill. Captain R. M. Watt (USN) was instructed by the Navy to report to Admiral Land and from the first of December until March he sat at Admiral Vickery's desk,[9] but in this period some vital matters, such as the award of contracts for Victory ships, were presented to the Commission by Admiral Land himself.[10]

When Admiral Land was away January 26 to February 12 on a trip with President Roosevelt to the Yalta conference, leaving his proxy with Captain Macauley, only three commissioners were available for meetings.[11]

As soon as Admiral Vickery was again at his office in Washington, Admiral Land spent March 4 to 30, 1945, on a trip to London in connection with his duties as Administrator of War Shipping,[12] and Vice Chairman Vickery presided over the meetings of the Commission. During this interval Commissioner Carmody began to register a persistent series of dissents on a matter of basic policy, namely, the terms of the new lump-sum contracts being concluded at that time. The Commission meeting on March 6, 1945 approved unanimously a lump-sum contract for tankers with Marinship,

[7] MC Minutes, p. 24701.
[8] Memoranda, in Land reading file, Sept. 25, and Sept. 28, 1944.
[9] Information from MC, Personnel Division; Land to Chief, Bureau of Ships, Feb. 26, 1945 in Land reading file; statements by Miss Van Valey and Miss Long, June 1949.
[10] See above, p. 680.
[11] MC Minutes and Land diary kept by his secretary.
[12] Diary kept by Miss Van Valey.

but when the minutes of that meeting were brought up for approval on March 13, Commissioners Carmody and Woodward said that when approving the contract they had been unaware of the content of its Article 10, and having discovered the content of that article they wished to be recorded as voting "nay."[13] The article which aroused this opposition provided that, when auditing costs for the purpose of recovering profits in excess of the maximum stated in the contract, the Commission would allow as costs even those resulting from the negligence of the corporate officers of the contractor. The justification made for the inclusion of this clause will be explained in the discussion of profits in the next chapter. Not only did Mr. Carmody object to this clause, he dissociated himself from the approval of a number of other lump-sum and selective-price contracts and was recorded as "not voting."[14] He did not oppose the main selective-price contracts for Victory ships, however.[15] It was not the return to lump-sum to which he objected, it was the terms of many of the new contracts —terms which Chairman Land later declared were necessary to obtain the consent of the contractors, but which Commissioner Carmody found objectionable. In this same general period he was also insistent about getting the East Coast Shipyards, Inc. out of the Bayonne yard,[16] and about pushing settlement of the claims against the Bethlehem Sparrows Point yard in connection with the welding frauds there.[17]

The opposition of Commissioner Carmody to Admirals Land and Vickery (who held each other's proxies on the numerous

[13] Approval of this contract (MCc 33546) in MC Minutes, pp. 31235-36 and the correction of minutes recorded on p. 31322. A contract with Kaiser-Swan Island (MC 29039) contains the same language in Article 10. The award of contract MC 29039 was approved by the Commission unanimously Feb. 15, 1945 (MC Minutes, pp. 31076-77, minutes approved p. 31130) but I have not found any record of the form of this contract being brought before the Commission for approval.

[14] MC Minutes, pp. 31394-95, 31409 (Mar. 20, 1945); pp. 31537-41 (Apr. 3, 1945); pp. 31697-98 (Apr. 12, 1945); pp. 31708-09 (May 17, 1945); pp. 32040-42 (May 17, 1945); pp. 32102 and 32182 (May 25, 1945). Carmody also noted "nay" in the 3 to 1 approval of an adjustment in Bethlehem Fairfield's contract for Liberty ships. Minutes, pp. 33501-06 (Oct. 16, 1945).

[15] Minutes, pp. 31184-86 (Feb. 27, 1945); pp. 31428-30 (Mar. 22, 1945); p. 31522 (Mar. 29, 1945); pp. 31229-30 (Mar. 1, 1945); pp. 31237-38 (Mar. 6, 1945).

[16] Minutes, pp. 30471-74, 30540-42.

[17] Minutes, pp. 31400, 31431, 31724, 31842-50 and compare Land to Lester A. Pratt and Co., July 13, 1945 in Land reading file, and final action on the case. Minutes, p. 33575 (Oct. 23, 1945).

DISRUPTION OF THE WARTIME COMMISSION 791

occasions when one or the other was absent) was expressed even more sharply when the facilities operated by Bethlehem-Fairfield and California Shipbuilding were sold to those companies. These two sales were approved by the Commission on November 27, 1945 despite Mr. Carmody's vote of "nay." On this occasion Commissioner Carmody submitted for record in the minutes a note explaining his "nay" vote, the note being submitted following Chairman Land's statement that the submission of such memoranda was a custom "going back to Moran's time," a pointed reference to a previous period, prior to 1941, in which there had been a split commission. In explaining his "nay" vote Mr. Carmody admitted the difficulties in disposing of the emergency shipyards but stated he thought these could best be overcome by "inventories by our own employees and independent appraisals and then by giving the widest possible publicity to the offering of these public properties. . . . I am unaware (1) of any appraisal made by an independent appraiser, (2) any inventory of materials not made by the purchaser or his employees, (3) of public offer through advertisement or otherwise." He concluded by saying he was "not persuaded a larger net sum could not be realized." [18]

Commissioner Carmody's opposition to the policies being carried through by the Chairman and Vice Chairman was the more serious, and the more galling, because it was paralleled by the criticism which was coming from certain members of Congress and from the General Accounting Office. Indeed, for some time those whom Admiral Land had reason to consider his political enemies had been making references to Mr. Carmody which dissociated him from their attacks on Admiral Land. In the furor over the cancellation of the Higgins contract, even though Mr. Carmody voted for cancellation and defended it when called to testify, he was referred to by Mr. Higgins as "sympathetic." [19] When the President's reappointment of Admiral Land as Chairman of the Commission was before the Senate for confirmation on March 30, 1943 there were five hours of blistering debate on charges of Senator George D. Aiken that Admiral Land had countenanced "illegal expenditures, incompetence, and wastefulness." In this and other attacks, Senator Aiken based most of his

[18] MC Minutes, pp. 33918-19, approvals on pp. 33916-23.
[19] House, *Higgins Contracts, Executive Hearings*, pt. 1, pp. 29, 34; pt. 3, p. 237.

charges on evidence and opinions obtained from the Comptroller General, and he quoted also from letters of Commissioner Carmody to the Commission, from which Senator Aiken concluded: "Carmody apparently is not in harmony with all the acts of the rest of the Commission."[20] Admiral Land was confirmed as chairman, being warmly endorsed by Senator Truman and many others, with only Senator Aiken and four other Republicans voting against him.[21]

In June 1944 the Senate again gave Land a vote of confidence by approving his promotion to the rank of Vice Admiral, although Senator Aiken repeated his attack.[22] During the next month, critics of the way the Maritime Commission was operating had a field day in the investigation of the mess at the Bayonne shipyard which had been taken from Marine Maintenance Corp. and put in the hands of Mr. Otterson's East Coast Shipyards, Inc. Knowledge of the letter which Mr. Carmody had written the Commission denouncing East Coast's management of the yard was evident in the embarrassing question which produced the above-quoted unfavorable comment by Admiral Vickery about Commissioner Carmody's judgment on shipbuilding.[23]

The same pattern of parallel action on the part of Admiral Land's opponents in Congress, the General Accounting Office, and Commissioner Carmody appeared in regard to the terms of the new lump-sum contracts. On April 25, 1945, about a month after Mr. Carmody registered in the Commission's meeting his objection to the article in these contracts which referred to "negligence," Admiral Land was critically questioned on that very matter in a hearing before the Appropriations Committee of the House by Congressmen Taber and Wigglesworth, Republican members. Admiral Land admitted that he did not "like

[20] *Congressional Record*, 78th Cong., 1st sess., vol. 89, p. 2, pp. 2686, 2721-23, Mar. 30, 1943.
[21] *Congressional Record*, vol. 89, pt. 2, pp. 2705, 2722.
[22] *Congressional Record*, 78th Cong., 2d sess., vol. 90, p. 6610. Since Admiral Land was on the retired list, a special bill was necessary for his promotion. It was introduced in 1942 and passed the House, but did not pass the Senate until the following session. House, *Hearings No. 283* on H. S. 7576 before the Committee on Naval Affairs, 77th Cong., 2d sess., Sept. 23, 1942; House, *Report No. 2487* of the same committee, Sept. 24, 1942; House, *Report No. 1204* to accompany H. R. 634 of the same committee, 78th Cong., 2d sess., Mar. 1, 1944; Senate, *Report No. 904* of the Committee on Naval Affairs, 78th Cong., 2d sess., May 25, 1944.
[23] House, *East Coast Shipyards, Inc., Hearings*, p. 384 and *passim*.

that phraseology," but he defended it as not really harmful and as necessary to secure the consent of the companies to a lump-sum form of contract.[24]

The General Accounting Office sharply criticized also the selective-price contracts concluded by the Commission in the first half of 1945.[25] In June the Commission recognized the seriousness of this criticism to the extent of referring a selective price contract with Bethlehem-Fairfield to the Comptroller General " for advice as to whether it meets with his approval." In light of the adverse opinion which it received from the Comptroller General, the Commission concluded with Bethlehem-Fairfield, instead of a selective-price contract, a fixed-price contract subject to renegotiation and not containing in Article 10 the clauses to which objection had been raised.[26]

In the fall of 1945 the composition of the Commission began to change. Commissioner Woodward's second term expired September 25, 1945. Chairman Land gave the President his recommendation concerning a new appointment, as he had done before when vacancies occurred.[27] He now recommended Captain Conway, who was Deputy Administrator under him in the WSA. Captain Conway was experienced both as a ship operator and a public administrator. He was a man in whose ability and loyalty Admiral Land had confidence, and to add him to the Maritime Commission would help in reuniting the WSA with it in the postwar demobilization.[28] But the President, now Harry S. Truman since Franklin D. Roosevelt had died April 12, 1945, decided to appoint Raymond S. McKeough, a former member of Congress from Illinois. McKeough's experience included connections in the banking field and rail transportation, as well as

[24] House, *National War Agencies Appropriation Bill for 1946, Hearings*, pp. 398-402.
[25] In House, *Investigation of Shipyard Profits, Hearings* (1946), pp. 685-92.
[26] MC Minutes, p. 32321 (June 19, 1945); pp. 32666-68 (July 26, 1945).
[27] For Land's recommendation of Woodward's appointment for his second term in 1940, see Land to the President, Sept. 5, 1939, in Land file. On his support of Vickery's first appointment, see Land to Bland Aug. 29, 1940 in same file. For his recommendations, which were not followed, in July 1941 when Dempsey retired, see Land to the President, July 1, 1941 in Land reading file.

Captain Macaulay's first appointment to the Commission was discussed by the President with Land (Land to the President, May 6, 1942, in Land file) and Land " earnestly recommended " his nomination for a second term (Land to the President, Aug. 16, 1944, in MC gf 107-1).

[28] Statements by Admiral Land in interview, Sept. 30, 1946.

national government, but not any connections with the merchant marine or shipbuilding. Because he had been affiliated with the Political Action Committee of the CIO his confirmation was opposed by the seamen's unions of the AFL, but he was confirmed by the Senate and took oath of office as a Commissioner October 16, 1945.[29]

The change in the White House affected also the kind of backing that Admiral Land could expect in his planning for a postwar merchant marine. Although postwar plans are outside the theme of this book, they were frequently in the minds of the Commissioners during the war. When Admiral Vickery made a visit to England in September 1943, just after the controversy over the Victory ship, a small storm of comment arose over a speech he made in which he emphasized the intention of the United States to build fast ships whether the British liked it or not, and to continue after the war to be a "maritime nation."[30] In 1945 postwar plans were pushing into the center of attention. Admiral Land was urging the immediate construction of new passenger vessels, and in November 1945 he wrote the Bureau of the Budget protesting against cuts, particularly those that eliminated passenger ships which had been approved by President Roosevelt.[31] But the construction of new passenger ships at that time was not approved.

Admiral Vickery, aggressive champion of a big American merchant marine, left the Commission at the end of 1945. The illness which had overcome him in the fall of 1944 resulted in his being placed on the Navy's retired list in October 1945, and he followed his physician's orders in resigning from the Commission November 29, 1945, effective December 31, 1945. He died March 21, 1946. In accepting Admiral Vickery's resignation President Truman wrote, "You have earned the thanks of the Nation."[32] Hackneyed words; but in this case they were the truth.

On January 2, 1946, Admiral Land sent President Truman his resignation as Chairman of the Maritime Commission and Administrator of the War Shipping Administration, recalling that he had

---

[29] Senate, *Nomination of Raymond S. McKeough, Hearing* before the Subcommittee of the Committee on Commerce, 79th Cong. 1st sess., Sept. 13, 1945. Dates of office furnished by MC Personnel Division.

[30] *New York Times*, Oct. 8, 1943, and Tyler, " The Victory Ship Design," pp. 82-85.

[31] Land to Harold D. Smith, Nov. 20, 1945, in Land reading file.

[32] Letters in MC, Personnel Division files.

DISRUPTION OF THE WARTIME COMMISSION 795

informed the President at conferences in May, September, and December 1945 of his desire to be relieved.[33] An attractive position in private industry, the presidency of the Air Transport Association of America, was awaiting him. He suggested, and the President approved pending the appointment of a new chairman, the designation of Commissioner Macauley as Acting Chairman of the Maritime Commission and Captain Conway as Acting Administrator of WSA.[34]

On November 19, 1945, shortly before the resignation of Admiral Land, the Comptroller General had sent to the Commission the General Accounting Office's highly critical audit of the Maritime Commission's accounts for the fiscal year 1943, and on January 15, 1946 sent the equally critical audit of the 1943 accounts of the War Shipping Administration.[35] Chairman Land's resignation was effective January 15, 1946. When the Commission, consisting of Commissioners Macauley, Carmody, and McKeough, met on Tuesday, January 22, they ordered the Director of Finance to furnish them on Thursday with a complete list of the unanswered letters received from the Comptroller General.[36] Next day, Wednesday, January 23, Congressman Wigglesworth placed in the *Congressional Record* the Comptroller General's audit reports.[37] Acting Chairman Macauley promptly wrote on behalf of the Commission informing Mr. Wigglesworth that the issuance of the *Congressional Record* of January 23, 1946 had " brought to the official notice of myself and my two colleagues for the first time that the Comptroller General had addressed to the Commission his letter of November 19, 1945, and his audit report of the Commission's balance sheet as of June 30, 1943." [38] On February 7, 1946 the Commission terminated the services of Mr. Anderson as Director of Finance.

Thus, while the appointment by the President of a new Chair-

---

[33] Land to the President, Jan. 2, 1946 in Land reading file.

[34] Land to the President, Jan. 8, 1946, with note by President Truman giving approval, in MC gf 107-1.

[35] Warren to Bland, Feb. 8, 1946, in House, *Accounting Practices, Hearings*, p. 454.

[36] MC Minutes, p. 34478.

[37] *Congressional Record*, 79th Cong., 2d sess., vol. 92, pt. 1, pp. 278-92.

[38] *Ibid.*, p. 1174 gives the text of the letter. The Commission approved on Jan. 28 the preparation of an " appropriate letter of acknowledgment to Congressman Wigglesworth in reply to his criticism appearing in the Congressional record, Jan. 23, 1946." MC Minutes, p. 34492.

man hung fire, the Commission of three began a reorganization and an upheaval in personnel, which continued for some time. The task of replying to the criticisms of the Comptroller General was entrusted by the Commission to William L. Slattery. In terminating Mr. Anderson's services, the Commission abolished the Division of Finance and created two new divisions: one for construction, the other for operations. Mr. Slattery became director of the "Division of Finance (Construction)"; Mr. Quinn the director of the "Division of Finance (Operations)." [39] Since Mr. Slattery had been during the war the General Auditor of Construction, he was ready to defend vigorously the work of the resident auditors whom he had supervised, but he felt no personal responsibility for the way the accounts had been handled, he condemned the basic method which had been used, and he characterized the general policy of the Commission during the war as "a shipbuilding program without regard to costs, rightly or wrongly, with the full knowledge and consent of the powers that be." [40] During the Congressional investigations and the public criticism touched off by the publication of the Comptroller's reports, the Commission was not represented by those who had been primarily responsible for its shipbuilding activities during the war.

In considering the Maritime Commission as an example of the commission-form of government, various periods in its history must be sharply distinguished. Although prior to the autumn of 1940 and again after 1944 there were times when the Commission was seriously divided over major policies, in the critical period of wartime shipbuilding the Commission was sufficiently united so that Chairman Land was able to act as its leader and the head of its administration. There were several reasons for the dominating position thus held by Chairman Land from 1940 through 1944. One was his personal energy and knowledge. Another was the extent to which his recommendations had been followed by the President in appointing the Commissioners serving in that period. Effective also in 1942-1944 was the fact that the functions of the Commission were under the war powers of the President, who

[39] MC Adm. Order No. 39, Suppl. 55 (revised); MC Minutes, p. 3456 (Feb. 7, 1946). Earlier, on Jan. 24, Slattery had been named Special Assistant to the Commission to investigate correspondence with the Comptroller General. MC Minutes, p. 34492.
[40] House, *Accounting Practices, Hearings*, p. 323.

could delegate them as he wished. Basically, it was Presidential support which enabled Admiral Land to use a commission-form of organization in directing ship construction for World War II.

An appropriate conclusion to these chapters on the administrative problems faced by the Commission under Admiral Land's chairmanship is one of the disarming disclaimers with which he so often won the confidence of Congressional committees: "As you gentlemen started us here in 1937 and 1938, we were a $300,000,000 organization. We have been blown up to about a $10,000,000,000 organization. And that there are faults, mistakes, errors of omission and commission is admitted. I would be a stupid ass to come up here and tell you that we are running this with any degree of perfection. We are not. We are just doing the very best we can with the tools we have." [41]

[41] House, *Independent Offices Appropriation Bill for 1944, Hearings*, p. 693.

CHAPTER 24

ADVENTURES IN HINDSIGHT

I. RENEGOTIATION AND INVESTIGATIONS OF PROFITS

IN THE PERIOD between the First and Second World Wars there was much talk about taking the profit out of war. In what was done about it, two approaches can be sharply distinguished. One was to limit the per cent of profit on contracts that were considered war contracts and thus attempt to prevent companies from receiving excessive profits. The other was to impose an excess profits tax on corporations and individuals and thus take away profits after they had been received. Limitations on profit were administered by procurement agencies such as the Maritime Commission; excess profits taxes were administered by the Treasury Department.

In addition to these two clearly distinguishable approaches to the problem Congress later authorized a third method—renegotiation, which was partly one thing, partly the other.[1] The story of the Renegotiation Act and its operation is an essential preliminary to any discussion of the amount of profit made in shipbuilding during World War II because it has shaped the facts and figures available.

In the spring of 1940 the legislative enactments to take the profit out of war were based on the first approach. The amount of profit a company could receive on contracts with the Navy or Army for ships or aircraft was limited to 10 per cent by the Vinson-Trammel Act. Builders of ships for the Maritime Commission were similarly limited to 10 per cent, first by the Merchant Marine Act of 1936 or later by Public Law 46 of May 2, 1941.[2]

---

[1] Miller, *Pricing of Military Procurement*, pp. 162-87, gives a general account and bibliography of renegotiation. See also, Connery, *The Navy and the Industrial Mobilization*, chap. XIII.

[2] Profit limitations were reviewed by Admiral Land in House, *Shipyard Profits Hearings*, before the Subcommittee on Shipyard Profits of the Committee on Merchant Marine and Fisheries, 78th Cong., 2d sess., Mar. 22, 1944, pp. 6-7. Cited hereafter as House, *Shipyard Profits, Hearings* (1944).

But the competition of British purchases made sweeping profit limitations impractical in the period before Pearl Harbor; manufacturers were free to take foreign contracts which contained no provisions limiting profits.³ Consequently Congress shifted from the first approach to the second. On October 8, 1940, Congress repealed the Vinson-Trammel Act and, in regard to the profit of subcontractors, lifted the restriction in the Merchant Marine Act of 1936.⁴ At the same time, the Second Revenue Act of 1940 enacted provisions designed to take away wartime profits by an excess profits tax.⁵

After Pearl Harbor, the question was reopened. Many members of Congress were unwilling to leave entirely to the tax laws the task of taking profit out of war. Congressman Case proposed the renegotiation of contract prices to eliminate all profits above 6 per cent. The Services objected to so inflexible a provision, which might remove incentives to efficiency by putting all contractors in the position of working for cost plus 6 per cent. The Renegotiation Act as passed in April 1942 and amended in February 1944 provided for much flexibility in its administration. Price Adjustment Boards were created to determine the reasonableness of profits. They were instructed to take into account (1) the efficiency of the contractor, (2) his volume of production and normal prewar earnings, (3) the amount of private capital employed, (4) the extent of risk, (5) his contribution to the war effort in new developments and technical aid to other contractors, (6) the character of the business, and (7) "such other factors the consideration of which the public interest and fair and equitable dealing may require." ⁶

The Price Adjustment Board of the Maritime Commission took advantage of the fact that so large a percentage of deliveries had been Liberty ships, which were built under manhour contracts. These contracts facilitated the comparison between performance and profits. Calculations for the renegotiation of these contracts consisted of two steps. The first step was the reduction of fees

---

³ Miller, *Pricing of Military Procurement*, p. 168.
⁴ Concerning MC subcontractors, see the Second Revenue Act of 1940 (54 Stat. 1003-04, Public Law 801, 76th Cong., 3rd sess., chap. 757). Concerning the legal limit on the profit of shipbuilders, see below, note 38 in this section.
⁵ 54 Stat. 975.
⁶ Miller, *Pricing of Military Procurement*, pp. 170-77; Truman Committee, *Hearings*, pt. 42, p. 25505.

according to a common time scale which was applied to all the yards that received their first contracts in the spring of 1941. For the ships delivered during the first fifteen months thereafter maximum fees were set in renegotiation at $120,000 and minimum fees at $60,000; for the ensuing six months $100,000 and $50,000 respectively; for the next six months $80,000 and $30,000; and for the final six months' period—namely, July to December 1943—$60,000 maximum and $20,000 minimum. This was a substantial reduction in fees, but it reduced least the fees of the companies that had begun production most rapidly and delivered most ships in the earlier periods.

The second step was to determine whether the company should receive the maximum or minimum fee or something in between. This was done by applying the " four-factor formula " which had been developed by the Commission for comparing the efficiency of the yards. The four factors which were given equal weight were (1) manhours per vessel, (2) dollar cost per vessel (3) number of ships delivered per way, and (4) dollars of government investment in facilities per vessel. When all four factors were averaged, the yard showing the best results for any period was given the maximum fees allowed for that period; the yard with the worst showing was given the minimum fees; and those in between received their percentage of the difference, subject to some adjustment for special cases.[7]

Although renegotiation with the big emergency yards received most public attention, it was only a small part of the total work of renegotiation performed by the Price Adjustment Board of the Maritime Commission. In renegotiation with the shipbuilding companies which had constructed the less standardized military, minor, and C-types, as also in negotiating with the vendors who had supplied the shipyards with components, the Price Adjustment Board could not use such a broad and precise system of rating to determine efficiency. It was forced to examine each case individually in the light of the seven considerations specified by Congress.

[7] House, *Shipyard Profits, Hearings* (1944), testimony of Commander Rydstrom, pp. 56-57. Admiral Land was responsible for the use of the four-factor plans—John R. Paull, Chairman, Price Adjustment Board to Capt. Edward Macauley, Mar. 29, 1946; T. M. Woodward and A. G. Rydstrom to Land, Jan. 20, 1944, and Rydstrom to Land, Dec. 9, 1943. On the organization of the Price Adjustment Board of the Maritime Commission, see above, chap. 23, Sect. IV.

Indeed, the Chairman of the Price Adjustment Board testified that all the seven factors were taken into consideration in all cases.[8]

When the process of renegotiation was 95 per cent completed in May 1947 its results were as summarized in Table 17. The coverage of these figures, and consequently their significance, was determined by the way in which the price adjustment boards organized their work. Since a contractor's whole business had to be taken into account to determine the amount of his excessive profits, a contractor who had done business with more than one procurement agency was generally assigned for renegotiation to the price adjustment board of the agency with which he had the largest volume of contracts. Consequently, not all builders of merchant ships had their contracts renegotiated by the Price Adjustment Board of the Maritime Commission. The Navy's Board renegotiated the contracts of the Consolidated Steel Corp., Gibbs & Cox, and Federal Shipbuilding and Dry Dock Co. The Price Adjustment Board of the Maritime Commission renegotiated the contracts of Bethlehem-Fairfield, of Sparrows Point, and of Bethlehem-Alameda, but the rest of Bethlehem's business was renegotiated by the Navy's Board. The Army's Board renegotiated the contracts of J. A. Jones, whose main business had been army installations. The profits and recoveries on the Maritime Commission contracts of these companies are not included in Table 17. On the other hand, it includes some profits and recoveries on the contracts with the Navy of companies, such as Moore Dry Dock Co., that had some Navy contracts but mainly Maritime Commission contracts.[9]

A second reason why the table does not give a record of all profits made from shipbuilding under the Maritime Commission is the fact that many contracts were exempt from renegotiation. As amended in 1944 the Renegotiation Act exempted all contracts when the company's total renegotiable sales for a year did not exceed $500,000.[10] Previously only those under $100,000 were exempt, and the Maritime Commission opposed in vain the raising of the lower limit to $500,000.[11] Many contracts between the Maritime Commission and the suppliers were excluded by this clause. Real estate transactions also were not renegotiable.

[8] Truman Committee, *Hearings*, pt. 42, pp. 25505-11, testimony of John R. Paull.
[9] Statements by E. Stanley Glynes in September and December 1949.
[10] Miller, *Pricing of Military Procurement*, pp. 172-74.
[11] MC Minutes, pp. 24601-03 (March 18, 1943).

TABLE 17

RESULTS OF RENEGOTIATION

TO MARCH 31, 1947

OF THOSE CONTRACTS WHICH WERE RENEGOTIATED BY THE U. S. MARITIME
COMMISSION PRICE ADJUSTMENT BOARD (EXCLUSIVE OF WSA)

	Before Renegotiation			After Renegotiation			Excessive Profits Recovered
	Sales	Profits	Percent of Sales	Adjusted Sales	Adjusted Profits	Percent of Sales	
Supply Contracts............	$ 4,070,023,205	$573,381,650	14.09	$ 3,803,058,365	$306,416,810	8.06	$266,964,840
Ship Construction Contracts*	6,523,844,725	372,334,323	5.71	6,473,750,922	322,240,520	4.98	50,093,803
No Profit Shipyard Facility Contracts.........	411,603,172	....	......	411,603,172	....	....	....
Materials Not Included in Amounts Shown Above†	2,789,967,770	....	......	2,789,967,770	....	....	....
Total, All Ship Construction Contracts.........	9,725,415,667	372,334,323	3.83	9,675,321,864	322,240,520	3.33	50,093,803
Grand Total	$13,795,438,872	$945,715,973	6.86	$13,478,380,229	$628,657,330	4.66	$317,058,643
Estimated savings effected through price reductions							$103,066,938
Total savings through renegotiation............							$420,125,581

* Does not include substantial amounts saved through the elimination in renegotiation of escalation and other claims under shipbuilding contracts.

† This represents amount of materials acquired by ship construction contractors from the Maritime Commission and not included in sales or costs by such contractors. It has been included in the total amount of ship construction contracts for the purpose of showing comparable results of the renegotiation of maritime ship construction contractors with other ship construction contractors who acquired similar materials under subcontracts and included them in sales and costs. Contracts for the acquisition of these materials by the Maritime Commission have been renegotiated and to the extent that such renegotiation was concluded by this Board the amounts have been included in the totals for supply contracts. The duplication resulting from the inclusion of the cost of the materials under ship construction contracts is a duplication present in all instances wherein prime contractors subcontract part of the work and the prime and subcontracts are negotiated separately.

Source: House, *Independent Offices Appropriation Bill for 1948, Hearings*, pt. 2, pp. 623, 904, and the reconciliation of the two sets of figures on p. 905. On coverage, see text.

Exempted likewise were selective-price shipbuilding contracts. They were exempted from renegotiation under the discretionary authority granted in the Act to exclude contracts when the profits could be determined with reasonable certainty at the time the contract price was specified.[12] These and various other exemptions make the totals of contracts subject to renegotiation by the Price Adjustment Board of the Maritime Commission less by two or three billion dollars than the total of contracts entered into by the Commission. The four chief reasons for the difference are (1) the limit of $500,000 on exemption of renegotiable sales, (2) exemption of selective price contracts, (3) the exclusion of contractors renegotiated by other price adjustment boards, and (4) about 300 unsettled cases on which a recovery of $50,000,000 for excessive profits was expected.[13]

Renegotiation proceeded contract by contract, and then the profits were allocated by fiscal years. Findings of excess profits were made with the proviso, "less tax credit, if any." This was a way of recognizing that a portion of what the Price Adjustment Board declared recoverable by the government had in many cases already been paid by the company in excess profits taxes. Although the Price Adjustment Board calculated the contractor's profits before taxes, and made its finding on that basis, it furnished Congress with some estimates of how much of what was recovered by renegotiation would have been recovered anyhow by taxes. On the total renegotiated contracts of the whole government, namely $191 billion, renegotiation recovered $10 billion of which about $7 billion would have been recovered in taxation.[14] If the proportion was the same on Maritime Commission contracts, seven-tenths of the $317 million recovered would have been taken by taxes if there had been no renegotiation.

Even the net recoveries (as are called recoveries in addition to what taxes would have taken) are a substantial figure, about $100 million. The biggest recoveries were made from supply contracts.

[12] MC Minutes, p. 25486 (June 22, 1943). House, *Independent Offices Appropriation Bill for 1948, Hearings*, pt. 2, p. 628; House, *Investigation of Shipyard Profits, Hearings* (1946), pp. 197, 689. On the exemption see also Truman Committee, *Hearings*, pt. 42, pp. 25494-504, 25887-89. Some "incentive contracts" were exempted from renegotiation by the Navy.—Connery, *The Navy and the Industrial Mobilization*, p. 218.

[13] House, *Independent Offices Appropriation Bill for 1948, Hearings*, p. 900.

[14] House, *Shipyard Profits, Hearings* (1944), pp. 61-63; Truman Committee, *Hearings*, pt. 42, pp. 25488, 25493.

Although the dollar volume of renegotiable shipbuilding contracts was about 50 per cent larger than that of the supply contracts, recoveries from supply contracts were more than five times those from shipbuilding contracts.

In addition there were savings through price reductions. On many supply contracts it was discovered before deliveries under the contract had been completed that the price had been set so high as to permit excessive profit. The contractor could then be told that he must reduce his prices below those in the contract. This repricing of contracts was backed by the threat that the profits would be taken away anyhow in renegotiation, and by the power conferred in the Revenue Act of 1943 to issue an order establishing fair and reasonable prices for future delivery in case fair and reasonable prices could not be negotiated voluntarily.[15] The repricing was administered for the Maritime Commission by its Procurement Division, which also was authorized to adjust prices upward when that was necessary to expedite the delivery of components.[16] The pricing of the government's large purchases of materials for the Maritime Commission's shipyards presented many problems comparable to those encountered in Army and Navy procurement, which it is not possible to go into here. But the large recoveries made from supply contracts by the Price Adjustment Board is noteworthy.

Some comparison of the profit margins (before taxes) of the shipbuilders and of manufacturing corporations as a whole is made in Table 18. It also emphasizes the savings accomplished by the Maritime Commission's Price Adjustment Board in renegotiating supply contracts.

Renegotiation did much, it is said, "to clear industry of a record of 'unconscionable profits.'"[17] But shipbuilders' profits, although low compared to the general rate of profit on sales (Table 18), were the subject of two Congressional investigations. The work of the Price Adjustment Board was criticized both directly and by implication on the grounds that it did not give sufficient weight to the relation of a contractor's earnings to his

---

[15] Title VIII, 58 Stat. 92.

[16] Explanation by Schell in House, *Navy Department Appropriation Bill for 1947, Hearings*, pp. 164-65; MC Minutes, pp. 30009-10 (Oct. 24, 1944); p. 30954 (Feb. 6, 1945).

[17] Miller, *Pricing of Military Procurement*, p. 178.

capital investment.[18] The same criticism was leveled against the whole work of the War Contracts Board.[19] Shipbuilders under the Maritime Commission were singled out for attack because so many of them operated plants belonging to the government. Their profits were relatively low compared to their output but were relatively high compared to their investment of capital.

It is difficult to determine exactly the amount of capital invested. At the hearing before the House Committee on the Merchant Marine and Fisheries in September 1946, the Maritime Commission, which at that time included only one member who had served through the war, namely Mr. Carmody, reported the following figures concerning corporations operating in plants fully owned by the government: [20]

Total capital investment of shipyard operators $23 million
Total estimated profits $356 million

In the hearings Henry J. Kaiser was charged with earning in one of his companies " something like 11,600 per cent on the original capital," [21] Bethlehem Fairfield was charged with earning 1,200 per cent,[22] and in the most extreme case, the Merrill interests in the St. Johns River Shipbuilding Company were charged with making 50,000 per cent after taxes on an initial investment of $400.[23] Representatives of the shipbuilders bitterly declared that this way of figuring and the figures themselves were incorrect. They claimed to have " invested " far more than was shown in the figures furnished by the Maritime Commission. In the hearings there were hot clashes over whether money borrowed by the shipbuilding companies or by their stockholders for use

---

[18] Both Congressional investigations were before the House Committee on Merchant Marine and Fisheries. At that of March 22 and 23, 1944, the only witnesses were Admiral Land and Commander Rydstrom of the MC Price Adjustment Board. —House, *Shipyard Profits, Hearings* (1944). Much longer and accompanied by voluminous exhibits were the hearings Sept. 23-26, cited as House, *Investigation of Shipyard Profits, Hearings* (1946). At that time Admiral W. W. Smith was Chairman of the Maritime Commission and Mr. Slattery was its Director of Finance, as explained above in chap. 23. Almost all the testimony was by shipbuilders. Although Admiral Land attended some of these hearings (I saw him there), he was not called to testify.

[19] Truman Committee, *Hearings*, pt. 42, pp. 25508-15.
[20] House, *Investigation of Shipyard Profits, Hearings* (1946), p. 372.
[21] *Ibid.*, p. 96.
[22] *Ibid.*, p. 369.
[23] *Ibid.*, p. 344.

TABLE 18

PROFITS BEFORE TAXES ON VARIOUS ENTERPRISES

Type of Enterprise	Profits As Per Cent of Sales 1936-39	1942-45
All Manufacturing*	5.6	9.4
War Industries	7.6	10.2
Non-war Industries	4.3	8.5
All Renegotiable Contracts†		
Before Renegotiation	...	12.7
After Renegotiation	...	8.4
Maritime Commission Renegotiable Contracts§		
Supply Contracts		
Before Renegotiation	...	14.1
After Renegotiation	...	8.1
Ship Construction Contracts‡		
Not including in "Sales" the cost of government furnished material		
Before Renegotiation	...	5.7
After Renegotiation	...	5.0
Including in "Sales" the cost of government furnished material		
Before Renegotiation	...	3.8
After Renegotiation	...	3.3

* Source: Miller, *Pricing of Military Procurement*, p. 213.

† Source: Truman Committee, *Hearings*, pt. 42, p. 25932. For those contracts from which recoveries were made by renegotiation the corresponding figures are 21.4 and 11.1 per cent. *Ibid.*, p. 25489.

§ Source: House, *Independent Offices Appropriation Bill for 1948, Hearings*, p. 904.

‡ Source: See Table 17 above. The figure which includes in "Sales" the cost of government-furnished material is the more significant for comparing the work of the Maritime Commission Price Adjustment Board with that of other Boards, since the cost of material was generally included in "Sales" and in many instances the material for Navy building, in addition to be being paid for by the government, was procured by a "lead" yard for all the other yards building the same type.

in the business should be counted part of the capital investment, or how much of the loans or debentures should be included. Old-line companies such as Bethlehem and newcomers such as Kaiser all maintained that their total resources stood back of the credit of their companies and that it was not fair to consider merely the paid-in capital. Moreover, all the shipbuilders also emphasized that they had put into their companies their " knowhow," their skilled engineering organizations—that this was the main justification for the profits received. We might say, although they did not phrase it exactly that way, that they wished to have the value of their engineering organizations as going concerns, and their reputations on which these organizations depended, counted as assets even if they were not shown as capital assets on the balance sheets.

But the main line of defense taken by the managers of shipbuilding corporations was to deny the validity of any attempt to figure profits as a percentage on capital investment. They considered the fair method to be that which was usual in the construction industry where profits were always figured as a percentage on the size of the job. Edwin I. Jones was able to put this argument especially clearly and forcefully because he represented a company, the J. A. Jones Construction Co., for which shipbuilding was definitely a minor part of its operations. He said: " No construction firm does business on the amount of money they have got invested. If you ask me to build you a house I will charge you a percentage of the total cost of the house, because the thing I would be furnishing you would be service, and that is what we furnished the Maritime Commission." [24]

That the contractors were selling services rather than products was emphasized also by Admirals Land and Vickery during the war. They contended that this point of view had received Congressional approval since all statutes limiting profits had stated the limit as a percentage of the contract price. Before the same committee of Congress at its earlier investigation Admiral Land roundly declared that fees and profits were paid, not for the use of a private firm's capital, but for management, skill, and organization.[25]

If the question is considered as one of social ethics, there is bound to be much difference of opinion. Probably most people will agree that corporations with war contracts should earn the going rate of interest on money invested and that their executives should receive salaries of the kind generally paid to men of their ability. The question is whether, aside from these payments, corporations should earn profits which enable some men to make millions. Should any man receive millions of dollars as a result of doing his part in the war effort? According to my own ideal of distributive justice, the answer is no. Neither in the armed forces nor among civilians does the distribution of awards in time of war correspond to ideal justice. But in the economic system operating in this country before the war—call it capitalism or

[24] *Ibid.*, p. 263. See also pp. 260, 263.
[25] House, *Shipyard Profits, Hearings* (1944), pp. 7-11.

free enterprise or by some other name—wealth was considered the just reward due to a man who by his enterprise, organizing ability, foresight, and readiness to run risks made an outstanding contribution to the production of needed goods. Much can be said for the justice of such a system of rewards in time of peace. The standard of rewards approved in the existing economic system was continued in the war economy, and as far as I can see they had to be used during the war if production was to be at a maximum. These rewards may be called "just" in a practical sense, that is, under the necessities of the given situation. If the legitimacy of profits is judged on this basis, the question whether a shipbuilder's personal profits represent 20 per cent or 2,000 per cent on his capital investment is irrelevant, for I do not see why a man is more entitled to make millions under wartime conditions just because he possesses many millions at the time a war starts. If anyone is entitled to make millions during a war, it is those who contribute most to victory. Huge profits, if they can be justified at all, can be justified only by their relation to the size of the job.

Kaiser-managed companies produced more ships than any other group, the dollar value of their fees was larger, and they operated entirely in government-owned shipyards. These were good reasons for the extended attention given by the Congressional committee to Henry J. Kaiser and his network of corporations. Figure 48 shows the Kaiser corporations concerned with shipbuilding and distinguishes between the yards which were Kaiser-managed and those in which Kaiser owned an interest although they were not Kaiser-managed.

The corporate connections between Kaiser's shipbuilding and his production of steel and magnesium made it difficult to determine his final profits. Kaiser Co., Inc., in addition to operating three big shipyards, built and operated a steel mill at Fontana, California. On steel, its books showed a $61 million loss due in part to having written off as depreciation 50 per cent of the cost of the steel mill, as was permitted by the tax laws, since the Navy Department certified that the steel plant was necessary for national defense. These losses in its steel division were larger than the fees or profits earned on the shipbuilding contracts of Kaiser Co., Inc. Consequently that company did not have to pay taxes

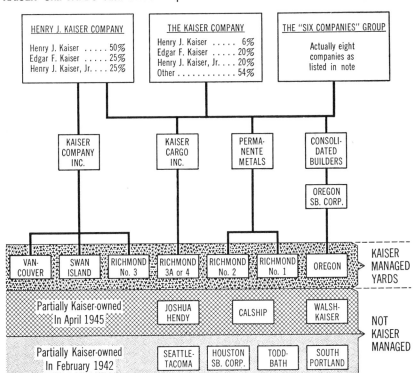

FIGURE 48. Source: Charts and reports in exhibits in House, *Investigation of Shipyard Profits, Hearings* (1946).

Note: The following companies, commonly associated with the name "Six Companies," owned shares in 1942 in the Permanente Metals Corp.: W. A. Bechtel Co., Bechtel-McCone-Parsons (Bechtel-McCone Corp.), General Construction Co., J. F. Shea Co., Inc., The Utah Construction Co., Morrison-Knudsen Co., MacDonald & Kahn, Inc., Pacific Bridge Co. On April 23, 1945, Henry J. Kaiser Co. and The Kaiser Co., which already owned a large share of Permanente Metals Corp., bought the shares in that company of W. A. Bechtel Co. and Bechtel-McCone Corp. and sold to W. A. Bechtel Co. and Bechtel-McCone Corp. the Kaiser shares in Calship on terms that about equalized the cash returns for each from the two ventures to that date. *Ibid.*, pp. 373-74.

On the exchange of equities in February 1942 between Todd Shipyards Corp. and the Kaiser interests, see *ibid.*, charts between pp. 374 and 375.

on profits.[26] Permanente Metals Corp. similarly was engaged in activities other than shipbuilding, mainly building magnesium plants and producing magnesium, and its profits on shipbuilding contracts were offset in the accounts of the company to the extent of $28 million by losses and depreciation recorded from its Magnesium Division. The taxable profits on the corporation were thereby reduced.[27] The use of the profits from shipbuilding contracts to pay for the steel plant was defended by Mr. Kaiser as not only entirely legal but also as basically right on the ground that he went into the steel business in order to get the steel needed for his ships after established steel companies had failed to make deliveries.[28]

In view of these corporate entanglements, the best that can be said about Kaiser's profits from shipbuilding is the statement finally elicited from Oscar Cox, Kaiser's lawyer, that for the four Kaiser-managed shipbuilding companies " theoretical earnings on shipbuilding, without regard to actual losses on other operations, and after theoretical Federal income taxes [what the taxes would have been if the corporations had not had losses on other operations] would have been about $40,000,000. Of that theoretical amount, the theoretical share of the Kaiser interests, based on their stockholding in those companies, would have been about $16,000,000." [29] Some addition to this figure would have to be made if one wished to include earnings on the shares owned by the Kaiser interests in companies which were not Kaiser-managed. But what Kaiser was actually worth in 1946 depended very largely on the value at that date of the steel, magnesium, and other plants which his corporations then owned.

Attacks on Kaiser in the 1946 hearings were interlarded with attacks on the Maritime Commission for having shown favoritism towards him. Pertinent to this charge are the figures which show the relation of the profits made by his companies in shipbuilding before taxes but after renegotiation, to the total volume of their contracts. The profits were 5.6 per cent of the total cost of ships delivered when the cost of government-furnished material is not counted part of the cost. But there are good arguments

---

[26] House, *Investigation of Shipyard Profits, Hearings* (1946), pp. 156, 387, 415.
[27] *Ibid.*, pp. 415, 436, 440.
[28] *Ibid.*, pp. 145, 404, and *passim*. Compare above, chap. 10, sect. I.
[29] House, *Investigation of Shipyard Profits, Hearings* (1946), p. 163.

for including in the total the cost of government furnished material. It was the general practice of construction companies and other corporations that sold primarily services to include the cost of materials, however paid for, in calculating the size of the job and the fee charged for doing it. Kaiser's own expediting organization was very active in seeing to it that the Kaiser yards were supplied. If materials and a number of other quite minor disputable items were included, the rate of profit before taxes was 3.3 per cent.[30] Exactly average! Compare the figures in Tables 17 and 18.

Against the charge of favoritism must be considered the terms of the contracts negotiated in 1942 and 1943. As told above in chapters 5 and 14, the story of contract negotiations during those years shows that the Kaiser companies were given terms harder than those given other companies. Kaiser was, as Admiral Vickery said, the lead horse used to set the pace. The only kind of "favoritism" shown Kaiser in these negotiations was the selection of his companies to build new yards which would be paid for by the government and in which the Kaiser corporations would have a chance to make large fees with very slight investment of the industrialist's capital. It seems to me clear that the reason Kaiser was thus "favored" in the winter and spring of 1942 was because he had at that time much the best record for rapid production.

Discussion of the profits of individual shipbuilders whose operations were smaller than Kaiser's does not seem practical here, especially since to be fair it would be necessary to consider in each case any special difficulties, such as the late start made by the 6-way yards in the South. But one conclusion reached in the discussion of the Commission's management of management seems to me reinforced by a study of renegotiation and of the profits retained by the shipbuilding corporations. The effort to draw contracts that would penalize the less productive contractors had little success. Inefficient yards as well as those relatively most efficient retained profits which were very large compared to the

[30] *Ibid.*, p. 418 These figures include not only the four Kaiser managed companies—Permanente Metals Corp., Kaiser Co., Inc., Oregon Shipbuilding Corp., and Kaiser Fleetwings, Inc. (successor of Kaiser Cargo, Inc.)—but also California Shipbuilding Corp., and Walsh-Kaiser Co. For Calship, on total contracts, not counting materials, of $601 million the profit was 6 per cent. The total government contracts for all six companies was $2,994 million.

amount of private capital invested. Even if inefficient yards made only the minimum fees assured by their contracts, their profits were in this sense high. In renegotiation there was a tendency to level off the rate of return. This has been criticized as failing to give adequate reward to the efficient. It can equally well be criticized as failing to punish the inefficient.[31] They too made millions.

Deservedly an object of critical examination were the terms on which the Commission had changed over to lump-sum contracts in the early months of 1945. It will be recalled that the Commission began trying to effect this change as early as June 1943 but that almost all the yards resisted the change and in December 1944 were still operating under cost-plus-variable-fee or price-minus. The contracts they had at that time would have been completed about August 1945. Not many awards of contracts for ship construction beyond that date were made until, in the last days of December 1944, the Joint Chiefs of Staff and the Director of the Office of War Mobilization and Reconversion authorized construction of additional tankers, Victorys, and some other types. The yards were then told that contracts for the additional vessels must be on a " fixed-price " basis. The new contracts, negotiated between January and May 1945, not only provided for the construction of the additional vessels recently authorized, but also converted to some form of lump-sum contracts many cost-plus-variable-fee contracts for ships already completed or partially completed.[32] In these negotiations the shipbuilders could use as a strong talking point their claims to complete reimbursement of all costs and to the fees earned under the existing contracts. The Commission on the other hand could threaten to close down first the yards making least satisfactory terms. The result of the negotiations was profitable, however, to some of the big shipbuilders.

Consider first the conversion to a fixed-price contract of a contract for ships already completed. In such a case the basis for the fixed price was a compromise between the figures of the company's accountants and those of the Commission's auditors as to what the cost had been, and an allowance for profit of

---

[31] On this general tendency in renegotiation see Miller, *Pricing of Military Procurement*, pp. 181-85, and Truman Committee, *Hearings*, pt. 42, pp. 25946, 25948.

[32] See above, chap. 19.

approximately the amount that would have been earned as fee under the original contract. For the shipbuilder the conversion of the contract had the disadvantage of preventing him from claiming reimbursement for items of cost that might turn up later.[33] The Commission on its side gave up some rights to further auditing of the costs of those ships. The General Accounting Office felt that it gave up too much auditing control.[34] It may be hard to prove who got the better bargain in these compromises, but one feature of the Comptroller General's attack on these settlements is clearly misleading. The General Accounting Office compiled a table showing that the price under the new lump-sum contracts was in many cases higher than the maximum set in the original cost-plus contracts. But these maxima had already been exceeded with the Commission's permission.[35] It was not the conversion of the contract that raised the price; the conversion merely recognized an increase in cost which the Commission would have been obligated to pay anyhow.

Second, there were fixed price contracts which covered in the same contract vessels already completed, vessels in process of construction, and vessels yet to be built—for example the new contracts for tankers with Marinship and Kaiser-Swan Island.[36] In these cases the price fixed for vessels already completed could be reached by reconciling the figures of the company's accountants and of the Commission's auditors as to what the cost had been, and adding an amount for profit which was figured on the basis of what the profit would have been under a cost-plus-variable-fee or price-minus contract; but in regard to the vessels not yet built, and for which the same price was fixed, the contractor was expected to reduce the cost of construction and so increase his profit. The profit he could earn was limited, however, to an amount specified

[33] House, *Investigation of Shipyard Profits, Hearings* (1946), pp. 643-52.
[34] House, *Independent Offices Appropriation Bill for 1948, Hearings*, pt. 1, p. 65.
[35] *Ibid.*, pt. 1, pp. 66-67. Note footnote 1 to the table. Also in *ibid.*, pt. 2, pp. 662-63.
[36] Contract MCc 33546, with Marinship dated Mar. 1, 1945 was for 41 T2-SE-A1's and four Navy type oilers (AO) of 10,000 horsepower. Nineteen of the T2's had been delivered. —Reynolds' Way Charts and *House, Investigation of Shipyard Profits, Hearings* (1946), p. 508. Contract MCc 29039, with Kaiser Co., Inc., dated Mar. 1, 1945 was for 55 T2's of which 11 were later cancelled. (*Ibid.*, p. 388, MC Minutes, pp. 31076-77.) The Company's statement does not show how many had been completed, but the number of tankers for which keels were laid in the yard after Mar. 1, 1944 was only 18.—Reynolds' Way Charts.

in the contract. For example, Marinship's contract set the price for T2's at $2,800,000 (of which Commission-furnished material amounted to $1,300,300) and limited the profit to $200,000 per vessel. Any profit above that amount would be recaptured by the Commission.[37] In the opinion of its General Counsel the Commission was not required by law to insert any clause in these lump-sum contracts that would limit the profit of the shipbuilder, but to do so was in accord with the general policy which Congress and the President had laid down, especially since the shipbuilder's risk was reduced by his knowledge at the time the contract was concluded of how his costs were running on that type of ship.[38]

Even under lump-sum contracts the contractor's costs had to be examined by the Commission's auditors in order to determine whether he had earned more than the maximum profit. Auditing to determine excess profits required determining cost, but the rules for determining cost were different from those governing the auditing of costs which were to be reimbursed by the Commission. The auditing would be done by many of the same persons who had previously audited costs for reimbursement, however, and the contractors were afraid that these auditors would carry over their old habits and exclude some of the companies' expenditures as not properly part of cost. To make clear that a new standard must now be applied, a standard that would allow the

[37] MC Minutes, pp. 31235-36.

[38] Authority granted by Public Law 46 (77th Cong. 1st sess.), did not extend beyond June 30, 1942. 55 Stat. 148. The appropriation of funds by Public Law 247 (the First Suppl. Nat. Defense Appropriation Act, 1942) approved Aug. 25, 1941 (55 Stat. 681) contained no profit limitation, and on that basis the General Counsel of the MC argued in regard to lump-sum contracts that none was obligatory. But Ex. Order No. 9001 of Dec. 27, 1941 (6 F. R. 6787) authorizing contracting activities under the war powers, provided in Title II par. 7, " Nothing herein shall be construed to authorize any contracts in violation of existing law relating to the limitation of profits, or the payment of a fee in excess of such limitation as may be specifically set forth in the act appropriating the funds obligated by the contract. In the absence of such limitation, the fixed fee to be paid the Contractor as a result of any cost-plus-a-fixed fee contract entered into under the authority of this Order shall not exceed seven per centum of the estimated cost of the contract (exclusive of the fee as determined by the Secretary of War, the Secretary of the Navy, or the United States Maritime Commission, as the case may be.) " Public Law 46 had also limited the "fixed fee" to 7 per cent while permitting bonuses to increase the total fee to a maximum of 10 per cent. Ex. Order 9001 makes no mention of lump-sum contracts. In view of the conditions under which the lump-sum and selective-price contracts were negotiated, however, the inclusion in these contracts of profit limitations may be considered required to carry out the intentions or policies expressed by Congress and the President.

contractors more freedom, including freedom to make mistakes, there was inserted in some of the lump-sum contracts awarded in the spring of 1945, including the contracts with Marinship and Kaiser-Swan Island for tankers, a sentence (in Article 10) which stated that the Commission " will allow . . . all costs, charges and liabilities incurred by the Contractor, including those resulting from the negligence of its corporate officers, agents or employees." [39]

When Commissioners Carmody and Woodward discovered this clause, allowing " costs . . . resulting from . . . negligence," they registered their opposition.[40] The Commission's lawyers defended it by pointing out that it was counterbalanced by other clauses of the contract which excluded costs incurred as a result of " reckless or willful misconduct or evasions," payments to affiliates, or where a director or officer had a pecuniary interest in the transaction. They said it " appeared to be necessary as the contracts were to be audited by representatives of the Commission who had previously been auditing the cost-plus-a-fee type of contract under which the contractor agrees to furnish a managerial service for a stipulated fee and is therefore chargeable with losses which are incurred as a result of his failure to render such service." [41] Of course it would have no practical importance unless the companies earned the maximum profit allowed by the contracts.

In fact, Marinship and Kaiser-Swan Island did earn the maximum. Marinship reported total cost of performance of the aforementioned contract (including cost of Commission-furnished material) as $116.9 million, profit as $7.6 million, and estimated recapture by the Maritime Commission as $2.5 to $2.9 million.[42] Kaiser-Swan Island reported cost as $61 million (not including the cost of Commission-furnished material, about $57 million), profit as $10.9 million, and recapture by the Maritime Commission of $1.8 million.[43] If sums had been excluded from cost on the grounds of negligence, the differences between contract price and cost would have been larger, and more would have been recaptured by the Commission.

[39] Contract, MCc 33546 and 29039, in MC, Finance Division.
[40] See above chap. 23, sect. V.
[41] House, *National War Agencies Appropriation Bill for 1946, Hearings*, p. 404.
[42] House, *Investigation of Shipyard Profits, Hearings* (1946), pp. 508-10.
[43] *Ibid.*, p. 388. The contract valued Commission furnished material as $1,300,300 per tanker.

Even when the costs were determined in accordance with Article 10, profits after recapture but before renegotiation were relatively high on some of these contracts, that is high compared to the general average for shipbuilding contracts. In renegotiation, the profits from Kaiser-Swan Island were of course balanced against the losses on the Kaiser steel mill.

Lump-sum contracts were at least subject to renegotiation. Selective-price contracts, on the other hand, were not subject to renegotiation. The award of new vessels under this form of contract gave the shipbuilders an opportunity—provided they reduced costs—to make and retain much larger profits per ship than they could have retained if they had been restricted to the general rate applied in renegotiation. Actually the drop in costs which occurred in the summer of 1945 can be considered a result of continuity in production of types already being built and a result of the elimination of the graveyard shift. The prices set in these contracts are open to criticism on the grounds that they did not require the contractor, in return for his higher retainable profit, to cut costs on successive rounds by a percentage higher than they had been cut in comparable rounds of the ways of Liberty ships. But at least it may be said that these higher retainable profits were in accord, theoretically, with the principle on which the selective-price contract was based: namely that the contractor should receive more profit in return for taking some risk and effecting economies.[44] Moreover, shipyards faced the prospect of closing down when these contracts were completed. Everyone in the shipyard was tempted to "string out the work" so as to make his job last as long as possible. Selective-price contracts counteracted this tendency, for they gave contractors larger profits if they finished promptly.

But, and this is the most questionable feature of these arrangements, the selective-price contracts concluded in 1945 were so written as to apply not only to the vessels remaining to be constructed but also to many which were partially built and even to many which were already delivered. Since, within the wide range stated in the contract, the builder could select his price for them, receiving a higher or lower profit according to the price selected, his only risk was that he might not have kept adequate account of

[44] See above, chap. 20.

his costs.[45] The risk hardly seems to justify exempting these contracts from renegotiation.

Profits exempt from renegotiation, on the three largest of the selective-price contracts and on those under which North Carolina

TABLE 19

PROFITS ON SELECTIVE-PRICE CONTRACTS
(Not subject to Renegotiation)

	CONTRACTORS			
	Permanente-Richmond No. 1 and No. 2	Oregon Ship	Cal Ship	North Carolina
Number of Ships				
Delivered Before Contract Made	53	15	15	..
Incomplete When Contract Made	57	59	64	..
Total	110	74	79	42
Contract Value				
Excluding Cost of Materials (in millions)	$142.9	$104.0	$ 89.6	.....
Cost of Materials (in millions)	$129.2	$ 95.0	$ 80.5	.....
Including Cost of Materials (in millions)	$272.1*	$199.0*	$170.1*	$96.1†
Profits (Before Taxes)				
Sum Retained by Contractor (in millions)	$ 12.5	$ 10.8	$ 8.4	$ 6.9
Rates of Profits Retained (Per Cent of Contract Value)				
Materials Excluded	8.7	10.4	9.4	.....
Materials Included	4.6*	5.4*	4.9*	7.2†
Recapture by MC (in millions)	$ 5.	$ 2.	$ 3.1	N.A.§

* Materials furnished by Commission.
† Materials furnished by company.
§ The company reported simply "refunds of excessive profits will be required by the terms of the contracts."
Source: House, *Investigation of Shipyard Profits, Hearings*, pp. 441, 476, 489, 528, 645, 649, 652, 654, 687.

Ship had been operating ever since June 1943, are shown in Table 19. The rates of profit can be compared with those in Table 18.

From the preceding discussion of renegotiation and of the various ways of figuring " profit," it should be clear that it is impossible from the figures available to state precisely what portion

[45] This risk was stressed by the contractors. See House, *Investigation of Shipyard Profits, Hearings* (1946), pp. 644, 651. The attack on such contracts by T. H. Reavis is in *ibid.*, pp. 685-92.

out of a total of about $14 billion spent for ships went to shipbuilding corporations as profit. In making a rough estimate, the profits after renegotiation reported by the Price Adjustment Board (Table 17) provide one starting point, $322 million, but this figure does not include the contracts not renegotiable or renegotiated by some other Board. A second starting point is the figure given by the Maritime Commission to the House Committee on Merchant Marine and Fisheries in July 1946, namely, $356 million, but this figure includes only those companies that operated plants that were fully owned by the government. It therefore excluded all the old-line shipyards, even those that had added extensively to their plants by constructing facilities paid for by the government. It does not include Sun Shipbuilding and Dry Dock Co. which, after the Commission had paid for a $27 million dollar addition to its $7 million prewar plant, had more building berths than any other single, contiguous shipyard.[46] And while the figure $356 million is incomplete in this very important respect, it derives from items nearly all of which were attacked by the shipbuilding companies as too large. Indeed it is hard to determine in these items how much renegotiation, nonreimbursable costs, and similar factors are excluded or included. Allowing for the inadequacies of this figure, and considering the figures reported by the Price Adjustment Board, I would guess that the profits (after renegotiation and before income and excess profits taxes) were: of suppliers about $310 million; of shipbuilding corporations about $350 to $400 million estimating conservatively.

This figure does not of course include the profits made in connection with the shipyards by individuals or corporations other than the shipbuilding corporations: it does not include the profits of those who sold real estate or services or supplies to the shipyards. While some instances have been referred to above in which high profits were made in such sales, there is no way of arriving at the total of such profits. My impression is that the total was not a large percentage of the total costs of shipbuilding, although the rate of profit was extremely high on some of these deals, such as real estate transactions, which were excluded from renegotiation.

[46] House, *Shipyard Profits, Hearings* (1944), p. 65; House, *Investigation of Shipyard Profits, Hearings* (1946), pp. 371-72.

ANALYSES OF COSTS

An indication of how much of the profits after renegotiation was paid as income tax and excess profits tax is the fact that for 1945 ship- and boat-building corporations that filed returns with net income paid 68 per cent of their net income as tax.[47]

II. ANALYSES OF COSTS

The cost of vessels built in World War II cannot in most cases be calculated for individual ships directly from the accounting records because the components and materials used in the shipbuilding

TABLE 20

AVERAGE COSTS OF SELECTED TYPES

(Explanation in text)

Type	Design	Column 1 "Direct Cost" (in thousands)	Column 2 "War Cost" (in thousands)	Column 3 "Pre-War Cost" (in thousands)	Displacement Tonnage	Direct Cost per Displacement Ton
Coastal Cargo..	C1-M-AV1	$2,136	$1,983	$1,280	2,391	$ 893.35
Standard Cargo.	C2-S-AJ1	2,292 ⎫	2,736	2,100	⎧ 4,285	534.89
	C2-S-B1	3,571 ⎭			⎩ 4,682	762.71
Standard Cargo.	C3-S-A2	3,523	3,659	2,460	5,386	654.10
Liberty Cargo..	EC2-S-C1	1,822	1,728	1,278	3,478	523.86
Victory Cargo..	VC2-S-AP2	2,494	2,512	1,958	4,442	561.46
	VC2-S-AP3	2,894	2,872	2,130	4,526	639.42
Tanker.........	T2-S-A1	3,068	3,011	2,316	5,458	562.11
Attack Transport	VC2-S-AP5	4,502	not available		5,839	771.02
Troop Transport	P2-SE2-R1	11,272	not available		9,676	1,164.94
	C4-S-A1	10,414	not available		8,442	1,233.59
Frigates........	S2-S2-AQ1	2,254	not available		1,240	1,817.74

were to a large extent bought and paid for by the Maritime Commission. For reasons already explained, the Commission's accounts concerning the actual distribution of this material to particular yards are not complete. The average cost of a particular type can be constructed, however, in such cases by adding to the builder's average cost, as determined in settling his contract (after recapture but before renegotiation), the average cost of a standard bill of Commission-furnished material for that type. Figures constructed in this way are shown in Table 20 as the " direct cost." The total " direct cost " of the 5,601 ships delivered on Maritime

[47] *Statistics of Income for 1945* (U. S. Treasury Dept., 1950), pt. 2, pp. 102-03.

TOTAL DIRECT COST OF PRINCIPAL TYPES
Of Vessels Delivered On Maritime Commission Contracts in 1939-1945

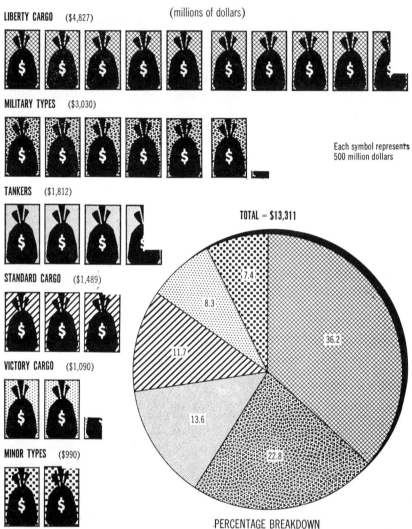

FIGURE 49. "Standard cargo" includes combination passenger-and-cargo ships. Source: Fischer, *Statistical Summary*, Table H-6.

ANALYSES OF COSTS 821

Commission contracts in 1939-1945 was $13.3 billion.[1] The total direct cost of principal types is shown in Figure 49.

On the other hand, the total disbursed by the Maritime Commission between October 1936 and June 1946 was about $18 billion. Part of the difference between $18 billion and $13.3 billion is easily accounted for and clearly assignable either to costs of ship construction in the broad sense or to activities of the Commission not connected with the ship construction of 1939-1945. For example, about $600 million was spent on facilities and about the same amount on conversions or on vessels still under construction in 1946. On the other hand, large sums were transferred from the Maritime Commission to the WSA. But the Maritime Commission's accounts contain so many different funds and so many transfers between them that I find it difficult to determine how much of the $4.7 billion, which is the difference between $18 billion and $13.3 billion, was spent because the Commission was building ships, and how much was spent because of one or another of the Commission's many other functions.[2]

In connection with determining the price at which ships should be sold after the war, the Finance Division prepared overall estimates of the cost of ships built and building as of June 30, 1946. After subtracting from "gross disbursements" many millions for recoveries, the remainder was broken down as shown in Table 21, Part A, in round figures. Although these figures are useful for the purpose for which they were constructed, they are not comparable with the "direct cost of ships" given above because different items are included and excluded. The $14.7 billion includes not only facilities and MC administration but also surplus material and some ships delivered after January 1, 1946. It excludes some costs which had been "recovered." It even excludes as "recovered" the cost of the four emergency yards which the Commission decided to keep as standby yards.[3]

Costs of ship construction are customarily broken down into three main categories: labor, material, and overhead. In attempting such a breakdown for the wartime shipbuilding as a whole, it seems best to start with the direct cost of the 5,601 ships delivered

[1] Fischer, *Statistical Summary*, Table H-5. Cost of defense features is included.
[2] MC *Report to Congress*, 1946, Appendix A, and above, chap. 22, sect. I.
[3] "Cost of War-Built Vessels from Inception, October 25, 1936 to June 30, 1946" in MC, files of Wesley Clark.

on Maritime Commission contracts in 1939-1945, namely $13.3 billion. For those ships the material cost can be estimated as $5.9 billion; the cost of labor as $5.8 billion; and profits, overhead and other costs of shipbuilding companies as $1.6 billion. The figure of $1.6 billion is derived by subtracting from the costs before renegotiation the estimates of material and labor costs.[4] Therefore

TABLE 21

TOTAL COST OF MARITIME COMMISSION SHIPBUILDING

(in billions)

A. Ships built and building as of June 30, 1946. Estimate of Finance Division allowing for recoveries and surplus materials.	
Costs and fees of shipbuilding companies	$ 9.4
Commission-procured material (including surplus)	4.7
Facilities paid for by Commission	.5
Administration in the Commission	.1
Total Cost	$14.7
B. Ships delivered, 1936-45. Estimate not allowing for recoveries or for all surplus materials.	
Labor	$ 5.8
Materials	5.9
Overhead of shipbuilding companies	1.6
Government-furnished facilities	.7
Maritime Commission administration	.2
Total Cost	$14.2

Sources explained in text.

it includes the $50 million or more recovered through renegotiation of shipbuilding contracts. Similarly, the estimated cost of material includes the $270 million, approximately, which was recovered by the renegotiation of supply contracts. The figure of $1.6 billion includes also, of course, the profits of shipbuilders, which I estimate as being, before renegotiation and taxes, about $400 to $450 million.

To estimate the total cost of constructing these 5,601 ships, certain items of overhead paid for by the government directly

[4] Costs of materials have been estimated on the basis of the average of actual costs of a standard bill of materials for each type for which the Commission furnished materials. From these data the material cost per displacement ton was calculated by types. This unit material cost was then applied to similar types of ships for which the Commission did not purchase materials. Labor costs were built up by using the figures on manhours and on wages described in Fischer, *Statistical Summary*, Section G.

must be added. The facilities, including housing and transportation facilities, paid for by the Maritime Commission cost $588 million. Other facilities paid for by the Defense Plant Corporation, mainly industrial equipment, cost $105 million.[5] The administrative cost in the Maritime Commission on account of ship construction has been estimated as $144 million,[6] but by

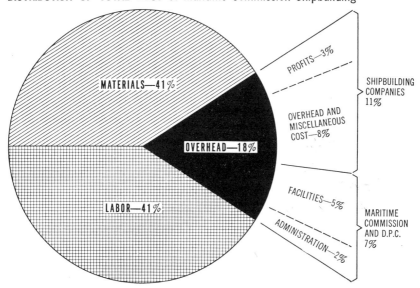

FIGURE 50. The bases for these estimates are described in the text.

including dubious items the figure could be raised to about $200 million. Adding these various estimates gives $14.2 billion as the total cost of ships delivered 1939-1945 (Table 21, Part B).

The distribution of costs for a shipbuilding company in normal peacetime operation has been estimated as: materials, 40 per cent; labor, 35 per cent; overhead, 25 per cent.[7] The distribution of costs in the Maritime Commission's shipbuilding program as a

[5] House, *Independent Offices Appropriation Bill for 1946, Hearings*, pp. 546-48; 584-85.

[6] This estimate was made, in order to determine costs for the Ship Sales Act, by going through payrolls and assigning appropriate salaries to ship construction. Statement by Wesley Clark, April 25, 1950.

[7] James W. Culliton, " Economics and Shipbuilding," in Fassett, ed., *The Shipbuilding Business*, p. 11.

whole is not so different as one might expect in view of the very different nature of the industrial process in a period of multiple production. (See Figure 50.) It shows relatively less overhead; but these figures are estimates open to doubt, and the Maritime Commission did not have to compute interest on money it invested in materials or facilities.

My doubts concerning the estimates I have given above relate particularly to the figure for materials and that for overhead. The estimate of $5.9 billion as the material cost may not include as much as it should of the cost of " surplus " material, that is, material which the Commission had paid for but which did not find its way into ships in 1939-1945. Much was left in the yards at the end of 1945. Some went into ships built later; much was disposed of as part of the settlements by which the Commission freed itself of the obligation to " restore " the shipyard property to its previous condition; some was sold as war surplus; and some went unaccounted for.[8] The estimate of the Commission's administrative cost for shipbuilding activities, also, may not be large enough in view of the difficulty of allocating funds among the many activities on which the Maritime Commission disbursed in all about $18 billion. Very likely the estimate fails to represent adequately the cost of having a large government agency directing the whole program and dispensing materials and money in billions.

Uncertainty in the figures is a less serious problem in comparing the costs of construction of different types of ships. So long as figures compared were constructed on the same basis, the comparison is of value. In Table 20 the figures in Column 1, " Direct Cost," are based on the amounts paid the shipbuilding companies for deliveries 1939-1945. In case all or part of the material was furnished by the Commission, the actual average cost of a standard

[8] In connection with figuring costs in order to determine sales price, Mr. Clark suggested that 15 per cent be added to the material cost derived from a standard bill of materials if it was desired to bring figures per ship up so that they would include surplus and would balance with the figure of $4,664 million given in Table 21, Part A, as the total cost of Commission Procurement. This method of distributing the cost of the surplus materials, by allocating a certain amount to each ship, was not applied, however, either in compiling figures for ship sales or in compiling the figures which are the basis of Table H-5 in Fischer, *Statistical Summary*. In the $5.9 billion shown estimated as material cost on Table 21, part B, about $4 billion was for Commission-procured material. Perhaps 15 per cent of $4 billion (namely $.6 billion) should be added to arrive at a figure including " surplus " material. This would raise the total cost of ships delivered, 1939-1945, from $14.2 billion to $14.8 billion.

ANALYSES OF COSTS 825

bill of Commission-furnished material for a ship of that design has been added. Cost of defense features is included but there is no allowance for the cost of facilities or of the Maritime Commission's administration.[9] Column 2, on the other hand, gives figures of cost calculated as directed by the Ship Sales Act (generally based on 1944 costs). They do not include the cost of defense features but do include an allocation to individual ships of the cost of facilities and of administration in the Maritime Commission.

Column 3 gives as " prewar cost " the estimates constructed by the Maritime Commission according to the directions in the Ship Sales Act. The Commission's staff was instructed to calculate what it would have cost to build ships of these types under average conditions of January 1941.[10] They used cost data concerning ship construction at that date, making adjustments for the differences between types in size, complexity, propulsion, and so on. Their " prewar " costs assumed an order for about 5 ships at a time. Therefore the figures in some cases permit comparing costs under the wartime conditions of high wages, high material costs and multiple production with costs under prewar conditions of relatively lower wages and material costs but without the savings of multiple production.

Not all the yards achieved the full benefits of multiple production during the war, however. Very similar standard cargo ships were built both at Moore Dry Dock (the C2-S-B1) and at North Carolina Ship (the C2-S-AJ1). Only in the latter yard were the equipment and the process of construction such as to give a clear example of multiple production. Figured according to the method used for the Ship Sales Act, the average cost of 18 standard freighters delivered from North Carolina June-July of 1944 was $2,330,000, whereas the prewar cost of a comparable ship was figured as $2,100,000.[11] Wage rates had gone up about 50 per cent between January 1941 and January 1944, and materials had risen also; but the savings of multiple production were such that the rise in the cost of building this ship was only 10 per cent.[12]

[9] Fischer, *Statistical Summary*, Table H-5.
[10] *Federal Register*, Aug. 17, 1946, pp. 8972-77. Cf. *ibid.*, Apr. 23, 1946, pp. 4459-74.
[11] Figures and explanations given me Apr. 25, 1950 by Joseph F. Barnes and L. M. Wuertele, MC, Bureau of Technical Activities, Division of Engineering Cost.
[12] Fischer, *Statistical Summary*, Table G-8, and above, chap. 7, section IV.

Comparison of the construction cost of the same type in various areas is worthwhile only for the Liberty ship (Table 22). Since these figures are direct cost, based on a standard bill of materials, the variations are mainly a reflection of differences in labor costs. Wage rates were highest on the West coast, but the productivity of labor was highest there and the average cost was the lowest.

TABLE 22

COST OF LIBERTY SHIPS

(E2-S-C1)

Sections and Yards	Number of Ships	Cost per Ship (in thousands)
Lower South		
Delta	128	$1938.9
Jones-Brunswick	85	1992.8
Jones-Panama	66	2020.4
St. Johns	82	2099.2
Southeastern	88	2043.0
Todd-Houston	208	1833.4
Average		1954.6
Pacific Coast		
California	306	1858.2
Permanente No. 1	138	1875.3
Permanente No. 2	351	1667.5
Oregon	330	1643.1
Average		1737.7
Other		
North Carolina	126	1543.6
Bethlehem-Fairfield	361	1754.9
New England	228	1892.0
Average		1761.6

In all calculations of the number of ships and the total cost the figures used are based on the number of ships delivered complete, in Fischer, *Statistical Summary*, Table H-5.

High productivity of labor depended above all on continuity in the production of the same type. Consequently comparison of the manhour costs in various regions is largely to be explained by examining how long the individual yards in the region had been in production without changing from the construction of one type of vessel to the construction of a different type. This is evident when we turn to consider manhour costs for the program as a whole in various regions at various time (Figure 42). In all regions the manhour cost rose in 1942 compared to 1941, that is, the number of light displacement tons produced per manhour was less in 1942 than in 1941. The drop in productivity was especially sharp in the Gulf region because the figures for 1941 are dominated by one low-cost yard, whereas the figures for 1942 include the first few rounds of ships to come off the ways of the new emergency yards. Productivity in the Gulf in 1942 was the lowest of all regions, and it increased very little in 1943 because many new yards were brought into production that year. It rose sharply in 1944, however. In averages for 1944 as a whole, the Gulf had the highest productivity per manhour of any region, as a result of the fact that there was more continuity in production through 1943 and 1944 on the Gulf than elsewhere (Table 23).

The West coast led all other regions in productivity per manhour from the time when new yards began to come into production until the time when West coast yards were changed over from Libertys to Victorys and Victory transports. Because of this change of types the West coast lost in 1944 about one third of the improvement in employee productivity made in 1942 and 1943.

The East coast maintained a fairly steady pace of increasing productivity, after a slight drop from 1941 to 1942, and on the whole ranked between the West coast and the Gulf. Productivity in yards on the Great Lakes varied widely, but performance there must be considered in a somewhat different light from that in the other three regions, inasmuch as they were without the elaborate facilities provided in the new yards elsewhere and were engaged in smaller scale production of particular types of vessels such as the frigates, which required many manhours but had little displacement.

Comparing the record of the four regions for the five years as a whole suggests some comment on the decisions made in 1941 in choosing locations. The Commission built many more new shipways on the West coast than in any other region (Figure 9). The high productivity on the West coast (Figure 42) proved that this was a wise decision. But, as expected, labor became scarce on the West coast. In order to draw on the labor supply of the South the Commission had decided to build new yards there in spite of obvious disadvantages. The Gulf lagged far behind the West coast, as expected, and was slow in reaching a comparable level of

TABLE 23

PRODUCTION, EMPLOYMENT, MANHOURS, AND PRODUCTIVITY
1941-1945

Region and Year	Production (Displacement tons, thousands)	Employment (thousands)	Manhours (millions)	Productivity	
				Per Employee (Displ. Tons)	Per Manhour (Displ. Tons per thousand Manhours)
All Regions					
1941	794	71	158	11.2	5.02
1942	3,466	341	822	10.2	4.21
1943	7,191	620	1,449	11.6	4.96
1944	6,892	574	1,347	12.0	5.12
1945	3,560	289	666	12.3	5.35
East Coast					
1941	434	39	91	11.0	4.77
1942	1,159	115	296	10.1	3.91
1943	2,363	206	510	11.5	4.63
1944	2,347	186	442	12.6	5.30
1945	1,242	96	225	13.0	5.52
Gulf Coast					
1941	91	7	13	13.8	7.40
1942	512	54	133	9.5	3.85
1943	1,056	105	246	10.1	4.29
1944	1,302	101	245	12.9	5.32
1945	627	53	128	11.8	4.90
West Coast					
1941	251	23	50	10.8	4.98
1942	1,673	163	369	10.3	4.54
1943	3,648	292	645	12.5	5.66
1944	3,121	269	612	11.6	5.10
1945	1,614	131	288	12.3	5.60
Great Lakes					
1941	18	2	4	9.9	4.77
1942	121	9	24	13.3	5.03
1943	124	18	48	7.0	2.60
1944	122	18	48	7.0	2.58
1945	78	10	24	8.2	3.21

Source: Fischer, "Labor Productivity," Table 1.

labor productivity. In 1943 the output per thousand manhours on the Gulf was only 4.29 displacement tons whereas on the West coast it was 5.66 displacement tons. In 1944 the output from the Gulf was 5.32, however. Given time and aided by continuity in production of the Liberty type, the Gulf yards attained in 1944 rates of productivity in line with the average on the East coast and West coast,[13] although for 1941-1945 as a whole the average cost of Libertys built in the Lower South was $200,000 more per ship than the average cost of these built elsewhere. All things considered, the decision to utilize the labor supply of the Deep South seems to have been well worth this extra cost.

As a whole an analysis of costs in shipbuilding during World War II emphasizes two conclusions already presented: the importance of adequate control of inventory, and the seriousness of decisions changing the types of ships being built. If a yard was devoted continuously to the production of a single type during the four-year period 1942-1945, it could double productivity of labor. Changes in the military situation and national needs prevented any such thorough concentration on the advantages of multiple production of standardized types.

III. Conclusions

In concluding, a historian may be tempted to sum up the lessons to be learned from this largest of all shipbuilding efforts. But the lessons that should be drawn from the experience of 1940-1945 will depend on the situation to which they are to be applied. In 1940 and 1941, the Maritime Commission and the leaders of the shipbuilding industry attempted to apply lessons learned from 1917-1919. Recalling the overconcentration in the Northeast, they wisely placed many new shipyards on the Gulf and Pacific coasts. Recalling the dangers of scamping, strikes, and an upward spiral of wages, they joined in the Stabilization Committee and through its agreements raised wages by a method that avoided the evils of inter-yard competition. Profiting from the achievements as well as the mistakes of World War I, they planned for multiple production and emphasized standardization through adoption of the Liberty ship design. Some of the lessons learned from World War I proved irrelevant, however, or wrong.

[13] Fischer, " Productivity."

Plans to have the steel fabricated outside the shipyards in structural steel plants had to be abandoned. The shipyards were equipped with fabricating facilities of their own. Direct experience with the new methods of cutting, welding, and subassembly was necessary before anyone realized fully how much space was needed in a yard devoted to multiple production. The success of the Maritime Commission's shipbuilding program was due in part to lessons learned from World War I, but it depended also on readiness to learn new lessons from day-by-day experience in new situations.

The making of technological innovations and of corresponding adjustments in the organization of labor was facilitated by the basic policy adopted by the Commission in regard to the management of management. Production methods within each yard were the responsibility of the private company managing the yard. Maritime Commission officials were at each yard to prevent fraud and to help the management, but were told that it was not their function to run the yard. Since the government was paying for everything in the new emergency yards, they were often spoken of as if they were government enterprises. Indeed, from the point of view of administering a national shipbuilding plan, the executives of the shipbuilding companies were subordinates of the Maritime Commission. Each contractor had freedom of action, however, within the terms of his contract, far more freedom of action than he would have had if he had been a government official subject to administrative orders. Contracts with private corporations thus provided a decentralized system of administering a national plan. This decentralization enabled managers of successful shipyards to go ahead without being hamstrung by interference from Washington. Difficulties occurred when a shipbuilding company failed to perform satisfactorily within the sphere of action allotted to it by its contract. In some cases it was judged necessary to change the management, and such a change was more difficult because the management could insist on its contractual rights. On the whole, however, this decentralized system, with the incentives provided by the variable-fee contracts and by the national need, was extremely effective in producing ships rapidly. It facilitated the innovations in technology and in labor utilization which were the essence of multiple production.

As a result of the new methods shipyards began to turn steel into ships at so rapid a rate in 1942 that there was not enough steel to keep them all fully occupied. Even after the contract with Higgins was cancelled, the shipyards could have used more steel than was allocated to them. The Maritime Commission fought for a larger share of the steel output, and at the same time struggled, not too successfully, to regulate the flow of steel efficiently, to "squeeze the pipe-line," and keep each of its yards supplied with the minimum inventory essential to operation. The outcome was that the shipbuilders met their goals brilliantly in 1943. The ships produced that year in American shipyards made possible the successful landing in Europe in June 1944.

In 1944 and 1945 the decentralized method of managing management was subjected to new strains. Standardized multiple production of Liberty ships was replaced by the construction of a variety of types of greater complexity. Pressure within the Maritime Commission for the building of better, faster merchant ships, and pressure from the Joint Chiefs of Staff for the construction of transports and other military types changed the character of the production problem. The WPB opposed the change but was overruled by the Joint Chiefs. In 1944 and 1945 the Maritime Commission was building Victory ships and military types which required more labor per ton delivered. At the time that the program of the Maritime Commission was thus changing, manpower, not steel, was coming to be regarded in the nation as a whole as the chief factor limiting production. In managing management it became important to stress not only speed but also economy in the use of labor.

To attain this result and at the same time keep the advantages of a decentralized system of managing management, the Maritime Commission attempted to put into effect new types of contracts with the shipbuilding companies. In making its initial contracts for operation of the emergency yards, the government had profited from the lessons of World War I to the extent of avoiding the kind of contract which had then acquired the most unsavory reputation, namely, that called cost-plus-a-percentage-of-cost. Instead, the Commission used variable-fee contracts which gave less profit to the least efficient producers. But for reasons that have been explained, these contracts provided insufficient incentives to economy in the use of manpower. Attempts were made to negotiate lump-sum contracts instead. Although the Maritime

Commission began urging the change in 1943, it was unable to effectuate the shift until 1945 after the emergency was passed.

Meanwhile there was considerable pressure for a more centralized management of ship construction. In regard to first one specific problem and then another, the Maritime Commission was urged to use its control over reimbursements under the variable-fee contracts in such a way as to dictate to management in detail, especially in regard to the training and utilization of labor. The Commission's staff of auditors and inspectors had many occasions to pass judgment on matters which a yard management considered within its sphere of decision. The auditors and engineers employed under Civil Service were fully subject to orders issued through the divisions in Washington or through the regional offices. Extending their authority would decrease the freedom of managements and thereby create a more centralized control over the shipbuilders.

To a certain extent Admiral Vickery and the regional directors did tell management how to manage and brought about improvements in the operation of the yards. But any more centralized, more "bureaucratic," method of managing management might place too heavy a burden on the administrative organization of the Commission. In the functions which it did perform through its Civil Service staff, the Maritime Commission had difficulties in coordinating the work of the regional offices and of the various divisions in Washington. Some difficulties of this kind were of course inevitable and without serious consequences. The most serious difficulty was in keeping track of materials. The very nature of shipbuilding with multiple production made this a huge problem anyhow; and the administrative organization was so fashioned as to put the emphasis on making sure that the shipyards received the materials and money they needed to turn out ships. It was less effective in making sure that all movement of materials was fully recorded. The attempts by the Production Division, the Finance Division, and the regional offices to keep track of materials did not mesh smoothly. One result was that inventories were larger than they might otherwise have been. Another result was that the accounting record was incomplete. Not all the difficulties in accounting were due to the nature of the records concerning the receipt and removal of material. That was one weakness; another was the deficient organization and staff for compiling a central accounting record. Inventory control and accounting were weak points in the Commission's functioning; its strong point was that it produced the ships and produced them fast.

In connection with the administrative organization the question naturally arises whether it would not have been better to have separated entirely the Maritime Commission from the War Shipping Administration, the administration of ship construction from the administration of ship operation. Without a thorough study of the WSA, no balanced judgment is possible, but I am inclined to think history can give no answer to guide future decision in such a matter. To be sure the Maritime Commission staff could have performed certain functions better, accounting for example, if it had not been busy also with the functions of the WSA. But how would the WSA have found its staff without drawing key men from the staff of the Commission? As is clearly symbolized by the position of Admiral Land as the head of both agencies, the problem was one of personnel or of personalities. If it arises in the future, it will then also very likely and correctly be decided on that basis.

Attempts to learn lessons from the past inevitably seek out mistakes, although it is perhaps as difficult to diagnose the causes of a particular failure as to diagnose the reasons for success. On the whole the Maritime Commission's shipbuilding in World War II was a success. The main demand on the Commission was that it deliver the ships. In spite of flaws in its administration, the organization headed by Admirals Land and Vickery had ships built with unparalleled speed. It is easy in retrospect to say that they should have done better in this or that; it is difficult to realize how desperately ships were needed in 1941-1944 and how unlikely it seemed then that means could be found to build 19 million tons in one year as was done in 1943. Compared with what had been done before or with what contemporaries thought it reasonable to expect, the achievement commands admiration.

BIBLIOGRAPHICAL NOTE

In addition to the official "Minutes" of the Maritime Commission, kept in the Office of the Secretary, the following files of the Commission were used.
1) Current files:
   (a) the General Files (gf), cited by file number.
   (b) the files of many administrative units. They are identified by the name of the unit at the time the files were used.
2) Files transferred from administrative units no longer in existence to the Maritime Commission's Record Administration, cited by the designation under which they were being kept by the Records Administration.
3) Reading files, cited as reading files, most of which are in the hands of whoever maintained the file originally. That of Admiral Land was used through the courtesy of Miss E. H. Van Valey.
4) Files in the "Historian's Collection," and so cited, namely:
   (a) the Research Collection, containing research reports written in the Office of the Historian, notes on interviews, and other papers collected in the process of preparing this volume.
   (b) the Land file, papers which had been collected from Admiral Land's office and were about to be distributed through the Commission's general files, but, instead, were as a unit placed at my disposal as Historical Officer, and consequently form part of the "Historian's Collection."
   (c) Other collections which came from the offices of Commissioners or from administrative units of the Commission and which were placed by the Records Administration in the "Historian's Collection."

I hope that most of this collection, at present 24 filing cabinets in the Y building with the other Maritime Commission records, will ultimately be placed as a unit in the National Archives.

ABBREVIATIONS USED IN FOOTNOTES

MC = Maritime Commission
gf = General files
WSA = War Shipping Administration
PR = Press release
Bland Committee = House, Committee on the Merchant Marine and Fisheries
Truman Committee = Senate, Special Committee Investigating the National Defense Program

Works cited are entered in the index under the abbreviations used in citing them, and the index gives reference to the first citation, which contains the full title and identification. Congressional documents are indexed under "House" or "Senate," followed by the title of the document.

INDEX

A

Absenteeism, 412-414, 440, 446, 484, 661, 673.
A. C. Stanley Co., 509, 520-521.
Accident rate in shipyards, 448.
Accounting Branch, Budget and Accounts Section (MC), *see* Finance Division.
Accounting profession, 732, 739, 744-745, 749.
Accounting Section (MC), *see* Finance Division.
"Accounting Sheets, Ship Construction Costs," 476 n.
Accounts:
 of Maritime Commission, 15, 739-749, 763, 767-773, 796, 819, 832-833; *see also* Auditing;
 of shipbuilders, 128, 472, 537, 729-730, 739, 816-817.
Accounts Section (MC), *see* Finance Division.
Acting Chairman (MC), 795.
Adams, Carl E., 323.
Administration, *see* Administrative conflicts; Administrative orders; Bureaucracy; Commission action; Commissioners; Decentralization; Personnel; Regional Offices.
Administrative and Processing Branch, Priorities Section (MC), *see* Production Division.
Administrative conflicts, 832;
 with other agencies, *see* General Accounting Office; Navy; War Manpower Commission; War Production Board; War Shipping Administration;
 within MC, *see* Accounts; Design changes; Inventories, control of; Material coordinators; Personnel of MC; Purchase controllers; Regional Directors; Resident auditors; Training supervisors.
Administrative expenses of MC, 712, 719, 724.
Administrative orders of MC, 17, 390, 698, 700, 705, 709, 723, 727, 767-769, 771 n, 778.

Administrative Section (MC), *see* Emergency Ship Construction Division; Production Division.
Administrative Unit, Construction Audit Section (MC), *see* Finance Division.
Admiralty metal, 409.
Advances to contractors, 401, 403-404, 480, 481 and n, 494, 780.
Advertising expenses, 737.
Advisory Committee on Official Production Statistics, 362.
AFL, *see* American Federation of Labor.
Agents of MC, by contract, 90, 97-99, 119, 401.
Agriculture, Department of, 19, 652.
Aiken, George D., 791-792.
Air pump, 75.
Air shipments, 648.
Air Transport Association of America, 795.
Aircraft carriers, 29, 37.
Aircraft carrier escorts, 4, 149, 177, 180 n, 324, 325, 329, 339, 360, 399, 574, 584, 608, 612-614, 720-721, 777, *facing* 362, 587;
 manhours per ship for, 234.
Aircraft industry, 10, 49, 107, 206, 235, 414, 656, 798.
AKA, 653. *See also*, Attack cargo ships.
Alabama Dry Dock and Shipbuilding Co., 35, 36, 50, 51, 59;
 change of management at, 537;
 construction methods at, 230, 563;
 cost of ships at, 682;
 fees, 476;
 labor at, 286, 294-295;
 layout and facilities, 51, 62, 223;
 productivity at, 209, 646, 683;
 race riot at, 253;
 speed of construction at, 208;
 and tanker program, 361, 583, 640;
 transportation to, 430.
Alameda, Cal., 117. *See also* Bethlehem-Alameda shipyard; Pacific Bridge Co.
Albany, N. Y., 152, 153.
Albina Engine and Machine Works, 276.
Alien Property Custodian, 585.
Allegheny Ludlum Steel Corp., 313.

835

Allen, Ethan P., *Policies Governing Private Financing*, 108 n.
Allen, T. R., 504 n.
Allis-Chalmers Manufacturing Co., 399, 595.
Allocation of funds to MC by President, 712, 713.
Allocation of labor, 662-663.
Allocation of materials, 97, 318, 354-355, 358, 366, 380-382, 390, 663;
in excess of capacity, 384, 393;
responsibility for, 370, 372, 376, 697.
See also, Components; Scheduling; Steel.
Allowance list, 18 n, 359, 360, 700.
Alloy steel, 339.
Aluminum, 339, 378, 409.
Amalgamated Clothing Workers of America, 270.
America, 18, 22, 73, 102, *facing* p. 234.
American Academy of Political and Social Science, *Annals*, 22 n.
American Bankers Association, *Special Bulletin*, 402 n.
American Bureau of Shipping, 2, 4, 86, 87, 93, 320, 408, 410, 511, 546-552, 556-560, 562, 563, 565, 566, 576-577, 591, 629, 630, 763;
Bulletin, 7 n.
American Federation of Labor, 240, 241, 271, 277, 495, 496, 794;
and Higgins yard, 183 n, 186-187, 195-197, 198;
and job classifications, 418;
rivalry with CIO, 196, 294-297, 522-525;
and Shipbuilding Stabilization Committee, 274-275;
and wages, 309.
See also, Metal Trades Dept., AFL.
American Hawaiian Steamship Co., 617.
American Hoist and Derrick Co., St. Paul, Minn., 405.
American Institute of Accountants, 732.
American Institute of Electrical Engineers, 408, 567.
American Iron and Steel Institute, 313.
American Magazine, 13 n.
American Merchant Marine Conference, *Proceedings*, 1944, 449 n.
American President Line, 624, 625.
American Shipbuilding Co., 52 n, 53, 92, 462, 584-585;
Cleveland yard, 544, 630;
and employees for Delta yard, 247;
and frigates, 614;
labor at, 295;
Lorraine yard, 630.
American Welding Society, 559, 567.

Ames, Amyas, 756, 770.
Amortization, 107, 400, 808, 810.
Analysis, Cost Studies and Reports Unit, Construction Audit Section (MC), *see* Finance Division.
Anchor chain, 85, 87, 358, 366, 397, 697.
Anchors, 85, 86.
Anderson, Col. P. M., 631.
Anderson, Marian, *facing* 75.
Anderson, Robert Earle, 16, 20, 156, 157, 762, 764, 766, 767, *facing* 362;
and contracts, 123, 124, 401, 403, 477, 482, 738;
and finances, 770;
dismissal of, 795, 796;
and managements, 130, 462, 467, 478-479, 499, 531, 535, 537, 538, 540-541, 681;
Merchant Marine and World Frontiers, The, 22 n;
on personnel, 136;
"'Price-Minus' Contracts Speed Shipbuilding," 124 n;
"United States Maritime Commission Procedure," 123 n.
ANMB, *see* Army and Navy Munitions Board.
Annapolis, *see* Naval Academy.
AP5's *see* Transports, Victory.
APA, 653. *See also*, Attack transports.
Appleton Iron Works, 405.
Apprentices, 237, 240, 246, 258, 262, 265, 418.
Appropriations, 24, 55, 341, 712-721, 741-742.
Archibald, Katherine F., *Wartime Shipyard*, 254 n.
Arctic, 36.
Area Manpower Priorities Committees, 664, 665.
Area Production Urgency Committees, 665.
Arizona, 249.
Arkies, 249, 442.
Armament of ships, 84, 89, 582, 619, 623, 695.
Army:
Chief of Engineers, 189;
personnel lost to, 740, 750-751; *see also*, Selective Service.
procurement by, 335, 356, 370, 397, 401, 432, 490, 798;
renegotiation by, 778, 801;
shipbuilding by, 57, 636;
shipbuilding by MC for, 4, 392, 605, 617-621;
shipping requirements of, 138, 143-144, 332, 605, 617, 634;

INDEX 837

steel requirements of, 200, 314, 322, 326, 331, 335, 337, 339;
Surgeon General of, 622.
Army and Navy Liaison Branch, Priorities Section (MC), *see* Production Division.
Army and Navy Munitions Board, 57, 314, 317, 324-327, 342, 355-358, 370.
Arnott, David, 560, 561, 566, 570, 581; "Some Observations on Ship Welding," 31 n.
Assembly lines in shipyards, 185, 186, 190-194, 224-230, 238-239.
Assignment of Claims Act, 479.
Assistant Director for Departmental Activities (MC), *see* East Coast Regional Office.
Assistant Directors of Finance (MC), *see* Finance Division.
Assistant Directors, Regional Offices (MC), *see* Regional Offices of Construction.
Assistant General Auditor (MC), *see* Finance Division.
Atlantic coast yards:
and efficiency experts, 675;
labor in, 250, 268, 282, 284-285, 293, 308, 412, 415, 418-420, 423-426, 661;
productivity in, 644, 827-829.
See also, North Atlantic coast.
Atlantic Coast Zone Conference, *see* Zone Conferences.
Atlantic coastal transportation, 635.
Attack cargo ships, 4, 605, 616-617, 625, 627, 638-639, 647, 652, *facing* 778.
Attack transports, 607, 625-627, 648, 652, 653, 678-680, 685, *facing* 586.
Attorney General, 498.
Audit Branch (MC), *see* Finance Division.
Audit Practice and Procedures Unit, Construction Audit Section (MC), *see* Finance Division.
Auditing:
of construction, 27, 105-106, 124-131, 168, 417, 682, 687, 689, 706-709, 722, 728-739, 766, 813-815, 832; *see also*, Resident auditors;
of disbursement, 691, 723-728;
of ship operations, 722, 743-744, 766-767;
of ship repair, 761.
See also, General Accounting Office.
Auditing and Financial Analysis Section (Operations) (MC), *see* Finance Division.
Audits Section, *see* Finance Division.
Automobile industry, 185, 206, 224-226, 237, 268, 315, 318, 319.

Auxiliaries, *see* Naval auxiliaries.
"Avalanche Hits Richmond," 442 n-443 n.
Avondale Marine Ways, Inc., 630.
Award of Merit by MC, 67, 451, 453-454.
Awards, Committee on (MC), 379, 681, 683.
Axis powers, 3.
Ayre, Sir Amos L., 81, 576-577;
"Merchant Shipbuilding during the War," 8 n.

B

B7-A1 design, 631.
See also, Barges, concrete.
B7-A2 design, 631.
See also, Barges, concrete.
Babcock & Wilcox Co., 95, 205.
Baby flattops, *see* Aircraft carrier escorts.
Ballast, 80, 84, 89, 547, 554, 555, 567, 582, 695.
Baltimore, Md., 33, 34, 50-52, 110, 247, 256, 399, 441, 442, 702.
See also, Bethlehem-Fairfield Shipyard, Inc.; Bethlehem-Sparrows Point, Shipyard, Inc.
Bananas, 35.
Bancroft & Martin, South Portland, Me., 507, 510.
Bankers Trust Co., 540.
Bankruptcy, 496, 542.
Banks, 107, 402-403, 482, 497.
See also, Loans.
Bard, Ralph A., 417.
Barges, 325;
concrete, 60, 63-64, 628-634, *facing* 778;
deliveries and use of, 633-636;
steel, 63-64, 635;
wooden, 634-636.
Barnes-Duluth Shipbuilding Co., 407, 537, 563.
Barrett and Hilp, 630, 633.
Bartlett-Hayward Division, Koppers Co., 399, 585.
Baruch, Bernard, 604.
Basic Principles to be Observed in Establishing Production Training Programs, 264 n, 676, 678.
Basin yards, 216, 220, 617-618, 631, 633.
Batcheller, H. G., 313.
Bates, James L., 19, 25, 63, 74, 77, 567, 568, 577, 578, 580, 612, 629, 788, 789, *facing* 362;
"The Victory Ship Design," 555 n.
Bath, Me., 34, 216, 247. *See also*, Bath Iron Works Corp.
Bath Iron Works Corp., 34, 35, 36 and

n, 55, 216, 503, 504, 508, 509, 513, 515, 528, 532, 533;
  labor relation at, 268, 283, 308, 522.
Bath Trust Co., 504, 507.
Battleships, 37.
Bay Cities Metal Trades Council, 288, 290.
Bayer Company, 95, 96.
Bayonne, N. J., 538, 539.
  See also, East Coast Shipyards, Inc.
Beaumont, Tex., 35, 49, 433, 436.
  See also, Pennsylvania Shipyards, Inc.
Bechtel, K. K., 249, 466.
Bechtel-McCone Corp., see Bechtel-McCone-Parsons.
Bechtel-McCone-Parsons, 54, 809.
Bechtel management, 222, 269, 463, 470.
Bechtel, S. D., 146, 147, 155.
Bedford, Clay, 54, 141, 146, 228, 238, 261, 463, 469-470.
Belle Isle, 544.
Beloit Iron Works, 405.
Bench work, 82.
Bending slab, 226.
Bennett, William, 80 n.
Bethlehem, Pa., 33.
Bethlehem-Alameda shipyard, 117, 142, 624, 625, 639, 801.
Bethlehem-Fairfield Shipyard, Inc., 51, 52, 68, 177, *facing* 331, 779;
  construction methods at, 84, 217-218, 558, 563, 581;
  contracts of, 140-141, 457-458, 475, 680-683, 793;
  cost of facilities, 116, 472;
  cost of Libertys at, 826;
  inventories at, 391;
  labor at, 247, 293, 294, 413, 430, 656-657;
  layout and facilities, 51, 59, 217-218, 224, 579;
  LST's at, 176, 468-469, 610;
  Negroes at, 256;
  productivity at, 209, 475, 645, 683-684;
  profits of, 801, 805-806;
  rentals for land at, 115;
  sale of, 115-116, 791;
  speed of construction at, 140, 141 n, 145, 168, 175, 208, 211, 364, 458, 468-469;
  Victory ships at, 586, 607, 648;
  Yard Staff Committee, 704.
Bethlehem management, 33, 53, 77, 92, 102, 103, 189, 195, 462, 470;
  and labor relations, 268, 269, 273, 278, 281-283, 293, 308, 421, 422;
  training program of, 260, 264.
Bethlehem Shipbuilding Corp., Ltd., 196, 274.

Bethlehem-Sparrows Point Shipyard, Inc., 33-37, 52, 58, 247;
  additions to, 62;
  labor relations at, 293;
  renegotiation, 801;
  welding frauds at, 735, 737, 790.
Bethlehem Steel Co., 52, 149, 204, 359;
  Fore River yard, 33-37; 58, 130, 308;
  Hoboken yard, 293;
  Introduction to Shipbuilding, 260 n;
  Key Highway (Baltimore) yard, 247, 293, 610;
  San Francisco Yard, 33-37, 58, 276, 288, 290, 297;
  San Pedro yard, 276;
  Shipbuilding Division, 33, 124, 464, 578, 580, 581, 801, and *see* under yard names above and below.
  Staten Island yard, 33-37, 58;
  Union Plant, *see* San Francisco Yard.
  See also, Bethlehem management.
" Bethlehem's record of Rudder Production," 204 n.
Bids, 91, 96, 102-104, 133, 361, 405, 495.
Big Five shipbuilders, 37, 51, 52, 57, 102.
Bilge keels, 548, 581.
Bills of lading, 725-727.
Bills of materials, 378, 410, 699, 748, 819, 822 n, 824 and n.
  See also, Requisitions.
Bing, Alexander M., *Wartime Strikes and Their Adjustment*, 246 n.
Birmingham, Ala., 35.
Black market in steel, 196.
Black-out, 300.
Blakely Island, Ala., 439-440.
  See also, Alabama Dry Dock and Shipbuilding Co.
Bland Committee, 526-529;
  Confidential Committee Print, Oct. 6, 1942, 527 n;
  Committee Print, No. 106, 60 n;
  Report No. 2653, 510 n.
  See also, Committee on Merchant Marine and Fisheries.
Bland, Schuyler Otis, 450, 498, 511 n, 525-527, 529, 530, 553.
Board of Investigation . . . Welded Steel Merchant Vessels, 566-571.
Board of School Commissioners of Baltimore City, *Annual Report*, 1942, 256 n.
Boat builders, 184.
Boeing Aircraft Co., 668.
Boggs, Thomas Hale, 194, 195.
Bogie (manhours), 120, 122, 126, 234, 474-475, 477.
  See also, Standard manhours.
Boilermakers, 242, 254-255, 296.

INDEX 839

Boilermakers, Iron Ship Builders and Helpers of America, International Brotherhood of, 241, 243, 276; Portland, Ore., local, 296, 297; *Proceedings of . . . Convention,* 244 n, 297 n; *Report* of President of, 1944, 243 n. segregation of Negroes in, 254-255.
Boilers, 45, 74, 75, 83, 95, 97, 205, 397.
Bonneville dam, 54, 248.
Bonuses:
 paid by contractors, 479, 737;
 for savings in manhours, 122, 458 and n, 473, 474, 476;
 for speed, 121, 458, 473, 474, 476, 525.
Booker T. Washington, facing 75.
Bookkeeping, see Accounts.
Booms, 31, 87, 410, 580, 581.
" Boss, The," 184 n.
Boston & Maine Railroad, 518.
Boston, Mass., 8, 28.
Boulder Dam, 277.
 See also, Hoover Dam.
Bowdoin College, 504.
Boykin, Frank W., 197.
Bradley, Fred, 627.
Brewster, Ralph Owen, 513, 519, 528, 529, 530.
Bridge cranes, 214.
 See also, Cranes.
Bridge fashion plate, 547, 549.
Bristol, Pa., 47.
Britain:
 aid to, 40, 41, 65-68, 164, 627, 714, 799; *see also,* Lend-lease.
 shipbuilding in, 6-8, 72, 81-83, 306 and n, 576-577, 600;
 shipbuilding in the U. S. by, 42-46, 47 n, 60, 66, 73-78, 138-139, 215, 216, 218, 248, 463-464, 504, 505 n, 515;
 steel allocation for, 339;
 strikes and manpower in, 306 and n, 687.
British Admiralty, 609.
British Chiefs of Staff, 603.
 See also, Combined Chiefs.
British Libertys, 508.
 See also, Ocean class ships.
British Merchant Shipping Mission, 39, 42 and n, 60, 75, 80-82, 93.
" British Prototype of the Liberty Ship," 81 n, 82 n.
British Technical Merchant Shipping Mission, 80, 82.
Brooklyn, N. Y., 33.
Brown, Harvey, 273, 288.
Brown, Walston S., 486, 537, 681, 751.
Brunswick, Ga., 148, 436, 539.
 See also, Jones-Brunswick shipyard.

Brunswick Marine Construction Co., 535.
Bryan, Comdr. Hamilton V., 165.
Buckling, 547, 554.
Budget and Accounts Section (MC), *see* Finance Division.
Budget of MC, 17, 638, 712, 716 n, 717, 748, 767.
Budget Officer, *see* Finance Division.
Building berths, 36 and n, 59, 143, and n, 149 n, 153, 341, 347, 608, 640, 690.
Bulkhead assemblies, 227.
Bulwarks, 84, 88.
Bureau of Applied Economics, Inc., *Wages in Various Industries and Occupations,* 272 n.
Bureau of Labor Statistics, 106, 413, 424, 763;
 " Basic Wage Rates in Private Shipyards, June 1943," 259 n.
Bulletin No. 283, 272 n.
Bulletin No. 727, 420 n.
Estimated Labor Requirements for the Shipbuilding Industry, 237 n.
 " Hourly Earnings in Private Shipyards, 1942," 259 n.
 " Labor Statistics for Private U. S. Shipyards," 658 n.
 " Monthly Labor Turnover in Commercial Shipyards with Maritime Commission Contracts," 656 n;
 " Wartime Employment, Production, and Conditions of Work in Shipyards," 236 n.
Bureau of Marine Inspection and Navigation, 2, 408, 556, 557, 566, 581, 629, 630.
Bureau of National Affairs, *see Labor Relations Reference Manual; War Labor Reports.*
Bureau of Standards, 408, 567.
Bureau of the Budget, 180, 341, 713, 718, 719, 741, 755, 767, 794;
 The United States at War, 10 n.
Bureau of the Census: *Population, Congested Production Areas,* 438 n.
Bureaucracy, 353, 688, 832. *See also,* Administrative conflicts; Inspection; Managements, control over.
Burkhardt, John E., 578.
Burning, 257, 320.
Burns, William R., 691.
Burntisland, Scotland, 80 n.
Burr, Kenneth T., 507, 509, 510, 526-529, 532.
Buses, 430, 431, 441, 443, 701, 718.
Bushey, Ira S., 126.
Business Historical Society, *Bulletin,* 743 n.
Business Week, 296 n.

840　　　　　　　　　　　　　　INDEX

Butler, Walter, 674, 709-710. See also, Walter Butler Shipbuilders, Inc.
Byrnes, James F., 603-604, 669, 681.

C

C-types, see Standard Cargo ships.
C1-B design, 577.
C1-M-AV1 design, 628, 640, 647, facing 619. See also, Coastal cargo vessels.
C2-S-AJ1 design, 825, facing 490.
C2-S-B1 design, 825.
C2-S1-A1 design, facing 490.
C3-S-A2 design, 612.
C4-S-A1 design, 234, 620.
C4-S-B1 design, 619.
C4-S-B2 design, 619. See also Transports, C4's.
C4-S-B5 design, 619. See also, Transports, C4's.
Cafeterias, 438, 449-450, 674, 704, 718, 731.
California, 9, 243, 407;
Governor of, 290;
Supreme Court of, 254.
California Shipbuilding Corp., 51, 54, 146, 147, 462, 557, 560, 809, facing 586, 746;
cost of facilities, 472 and n;
cost of Liberty ships at, 826;
labor at, 249, 430, 451, 455 n, 664, 669 and n;
layout and facilities, 51, 59, 222, 340;
lease of land for, 115;
Liberty tankers, 458;
productivity at, 209, 475, 645, 685;
profits of, 817;
sale of, 115, 791;
speed of construction at, 140 n, 175, 208, 211, 224, 333, 350, 364, 468, 471;
steel shortage at, 329;
Victorys at, 586, 607, 626, 638, 685.
Calship, see California Shipbuilding Corp.
Cambria, Pa., 33.
Camden, N. J., 33.
Canada, 83.
Cancellation of contracts, 108, 182, 183, 190-200, 330, 332, 335, 391, 493-494, 529-531, 534, 635, 686, 687, 699, 777, 787, 791.
Canon City, Col., 14.
Capital:
invested by shipbuilding contractors, 805-808, 811.
See also, Financing new plants; Working capital.
Cardinal Canal, 615.
Cargill, Inc., shipyard of, 152-154.

Cargo ships, see Attack cargo ships; Dry cargo vessels, etc.
Caribbean Sea, 138.
Carmody canal, 185 n.
Carmody, John M., 16, 70 n, 163, 774, 779, 805, facing 106;
on Commission form, 785-786, 787;
and contracts, 403, 781-783;
dissents, 789-792;
and finances, 795;
on fractures in ships, 551-552;
and labor, 446, 677, 678;
and management, 189-190, 199, 462, 484-485, 486-487, 518, 541-542, 675, 783-784;
and personnel, 136, 664 n, 784-785;
relations with Admiral Land, 785-786, 791-792;
relations with Admiral Vickery, 785-786, 788, 792.
Carnegie-Illinois Steel Corp., 550-552.
Carpenters, 246.
Case, Francis H., 799.
Castings, 353.
Catton, Bruce, The War Lords of Washington, 452 n.
Certificate of availability, 663, 664.
Certificate of necessity, 400, 808.
Chaikin, William and Charles H. Coleman, Shipbuilding Policies, 9 n.
Chain rails, 84, 88.
Chairman of MC, 11-12, 271, 777, 786. See also, Land, Vice Admiral Emery Scott.
Challenge, 28.
Change-overs to new types, 344, 458 n, 468-469, 491, 575, 583, 591, 599-601, 607, 610-612, 626, 627, 637, 644-646, 648, 657, 676, 683, 827, 829. See also, Continuity of production.
Chappell, P. W., 279.
"Characteristics of Recently Hired Shipbuilding Labor," 250 n.
"Characteristics of Shipbuilding Labor Hired During First Six Months of 1941," 250 n.
Charleston Shipbuilding and Dry Dock Co., 501.
Chart room in MC, 460-461, 696.
Chase, R. K., 154.
Chesapeake Bay, 9.
Chester, Pa., 33, 36, 57, 169, 621. See also, Sun Shipbuilding and Dry Dock Co.
Chicago, Ill., 170, 390, 689. See also, American Shipbuilding Co.; Zone Conferences.
Chicago Drainage Canal, 615.

INDEX

Chickasaw, Ala., 36, 49, 50, 438. *See also*, Gulf Shipbuilding Corp.
Chief Health Consultant, *see* Shipyard Labor Relations, Division of.
Chief Safety Consultant, *see* Shipyard Labor Relations, Division of.
Chipping, 562.
"Chronological Statement Relating to Labor Situation at Federal Shipbuilding and Dry Dock Company," 291 n.
Chubb, Percy, 765, 768, 770.
Churchill, Chester L., 533.
Churchill, Winston S., 43, 139, 160, 177, 342;
 Their Finest Hour, 43 n.
Cimarron class, 22, 30, 56.
CIO, *see* Congress of Industrial Organizations.
Cissel, C. Wilbur, 749.
Civil Service, 129, 132 and n, 136, 740, 750, 751, 754, 832.
Civil Service Commission, 129, 132, 136, 693.
Civilian Production Administration, 642;
 Industrial Mobilization for War, 10 n;
 Minutes of the Supply Priorities and Allocations Board, 319 n;
 Minutes of War Production Board, 165 n;
 The Production Statement, 10 n.
Class I fractures, 554.
Class I yards defined, 35.
Classification of jobs, 263, 281, 416-418, 420-424, 523-524, 731.
Clawson, Augusta H., *Shipyard Diary*, 257 and n, 258.
Clerical Section (MC), *see* Procurement Division; Technical Division.
Cleveland, Paul, 402 n.
Closed shop, 241, 254-255, 270, 278, 281, 287, 290-292, 296-299, 495.
Coal, 74, 84, 634.
Coast Guard, 408, 550 n, 557, 565, 566, 616, 621, 622. *See also*, Bureau of Marine Inspection and Navigation.
Coastal cargo vessels, 60, 325, 615, 627, 628, 630, 636, 640, 710;
 cost of, 819;
 procurement for, 407.
Coastal tankers, 60, 225, 538, 539, 627, 636.
Cochrane, Rear Admiral Edward L., 173, 566, 570;
 "Shipbuilders Meet the War Challenge," 173 n.
Coding of vouchers, 741-742.
Cohen, Frank, 155-158, 498, 502.
Cohen, Hymen Ezra:

"Budgeting Process," 713 n;
"Commissioner John Michael Carmody," 16 n.
Colby College, 504.
Coleman, Charles H., *Shipbuilding Activities*, 38 n.
 See also, Chaikin, William.
Coll, Blanche D., "Administration of Functions Assigned to the Technical Division," 26 n;
 "1937 Audit of Maritime Commission Accounts," 744 n;
 "Division of Information," 760 n;
 "Labor in South Portland," 392 n;
 "The Sea Otter," 164 n;
 "Stabilization," 240 n.
Collection and Securities Unit (MC), *see* Finance Division.
Collective bargaining, 268-287, 306-310;
 contracts, 281, 284, 286-290, 292-297, 309, 522 n;
 approval of, 284, 416-418, 523-525;
 Maritime Commission's role in, 278-279, 281, 283, 285, 287, 291-292, 308-310, 411, 416, 422, 423; *see also*, Collective bargaining, contracts, approval of;
 representation elections, 269, 293-296, 522, 525.
Colorado, 20.
Columbia River, 9, 146.
Columbia River Shipbuilding Corp., 210.
Combat loaders, 605, 652, 668. *See also*, Attack cargo ships; Attack transports.
Combined Chiefs of Staff, 327, 332, 337, 592, 602, 603, 605;
 Combined Steel Plate Committee of, 331.
Combined Production and Resources Board, 337, 592.
Combined Shipbuilding Committee (Standardization of Design), 592, 593, 602, 604.
Combined Shipping Adjustment Board, 337.
Combustion Engineering Co., Inc., 95, 205.
Commerce, Dept of, *see* Bureau of Standards; Bureau of Marine Inspection and Navigation; Bureau of the Census; Shipping Board Bureau.
Commission action (MC), 10-12, 20-21, 154, 159, 161, 162, 192, 199, 403, 406, 519, 527, 530, 539, 598, 737, 769, 773-778, 783, 786-787, 788, 796-797.
Commission-procured materials, 89-98, 359-362, 365-366, 388, 479, 489, 700,

842 INDEX

726, 733, 734, 747, 802, 806, 811, 815, 817, 819, 822 and n, 824.
Commissioners (MC):
dissents of, 20, 23, 112, 501, 774, 779-780, 783, 789-792;
and labor, 262-263, 270 ff., 783;
meetings of, 20, 774, 787-788, 789;
office of, 18, 758;
political affiliations of, 16, 20, 791-792, 793;
responsibilities of, 161-163, 774-780, 785-786;
review function of, 774, 785-786, 787;
terms of, 15-16, 20, 162, 163, 774 and n, 792-795.
See also, under names of Commissioners.
Committee Investigating the National Defense Program (Senate), see Truman Committee.
Committee on Congested Production Areas, The President's, 436.
Committee on Merchant Marine and Fisheries (House), 103, 104, 195, 197, 242, 450, 451, 493, 498, 526-529, 534, 536, 540, 553, 702, 804-805, 818;
Committee Prints, Hearings and *Reports, see* Bland Committee; House of Representatives.
Committee on Standard Merchant Vessels, 25 n.
Committees of Congress, see Congress, committees of.
Committees of MC, see the committee's name.
Communists, 288.
Community planning, 132, 187. See also, Shipyards, living conditions near; Shipyards, transportation of workers to; Shipyards, railroads to.
Company unions, 269, 283. See also, Independent unions.
Compasses, magnetic and gyro, 85, 86, 579.
Competitive bidding, 102-107, 117-119, 411, 510, 528. See also, Bids.
Components, 230:
scheduling of, 205, 233, 365-369, 381-382, 647-649, 697;
shortages of, 237, 343-344, 353, 481, 637, 647, 653, 705; *see also,* Turbines; Valves, etc.
See also, Procurement; Suppliers.
Comptroller General, U. S., 497, 532, 739, 791-793;
Annual Report (1941, 1942, 1943, 1944), 744 n, 745 n;
Audit Report on MC and WSA for 1943 and 1944, 741, 743, 745 n, 795-796.
See also, General Accounting Office.
Concrete Control Subsection (MC), see Technical Division.
Concrete Ship Constructors, 630, 633.
Concrete vessels, 60, 63-64, 627-633, *facing* 619, 778;
cost of, 633-634.
Condemnation, see Eminent domain.
Condenser tubes, 409.
Condensers, 79.
"Conference between Representatives of East Coast Independent Unions and Representatives of East Coast Shipbuilders, May 20, 1941," 283 n.
Congress of Industrial Organizations, 240, 270;
and job classifications, 418;
organizational drives, 268-269;
Political Action Committee, 794;
and Shipbuilding Stabilization Committee, 274-275;
and wages, 309.
See also, American Federation of Labor, rivalry with CIO; Industrial Union of Marine and Shipbuilding Workers of America.
Congress, U. S., 2, 10, 12, 15, 38, 40, 47, 61, 63, 68, 70, 96, 103, 104, 151, 158, 161, 194, 195, 198, 297, 299, 413, 553, 663, 702, 703, 719, 720, 739, 744, 748, 784, 786, 791, 792, 794;
committees of, 13, 120, 165, 195-199, 329, 406, 464, 493, 503, 623, 628, 662, 663, 718, 749, 788, 792, 796-797.
See also, Committee on Merchant Marine and Fisheries; Truman Committee.
Congressional Record, 102 n, 795.
Connery, Robert H., *The Navy and the Industrial Mobilization*, 108 n.
Conservation Branch, Priorities Section (MC), *see* Production Division.
Conservation Committee (MC,) see Technical Division.
Conservation of materials, 409-410.
Conservation Section (MC), *see* Production Division.
Consolidated Builders, 809.
Consolidated Steel Corp., Ltd., 38, 143 n, 476, 640, 669, 801;
Annual Report for 1944, 35 n;
Craig (Long Beach) yard, 35, 58, 640, *facing* 42;
Orange yard, 170, 264;
Wilmington yard, 58, 62, 115, 117, 451, 614, 616, 625, 640.

INDEX 843

Construction Audit Section (MC), see Finance Division.
Construction auditors, see Auditing, of construction.
Construction Cost Committee (Washington) (MC), 736, 738, 739.
Construction Division (MC), 169-171, 511, 691, 694;
  Inspection Section, 135, 512;
  Plant Engineering Section, 499;
  Production Engineering Section, 355;
  Purchasing Section, 359.
Construction, Division of (MC in 1945), 710.
Construction Fund of MC, 712, 713-717, 744, 748.
Construction industry, 53-54, 148, 238, 244, 248, 250, 463, 471, 509-510, 513, 535, 536.
Construction Review and Analysis Section (MC), see Finance Division.
Construction Section (MC), see Technical Division.
Construction time, 175, 211-212. See also, Outfitting; Speed of Construction; Way time; and under names of ship types.
Consulting engineers, 483, 486-487, 561-564, 675-676, 678, 781.
Continuity of Production, 202, 204, 231-232, 234, 687, 816, 826, 829.
  See also, Change-overs.
Contract Settlement and Surplus Property Division (MC), 758.
Contractors, see Capital; Managements; Shipyards.
Contracts, 3, 127, 456-457, 542-543, 830-831;
  adjusted-price, 106, 126;
  adjustment of prices in, 490, 653, 804;
    see also, Escalator clauses.
  conversions of, 682-687, 812-817;
  cost-plus-fixed-fee, 98, 101, 123, 127, 289, 457, 501, 815;
  cost-plus-percentage-of-cost, 101, 117, 119, 123;
  cost-plus-variable-fee, 101, 102, 117-123, 473, 477, 487, 679-689, 812, 813; see also, Contracts, manhour;
  EPF, 107, 111;
  facilities, 107-117, 432, 472-473, 779;
  fixed-price, 106, 126; see also, Contracts, lump-sum;
  incentive, 490; see also, Contracts, cost-plus-variable-fee, price-minus, selective-price;
  labor productivity affected by, 683-687;
  lump-sum, 101-107, 117, 125, 126, 128-129, 354, 456-457, 478, 487, 488, 490, 491, 635, 679-683, 687, 688, 710, 723, 731, 789-790, 792, 793, 812-816;
  manhour, 117-123, 124, 128, 365, 456-459, 543, 723, 730, 734, 799-800;
  predated, 127, 682;
  price-minus, 124-128, 456-457, 476, 477, 480, 543, 635, 636 n, 730, 812, 813;
  selective-price, 127, 488-490, 679, 681-683, 685-686, 731, 790, 793, 803, 816-817.
  See also, Renegotiation; Cancellation of Contracts; Defaults on Contracts.
Controlled Materials Branch, Production Engineering Section (MC), see Production Division.
Controlled Materials Plan, 376-378, 382-384.
Controller of Shipbuilding, 587-607. See also, under War Production Board.
Conversions of ships, 229, 230, 324, 652, 821. See also, Design changes; Standard cargo ships, converted.
Convoys, 576, 600.
Conway, Capt. Granville, 762, 770, 793, 795.
Cooke, Morris L., 273.
Cooperative shipyard, 159-160.
Coordinator of Ship Repair, 539.
Copper, 339, 378, 390.
Coppersmiths, 622.
Corcoran, Thomas, 158.
Cork, Ireland, 28.
Corvettes, 177, 178, 180, 181, 360, 614-615.
  See also, Frigates.
Cost of facilities, see Facilities.
Cost of ships, 468, 819-829;
  administrative expense of MC in, 822 and n, 823, 825;
  direct costs, 819-821;
  efforts to reduce, 471-477, 487-492, 659, 661, 675, 680, 682, 687, 733, 785, 812, 816;
  estimates of, 103, 117, 118, 120, 124, 125, 128, 172;
  labor costs, 421, 822-829; see also, Labor Productivity.
  material costs, 106, 120, 642, 819, 822-824, 825;
  prewar cost, 825;
  records of, 747-748, 819;
  regional differences in, 102, 285, 826-827;
  savings in from suggestion program, 454-455;
  war costs, 819, 821, 825.
  See also, Standard manhours; under names of ship types.

Cost of steel, 322.
"Cost of War-Built Vessels from Inception, October 25, 1936 to June 30, 1946," 821 n.
Cost-per-ton-mile, 599-600.
Cost-plus, 101, 105, 107, 119, 124, 458, 476-478, 485, 489, 491, 687, 799.
Cost Review and Material Control Section (MC), *see* East Coast Regional Office.
Cost Section (MC), *see* Technical Division.
Cost Valuation Unit, Construction Audit Section (MC), *see* Finance Division.
Court of Claims, 737.
Coy, Wayne, 182.
CPA, *see* Civilian Production Administration.
Crack arrestors, 548-550, 555, 571, 573.
Craft rules, 241-243.
Craft skills, 237, 238.
Craft unions, 240-244, 254-255;
fees of, 243, 254-255, 296.
Craig yard, *see* Consolidated Steel Corp.
Cramp Shipbuilding Co., 37.
Crane operators, 257.
Cranes, 2, 38, 185 n, 210, 214-216, 222-224, 229-233, 353-355, 358, 500, 514, 542, 615, 619.
Crankshafts, 94-95.
Crap games, 301, 664.
Crew accommodations, 30, 76, 84, 85, 88, 582, 615, 620.
Crews, *see* Seamen.
Cromwell, James, 159.
Cruisers, 37, 572.
Culliton, James W., "Economics and Shipbuilding," 823 n.
Cupro-nickel, 409.
Curtailment orders, 319.
Cushing Point, Me., 216, 504, 507, 515-517, *facing* 522. *See also*, South Portland Shipbuilding Corp.
Customs Bureau, 763.
Czar of shipbuilding, 163, 166-168, 589, 592.

D

Daniel Construction Co., 500-501.
Daniel Guggenheim Fund, 12.
Davis, William H., 269 n, 292, 309.
Davits, 87.
Deadweight tonnage, 4-6.
Decatur, Ala., 225, 630.
Decentralization, 168-172, 355, 689, 690, 693-695, 700, 711, 722, 764, 767, 830-832.
Deck assemblies, 227.

Deck covering, 92.
Deck loads, 80, 544, 579 and n.
Deck machinery, 647. *See also*, Winches.
Deck strap, 549, 555, 573.
Deckhouse assemblies, 228-229, *facing* 330.
Deckhouse slip, 228-229.
Declaration of war, 138, 240, 498, 508. *See also*, Pearl Harbor.
Defaults on contracts, 108, 477, 481, 491, 494, 496, 499, 527, 530, 531, 533, 534.
Defense Aid Fund, 714-716, 741.
Defense features on ships, 22, 23, 56, 102, 104 n, 629, 825. *See also*, Armament.
Defense Plant Corporation, 109, 111, 155, 400-401, 479, 823.
Degaussing, 86.
Delair, N. J., 159.
Delano, Frederic A., 49.
Delano, Sara, 164.
DeLaval Steam Turbine Co., 399, 595, 598.
Delaware River, 9, 58, 146.
Deliveries, totals:
by managerial groups, 470;
by months, 203, 641;
by quarters, 64;
by regions, 9, 690;
by types, 4-6, 820;
by years, 7, 8, 343, 637-638, 640-641, 833;
by years by types, 5, 336, 651.
See also, under names of ship types and yards.
Delta Shipbuilding Co., 51, 53, 59, 462, *facing* 363;
cost per ship at, 209, 826;
labor at, 247, 295;
layout of, 62, 223;
Liberty tankers, 458;
productivity at, 209;
speed at, 208, 232, 637
Victory ships planned at, 586.
Demobilization, *see* Postwar planning.
Democratic Party, 710.
Dempsey, John J., 16, 20, 21, 34, 152, 153, 163.
Design changes:
approval of, 170-171, 581, 694-695;
effects of on construction, 229, 391, 392, 394, 542, 609, 611, 616, 619, 621, 627, 679;
effects of on contracts, 457, 481.
Design of ships, *see* Technical Division and under names of ship types.
Destroyer Tender, 38, 51.
Destroyers, 37, 40, 100, 509, 513, 514.
Detroit, Mich., 390.

INDEX

Deutsch, Herman B., "Shipyard Bunyan," 184 n, 195.
Diamond Co., 95.
Diesel engines, 28, 29, 85, 104, 397, 399, 577, 584, 586, 609, 610, 636.
Diesel generator sets, 647.
Di-lok process, 358.
Direct manhours, 121 and n.
Direction finders, 87.
Directives for shipbuilding:
  from the Joint Chiefs, 335, 339-340, 342, 369, 379, 383, 603-606, 626, 628, 647, 650, 681, 775, 780, 788, 812;
  from the President, 43, 45, 55, 57, 61, 139, 143-145, 147, 149, 160, 173, 176, 179 and n, 180, 182-185, 193, 199, 323, 329, 330, 335, 340-342, 602, 637-638, 650, 775.
Disallowances, 707-708, 735-739.
Disbursement Section (MC), see Finance Division.
Disbursements, 168, 717, 721-728, 734.
Discrimination, 186, 252-257, 495.
Displacement tonnage, 4-6.
Dividends, 402-403, 481.
Division of Labor Standards (Labor Dept.), *Report on Apprenticeship System*, 237 n.
Dollar-a-year men, 752 and n.
Dombrowski, James A., 188.
Donnels, W. F., 187.
Dooley, C. R., 259.
Double bottom, 84.
Double fillet weld, 561.
Douglas, Lewis W., 162, 178, 182, 183, 337-339, 755-756, 764, 768-769, 770.
DPC, see Defense Plant Corporation.
Drawings for engines, 93.
Dreadnaught, 216.
Dreesen, Waldemar C., et al., "Health of Arc Welders in Steel Ship Construction," 448 n.
Drilling steel, 206, 218.
Drinker, Philip, 446, 447; "Health and Safety in Shipyards," 449 n.
Drummond, Wadleigh, 510.
Dry cargo vessels, 3, 27, 324. *See also*, Coastal cargo vessels; Liberty ships; Standard cargo ships; Victory ships.
Duff., P. J., 154, 157.
Duluth, Minn., 407, 614, 673.
Dunn, Gano, Report of, 63 n, 315, 318.
Dutch motor coasters, 636.

E

East Bay Union of Machinists, 288-289.
East Boston, Mass., 33.

East Coast Regional Office (MC), 170, 364, 564, 690, 697-698;
  Assistant Director for Departmental Activities, 691;
  Industrial Adviser, 524, 691;
  Inspection Section, 674;
  location of, 702;
  Manpower Survey Board, 674;
  Production Section, 698.
  See also, Yard staff committees.
East Coast Shipyards, Inc., 430, 538-542, 630, 640, 729-730, 788, 790, 792.
  See also, Marine Maintenance Corp.
Eberstadt, Ferdinand, 313, 342, 376.
EC2-S-A1 (design), 67, 83. *See also*, Liberty ships.
EC2-S-AP1 (design), 585.
EC2-S-AP2 (design), 585.
EC2-S-AP3 (design), 585.
EC2-S-AP4 (design), 586.
Economic Survey of the American Merchant Marine, 22 n.
Economics and Statistics, Division of (MC), 18 n, 762, 763.
"Economy Ship," 82.
Eddystone property, 112.
Education, U. S. Office of, 261, 556 n.
"Effects of Unannounced Quits on Absenteeism in Shipbuilding," 413 n.
Eisenhower, Dwight D., *Crusade in Europe*, 609 n.
Electric Boat Co., 34, 152, 283.
Electric drive, 624. *See also*, Turboelectric units.
Electric motors, 368, 408, 647, 648, 650, 652.
Electric power for shipyards, 154, 163, 190, 741, 780-783.
Electricians, 242, 246, 257, 622, 657.
Electronic equipment, 390.
Electronics Branch, Priorities Section (MC), *see* Production Division.
Eleven-Percent Expansion Act, 36.
Elliott Co. (Pa.), 399.
Emergency agencies, 24.
Emergency Construction Audit Branch, Construction Audit Section (MC), *see* Finance Division.
Emergency Construction Audit Section (MC), *see* Finance Division.
Emergency Fund for the President, 45 n.
Emergency purchases, 350 n, 375-376, 380, 381, 697-699.
Emergency Ship Construction Act, 132.
Emergency Ship Construction Division (MC), 78, 83, 91-93, 127, 131-135;
  Administrative Section, 131;
  Engineering Plan Approval Section, 78, 131;

Hull Plan Approval Section, 78, 131; Plant Engineering Section, 131, 133, 215; Production Engineering Section, 131, 135, 355.
Emergency Ship Construction Fund, 45 n, 713-716, 741.
Emergency Shipping, Division of (MC), 754.
Emergency ships, 40, 44, 46-58, 67, 68, 81, 83, 84, 88, 166. *See also,* Liberty ships.
Emergency yards, 42-43, 46-55, 169, 344, 350-352, 355, 360, 371, 661. *See also,* Liberty ships, yards building.
Emerson Engineers, 675 n.
Emeryville, Cal., warehouse, 743.
Eminent domain, power of, 112-117, 494, 498, 502-503, 516, 521, 532, 538, 539, 741, 779.
Empire Liberty, 81, 83.
Empire Ordnance, 155, 158, 498.
Employee-representation plans, 282. *See also,* Company unions.
Employment, *see* Labor force, size of.
Employment Service, U. S., 249, 435, 663, 666-668, 670;
" Short History of the War Manpower Commission," 663 n.
Engineering Design Specifications and Priorities Section (MC), *see* Technical Division.
Engineering Design and Specifications Section (MC), *see* Technical Division.
Engineering Plan Approval Section (MC), *see* Emergency Ship Construction Division.
Engineering Scheduling and Planning Branch, Production Engineering Section (MC), *see* Production Division.
Engineering Section (MC), *see* Technical Division.
Engines, 149-150, 167, 344, 404. *See also,* Diesel engines; Reciprocating engines; Turbines.
Equipment, *see* Shipyards, equipment of.
Erection on shipways, 210-212, 224, 229, 347-349, 351. *See also,* Way time.
Erection sequences, 320, 559-560, 567, 569, 614.
Escalator clauses, 106, 125, 489, 802.
Escort aircraft carrier, *see* Aircraft carrier escorts.
Escort vessels, 177-180, 323, 338, 339, 588, 590, 591, 594, 605, 611, 612. *See also,* Aircraft carrier escorts; Corvettes.

Esmond, William G., 19, 24, 25, 74, 76, 78, 79, 83, 86, 87, 171, 511;
" Liberty Ships. Historical Data," 86 n.
Esso Manhattan, 545 and n, 565, 566, 571, 572, *facing* 523.
Evening Express, Portland, 506 n.
Evening Star, Washington, 548 n.
Examiner, San Francisco, 289 n.
Examining Division (MC), 18, 722 n, 762.
Examining Section (MC), *see* Finance Division.
Excess materials, *see* Surplus materials; Steel inventories.
Executive Director (MC), 16-17, 20, 67. *See also,* Schell, Samuel Duvall.
Exchequer, 31.
Expediters:
of MC, 92, 168, 354, 355, 387-390, 396, 649;
of Navy, 388, 649-650, 652;
of shipbuilding contractors, 364, 380, 381, 649 and n, 697, 699, 811.
Expediting Branch, Production Engineering Section (MC), *see* Production Division.
Eye diseases in shipyards, 447-449.
Ezekiel, Mordecai, report of on steel, 166 n, 197-198, 346 and n, 373-374, 377, 393.

F

Fabrication buildings, 186, 210, 213, 214, 217-223, 226, 258, 516.
Fabrication of steel, 830;
in shipyard, 111, 206-207, 210-212, 226-227, 320, 345, 347-349, 351-352, 374, 472, 516, 564, 830;
outside of shipyard, 46 n, 147, 166, 204, 206-207, 217, 218, 508-509, 516, 539, 830.
Facilities:
cost of, 108, 109 n, 111-112, 190, 191, 397, 472-473, 536, 618-619, 628, 633, 821-825;
expansion of, 150, 167, 172, 173, 340-341, 397, 575; *see also,* Building Berths; Emergency yards; Shipyards, additions to;
funds for, 741;
option to purchase, 109;
overexpansion of, 150, 176, 184, 191, 201, 343;
regional distribution of, 7, 46-51, 153, 827, 829
removal of, 110, 111, 115-117, 823;

INDEX 847

shortage of, 32, 39, 42, 237, 343.
See also, Housing; Shipyards.
Fair Employment Practice Committee, 252-255.
Falk Corp., 399.
Fans, 648.
Farm Security Administration, 444.
Farmer, Norman R., 180.
Farrel Birmingham, 595.
Farrell, J. A., 67.
Fassett, F. G., Jr., ed., *Shipbuilding Business*, 123 n.
" Fatal Work Injuries in Shipyards, 1943 and 1944," 448 n.
Fathometer, 85-86.
Favoritism, charges of, 152-154, 538, 542, 810-811.
FBI, see Federal Bureau of Investigation.
Federal Bureau of Investigation, 735.
See also, Justice, Dept. of.
Federal Public Housing Authority, 428, 432-435, 444.
Federal Register, 421 n.
Federal Reserve System, 402, 403.
Federal Security Agency, *Preliminary Job Study*, 239 n.
Federal Shipbuilding and Dry Dock Company, 33-37, 52, 92, 121, 128, 142-143, 789, 801;
 contracts of, 102-104, 126, 128, 679;
 labor at, 268, 282, 283, 290-292;
 P2's at, 624, 625, 639.
Federal Works Agency, 16.
Fees, 108, 122-123, 506, 527, 531, 679, 781;
 advance payment of, 481;
 to consulting engineers, 675-676, 737;
 of insurance advisers, 782;
 launching and delivery, 125, 481;
 maximum and minimum, 101, 120, 122, 125, 457-458, 474-477, 532, 633, 799-800;
 of naval architects, 98;
 reduction proportionate to facility cost, 190, 473, 502, 521-522, 527, 531-532;
 standard, 120, 122-123; see also, Renegotiation.
 to subcontractors, 478 n.
 See also, Profits.
Feis, Herbert, *The Spanish Story*, 501 n.
FEPC, see Fair Employment Practice Committee.
Ferguson, Homer L., 52.
Ferries, 430, 431, 438, 439, 441, 443, 701.
" Few Accomplishments in Shipbuilding, A," 675 n.

Field Coordinating Section (MC), *see* Production Division.
Field offices, *see* Expediters of MC; Regional offices (MC); War Shipping Administration, district offices.
Fife county, Scotland, 80 n.
Filer & Stowell Co., 585.
Final Design Section (MC), *see* Technical Division.
Final Report of Board of Investigation (on Welded Vessels), 566 and n, 570 n.
Finance Division (MC), 16-18, 27, 487, 716, 723, 733, 738, 748, 750, 757, 758, 760, 762, 772, 775, 796;
 Accounting Branch, Budget and Accounts Section, 742;
 Accounting Section, 722, 723;
 Accounts Section, 767;
 Administrative Unit, Construction Audit Section, 131, 728;
 Analysis, Cost Studies and Reports Unit, Construction Audit Section, 731;
 Assistant Director of Finance (Construction), 722; see also, Honsick, J. A.;
 Assistant Director of Finance (Operations), 722, 767; see also, Quinn, J. M.;
 Assistant Director of Finance (West Coast), 708;
 Assistant General Auditor, 731;
 Audit Branch, 79;
 Audit Practice and Procedures Unit, Construction Audit Section, 731;
 Auditing and Financial Analysis Section (Operations), 722, 723, 764, 767, 771;
 Audits Section, 771;
 Budget and Accounts Section, 722, 740, 742-743, 764, 767-769;
 Budget Officer, 17, 757; *see also*, Kirsch, William U.;
 Collection and Securities Unit, 722 n;
 Construction Audit Section, 27, 128-131, 722-724, 728, 730;
 Construction Review and Analysis Section, 728, 737, 778;
 Cost Valuation Unit, Construction Audit Section, 131;
 decentralization of, 171;
 Director of, 129, 706, 736-737; *see also*, Anderson, Robert Earle; Houlihan, D. F.;
 Director of Finance (Construction), 738;

848 INDEX

Disbursement Section, 722-724, 726, 727, 740, 742, 764, 767, 771;
Emergency Construction Audit Branch, Construction Audit Section, 130 and n, 322;
Emergency Construction Audit Section, 127; see also, Emergency Construction Audit Branch;
Examining Section, 722 n;
General Accountant, 767, 772;
General Auditor, 771;
General Auditor of Construction, 706, 727, 728, 730-732, 737-738, 742, 764, 796; see also, Slattery, W. L.;
General Auditor of Disbursements, 727;
Home Office Shipyard Unit, Construction Audit Section, 130;
Labor Unit, Construction Audit Section, 731;
Materials Unit, Construction Audit Section, 731;
Overhead Unit, Construction Audit Section, 731;
Report Review Unit, Construction Audit Section, 131;
Special Assignment Unit, 130-131;
Special Studies Branch, 467;
Special Studies Unit, Assistant Director of Finance (Construction), 130-131, 728;
Subcontract Audit Unit, Construction Audit Section, 130;
Voucher Examining Unit, Construction Audit Section, 131.
Finance, Division of (Construction), 796.
Finance, Division of (Operations), 796.
Financing of new plants, 103, 107-109, 151, 154, 159, 399-401, 404. See also, Working capital.
Fire-proofing, 22, 30, 84, 85.
Fischer, Gerald J., 180 n;
"Labor Productivity," 230 n;
"Organization and Procedures," 375 n;
"Problems of Steel Production," 317 n-318 n;
"Productivity," see "Labor Productivity";
"Programming of Ship Construction," 23 n;
"Statistical Analysis, Steel," 210 n-212 n;
Statistical Summary of Shipbuilding, 3 n;
"Wages," 245 n.
Flame-cutting, 226.
Fleet Fillet weld, 562, 564.
Flesher, Carl W., 170, 171, 355, 450, 564, 621, 678, 692, 693, 697-699, 702, 708-710, *facing* 746;
"Draft is Retarding Ship Production," 656 n.
Floyd, William, 383.
Fontana, Cal., 338, 808.
Ford Motor Co., 238.
Foreign contracts for shipbuilding, 4, 5, 6. See also, Britain, shipbuilding in U. S. by.
Forelady, 257.
Foreman's aides, 484.
Forepeak assembly, 218.
Forgings, 353, 354, 366.
Forrestal, James, 570, 653, 662.
Fort class ships, 82, 88, 90.
Fortune, 53, 54 n, 442, 445.
Foster-Wheeler Corp., 95.
Four factor formula, 468, 800.
FPHA, see Federal Public Housing Authority.
F.R. see *Federal Register*.
Fractures:
in riveted ships, 544, 554 and n;
in welded ships, 80, 267, 544-546, 553-554, 569-573.
See also, Crack arrestors; Girder reinforcements; Welding.
France, 36, 40.
Franklin, Brig. Gen. John M., 604.
Frauds, 509-511, 538, 542, 558, 632, 633, 735, 737, 739, 748, 784 and n, 790, 830.
Free enterprise, 1, 808. See also, Managements.
Freeman, Ward B., 781.
Freight charges, 163, 322, 747, 780.
Freighters, see Dry Cargo vessels.
French, E. S., 518 n.
Frey amendment, 297, 303.
Frey, John P. 254, 271, 273, 276-278, 281, 288, 289-290, 296-297.
Frick, Henry E., 462, 486, 538, 540, 752 n.
Friedman, Jesse J., see Rogers, H. O.
Frigates, 574, 589, 614-617, 639, 695, 710, *facing* 618;
cost of, 819;
manhour cost of, 642;
procurement for, 407;
yards building, 674, 827.
Froemming Brothers, Inc., Shipbuilding Division, 614, 630.
Fuel consumption, 28, 29, 81.
Fuel tanks, 84, 582.
Furer, Capt. J. A., 279.
Furnaced plates, 76, 77, 218, 226, 582.
Furniture, 85, 359.

INDEX 849

G

G. S. May Co., 562.
Gallagher, J. T., 135, 362, 378, 384, 388.
Galley equipment, 85, 359.
Gallup poll, 299 n.
Gantry cranes, 214, 228. *See also,* Cranes.
GAO, *see* General Accounting Office.
Garbage cans, 434, 437, 444.
Gasoline shortage, 428, 437, 508, 516, 628, 633.
Gauges of steel plate, 87, 320.
Gawne, Capt. James O., 366-367, 588.
Gear hobbers, 38, 398, 595.
Gears, 28, 38, 42, 56, 61, 353, 355, 367, 397, 398, 395.
General Accountant (MC and WSA), 767, 772.
General Accounting Office, 98, 485, 731, 735-736, 738, 739, 741, 743-746, 748, 768, 813. *See also,* Comptroller General (U. S.).
General Auditor (MC), *see* Finance Division.
General Auditor of Construction (MC), *see* Finance Division.
General Auditor of Disbursements (MC), *see* Finance Division.
General Construction Co., 54 n, 809.
General Counsel (MC), *see* Legal Division.
General Electric Co., 398-399, 401, 583, 587, 595, 598.
General Engineering Works, 276.
General Machinery Co., 93-94, 397, 501, 585, 586.
General Production Urgency List, 652 n.
General Scheduling Order M-293, 368.
Generators, 360, 397, 647.
Geoghegan, T. D., 429, 430, 517, 518, 780.
Gerhauser, W. M., 53 n.
Germain, E. B., 164.
Germany, 36, 587, 687. *See also,* Nazis.
Gibbons, C. D., 770, 771.
Gibbs & Cox, Inc., 73-79, 83, 90-100, 135, 205, 354, 356, 365, 401, 503, 559, 560, 586, 588, 606, 801.
Gibbs, William Francis, 73, 74, 76, 77, 83, 92, 99, 586. *See also,* War Production Board, Controller of Shipbuilding.
Gilbert Partnership, The, 510.
Girder reinforcements, 548, 554, 573.
Gland sealing ring, 82.
Globe Shipbuilding Co., 407, 614, 630.
Gordon, Lincoln, "An Official Appraisal of the War Economy and its Administration," 370 n.
Goudy, Clinton T., 519.
Goverment Aids Division (MC), 758.
Government Manual (1944), 429 n.
Government ownership, 107-117, 400-401, 484, 493, 805, 808.
Gowing, M. M., *see* under Hancock, W. K.
Grace Line, 624.
Graf, S. H., 550.
Graham, Frank P., 269 n, 292.
Grain carrier, 153.
Grand Coulee dam, 54, 277.
Grant, John E. P., 78, 79, 83, 86, 94, 585.
Gray's Iron Works, 537. *See also,* Todd Galveston Dry Docks.
Great Britain, House of Commons Sessional Papers, 1940-1941, 601 n. *See also,* Britain.
Great Lakes, 49, 53, 60, 65.
Great Lakes Engineering Co., 295.
Great Lakes Regional Office (MC), 170, 364, 690.
  combined with Gulf, 674;
  frigate program, 614-615;
  Manpower Survey Board, 673-674;
  See also, Gulf-Great Lakes Regional Office.
Great Lakes yards:
  and CI's, 640;
  and frigates, 614, 615, 639;
  labor in, 250, 268, 286, 294-295, 308, 412, 415, 419, 420, 423, 424, 425, 426, 660, 661;
  and N3's, 628;
  productivity in, 644, 827-828;
  and tugs, 636.
Great Lakes Zone Conference, *see* Zone Conferences.
Green, John, 273, 303, 308, 551.
Green, W. L., 533.
Green, William, 195, 288.
Greylock, 75 n.
Griffith, Sanford, *Where Can We Get War Workers?* 256 n.
Gross tonnage, 4-6.
Groton, Conn., 48, 152.
Groton Iron Works, 152.
Gulf Coast Regional Office (MC), 170, 364, 690;
  combined with Great Lakes, 674;
  and housing, 434;
  inspector's manual, 704;
  Manpower Survey Board, 672-673;
  Material Coordinator Section, 697;
  purchase controller, 700;
  Staff conferences, 705-706;

850 INDEX

transfers of components and materials, 697-698.
See also, Yard Staff Committees.
Gulf Coast Zone Conference, see Zone Conferences.
Gulf coastal transportation, 635.
Gulf-Great Lakes Regional Office, formation of, 710.
Gulf of Mexico, 138, 184.
Gulf Shipbuilding Corp., 35, 36, 37, 49, 50.
Gulf yards, 9, 34, 48, 49, 145, 223;
 community facilities near, 430-431, 433;
 and efficiency experts, 675;
 and frigates, 615;
 hiring of women in, 257;
 labor in, 240, 250, 261, 268, 285-286, 294-295, 308, 309, 412, 415, 418, 419, 420, 423, 424, 425, 526, 660;
 productivity in, 644, 827-829;
 and Victory ships, 607.
Gunwale bar (arrestor), 548, 549, 555, 573.
Gyrocompasses, 85, 579.

H

Hamilton, Ohio, 93.
Hancock, W. K. and M. M. Gowing, *British War Economy*, 203 n.
Handyman, 239, 258, 418.
Hanna, Hugh S. and W. Jett Lauck, *Wages and the War*, 272 n.
Hannay, Edwin W., Sr., 248, 464, 622.
Hansbury, John E., 517-518, 519, 520.
Harding, Me., 216, 217, 508-509. *See also*, South Portland Shipbuilding Corp.
Harris, Mortier, 510.
Harris Structural Steel Co., 539.
Harrisburg Machinery Co., 403.
Harrison, Gregory, 274.
Harrison, W. H., 148, 165, 514.
Hartline, E. E., 731.
Harvard School of Public Health, 446.
Harvard Report, 22 n.
Hatch corner reinforcement, 80, 549-550, 555, 570, 573.
Hawsepipe, 86.
Health consultants, see under Regional Offices of Construction.
Health of shipyard workers, 413, 415, 446-449, 701.
Hebert, F. Edward, 194.
Heck, J. S., 80 n.
Helpers, 258, 281, 416, 418.
Henry J. Kaiser Co., 54 n, 809.
Henry Wynkoop, 544.

Herald-Tribune, New York, 525 n.
Herrick, Mrs. Elinore, 533.
Hess Company, 103.
Hicks, Frederick E., 772.
Higgins, Andrew J., 148, 155, 183, 184-192, 195-197, 199, 225, 252, 338, 373.
Higgins, Frank, 189, 196.
Higgins yard, 142, 148, 184-191, 223, 225-226, 328-340, 346, 377, 430, 720, 782, 791.
Hill, R. F., 504 n.
Hillman, Sidney, 262, 263, 270, 273, 283-284, 289-290, 295, 308, 312, 428.
Hitler, Adolph, 40, 188, 202.
Hoboken, N. J., 33.
Hod Carriers, Building and Common Laborer's Union, International, 187, 241.
Hog Island, N. J., 43, 46, 206, 213, 219.
Hog Islanders, 18, 75, 206.
Hogan, Newell A., 264.
Hogging tests, 547, 554.
Holton, Charles R., "Procurement and Storekeeping," 388 n.
Home Office Shipyard Unit (MC), Construction Audit Section, see Finance Division.
Home Owners Loan Corporation, 428.
Homestead plant, see Carnegie-Illinois Steel Corp.
Honsick, Joseph A., 722, 723, 728, 731 and n, 736, 764.
Hooker, R. G., 264.
Hoover Dam, 54. *See also*, Boulder Dam.
Hopkins, Harry, 178, 183, 193, 315, 328, 329-330, 331, 337, 338, 589.
Horizontal yards, 213, 222.
Horton, Robert, 760.
Hospital ships, 619, 623, 627.
Hospital space on transports, 622.
Hospitals, 444, 449.
Houlihan, D. F., 744-745, 756.
Hours worked in shipyards:
 eight-hour day, 280, 286, 427;
 forty-hour week, 280, 286, 307;
 Sunday work, 286, 307, 309-310, 659, 661, 777;
 weekly hours, 244-245, 418, 419, 424, 659-661.
House of Representatives:
 Accounting Practices, Hearings, 725 n;
 Appropriation Bill for 1943, Heargency Cargo Ship Construction, Report 10, 83 n;
 Correspondence Between Comptroller General and Attorney General Relative to Tampa Shipbuilding Co., Document No. 45, 498 n;

INDEX 851

Department of Labor—Federal Security Agency Appropriation Bill for 1944, Hearings, 296 n;
East Coast Shipyards, Inc., Hearings, 127 n;
Emergency Cargo Ship Construction, Hearings, 52 n;
Fifth Supplemental National Defense Appropriation Bill for 1942, Hearings, 120 n;
First Supplemental National Defense Appropriation Bill for 1942, Hearings, 62 n;
First Supplemental National Defense Appropriation Bill for 1943 Hearings, 715 n;
Foreign Construction Costs, Executive Hearings, 118 n;
Full Utilization of Manpower, Hearings, 656 n;
Hearing before the Committee on Naval Affairs on H.R. 1876 . . . , 412 n;
Hearings No. 283 on H.R. 7576, 459 n;
Hearings on H.R. 6790, 163 n;
Hearings on H.R. 8532, 104 n;
Hearings on S. Res. 646, 779 n;
Higgins Contracts, Executive Hearings, 140 n;
Higgins Contracts, Hearings, 140 n;
Higgins Contract, Interim Report, 179 n;
Independent Offices Appropriation Bill for 1942, Hearings, 34 n;
Independent Offices Appropriation Bill for 1943, Hearings, 70 n;
Independent Offices Appropriation Bill for 1944, Hearings, 147 n;
Independent Offices Appropriation Bill for 1945, Hearings, 548 n;
Independent Offices Appropriation Bill for 1946, Hearings, 70 n;
Independent Offices Appropriation Bill for 1948, Hearings, 109 n;
Interim Report No. 2057, 538 n;
Investigation of Certain Transactions of the Tampa Shipbuilding Co., Report No. 938, 495 n;
Investigation of East Coast Shipyards, Inc., Interim Report, No. 2075, 538 n;
Investigation of Rheem, Hearings, 144 n;
Investigation of Rheem, Interim Report (No. 2057), 148 n;
Investigation of Shipyard Profits, Hearings (1946), 53 n;
Investigation of South Portland Shipbuilding Corporation, Interim Report, 465 n;
Investigation of the Progress of the War Effort, Hearings, 73 n;
Investigation of the Progress of the War Effort, Report No. 733, 778 n;
National Defense Migration, Hearings, 441 n;
National War Agencies Appropriation Bill for 1946, Hearings, 686 n;
Navy Department Appropriation Bill for 1942, Hearings, 37 n;
Navy Department Appropriation Bill for 1945, Hearings, 91 n;
Navy Department Appropriation Bill for 1947, Hearings, 115 n;
Production in Shipbuilding Plants, Hearings, 112 n;
Renegotiation of War Contracts, Hearings, 753 n;
Report No. 567, 779 n;
Report No. 938, 495 n;
Report No. 1204 on H.R. 634, 459 n;
Report No. 2168, 104 n;
Report No. 2487, 459 n;
Report No. 2582, 104 n;
Report of Activities of Investigating Committee for the Year 1945. Committee Doc. No. 82, 536 n-537 n;
Second Supplemental National Defense Appropriation Bill for 1942, Hearings, 106 n;
Shipyard Profits, Hearings (1944), 798 n;
Shortage of Steel in Shipbuilding, Executive Hearings, 329 n;
Tampa Hearings, 103 n;
Tampa Shipbuilding & Engineering Co., Hearings, 103 n;
Third Deficiency Appropriation Bill for 1937, Hearings, 18 n;
Walsh-Kaiser Co., Inc., Hearings, 147 n;
War Housing, Hearings, 432 n;
Wartime Accounting, Report, 739 n;
See also, Congress; Bland Committee.
Housing, 114, 253, 333, 428, 429, 433-444, 472, 482, 516, 633, 701, 718, 731, 780, 783, 823.
"Housing for War," 438 n;
Houston Shipbuilding Corp., 51, 59. See also, Todd-Houston.
Houston, Tex., 48, 50, 51, 55, 97, 630, 631, 632. See also, Houston Shipbuilding Corp.; McCloskey and Co., San Jacinto; Todd-Houston Shipbuilding Corp.
Howell, George B., 497.

852   INDEX

Hull Final Design Section (MC), see Technical Division.
Hull Inspector Coordinator, 135, 500.
Hull Inspectors, 133-135. See also, Inspection in shipyards.
Hull Plan Approval Section (MC), see Emergency Ship Construction Division; Technical Division.
Hull Scheduling and Planning Branch, Production Engineering Section (MC), see Production Division.
Hull Section (MC), see Technical Division.
Human Engineering, Inc., 737-738.
Hunter, Harry, 80 n, 81, 86, 93, 94. See also, Thompson, Robert C.
Hunter, James B., 578, 581.
Hurley, Edward N.: *The Bridge to France*, 75 n;
  The New Merchant Marine, 206 n.
Hurricane, 392.

I

Ice service, 438.
Illinois River, 615.
Independent unions, 268, 274, 282, 283, 284, 308. See also, Company unions; Sparrows Point, Independent Shipbuilders Association of.
Indirect manhours, 121 and n, 474.
Industrial Mobilization for War, 10 n.
Industrial Relations Advisor (MC), see Regional Offices of Construction.
Industrial Union of Marine and Shipbuilding Workers of America, 551;
  and job classifications, 420, 422;
  membership in, 268, 276, 282, 286;
  Officers' Report, 293 n, 294 n;
  organizing drives, 282-283, 290-295, 303;
  and Shipbuilding Stabilization Committee, 273, 275, 278 and n, 282, 284, 285, 308;
  strikes of, 290-292, 294-295.
Inflation, 302, 306-310, 421, 422, 662.
Information, Division of (MC), 18, 66-67, 70, 172, 453, 454, 715-716, 757, 758, 760.
Ingalls Iron Works, 35, 437.
Ingalls Shipbuilding Corp.:
  Decatur yard, 225, 630;
  Pascagoula yard, 31, 35, 36, 38, 48, 50, 58, 62, 126, 149, 285, 294, 638, 669 n, 679, *facing* p. 234.
In-migrants, 247, 248, 249, 427, 435, 442, 667.
In-plant feeding, 449-451, 701, 780.
Insignia of MC, 67.

Inspection:
  by contract, 127;
  in shipyards, 131-137, 168-171; 460-461; 486, 511, 512 and n, 556-557, 561, 563, 567, 608, 689, 691, 702, 704, 707-709, 729, 733, 830, 832;
  of suppliers, 408-409, 551, 567, 726.
  See also, Safety at sea.
Inspection Section (MC), see Construction Division.
Inspectors, see Inspection, in shipyards.
Inspector's manual, 704.
Institute on Human Science, see Human Engineering, Inc.
Institution of Naval Architects, *Transactions*, 8 n.
Insulation, 92.
Insurance, 731, 765, 782-783.
Insurance fund of MC, 718.
Insurance Division (MC), 18, 758, 782, 783.
Interchangeability of parts, 93-96, 205, 405. See also, Standardization.
Interest, 401, 479-480, 824.
Interstate Commerce Commission, 11, 15, 114, 429, 517, 518, 520, 752 n.
Inventories:
  control of, 131, 197, 370-371, 374-375, 390, 391, 395, 699, 732 and n, 733, 734, 739, 743, 747, 791, 829, 831-833;
  size of, 197, 344-352, 370, 373, 390-395, 609, 699.
Investigation, Division of (MC), 784.
Investigation of the National Defense Program, see Truman Committee; Congress, committees of.
Investigation Section (MC), see Technical Division.
Investment Bankers Association, 303.
Invoices, 728, 729.
Iron Age, 747 n.
Iron and Steel Advisory Committee, 313.
Irvin Works, Carnegie-Illinois Steel Co., see Carnegie-Illinois Steel Co.
Isaacson, Henry D., 664 n.

J

J. A. Jones Construction Co., 148, 250, 463, 535, 801, 807. See also, Jones-Brunswick shipyard; Jones-Panama City shipyard.
J. F. Shea Co., Inc., 809.
J. G. White, 52.
J. L. M. Curry, 553.
J. W. Van Dyke, 572.
Jackson, Henry M., 553.
Jacksonville, Fla., 148, 250. See also, St. Johns River Shipbuilding Co.

INDEX 853

Jigs, 213, 239.
Job breakdown, 238, 240, 243, 244, 264, 267. *See also*, Labor force, skill in.
Johnson, C. H., 789.
Johnson, Rear Admiral Harvey F. M., 566, 570.
Joiner work, 204, 694.
Joint Chiefs of Staff, 335-342, 491, 574, 602-605, 619, 658, 718, 831. *See also*, Directives for shipbuilding.
Joint Instructions to auditors of MC, 727.
Joint Military Transportation Committee, 603-604.
Joint Service Organization of MC, 172, 722, 723, 728, 736, 756-765, 767, 771.
Jones-Brunswick shipyard, 142, 208, 209, 223, 250, 475, 535, 557, 563, 640, 826, *facing* 235.
Jones, Edwin I., 807.
Jones, J. A., 148. *See also*, J. A. Jones Construction Co.
Jones-Panama City shipyard, 142, 208, 209, 250, 475, 535, 563, 826.
Joseph L. Thompson & Sons, 81.
Joshua Hendy Iron Works, 397, 399, 404, 585, 595, 809.
Journal of Accountancy, 732 n.
Journal of the American Welding Society, 570 n.
Journal of Commerce (Liverpool), 46 n.
Journal of Commerce (New York), 42 n, 166, 167, 546 n, 548 n, 576 n, 662 n.
Journal of Economic History, 107 n.
Journeymen, 187.
Junior Construction Cost Auditor, 27.
Jurisdictional disputes, 241-243, 276.
Justice, Dept. of, 112, 113, 189, 498, 502-503, 735, 784.
" Justification and Schedules for Budget Estimates, 1944," 728 n.

K

Kahn, A. D.: " History of the Concrete Ship and Barge Program, 1941-1944," 629 n;
" History of the Wood Ship and Barge Program, 1941-1944," 635 n.
Kahn, Felix, 664 n.
Kaiser Cargo, Inc., 809;
frigates at, 614;
LST's at, 177;
procurement by, 360.
See also, Richmond No. 4.
Kaiser Company, 54 n, 809.
Kaiser Co., Inc., 407, 809;
LST's at, 176, 611;
procurement by, 733;
profits of, 808.
See also, Richmond No. 3; Swan Island; Vancouver.
Kaiser, Edgar, 54, 142, 146, 219 n, 220, 248, 254, 342, 463, 469-470, 471, 611, 664 n, 668, 707;
on freedom of management, 484-486.
Kaiser, Henry J., 57, 69, 73, 113, 196, 442, 464, 465, *facing* 74.
business connections of, 54, 141, 469, 808-809, 810;
on fractures in ships, 551;
and labor, 277, 413-414, 428;
and military types, 611-613, 622;
profits of, 805, 808-811;
and steel shortage, 315, 316, 338.
See also, Kaiser management; and under names of companies and yards.
Kaiser management, 53-54, 140-142, 145-146, 189, 220, 227, 228, 238, 248, 463, 470;
and continuity of production, 468;
corporate organization, 809;
and escort aircraft carriers, 612-613;
and housing, 434;
labor recruitment by, 248-249, 254, 667;
labor relations of, 269, 277-278, 277 n, 296-297;
and priorities, 357;
procurement by, 360;
training of workers, 678.
Kaiser-Richmond, *see* Richmond No. 3; Richmond No. 4.
Kaiser-Swan Island, *see* Swan Island shipyard.
Kaiser-Vancouver, *see* Vancouver shipyard.
Kaiser yards:
achievements of, 469-471;
and fractures, 545, 546, 553;
speed of, 513.
Kearny, N. J., 33, 143. *See also*, Federal Shipbuilding and Dry Dock Co.
Keel laying, in relation to fabrication and assembly, 210-212.
Keller, George M., Jr., " Shipbuilding Stabilization History," 289 n.
Kennedy, Eleanor V., " Absenteeism in Commercial Shipyards," 412 n.
Kennedy, Joseph P., 12, 13, 15, 16, 17, 103, 166-168, 168 n.
Kerr, Clark, " Collective Bargaining on the Pacific Coast," 277 n.
Keyhoe, Donald E., " You Can't Sink This Admiral," 13 n.
" Key Men of the Maritime Commission," 171 n.

Kickbacks, 784.
Kiely, J. R., and W. C. Ryan, *Construction Procedure used by California Shipbuilding Corporation*, 226 n.
Kilgore, Harley M., 528, 529, 530.
Kilgore, Mrs. Harley M., 528.
King, Admiral Ernest, 177, 338.
King posts, 31, 84, 88.
Kirkpatrick, James, 500.
Kirsch, William U., 717, 718, 719, 722, 723, 741, 764, 767, 769, 772.
Klitgaard, Karl E., 514, 515, 518, 521, 522, 525-526, 528, 530-531.
Knowlson, James S., 327, 331.
Knox, Frank, 51, 165, 338, 662.
Knudsen, William S., 42, 47, 315, 316, 317.
Knuth, Helen E.: "The Building of LST's by the Maritime Commission," 391 n; "The C-4 Shipbuilding Program," 256 n.
Kohler, Eric L., 768-769.
Koppers Co., Baltimore, 399, 585.
Korndorff, L. H., 291.
Kossoris, Max D., "Work Injuries in the United States During 1944" 448 n.
Kreher, Ernest, 495-496.
Krock, Arthur, 166, 299.
Ku Klux Klan, 49, 495.

L

Labor density, 233.
Labor, Department of, 237, 279;
Apprentice Training Service, 262;
Women's Bureau, 257.
See also, Bureau of Labor Statistics; Division of Labor Standards; Employment Service, U. S.
Labor draft, 263, 272, 299, 662, 663, 687, 688.
Labor force, 413;
ceilings on, 664, 665, 666, 670, 671, 674, 710;
efficient utilization of, 244, 671-672, 679, 683-684, 687-688; see also, Job breakdown; Labor productivity; Manpower surveys;
employment per way, 212, 233;
hoarding by shipyards, 302, 488, 609, 617, 668, 669, 673-674;
lay-offs, 329, 656, 657;
morale, 411-412, 452-455, 627, 657-658, 662, see also, Loafing; Stabilization of labor;
size of, 65, 236, 240, 250-251, 258, 267, 642-643, 654, 669, 671, 673, 683-684, 688, 828;
skill in, 34, 38, 186, 235-240, 244, 246-250, 255-256, 258, 264, 267, 310, 343-344, 423-424, 456, 471, 547, 642, 654, 656, 657, 690; see also, Training;
specialization in, 226-227, 237, 238, 264, 267; see also, Job breakdown;
suggestions from, 454-455;
utilization of, 830, 831, 832.
See also, Absenteeism; Classification of jobs; Hours; Labor supply; Loafing; Recruitment; Shipyards, living conditions near; Supervisors; Training; Turnover; Wages.
Labor leaders, 151, 196-198, 241-242, 254, 420, 435, 436, 450, 657. See also, Collective Bargaining; No-strike pledge; Strikes; Union organization; Working rules.
Labor productivity, 212, 230-236, 298, 301, 302, 310, 411-412, 642-646, 683-685, 826-829.
Labor relations, 15, 16, 49, 166, 411;
grievance machinery, 275, 279.
See also, Collective bargaining; Seamen; Stabilization.
Labor Relations Reference Manual, 291 n.
Labor supply, 1, 34, 37, 47-50, 52, 61, 66, 147, 152, 154, 184, 246-250, 310, 343, 361, 614 n, 637, 763, 827;
competition between yards for, 246, 416, 426, 658; see also, Scamping;
shortage predicted, 236-237.
See also, Manpower shortage.
Labor temple at Portland, 296-297.
Labor unions, 240;
protection of incompetents by, 301-302;
and upgrading, 677.
See also, American Federation of Labor; Collective Bargaining; Congress of Industrial Organizations; Craft unions; Independent Unions; Industrial Union of Marine and Shipbuilding Workers of America; Metal Trades Dept., AFL; National unions; Working rules.
Labor Unit, Construction Audit Section (MC), see Finance Division.
Laborers (classification), 418.
Laborer's Union, see Hod Carriers.
Labor-management committees, 451-454, 465, 523.
LaChine canal, 615.
Lackawanna, Pa., 33.
Lake Carriers Association, 63.
Lake Erie, 615.
Lake Michigan, 615.

INDEX 855

Lake Superior, 63, 615, 710.
Lakeside Engineering Co., 584-585.
Lalley, W. H., 429, 517, 518, 520, 530, 531.
Lancaster Iron Works, Inc., 630.
Land canal, 185 n.
Land file, 55 n, 834.
Land, Mrs. Emory Scott, 81.
Land ownership by MC, see Real Estate.
Land, Vice Admiral Emory Scott, 2, 13-21, 720, 774, 779, *facing* 42, 106;
 as Administrator, War Shipping Administration, 161-162, 754, 755, 757, 760, 764, 833;
 and appointment of Commissioners, 15, 20, 541, 793 and n, 796;
 and British shipbuilding, 42, 60, 576;
 "Building an American Merchant Marine," 22 n;
 as Chairman, 13-14, 20, 21, 159-161, 774-777, 785-789, 791-792, 794-796;
 on Combined Shipbuilding Committee, 592;
 on Commission form, 14;
 and concrete ships, 628, 631;
 and contracts, 99, 104, 105-106, 118, 680, 790, 792-793, 800 n, 805 n, 807;
 on Diesel engine, 29;
 on Ezekiel report, 373;
 and facilities, 46, 47, 49, 111, 147, 449;
 and finances, 719, 744, 746, 749, 768, 770, 771;
 and fractures in ships, 548, 552;
 and Higgins contract, 188, 192-193, 198-200;
 and labor, 61, 166, 243, 263, 270-273, 283, 284, 285, 287, 289, 297-304, 658, 661, 662, 668;
 on launching gifts, 70;
 and Liberty ships, 43, 45, 68, 77;
 and long-range program, 21, 23, 43, 606;
 and management, 52, 151, 160, 236, 300, 456, 462, 478, 537, 670, 671-672, 783;
 and military types, 178, 610, 612, 613;
 and Navy shipbuilding, 51;
 in NDAC, 42, 366, 588, 591;
 in OPM, 366;
 and personnel, 752;
 pleas for shipping, 182, 183;
 and political influence, 703;
 and production, 139, 179, 206-207, 224;
 and programs, 55, 57, 61, 66, 143-145, 165-166, 341, 342, 650-651;
 promotion to Vice Admiral, 459, 792;
 relations with Robert Earle Anderson, 766;
 relations with John M. Carmody, 791-792;
 relations with Capt. Granville Conway, 770;
 relations with William Francis Gibbs, 99-100, 589-592;
 relations with Joseph P. Kennedy, 167-168;
 relations with W. S. Newell, 503, 514 and n;
 relations with John E. Otterson, 541;
 relations with Joseph W. Powell, 200;
 relations with Franklin D. Roosevelt, 12-13, 14, 777, 779;
 relations with Admiral Vickery, 15, 20, 168, 775-777, 788, 790-791;
 and *Sea Otter*, 164-165;
 and shipbuilding czar, 163, 167-168, 589-592;
 on small business, 406-407;
 "Some Policies of the U. S. Maritime Commission," 29 n;
 and South Portland Shipbuilding Corp., 513, 518, 520, 525, 527, 529, 530;
 on steel, 181, 193, 316-317, 327-328, 330-332, 335, 338, 340;
 and Victory ships, 590, 594, 599 and n, 601-605, 606.
Landing craft, 225, 323, 650, 651, 652.
Landing ships, 4, 177, 225, 323, 357, 391, 468-469, 574, 608, 609-612, 613, 645, 777, *facing* 587;
 yards building, 777.
Lane, F. C.: *Andrea Barbarigo, Merchant of Venice*, 743 n.
"Venture Accounting in Medieval Business Management," 743 n.
Lank, Cmdr. R. B., Jr., 567.
Lapland, 40.
Large Vessel Sales Division (MC), 758.
Larsen, A. W., 390.
Lauck, W. Jett, see Hanna, Hugh S.
Launching ceremonies, 68-71, 81, 454, 528, 736.
Launchings, 68. See also, Side launchings.
Lawrence, D. E., 429, 430, 431, 432, 433, 439.
Lay-offs, 329, 333, 656, 657.
Lead times, 211, 349-351, 361, 379-380.
Leaderman, 258, 557.
Leaderwoman, 257.
Leahy, Admiral W. D., 335, 338, 339, 341, 342, 589, 592, 603, 604, 605.
Leases:
 of DPC, see Defense Plant Corporation;

856 INDEX

of land for facilities, 108, 110, 111, 114-115, 511.
Leathem D. Smith Shipbuilding Co., 155, 563, 614, 630.
Legal Division (MC), 18, 171-172, 486, 691, 758, 765;
  General Counsel, 194 n, 552, 598, 719, 736, 741, 762;
  Legislative Counsel, 719;
  Solicitor, 112, 116; see also, Page, Paul D., Jr.
Legislation, Committee on (MC), 119.
Legislative Counsel (MC), see Legal Division.
Leiserson, William L., 299.
Lend-lease, 40, 41, 45, 55, 60, 62, 66, 634, 713.
Lentz engine, 584-587, 590, 593-596, 598, 601.
Lester, Pa., 398, 399.
Letters of intent, 127, 500, 680.
Liberty fleet, 43, 68.
Liberty Fleet Day, 68.
Liberty ships, 28, 41, 55, 60, 61, 62, 68, 70, 131, *facing* 74, 779;
  accommodation ladder on, 80, 548, 569, 573, 581;
  booms for, 87, 410;
  British prototypes of, 80-83;
  cargo holds of, 84, 88-89;
  as colliers, *facing* 363;
  compared to other types, 28, 164-165, 513, 591 and n, 633-634;
  construction of favored by WPB, 590-594, 596, 599-601;
  construction time of, 121-122, 140-141, 174-177, 207, 208, 210-212, 231, 457-458;
  cost of, 122, 491, 819, 820, 826, 829; see also, manhour cost of;
  cubic of, 28, 89;
  deadweight of, 86, 89;
  deliveries of, 4-5, 6, 336, 574, 651, 820;
  design, origin and changes, 74-77, 80, 83-88, 320-321, 548-549, 554-555;
  draft of, 28, 86;
  engine room, 87-89;
  form of contract for, 119-123, 456, 491;
  fractures and crack arrestors on, 548-550, 553-555, 561, 571, 574;
  fuel tanks of, 84;
  funds for, 713-714, 716 and n;
  hatches on, 84, 89, 549; see also, Hatch corner reinforcements;
  lines of, 86;
  manhour cost of, 77, 121, 209, 212, 231, 234-235, 320, 475, 642, 675, 685, 686;
  modified forms of, 89, 208, 209; see also, as tankers.
  name of type, 67-68;
  names for, 70, 779;
  opposition to, 43, 553, 576, 599, 720;
  plate model, 77, 320, 321;
  propeller troubles of, 82 n;
  reason for building, 42-45, 72-77, 575, 829;
  rudders for, 204;
  specifications for, 78-80; see also, Requisitions;
  stability of, 695;
  supplies and components for, 396;
  as tankers, 208, 209, 458, 588, 640, 645, *facing* 363;
  as transports, 554;
  weight of steel plate in, 320, 321;
  yards building, 51-55, 59, 62-63, 138-142, 145-148, 150, 175-177, 204, 208-209, 472, 583, 610, 617, 626, 639, 640, 645, 777, 827, 829; see also, Emergency yards; Higgins yard; and other yard names.
  See also, under Standardization; Steel, plate; Working plans.
Liddell, F. A., 274.
Life, 501 n.
Life saving appliances, 629.
Lifeboats, 85, 366, 407, 408, 622.
Lighters, concrete, 631, 633.
Lincoln Electric Co., 558, 561-564.
Lincoln, James P., see Lincoln Electric Co.
Lippmann, Walter, 164.
Little Steel formula, 421, 423.
Lloyd's Register of Shipping, 80 n, 86, 93.
Loafing, 67, 166, 255, 297, 299-302, 412, 414, 542, 558, 664.
Loans, 479-480, 494, 495, 498, 735, 805-806. See also, V-loans.
Local governments, 114-116, 428, 436.
Log, The, 220 n.
Long Beach, Cal., 35. See also, Consolidated Steel Corp.
Long, Irene, 460, 466.
Long range program, 23-27, 36, 39, 40, 42, 44, 55, 58, 355, 359, 572, 713, 716;
  fiscal meaning of, 716 n;
  yards in, 34-36, 57-58, 125-126, 169, 320, 344, 350, 354, 355, 373, 386-387, 392, 394, 406-407; see also, Old-line shipyards.
Los Angeles, Cal., 33, 38, 48, 49, 51, 54, 58, 110, 115, 249, 431, 432, 436. See also, California Shipbuilding Corp.; Consolidated Steel Corp.
Los Angeles class, 75.

INDEX

Los Angeles Metal Trades Council, 329.
Los Angeles Shipbuilding and Dry Dock Co., 75.
Louisiana, 198. See also, New Orleans.
Louisiana Shipyards, Inc., 472 n.
LST's, see Landing ships.
Luckenbach, J. Lewis, 576, 577.
Lumber, 410, 636.
Lunch period pay, 524, 532.
Lykes Bros., 286.

M

" M " pennants, 453-454, 469.
Macauley, Capt. Edward, 15, 20, 68 n, 774, 779, 789, 795, facing 106;
 and contracts, 403, 783;
 dissents, 501;
 and labor, 21, 263, 264, 271 and n, 302, 309;
 and management, 189, 530;
 and War Shipping Administration, 162, 754, 764.
McCloskey and Co.:
 San Jacinto shipyard, 630, 632;
 Tampa shipyard, 630, 633, facing 619.
McCone, J. A., 55, 219 n.
McCone management, 269, 463, 470.
MacCutcheon, Lt. Comdr. E. M., 570.
MacDonald & Kahn Inc., 54 n. 809.
McElroy, Frank S. and Arthur L. Svenson, " Shipyard Injuries During 1943," 448 n.
MacEvoy, Clifford, 632.
MacEvoy Shipbuilding Corp., 453, 630, 632.
McGinn, B. A., 98.
MacGowan, Charles J., 255.
Machine tools, 38, 46, 47 n, 153, 207, 353, 354, 355, 398.
Machine Tools Branch, Priorities Section (MC), see Production Division.
Machinery inspector coordinator, 135.
Machinery inspectors, 133-135. See also, Inspection in shipyards.
Machinery and Scientific Specifications Section (MC), see Technical Division.
Machinists, 288, 295.
Machinists, International Association of, 241, 273;
 Local 68, 288-290.
Machinists strike, see San Francisco Bay strike.
McInnis, J. F., 170, 171, 537, 564, 705-706, 710, facing 746.
McIntyre, Marvin, 192, 193, 331, 613.
McKeough, Raymond S., 793-794, 795.
McKesson and Robbins, 732, 744.

Mackusick, W. H., 557.
MacLean, Allen D., 367, 378-379, 383, 384, 752, facing 362;
 " Functions of the Maritime Commission Production Division," 205 n.
McNutt, Paul V., 263.
MacPhedran, A., 74.
Macy Board, 167.
Magnuson, Warren G., 554.
Mail, 169, 760.
Main, A. M., 504 n.
Maine, 48, 635, 636 n.
Maine Central Railroad Co., 504, 506, 510, 517.
Maintenance of membership, 282, 290-294, 299 n.
Maintenance and Repair Division (MC and WSA), 19 n, 758, 761.
Managements, 830-832;
 ability in, 46, 47, 49, 51, 66, 145, 151-152, 167, 168, 233, 236, 300, 302, 343-344, 390, 391, 392, 394, 456, 462-464, 498, 534, 542, 609, 616, 617, 632, 642, 690, 806-807;
 changes in Class I and Class II yards, 493, 537;
 control over, 1-2, 108, 128, 132, 136-137, 404, 485, 496, 516, 522, 542-543, 598, 689, 739, 783;
 exchange of ideas among, 77, 92, 189, 215, 223, 460, 464-465, 705;
 functions of, 1, 215, 246, 259-263, 265, 266, 268, 365, 436, 446, 666, 670-671, 679, 688, 689, 734, 783;
 initiative in, 136, 215, 238, 371, 456, 483, 485-487, 556, 559;
 labor policies of, 241-242, 269, 274, 291, 302, 416, 420-421;
 needling of, 300, 384-385, 461, 465-468, 471, 492, 513, 622-624, 672;
 new to shipbuilding, 53-55, 224, 269, 277, 420-421, 463, 471, 513, 545, 571, 806;
 old-line, see Old-line shipyards;
 rating competitively, 467-471, 513, 748, 800;
 selection of, 48-55, 160, 184, 456, 497-498, 501-502, 514.
Manhattan project, 650, 668.
Manhours, see Cost of ships, labor costs; Direct manhours; Indirect manhours; Labor productivity; Standard manhours; and under names of ship types.
Mann, Charles F. A.:
 article on Delta, 215 n;
 " Emergency Shipyard of the Pacific Northwest," 219 n;

"How They Break Records in the Oregon Shipyard," 219 n;
"Oregon Shipyard, a Record Breaking Yard," 220 n.
Manpower shortage, 252, 263, 343-344, 647, 654, 656- 657, 661, 671, 680, 687-688, 831. See also, War Manpower Commission.
Manpower Survey Board reports, 244 n.
Manpower surveys, 243;
by MC, 669, 672-674, 678, 688, 777;
by War Manpower Commission, 669-670; 673.
Mansfield, Harvey C., *The Comptroller General*, 746 n.
Marin County, Cal., 147. See also, Marinship Corp.
Marine Age, 214 n.
Marine Engineer, 576.
Marine Engineering and Shipping Review, 22 n.
Marine instruments, 397.
Marine Maintenance Corp., 538-539. See also, East Coast Shipyards, Inc.
Marinship Corp., 142, 146-147, *facing* 138, 235, 330, 778;
contracts of, 333, 789, 813, 814, 815;
cost of ships at, 682;
labor at, 249, 430, 669 n, 674;
layout and facilities at, 222, 340;
speed of construction at, 175, 466, 471;
and tankers, 333, 469, 583, 640;
women at, 257.
Maritime Commission:
chairman, 11, 12, 271, 777, 786, 795; see also, Land, Vice Admiral Emory Scott;
committees of, *see* under names of committees;
"Cost of War-Built Vessels," 820 n;
Division Directors, 483; see also, under names of Divisions;
Divisions of, 17, 18, 20, 788-789; see also, under names of divisions;
Docket of, 20;
Executive Director of, see Executive Director;
Historian's Collection, Land file, 55, 834;
Historian's Collection, Research file, 117, Statistics, 61 n;
Maritime News Digest, 166 n, 760;
Minimum Requirement for Safety and Industrial Health in Contract Shipyards, 447 n;
Minutes, 774;
Official Construction Record, 60 n;
Officials of, *see* under names or title of official;
"Permanent Report of Completed Ship Construction Contracts," 121 n;
Program Collection, 60 n;
Report to Congress, 1936, 18 n;
Report to Congress, 1938, 103 n;
Report to Congress, 1939, 12 n;
Report to Congress, 1940, 24 n;
Report to Congress, 1941, 716 n;
"Report to Congress, 1943," 179 n;
Report to Congress, 1944, 763 n;
Report to Congress, 1946, 721 n;
Report to Congress, 1947, 117 n;
Report to Congress, 1948, 3 n;
Vice Chairman, 160, 776; see also, Vickery, Vice Admiral Howard L.
See also, Commissioners of MC; Commission action; War Shipping Administration, relations to MC.
Maritime Day, 174, 189.
"Maritime 'M' Award Record," 454 n.
Maritime News Digest, 166 n, 760.
Maritime Personnel, Division of (MC), 18, 271, 418. See also, Shipyard Labor Relations, Division of.
Maritime Promotion and Information, Division of (MC), see Information Division.
Marshall, General George C., 144, 338.
Marshall, W. L., 129.
Martin Joseph W. Jr., 166.
Martinsky, E. E., 559, 567.
Maryland Drydock Co., 285, 610.
"Mass Produced 'Landing Craft Infantry' at Todd Managed Shipyard," 225 n.
Mass production, 45, 76, 77, 205-206, 227, 235, 238, 513, 590. See also, Multiple Production.
Massachusetts Institute of Technology, 14, 504.
Masts, 84, 88.
Material Control Section (MC), see West Coast Regional Office.
Material Coordinator Section (MC), Gulf Regional Office.
Material coordinators, 171-172, 363, 381, 394, 696-700, 703-704, 711, 727, 733.
Material inspectors, 408-410, 551.
Material Receiving Reports, 726-728, 734, 747, 832.
Material requirements schedule, 380.
Materials, *see* Commission-procured materials; Costs, material; Emergency purchases; Procurement; Scheduling; Steel; Surplus materials; Transfers.
Materials Handling Corp., 510, 527.

INDEX

Materials Requirement Branch, Production Engineering Section (MC), *see* Production Division.
Materials Section (MC), *see* Technical Division.
Materials Unit, Construction Audit Section (MC), *see* Finance Division.
Mattresses, 409.
May, Stacy, 165.
MC gf (general files of Martime Commission), 834.
Mechanics (classification):
 first class, 240, 245, 256, 258, 264, 275, 279, 280, 281, 285, 309, 416-418, 420-424, 425, 524, 525, 556, 676;
 second class, 281, 420;
 third class, 420.
Meidlich, Paul, 584.
Merchant and Miners Transportation Co., 164.
Merchant Fleet Corporation of U.S. Shipping Board, 767.
Merchant Marine Act of 1928, 21.
Merchant Marine Act of 1936, 10-12, 22, 27, 56, 101, 102, 105, 498, 712, 714, 716, 762, 775, 776, 798, 799.
Merchant marine of U.S., 3, 21, 29, 44, 56, 74, 763, 766;
 postwar, 575-577, 600-601, 626.
 See also, Dry cargo carriers, Long-range program; Tankers.
Merrill, James C., 805.
Merrill, R. T., 164, 805.
Metal Trades Councils, 297.
Metal Trades Dept., AFL:
 and craft rules, 240, 241, 242;
 and jurisdictional disputes, 276-277;
 membership in, 268, 276, 277, 285-286;
 and Negroes, 254;
 and organizing drives, 295;
 and Pacific coast master agreement, 281-282;
 Proceedings of . . . Convention, 240 n;
 and recruitment of labor, 277;
 and Shipbuilding Stabilization Committee, 273, 282 and n, 285, 286;
 and strikes, 294, 303;
 and wages, 308-309;
 and welders secession, 243.
Metten, John F., "The New York Shipbuilding Corporation," 214 n.
Meyer, Henry C. E., 83, 93, 94.
Midship house, 30, 76, 83, 88, 582.
Mikhalapov, G. S., 567, 568.
Military service, *see* Selective service.
Military types, 4, 27, 199, 235, 329, 392, 491, 575, 607, 644, 646, 666, 671, 679, 716 n, 777, 788, 831;
 cost of, 819, 820;

deliveries, 4, 5, 336, 651, 820;
form of contracts for, 457, 800;
manhour costs, 642;
reason for MC building, 176, 608-609, 613;
yards building, 344, 360, 392, 394, 630, 638-640, 645.
 See also, Aircraft carrier escorts; Landing ships; Transports, etc.
Miller, John Perry, *Pricing of Military Procurement*, 101 n.
Mills, R. P., 681.
Milwaukee, Wisc. 399. *See also*, Froemming Bros., Inc.
Minimum inventory, steel, 349-352.
"Minimum Requirements for Safety and Industrial Health in Contract Shipyards," 447.
Ministry of Labour Gazette, 306 n.
Minnesota, 710.
Minor types, 27, 60, 62, 329, 627-636, 716 n;
 cost of, 820;
 deliveries of, 4, 5, 336, 651, 820;
 form of contract for, 457, 800;
 yards building, 360.
 See also, Barges; Coastal cargo vessels, etc.
Minutes of the Advisory Committee to the Council of National Defense, 111 n.
Minutes of MC, 760.
Minutes of the Planning Committee of the War Production Board, 325 n.
Minutes of the War Production Board, 165 n.
Mississippi River, 615, 635.
Mobile, Ala., 48, 50, 51, 53, 62, 144, 146, 432, 433, 436, 438-440;
 race riot at, 253.
 See also: Alabama Dry Dock and Shipbuilding Co.; Gulf Shipbuilding Corp.
Mohler, R. H., 130, 322.
Mold loft, 2, 86, 221, 500, 564, *facing* 107.
Monotube cargo booms, 410.
Monthly Labor Review, 249 n;
Monthly Report of Progress, 381, 384.
Moore Dry Dock Co., 35, 38, 58, 104, 129, 149, 464;
 and attack cargo vessels, 638;
 construction methods at, 825;
 labor at, 276, 290, 430, 451, 669 and n;
 Negroes at, 254;
 profits of, 801;
 Victory ships at, 626 n.
Moore, Joseph A., 254.
Moore-McCormack Lines, 624.

Moorehead City, N. C., 50.
Morale, *see* under Labor force.
Moran, Edward C., Jr., 20, 791.
Morison, Samuel Eliot, *History of United States Naval Operations*, 10 n;
The Battle of the Atlantic, 10 n.
Morpurgo, Henry von, 664 n.
Morrison-Knudsen Co., 54 n, 809.
Morro Castle, 30.
Morton, Harry F., 277.
Moses, Walter B., 185, 189.
Motorship, 576.
Mowry, George E., *Landing Craft and the War Production Board*, 609 n.
Multiple production, 32, 73, 169, 205-207, 214-216, 219, 222, 224, 230-235, 237, 243, 456, 462, 463, 471, 572, 637, 824, 829, 830;
costs under, compared, 825;
of military types, 613, 626.
Munitions production, total, 10, 180 n, 202, 323, 335-336.
Murray, James E., 406.
Murray, Philip, 288, 296, 303.
Musser, W. Daniel, "Vocational Training for War Production Workers," 259 n.
" Must " programs, 650, 652, 668.
Mutual Life Insurance Co., 755.

N

N3's, 225, 628. *See also,* Coastal cargo vessels.
Names of ships, selection, 70, 779.
Narragansett Bay, 148.
Nathan, Robert R., 331, 339, 340, 656.
National Academy of Sciences, 567.
National Archives, 44 n, 834.
National Association of Manufacturers, 299 n.
National City, Cal., 630. *See also,* Concrete Ship Constructors.
National Council of American Shipbuilders, 2, 274, 591;
Annual Report, 206 n.
National Defense Advisory Commission, 41-42, 56, 107, 111, 236, 270, 273, 274, 428;
Shipbuilding Section of, 42, 66.
National Defense Advisory Council, *see* National Defense Advisory Commission.
National Defense Mediation Board, 269 and n, 270, 289-292, 294-295, 297, 298.
National Emergency, 61, 119, 315.

National Housing Agency, 428, 429, 432, 433, 434, 435, 437, 439, 440, 441;
Third Annual Report (1945), 429 n.
National Labor Relations Act, 269, 297, 495.
National Labor Relations Board, 269, 283, 296, 297, 522, 524;
Decisions and Orders, 522 n.
National Manpower Priorities system, 668. *See also,* Urgency ratings, for labor.
National Resources Planning Board, 49-50.
National Safety Council, 446.
National Service Act, *see* Labor draft.
National unions, 269, 273, 287, 308. *See also,* American Federation of Labor; Congress of Industrial Organizations; Industrial Union of Marine and Shipbuilding Workers.
National Urban League, *Summary of a Report on the Race Riots*, 253 n.
National War Labor Board, *see* War Labor Board.
National Youth Administration, 262.
Nautical Gazette, 214 n.
Naval Academy, U. S., 13-14, 171, 541.
Naval architects, firms of, 73, 78, 99-100, 354, 359-361, 380, 465, 580-581, 586, 587, 606, 649.
Naval Architects, Society of, *see* Society of Naval Architects and Marine Engineers.
Naval auxiliaries, 4, 22, 28, 29, 37, 394, 626, 647, 652, 684, 720. *See also,* Military types; Standard cargo ships.
Naval Boiler and Turbine Laboratory, 586.
Naval Transport Service, 621.
Navy, Department of the:
Bureau of Construction and Repair, 12, 13, 15, 16, 766;
Bureau of Ordnance, 652;
Bureau of Personnel, 652;
Bureau of Ships, 143, 324, 541, 610, 612, 621, 652;
Chief of Naval Operations, 652;
Construction Corps, 14;
contracts used by, 111, 119, 490, 803 n;
conversions for, *see under*, Standard cargo ships.
crews on frigates, 616;
" E " pennants, 453, 509;
and fractures in ships, 549, 565; *see also* Board of Investigation;
and Higgins, 184, 195-196, 199-200;
Inspectors, 652, 711;
labor policies of, 264, 270, 279, 291,

INDEX 861

309, 416-418, 422, 423, 427, 446, 452 and n, 659, 677;
Material Coordinating Agency, 711;
and merchant shipping, 11, 22, 24, 51, 200, 332, 575, 581, 605, 620-622;
Office of Naval Intelligence, *Fuehrer Conference*, 202 n;
Office of Procurement and Material, 196, 200;
operation of ships by, 620-621;
personnel from in MC, 19, 170, 750-751, 773, 789;
personnel drawn from MC by, 740;
priorities of, 38, 118, 323, 335, 356, 358, 432, 652; *see also*, Navy, steel requirements;
procurements by, 396-399, 401, 407, 408, 409, 611;
renegotiation by, 778, 801, 803 n, 806;
scheduling by, 342, 370, 391;
Secretary of, 183; *see also*, Knox, Frank;
shipbuilding by, 3, 8, 9, 33-34, 36-39, 42, 48, 49, 51, 52, 55, 56, 57, 58, 65 and n, 73, 92, 99, 100, 104, 105, 128, 138, 143, 149, 152, 164-165, 167, 178, 181-183, 225, 236, 292, 478, 497-498, 504, 509, 539, 588, 592, 611, 636, 658, 798;
shipbuilding by MC for, *see* Military types;
steel requirements of, 200, 314, 322, 331, 335-337, 339, 342, 344-346;
transfers of funds to MC, 720, 721 and n;
turbines for, 367, 398, 588-590; *see also*, Turbines shortage of, scheduling of.
Navy Liaison Group (MC), *see* Production Division.
Navy oilers, 4, 56, 627, 682.
Nazis, 40, 42, 66, 138, 300, 501.
NDAC, *see* National Defense Advisory Commission.
NDMB, *see* National Defense Mediation Board.
Negligence of contractor (contract clause), 790, 792, 815.
Negotiation of contracts, 119, 127, 130, 379, 477, 766;
before the war, 102, 105;
in January 1942, 140-142, 811;
in January 1945, 681-683, 686-687, 812-817.
Negroes, 186, 188, 189, 252-256, 258, 442, 495, 563, 622, 740.
Nelson, Donald, 182, 183, 192, 193, 198, 313, 325, 329-330, 335, 338-340, 346, 451, 592, 603-604, 605;
Arsenal of Democracy, 42 n.
Neutrality Act, 40.
New Deal, 16, 42, 269, 428, 783.
New England, 147, 148.
New England Shipbuilding Corp., 458, 504, 505, 507, 533-534;
cost of ships at, 826;
east area, 208, 209, 216, 504, 507; *see also*, Todd-Bath Iron Shipbuilding Corp.,
layout and facilities at, 675;
north area, 507, 534;
and power rates, 781;
productivity at, 475;
speed at, 457;
and Thompson & Lichtner, 675, 676;
welding at, 563;
west area, 208, 209, 504, 507; *see also*, South Portland Shipbuilding Corp.
New Jersey Shipbuilding Corp., 539.
New London, Conn., 34.
New Mexico, 20, 249.
New Orleans, La., 48, 50, 51, 52, 53, 62, 170, 184, 187, 189, 190, 195, 196, 689. *See also*, Delta Shipbuilding Co., Inc.; Higgins yard.
New York, N. Y., 9, 78, 147, 667.
New York Shipbuilding Corp., 33, 37, 52, 57, 214, 268, 282, 283, 285.
New York Times, 54 n, 303.
Newburgh, N. Y., 48, 152.
Newell, William S., 53, 55, 73, 216, 217, 465, 503-504, 506, 507, 509, 510, 513, 514, 515 and n, 516, 519, 522, 526-529, 531, *facing* 74;
"A Basin Type Shipyard," 217 n.
Newfoundland, 40.
Newman, Dorothy K., "Employing Women in Shipyards," 257 n.
Newport News, Va., 33, 526.
Newport News Shipbuilding and Dry Dock Co., 33, 34, 35-37, 50, 52, 53, 57, 77, 86, 92, 102, 103, 129, 462, 463, 470;
labor relations at, 268, 269, 273, 282-283, 294, 308;
and Liberty ship program, 464-465;
and North Carolina yard, 218, 246;
training of workers at, 237, 238.
Newspapers (plant organs), 483, 707, 731.
NHA, *see* National Housing Agency.
Nicholls, Maurice, 219, 464.
NLRB, *see* National Labor Relations Board.

No-strike pledge, 274, 275, 279, 287, 288, 289, 290, 292, 298.
Nordberg Manufacturing Co., Milwaukee, 399.
Norfolk, Va., 36, 50.
Normandie, 22.
Norris, George W., 199.
North Atlantic Coast, 34, 48, 49, 97, 257, 660. *See also*, Atlantic Coast yards.
North Carolina, 148, 247.
North Carolina Shipbuilding Co., 51, 129, 462, 463, 470, 512, *facing* p. 267;
and attack cargo vessels, 639;
change to C2's, 333, 583;
construction methods at, 230, 825;
contracts of, 490 and n, 679, 817;
cost of facilities at, 472;
cost of ships at, 825, 826;
Five Years of North Carolina Shipbuilding, 227 n;
labor at, 246-247, 294, 308, 413, 452;
layout and facilities at, 59, 218 and n, 222, 224, 227;
procurement by, 360;
productivity at, 209, 475, 645, 646;
profits of, 458 n, 817;
speed of construction at, 140 n, 141 n, 145, 175, 208, 234, 458, 468-469, 471, 671;
as standby yard, 117.
North East Coast Institution of Engineers and Shipbuilders, *Transactions*, 72 n.
North Pacific, fractures in, 553.
North Sands, England, 83.
North Sands Shipbuilding Yard, *see* Joseph L. Thompson & Sons.
Northeast, *see* North Atlantic coast.
Northwest, *see* Pacific Coast.
Norton, T. L., 279.
Notch sensitivity of steel, 547, 569, 570, 573.
' NYA Trains Shipyard Workers," 262 n.

O

Oakies, 249, 442.
Oakland, Cal., 35, 104, 276, 289, 390, 407, 433, 689, 702, 762; *See also*, Moore Dry Dock Co.
Ocean class ships, 81-82, 84, 85, 88, 90, *facing* 74.
Ocean Courage, *facing* 74.
Ocean Vanguard, 81.
O'Dea, Mark, 760.
ODT, *see* Office of Defense Transportation.
Office of Defense Transportation, 428-431, 517, 635, 752 n;
Civilian War Transport, 428 n.

Office of Emergency Management, 161.
Office of Production Management, 56, 57, 93, 143, 178, 259, 262, 270, 279, 283, 285, 289, 291, 368;
priorities of, 314, 356, 358; *see also*, SPAB;
Shipbuilding Section, 366;
steel under, 312-313, 316-319.
Office of Production Research and Development, 567.
Office of War Mobilization and Reconversion, 602-603, 681, 686.
Official Construction Record, 60 n.
Official Production Statistics, Advisory Committee on, 362.
Ohio, 20.
Oil, 74, 83, 84, 325.
Old-line shipyards, 32-34, 53, 195, 200, 268-269, 308, 371, 420-421, 462, 465-466, 471, 545, 661, 806, 818.
Oliver, Eli, 289.
Oliver, James C., 517, 518, 526, 530.
Open shop, 270, 278, 291, 298.
Operation of ships, 18, 620, 713, 746, 767, 833. *See also*, Ship Operators; War Shipping Administration.
Operations Division (MC), *see* Ship Operations Division (MC).
OPM, *see* Office of Production Management.
Orange, Tex., 38, 170, 436.
Ore carriers, 60, 63, 325, 544, 551, 574, 584, 624, 636.
Oregon Shipbuilding Corp., 51, 69, 123, 685, 686, 809, *facing* 75, 138, 139, 330, 490;
cost of facilities, 472;
cost of ships at, 826;
financing of, 480, 481;
Hull No. MCE 581;
inventories at, 391, 392;
labor at, 212, 248, 282, 296, 430, 668, 669;
layout and facilities, 59, 150, 218-219, 220, 223-224, 227, 340;
Negroes at, 254;
productivity at, 209, 475, 645;
profits of, 476, 817;
speed at, 145, 175, 176, 207, 208, 210, 224, 232, 333, 343, 350, 458, 467, 468;
steel shortage at, 316;
Victory ships at, 586, 607;
Victory transports at, 626, 638;
women at, 257.
Oregon State College, 550.
Organization charts, 15 n, 132 n, 367, 764-765.
Osbourne, Alan, 75, 583 n, 587 n.
Otterson, John E., 539-541, 792.

INDEX 863

Outfitting, 207, 347-349;
docks, 221-222, 224, 228-229;
time in, 140, 141 and n, 174, 212.
of Victory transports, 626.
Output, 640-643. *See also*, Deliveries; Productivity.
Ovenden, Cmdr. Philip A., 566.
Over, short, and damage reports, *see* Material Receiving Reports.
Overhead:
governmental, 822-824;
of shipbuilding companies, 105, 474, 729, 731, 735, 822-824.
Overhead Unit, Construction Audit Section (MC), *see* Finance Division.
Overpayments by MC, 746.
Overscheduling, 202, 364, 365, 392.
Overstaffing in MC, 753-754.
Overtime, 275, 279, 280, 286, 307, 309-310, 415, 418, 427, 637, 653, 658, 659, 661, 674.

P

P1 (design), 612.
P2-SE2-R1 design, 624.
P2-S2-R2 design, 624, 625 n.
P2-SE2-R3 design, 625.
P2-S2-R4 design, 625.
P4-P design, 29.
Pach, G. V., 504 n.
Pacific Bridge Co., 54 n, 630, 809.
Pacific Coast Metal Trades District Council, 276.
Pacific Coast yards, 34, 48, 54, 148.
hiring of women in, 257;
labor in, 240, 248, 250, 259, 261, 268, 276-277, 280, 281-282, 295-296, 308, 412, 415, 418, 419, 420, 423, 424, 425, 426, 450, 657, 659, 660, 664;
and military types, 638;
productivity in, 644, 827-829;
transfers of material between, 648.
See also, Zone Conferences, Pacific Coast.
Pacific Shipper, 20 n.
" Pacific Yards Lead U. S. Shipbuilders," 468 n.
Packard Motor Co., 111.
Page, Paul D., Jr., 112, 113, 154, 499, 779.
Paints, 408.
Palmer, M. B., *We Fight with Merchant Ships*, 196 n.
Panama City, Fla., 148, 433-434, 437-438.
See also, Jones-Panama City shipyard.
Paramount Pictures, 16, 541.
Parking lots at shipyards, 114, 440.

Pascagoula, Miss., 31, 35, 50, 62, 433, 436, 437-438. *See also*, Ingalls Shipbuilding Corp.
Passenger-and-cargo ships, 4, 21, 22, 29, 605, 617, 624, 625.
Passenger ships, 4, 29-30, 794. *See also*, Passenger-and-cargo ships.
Patents, 358, 584, 585.
Patric, John, " Remove Union Restrictions," 242 n.
Patriotism, 120, 244, 246, 412, 414, 415, 429, 445, 451, 453-455, 456, 459, 477, 592.
Patterson, Robert P., 662.
Paul, Col. C. M., 453.
Paull, John R., 778.
Payrolls, 417, 479, 536, 564, 677, 708, 728, 734-735, 737.
Pearl Harbor, 40, 66, 70-72, 112, 127, 138, 150, 168, 220, 227, 243, 250, 295, 297, 298, 319, 321, 322, 399, 428, 457, 499, 505, 508, 513, 750, 799.
Pearson, Drew, 199 n.
Peet, W. C., Jr., 760.
Penalties for delayed deliveries, 121.
Penalty for excessive manhours, 122.
Pendleton Shipyards Co., Inc., 630.
Pennsylvania Shipyards, Inc., 35, 49, 58, 285.
Performance Section (MC), *see* Technical Division.
Perkins, Frances, 289.
Perlman, J., *et al.*, " Earnings and Hours in Private Shipyards, 1936 and 1937," 236 n.
" Permanent Report of Completed Ship Construction Contracts," 121 n.
Permanente Metals Corp., 59, 664 n, 809;
expeditors of, 649;
financing of, 481;
labor at, 277, 454-455 n;
profits of, 810, 817.
See also, Richmond No. 1; Richmond No. 2.
Perry, Carl Carroll, 154.
Personnel Division (MC), 18, 129, 172, 693, 698, 715, 753, 756, 764, 785.
Personnel Management, Division of (MC), *see* Personnel Division.
Personnel of MC, 15, 17, 18, 24, 26, 127, 690, 715, 725, 740, 741, 749-773, 833;
classification and appointment of, 693-694;
management of, 163, 689, 784, 785, 796;
training of, 26.
Peterson, J. Hardin, 197, 199.
Pew, John, 300, 621.
Pew, Joseph N., Jr., 51.

864 INDEX

Philadelphia, Pa., 20, 170, 689, 702.
Photographs used by Vickery, 460.
Piez, Charles, *Report* (1919), 119 n.
Pile driving, 189.
Pillows, 409.
Pinto Island, Ala., 223, 438-439. *See also,* Alabama Dry Dock and Shipbuilding Co.
Pipefitters, 242, 257, 622, 657.
Piping, 92, 121, 366.
Piping Branch, Production Engineering Section (MC), *see* Production Division.
Pittsburgh Equitable Meter Co., 379.
Pittsburgh, Pa., 390.
Plan approval, *see* Design changes; Shipyards; Working plans.
Plancors, 400 and n, 401.
Planning, *see* Directives for shipbuilding; Programs of MC.
Plant Engineering Section (MC), *see* Construction Division; Emergency Ship Construction, Division of; Regional Offices of Construction; Resident Plant Engineers.
Plastic armor, 87.
Plate mills, *see* Steel.
Plate shops, *see* Fabrication buildings.
Platens, *see* Shipyards, assembly areas in.
Plumbers, 246.
Pneumatic tools, 465.
Politics, 47, 96-98, 112, 113, 150-152, 154-155, 173, 192, 195, 198, 518-519, 531, 532, 629, 702-703, 720, 723, 739, 791-796.
Pool orders, 205, 382, 747.
Porter, Paul R., 308, 422; "Labor in the Shipbuilding Industry," 242 n.
Portland, Me., 247, 507.
Portland, Ore., 48, 49, 51, 54, 115, 117, 146, 210, 220, 254, 432, 436, 442. *See also,* Oregon Shipbuilding Corp.; Swan Island; Vancouver.
Portland Terminal Co., 510, 517.
Portugal, 37.
Post (Washington), 296 n.
Posters, 67, 451, 453, 760.
Postwar planning, 763, 787, 794, 816. *See also,* under Merchant marine of U. S.
Postwar shipbuilding, 58, 107, 109, 117, 658.
Powell, Joseph W., 124, 164-165, 195-196, 200, 273, 284, 285.
Powell, R. R., 80 n.
Power, *see* Electric power.
Power Procurement Officer of MC, 781.

PR (Press Releases), 17, 66.
Preassemblies, 207, 208, 212, 213, 214, 217-218, 220-221, 226-230, 237, 239, 345, 349, 351, 539, 547, 548, 560, 582, 830.
Prefabricated ships, 44 n, 206-207. *See also,* Fabrication of steel, outside shipyard.
Prefabrication, *see* Fabrication; Preassembly.
Preference ratings, *see* Priorities.
Preliminary Design Branch (MC), *see* Technical Division.
Premium men, 417, 420. *See also,* Supervisors of shipyard labor.
President of the U. S.:
    approval of contracts by, 12, 56 n, 104;
    funds allocated to MC by, 712, 713, 714;
    war powers of, 161, 162, 776, 787, 796. *See also,* Directives, from the President; Roosevelt, F. D.; Truman, H. S.
President's Committee for Congested Production Areas, *Final Report,* 436 n.
President's Scientific Research Board, John R. Steelman, Chairman, *The Federal Research Program,* 571 n.
Press Herald (Portland), 465 n.
Price adjustment, *see under* Contracts.
Price Adjustment Board (MC), 123, 758, 778, 799. *See also,* Renegotiation.
Price, Waterhouse & Co., 744-745.
Prices, *see* Costs.
Pricing, *see* Contracts; Cost-plus; Procurement.
Principal Construction Cost Auditor, 27, 129, 130.
Principal Hull Inspector, 132 n, 700, 704, 729.
Principal Machinery Inspector, 132 n, 700, 703, 733.
Principal Material Coordinator, 375.
Principal Types, 5.
Priorities, 38, 44, 51, 52, 57, 97, 127, 314, 318, 342, 353-359, 361, 368, 372, 380, 388, 396, 521, 589, 593-594, 601, 606-607;
    blanket orders, 354-355;
    certificates, 38, 118, 314, 354, 356, 357;
    Critical List, 314;
    for housing, 431-433;
    overissue of, 326, 327-328;
    rating AAA, 258, 325, 389, 650-652;
    ratings at various dates, 314, 317, 323-327.
See also, Urgency ratings.

INDEX 865

Priorities Section (MC), see Procurement Division; Production Division.
Private contracts for shipbuilding, 2, 4-6, 23, 34, 36, 58, 63, 66, 68, 69, 128, 138.
Procurement, 3, 72, 73, 89, 97, 100, 129, 205, 353, 354, 359-361, 401, 404, 405, 409, 410;
  centralization of, 170-172, 407, 695;
  by Kaiser Cargo, Inc., 614.
  See also, Commission-procured materials.
Procurement Division (MC), 89, 359-362, 375, 380, 385, 386, 388, 389, 400, 406-409, 649, 715, 750, 758, 761, 762, 764;
  buying sections, 361;
  Clerical Section, 362;
  conflicts with Regional Offices, 698-700;
  Director, 653, 789; see also, Walsh, Charles E.;
  liaison with the OPA, WPB, etc. Section, 362;
  Priorities Section, 362;
  Purchasing Section, 362;
  Rates and Routing Section, 362;
  Records Section, 361;
  Small Business Section, 362, 407;
  Surplus Material Section, 362.
Procurement Section (MC), see Production Division.
Production (index), 643. See also, Output.
Production certificate, 402.
Production Division (MC), 89, 360-369, 374-379, 381, 385, 388, 400, 408, 648-649, 651, 652, 655, 733, 750, 759, 762-764, 775;
  Administrative and Processing Branch, Priorities Section, 367;
  Administrative Section, 367;
  Army and Navy Liaison Branch, Priorities Section, 367;
  conflicts with Regional Offices, 695-699;
  Conservation Branch, Priorities Section, 367;
  Conservation Section, 409-410;
  Controlled Materials Branch, Production Engineering Section, 367, 378, 382, 383, 753;
  Director, 326, 360, 789; see also, MacLean, Allen D.; Rockwell, Col. W. F.;
  Electronics Branch, Priorities Section, 390;
  Engineering Scheduling and Planning Branch, Production Engineering Section, 366, 367, 380, 382;
  Expediting Branch, Production Engineering Section, 367, 388, 390, 649;
  Field Coordinating Section, 367, 375;
  Hull Scheduling and Planning Branch, Production Engineering Section, 367, 369, 380, 381-382, 388, 389;
  Machine Tools Branch, Priorities Section, 367;
  Machinery Branch, 388;
  Materials Requirement Branch, Production Engineering Section, 369, 374, 386;
  Navy Liaison Group, 652;
  Piping Branch, Production Engineering Section, 366;
  Priorities Section 358, 367, 377, 389, 390;
  Procurement Section, 359, 724;
  Production Engineering Section, 349, 362-363, 365, 367, 378, 384, 648;
  Statistical Analysis Branch, Production Engineering Section, 363, 364-365, 367, 374-376, 379-384, 386, 665, 696, 763;
  Steel Requirements Branch, Production Engineering Section, 367, 369, 374, 376, 385, 386, 390;
  Survey and Scheduling Branch, Production Engineering Section, 367-369;
  Survey Branch, 368-369, 379, 390;
  Voucher Certification Section, 724;
  WPB Branch, Priorities Section, 367.
Production Engineering Section (MC), see Construction Division; Emergency Ship Construction, Division of; Production Division.
Production Requirements Plan, 376, 377.
Production Urgency List, 652 n, 668.
Productivity, see Labor Productivity; Way time.
Profit motive, 1, 456, 459, 492, 526, 531, 542, 688, 807-808.
Profits, 101, 105-108, 527, 528, 532, 533, 684, 687, 807-819;
  limits on, 27, 102, 122, 125, 126, 299 n, 798, 799, 807, 813, 814 and n; see also, Recapture of profits;
  rates of, 120, 122, 124-125, 488 n, 682, 802, 805-807, 816;
  total, 802, 806, 818, 822-824.
  See also, Fees; Renegotiation.
Program Collection, 60 n.
Programs of MC:
  by dates
    December 1, 1940, 40;

January-March 1941, 40-41, 44-45, 177;
April 19, 1941, 41, 56 and n, 57-59, 60;
June 1, 1941, 61, 315;
December 1, 1941, 41, 64-66, 138, 157;
January 1, 1942, 138-139, 181;
February 1, 1942, 139-143;
February-April, 1942, 144-149, 173, 177, 181;
June 19, 1942, 180 n;
June 23, 1942, 180 n, 181, 336;
July 25, 1942, 181, 336, 575 and n;
August 1942, 334;
December 1942, 336, 340, 575-576;
April 15, 1943, 596, 597;
May 15, 1943, 596, 597;
July 15, 1943, 596, 597, 605;
November 15, 1943, 597;
November 20, 1943, 597;
January 15, 1944, 651, 655;
April 15, 1944, 646;
September 15, 1944, 681;
December 1944, 681;
August 1945, 686;
exceeded, 176, 202, 625;
formulation of, 1-6, 40-42, 62, 180, 379, 638, 696, 780; *see also*, Directives; Scheduling ship construction;
reduced or stabilized, 191, 193, 263, 329, 332-334, 343, 344, 546, 575, 599;
slippage behind, 165-166, 176, 202, 300, 302, 452, 610-611, 613, 616, 623, 637, 646, 653, 654-656, 658, 788.
Progress payments, 488, 496, 723.
Propeller shaft, 82 and n.
Propeller speed, 595.
Propellers, 397.
Property Removal Notices, *see* Material Receiving Reports; Transfers of materials.
Propulsion machinery, *see* Engines.
Prout, Charles H., 519.
Providence, R. I., 147, 148, 223, 536, 539. *See also*, Rheem Manufacturing Co.; Walsh-Kaiser Co., Inc.
PRP, *see* Production Requirements Plan.
Public Health Bulletin No. 298, 488 n.
Public Health Service, U. S., 448.
Public Information, Division of, (MC), *see* Information, Division of.
Public Law 46, 119, 123, 798.
Public Papers and Addresses of Franklin D. Roosevelt, 43 n.
Public Relations, Division of (MC), *see* Information, Division of.
Public Relations of MC, 15, 65, 67-71, 132, 151-152, 163-166, 198, 297-301, 453, 493, 509, 515, 545-546, 553-554, 691, 701, 719, 720, 760.

Public Schools, 256, 259, 261, 262, 265, 266, 678.
Puget Sound, 9, 243.
Pullman plant, Baltimore, 115, 217-218.
Pumps, 647, 648, 650.
Purcell, Richard E., *Labor Policies*, 270 n.
Purchase and Supply, Division of (MC), 18, 359, 360, 700, 761;
Director, 359-360.
Purchase controllers, 135, 699-700, 703, 704 n, 729, 733.
Purchase orders, 91, 93, 96, 380, 385, 387, 725;
approval of, 129, 131-132, 133.
Purchases of land, 114.
Purchasing, 124 n.
Purchasing agent of MC, 89-100.
Purchasing Section (MC), *see* Construction Division; Procurement Division.
Purissima, 466.
Pusey and Jones Corp., 35, 36, 58, 143 n, 283, 285, 562.
PWA (Public Works Administration), 133.
Pyrenees, 40.

Q

Quarg, William H., 129, 708.
Queen Mary, 22.
Quincy, Mass., 33. *See also*, Bethlehem Steel Co., Fore River yard.
Quinn, Joseph M., 722, 723, 760, 764, 765, 767, 768, 770-772.
Quitting early, 485.
Quits, 413, 414, 427, 440, 442, 446, 655, 657, 661.
Quotas:
of Regions, 333, 461, 690, 692-693, 706, 710;
of yards, 461, 637.

R

Radiagraph, 226.
Radio apparatus, 359.
Radio direction finder, 85.
Rat proofing, 30, 84.
Rates and Routing Section, *see* Procurement Division (MC).
Rationing of tires, 428 n.
Ray, Tom, 296, 297.
Reader's Digest, 242 n.
Real Estate, 108-117, 163, 779 and n, 780, 783, 801, 818. *See also*, Leases.
Reavis, T. H., 489, 744.
Recapture of profits, 105, 478, 488, 682-684, 790, 814, 815. *See also*, Renegotiation.

INDEX 867

Reciprocating engines, 45, 74, 75, 81, 90, 93, 205, 353, 366, 396-398, 577, 578, 583-587, 615, 633. *See also,* Skinner Uniflow engine.
Reconstruction Finance Corporation, 103, 104, 109, 164, 495, 745.
Recruitment of labor, 244-258, 482;
 helped by WMC, 665-669;
 MC's role in, 244, 334;
 restricted by WMC, 665-668;
 unions' role in, 240, 246, 249, 277, 666 and n.
Redmond, Roland, 164.
Reed, F. E., 75.
Reeves, Admiral J. M., 331.
Refrigeration on ships, 30, 84, 204.
Refrigerator ships, 627, 636.
Regional Attorney, *see* Regional Offices of Construction.
Regional Construction Auditors, *see* Regional Offices of Construction.
Regional Construction Cost Committee (MC), 736, 739.
Regional Directors of Construction (MC), 169-170, 363, 364, 375, 381, 460, 462, 478-479, 482, 483, 664, 678, 700, 705, 736, 738, 784;
 authority of, 700-702;
 conflicts with Production Division, 695-699;
 conflicts with Technical Division, 694-695.
 See also, Flesher, C. W.; McInnis, J. F.; Sanford, L. R.; Spofford, W. E.
Regional Disbursements Auditors, *see* Regional Offices of Construction.
Regional Industrial Adviser, *see* Regional Offices of Construction.
Regional Offices of Construction (MC), 169-172, 376, 429, 550, 665, 689-711, 733, 750, 759, 760-764, 775;
 accounting, 743;
 Assistant Directors, 691;
 Assistants to the Director of Finance, 482;
 Attorneys, 171, 691, 736;
 Directors, *see* Regional Directors;
 functional responsibility in, 691, 693, 701, 703;
 health and safety consultants, 446-449, 691, 701;
 Manpower Survey Boards, 669;
 Plant Engineering Section, 215;
 Regional Attorney, 736;
 Regional Construction Auditors, 171, 482, 706, 724, 728, 736, 738-739, 764;
 Regional Disbursement Auditors, 171, 724;

Regional Industrial Adviser, 171, 691, 701;
Regional Trial Boards, 691;
Small Business Sections, 407, 691.
 See also, East Coast, Great Lakes, Gulf, and West Coast Regional Offices.
Regulations Division (MC), 18, 758, 762.
Regulations Prescribing Methods of Determining Profit, 27 n, 730.
Reilly, John D., 53, 514, 528, 529, 531, 532.
"Reinforcing the Liberty Ship," 548 n.
Removal notices, *see* Material Receiving Reports; Transfers.
Renegotiation, 123, 488, 490, 777-778, 793, 798-806, 812, 816-819, 822.
Renegotiation Act, 753.
Rentals:
 for facilities, 109, 115, 157, 511, 532;
 of shipyard equipment, 732.
Rentschler, G. A., 501.
Repair yards, 53, 109, 216, 236, 311 n, 427, 463, 537, 538, 658, 761.
"Report of the American Federation of Labor Committee . . . to investigate . . . Higgins . . . ," 183 n.
Report on Abuses of Merit System, 733 n.
Report Review Unit, Construction Audit Section (MC), *see* Finance Division.
Repricing of contracts, 804. *See also,* Contracts.
Requisitioning of ships, 746.
Requisitions from naval architects, 90, 91, 361, 380, 614, 649.
Rescissions, 717, 721.
Research Division, (MC), 18, 758.
Reserve Fleet, 18, 759.
Resident auditors, 27, 121, 128-131, 132, 137, 255, 472, 482-483, 485, 486, 499, 501, 699 n, 701, 703-704, 706-709, 723-739, 748, 764, 796.
Resident plant engineers, 132, 133, 135, 446, 501, 506, 703, 707-708, 729, 733, 781.
Respiratory diseases, 446, 447, 448.
Review Committee (Construction Cost) (MC), 736 n.
Review of Economic Statistics, 370 n.
Reynolds' Way Charts, 34 n.
Rheem Manufacturing Co., 142, 148, 453, 536. *See also,* Walsh-Kaiser Co., Inc.
Rheem, Richard S., 148, 223, 536.
Richfield Oil Co., 113.
Richmond, Cal., 48, 51, 54, 59, 249, 432,

433, 442-445. *See also*, Richmond shipyards.
Richmond Shipbuilding Corp. (of Calif.), 59, 139.
Richmond No. 1 (shipyard), 54, 138, 265, 463-464, 809;
cost of ships at, 826;
layout and facilities, 218-219, 227;
productivity at, 209, 283-284, 475;
profits of, 473;
speed at, 145-146, 175, 176, 208, 350, 457, 458, 468;
Victory ships at, 586, 607.
See also, Richmond shipyards.
Richmond No. 2 (shipyard), 59, 809;
cost of ships at, 826;
Full Ahead (1942) [Kaiser Manual], 260 n;
Hull No. MCE 440, 210 n;
layout and facilities, 62, 146;
productivity at, 209, 475, 683-684;
profits of, 473;
speed at, 208, 210, 350, 458, 468;
Victory ships at, 586, 607;
Victory transports at, 626, 638.
See also, Richmond shipyards.
Richmond No. 3 (shipyard), 141, 142, 611, 809;
C4's at, 617-618, 619, 622, 623, 639;
construction methods at, 563;
inventories at, 392, 733-734;
land for, 112-113;
layout and facilities at, 220-222;
as standby yard, 117.
See also, Richmond shipyards.
Richmond No. 3A, *see* Richmond No. 4 (shipyard).
Richmond No. 4 (shipyard), 146, 622, 630, 809;
C1's at, 640;
frigates at, 614, 616;
LST's at, 468, 611.
See also, Richmond shipyards.
Richmond shipyards, 129, 141, 189, 264, 464, 562, *facing* 106, 267, 491;
deckhouse assembly area at, 228;
expeditors of, 649;
labor at, 248, 257, 260-261, 430, 668, 669;
speed at, 224, 637;
steel shortage at, 315, 316.
See also, Kaiser Cargo, Inc.; Kaiser Co.. Inc.; Permanente Metals Corp.; Richmond Shipbuilding Corp.; Todd-California Shipbuilding Corp.
" Richmond Took a Beating," 249 n.
Rieber, Capt. T., 501.
Riegelman, Carol, *Labour-Management Co-operation in United States War Production*, 453 n.
Ring, Daniel S., 254, 264, 271 and n, 272-273, 279, 309, 418, 445-456, 451, 484, 495, 559, 665, 788.
Riveting, 31, 47 n, 81-84, 88, 206, 213, 214, 218, 239, 469, 548, 573, 581.
Roach, Alden, 664 n.
Robert, Admiral W. P., 103.
Robert W. Hunt Co., 389.
Robinson, Admiral S. M., 143, 592.
Robson, H. H., 754, 755.
Roche, John M., 446, 447.
Rockwell, Col. Willard F., 326, 330, 357, 367, 373, 378-379, 389, 393, 752-753, *facing* 362.
Rocky Mountains, 160.
Rogers, H. O. and Jesse J. Friedman, *Preliminary Survey of Shipbuilding Facilities in the United States*, 236 n.
Roosevelt, Eleanor, 188, 520.
Roosevelt, Franklin D., 3, 10, 166;
and aid to Britain, 40, 66;
bedroom conference, 144, 618;
and Higgins contract, 188, 192-193, 195, 196, 198-199;
and labor, 273, 284, 292, 297-299, 301-302, 308-309, 451;
and Liberty ships, 43, 45, 68, 74;
and military types, 177-178, 613, 653;
and programs, *see* Directives;
relations with Commissioners, 12-16, 160, 167-168, 271, 302, 776, 777-779;
and Sea Otter, 164;
and shipyards in South, 48, 50, 184;
and South Portland Shipbuilding Corp., 520, 527, 529;
and steel, 178, 193, 317, 328, 330, 332, 335, 337, 338, 339-340;
and Victory ships, 589, 601, 602 and n, 604;
and War Shipping Administration, 754-755;
and wooden ships, 629, 635.
See also, President of the U. S.
Roseman, Samuel I., ed., *Public Papers and Addresses of Franklin D. Roosevelt*, 43 n.
Ross, Holt, 187.
Ross, Malcolm, *All Manner of Men*, 254 n.
Rossell, H. E., *et al.*, *The Training of Shipyard Personnel*, 259 n.
Rotherwick, Lord, 576.
Rounds of the ways, 139, 202, 208, 231-235.
Rowell, E. P., 74.
Rubber shortage, 428, 429-430, 443, 508, 516.

INDEX 869

Rudders, 204.
Rural Electrification Administration, 781.
Ryan, W. C., *see* under Kiely, J. R.
Rydstrom, Comdr. Arthur G., 778, 805 n.

S

S2-S2-AQ1 design, 615. *See also*, Frigates.
S4-S2-BB3 design, 234, 612-613. *See also*, Aircraft carrier escorts.
S4-SE2-BD1 design, 625
S4-SE2-BE1 design, 616, 625, *facing* 778.
Safety:
  at sea, 2, 30, 85;
  in shipyards, 261, 447-449, 674, 701.
Safety consultants, *see* under Regional Offices of Construction.
St. Johns River Shipbuilding Co., 142, 175, 208, 209, 223, 250, 455, 475, 805, 826.
St. Lawrence River, 138, 615.
St. Paul, Minn., 405.
Salaries, 99, 135-136, 402, 403, 478-479, 481, 708, 751, 754, 782, 807.
Sale of ships, 713, 716, 821.
Salter, Sir Arthur, 60, 64, 178.
San Diego, Cal., community facilities at, 436.
San Francisco, Cal., 35, 49, 113, 147, 170, 189, 431, 436, 702. *See also*, Barrett and Hilp; Western Pipe and Steel; Bethlehem Steel Co., San Francisco Yard; Richmond, Cal.; Sausalito, Cal.; Oakland, Cal.
San Francisco Bay, 9, 33, 34, 117, 142, 146, 249, 442.
San Francisco Bay strike, 287-290.
San Jacinto Shipbuilders, Inc., *see* McCloskey and Co., San Jacinto shipyard.
San Pedro, Cal., 33, 37.
Sandhogs, 463-464, 622.
Sanford, L. R., 135, 154, 170, 172, 194, 535, 537, 672, 674, 692, 706, 709-710, 738, *facing* 746;
  and housing, 434-435;
  on material coordinators, 696;
  on Regional Director's authority, 701-702.
Sanitary facilities, 449.
Sault Ste. Marie, 615.
Sausalito, Cal., 147, 222, 249, 432. *See also*, Marinship Corp.
Savannah, Ga., 142, 630, 631. *See also*, Savannah Shipyards; Southeastern Shipbuilding Corp.; MacEvoy Shipbuilding Corp.
Savannah Shipyards Inc., 142, 155-158, 498-503, 530, 542, 720. *See also*, Southeastern Shipbuilding Corp.
Scamping, 49, 272, 277-278, 287, 418, 829.
Scantlings, 694.
Scheduling:
  of components, *see* Components;
  of materials in shipyards, 233, 238;
  of materials to shipyards, 90, 91, 171-172, 350, 353-354, 359, 365-387, 389-391, 394, 695-699, 733, 734;
  of ship construction, 140 n, 177, 353, 362-365, 512; *see also*, Programs of MC.
Scheduling and Planning Section (MC), *see* Production Division, Production Engineering Section.
Schell, Samuel, Duvall, 17-18, 20, 136, 693-694, 698, 718, 719, 752, 754, 764, 765, 766-767, 771, 772, 784, 785.
Schenectady, 539, 544-547, 550-552, 555, 558, 562, 565, 571, 572, 581, 590, *facing* 523.
Schmeltzer, J. E., 19, 24, 25, 74, 83, 96, 131, 156, 169, 171, 355, 462, 505, 583, 789, *facing* 107;
  "The Commission's P4-P Design," 30 n.
Schumacher, Capt. T. L., 570.
Scoll, David E., 156 n, 352.
Score boards, 461.
Scrap steel, 322, 330, 716, 718.
Sea Fox, 29.
Sea Otter, 164-165.
Sea Plane Tender, 37-38.
Sea-trains, 145, 618.
Seamen, 11, 15, 88, 162, 718, 754, 757, 779.
Searchlights, 84, 85, 579.
Searls, Fred, Jr., 337, 338.
Seattle-Tacoma Shipbuilding Corp., 35, 36, 38, 53, 54, 58, 276, 533, 612, 809; transfer to Navy, 149, 613.
Seattle, Wash., 33, 49. *See also*, Seattle-Tacoma Shipbuilding Corp.
Secretary of MC, Office of, 172, 716, 757-758, 760.
Selective Service, 40, 252, 263, 266, 413, 656, 751.
Senate, U. S.:
  Emergency Cargo Ship Construction, Hearings, 45 n;
  First Supplemental National Defense Appropriation Bill for 1943, Hearings, 302 n;
  Government's Wartime Research and Development, 1940-44, Report, 567 n;

Independent Offices Appropriation Bill for 1945, Hearings, 449 n;
Investigation of the National Defense Program, Hearings, 33 n;
Investigation of the National Defense Program Reports, 155 n; *see also,* Truman Committee;
Manpower, Hearings, 253 n, 256 n;
Maritime Commission Nominations, Hearings, 13 n;
Munitions Industry, Preliminary Report, 111 n;
National War Service Bill, Hearings, 663 n;
Nomination of Raymond S. McKeough, Hearing, 794 n;
Problems of American Small Business, Hearings, 205 n;
Report No. 80, 779 n;
Report No. 904, 459 n, 792 n;
Report No. 2076, 459 n;
Wartime Health and Education, Hearings, 437 n.
See also, Congress; Truman Committee.
Sessional Papers, 601.
Seventy-Percent Expansion Act, 37.
Seward, H. L., 274.
Sewerage, 437, 439, 441.
Shaft tunnel, 85, 87-88, 89.
Shafting for ships, 397.
Sharp, George, G., 24, 73, 76, 580, 586, 612, 617.
Shaw, T. E., 704.
Sheehan, Joseph R., 16, 17.
Sheer strake:
 cut in, 80, 548, 569, 573, 581;
 slot, 555, 548, 573.
Sheet metal workers, 622, 657.
Shift premiums, *see* Overtime; Shifts.
Shifts, 212, 232, 233, 286, 307, 414, 421, 427, 659, 674, 684, 687, 816.
Ship Naming Committee (MC), 70.
Ship Operations, Division of (MC), 18, 757, 758, 761.
Ship operators, 2, 23, 31, 68, 576, 762, 763, 765, 766. *See also,* under company names.
Ship Sales Act, 825.
Ship Structure Committee, 571.
" Shipbuilders Meet the War Challenge," 65 n.
" Shipbuilding Activity in Portland Oregon," 248 n.
Shipbuilding and Marine Engineering in Germany, British Intelligence Report No. 2, 587 n.
Shipbuilding and Shipping Record, 82 n.
Shipbuilding brains, 57, 66, 236, 503, 513. *See also,* Management, ability in.
Shipbuilding Commission of the WLB, 421-423.
Shipbuilding Stabilization Committee, 273-275, 279, 284, 285, 287, 302, 306-308, 411, 415, 420-423, 829;
 Minutes, 274 n.
 See also, Zone Conferences.
" Shipbuilding Stabilization Committee," *see* Shulman, Rosalind S.
Shipbuilding Stabilization Zone Standards Committee, 677.
Shipfitter, 239, 242, 257, 547, 564.
Shipping:
 losses of, *see* Sinkings;
 need for, 337, 339;
 See also, Merchant marine of U. S.
Shipping Board, 7, 8, 15, 17, 76, 170, 210, 744, 767;
 Report of Director General Charles Piez (1919), 119 n;
 Sixth Annual Report, 7 n.
Shipping Board Bureau, 17, 19, 24, 25, 712, 771.
Shipping rates, 11, 746, 762.
Shipping World, 72 n.
Ships, Inc., 164-165.
Shipsworth produced, 641-642.
Shipways, *see* Building berths; Erection; Shipyards.
Shipyard Efficiency Awards Committee, 453, 465.
" Shipyard Facilities Data," 217 n.
Shipyard Labor Relations, Division of (MC), 162, 271 and n, 429, 559, 564, 676, 759, 763;
 Assistant Training Supervisors, 266;
 Chief Health Consultant, 701 n;
 Chief Safety Consultant, 701 n;
 Director, *see* Ring, Daniel S.;
 Supervisor of Training, 263; *see also,* Wolff, Jack.
 See also, Maritime Personnel, Division of.
Shipyard Site Planning Committee (MC), 154-156, 158-163, 779.
Shipyard Worker, 279 n, 283, 301.
Shipyards:
 additions to, 58-59, 62-63, 145-146, 443, 445, 482, 505; *see also,* under yard names;
 assembly areas in, 206, 210, 213-214, 219, 220-223, 226-227;
 assembly buildings in, 214, 217-223, 227, 228;
 as assembly points, 206, 224, 230;
 basin type, 216, 220;
 cost of, *see* Facilities, cost of;

INDEX 871

disposal of, 111, 115-117, 791, 824;
equipment of, 108, 111, 233, 355, 472, 473, 642, 671, 673, 827, 830;
land ownership in, see Real Estate;
layouts of, 185, 214-224, 232, 233, 294, 390, 391, 671;
approval of, 78, 108, 131-133, 170-171, 215, 508, 509;
listed and located, 33-36, 59, 142, 630;
living conditions near, 411, 414, 415, 436-446, 657; see also, Housing;
MC staff at, 132-137, 703-704; see also, Inspection in shipyards; and under titles of officials;
productive capacity of in 1943, 340;
railroads to, 50, 48, 152, 154, 163, 217, 221, 230, 472, 506-508, 516-521;
records of compared, 175, 208-209, 467-471, 826-827;
selection of sites for, 46-51, 71, 147, 150-160, 536 and n, 670;
size of, 46-47, 145-147, 186, 191-192;
space in, 213, 215-224, 227, 233, 505, 516, 830;
storage space in, 213, 218-221, 508, 521;
transportation of workers to, 114, 152, 333, 428, 429-431, 442, 701, 823;
warehouses in, 213, 221.
See also, Assembly lines; Emergency yards; Fabrication of steel; Facilities; Management; Old-line yards; and under names of yards and of ship types.
" Short History of War Manpower Commission," 663 n.
Short tons, 311.
Shulman, Rosalind S., *The Shipbuilding Stabilization Committee*, 283 n.
Side launchings, 35, 36, 223, 225.
Side shell assembly, 227.
Siderosis, 448.
Sides, A. B., 533.
Sill, Van Rensselaer, *American Miracle*, 471 n.
Sims, Admiral William S., 14.
Sinkings, 3, 40, 60-62, 64-66, 138, 143, 173, 177, 182, 202-203, 300, 332, 428, 508, 605, 630, 631, 638.
Sirrine and Co., 501.
Six Companies, 54 and n, 146, 809.
Six-way yards, 47, 192, 221, 223, 466, 502.
Skeg, 75, 86.
Skids, see Shipyards, assembly areas.
Skinner uniflow engine, 399, 583, 584, 612.
Skinner, Wade, 194 n. See also, Legal Division, General Counsel, MC.

Slacks, John, 103, 154.
Slattery, W. L., 486, 733, 738-739, 796, 805 n.
Slottman, G. V.: " Organizing Facilities to Promote Production Welding in Shipyards," 214 n;
" Subassembly Welding and Erection," 227 n.
Slowdowns, 270, 301-302.
Siugging, 558.
Small business, 361, 404-408, 691. See also, Subcontracting.
Small Business Act, 648.
Small Business Committee, 614.
Small Business Section, see Procurement Division; Regional Offices of Construction.
Smaller War Plants Corporation, 406, 408.
Smith, H. Gerrish, 274, 278;
" American Shipbuilding Industry at War," 225 n;
" Events of 1939 Prove Wisdom of U. S. Shipbuilding Policy," 28 n.
Smith, William H., 501.
Smith, Admiral W. W., 805 n.
Smokestack assembly, 228.
Social Security Board, Bureau of Employment Security, *Labor Supply Available at Public Employment Offices in Selected Defense Occupations*, 236 n.
Society of Naval Architects and Marine Engineers, 259;
Historical Transactions, 214 n;
presidency of, 14;
Transactions, 14 n.
Solicitor of MC, 499, 502. See also, Legal Division.
Somervell, General Brehon B., 331, 335.
Sootblowers, 95-96.
Sounding machines, 85.
" Sources of Labor Supply in West Coast Shipyards and Aircraft Parts Plants," 248 n.
South, facilities in, 9, 48, 49, 184, 187, 191, 195, 247, 250, 252, 827-829. See also, Gulf yards.
South Carolina, 247, 255.
South Portland Citizens Club, 519.
South Portland, Me., 48, 51, 55, 59, 63, 148, 177, 433, 435, 436, 503-527, 534. See also, South Portland Shipbuilding Corp.
South Portland Shipbuilding Corp., 59, 136, 168, 809, *facing* 522;
auditing at, 473 n;
cafeteria at, 509;
capital ownership of, 527;

872 INDEX

change of management at, 539, 542, 720;
construction methods at, 511, 512, 514;
deliveries from, 526, 527;
inventories at, 392;
labor at, 212, 247, 294, 303, 413, 430, 522-525;
land at, 516, 521;
layout and facilities, 62, 217, 505-507, 511, 516, 521, 522 n, 526, 534;
lease of Thompson's point, 510, 511 and n, 527, 528, 532;
managers of, 504 and n, 514, 526, 527, 529, 531, 533;
operation of Todd-Bath, 505 and n, 516;
profits of, 506, 521, 527, 531-532;
railroad to, 506, 507, 516-521;
speed at, 208, 333, 513;
steel storage at, 508-509, 516, 521;
supervision at, 514, 515, 523, 530;
termination of contract, 527, 529-533;
trucking to, 507-510, 516, 520, 526-528, 532;
women at, 257.
See also, New England Shipbuilding Corp.
South San Francisco, Cal., 33.
Southeastern Shipbuilding Corp., 142, 175, 208, 209, 295, 475, 501-502, 586, 640, 697, 826.
Southern Conference for Human Welfare, 188.
"Southern Shipyard Produces EC-2 Ships," 223 n.
SPAB, *see* Supply Priorities and Allocation Board.
Sparrows Point, Independent Shipbuilders Association of, 293.
Sparrows Point, Md., 33, 52. *See also,* Bethlehem-Sparrows Point Shipyard, Inc.
Speakers Bureau of MC, 453.
Special Assignment Unit (MC), *see* Finance Division.
Special Studies Branch (MC), *see* Finance Division.
Special Studies Section (MC), 18 n.
Special Studies Unit (MC), *see* Finance Division.
Specifications, 23, 25, 103, 106, 134, 171, 408-410, 694. *See also,* Requisitions.
Speed of construction, 61, 121-122, 138-141, 143, 149, 174-177, 201, 202, 207-210, 230-232, 235, 455-458, 546, 562, 573, 624, 684, 687;
of concrete barges, 631 and n;
relation to labor productivity, 232;
the speed record, 210.

Speed of ships, 21-22, 28-30, 42-44, 574-577, 599-601, 605, 794, 831. *See also,* under names of ship types.
Spencer, White and Prentice, 537.
Spofford, W. E., 170, 673, 674, 709, *facing* 746;
"History of the Design, Construction and Operation of the EC-2 Emergency (Liberty) Ships," 86 n.
Spokes for steering wheels, 454.
Spot allocations, 382, 747.
Stability of ships, 582, 620, 695.
Stabilization of labor, 246, 272, 274, 276, 298, 303-304, 306-307, 309, 411-412, 427, 445-446. *See also,* Scamping; Shipbuilding Stabilization Committee.
Staff conferences at regional offices, 705-706, 708 and n, 709.
Staff meetings, of Finance Division, 723, 766.
Stakem, T. E., 784.
Standard 1945 Munitions Dollars, 10.
Standard cargo ships, 27-32, 43-45, 50, 55, 58, 60-62, 68, 166, 336, 355, 574, 583, 605;
C1's, 28-29, 35, 36, 56, 58, 590, 598, *facing* 42, 43;
C2's, 28-29, 34, 36, 56, 104, 121, 234, 590, 597-600, *facing* 42, 490;
C3's, 28-29, 31, 35, 56, 121-122, 149, 324, 590, 612, *facing* 43;
C4's, 619, 623;
converted into naval auxiliaries, 28, 230, 324, 394, 497, 575, 601, 605, 608, 612, 626, 639;
costs of, 495, 642, 819, 820, 825;
deliveries of, 4, 5, 336, 651, 820;
form of contract for, 102-107, 126, 128, 456-457, 490, 800;
manhour costs, 642;
multiple production of, 230, 234-235;
procurement for, 125, 320, 354, 359, 360, 369, 373, 374, 386-387, 394, 406-407;
yards building, 638-640; *see also,* Long range program, yards in; North Carolina Shipbuilding Co.
Standard manhours, 234-235, 640 n, 641-643, 646.
Standard Oil Company of N. J., 30, 56, 102 n, 127, 514.
Standard types, 27-32, 169.
Standard-Vacuum Co., 584.
Standardization, 45-46, 55, 72, 90, 92-96, 202-206. 574, 575, 577, 587, 590-591, 636, 829, 831.
Standby yards, 117, 821.
Stanley Co., *see* A. C. Stanley Co.

INDEX 873

Star of Oregon, 69.
Star (Washington), 576 n.
Stat. (*United States Statutes at Large*), 10 n.
Statement by the Higgins Corporation, A, 199 n.
Staten Island, N. Y., 33, 147.
Statistical Analysis Branch, Production Engineering Section (MC), *see* Production Division.
Statistical units in MC, 18 and n, 140 n, 362-363, 374, 377, 395, 467, 763-764.
Statistics of Income for 1945, 819 n.
Steam auxiliaries, 45.
Steel (carbon steel):
 allocations to MC, 180-182, 194 n, 197, 198, 318-319, 323, 326-335, 338, 342-343, 344, 348, 365, 369-378, 382-384;
 allocations to yards, 197, 346, 350 n, 365, 369, 370, 371-379, 382, 383, 384-387, 394;
 expediters for, 389;
 freight on, 747, 780;
 inventories of, *see* Inventories;
 plate: increase in output, 312, 318-322;
  requirements per vessel, 320, 347 n;
  specifications of MC, 87, 319-322, 552-553, 567, 570;
  see also, Inventories; Steel, allocations; Steel, shipments; Steel, supply and shortage;
 reports from shipyard, 370-371, 374-375, 696, 697;
 requirements of MC, 6, 311, 312, 328, 329-330, 334, 374, 382-383;
 rolling mill schedules, 313, 314, 318-319, 327, 346, 372, 383, 384-387, 389, 393 and n;
 shapes, 312, 330, 339, 380, 385, 579 and n;
 shipments compared to shipyard needs, 311-312, 316, 323, 328, 330, 333, 334, 342, 343, 387, 393, 394;
 specifications for, 570, 615; *see also*, plate;
 substandard, 547, 550-553, 564, 566, 569;
 supply and shortage of, 38, 61, 63-64, 87, 149-150, 166, 167 n, 178-183, 193, 197, 199, 200, 263, 301, 302, 310, 311, 312, 315-317, 328-335, 338-340, 343-352, 383, 410, 575, 591, 599, 628, 629, 630, 631 n, 831;
 in transit, 347-348;
 vouchers, 724, 747.
 See also, Alloy steel; Fabrication; Scrap Strip mills.
Steel producers, 49, 50, 182, 268, 305, 318, 343, 344, 346, 352, 372, 376, 550-552, 579 n. *See also*, Steel, rolling mill schedules.
Steel Requirements Branch, Production Engineering Section (MC), *see* Production Division.
Steel Valve Advisory Committee, 368.
Steel Workers Organizing Committee, CIO, 276, 288.
Steelman, John R., *The Federal Research Program*, 571 n.
Steelton, Pa., 33.
Steering gear, 358.
Stephan, Herman R., 511-512, 527.
Stern frames, 366.
Sternposts, 204.
Stettinius, Edward R., Jr., 316.
Stevedores, 271.
Stimson, Secretary of War Henry L., 338, 662.
Stockton, Cal., 148.
Stokes, Richard L., 576.
Stone, J. F., 769, 770, 771-772.
Straight-line flow, 217, 219-223.
Stresses in steel, 546, 547, 548, 559, 562, 563, 568, 569, 570, 572.
Strikes, 242-243, 270, 272, 276, 287-299, 391, 392, 394, 398, 411, 412, 440, 446, 452, 495, 522, 524, 525, 829;
 time lost due to, 304 and n, 305, 306.
Strip mills, 318-322, 550.
Structural Alterations on Liberty Ships, 550 n.
Structural failure of ships, *see* Fractures.
"Structural Reinforcement of Liberty Ships, The," 548 n.
Structural steel plants, *see* Fabrication of steel, outside shipyard; Steel producers.
Stunt ships, 210.
Sturgeon Bay, Wisc., 155, 433. *See also*, Leatham D. Smith Shipbuilding Co.
Subassembly, *see* Preassembly.
Subcontract Audit Unit, Construction Audit Section (MC), *see* Finance Division.
Subcontracting, 92-93, 95, 166, 388, 401, 454, 472, 500, 509, 728, 799, 802.
Submarine tenders, 51.
Submarines, *see* Sinkings.
Subsidies, 2, 10, 11, 12, 21, 22, 104 n, 712, 762, 763.
Suggestion Committee (MC), 784.
Suggestions committees in shipyards, 454.
Sughrue, T. G., 517, 518 n.
Sun (Baltimore), 13 n.
Sun Doxford engines, 29 n, 584.
Sun Oil Co., 33.

874   INDEX

Sun Shipbuilding and Dry Dock Co., 27, 33, 34, 35-36, 37, 51, 56, 57, 102, 103, 112, 122, 127, 146, 300, 462, 470, 640;
    construction methods at, 230, 563;
    cost of ships at, 682;
    and fractures, 545;
    labor at, 268, 269, 273, 282-283, 294, 308, 669 n, 670;
    layout and facilities at, 58, 145, 159, 818;
    Negroes at, 255;
    No. 4 yard, 255-256, 618-619, 621-623, 639;
    profits of, 476.
Sunday Telegram (Portland), 521 n.
Sunday work, see Hours.
Sunderland, England, 81, 83.
Sunship, see Sun Shipbuilding and Dry Dock Co.
Supervisors of shipyard labor, 233, 239, 255, 258-261, 265, 266, 417, 420, 483-485, 515, 523, 545, 557-558, 562-565, 567, 673, 708.
Supplement to Preliminary Report on Comptroller General's Audit Reports, 1943 and 1944, 742 n.
Suppliers, 93-97, 213, 396-408, 481, 723-724, 726-727;
    profits of, 800, 801, 803-804, 806, 818, 822.
    See also, Components; Procurement.
Supply Officers (field), see Procurement Division.
Supply Officers of MC at shipyards, 700, 703.
Supply Priorities and Allocations Board, 314 n, 317, 428.
Supreme Court of U. S., 282.
Supreme Court Reporter, 124 n.
Surgeon General, U. S. Army, 622.
Surplus Material Section (MC), see Procurement Division.
Surplus materials, 115, 116, 376, 381, 698, 699 and n, 821-824 and n.
Survey Branch, Production Engineering Section (MC), see Production Division.
Survey and Scheduling Branch, Production Engineering Section (MC), see Production Division.
Svenson, Arthur L., see under McElroy, Frank S.
" Swan Island Building Large Tankers," 227 n.
Swan Island shipyard, 142, 146, 469, 583, 640, 809, facing 139, 491;
    construction methods at, 228, 230, 563;
    contracts of, 476, 813, 815;
    cost of ships at, 682;
    labor at, 248, 296, 430, 669;
    land for, 114-115;
    layout and facilities at, 220, 227;
    management of, 484;
    procurement by, 361;
    productivity at, 646, 683;
    speed at, 204;
    splitting of Schenectady, 544, 551.
    See also, Kaiser Co., Inc.
Swope, Gerard M., 285.
Sword Line, 777 n.

T

T2-SE-A1 design, 56-57, facing p. 363.
    See also, Tankers.
T3-S2-A1 design, 30.
Taber, John, 792.
Tacoma, Wash., 36.
Tampa, Fla., 34, 49, 433. See also, Tampa Shipbuilding Co., Inc.
Tampa Shipbuilding Co., Inc., see Tampa Shipbuilding and Engineering Co.
Tampa Shipbuilding and Engineering Co., 34, 35, 38, 48, 73, 103, 105, 285, 495-498, 720, 745-746.
Tank carrying ship (C4), 619, 623.
Tank carrying landing ships, see Landing ships.
Tank top assembly, 227.
Tankers, 3, 22, 30, 55-57, 76, 324, 409, 574, 583, 623, 626, 652, 668, 686, facing 43, 330, 363;
    construction time of, 204, 466;
    cost of, 682, 814, 819, 820;
    deliveries of, 4, 5, 336, 651, 820;
    form of contract for, 102-107, 123-124, 126, 456, 476, 682-683, 813-815;
    fractures of, 544-545, 561;
    Liberty as, see Liberty, tankers;
    losses of, 203, 332;
    manhour costs of, 234, 642;
    military, 639, 640;
    multiple production of, 230, 235;
    reinforcements on, 548;
    way time for, 683;
    yards building, 35-36, 62-63, 102, 145, 333, 344, 350, 351, 360, 361, 394, 469, 583, 639, 640, 645, 681-683.
Tartter, Edwin E., 729-730.
Tavares, Carlos, 664 n.
Taxes, 110, 481, 488, 491, 798-799, 803, 810, 818, 819.
Taylor, C. M., " The LST, the Kingpin of Invasion Fleet," 225 n.
Taylor, Frank J., " Builder No. 1," 249 n.
Technical Division (MC), 19, 25, 46, 60, 74, 75, 78, 361, 378, 380, 409,

577, 585, 606, 612, 620, 628, 629, 641, 646, 682, 691, 750, 759, 762, 764, 775;
Assistant Director, 736;
Associate Director, 25, 131;
Clerical Section, 25;
conflicts with Regional Offices, 694-695;
Conservation Committee, 409;
Construction Section, 25, 170;
Cost Section, 717, 718;
and decentralization, 169-172;
design development by, 74; *see also* under names of types;
Director, *see* Bates, James L.
Engineering Design, Specifications and Priorities Section, 355;
Engineering Design and Specifications Section, 171, 409;
Engineering Plan Approval Section, 78;
Engineering Section, 25, 583;
Hull Plan Approval Section, 78, 171, 409;
Hull Section, 25;
inspectors in, 134, 135;
Interior and Styling Group, 25;
Investigation Section, 25;
Materials Section, 25, 408;
Performance Section, 26;
Preliminary Design Branch, 25;
Preliminary Design Section, 610.
Telephone calls, 731.
Templates, 2, 226, 500, *facing* 107.
Tennessee Valley Authority, 768, 769, 773.
Terminal Island, Cal., 51. *See also*, California Shipbuilding Corp.
Terminals, 759.
Termination of contract, 536, 543 and n. *See also*, Default.
Termination Report of the National War Labor Board, 288 n.
Testing models, 25, 76, 77.
Texas coast, 9.
Texas Co., 501.
Texies, 442.
Textiles, 359, 408.
Thomas, Albert, 97, 116, 299.
Thomas Hooker, 553.
Thompson & Lichtner, 675, 676.
Thompson, Robert C., 80 n, 81.
Thompson, Robert C. and Harry Hunter, "The British Merchant Shipbuilding Programme in North America, 1940-1942," 72 n, 90 n.
Thompson's Point, Portland, Me., 507, 508, 509, 510, 516, 527, 532, 534.
Tigris River, 634.
Time Magazine, 67 n.

Times (London), 8 n.
Times Record of British War Production, 72 n.
Timken-Detroit Axel Co., 753.
Titusville, Pa., 358.
Todd-Bath Iron Shipbuilding Corp., 51, 59, 81, 82, 90, 138, 168, 216, 504, 512, 515, 522, 526, 528, 809, *facing* 522;
layout and facilities at, 217;
purchase of yard by MC, 505 and n, 516;
trucking to, 508, 510.
See also, New England Shipbuilding Corp.
Todd-California Shipbuilding Corp., 51, 54, 59, 81, 82, 90, 139;
Full Ahead (1942) [Kaiser Manual], 260 n.
See also, Richmond No. 1.
Todd-Galveston Dry Docks, Inc., 286, 537, 630.
Todd-Houston Shipbuilding Corp., 51, 59, 430, 431, 563, 586, *facing* 75, 331;
cost of facilities at, 472, and n;
cost of ships at, 826;
productivity at, 209, 475;
speed of construction at, 141, 175, 208, 364, 457, 458.
See also, Houston Shipbuilding Corp.
Todd Shipyards Corp., 48, 50, 53-55, 73, 77, 98, 141, 158, 168, 268, 283, 470, 501, 514, 532, 533, 535, 537, 539.
Toele, Utah, 607.
Tolerances, 93, 95, 322.
Tons, kinds of explained, 4-6, 311 n, 640.
Torches, *see* Flame-cutting.
Tower, Walter S., 313.
Tracy, Edward J., 271, 418.
Trailer Camps, 440-441, 444.
Trainees, 258, 263-265, 333, 416, 418.
Training of shipyard labor, 186, 189, 233, 237, 239, 240, 244, 256, 258-267, 548, 556-559, 562, 564-565, 671, 673, 676-679, 701, 737, 783, 832.
Training Organization (MC and WSA), 759, 761. *See also*, Seamen.
Training, Supervisor of (MC), *see* Shipyard Labor Relations, Division of.
Training Within Industry Service, OPM, 259, 262, 265.
Tramp ships, 80, 81, 82 n, 83.
Transfers of funds between agencies, 718, 720-721, 821.
Transfers of material between shipyards, 375-376, 380, 381, 697-699, 742, 747, 763.

Transportation, 779, 780, 823. *See also,*
  Freight charges; Shipyards.
Transportation Advisor of MC, 517.
Transports, 4, 138, 142;
  C3's as transports, 638;
  C4 transports, 234, 360, 392, 590, 605, 617-624, 639, *facing* 618;
  cost of, 819;
  manhour costs, 234, 642;
  Navy transports, 625;
  P2 transports, 590, 624-625, 639, *facing* 618;
  Victory transports, 607, 625-626, 648, 653, 679, 680, 685, *facing* 586;
  yards building, 142, 638, 645, 777.
Treasury, Department of the, 712, 721, 723, 729, 741, 748, 798. *See also,* Coast Guard; *Statistics of Income.*
Trial Board (MC), 25. *See also,* under Regional Offices of Construction.
Trials and Surveys Division (MC), 758.
Triatic Stay, 85.
Tribune (Chicago), 554 n.
Truck horse of the fleet, 554.
Trucking in shipyards, 632. *See also,* South Portland Shipbuilding Corp.
Truman, Harry S., 96, 163, 197, 198, 467, 509, 513, 528, 792, 793, 794.
Truman Committee, 33 n, 152, 155 n, 158, 163, 166, 182, 197, 198, 289, 328, 467-468, 493, 513, 528, 529, 550, 551, 554, 601, 602, 630, 636, 657;
  Additional Report, 155 n;
  Hearings, 33 n;
  Interim Report on Shipbuilding at the South Portland Shipbuilding Corporation, 529 n;
  Report, 155 n.
Tuberculosis, 448.
Tugs, 126, 225, 627;
  harbor, 60, 634;
  sea-going (V4), 60, 63-64, 325, 615, 634, 636, *facing* 619.
Tulloss, S. B., 745.
Turbines, 28, 29, 577, 704;
  C1 turbine, 594;
  C2 turbine, 585, 593, 596, 597, 598, 600, 601, 605;
  C3 turbine, 583, 586, 593, 596, 598, 607, 624;
  exhaust turbine, 585;
  procurement of, 360, 398, 747;
  producers of, 397-399, 401, 583, 587, 594-599;
  scheduling of, 353, 355, 367, 369, 390, 587-589;
  shortage of, 38, 42, 45, 373, 575, 583;
  standardization of, 590, 594-596;
  Victory turbine, 595, 596, 599-601.

Turbines and Gears Defense Industry Advisory Committee, 367-368.
Turbo-electric units, 56, 398, 399.
Turbo-generators, 360.
Turner Construction Co., 537.
Turning flow, 221-222.
Turnover of labor, 247, 412, 414-415, 426, 427, 446, 654-655, 656, 657, 664, 669, 673, 688. *See also,* Quits.
Tutin, John, 225.
Twenty-Percent Expansion Act, 36.
Tyler, David B.:
  "Construction and Operation of Liberty Ships," 74 n;
  "The Victory Ship Design," 578 n;
  "Welding," 544 n;
  "Welding Before World War II," 31 n.

U

"Ugly duckling," 67, 74.
Unemployment, 102, 104, 147, 244, 252, 262, 654.
Uniflow engines, 577. *See also,* Skinner uniflow engine.
"Uniform Classification of Accounts for Shipbuilders," 128 n, 730.
"Union Agreements in Shipbuilding," 276 n.
Union Metal Manufacturing Co., 410.
Union organization, 273, 278, 287-298, 304, 411, 452, 522, 524, 525.
Union Ship Co., 115.
Union shop, 196, 241, 269, 270, 291, 293, 295, 308.
Union status, *see* Closed shop; Open shop; Union organization; Union shop.
Unionmelt automatic welding, 547, 560.
Unions, *see* Labor unions.
United Nations, 190.
United States at War, 10 n.
U. S. Departments, *see* under name, e. g., Treasury, Dept. of.
U. S. Government Manual, 1944, 429 n.
United States Lines, 22, 604, 625.
U. S. Maritime Commission, *see* Maritime Commission.
U. S. Shipping Board, *see* Shipping Board.
United States Steel Corp., 33, 552.
Use and Disposition of Ships and Shipyards at the End of World War II, 22 n.
University of Maine, 504.
University of Pennsylvania, 250.
Unskilled labor, *see* Job breakdown; Labor force.

INDEX

Upgrading, 263-266, 423-424, 427, 564, 676, 677 and n, 753;
of Negroes, 252, 253.
Urgency ratings:
for components, 650-653, 671;
for labor, 663, 665-668, 670.
Utah Construction Co., 809.

V

V-loans, 402-404, 480, 481, 482, 780.
Valery Chkalov, facing 523.
Value in place, 642.
Valves, 312, 353, 354, 355, 366, 368, 397, 647, 650, 697.
Van Gelder, Philip H., 273.
Van Riper, Francis H., 367.
Vancouver (shipyard), 142, 391, 809, *facing* 362, 587;
C4's at, 623 and n;
escort aircraft carriers at, 177, 180 n, 399, 613;
labor at, 248, 296, 430-431, 669;
land for, 114-115;
layout and facilities at, 146, 220, 227, 228, 613-614;
LST's at, 176, 611, 777;
productivity at, 645;
speed at, 613, 671;
as standby yard, 117;
Victory ships at, 607;
Victory transports at, 626 and n, 638.
See also, Kaiser Co., Inc.
Vancouver, Wash., 114, 432, 435, 436. *See also*, Vancouver (shipyard).
Vanport, Ore., 445.
Vassar, C. D., "Defense Plant Corporation Projects," 109 n.
Vasta, John, 567, 568;
"History of Structural Design of Concrete Ships," 629 n.
VC2-S-AP1 design, 586.
VC2-S-AP2 design, 585-586, 685. *See also*, Victory ships.
VC2-S-AP3 design, 586. *See also*, Victory ships.
VC2-S-AP5 design, *see* Transports (Victorys).
Ventilation, 30, 85, 88, 694.
Vernon, Cal., 33.
Vertical yards, 213, 221-223.
Vestibule schools, 261.
Via, G. Guy, 237, 264;
"The Wartime Training of Shipbuilders," 238 n.
Vice Chairman of MC, 160, 776. *See also*, Vickery, Vice Admiral Howard L.
Vickery canal, 185 n.

Vickery, Vice Admiral Howard L., 15, 19-21, 68 n, 163, 168, 180, 188, 379, 767, 774, *facing* 106, 491;
and British shipbuilding, 42, 60, 576, 794;
and Combined Shipbuilding Committee, 592;
"Commission Ships for Postwar Service," 624 n;
as Commissioner, terms of office, 19, 774 n-775 n, 794;
and contracts, 125, 471, 477, 490-491, 807;
as Deputy Administrator, WSA, 162, 754, 764;
dissents, 780;
and facilities, 146, 219, 473, 505;
and finances, 717;
and fractures in ships, 546, 548, 552, 553;
and Higgins contract, 185-186, 191-192, 198, 377;
illness and death of, 169, 672, 680, 681, 788, 794;
and labor, 253, 263, 266, 270-271, 271 n, 417, 659, 661, 662, 665, 668, 677;
and Liberty ships, 74, 76-77, 575, 580;
and long-range program, 25;
and managements, 147, 148, 151, 153, 159, 160, 215, 384-385, 456, 459-462, 465-467, 478, 487, 493, 535, 537, 538, 539, 627, 650, 669, 671-672, 675, 676, 733 and n, 783, 786;
and postwar merchant marine, 575-577, 626, 794;
and military types, 610, 620;
and ore carriers, 63;
and personnel, 136, 752;
and production, 62, 139, 140, 141 n, 144, 149, 174, 176, 177, 206-207, 224, 696;
promotion to Rear Admiral and Vice Admiral, 162, 459;
relations with John M. Carmody, 785-786, 788, 792;
relations with William Francis Gibbs, 588; *see also*, and Victory ships;
relations with Admiral Land, 15, 20, 168, 775-777, 788, 790-791;
relations with Regional Offices, 172, 429, 692-693, 701, 708, 709;
relations with Franklin D. Roosevelt, 160, 776;
and Savannah Shipyards, Inc., 155-158, 499, 501, 502-503;
"Shipbuilding in World War II," 205 n;

878 INDEX

and South Portland Shipbuilding Corp,. 513, 514, 503;
and steel, 178, 315, 324, 325, 328, 338, 343, 384;
as Vice-Chairman, MC, 160, 775-777, 778, 785, 787;
and Victory ships, 577, 578, 590, 593 n, 594, 597, 605, 606;
and welding, 563, 566.
Vickery's memo on Victory ship, 593 n.
Victory Fleet Drive, 453.
Victory ships, 28, 235, 491, 831, *facing* 586;
award of contracts for, 586, 593, 679 n, 680, 789, 790;
cancellations of, 687;
change in beam, 579;
compared to C2, 580;
compared to Liberty ship, 28, 577-579, 581-582;
components for, 647-653;
cost of, 686, 819, 820;
deliveries of, 4-6, 607, 651, 820;
deliveries planned, 574, 580, 583, 593 596, 607;
design development, 578-583;
features to prevent fractures, 555, 571, 581;
form of contract for, 456, 679-682;
machinery space in, 580;
priorities for, 593-594, 601, 606-607;
propulsion machinery for, 580, 583-587, 593-596;
reasons for building, 574-577, 599-601, 626, 794;
way time for, 683-684;
WPB's opposition to, 590, 593-601, 607;
yards building, 344, 351, 360, 593, 605, 626, 638-639, 645, 681, 684, 685, 777, 827.
See also, Transports, Victory.
Vinson, Carl, 720.
Vinson-Trammell Act, 798, 799.
Voucher Certification Section (MC), *see* Finance Division.
Voucher Examining Unit, Construction Audit Section (MC), *see* Finance Division.
Vouchers, 128, 130, 483, 721-729, 739-742, 745, 772;
volume of, 724, 725, 742.
Vulcan sootblowers, 95.

W

W. A. Bechtel Co., 54 n, 147 n, 809.
Waesche, Admiral R. R., 566.
Wages of shipyard labor, 106, 120, 241, 244-246, 267, 272, 306, 308-310, 411, 412, 415-427, 444, 445, 658-659, 662, 825, 826, 829;
approval and enforcement of rates, 266, 417-418, 421-422, 427;
cost of living in relation to, 286, 306 and n, 307-308, 420-422;
disputes over, 288, 295, 304, 523;
fringe issues, 421;
incentive systems of, 302, 412, 415, 418, 451, 675, 737;
zone standards, 275, 279, 280, 284-287, 306 and n, 308-309, 415, 420, 423-425, 482.
See also, Collective bargaining; Overtime; Shifts.
Wall, Frank E., 506, 509, 517, 520, 521, 527, 528-529.
Waller, A. C., 764, 767, 769.
Wallgren, Monrad C., 97.
Walsh, Charles E., Jr., 359, 360, 406, 752.
Walsh-Kaiser Co., Inc., 142, 536, 809;
attack cargo vessels at, 625, 639;
continuity of production at, 616-617;
efficiency experts at, 676;
layout and facilities, 223;
power rates at, 781;
speed at, 333, 671;
welding at, 563.
Walter Butler Shipbuilders, Inc., 407, 537, 614, 710;
Duluth yard, 630;
Superior yard, 630, 673.
Wanless, Ivan J., 74, 579;
"History of Emergency Ship Design," 77 n.
War Contracts Board, 805.
War Department, *see* Army.
War Foods Administration, 752 n.
War Labor Board, 242, 269 and n, 270, 292, 293, 298, 299 and n, 306, 309, 421-423, 523-525.
War Labor Reports, 242 n.
War Loans Administrator, 403.
War Manpower Commission, 2, 10, 253 n, 263, 264, 659, 662-670, 673, 751;
Basic Principles to be Observed in Establishing Production Training Programs in Shipyards, 259 n;
"Report on Training at the South Portland Shipbuilding Corporation," 257 n;
"Short History of," 663 n;
The Training Within Industry Report, 1940-1945, 259 n.
War Metallurgy Committee, 567, 568.
War Powers Act, 161, 776.

INDEX 879

War Production Board, 2, 10, 148, 178, 195, 514, 642, 751, 775, 778;
authority of challenged by MC, 594, 597-599, 601, 604;
Controller of Shipbuilding, 587-607;
Conservation and Limitation Orders, 410;
criticism of MC by, 166, 168, 345-346, 373-374, 376, 383, 386, 574, 659, 661, 667, 670, 831; see also, Controller of Shipbuilding;
files of, 44 n;
Facilities Clearance Board, 433, 605;
Historical Reports on War Administration, 9 n-10 n;
Industrial Facilities Committee, 606;
Iron and Steel Industry Advisory Committee, 552;
Iron and Steel Branch, see Iron and Steel Division;
Iron and Steel Division, 313, 318, 319, 323, 327, 371-372, 374, 384-388;
Labor Division, 661;
labor-management committees of, 451-453;
Labor problems, 263, 656, 659, 663;
Planning Committee, 373;
Official Munitions Production of the United States . . . , 180 n;
Official Plan Book, War Production Drive, 451 n;
Program Adjustment Committee, 383;
Program Vice Chairman, 313, 376;
Production Executive Committee, 594, 650, 652 n, 668;
production scheduling by, 389, 587, 589, 597-598, 649-650;
Production Vice Chairman, 368, 587;
Requirements Committee, 327, 342, 383-385;
ship delivery estimates of, 165-166, 323;
Shipbuilding Branch, 366; see also, Shipbuilding Division;
Shipbuilding Division, 366-368, 588, 597;
Smaller War Plants Division, 406;
Special Rating Branch, 389, 651;
Special Rating Committee, 652;
Steel Division Requirements Committee, 382-83;
Vice Chairman on Program Determination, 342;
Victory ship opposed by, 831; see also, Controller of Shipbuilding;
War Production Drive of, 453.
Wartime Production Achievements, 202 n;
See also, Civilian Production Administration; Components; Priorities; Scheduling; Steel, allocation.
War Production Drive, 453.
War Service appointees, 750-751.
War Shipping Administration, 161-162, 538, 740;
accounts of, 724, 725, 726, 742, 748, 762, 764, 795;
Administrative Officer, 756, 768;
Administrator, 758, 764, 777; see also, Land, Vice Admiral Emory Scott;
Assistant Deputy Administrators, see names of their divisions;
Budget and Accounts, Division of, 769;
budget of, 715-716, 758, 768;
Comptroller, 764, 768, 771; see also, Quinn, Joseph M.;
Deputy Administrator for Recruitment and Manning, Training and Labor Relations, 162, 764; see also, Macauley, Capt. Edward;
Deputy Administrator for Ship Construction, 162, 754, 764;
Deputy Administrator for Vessel Utilization, Planning and Policies, 755, 764; see also, Douglas, Lewis;
and design of ships, 555, 579, 620;
District Directors, 758;
District Offices, 760, 761-762, 764;
Executive Deputy Administrator, 770;
Executive Officer, 764, 765;
Finance Division, 757;
Fiscal Affairs, Assistant Deputy Administrator for, 745, 756, 758, 764, 765, 768; see also, Chubb, Percy; Houlihan, D. F.;
Fiscal Affairs, Deputy Director of, see Quinn, Joseph M.;
General Accountant, 772;
General Auditor, 771;
Information, Division of, 757;
Insurance Division, 758;
Legal Division, 758;
Maintenance and Repair Division, 758;
Maritime Labor Relations, Assistant Deputy Administrator for, 764;
and operators of ships, 719-720, 745-746;
personnel in, 753, 756-762, 771;
Personnel Management Division, 756, 758, 768, 770;
Price Adjustment Board, 758;
Recruitment and Manning, Assistant Deputy Administrator for, 759, 764;
Relations to MC, 161-162, 168, 172, 754-765, 776, 833;
Research Division, 758;

Ship Control, Assistant Deputy Administrator for, 756, 763, 764;
Ship Operations, Assistant Deputy Administrator for, 756-758, 764;
Shipping Summary, 65 n;
Small Vessel Sales Division, 758;
Statistics and Research, Division of, 763;
Tanker Operations, Assistant Deputy Administrator for, 756, 764;
Training, Assistant Deputy Administrator for, 764.
Ward, C. A., 83, 86.
Ward, Emma F., Report, 257 n.
Warehouses of MC and WSA, 743, 759, 761, 763.
Wartime Production Achievements and the Reconversion Outlook, 322 n.
Washington (state of), 97, 160, 635.
Washington, D. C., 194, 196. See also, Decentralization.
Washington Post, 296 n.
Watt, Capt. R. M., 789.
Way charts, 104 n, 364, 461;
work sheets of, 140 n, 141 n.
See also, Reynolds' Way Charts.
Way time, 140-141, 174-177, 207, 208, 210, 231, 457-458.
Weber, Lt. Comdr. William A., 460, 467, 537, 681, 751, 789.
Wedeberg, Sivert M., 749.
Welch, Richard J., 450, 702 n.
Welders:
classification of, 416, 556, 564, 565;
frauds by, 558, 735, 737, 790;
strike of, 295-296;
tack welders, 239, 242, 243, 266;
training of, 239, 243, 260, 266, 556-558, 562-565;
women as, 239, 257-258.
Welding, 47 n, 75, 80, 81, 84, 88, 213-214, 218, 321, 675, 695, 830;
before the war, 31, 35, 572;
defective welds, 547, 551, 556-558, 566, 569;
downhand, 213, 239;
fumes, 447-448;
machines, 353, 357, 473, 547, 548, 556, 560-562;
manual, 559, 565;
overhead welding, 511, 557;
preparation of steel for, 320, 562;
procedures, 556-565, 568, 572;
research on, 567-568, 572;
rod, 322, 353, 557, 562, 563;
sequences, 511, 547, 559-560, 581.
Welding Advisory Committee, 565, 567, 568.
Welding Research Committee, 567.

Welding Research Council, 568.
Welding Shipyards, Inc., 35, 36.
Werner, Marie D., "Administrative Study of the Division of Economics and Statistics," 764 n.
West Coast, see Pacific Coast.
West Coast Regional Office, 170, 364, 486, 561, 564, 690;
Assistant Director of Finance, 708;
Assistant Supervisor of Training, 678;
Auditor, 707-709;
conferences with management, 705-706;
Director, 656, 664; see also, Flesher, C. W.
frigate program, 614;
Industrial Relations Advisor, 678;
location of, 702;
manpower survey, 674;
Material Control Section, 698;
purchase controller, 700;
relations with auditors, 707-708.
West Lynn, Mass., 398.
West Virginia, 247.
Western Pipe and Steel Co., 35, 58, 276, 281, 451, 638.
Westinghouse Electric Co., 171, 398, 399, 595.
Weyforth, William O., *Manpower Problems*, 247 n.
Wheeler, E. W., 517.
Wheelock, Capt. C. D., 567.
White, E. E., 764, 767, 771, 772.
White, Gerald T., "Financing Industrial Expansion for War," 107 n.
White, Wallace H., Jr., 519.
Whiteside, Arthur D., 313 n, 319.
Wigglesworth, Richard B., 719, 792, 795.
Wiley, Admiral H. A., 15, 20.
Willamette Iron and Steel Corp., 585.
Willamette River, 146.
Williams, A. J., 760.
Williams, E. B., "The Delta Shipyard," 53 n.
Williams, Capt. Roger, 52, 141 n, 246, 413, 463, 471.
Willis, Horace D., 450.
Wilmington, Cal., 58, 62. See also, Consolidated Steel Corp., Ltd.
Wilmington, Del., 36.
Wilmington, N. C., 50, 51, 52, 117, 440-441. See also, North Carolina Shipbuilding Co.
Wilson, C. E., 368, 587, 589-590, 593, 594, 597, 599, 602, 603, 605, 606.
Wilson, J. S., 788.
Wilson, James W., 512, 556-559, 563, 567.
Winches, 31, 358, 405, 580.
Winchester Repeating Arms Co., 16, 541.

INDEX 881

Windlasses, 358.
WLB, see War Labor Board.
Wolfe, George F., "Production Line Welding Plant Speeds the War Program," 225 n.
Wolff, Jack, 264, 265-266, 559, 562, 564-565, 676-678.
Women in shipyards, 239, 257-258, 415, 449, 656.
Women's State Republican Club of N. J., 166.
Wooden decks, 84.
Wooden vessels, 629, 634-636.
Woodrum, Clifton A., 719, 720.
Woodward, Thomas M., 15, 20 n, 21, 160, 163, 774, 793, *facing* 106;
and Commission form, 786;
and contracts, 402-403, 778, 782-783;
dissents, 116, 158, 501, 780, 790;
and managements, 154, 779;
and South Portland Shipbuilding Corp., 517, 518, 519, 520, 530;
and transportation of workers, 429.
"Work Injuries in the United States During 1943," 448 n.
Working capital, 401-404, 479, 480, 487, 488, 491, 496, 499, 538, 539, 541, 735.
Working conditions, see Collective bargaining.
Working funds, see Transfers of funds.
Working plans, 705, 741;
for aircraft carrier escort, 612;
approval of, 649, 652, 689, 691, 694;
for British Libertys, 72-79, 82, 93, 98-99;
for frigates, 614, 695;
for Liberty ships, 72-80, 82, 83-89, 91, 93, 98-99, 205;
for Victory ships, 580, 586, 625.
Working rules of unions, 240-244, 278, 302, 465, 485.
Works Progress Administration, 133, 262.
World War I:
boom and slump after, 21, 32, 81;
concrete ships in, 628, 631-632;
construction methods and records, 76, 151, 166, 206-207, 210, 213, 239, 829-830;
contracts used in, 101, 111, 119 n. 123, 124;

deliveries of ships in, 3, 7, 8;
labor in, 126, 246, 272, 273, 412, 829;
shipyards of, 47-50, 103, 111, 118, 133, 147, 213, 248;
wooden ships in, 629.
See also, Shipping Board.
World War II, 203;
battle of the Atlantic, 612; *see also,* Sinkings;
beginning of, *see* Declaration of War; France; Pearl Harbor;
defeat of Germany, 681;
end of, 658, 686-687, 787; *see also,* Postwar planning.
landings in Europe, 176, 203, 634, 636, 647, 650, 652, 657-658, 718;
Pacific operations, 612, 634;
Worthington Pump & Machinery Corp., 94.
WPB, *see* War Production Board.
WSA, *see* War Shipping Administration.
WSA Shipping Summary, 65 n.
Wyoming, 14.

Y

Yalta conference, 789.
Yancy, John A., "Training a Democracy to Build Liberty Ships," 250 n.
Yard staff committees of MC, 703.
Yates, J. H., 771 n, 772.
Yearbook of American Labor, 1944, 242 n.

Z

Z-ET1-S-C3, *facing* p. 363.
Zimmerman, Dr. Rufus E., 552.
Zone Conferences, 275, 416, 423;
Atlantic Coast, 282-285, 308;
Chicago, 308-310;
Great Lakes, 286;
Gulf Coast, 285-286;
Pacific Coast, 276-282, 309.
Zone standards, 279-280;
Atlantic Coast, 290, 306 n;
Chicago amendments, 309-310;
Great Lakes, 306 n;
Gulf, 306 n;
Pacific Coast, 287, 306 n;
See also, Overtime; Shifts; Wages.